S. Axler
F.W. Gehring
P.R. Halmos

Springer Science+Business Media, LLC

Undergraduate Texts in Mathematics

Anglin: Mathematics: A Concise History and Philosophy.
Readings in Mathematics.

Anglin/Lambek: The Heritage of Thales.
Readings in Mathematics.

Apostol: Introduction to Analytic Number Theory. Second edition.

Armstrong: Basic Topology.

Armstrong: Groups and Symmetry.

Bak/Newman: Complex Analysis.

Banchoff/Wermer: Linear Algebra Through Geometry. Second edition.

Berberian: A First Course in Real Analysis.

Brémaud: An Introduction to Probabilistic Modeling.

Bressoud: Factorization and Primality Testing.

Bressoud: Second Year Calculus.
Readings in Mathematics.

Brickman: Mathematical Introduction to Linear Programming and Game Theory.

Cederberg: A Course in Modern Geometries.

Childs: A Concrete Introduction to Higher Algebra. Second edition.

Chung: Elementary Probability Theory with Stochastic Processes. Third edition.

Cox/Little/O'Shea: Ideals, Varieties, and Algorithms.

Croom: Basic Concepts of Algebraic Topology.

Curtis: Linear Algebra: An Introductory Approach. Fourth edition.

Devlin: The Joy of Sets: Fundamentals of Contemporary Set Theory. Second edition.

Dixmier: General Topology.

Driver: Why Math?

Ebbinghaus/Flum/Thomas: Mathematical Logic. Second edition.

Edgar: Measure, Topology, and Fractal Geometry.

Elaydi: Introduction to Difference Equations.

Fischer: Intermediate Real Analysis.

Flanigan/Kazdan: Calculus Two: Linear and Nonlinear Functions. Second edition.

Fleming: Functions of Several Variables. Second edition.

Foulds: Combinatorial Optimization for Undergraduates.

Foulds: Optimization Techniques: An Introduction.

Franklin: Methods of Mathematical Economics.

Hairer/Wanner: Analysis by Its History.
Readings in Mathematics.

Halmos: Finite-Dimensional Vector Spaces. Second edition.

Halmos: Naive Set Theory.

Hämmerlin/Hoffmann: Numerical Mathematics.
Readings in Mathematics.

Iooss/Joseph: Elementary Stability and Bifurcation Theory. Second edition.

Isaac: The Pleasures of Probability.
Readings in Mathematics.

James: Topological and Uniform Spaces.

Jänich: Linear Algebra.

Jänich: Topology.

Kemeny/Snell: Finite Markov Chains.

Kinsey: Topology of Surfaces.

Klambauer: Aspects of Calculus.

Lang: A First Course in Calculus. Fifth edition.

Lang: Calculus of Several Variables. Third edition.

Lang: Introduction to Linear Algebra. Second edition.

Lang: Linear Algebra. Third edition.

Lang: Undergraduate Algebra. Second edition.

Lang: Undergraduate Analysis.

Lax/Burstein/Lax: Calculus with Applications and Computing. Volume 1.

LeCuyer: College Mathematics with APL.

Lidl/Pilz: Applied Abstract Algebra.

Macki-Strauss: Introduction to Optimal Control Theory.

(continued after index)

Bruce P. Palka

An Introduction to Complex Function Theory

With 138 Illustrations

Springer

Bruce P. Palka
Department of Mathematics
University of Texas at Austin
Austin, TX 78712-1082
USA

Editorial Board:

S. Axler
Department of Mathematics
Michigan State University
East Lansing, MI 48824
USA

Frederick W. Gehring
Department of Mathematics
University of Michigan
Ann Arbor, MI 48109
USA

Paul R. Halmos
Department of Mathematics
Santa Clara University
Santa Clara, CA 95053
USA

Mathematics Subject Classifications (1991): 30-01

Library of Congress Cataloging-in-Publication Data
Palka, Bruce P.
   An introduction to complex function theory / Bruce P. Palka.
      p.   cm.
   Includes index.
   ISBN 978-1-4612-6967-0   ISBN 978-1-4612-0975-1 (eBook)
   DOI 10.1007/978-1-4612-0975-1
   1. Functions of complex variables.   I. Title.
QA331.7.P35   1990
515'.9—dc20   90-47375
                                                            CIP

© 1991 Springer Science+Business Media New York
Originally published by Springer-Verlag New York Inc. in 1991
Softcover reprint of the hardcover 1st edition 1991
All rights reserved. This work may not be translated or copied in whole or in part without the written permission of the publisher (Springer Science+Business Media, LLC), except for brief excerpts in connection with reviews or scholarly analysis. Use in connection with any form of information storage and retrieval, electronic adaptation, computer software, or by similar or dissimilar methodology now or hereafter developed is forbidden.
The use of general descriptive names, trade names, trademarks etc., in this publication, even if the former are not especially identified, is not to be taken as a sign that such names, as understood by the Trade Marks and Merchandise Marks Act, may accordingly be used freely by anyone.

Camera-ready copy prepared by TeXniques, Austin, Texas, using LaTeX.

9 8 7 6 5 4 3 2 (Corrected second printing, 1995)

ISBN 978-1-4612-6967-0

To my parents,
Leonard Palka and Charlotte Fogarty Palka

Preface

The book at hand has its origins in and reflects the structure of a course that I have given regularly over the years at the University of Texas. The course in question is an undergraduate honors course in complex analysis. Its subscribers are for the most part math and physics majors, but a smattering of engineering students, those interested in a more substantial and more theoretically oriented introduction to the subject than our normal undergraduate complex variables course offers, can usually be found in the class. My approach to the course has been from its inception to teach it in everything save scope like a beginning graduate course in complex function theory. (To be honest, I have included some material in the book that I do not ordinarily cover in the course, this with the admitted purpose of making the book a suitable text for a first course in complex analysis at the graduate level.) Thus, the tone of the course is quite rigorous, while its pace is rather deliberate. Faced with a clientele that is bright, but mathematically less sophisticated than, say, a class of mathematics graduate students would be, I considered it imperative to give students access to a complete written record of the goings-on in my lectures, one containing full details of proofs that I might only sketch in class, the accent there being on the central idea involved in an argument rather than on the nitty-gritty technicalities of the proof. I also deemed it wise to provide the students with a generous supply of worked-out examples appropriate to the lecture material. Since none of the textbooks available when I started teaching the course had exactly the emphasis I was looking for, I began to compile my own set of lecture notes. It is these notes that have evolved into the present book.

In rough terms the course I have been describing comprises Chapters I, III, IV, V, VII, and VIII of the book, together with the first three sections of Chapter IX. Chapter II, a resume of information from plane topology, is a reference chapter. It would never occur to me — nor would I recommend to anyone else — to go systematically through this chapter in teaching a complex variables course. Instead, the ideas from Chapter II get dispersed throughout my lectures, each topological notion being brought up as it becomes germane to the development of complex function theory. While this system works fine in the setting of a lecture, I find it disruptive to the ongoing narrative of a book. Therefore, just as other authors have done before me, I have chosen to assemble all the background material from elementary topology in a single place for ease of reference. Chapters VI and X, and with them the last two sections of Chapter IX, furnish "enrichment topics" to those who wish to proceed slightly beyond the essential core of basic complex analysis. The subject matter in Chapters VI and X would, I think, be regarded as standard in most beginning graduate courses.

Located at the end of each chapter is a collection of exercises. Though some of these are intended to foster the development of the computational skills pertinent to complex analysis, most have a pronounced theoretical flavor to them, in keeping with the course for which they were designed. Many of the "classic" exercises in function theory turn up among these problems. Quite a few of the exercises, on the other hand, are original to this book (or they are, at least, to the best of my knowledge).

It is high time that I expressed my gratitude to everyone who has had a hand in the creation of this book. These individuals include a number of graduate students at Texas — Michael Pearson, Michael Westmoreland, and Edward Burger are three that spring immediately to mind — who carefully read through early versions of the manuscript, helped rid it of numerous errors, and, most importantly, identified places that from a student's perspective were badly in need of change. I am grateful to colleagues (in particular, to Barbara Flinn and Jean McKemie) who agreed to "field test" portions of the manuscript in their own classes. Their input has greatly improved the finished product. My special thanks go to Aimo Hinkkanen, with whom I've had many useful conversations during the final stages of preparation of the book and who has been an invaluable source of suggestions for problems. This book would have remained a pipedream were it not for the diligent efforts of Suzy Crumley, who typed it, and Buff Miner, who did the graphics and generally oversaw the production of the manuscript. Both patiently bore the brunt of my revisionist tendencies. Needless to say, they share none of the blame for the inevitable errors that have crept into the text and managed to escape detection under my proofreading. The editorial staff at Springer-Verlag (notably, Rob Torop and his successor, Ulrike Schmickler-Hirzebruch) have been extremely helpful and understanding. Above all, I appreciate the fact that they did not pressure me with deadlines during my stint as graduate advisor, when my literary output slowed to a trickle. A "tusen tack" goes to the Mittag-Leffler Institute in Djursholm, Sweden, where some finishing touches were applied to the manuscript in the course of my stay there during the academic year 1989-90 (and where Kari Hag and David Herron obliged with some greatly valued proofreading). My teacher, Fred Gehring, has been a source of both inspiration and encouragement for the undertaking. Finally, I would like to

Preface   ix

acknowledge the support of my wife, Mary Ann, and my sons, Kevin and Sean. Despite being innocent bystanders, they were often in perfect position to catch the flak of my frustration when things did not go as planned with this project. To them I say: the struggle is over and dad is a happy camper again.

Contents

	Preface	vii
I	**The Complex Number System**	**1**
	1 The Algebra and Geometry of Complex Numbers	1
	1.1 The Field of Complex Numbers	1
	1.2 Conjugate, Modulus, and Argument	5
	2 Exponentials and Logarithms of Complex Numbers	13
	2.1 Raising e to Complex Powers	13
	2.2 Logarithms of Complex Numbers	15
	2.3 Raising Complex Numbers to Complex Powers	16
	3 Functions of a Complex Variable	17
	3.1 Complex Functions	17
	3.2 Combining Functions	19
	3.3 Functions as Mappings	20
	4 Exercises for Chapter I	25
II	**The Rudiments of Plane Topology**	**33**
	1 Basic Notation and Terminology	33
	1.1 Disks	33
	1.2 Interior Points, Open Sets	34
	1.3 Closed Sets	34
	1.4 Boundary, Closure, Interior	35
	1.5 Sequences	35
	1.6 Convergence of Complex Sequences	36
	1.7 Accumulation Points of Complex Sequences	37
	2 Continuity and Limits of Functions	39
	2.1 Continuity	39
	2.2 Limits of Functions	43
	3 Connected Sets	47

		3.1 Disconnected Sets	47
		3.2 Connected Sets	48
		3.3 Domains	50
		3.4 Components of Open Sets	50
	4	Compact Sets	52
		4.1 Bounded Sets and Sequences	52
		4.2 Cauchy Sequences	53
		4.3 Compact Sets	54
		4.4 Uniform Continuity	57
	5	Exercises for Chapter II	58

III Analytic Functions 62

	1	Complex Derivatives	62
		1.1 Differentiability	62
		1.2 Differentiation Rules	64
		1.3 Analytic Functions	67
	2	The Cauchy-Riemann Equations	68
		2.1 The Cauchy-Riemann System of Equations	68
		2.2 Consequences of the Cauchy-Riemann Relations	73
	3	Exponential and Trigonometric Functions	75
		3.1 Entire Functions	75
		3.2 Trigonometric Functions	77
		3.3 The Principal Arcsine and Arctangent Functions	81
	4	Branches of Inverse Functions	85
		4.1 Branches of Inverse Functions	85
		4.2 Branches of the pth-root Function	87
		4.3 Branches of the Logarithm Function	91
		4.4 Branches of the λ-power Function	92
	5	Differentiability in the Real Sense	96
		5.1 Real Differentiability	96
		5.2 The Functions f_z and $f_{\bar{z}}$	98
	6	Exercises for Chapter III	101

IV Complex Integration 109

	1	Paths in the Complex Plane	109
		1.1 Paths	109
		1.2 Smooth and Piecewise Smooth Paths	112
		1.3 Parametrizing Line Segments	114
		1.4 Reverse Paths, Path Sums	115

		1.5 Change of Parameter	116
	2	Integrals Along Paths	118
		2.1 Complex Line Integrals	118
		2.2 Properties of Contour Integrals	122
		2.3 Primitives	125
		2.4 Some Notation	129
	3	Rectifiable Paths	131
		3.1 Rectifiable Paths	131
		3.2 Integrals Along Rectifiable Paths	133
	4	Exercises for Chapter IV	136
V	**Cauchy's Theorem and its Consequences**		**140**
	1	The Local Cauchy Theorem	140
		1.1 Cauchy's Theorem For Rectangles	140
		1.2 Integrals and Primitives	144
		1.3 The Local Cauchy Theorem	148
	2	Winding Numbers and the Local Cauchy Integral Formula	153
		2.1 Winding Numbers	153
		2.2 Oriented Paths, Jordan Contours	160
		2.3 The Local Integral Formula	161
	3	Consequences of the Local Cauchy Integral Formula	164
		3.1 Analyticity of Derivatives	164
		3.2 Derivative Estimates	167
		3.3 The Maximum Principle	170
	4	More About Logarithm and Power Functions	175
		4.1 Branches of Logarithms of Functions	175
		4.2 Logarithms of Rational Functions	178
		4.3 Branches of Powers of Functions	182
	5	The Global Cauchy Theorems	185
		5.1 Iterated Line Integrals	185
		5.2 Cycles	186
		5.3 Cauchy's Theorem and Integral Formula	188
	6	Simply Connected Domains	194
		6.1 Simply Connected Domains	194
		6.2 Simple Connectivity, Primitives, and Logarithms	195
	7	Homotopy and Winding Numbers	197
		7.1 Homotopic Paths	197

		7.2 Contractible Paths	203
	8	Exercises for Chapter V	204

VI Harmonic Functions 214

1. Harmonic Functions 215
   1.1 Harmonic Conjugates 215
2. The Mean Value Property 219
   2.1 The Mean Value Property 219
   2.2 Functions Harmonic in Annuli 221
3. The Dirichlet Problem for a Disk 226
   3.1 A Heat Flow Problem 226
   3.2 Poisson Integrals 228
4. Exercises for Chapter VI 238

VII Sequences and Series of Analytic Functions 243

1. Sequences of Functions 243
   1.1 Uniform Convergence 243
   1.2 Normal Convergence 246
2. Infinite Series 248
   2.1 Complex Series 248
   2.2 Series of Functions 253
3. Sequences and Series of Analytic Functions 256
   3.1 General Results 256
   3.2 Limit Superior of a Sequence 259
   3.3 Taylor Series 260
   3.4 Laurent Series 269
4. Normal Families 278
   4.1 Normal Subfamilies of $C(U)$ 278
   4.2 Equicontinuity 279
   4.3 The Arzelà-Ascoli and Montel Theorems 282
5. Exercises for Chapter VII 286

VIII Isolated Singularities of Analytic Functions 300

1. Zeros of Analytic Functions 300
   1.1 The Factor Theorem for Analytic Functions 300
   1.2 Multiplicity 303
   1.3 Discrete Sets, Discrete Mappings 306
2. Isolated Singularities 309

	2.1	Definition and Classification of Isolated Singularities	309
	2.2	Removable Singularities	310
	2.3	Poles	311
	2.4	Meromorphic Functions	318
	2.5	Essential Singularities	319
	2.6	Isolated Singularities at Infinity	322

3 The Residue Theorem and its Consequences 323
   3.1 The Residue Theorem 323
   3.2 Evaluating Integrals with the Residue Theorem 326
   3.3 Consequences of the Residue Theorem 339

4 Function Theory on the Extended Plane 349
   4.1 The Extended Complex Plane 349
   4.2 The Extended Plane and Stereographic Projection . . 350
   4.3 Functions in the Extended Setting 352
   4.4 Topology in the Extended Plane 354
   4.5 Meromorphic Functions and the Extended Plane . . . 356

5 Exercises for Chapter VIII 362

IX   Conformal Mapping   374

1 Conformal Mappings 375
   1.1 Curvilinear Angles 375
   1.2 Diffeomorphisms 377
   1.3 Conformal Mappings 379
   1.4 Some Standard Conformal Mappings 383
   1.5 Self-Mappings of the Plane and Unit Disk 388
   1.6 Conformal Mappings in the Extended Plane 389

2 Möbius Transformations 391
   2.1 Elementary Möbius Transformations 391
   2.2 Möbius Transformations and Matrices 392
   2.3 Fixed Points 394
   2.4 Cross-ratios 396
   2.5 Circles in the Extended Plane 398
   2.6 Reflection and Symmetry 399
   2.7 Classification of Möbius Transformations 402
   2.8 Invariant Circles 408

3 Riemann's Mapping Theorem 416
   3.1 Preparations 416
   3.2 The Mapping Theorem 419

4 The Carathéodory-Osgood Theorem 423
   4.1 Topological Preliminaries 423
   4.2 Double Integrals 426

	4.3	Conformal Modulus	427
	4.4	Extending Conformal Mappings of the Unit Disk	440
	4.5	Jordan Domains	445
	4.6	Oriented Boundaries	447
5	Conformal Mappings onto Polygons		450
	5.1	Polygons	450
	5.2	The Reflection Principle	451
	5.3	The Schwarz-Christoffel Formula	454
6	Exercises for Chapter IX		466

X Constructing Analytic Functions — 477

1	The Theorem of Mittag-Leffler		477
	1.1	Series of Meromorphic Functions	477
	1.2	Constructing Meromorphic Functions	479
	1.3	The Weierstrass $\wp$-function	486
2	The Theorem of Weierstrass		490
	2.1	Infinite Products	490
	2.2	Infinite Products of Functions	493
	2.3	Infinite Products and Analytic Functions	495
	2.4	The Gamma Function	504
3	Analytic Continuation		507
	3.1	Extending Functions by Means of Taylor Series	507
	3.2	Analytic Continuation	510
	3.3	Analytic Continuation Along Paths	512
	3.4	Analytic Continuation and Homotopy	517
	3.5	Algebraic Function Elements	520
	3.6	Global Analytic Functions	527
4	Exercises for Chapter X		535

Appendix A Background on Fields — 543

1	Fields		543
	1.1	The Field Axioms	543
	1.2	Subfields	544
	1.3	Isomorphic Fields	544
2	Order in Fields		545
	2.1	Ordered Fields	545
	2.2	Complete Ordered Fields	546
	2.3	Implications for Real Sequences	546

Appendix B Winding Numbers Revisited   **548**

    1   Technical Facts About Winding Numbers 548
         1.1   The Geometric Interpretation 548
         1.2   Winding Numbers and Jordan Curves 550

Index   **556**

Chapter I

The Complex Number System

Introduction

Most of us met complex numbers for the first time in high school algebra, where the imaginary unit $i = \sqrt{-1}$ and expressions of the type $x + iy$ arose quite naturally in the study of quadratic equations. Back then we didn't balk for an instant at the prospect of extracting square roots of negative numbers. We simply learned to deal with complex numbers on a formal level by mastering the few innovations that caused the algebra of the complex number system to differ from that of the real numbers — and we managed just fine. It is in much the same spirit that we wish to begin the study of complex analysis in this chapter. After a nod in the direction of more rigorous mathematics in which we outline a proper definition of the complex number system, we review for the reader some of the standard elementary facts of complex arithmetic. By then embellishing these facts with detail of a perhaps less familiar character, we lay the groundwork for a careful treatment of the theory of analytic functions of a single complex variable.

1 The Algebra and Geometry of Complex Numbers

1.1 The Field of Complex Numbers

We assume on the part of the reader a knowledge of calculus, including a basic familiarity with the structure of the real number system $\mathbb{R}$ and with analytic geometry in two-dimensional real euclidean space $\mathbb{R}^2$. An element z of $\mathbb{R}^2$ is nothing but an ordered pair $z = (x, y)$ of real numbers x and y. For elements $z = (x, y)$ and $w = (u, v)$ of $\mathbb{R}^2$ to be declared equal it is

required that $x = u$ and $y = v$. The addition of such pairs is carried out in the most straightforward fashion,
$$z + w = (x + u, \; y + v).$$
So-called "imaginary numbers" involving the quantity $\sqrt{-1}$ were in reasonably wide-scale use long before the time of the Irish mathematician William Rowan Hamilton (1805–1865). That use amounted in large part to formal — which is not to suggest uninventive — algebraic manipulation, not unlike what most of us experience today in our initial contact with complex numbers. It was Hamilton, however, who in an 1833 paper finally "demystified" these numbers and placed them on a firm logical footing. He did so by observing that there is an advantageous way to define a product zw of the elements z and w; namely,
$$zw = (xu - yv, \; xv + yu).$$
The term "advantageous" alludes to the fact that, with the operations of addition and multiplication in $\mathbb{R}^2$ so defined, the ordinary rules governing the arithmetic of real numbers — e.g., the associative laws for addition and multiplication, the commutative laws for these operations, the distributive law — see their validity extended to $\mathbb{R}^2$. In standard mathematical parlance, when endowed with these two algebraic operations $\mathbb{R}^2$ acquires the structure of a *field*. (See Appendix A for the exact definition of this concept and for other material pertaining to fields.) It has become traditional to employ the designation $\mathbb{C}$ for $\mathbb{R}^2$ in its manifestation as a field and to refer to the elements of $\mathbb{R}^2$, in this context, as *complex numbers*. The role of zero in the field $\mathbb{C}$ is played by $(0,0)$ and the element $(1,0)$ serves as its multiplicative identity. The additive inverse of the complex number $z = (x,y)$ is the element $-z = (-x,-y)$; the multiplicative inverse z^{-1} (or $1/z$) of z is given by $z^{-1} = (x[x^2+y^2]^{-1}, \; -y[x^2+y^2]^{-1})$ — provided, of course, that z is different from zero.

Just how does the mysterious number i fit into this scheme? In order to answer this question, we first remark that the set of complex numbers of the form $z = (x,0)$ constitutes a subfield of $\mathbb{C}$ that is structurally indistinguishable from the field of real numbers. For this reason it is customary to blur any distinction between the real number x and the complex number $(x,0)$ and actually to employ x as a convenient abbreviation for $(x,0)$. Under this convention $\mathbb{R}$ itself is regarded as a subfield of $\mathbb{C}$. If we introduce the symbol i as a special notation for the complex number $(0,1)$, we observe that
$$i^2 = (0,1)(0,1) = (-1,0) = -1,$$
in accordance with the convention just established. In the field $\mathbb{C}$ the real number -1 does, indeed, have i as a square root! Furthermore, an arbitrary complex number $z = (x,y)$ can be rewritten as follows:
$$z = (x,y) = (x,0) + (0,y) = (x,0) + (0,1)(y,0) = x + iy.$$

1. The Algebra and Geometry of Complex Numbers

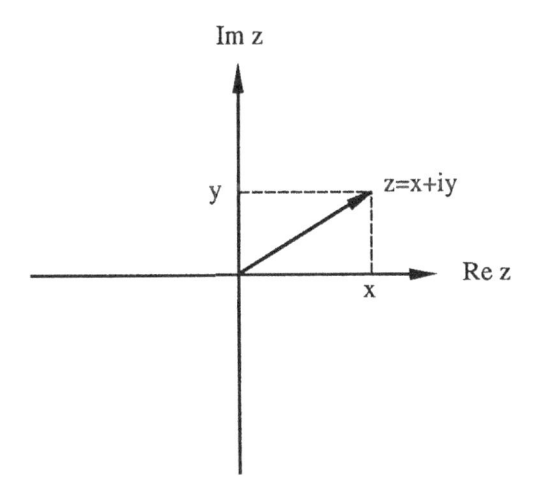

Figure 1.

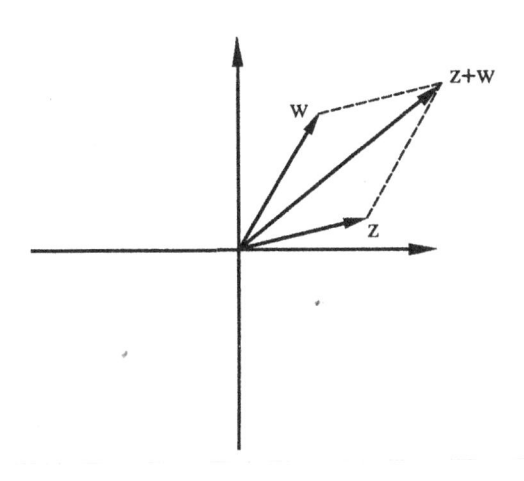

Figure 2.

The upshot of the preceding comments is that we are justified in thinking of a complex number z as a quantity of the type $z = x + iy$, where x and y are real numbers and where $i^2 = -1$. In fact, the representation of complex numbers in the form $z = x + iy$ — as opposed to the ordered pair notation $z = (x, y)$ — is the universally preferred one. (On aesthetic grounds we may occasionally opt to write $x + yi$. Somehow $1 + i5$ looks peculiar, while $1 + 5i$ doesn't.) For reasons chiefly of historical interest x is called the *real part* of z, and y — not iy, as one might expect — is termed its *imaginary part*. Use of the notations $\operatorname{Re} z$ and $\operatorname{Im} z$ to signify the real and imaginary parts of z is quite common. The formulas that define the sum and product of complex numbers $z = x + iy$ and $w = u + iv$ now take the forms

$$z + w = x + u + i(y + v) \quad , \quad zw = xu - yv + i(xv + yu) \ .$$

For example, we compute

$$(1 + 4i) + (-2 - i) = -1 + 3i \ ,$$

$$(2 - 6i)(1 + i) = 8 - 4i \ ,$$

$$\frac{2 + 5i}{3 - 4i} = (2 + 5i)(3 - 4i)^{-1} = (2 + 5i)\left(\frac{3}{25} + \frac{4i}{25}\right) = -\frac{14}{25} + \frac{23i}{25} \ .$$

One big difference between $\mathbb{R}$ and $\mathbb{C}$ must be stressed from the outset. The complex number field is an "unorderable field" — we once again refer the reader to Appendix A for a precise definition of the term — so that expressions like $z \leq w$ are not generally meaningful in $\mathbb{C}$. When inequalities appear in this book it will always be tacitly assumed that the quantities under comparison are real numbers.

While few mathematical illustrations can claim to be worth the thousand words promised by the adage, there is often genuine benefit to be gained by studying a mathematical concept from a geometric viewpoint. The usual way of representing complex numbers graphically is to depict $z = x + iy$, depending on the context, either as the point with coordinates (x, y) in the cartesian plane or as the vector from the origin to that point (Figure 1). In this setting the cartesian plane is spoken of as the *complex plane*, the x-axis as the *real axis*, and the y-axis as the *imaginary axis*. The graphical interpretation of complex addition is indicated in Figure 2. The geometric meaning of multiplication is less transparent and will be described shortly, following some additional preparation. Mathematicians tend to use the expressions "complex numbers" and "complex plane" interchangeably, despite the geometric overtones of the latter. We join with the crowd in our usage of these terms, drawing little or no distinction between them.

1.2 Conjugate, Modulus, and Argument

The *conjugate* of the complex number $z = x + iy$ is the complex number $\bar{z} = x - iy$. For instance, $\bar{i} = -i$ and $\overline{3 - i\sqrt{2}} = 3 + i\sqrt{2}$. Graphically z and $\bar{z}$ correspond to points that are mirror images of one another with respect to the real axis (Figure 3).

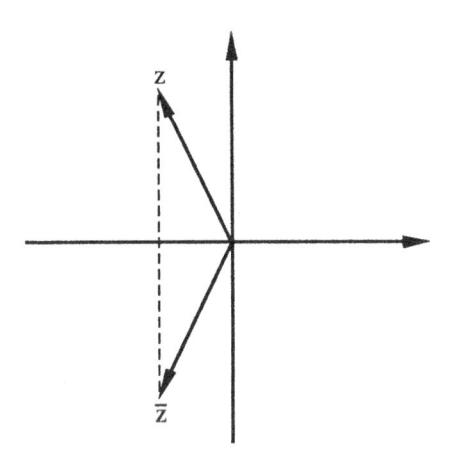

Figure 3.

One verifies with little effort the following elementary identities:

(1.1)
$$\begin{cases} \bar{\bar{z}} = z \, ; \\ \overline{z \pm w} = \bar{z} \pm \bar{w} \quad , \quad \overline{zw} = \bar{z}\bar{w} \quad , \quad \overline{z/w} = \bar{z}/\bar{w} \, ; \\ \operatorname{Re} z = (z + \bar{z})/2 \quad , \quad \operatorname{Im} z = (z - \bar{z})/2i \, . \end{cases}$$

A real number z in $\mathbb{C}$ is characterized by the property that $\bar{z} = z$; an element z of $\mathbb{C}$ is *purely imaginary*, meaning that $\operatorname{Re} z = 0$, precisely when $\bar{z} = -z$.

The *modulus* $|z|$ of $z = x + iy$ is defined by $|z| = \sqrt{x^2 + y^2}$. (The terms *magnitude* and *absolute value* are commonly used synonyms for "modulus.") As an example, $|-3 + 4i| = 5$. Clearly $|z|$ gives the length of the vector corresponding to z in the complex plane (Figure 4). More generally, $|z - w|$ is the ordinary distance between the points of the plane that represent z and w.

Some basic properties of the modulus are summarized in (1.2). Only the verifications of the final two assertions entail even the slightest complication. For these we refer the reader to Examples 1.1 and 1.2. The checking of the other statements is left as an exercise.

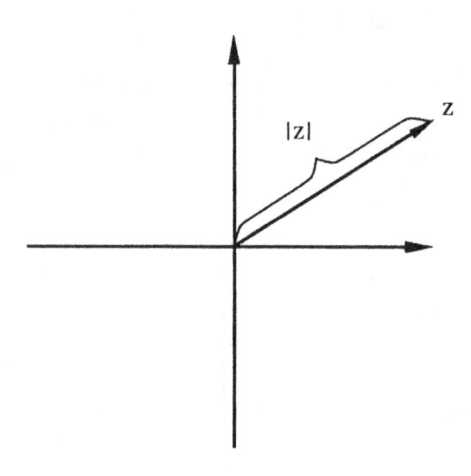

Figure 4.

(1.2) $$\begin{cases} |zw| = |z||w| \ , \quad |z/w| = |z|/|w| \ ; \\ |z| = |\bar{z}| \ , \quad z\bar{z} = |z|^2 \ ; \\ |\operatorname{Re} z| \leq |z| \ , \quad |\operatorname{Im} z| \leq |z| \ ; \\ |z+w| \leq |z| + |w| \ ; \\ |z+w| \geq ||z| - |w|| \ . \end{cases}$$

If $z \neq 0$, we remark that
(1.3) $$z^{-1} = \frac{\bar{z}}{|z|^2} \ .$$

In particular, $z^{-1} = \bar{z}$ if $|z| = 1$. Identity (1.3) shows how z and z^{-1} compare graphically: z^{-1} points in the direction of $\bar{z}$ and has modulus $1/|z|$ (Figure 5).

Consider next a complex number z other than zero. It is always possible to express z in *polar form*,

(1.4) $$z = |z|(\cos\theta + i\sin\theta) \ ,$$

where θ is a real number: if $z = x + iy$, we merely choose any θ that satisfies $\cos\theta = x/|z|$ and $\sin\theta = y/|z|$. For instance, $1 + i$ has a polar representation
$$1 + i = \sqrt{2}\left(\cos\frac{\pi}{4} + i\sin\frac{\pi}{4}\right) \ .$$

Each real number θ for which (1.4) holds is termed an *argument* (or an *amplitude*) of z. Geometrically, θ simply provides a measurement in radians

1. The Algebra and Geometry of Complex Numbers

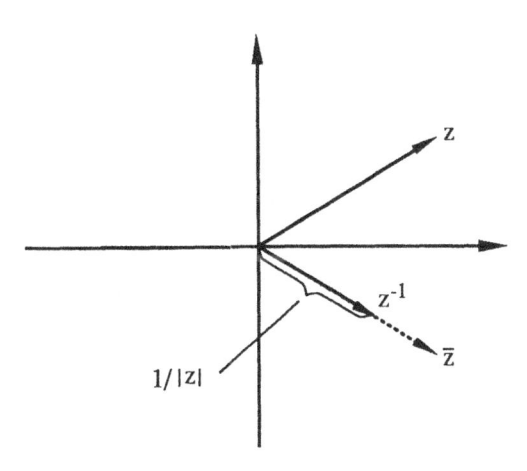

Figure 5.

of the angle from the positive real axis to the vector depicting z in the complex plane. The usual sign conventions for angles in polar coordinates are to be observed. In Figure 6 we indicate with θ_1, θ_2, and θ_3 three arguments of z; θ_1 and θ_2 are positive arguments, whereas θ_3 is a negative argument.

The notation $\arg z$ will be used in this text to designate the set of all arguments of z. Assuming that one such argument θ_0 is known, this set is readily described: $\arg z$ consists of all real numbers θ having the

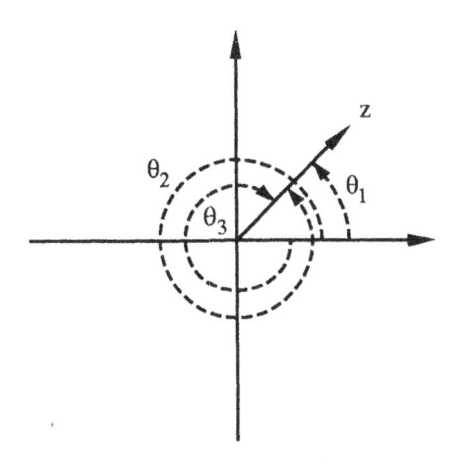

Figure 6.

form $\theta = \theta_0 + 2k\pi$, where k is an integer. To give an illustration of this, $\arg(-1) = \{(2k+1)\pi : k = 0, \pm 1, \cdots\}$. (N.B. It is customary in complex analysis to write $\arg z = \theta$ — rather than $\theta \in \arg z$ — to express the fact that θ is an argument of z. This mild abuse of notation is so well established that any attempt to avoid it would ultimately do the reader a disservice.) We select one special member of $\arg z$, the unique argument θ of z in the interval $(-\pi, \pi]$, for "preferential treatment" and employ it, circumstances permitting, in situations where a definite argument is called for. This argument is referred to as the *principal argument* of z. It is denoted by Arg z. For example,

$$\text{Arg}\, i = \frac{\pi}{2} \quad, \quad \text{Arg}(1-i) = -\frac{\pi}{4} \quad, \quad \text{Arg}(-\pi) = \pi \ .$$

In general, the principal argument of $z = x + iy\ (\neq 0)$ is given by

$$\text{Arg}\, z = \begin{cases} \text{Arcsin}(y/|z|) & \text{if } x \geq 0\ , \\ \pi - \text{Arcsin}(y/|z|) & \text{if } x < 0 \text{ and } y \geq 0\ , \\ -\pi - \text{Arcsin}(y/|z|) & \text{if } x < 0 \text{ and } y < 0\ . \end{cases}$$

Here, for t in the interval $[-1, 1]$, Arcsin t signifies the unique number u in $[-\pi/2, \pi/2]$ satisfying $\sin u = t$. The distinction between $\arg z$, which is a whole set of numbers, and Arg z, which is a particular element of $\arg z$, is an important one to keep straight. Thus, our notational conventions permit us to say that $\arg(-1) = -\pi$, for $-\pi$ is certainly an argument of -1, but not $\text{Arg}(-1) = -\pi$, since $-\pi$ is not the argument of -1 in the interval $(-\pi, \pi]$. Let it be emphasized one last time that no assignment of arguments is made to the complex number zero.

Given non-zero complex numbers z and w written in polar form, $z = |z|(\cos\theta + i\sin\theta)$ and $w = |w|(\cos\psi + i\sin\psi)$, we can without difficulty derive polar representations for z^{-1} and zw. First, it follows from (1.3) that

(1.5) $\quad z^{-1} = |z|^{-1}(\cos\theta - i\sin\theta) = |z|^{-1}[\cos(-\theta) + i\sin(-\theta)]\ .$

As for the product zw,

$$zw = |z||w|(\cos\theta + i\sin\theta)(\cos\psi + i\sin\psi)$$

$$= |z||w|[\cos\theta\cos\psi - \sin\theta\sin\psi + i(\cos\theta\sin\psi + \sin\theta\cos\psi)]\ ,$$

from which we infer that

(1.6) $\quad zw = |z||w|[\cos(\theta + \psi) + i\sin(\theta + \psi)]\ .$

Identity (1.6) leads directly to the geometric interpretation of complex multiplication promised earlier: zw is the complex number with magnitude

1. The Algebra and Geometry of Complex Numbers

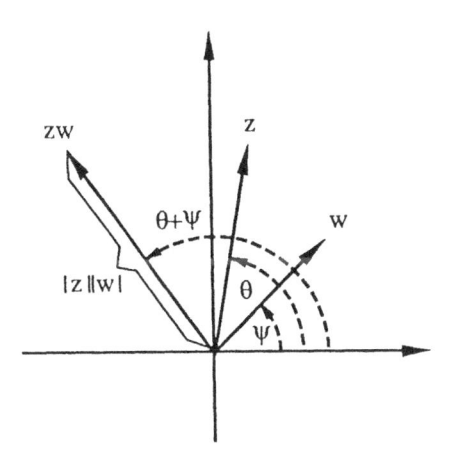

Figure 7.

$|z||w|$ and with $\theta+\psi$ as one of its arguments. (See Figure 7.) As implications of formulas (1.5) and (1.6) we observe that

$$\arg(z^{-1}) = -\arg z \quad , \quad \arg(zw) = \arg z + \arg w$$

— understood in the sense that the set of arguments of z^{-1} is made up of all negatives of arguments of z and that all sums of arguments of z and w constitute the set of arguments of zw. It is not always the case, by the way, that $\text{Arg}(z^{-1}) = -\text{Arg}\, z$ or that $\text{Arg}(zw) = \text{Arg}\, z + \text{Arg}\, w$. For instance, $\text{Arg}(-1/2) = \pi \neq -\text{Arg}(-2)$ and $\text{Arg}(-1) = \pi \neq \text{Arg}(-i) + \text{Arg}(-i)$.

A further implication of (1.5) and (1.6), in conjunction with a straightforward induction argument, is the relation

(1.7) $$(\cos\theta + i\sin\theta)^n = \cos(n\theta) + i\sin(n\theta) ,$$

valid for any real number θ and for any integer n. Identity (1.7) bears the name *de Moivre's formula* in honor of Abraham de Moivre (1667-1754), to whom its discovery is ascribed. An important consequence of de Moivre's formula, as we shall now see, is a simple procedure for finding roots of complex numbers.

Theorem 1.1. *Suppose that z is a non-zero complex number and that n is a positive integer. Then z has exactly n distinct complex n^{th}-roots. These roots are given in polar form by*

(1.8) $$\sqrt[n]{|z|}\left[\cos\left(\frac{\text{Arg}\, z + 2k\pi}{n}\right) + i\sin\left(\frac{\text{Arg}\, z + 2k\pi}{n}\right)\right]$$

for $k = 0, 1, \ldots, n-1$.

Proof. Write $z = |z|(\cos\theta + i\sin\theta)$, with $\theta = \text{Arg } z$. We seek all complex numbers $w = |w|(\cos\psi + i\sin\psi)$ for which it is true that $w^n = z$. An appeal to de Moivre's formula transforms this equation into

$$|w|^n [\cos(n\psi) + i\sin(n\psi)] = |z|(\cos\theta + i\sin\theta) ,$$

from which it is apparent that we are imposing on w the requirements $|w|^n = |z|$, $\cos(n\psi) = \cos\theta$, and $\sin(n\psi) = \sin\theta$. The first condition is met precisely when $|w| = \sqrt[n]{|z|}$, while the latter two can be satisfied if and only if $n\psi = \theta + 2k\pi$ for some integer k; i.e., $\psi = (\theta + 2k\pi)/n$ for some integer k. The choices $k = 0, 1, \ldots, n-1$ produce distinct n^{th}-roots of z. Any other choice for k merely leads to a duplicate of one of these roots. Accordingly, (1.8) describes the complete set of nth-roots of z. ∎

The geometric content of Theorem 1.1 is that the nth-roots of z are represented by a certain set of n points located on and equally spaced along the circle of radius $\sqrt[n]{|z|}$ centered at the origin of the complex plane. (The case $z = 4i$ and $n = 4$ is illustrated in Figure 8.) The nth-root of z

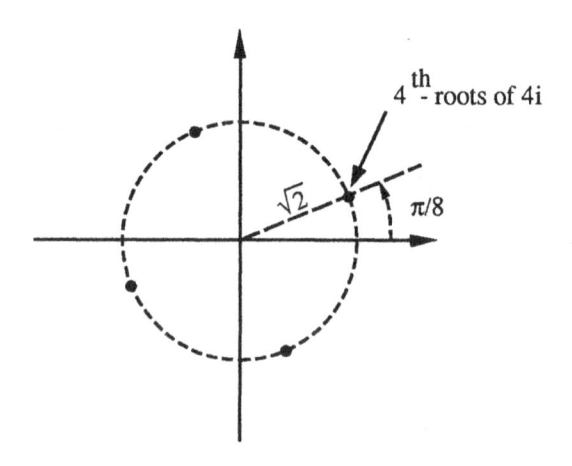

Figure 8.

obtained by taking $k = 0$ in (1.8) will be called the *principal n^{th}-root of z*. The notation $\sqrt[n]{z}$ is reserved for this distinguished root, a usage consistent with (1.8), where it is implicit that $\sqrt[n]{|z|}$ indicates the unique real n^{th}-root of $|z|$. We repeat for the sake of emphasis: *in this book $\sqrt[n]{z}$ always means the principal n^{th}-root of z*; i.e.,

$$\sqrt[n]{z} = \sqrt[n]{|z|}\left[\cos\left(\frac{\text{Arg } z}{n}\right) + i\sin\left(\frac{\text{Arg } z}{n}\right)\right] .$$

Technically speaking this formula does not apply to $z = 0$, since $\text{Arg } 0$ is undefined. The unique n^{th}-root of zero is, of course, zero itself. For this

1. The Algebra and Geometry of Complex Numbers

reason it is sensible to include zero in the above notation by agreeing that $\sqrt[n]{0} = 0$. As usual, we abbreviate $\sqrt[2]{z}$ to $\sqrt{z}$.

The following set of examples is intended to demonstrate techniques for working with the various concepts introduced above.

EXAMPLE 1.1. Verify the "triangle inequality": $|z + w| \leq |z| + |w|$.

We first use the properties listed in (1.1) and (1.2) to derive the identity

$$(1.9) \qquad |z + w|^2 = |z|^2 + 2\operatorname{Re}(z\overline{w}) + |w|^2 .$$

Direct computation produces

$$|z + w|^2 = (z + w)\overline{(z + w)} = (z + w)(\overline{z} + \overline{w}) = z\overline{z} + z\overline{w} + w\overline{z} + w\overline{w}$$

$$= z\overline{z} + z\overline{w} + \overline{z\overline{w}} + w\overline{w} = |z|^2 + 2\operatorname{Re}(z\overline{w}) + |w|^2 ,$$

and (1.9) follows. Next,

$$|z|^2 + 2\operatorname{Re}(z\overline{w}) + |w|^2 \leq |z|^2 + 2|\operatorname{Re}(z\overline{w})| + |w|^2 \leq |z|^2 + 2|z\overline{w}| + |w|^2$$

$$= |z|^2 + 2|z||w| + |w|^2 = (|z| + |w|)^2 .$$

Referring to (1.9) we conclude that

$$|z + w|^2 \leq (|z| + |w|)^2 .$$

Taking square roots on both sides results in the desired inequality.

EXAMPLE 1.2. Prove that $|z + w| \geq ||z| - |w||$.

The trick is to write $z = (z+w)-w$ and to apply the triangle inequality,

$$|z| = |(z + w) - w| \leq |z + w| + |-w| = |z + w| + |w| .$$

This leads to the inequality

$$|z + w| \geq |z| - |w| .$$

Repetition of the preceding computation with the roles of z and w interchanged gives

$$|z + w| \geq |w| - |z| .$$

Now $||z| - |w|| = |z| - |w|$ if $|z| \geq |w|$ and $||z| - |w|| = |w| - |z|$ if $|w| \geq |z|$. In all cases, therefore, we can assert that $|z + w| \geq ||z| - |w||$.

EXAMPLE 1.3. Describe geometrically the set S of complex numbers z that obey the condition $|z - 1| = 2|z + 1|$.

Write $z = x + iy$. Recalling (1.9), we remark that

$$z \in S \Leftrightarrow |z-1| = 2|z+1|$$
$$\Leftrightarrow |z-1|^2 = 4|z+1|^2$$
$$\Leftrightarrow |z|^2 - 2\operatorname{Re} z + 1 = 4|z|^2 + 8\operatorname{Re} z + 4$$
$$\Leftrightarrow 3|z|^2 + 10\operatorname{Re} z + 3 = 0$$
$$\Leftrightarrow 3(x^2 + y^2) + 10x + 3 = 0$$
$$\Leftrightarrow x^2 + \frac{10}{3}x + y^2 = -1$$
$$\Leftrightarrow x^2 + \frac{10}{3}x + \frac{25}{9} + y^2 = \frac{16}{9}$$
$$\Leftrightarrow \left(x + \frac{5}{3}\right)^2 + y^2 = \frac{16}{9}.$$

(N.B. The symbol $\Leftrightarrow$ stands for "if and only if" or "is equivalent to." The notation $\Rightarrow$ is read "implies" and $\Leftarrow$ means "is implied by.") The set S is now easily seen to be a circle of radius 4/3 centered at the point $-5/3$ on the real axis.

EXAMPLE 1.4. Find a "complex" equation for the hyperbola with "real" equation $x^2 - y^2 = 1$.

Make the substitutions $x = (z + \bar{z})/2$, $y = (z - \bar{z})/2i$. Then

$$x^2 - y^2 = \frac{(z+\bar{z})^2}{4} + \frac{(z-\bar{z})^2}{4} = \frac{z^2 + \bar{z}^2}{2}.$$

It follows that the given hyperbola is described by the complex equation $z^2 + \bar{z}^2 = 2$.

EXAMPLE 1.5. Determine all solutions of the equation $z^4 + 16 = 0$.

It is required to find all 4th-roots of -16. Noting that $\sqrt[4]{16} = 2$ and that $\operatorname{Arg}(-16) = \pi$, we invoke Theorem 1.1 to identify these roots:

$$2[\cos(\pi/4) + i\sin(\pi/4)] \quad , \quad 2[\cos(3\pi/4) + i\sin(3\pi/4)],$$

$$2[\cos(5\pi/4) + i\sin(5\pi/4)] \quad , \quad 2[\cos(7\pi/4) + i\sin(7\pi/4)].$$

After evaluation of the assorted trigonometric expressions, this list reduces to $\sqrt{2} + i\sqrt{2}$, $-\sqrt{2} + i\sqrt{2}$, $-\sqrt{2} - i\sqrt{2}$, and $\sqrt{2} - i\sqrt{2}$.

EXAMPLE 1.6. Solve the quadratic equation $z^2 - 3z + 3 + i = 0$.

2. Exponentials and Logarithms of Complex Numbers 13

The quadratic formula works just as efficiently in solving quadratic equations with complex coefficients as it does in solving the corresponding equations with real coefficients. (See Exercise 4.13.) As a consequence, the desired roots are given by

$$z = \frac{3 \pm \sqrt{9 - 4(1)(3+i)}}{2} = \frac{3 \pm \sqrt{-3-4i}}{2}.$$

To compute $\sqrt{-3-4i}$, set $\theta = \text{Arg}(-3-4i)$. Then $-\pi < \theta < -\pi/2$, $\cos\theta = -3/5$, and $\sin\theta = -4/5$. In view of Theorem 1.1 and the definition of the principal square root,

$$\sqrt{-3-4i} = \sqrt{5}\left[\cos(\theta/2) + i\sin(\theta/2)\right].$$

Appealing to the half-angle formulas from trigonometry — note that $\cos(\theta/2) > 0$ and $\sin(\theta/2) < 0$ — we compute

$$\cos(\theta/2) = \sqrt{\frac{1+\cos\theta}{2}} = \sqrt{\frac{1-(3/5)}{2}} = \frac{1}{\sqrt{5}}$$

and

$$\sin(\theta/2) = -\sqrt{\frac{1-\cos\theta}{2}} = -\sqrt{\frac{1+(3/5)}{2}} = -\frac{2}{\sqrt{5}},$$

whence $\sqrt{-3-4i} = 1 - 2i$. The solutions of the given equation are, therefore, $2 - i$ and $1 + i$.

2 Exponentials and Logarithms of Complex Numbers

2.1 Raising e to Complex Powers

It is our goal in this section to define the quantity e^z for complex z and to derive its elementary properties. Recognizing that from a strictly rigorous outlook the introduction of this notion here might be judged premature, we feel that this point is far outweighed by the enormous convenience of having complex exponentials at our disposal early on. There is, after all, no great problem in deciding what e^z ought to mean, as we now discover.

Recall from calculus the Taylor series expansion of e^t for real t,

$$e^t = 1 + t + \frac{t^2}{2!} + \frac{t^3}{3!} + \frac{t^4}{4!} + \cdots.$$

If we substitute iy for t in this series and compute formally, fighting off any qualms about the precise meaning of convergence, we arrive at

$$e^{iy} = 1 + iy - \frac{y^2}{2!} - i\frac{y^3}{3!} + \frac{y^4}{4!} + \cdots$$

$$= \left(1 - \frac{y^2}{2!} + \frac{y^4}{4!} - \frac{y^6}{6!} + \cdots\right) + i\left(y - \frac{y^3}{3!} + \frac{y^5}{5!} - \frac{y^7}{7!} + \cdots\right).$$

The two series in the last line should again evoke memories from calculus — they are Taylor expansions of $\cos y$ and $\sin y$, respectively. In other words, the suspicion arises that $e^{iy} = \cos y + i \sin y$ represents the proper interpretation of e^{iy}. Next, since $e^{s+t} = e^s e^t$ for real numbers s and t, in an attempt to assign meaning to the quantity e^{x+iy} it is difficult to avoid the suggestion that $e^{x+iy} = e^x e^{iy}$. In fact, motivated by the preceding considerations, we now actually take the step of defining e^z for $z = x + iy$ via the formula

(1.10) $$e^z = e^x(\cos y + i \sin y).$$

The notation $\exp(z)$ is a frequently employed substitute for e^z, used especially in situations where z is replaced by a more complicated expression. For instance, we might write $\exp[(z+1)/(z^2+4)]$ in preference to the corresponding expression involving e. Here are a few sample computations of e^z:

$$e^0 = 1, \quad e^{\pi i} = -1, \quad e^{1+(\pi i/2)} = ei, \quad e^i = \cos 1 + i \sin 1.$$

The definition (1.10) makes it obvious that $e^z \neq 0$ holds for every complex number z and that

(1.11) $$|e^z| = e^{\operatorname{Re} z}, \quad \arg(e^z) = \operatorname{Im} z.$$

Furthermore, de Moivre's formula implies directly that

(1.12) $$(e^z)^n = e^{nz}$$

for every integer n. In particular, $(e^z)^{-1} = e^{-z}$.

Let $z = x + iy$ and $w = u + iv$. Bearing in mind formula (1.6), we calculate

$$e^z e^w = e^x(\cos y + i \sin y)\, e^u(\cos v + i \sin v)$$
$$= e^{x+u}[\cos(y+v) + i \sin(y+v)] = e^{z+w}$$

and so verify a fundamental law of exponents:

(1.13) $$e^z e^w = e^{z+w}.$$

For which values of z is it the case that $e^z = 1$? Certainly $z = 0$ has this property, but many other complex numbers share it. For $e^z = 1$ to hold, it is essential first of all that $e^{\operatorname{Re} z} = |e^z| = 1$. It follows that $\operatorname{Re} z = 0$, so only purely imaginary numbers $z = iy$ fall under consideration. But $e^{iy} = \cos y + i \sin y = 1$ is true if and only if $\cos y = 1$ and $\sin y = 0$, conditions

2. Exponentials and Logarithms of Complex Numbers

satisfied precisely when $y = 2k\pi$ for some integer k. The conclusion: $e^z = 1$ if and only if $z = 2k\pi i$ for some integer k. More generally, since by (1.12) and (1.13) the equation $e^z = e^w$ is equivalent to $e^{w-z} = 1$, we remark that

(1.14) $\qquad e^z = e^w \Leftrightarrow w = z + 2k\pi i$ for some integer k .

We have earlier represented a non-zero complex number z in the polar form
$$z = r(\cos\theta + i\sin\theta) ,$$
where $r = |z|$ and θ is any argument of z. This polar description of z can now be expressed more economically as

(1.15) $$z = re^{i\theta} .$$

In most situations (1.15) will henceforth be our preferred mode of polar representation. Notice that if we set $r = 0$ in (1.15) we obtain what amounts to a polar representation of $z = 0$ — θ can be chosen arbitrarily — even though we assign no argument to 0.

2.2 Logarithms of Complex Numbers

What does it mean to state that $s = \ln t$, the (natural or base e) logarithm of the positive real number t? The answer confided to us by our high school math teachers is, of course, that $e^s = t$. We would like to mimic this idea and say what it means for a complex number w to be the logarithm of a non-zero complex number z. The analogous requirement would be that $e^w = z$ and, indeed, any w enjoying this property is pronounced a *logarithm* of z. The difference between the real and complex situations is that, whereas $e^s = t$ is satisfied for a unique real value of s, there are infinitely many complex values of w for which $e^w = z$. For example, $e^w = 1$ is satisfied by $w = 2k\pi i$ for $k = 0, \pm 1, \cdots$. Each such w is to qualify as a complex logarithm of 1! As happened with arguments, one of the logarithms of a given non-zero complex number z will be singled out for distinction, that being $w = \ln|z| + i\operatorname{Arg} z$. Owing to (1.13) and (1.15), the calculation
$$e^w = e^{\ln|z| + i\operatorname{Arg} z} = e^{\ln|z|}e^{i\operatorname{Arg} z} = |z|e^{i\operatorname{Arg} z} = z$$
confirms that w definitely is a logarithm of z. It is termed the *principal logarithm* of z and is denoted by $\operatorname{Log} z$. (Our use of the notation $\log z$, on the other hand, parallels the usage we've established for $\arg z$; i.e., $\log z$ indicates the set of all logarithms of z. To write $\log z = w$ merely signifies that w is one of the possible logarithms of z.) In order to reinforce it, we restate this notational convention: *throughout this text* $\operatorname{Log} z$ *designates the principal logarithm of* z, *the one defined by*

$$\operatorname{Log} z = \ln|z| + i\operatorname{Arg} z .$$

This definition insures that $\text{Log } x = \ln x$ for a positive real x. (From now on we shall write $\text{Log } x$ instead of $\ln x$ when $x > 0$.) As examples, we record

$$\text{Log}(-1) = \pi i \quad , \quad \text{Log}(ei) = 1 + (\pi/2)i \quad , \quad \text{Log}(1-i) = \text{Log } \sqrt{2} - (\pi/4)i \ .$$

Because an arbitrary logarithm w of z satisfies $e^w = z = e^{\text{Log } z}$, it is a consequence of (1.14) that w acquires the form

$$w = \text{Log } z + 2k\pi i = \text{Log } |z| + i(\text{Arg } z + 2k\pi)$$

for some integer k. Furthermore, any complex number of this type is a logarithm of z. In other words, the complete set $\log z$ of logarithms of z is made up of all complex numbers $\text{Log } |z| + i\theta$, where θ ranges over the set of arguments of z. For $z = -1$, to give one example, we have $\log(-1) = \{(2k+1)\pi i : k = 0, \pm 1, \cdots\}$.

If w_1 is a logarithm of z_1 and w_2 a logarithm of z_2, then

$$e^{w_1 + w_2} = e^{w_1} e^{w_2} = z_1 z_2 \ ,$$

which demonstrates that $w_1 + w_2$ is a logarithm of $z_1 z_2$. This fact leads to the statement

$$\log(z_1 z_2) = \log z_1 + \log z_2 \ ,$$

interpreted as follows: the set of logarithms of $z_1 z_2$ consists of all sums of logarithms of z_1 with logarithms of z_2. It is not, in general, to be expected that $\text{Log}(z_1 z_2) = \text{Log } z_1 + \text{Log } z_2$. (For instance, $\text{Log}(-i) = (-\pi/2)i \neq (3\pi/2)i = \text{Log}(-1) + \text{Log } i$.) Certainly $\text{Log } z_1 + \text{Log } z_2$ is a logarithm of $z_1 z_2$, just not necessarily the principal one.

2.3 Raising Complex Numbers to Complex Powers

One standard way to define t^a, where t is a positive real number and a is an arbitrary real number, is by means of the formula $t^a = e^{a \, \text{Log } t}$. If we attempt to carry this formula over to the complex setting with a mind toward raising a non-zero complex number z to a complex power λ, we immediately encounter an apparent obstacle: z has many different logarithms. Which one should we use? The answer: all of them! Just as we must accept the fact that z has an infinite number of logarithms, we must be prepared for there to be many different ways — but not always infinitely many, as it turns out — to raise z to the power λ. To be precise, for any logarithm w of z the complex number $e^{\lambda w}$ is called *the λ-power of z associated with w*. The choice $w = \text{Log } z$ gives rise to the *principal λ-power* of z. The familiar notation z^λ will see its use restricted to this special λ-power of z. To repeat, lest there be any confusion later on: z^λ *will be employed in this book exclusively to symbolize the principal λ-power of z, the one given by*

$$z^\lambda = e^{\lambda \, \text{Log } z} \ .$$

3. Functions of a Complex Variable

Here are two simple examples:
$$i^{2\pi} = e^{2\pi \operatorname{Log} i} = e^{\pi^2 i} = \cos(\pi^2) + i\sin(\pi^2),$$

$$i^{2\pi i} = e^{2\pi i \operatorname{Log} i} = e^{-\pi^2}.$$

Since an arbitrary logarithm of z has the structure $\operatorname{Log} z + 2k\pi i$ for a suitable integer k, it follows easily that the general λ-power of z inherits the form $e^{2k\pi\lambda i} z^\lambda$, where k is an integer. (By way of illustration, if $z = i$ and $\lambda = 2\pi i$, then $z^\lambda = e^{-\pi^2}$ and $e^{2k\pi\lambda i} = e^{-4k\pi^2}$, so the collection of λ-powers of z in this instance is seen to consist of the numbers $e^{-(4k+1)\pi^2}$ for $k = 0, \pm 1, \cdots$.) There are two cases meriting special commentary. Should λ itself be an integer, say $\lambda = n$, then quite clearly $e^{2k\pi\lambda i} = 1$ whenever k is an integer, in which event the sundry λ-powers of z all reduce to z^n, the "usual" n^{th}-power of z. Secondly, if $\lambda = 1/n$ for a positive integer n, we observe that

$$e^{2k\pi\lambda i} z^\lambda = \exp\left(\frac{2k\pi i}{n}\right) \exp\left(\frac{\operatorname{Log} z}{n}\right)$$

$$= \exp\left[\frac{\operatorname{Log}|z|}{n} + i\left(\frac{\operatorname{Arg} z + 2k\pi}{n}\right)\right]$$

$$= \exp\left(\operatorname{Log} \sqrt[n]{|z|}\right) \exp\left[i\left(\frac{\operatorname{Arg} z + 2k\pi}{n}\right)\right]$$

$$= \sqrt[n]{|z|} \exp\left[i\left(\frac{\operatorname{Arg} z + 2k\pi}{n}\right)\right]$$

and conclude, as might be anticipated, that the set of λ-powers of z coincides with the set of nth-roots of z listed in Theorem 1.1. In particular, $z^{1/n} = \sqrt[n]{z}$.

Although it is true that $z^{\lambda+\mu} = z^\lambda z^\mu$, other "laws of exponents" have only limited validity for principal powers. To single out one such law, $(zw)^\lambda$ is not generally the same as $z^\lambda w^\lambda$; e.g., $(-i)^{1/2} = (\sqrt{2} - i\sqrt{2})/2$, which is different from $(-1)^{1/2} i^{1/2} = (-\sqrt{2} + i\sqrt{2})/2$. Similarly, $(z^\lambda)^\mu$ can differ from $z^{\lambda\mu}$. As a rule it is advisable to exercise due caution in carrying out algebraic manipulations that involve complex exponents.

3 Functions of a Complex Variable

3.1 Complex Functions

Having reviewed the algebra of complex numbers and established some convenient notation and terminology, we conclude this chapter by casting a preliminary glance at the ultimate objects of study in complex analysis, complex-valued functions of a complex variable. (In truth, it is only a special class of these functions that comes under scrutiny in the subject, but this fact need not concern us for the time being.) To begin, we recall some pertinent background material concerning functions in general.

If A and B are sets, a *function f from A to B* is a rule of correspondence that assigns to each element x of A an element $f(x)$ of B, the *value of f at x*. The notation $f : A \to B$ is used to indicate that f is such a function. We refer to the set A as the *domain-set* of f and to B as its *target-set*. For a subset S of A the *image of S under f*, denoted $f(S)$, is the set of values that f attains on S; i.e., $f(S) = \{f(x) : x \in S\}$. The particular set $f(A)$ is called the *range of f*. For the most part the functions of interest in this text are those of the kind $f : A \to \mathbb{C}$, where A is a subset of $\mathbb{C}$. In fact, we go so far as to lay down the following guidelines: *in the absence of any statement to the contrary, $f : A \to \mathbb{C}$ will always signify a function whose domain-set A is a subset of $\mathbb{C}$; unless otherwise stipulated, any function referred to in the text is presumed to be of this type.* (The chief situation covered by "otherwise stipulated" occurs in Chapter VIII, where functions involving the "extended complex plane" are discussed.) Frequently such a function will be defined by giving a formula that prescribes its values, something like $f(z) = z^2$ or $f(z) = e^z$, with no specification made regarding its domain-set. When this is the case, it will be our normal convention to take as the domain-set of the function in question the set of all complex numbers for which the defining formula "makes sense." For example, $f(z) = (z^2 - 1)/(z^2 + 1)$ represents a function f with domain-set $A = \{z : z \neq \pm i\} = \mathbb{C} \sim \{\pm i\}$, since the formula defining f only ceases to be meaningful at complex numbers where its denominator vanishes. (N.B. The notation $S \sim T$ is used for $\{z : z \in S, z \notin T\}$.) By the same convention, $f(z) = \operatorname{Log} z$ determines a function whose domain-set is $A = \mathbb{C} \sim \{0\}$.

The class of *rational functions of z* is going to be a prominent source of examples for this book. We are thinking here of functions of the type

$$f(z) = \frac{a_0 + a_1 z + \cdots + a_n z^n}{b_0 + b_1 z + \cdots + b_m z^m},$$

where the coefficients $a_0, a_1, \ldots, a_n$ and $b_0, b_1, \ldots, b_m$ are complex numbers. Included in this class are the *polynomial functions of z*, the rational functions having the special form

(1.16) $$f(z) = a_0 + a_1 z + \cdots + a_n z^n .$$

(Assuming that $a_n \neq 0$ in (1.16), we say that f is a polynomial function of *degree n*.) Sometimes we shall also want to draw examples from a broader class of rational functions, the *rational functions of x and y*. This class consists of *polynomial functions of x and y* — the value of such a function at $z = x + iy$ is given by an expression of the sort

$$f(z) = a_{00} + a_{10} x + a_{01} y + \cdots + a_{mn} x^m y^n ,$$

3. Functions of a Complex Variable

again with complex coefficients $a_{00}, a_{10}, \ldots, a_{mn}$ — and their quotients. Every rational function of z is clearly a rational function of x and y, but not the other way around. By our convention polynomials are defined everywhere in $\mathbb{C}$, whereas to obtain the domain-set of a general rational function we must remove from $\mathbb{C}$ all points at which its denominator takes the value zero.

It is occasionally instructive to look at a function $f: A \to \mathbb{C}$ in terms of its real and imaginary parts; i.e., to represent f in the form $f = u + iv$, where $u(z) = \text{Re}[f(z)]$ and $v(z) = \text{Im}[f(z)]$. The functions u and v are, naturally, real-valued functions on A. It is even possible to carry this a step further, by identifying $z = x + iy$ with the ordered pair (x, y) of real numbers and by writing $u(z) = u(x, y)$ and $v(z) = v(x, y)$. In other words, we can regard u and v as real-valued functions of the two real variables x and y. Applied to the function $f(z) = z^2$ this procedure yields

$$u(z) = u(x, y) = x^2 - y^2 \quad, \quad v(z) = v(x, y) = 2xy.$$

In case $f(z) = e^z$ the corresponding functions are

$$u(z) = u(x, y) = e^x \cos y \quad, \quad v(z) = v(x, y) = e^x \sin y.$$

3.2 Combining Functions

The elementary algebra of complex-valued functions of a complex variable does not differ radically from that of real-valued functions of a real variable. Presented with a pair of functions $f: A \to \mathbb{C}$ and $g: B \to \mathbb{C}$, we can multiply f by a complex number c to obtain the function cf, add f and g to form $f + g$, multiply these functions to produce fg, and take their quotient f/g. Both $f + g$ and fg have domain-set $A \cap B$, and the domain-set of f/g is $\{z \in A \cap B : g(z) \neq 0\}$, provided these sets are non-empty. Two other useful functions associated with a given function $f = u + iv$ are its modulus $|f| = (u^2 + v^2)^{1/2}$ and its conjugate $\overline{f} = u - iv$.

A more interesting way of combining functions $f: A \to \mathbb{C}$ and $g: B \to \mathbb{C}$ to manufacture a new function arises when the range of f is contained in B. In this situation it is possible to form a function $g \circ f: A \to \mathbb{C}$, the *composition* of g with f, by setting $g \circ f(z) = g[f(z)]$. In the case of the functions $f(z) = z^2$ and $g(z) = e^z$, for instance, we have $g \circ f(z) = e^{z^2}$ and $f \circ g(z) = e^{2z}$.

An idea of major importance associated with the composition of functions is that of an "inverse function." Suppose that a function $f: A \to \mathbb{C}$ is *one-to-one* (or, to use a synonym favored by complex analysts, *univalent*), which means that $f(z_1) = f(z_2)$ only when $z_1 = z_2$. Under this assumption we can define a function $g: f(A) \to A$ as follows: for w in $f(A)$, $g(w)$ is the unique element of A such that $f[g(w)] = w$. Thus, to find $z = g(w)$ we must solve the equation $f(z) = w$ for z in terms of w. The function g is called the *inverse* of f and is usually denoted by f^{-1}. (Care must be exercised lest the inverse f^{-1} of f be confused with the reciprocal of f, that is, with $1/f$. The context will ordinarily rule out any possible ambiguity along these lines.) Setting $B = f(A)$, one readily checks that $f^{-1}(B) = A$, that $f^{-1} \circ f(z) = z$ for all z in A, and that $f \circ f^{-1}(w) = w$ for all w in B. Consider, as an illustration, the exponential function $f(z) = e^z$. This function fails to be univalent on its full domain-set $\mathbb{C}$, and, consequently, it does not have an inverse function. If, however, we modify the exponential function by suitably restricting its domain-set, we do arrive at functions that admit inverses. (This sounds a theme that we shall take up in earnest in Chapter III under the heading "branches of the logarithm.") For example, it follows from (1.14) that the function $f: A \to \mathbb{C}$, where $A = \{z: -\pi < \mathrm{Im}\, z \leq \pi\}$ and $f(z) = e^z$, is univalent, and it is easy to see that $B = f(A) = \mathbb{C} \sim \{0\}$. We conclude that f possesses an inverse function $f^{-1}: B \to A$. In order to compute $z = f^{-1}(w)$ for w in B we must find the solution z of $e^z = w$ belonging to A, a task quickly accomplished: $z = \mathrm{Log}\, w$, the principal logarithm of w. As a result, $f^{-1}(w) = \mathrm{Log}\, w$ gives us a formula for the inverse of f. Of course, once the inverse of a function f has been determined, there is no real need to use different letters for the variables in f and f^{-1}, as long as the domain-set of each function is kept in mind. We are thus free to write $f^{-1}(z) = \mathrm{Log}\, z$ in the present situation without fear of confusion.

3.3 Functions as Mappings

We have yet to mention a view of complex functions that occupies a central place in the conception of this book, a geometric view. From its perspective attention focuses on the role of a complex-valued function f of a complex variable as a mapping (or transformation) of the complex plane. In practice one often considers two separate copies of the complex plane — typically labeled the z-plane and the w-plane — and studies the manner in which geometric entities (lines, circles, etc.) in the z-plane are transformed into the w-plane under the correspondence $w = f(z)$. Such input can often contribute appreciably to one's understanding of a function. The following examples are meant to dramatize the geometric aspect of complex functions.

EXAMPLE 3.1. Discuss the geometric effect on the complex plane of the transformation $f(z) = az + b$, where $a \neq 0$ and b are complex numbers.

Writing $a = |a|e^{i\theta}$ with $\theta = \mathrm{Arg}\, a$, we realize f as the composition $f = k \circ h \circ g$ of three extremely simple transformations:

$$g(z) = e^{i\theta} z \quad , \quad h(z) = |a|z \quad , \quad k(z) = z + b \, .$$

Based on our geometric understanding of the algebraic operations in $\mathbb{C}$, the action on the complex plane of each of the individual components of f is

3. Functions of a Complex Variable

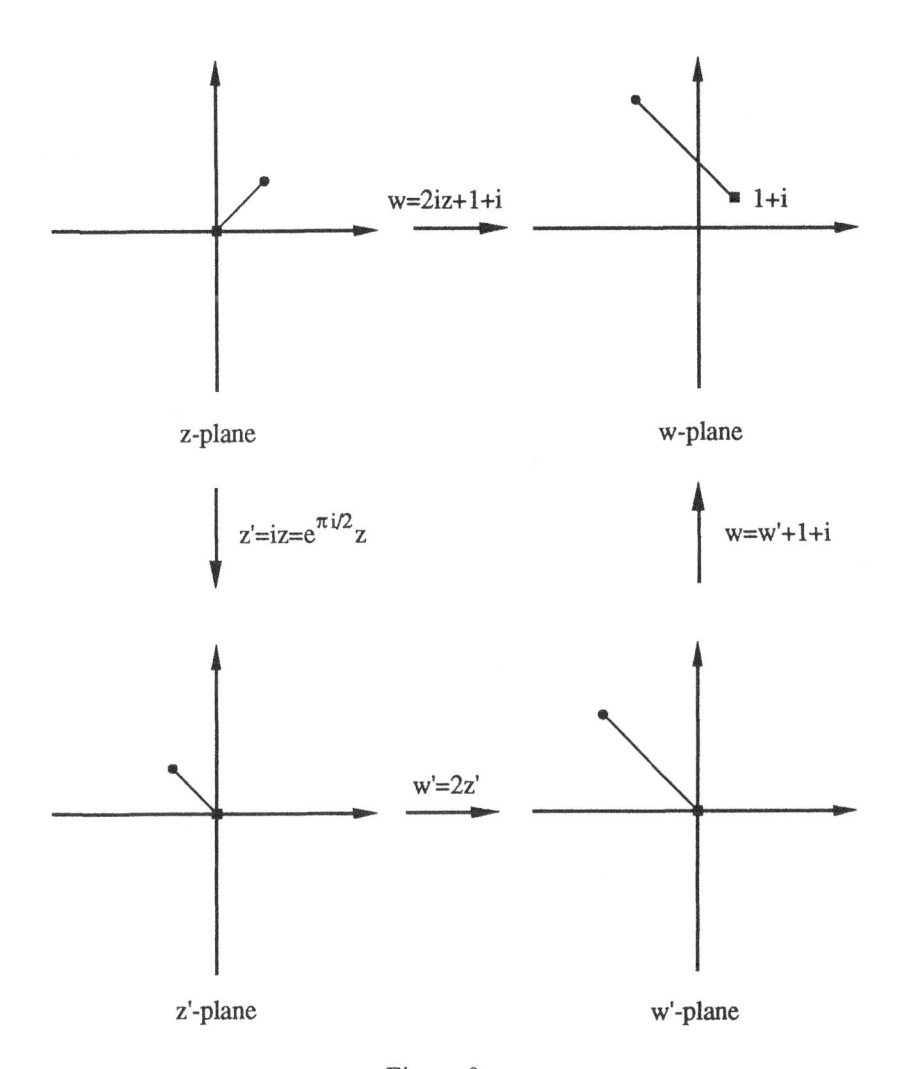

Figure 9.

very easy to describe. The geometric effect of f is just the cumulative effect produced when the transformations g, h, and k operate in sequence, as follows. First, g rotates the plane about the origin through the angle θ, with positive θ giving rise to a rotation in the counter-clockwise direction and negative θ to a clockwise rotation; next, the mapping h dilates (= stretches or shrinks) the plane so that each ray issuing from the origin is mapped to itself and all distances get multiplied by the factor $|a|$; finally, k causes each point of the plane to be translated a distance $|b|$ in the direction that the vector representing b points. (Figure 9 traces the successive images of a particular line segment under the elementary transformations that compose to give $f(z) = 2iz + 1 + i$.) Transformations of the type treated in this example are called *sense-preserving similarity transformations*. Such transformations map any geometric figure in the complex plane — a triangle, for instance — onto a figure similar, in the strict geometric meaning of the word, to the given one.

EXAMPLE 3.2. Let n be a positive integer. Discuss the behavior of the mapping $w = z^n$ on circles centered at the origin and on rays emanating from the origin.

Observe that $w = 0$ if and only if $z = 0$. For $z \neq 0$ we express z in the polar form $z = |z|e^{\text{Arg } z}$. Then $w = z^n$ has a polar representation $w = |z|^n e^{n \text{ Arg } z}$, which permits us to write $|w| = |z|^n$ and $\arg w = n \text{ Arg } z$. (But not necessarily $\text{Arg } w = n \text{ Arg } z$!) It follows that for $r > 0$ the circle $K = \{z : |z| = r\}$ is transformed to the circle $K^* = \{w : |w| = r^n\}$. As a matter of fact, in view of Theorem 1.1 there are exactly n points on K that are carried under $w = z^n$ to any given point of K^*. In effect, this transformation takes K and "wraps" it n times around K^*. For $-\pi < \theta \leq \pi$ the ray $R = \{re^{i\theta} : r \geq 0\}$ is transformed to the ray $R^* = \{se^{in\theta} : s \geq 0\}$, this in a one-to-one fashion. Two rays forming an angle ψ at the origin of the z-plane have as their images a pair of rays meeting at the origin of the w-plane at an angle $n\psi$. The geometry of this mapping is portrayed in Figure 10.

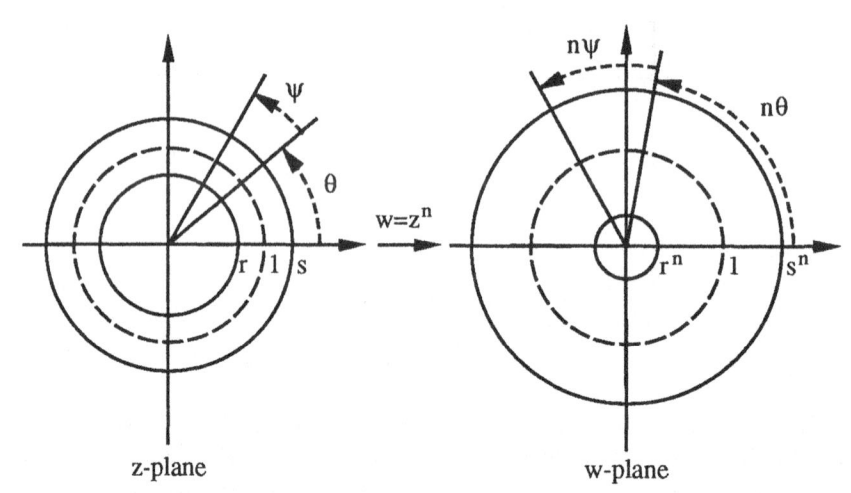

Figure 10.

EXAMPLE 3.3. Describe the image of an arbitrary line L in the complex plane under the function $f(z) = e^z$.

We first deal with the case of a vertical line $L = \{z : \text{Re } z = x_0\}$. A

3. Functions of a Complex Variable

typical point z of L has the form $z = x_0 + iy$; its image $w = f(z)$ is given by $w = e^{x_0+iy} = e^{x_0}e^{iy}$. Accordingly, $|w| = e^{x_0}$, implying that $f(L)$ is contained in the circle K of radius e^{x_0} centered at the origin of the w-plane. Due to the relationship $\arg w = y$, w is seen to visit every point of K as z varies over any half-open interval of length 2π on L. The image of L is, as a consequence, the complete circle K, each point of which has infinitely many pre-images on L.

The general non-vertical L line in the complex plane is described by another linear equation, $y = mx + b$. A point z of L has the representation $z = x + i(mx + b)$, with the result that $w = e^z = e^x e^{i(mx+b)}$. In particular, w has modulus e^x. We can also specify an argument $\theta(w)$ of w by taking $\theta(w) = mx + b$. When $m = 0$ — meaning that L is a horizontal line — we see that w traverses the open ray $R = \{re^{ib} : r > 0\}$ as z ranges over L. If $m > 0$, both $|w|$ and $\theta(w)$ increase steadily — $|w|$ covering the interval $(0, \infty)$ and $\theta(w)$ the interval $(-\infty, \infty)$ — as x increases from $-\infty$ to ∞. The image produced is a curve that winds infinitely often around the origin and is known as a *logarithmic spiral*. The spiral in question here is the one with polar equation $r = e^{(\theta-b)/m}$. When $m < 0$ the image of L is another logarithmic spiral. (See Figure 11.)

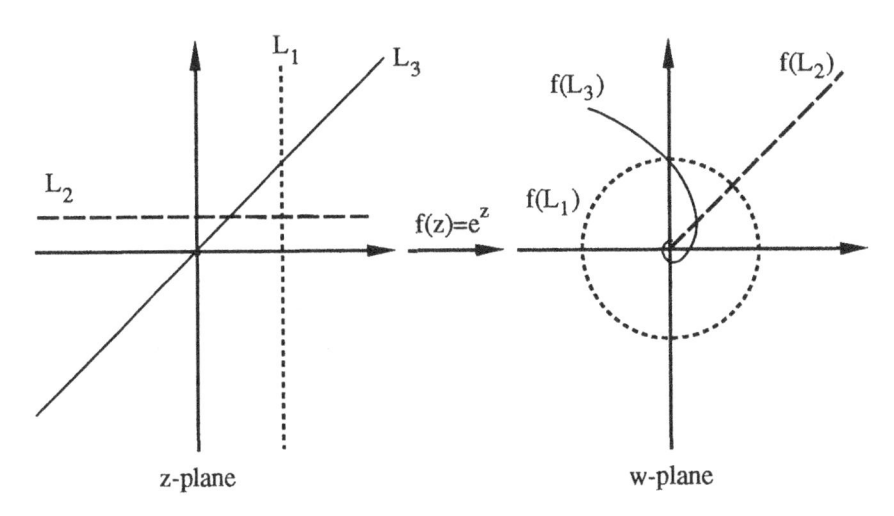

Figure 11.

From the preceding discussion we distill Figure 12, which is recorded for future reference. The mapping depicted is again $w = e^z$, but with z restricted to satisfy $-\pi < \operatorname{Im} z \le \pi$. In diagrams of this kind the use of the same symbol to mark points in the z-plane and w-plane serves to indicate that such points correspond to one another under a mapping and its inverse. In this figure, for instance, the origin of the z-plane corresponds to the point 1 in the w-plane.

24 I. The Complex Number System

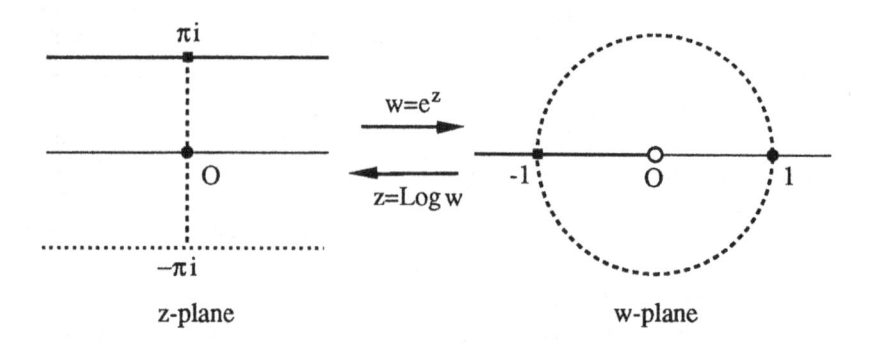

Figure 12.

EXAMPLE 3.4. Show that the function $f(z) = (1-z)/(1+z)$ maps the disk $D = \{z \colon |z| < 1\}$ onto the half-plane $H = \{w \colon \operatorname{Re} w > 0\}$.

Set $w = f(z)$ and rewrite w in the following manner: for $z = x + iy$,

$$w = \frac{1-z}{1+z} = \frac{1-z}{1+z} \cdot \frac{1+\bar{z}}{1+\bar{z}} = \frac{1-z+\bar{z}-z\bar{z}}{|1+z|^2} = \frac{(1-|z|^2) - 2yi}{|1+z|^2}.$$

From this we read off the relationship

$$\operatorname{Re} w = \frac{1-|z|^2}{|1+z|^2},$$

which makes it evident that w belongs to H for z in D and for no other z. We infer that $f(D)$ is contained in H. Does every point of H belong to $f(D)$? To answer this, fix w in H and solve the equation $w = (1-z)/(1+z)$ for z in terms of w. The solution is $z = (1-w)/(1+w)$. By construction $w = f(z)$, so the above remark shows that z lies in D. Therefore, $f(D) = H$. (See Figure 13.)

The mapping in Example 3.4 harbors several other noteworthy properties of which we will later have occasion to take advantage. The first is that f is a univalent function whose domain-set $A = \mathbb{C} \sim \{-1\}$ coincides with its range and whose inverse is f itself. This is so because the equation $w = (1-z)/(1+z)$ can be solved for z in terms of w precisely when $w \neq -1$, in which event the unique solution z is given by $z = (1-w)/(1+w) = f(w)$. Considerations similar to those in Example 3.4 reveal that the image under f of the unit circle $K = \{z \colon |z| = 1\}$ — or, more accurately, of the "punctured" circle $K \sim \{-1\}$ — is the imaginary axis in the w-plane. Since $f = f^{-1}$, the imaginary axis in the z-plane must then transform to the unit circle, less the point $w = -1$, in the w-plane. (There is a standard way to "fill in" the "punctures" occurring here, that being to adjoin an

4. Exercises for Chapter I

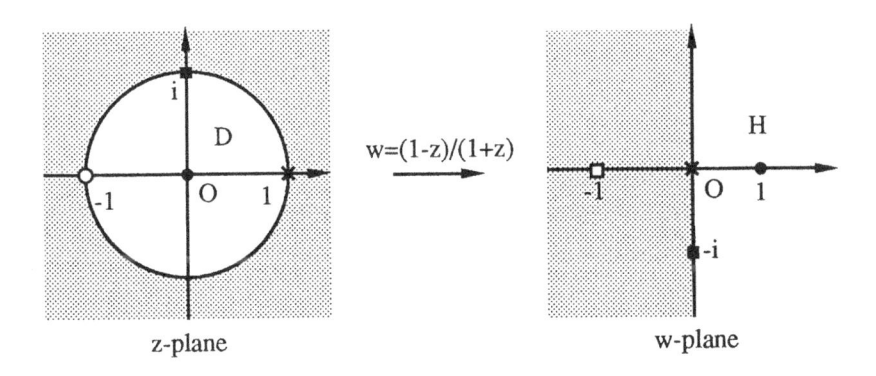

Figure 13.

ideal "point at infinity" to the complex plane and then to regard f as a function on the so-called *extended complex plane* $\widehat{\mathbb{C}} = \mathbb{C} \cup \{\infty\}$ satisfying $f(-1) = \infty$ and $f(\infty) = -1$. So regarded, f maps the full circle K onto the extended imaginary axis, meaning the imaginary axis together with the point ∞, and vice versa. These matters will be discussed more carefully in Chapter VIII.) Under f the set $\{z: |z| > 1\}$ is sent to the left half-plane $\{w: \operatorname{Re} w < 0\}$ with the point -1 removed. The punctured real axis $\mathbb{R} \sim \{-1\}$ is mapped by f to itself, the intervals $(-\infty, -1)$ and $(-1, \infty)$ each going to itself. The upper half-plane $\{z: \operatorname{Im} z > 0\}$ and lower half-plane $\{z: \operatorname{Im} z < 0\}$ are interchanged by f; i.e., they are carried to $\{w: \operatorname{Im} w < 0\}$ and to $\{w: \operatorname{Im} w > 0\}$, respectively.

The function $f(z) = (1-z)/(1+z)$ is just one member of a very important class of complex functions, the class of *Möbius transformations* (also called *linear fractional transformations*). This class consists of all functions of the type $f(z) = (az+b)/(cz+d)$, where $a, b, c,$ and d are complex numbers for which $ad - bc \neq 0$. Such functions will be subjected to a systematic examination in Chapter IX. Notice that every similarity transformation $f(z) = az + b$ qualifies as a Möbius transformation, one for which $c = 0$ and $d = 1$.

4 Exercises for Chapter I

4.1 Exercises for Section I.1

4.1. If $z = 1 + 2i$ and $w = 3 + 4i$, express the following in the form $x + iy$:
(i) $3z + iw$; (ii) $2z^2 - z\overline{w}$; (iii) $2|w| + (1-i)z^2$; (iv) $(w+z)/(w-z)$;
(v) $(1-iz)/(1+iz)$; (vi) $(z + 5z^{-1})^{-1}$; (vii) $\operatorname{Im}(\overline{z}w^2) + 25i\operatorname{Re}(zw^{-1})$;

(viii) $5\cos[\text{Arg}(z^2)] + 5i\sin[\text{Arg } w]$.

4.2. Verify that $\overline{z+w} = \overline{z}+\overline{w}$, $\overline{zw} = \overline{z}\,\overline{w}$, and $|zw| = |z||w|$ for all complex numbers z and w. Assuming that $w \neq 0$, show also that $\overline{(z/w)} = \overline{z}/\overline{w}$ and $|z/w| = |z|/|w|$.

4.3. Confirm that the identity $1+z+\cdots+z^n = (1-z^{n+1})/(1-z)$ holds for every non-negative integer n and every complex number z, save for $z = 1$.

4.4. Establish the so-called "parallelogram law" for complex numbers z and w: $|z+w|^2 + |z-w|^2 = 2|z|^2 + 2|w|^2$.

4.5. Given that $z \neq 0$ and $w \neq 0$, demonstrate that $|z+w| = |z|+|w|$ is true if and only if $w = tz$ for some $t > 0$.

4.6. Provide geometric descriptions for the following subsets of $\mathbb{C}$:
(i) $\{z : |z-1| = |z-i|\}$; (ii) $\{z : |z-1| = 2|z-i|\}$; (iii) $\{z : |z-1| = x\}$;
(iv) $\{z : (1+i)z + (1-i)\overline{z} = 1\}$; (v) $\{z : z\overline{z} + iz - i\overline{z} - 3 = 0\}$;
(vi) $\{z : |z-i| + |z+i| = 4\}$; (vii) $\{z : |z-i|^2 + |z+i|^2 = 4\}$;
(viii) $\{z : 2z^2 - \overline{z}^2 = 1\}$; (ix) $\{z : z^2 - \overline{z}^2 = i\}$. In each instance, sketch a rough graph of the set.

4.7. Depict each of the following sets in the complex plane graphically:
(i) $\{z : z = 0 \text{ or } \text{Arg } z = \pi/4\}$; (ii) $\{z : |\text{Arg } z - \text{Arg } i| < \pi/6\}$;
(iii) $\{z : |\text{Arg}(z-i)| < \pi/6\}$; (iv) $\{z : |\text{Arg}(iz+1)| = \pi/3\}$;
(v) $\{z : \text{Arg } z + \text{Arg}(z-1) = \pi\}$; (vi) $\{z : \text{Arg}(z-1) - \text{Arg } z = \pi/2\}$.

4.8. Verify that $2\text{Arg}(1+z) = \text{Arg } z$ when $|z| = 1$, but $z \neq -1$. (*Hint.* What is the set $K = \{1 + z : |z| = 1\}$?)

4.9. Establish de Moivre's formula (1.7). (*Hint.* Deal with $n = 0$ directly, use induction for $n \geq 1$, and reduce the case of negative n back to the case $n \geq 1$.)

4.10. Compute: (i) the square roots of $-1 + i\sqrt{3}$; (ii) the cube roots of -8; (iii) the fourth roots of i ; (iv) the square roots of $5 + 12i$; (v) the fourth roots of $7 - 24i$. The answers may involve radicals, but should not contain any trigonometric expressions.

4.11. Show by example that $\sqrt{zw} = \sqrt{z}\sqrt{w}$ need not be true for arbitrary complex numbers z and w. Confirm that this formula is valid, however, if either z or w is a non-negative real number. (N.B. Similar comments apply to the principal n^{th}-root.)

4.12. Knowing that $\text{Re } c > 0$, demonstrate that $|c + \sqrt{c^2 - 1}| \geq 1$, with equality holding only when c is a real number in the interval $(0, 1]$. (*Hint.* First check that $\text{Re}(\overline{c}\sqrt{c^2-1}) \geq 0$ for every c under consideration.)

4.13. Prove by completing the square that the solutions of a quadratic equation $az^2 + bz + c = 0$, in which a, b, c are complex numbers and $a \neq 0$, are

4. Exercises for Chapter I

given by the usual quadratic formula; i.e., by $z = (-b \pm \sqrt{b^2 - 4ac})/(2a)$.

4.14. Find all the solutions of the equations: (i) $z^2 - 4iz - 4 - 2i = 0$; (ii) $2z^2 - (2 + 5i)z - 2 + i = 0$; (iii) $z^4 + (-4 + 2i)z^2 - 1 = 0$.

4.15. When c is real and in the interval $[-1,1]$, the roots z of $z^2 - 2cz + 1 = 0$ have $|z| = 1$; when c is any other complex number, this equation has one root z_1 with the property that $|z_1| > 1$ and a second root z_2 having $|z_2| < 1$. Corroborate these statements and, in the latter case, identify z_1 and z_2 explicitly. (*Hint*: Recall Exercise 4.12. The determinations of z_1 and z_2 will vary, depending on whether $\operatorname{Re} c > 0$, $\operatorname{Re} c = 0$, or $\operatorname{Re} c < 0$.)

4.16. Determine all solutions of $z^{n-1} + z^{n-2} + \cdots + z + 1 = 0$. Here n is an integer greater than one.

4.17. It is plainly true of any complex number z that either $\sqrt{z^2} = z$ or $\sqrt{z^2} = -z$. Identify the set of z for which the minus-sign represents the proper choice.

4.18. For which complex numbers z is it the case that $\sqrt{z/\bar{z}} = z/|z|$?

4.19. Show that the locus of points z in the complex plane satisfying an equation of the type $Az\bar{z} + Bz + \overline{B}\bar{z} + C = 0$, in which both A and C are real numbers and $|B|^2 - AC > 0$, is a circle when $A \neq 0$ and a line otherwise. Conversely, show that any circle or line in the complex plane admits an equation of the kind just described.

4.20. Show that the locus S of points z in the complex plane subject to the condition $|z + 1||z - 1| = 1$ is the curve described in polar coordinates (r, θ) by the equation $r^2 = 2\cos(2\theta)$. (The curve in question is known as a *lemniscate*.) Sketch a rough graph of this curve in order to obtain an approximate picture of the set S. (N.B. In describing a point of $\mathbb{R}^2$ in terms of polar coordinates (r, θ) we always assume that $r \geq 0$. Naturally, $r = 0$ holds only for the origin. Exercise 4.8 may be of help in establishing that every point of S lies on the given lemniscate.)

4.21. Let c be a complex number satisfying $|c| < 1$. Demonstrate that $|z + c| \leq |1 + \bar{c}z|$ if and only if $|z| \leq 1$, with equality holding if and only if $|z| = 1$.

4.22. Prove *Lagrange's identity*: for $z_1, z_2, \ldots, z_n$ and $w_1, w_2, \ldots, w_n$ in $\mathbb{C}$,

$$\left| \sum_{k=1}^{n} z_k w_k \right|^2 = \left(\sum_{k=1}^{n} |z_k|^2 \right) \left(\sum_{k=1}^{n} |w_k|^2 \right) - \sum_{1 \leq k < j \leq n} |z_k \overline{w}_j - z_j \overline{w}_k|^2.$$

From it deduce *Cauchy's inequality*:

$$\left| \sum_{k=1}^{n} z_k w_k \right|^2 \leq \left(\sum_{k=1}^{n} |z_k|^2 \right) \left(\sum_{k=1}^{n} |w_k|^2 \right).$$

4.2 Exercises for Section I.2

4.23. Certify that $\overline{e^z} = e^{\bar{z}}$ and that, except for non-positive real values of z, $\overline{\text{Log } z} = \text{Log } \bar{z}$ and $\overline{z^\lambda} = \bar{z}^{\bar{\lambda}}$. What happens to the latter formulas when z is both real and negative?

4.24. If n is a positive integer, establish the truth of the identities

$$1 + \cos\theta + \cos(2\theta) + \cdots + \cos(n\theta) = \frac{1}{2} + \frac{\sin[(2n+1)\theta/2]}{2\sin(\theta/2)}$$

and

$$\sin\theta + \sin(2\theta) + \cdots + \sin(n\theta) = \frac{\cot(\theta/2)}{2} - \frac{\cos[(2n+1)\theta/2]}{2\sin(\theta/2)}$$

for all real θ other than integral multiples of 2π. (*Hint.* Make use of Exercise 4.3 with $z = e^{i\theta}$.)

4.25. Express in the form $x + iy$: (i) $\text{Log}(-e^2)$; (ii) $\text{Log}(1 - i\sqrt{3})$; (iii) $(-1)^i$; (iv) $i^{\text{Log } i}$; (v) $i^{\sqrt{2i}}$; (vi) i^{i^i}; (vii) $(\sqrt{3} + i)^{6-i}$.

4.26. Determine the collection of all λ-powers of z when (i) $z = -e$ and $\lambda = \pi i$; (ii) $z = ie^{\pi/2}$ and $\lambda = i$; (iii) $z = i$ and $\lambda = \text{Log } i$; (iv) $z = 1$ and $\lambda = 1 - i$.

4.27. As pointed out in the text, it is not always so that $\text{Log}(z_1 z_2 \cdots z_n) = \text{Log } z_1 + \text{Log } z_2 + \cdots + \text{Log } z_n$. The best one can usually say is that these quantities differ by an integral multiple of $2\pi i$, a state of affairs often indicated by writing $\text{Log}(z_1 z_2 \cdots z_n) = \text{Log } z_1 + \text{Log } z_2 + \cdots + \text{Log } z_n \pmod{2\pi i}$. Show that actual equality $\text{Log}(z_1 z_2 \cdots z_n) = \text{Log } z_1 + \text{Log } z_2 + \cdots + \text{Log } z_n$ occurs when and only when $-\pi < \text{Arg } z_1 + \text{Arg } z_2 + \cdots + \text{Arg } z_n \leq \pi$.

4.28. Verify that $\text{Log}(1 - z^2) = \text{Log}(1 - z) + \text{Log}(1 + z)$ when $|z| < 1$. What can be said about $\text{Log}[(1 - z)/(1 + z)]$ for such z?

4.29. Establish that $\text{Log}(z^\lambda) = \lambda \text{Log } z \pmod{2\pi i}$, but give an example of z and λ for which $\text{Log}(z^\lambda) \neq \lambda \text{Log } z$. Assuming that λ is both real and positive, determine the set of all z such that $\text{Log}(z^\lambda) = \lambda \text{Log } z$. (*Hint.* For the last part check first that $\lambda \text{Arg } z$ is an argument of z^λ when λ is real.)

4.30. Confirm that the law of exponents $z^\lambda z^\mu = z^{\lambda+\mu}$ is valid for all non-zero complex numbers z and all complex exponents λ and μ. Give an example of complex numbers z, λ, and μ for which $(z^\lambda)^\mu \neq z^{\lambda\mu}$.

4.31. Prove that the only circumstances under which $|z^\lambda| = |z|^\lambda$ are (a) when λ is a real number or (b) when z is a positive real number and the quantity $(\text{Im } \lambda) \text{Log } |z|$ is an integral multiple of 2π.

4.3 Exercises for Section I.3

4.32. Determine the domain-sets of the following functions: (i) $f(z) = (iz^2 + 2z + 5)/(z^4 + 3z^2 - 4)$; (ii) $g(z) = z\bar{z}/(z^2 + \bar{z}^2)$; (iii) $h(z) = (z^2 - \bar{z}^2 + 1)^{-1}$; (iv) $k(z) = (e^z + 1)/(e^z - 1)$; (v) $\ell(z) = \text{Log}(e^z - e^{-z})$; (vi) $m(z) = (x + y^3)/(i - x^2 - ix^2y^2)$.

4.33. Express the real and imaginary parts of the following functions as functions of the two real variables x and y: (i) $f(z) = z^3 - iz$; (ii) $g(z) = z^2 - 2i\bar{z}^2 + 1$; (iii) $h(z) = ze^z + z^{-1}e^{-z}$; (iv) $k(z) = \bar{z}e^z - ze^{\bar{z}}$; (v) $\ell(z) = z \text{ Log } z$ for $\text{Re } z > 0$.

4.34. If $f(z) = c_0 + c_1 z + \cdots + c_n z^n$, derive the ensuing polar coordinate representations for $u = \text{Re } f$ and $v = \text{Im } f$:

$$u(re^{i\theta}) = a_0 + \sum_{k=1}^{n} r^k \left[a_k \cos(k\theta) - b_k \sin(k\theta)\right]$$

and

$$v(re^{i\theta}) = b_0 + \sum_{k=1}^{n} r^k \left[a_k \sin(k\theta) + b_k \cos(k\theta)\right] ,$$

where $c_k = a_k + ib_k$ for $k = 0, 1, \ldots, n$.

4.35. Verify that the composition $g \circ f$ of Möbius transformations g and f is again a Möbius transformation. (N.B. It will follow from Exercises 4.36 and 4.48 that the inverse of a Möbius transformation is likewise a mapping of this type, so the class of Möbius transformations forms what is known as a "group" under the operation of composition.)

4.36. Suppose that $f(z) = (az + b)/(cz + d)$ is a Möbius transformation with the property that $c \neq 0$. Through the process of actually computing f^{-1}, which turns out to be another Möbius transformation, certify that f is a univalent function whose domain-set is $\mathbb{C} \sim \{-d/c\}$ and whose range is $\mathbb{C} \sim \{a/c\}$.

4.37. Let $A = \{z : \text{Re } z > 0\}$, and let $f : A \to \mathbb{C}$ be the function given by $f(z) = \text{Log}(z^2 + 1)$. Show that f is a univalent function. Find its range $B = f(A)$ and its inverse function f^{-1}. (*Hint.* Perhaps the most efficient way of identifying B is by geometrically tracking the image of A under successive application of the elementary functions $g(z) = z^2$, $h(z) = z + 1$, and $k(z) = \text{Log } z$ that compose to produce f.)

4.38. Define $f : A \to \mathbb{C}$, where $A = \{z : |z| > 1\}$, via the formula $f(z) = (z + z^{-1})/2$. Verify that f is a univalent function with range $\mathbb{C} \sim [-1, 1]$ and compute f^{-1}. (*Hint.* Exercise 4.15 is relevant here. The rule of correspondence giving $f^{-1}(w)$ involves two different expressions, the applicable one varying with the location of w.)

4.39. Let $D=\{z : |z| < 1\}$. Under the assumption that $|c| < 1$, show that $f(z) = (z+c)/(1+\bar{c}z)$ satisfies $f(D) = D$. (*Hint.* Recall Exercise 4.21.)

4.40. Determine $f(S)$, where $f(z) = e^{1/z}$ and $S = \{z : 0 < |z| < r\}$.

4.41. If z_0 is a root of a polynomial function $f(z) = a_0 + a_1 z + \cdots + a_n z^n$ with real coefficients $a_0, a_1, \ldots, a_n$, then $\bar{z}_0$ is also a root of f. Prove this assertion.

4.42. Suppose that z_0 is a root of a polynomial $f(z) = a_0 + a_1 z + \cdots + a_n z^n$. Demonstrate that $z - z_0$ is a factor of f; i.e., $f(z) = (z - z_0)g(z)$ for all z in $\mathbb{C}$, where g is also a polynomial function of z. (*Hint.* Look at $f(z) - f(z_0)$ in two different ways.)

4.43. Prove that a polynomial function $f(z) = a_0 + a_1 z + \cdots + a_n z^n$ of degree $n \geq 1$ has at most n roots. (*Hint.* Make use of the previous exercise.)

4.44. A point z is termed a *fixed point* of a function f provided $f(z) = z$. Locate all fixed points of the following functions: (i) $f(z) = (1+i)z + 1$; (ii) $g(z) = (1+i)\bar{z} + 1$; (iii) $h(z) = \bar{z} + i$; (iv) $k(z) = ze^z$; (v) $\ell(z) = (3z - 4)/(z - 1)$; (vi) $m(z) = (2z + 1)/(z + 1)$; (vii) $p(z) = 1 + z^2 - z^3$; (viii) $q(z) = z^2 + \bar{z}^2 + z - 2$; (ix) $r(z) = (z^2 + 2z)/(z^2 + 1)$.

4.45. What is the largest number of fixed points a polynomial $f(z) = a_0 + a_1 z + \cdots + a_n z^n$ of degree $n \geq 2$ could have? What about a proper rational function $f(z) = (a_0 + a_1 z + \cdots + a_n z^n)/(b_0 + b_1 z + \cdots + b_m z^m)$, where $a_n \neq 0$, $b_m \neq 0$, and $m \geq 1$? (The word "proper" here signifies that $n < m$.)

4.46. Suppose that $f(z) = (az + b)/(cz + d)$, where $ad - bc = 1$ and $c \neq 0$. Show that f has only one fixed point when $a + d = \pm 2$, but two such points in all other cases.

4.47. Describe the geometric effect on the complex plane of the transformation $f(z) = a\bar{z} + b$, where $a \neq 0$ and b are complex numbers. (Such a mapping is called a *sense-reversing similarity transformation*.)

4.48. If f and g are sense-preserving similarity transformations, show that the composition $g \circ f$ and the inverse function f^{-1} are transformations of that same type. What happens to $g \circ f$ when either f or g is changed to a sense-reversing similarity? When both f and g are so changed? What is the character of the inverse of a sense-reversing similarity transformation?

4.49. A similarity transformation $f(z) = az + b$ is devoid of fixed points in the complex plane precisely when $a = 1$ and $b \neq 0$; otherwise, f has exactly one fixed point or fixes every point. Support this claim.

4.50. Let the similarity transformation $f(z) = az + b$ have z_0 as a fixed point. Show that f can be rewritten in the form $f(z) = z_0 + a(z - z_0)$. Deduce from this representation that the geometric effect of f on the complex

4. Exercises for Chapter I   31

plane amounts to a rotation about the point z_0 through an angle $\theta = \operatorname{Arg} a$, followed by a dilation with respect to z_0 in which every ray issuing from z_0 is mapped to itself and all distances get multiplied by the factor $|a|$.

4.51. Prove that a sense-reversing similarity transformation $f(z) = a\bar{z} + b$ has a unique fixed point if $|a| \neq 1$, has a line as its set of fixed points if $|a| = 1$ and $a\bar{b} = -b$, and has no fixed points in the complex plane if $|a| = 1$ and $a\bar{b} \neq -b$. (*Hint.* Remember Exercise 4.19.)

4.52. Consider the transformation $w = (1-z)/(1+z)$ from Example 3.4. Check that under this mapping the following sets transform as indicated. (i) $\{z : \operatorname{Im} z > 0\} \to \{w : \operatorname{Im} w < 0\}$; (ii) $\{z : \operatorname{Re} z = -1, z \neq -1\} \to \{w : \operatorname{Re} w = -1, w \neq -1\}$; (iii) $\{z : |z+2| = 1, z \neq -1\} \to \{w : u = -2\}$; (iv) $\{z : x = 1\} \to \{w : |w + (1/2)| = 1/2, w \neq -1\}$.

4.53. Demonstrate that under the correspondence $w = z^2$ any horizontal line in the z-plane, apart from the real axis, is transformed to a parabola in the w-plane. To what does the real axis get mapped?

4.54. Extend the previous exercise as follows: under the correspondence $w = z^2$ every line in the z-plane, except for lines passing through the origin, is carried to a parabola in the w-plane. What are the images of lines containing the origin? (*Hint.* Use Exercise 4.53 and the identity $z^2 = e^{-2i\theta}(e^{i\theta}z)^2$.)

4.55. By a *cardioid* is meant a plane curve that is similar, in the technical geometric meaning of the term, to the curve whose polar equation is $r = 1 + \cos\theta$. Show that the transformation $w = z^2$ maps an arbitrary circle K passing through the origin of the z-plane to a cardioid in the w-plane. (*Hint.* If K has center z_0, then this circle admits the description $K = \{z_0 + z_0 e^{i\theta} : -\pi < \theta \leq \pi\}$. Begin by looking at the case $z_0 = 1$.)

4.56. If L is a line in the z-plane, determine its image under the mapping $w = \sqrt{z}$ when (i) L is horizontal, (ii) L is vertical, and (iii) L has equation $y = x\sqrt{3}$. (N.B. The character of the image in (i) and (ii) will vary, depending on whether L intersects the negative real axis or not. Keep in mind that $\operatorname{Re} \sqrt{z} \geq 0$.)

4.57. Let K be a circle in the z-plane. Show that the image of K under the mapping $w = z^{-1}$ is a circle in the w-plane if K does not pass through the origin, whereas its image — or, more correctly, the image of $K \sim \{0\}$ — is a line when the origin is a point of K. If L is a line in the z-plane, prove that $w = z^{-1}$ maps L to a circle punctured at the origin when the origin does not lie on L and transforms L — or, rather, $L \sim \{0\}$ — to a line with the origin deleted when L contains the origin. (*Hint.* Put Exercise 4.19 to work.)

4.58. Assuming that $r \neq 1$, check that the circle $K = \{z : |z| = r\}$ is transformed under the correspondence $w = (z + z^{-1})/2$ to an ellipse in the w-plane. What is the image of K when $r = 1$? (*Hint.* Observe that $K = \{re^{i\theta} : 0 \leq \theta \leq 2\pi\}$. Start by expressing $u = \operatorname{Re} w$ and $v = \operatorname{Im} w$ for w on the image of K as functions of the parameter θ.)

4.59. Let $w = (az+b)/(cz+d)$, where $ad - bc = 1$ and $c \neq 0$. Show that the circle $K = \{z : |z + (d/c)| = 1/|c|\}$ in the z-plane is mapped by this transformation to the circle $K^* = \{w : |w - (a/c)| = 1/|c|\}$ in the w-plane.

4.60. Given that z_1, z_2, and z_3 are distinct points in the complex plane, verify that these points are the vertices of an equilateral triangle if and only if

(1.17) $$z_1^2 + z_2^2 + z_3^2 = z_1 z_2 + z_1 z_3 + z_2 z_3 \ .$$

(*Hint.* Exploit the fact that both the property of being the set of vertices of an equilateral triangle and property (1.17) are invariant under similarity — i.e., if z_1, z_2, and z_3 enjoy one of these properties and if $w_j = az_j + b$, where $a \neq 0$, then w_1, w_2, and w_3 exhibit the same property — to reduce the problem to a simple case.)

4.61. Let z_1, z_2, z_3, and z_4 be distinct points of $\mathbb{C}$. If these points list in consecutive order the vertices of a square, demonstrate that

(1.18) $$\frac{(z_1 - z_3)(z_2 - z_4)}{(z_1 - z_2)(z_3 - z_4)} = 2 \ .$$

Provide an example to show that the validity of (1.18) does not, in general, imply that z_1, z_2, z_3, and z_4 are the consecutive vertices of a square. Under the assumption that these points are the consecutive vertices of a parallelogram, however, and also satisfy (1.18), establish that the parallelogram in question is necessarily a square. (*Hint.* Appeal to similarity invariants.)

Chapter II

The Rudiments of Plane Topology

Introduction

The present chapter is devoted to a survey of the basic definitions and theorems from plane point-set topology that are prerequisite to a careful discussion of complex function theory. (Additional topological concepts will be introduced later in the text, as they become pertinent to developments.) We warn the reader that a number of the statements found in this chapter are just that, statements made without any attempt to justify them. Our objective here is simply to assemble in one place for handy reference the relevant background material from topology, not to provide an in-depth account of the subject. Those desirous of a more detailed treatment are encouraged to consult one of the many available textbooks on elementary topology, e.g., *Topology – A First Course* by J.R. Munkres (Prentice-Hall, Englewood Cliffs, N.J., 1975). Having said this, we hasten to add that the bulk of the results offered here without proof are really nothing more than exercises in sorting through the definitions of the concepts involved. They can be treated as such by the ambitious reader.

1 Basic Notation and Terminology

1.1 Disks

We start by fixing notation for a number of standard sets that will be in constant use throughout the book. If z_0 is a point in the complex plane and if $0 < r < \infty$, we write

$$\Delta(z_0, r) = \{z : |z - z_0| < r\},$$

$$\overline{\Delta}(z_0, r) = \{z : |z - z_0| \leq r\},$$

$$\Delta^*(z_0, r) = \{z : 0 < |z - z_0| < r\} = \Delta(z_0, r) \sim \{z_0\},$$

$$K(z_0, r) = \{z : |z - z_0| = r\} .$$

The sets $\Delta(z_0, r)$, $\overline{\Delta}(z_0, r)$, and $\Delta^*(z_0, r)$ are called, respectively, the *open disk*, the *closed disk*, and the *punctured open disk* of radius r centered at z_0, while $K(z_0, r)$ is nothing but the circle with center z_0 and radius r.

1.2 Interior Points, Open Sets

Suppose that A is a subset of $\mathbb{C}$ and that z is an element of A. We say that z is an *interior point* of A if there exists an $r > 0$ such that the open disk $\Delta(z, r)$ is contained in A. A set U in the complex plane with the property that every point of U is an interior point of U is known as an *open set*. (N.B. The letters U, V, and W, when employed in this book to represent sets, will invariably designate open sets.) Some elementary examples of open subsets of $\mathbb{C}$ are the empty set ϕ, the entire complex plane $\mathbb{C}$, the open disk $\Delta(z_0, r)$, the punctured disk $\Delta^*(z_0, r)$, the set $\mathbb{C} \sim \overline{\Delta}(z_0, r) = \{z : |z - z_0| > r\}$, and the open upper half-plane $\{z : \operatorname{Im} z > 0\}$. Two fundamental properties of open sets are described in:

Theorem 1.1. *The union of an arbitrary collection of open subsets of $\mathbb{C}$ is an open set. The intersection of a finite collection of open subsets of $\mathbb{C}$ is an open set.*

1.3 Closed Sets

A subset A of $\mathbb{C}$ is pronounced a *closed set* provided its complement $\mathbb{C} \sim A$ is open. Examples of such sets include ϕ, $\mathbb{C}$, the closed disk $\overline{\Delta}(z_0, r)$, the circle $K(z_0, r)$, the closed upper half-plane $\{z : \operatorname{Im} z \geq 0\}$, and the real line $\mathbb{R}$. The sets ϕ and $\mathbb{C}$ are the only subsets of the plane that are simultaneously open and closed, although this fact is not immediately obvious just from the definitions of open and closed sets. It goes without saying that most subsets of $\mathbb{C}$ are neither open nor closed. There is a proposition dual to Theorem 1.1 dealing with closed sets. It can be inferred from the earlier theorem by application of de Morgan's rules for taking complements of unions and complements of intersections.

Theorem 1.2. *The intersection of an arbitrary collection of closed subsets of $\mathbb{C}$ is a closed set. The union of a finite collection of closed subsets of $\mathbb{C}$ is a closed set.*

1. Basic Notation and Terminology

1.4 Boundary, Closure, Interior

A point z in $\mathbb{C}$ is termed a *boundary point* of the plane set A if for every $r > 0$ the open disk $\Delta(z,r)$ has non-empty intersection with both A and $\mathbb{C} \sim A$. The set consisting of all such boundary points is known as the *boundary of A*. We use the notation ∂A to signify this set. As examples, observe that $\partial \Delta(z_0, r) = \partial \overline{\Delta}(z_0, r) = \partial K(z_0, r) = K(z_0, r)$, $\partial \Delta^*(z_0, r) = K(z_0, r) \cup \{z_0\}$, and $\partial\{z : \operatorname{Im} z > 0\} = \mathbb{R}$. An open set in the complex plane is characterized by the property that it contains none of its boundary points, whereas a closed set is distinguished by the feature of containing all of its boundary points.

The set $\overline{A} = A \cup \partial A$ associated with a subset A of $\mathbb{C}$ is referred to as the *closure of A*. (We rely on the context to make evident whether the symbol — above a letter stands for "conjugate" or for "closure.") By way of illustration, the closure of the open disk $\Delta(z_0, r)$ is the closed disk $\overline{\Delta}(z_0, r)$. Clearly a point z belongs to $\overline{A}$ if and only if $A \cap \Delta(z,r)$ is non-empty for every $r > 0$.

We record the following observation.

Theorem 1.3. *The boundary ∂A and closure $\overline{A}$ of a subset A of $\mathbb{C}$ are closed sets.*

Notice that a plane set A is closed precisely when $A = \overline{A}$. To add one last bit of basic terminology, the set of all interior points of a set A is called the *interior* of A. The notation A° is commonly utilized to designate this set, which is easily seen to be open.

1.5 Sequences

A *sequence σ in a set A* — here we are not necessarily restricting A to be a subset of $\mathbb{C}$ — is usually defined to be a function $\sigma : \mathbb{N} \to A$, where $\mathbb{N}$ is the set of positive integers. Writing $a_n = \sigma(n)$ we can view σ as a rule for listing in a specified order — namely, $a_1, a_2, a_3, \cdots$ — the elements of its range. Since σ may fail to be one-to-one, we must allow for duplications to crop up in this list. In other words, we are free to regard a sequence σ in A as an ordered, but possibly repetitious list $a_1, a_2, a_3, \cdots$ of elements of A, indexed by the positive integers. The nth entry a_n in the list is then called the nth *term* of the sequence. In this book the symbol $\langle a_n \rangle$ will be employed as shorthand notation for the sequence $a_1, a_2, a_3, \cdots$. (Sometimes we shall run into sequences indexed by sets of integers other than $\mathbb{N}$. Sequences $a_0, a_1, a_2, \cdots$ will, for instance, occur quite regularly. Such minor deviations from the discussion at hand should cause no severe problems of adjustment. We may occasionally write $\langle a_n \rangle_{n=1}^\infty$ or $\langle a_n \rangle_{n=0}^\infty$ in order to emphasize the index set, but the context will ordinarily make even this unnecessary.)

There are several options available for presenting sequences in the concrete. First, if there is a simple pattern to its terms, a sequence might be described by writing out enough of the terms so that the pattern becomes recognizable; a second option is simply to give the rule of correspondence that expresses a_n as a function of n; a third way to define a sequence is to do so *recursively* by identifying a_1 and by detailing the procedure for obtaining a_n once $a_1, a_2, \ldots, a_{n-1}$ are known. We can exemplify all three methods using the sequence $a_1 = 1, a_2 = -1, a_3 = 1, a_4 = -1, a_5 = 1, a_6 = -1, \cdots$. The pattern here is evident. Also, the general term a_n is readily expressed: $a_n = (-1)^{n+1}$. Lastly, the same sequence is presented recursively as follows: $a_1 = 1$ and $a_n = -a_{n-1}$ for $n \geq 2$.

If $\langle a_n \rangle$ is a sequence in a set A, then by a *subsequence of* $\langle a_n \rangle$ is meant a sequence, call it $\langle b_n \rangle$, related to the given one in the following manner: there is a sequence $n_1 < n_2 < n_3 \cdots$ of positive integers such that $b_1 = a_{n_1}$, $b_2 = a_{n_2}, b_3 = a_{n_3}, \cdots$. The notation $\langle a_{n_k} \rangle$ will serve to indicate such a subsequence of $\langle a_n \rangle$.

We stress again that the notion of a sequence is not confined to sequences of complex numbers. At various times in this book we shall have occasion to consider sequences of sets $A_1, A_2, A_3, \cdots$, sequences of complex functions $f_1, f_2, f_3, \cdots$, and, yes, even sequences of sequences!

1.6 Convergence of Complex Sequences

We now focus our attention on a sequence $\langle z_n \rangle$ of complex numbers. It may happen that, as n increases, the term z_n develops an overwhelming attraction for and finds itself inescapably drawn toward some particular complex number z_0. When this behavior is exhibited, we characterize $\langle z_n \rangle$ as a "convergent" sequence with z_0 as its "limit." To make this precise, we define a sequence $\langle z_n \rangle$ in $\mathbb{C}$ to be *convergent* if there exists a complex number z_0 of which the following is true: corresponding to each $\epsilon > 0$ there is an index $N = N(\epsilon)$ such that z_n lies in the disk $\Delta(z_0, \epsilon)$ for every $n \geq N$. (See Figure 1.) This being so, there is exactly one such number z_0 and it is called the *limit* of $\langle z_n \rangle$. We write $z_0 = \lim_{n \to \infty} z_n$ or $z_n \to z_0$ to express the fact that $\langle z_n \rangle$ is a convergent sequence with limit z_0. In symbol-free language one speaks of z_n "tending to" or "converging to" or "approaching" z_0. We point out that $z_n \to z_0$ is equivalent to $|z_n - z_0| \to 0$. This observation can be quite helpful in confirming that the suspected limit of a complex sequence actually is the limit, for the reason that $\langle |z_n - z_0| \rangle$, being a real sequence, is subject to analysis by techniques — the "squeeze rule," for one — without counterparts in $\mathbb{C}$. The inequalities

$$|\operatorname{Re} z_n - \operatorname{Re} z_0| \leq |z_n - z_0| \quad , \quad |\operatorname{Im} z_n - \operatorname{Im} z_0| \leq |z_n - z_0|,$$

and

$$|z_n - z_0| \leq |\operatorname{Re} z_n - \operatorname{Re} z_0| + |\operatorname{Im} z_n - \operatorname{Im} z_0|,$$

when combined with the squeeze rule, lead to a useful remark.

1. Basic Notation and Terminology

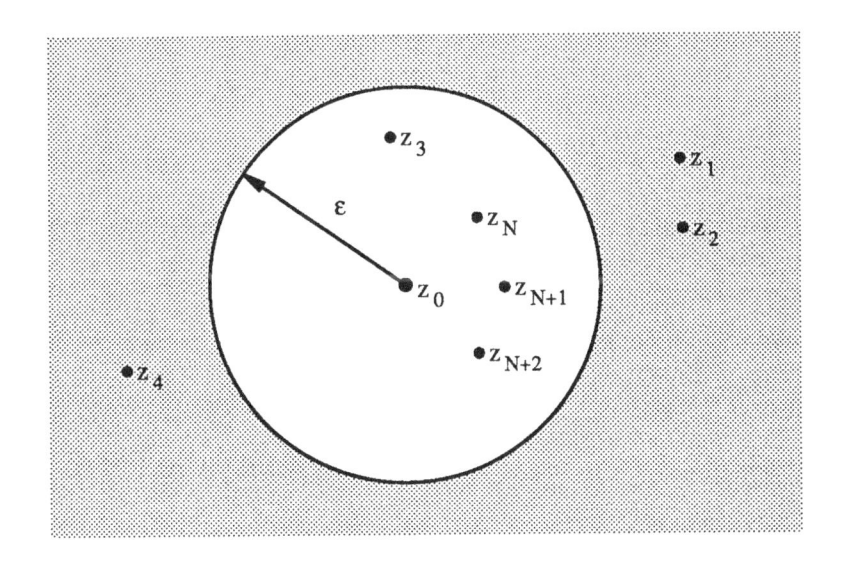

Figure 1.

Lemma 1.4. *A complex sequence $\langle z_n \rangle$ is convergent and has limit z_0 if and only if both of the real sequences $\langle \operatorname{Re} z_n \rangle$ and $\langle \operatorname{Im} z_n \rangle$ are convergent and have $\operatorname{Re} z_0$ and $\operatorname{Im} z_0$ as their respective limits.*

The algebra of limits for complex sequences is essentially the same as it is for the real sequences studied in beginning calculus. With the aid of Lemma 1.4 one can easily deduce the contents of the following result from the limit theorems of calculus.

Theorem 1.5. *Let $\langle z_n \rangle$ and $\langle w_n \rangle$ be sequences in $\mathbb{C}$ such that $z_n \to z_0$ and $w_n \to w_0$. Then $cz_n \to cz_0$ for any complex number c, $\bar{z}_n \to \bar{z}_0$, $|z_n| \to |z_0|$, $z_n + w_n \to z_0 + w_0$, $z_n w_n \to z_0 w_0$, and, if $w_0 \neq 0$, $z_n/w_n \to z_0/w_0$.*

The quotient z_n/w_n is undefined, naturally, for any n with $w_n = 0$. However, under the assumption that $w_0 \neq 0$, this creates a problem for at most finitely many terms z_n/w_n. We just disregard those terms in dealing with the quotient sequence.

1.7 Accumulation Points of Complex Sequences

Even a non-convergent complex sequence $\langle z_n \rangle$ may have associated with it certain attention-grabbing points. In particular, there may exist one or more points z_0 distinguished by the fact that, as n grows, z_n returns arbitrarily often to each open disk centered at z_0. Such a point is known as an "accumulation point" of $\langle z_n \rangle$. An exact definition of this concept reads: a sequence $\langle z_n \rangle$ in $\mathbb{C}$ has the complex number z_0 as an *accumulation point* if for each $\epsilon > 0$ the disk $\Delta(z_0, \epsilon)$ contains z_n for infinitely many values of n. A given sequence $\langle z_n \rangle$ may well fail to have any accumulation points. At the opposite extreme it is possible for a single sequence $\langle z_n \rangle$ to have every point of $\mathbb{C}$ as an accumulation point! If $\langle z_n \rangle$ is a convergent sequence with limit z_0, then z_0 is the unique accumulation point of $\langle z_n \rangle$. On the other hand, there exist non-convergent sequences having one and only one accumulation point. (See (2.2) below.)

We examine how the ideas just introduced apply to some specific sequences. The constant sequence

$$z_1 = c, \; z_2 = c, \ldots, z_n = c, \ldots$$

is obviously convergent and has limit c. A less trivial example of a convergent sequence is given by

$$z_1 = i, \; z_2 = -\frac{1}{2}, \ldots, z_n = \frac{i^n}{n}, \ldots.$$

Here $z_n \to 0$. The sequence

(2.1) $$z_1 = i, \; z_2 = -1, \ldots, z_n = i^n, \ldots$$

does not have a limit, although it does have four accumulation points — namely, the points $1, -1, i,$ and $-i$. A non-convergent sequence with exactly one accumulation point is

(2.2) $$z_1 = 1, \; z_2 = \frac{9}{2}, \ldots, z_n = 2^{1-n} + n + (-1)^n n, \ldots.$$

The unique accumulation point here is 0, but $z_n \not\to 0$ (i.e., z_n does not tend to 0), since the even-numbered terms of this sequence all lie outside the disk $\Delta(0, 1)$. Finally, it can be shown that for z in $\mathbb{C}$ the sequence

(2.3) $$z_1 = z, \; z_2 = z^2, \ldots, z_n = z^n, \ldots$$

displays the following behavior: if $|z| < 1$, $z_n \to 0$; if $|z| > 1$, $\langle z_n \rangle$ has no accumulation points in $\mathbb{C}$; if $z = e^{i\theta}$ with $\theta/2\pi$ a rational number, the accumulation points of $\langle z_n \rangle$ are $1, z, z^2, \ldots, z^{q-2}$, and z^{q-1}, where $\theta/2\pi = p/q$ in lowest terms; if $z = e^{i\theta}$ with $\theta/2\pi$ irrational, the set of accumulation points of $\langle z_n \rangle$ is the entire circle $K(0, 1)$. (Only the last assertion is hard to verify — and that quite tricky indeed.)

Assume that z_0 is an accumulation point of a complex sequence $\langle z_n \rangle$. We inductively construct a sequence $n_1 < n_2 < n_3 \cdots$ of positive integers as follows: we choose n_1 such that z_{n_1} belongs to $\Delta(z_0, 1)$ and, having constructed n_k, we choose $n_{k+1} > n_k$ such that $z_{n_{k+1}}$ lies in $\Delta\left(z_0, (k+1)^{-1}\right)$.

The fact that z_0 is an accumulation point of $\langle z_n \rangle$ makes the above choices possible. In this way we arrive at a subsequence $\langle z_{n_k} \rangle$ of $\langle z_n \rangle$ with the property that $|z_{n_k} - z_0| < 1/k$ for all k, which implies that $z_{n_k} \to z_0$ as $k \to \infty$. This establishes the less obvious direction in the next result.

Theorem 1.6. *A sequence $\langle z_n \rangle$ in $\mathbb{C}$ has the complex number z_0 as an accumulation point if and only if there is a subsequence $\langle z_{n_k} \rangle$ of $\langle z_n \rangle$ such that $\lim_{k \to \infty} z_{n_k} = z_0$.*

The sequence (2.1) lists i among its accumulation points. One choice of subsequence converging to i in that example is $z_4, z_9, z_{13}, \ldots, z_{4k+1}, \cdots$; another is $z_2, z_3, z_7, \ldots, z_{k!+1}, \cdots$. When a sequence $\langle z_n \rangle$ is actually convergent the behavior of its subsequences simplifies greatly, as the following theorem shows.

Theorem 1.7. *Suppose that a complex sequence $\langle z_n \rangle$ is convergent and has limit z_0. Then $\lim_{k \to \infty} z_{n_k} = z_0$ for every subsequence $\langle z_{n_k} \rangle$ of $\langle z_n \rangle$.*

Let A be a set in the complex plane and let z_0 be an element of $\overline{A}$. For each integer $n \geq 1$ we can pick a point z_n belonging to $A \cap \Delta(z_0, 1/n)$. By this process we manufacture a sequence $\langle z_n \rangle$ in A with the property that $z_n \to z_0$. Conversely, the existence of a sequence $\langle z_n \rangle$ in A with z_0 as its limit clearly places z_0 in $\overline{A}$. We infer:

Theorem 1.8. *The point z_0 belongs to the closure of a plane set A if and only if there exists a sequence $\langle z_n \rangle$ in A such that $z_n \to z_0$.*

Combining Theorems 1.7 and 1.8 with the fact that closed sets A in $\mathbb{C}$ are characterized by the condition $\overline{A} = A$, we can give a new description of such sets.

Theorem 1.9. *The subset A of $\mathbb{C}$ is closed if and only if A contains every accumulation point of every sequence in A.*

2 Continuity and Limits of Functions

2.1 Continuity

To say that a function $f: A \to \mathbb{C}$ is continuous at a point z_0 of A means intuitively that, when f is evaluated at a point z near z_0, the value $f(z)$ lies in close proximity to $f(z_0)$ and, even more, that the distance between $f(z)$ and $f(z_0)$ shrinks to zero as z is moved closer and closer to z_0. By the same token, to declare that f is discontinuous at z_0 conveys the information that there exist points z in A ever so slightly removed from z_0 for which $f(z)$ lies at a distance from $f(z_0)$ larger than some fixed positive number. The intuitive notion of continuity is incorporated into a formal definition

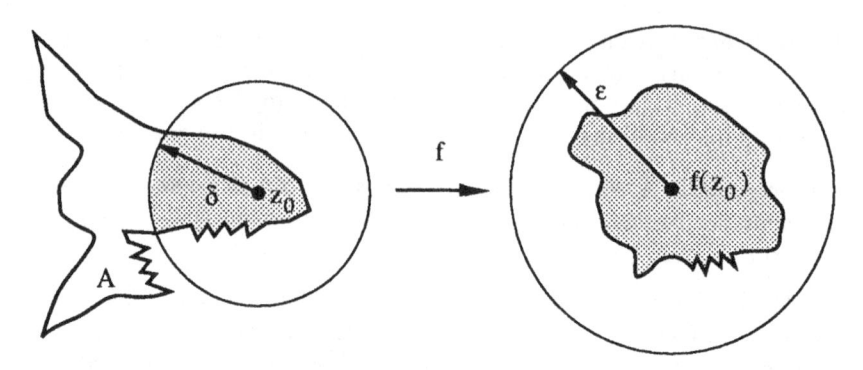

Figure 2.

as follows: a function $f: A \to \mathbb{C}$ is *continuous at a point* z_0 if z_0 belongs to A and if corresponding to each $\epsilon > 0$ there exists a $\delta = \delta(\epsilon, z_0) > 0$ with the property that

(2.4) $$f[A \cap \Delta(z_0, \delta)] \subset \Delta(f(z_0), \epsilon) .$$

(See Figure 2.) We have written $\delta = \delta(\epsilon, z_0)$ here in order to underscore the fact that δ will normally depend on z_0, not just on ϵ. (N.B. Use of (2.4) in place of the equivalent statement involving inequalities — i.e., $|f(z) - f(z_0)| < \epsilon$ for every z in A satisfying $|z - z_0| < \delta$ — is made with some foresight: as formulated, the definition of continuity carries over almost verbatim to a definition of continuity for functions whose domain-sets and ranges lie in the extended complex plane $\widehat{\mathbb{C}} = \mathbb{C} \cup \{\infty\}$, which will be discussed in Chapter VIII, whereas the corresponding definition in terms of inequalities is somewhat awkward in the extended setting.) A function $f: A \to \mathbb{C}$ is *discontinuous* (or *has a discontinuity*) *at a point* z_0 if z_0 belongs to A, yet f is not continuous at z_0. Negating the definition of continuity, we see that the latter condition reduces to the existence of some $\epsilon > 0$ with the property that

(2.5) $$f[A \cap \Delta(z_0, \delta)] \sim \Delta(f(z_0), \epsilon) \neq \phi$$

for every $\delta > 0$. Note that the expressions "f is continuous at z_0" and "f is discontinuous at z_0" are employed only in relation to points z_0 belonging to the domain-set of f. Thus, we do not speak of the function $f(z) = 1/z$ as discontinuous at the origin, for f is not defined there. (In Chapter VIII we shall introduce the terminology "isolated singularity" in conjunction with behavior akin to that of $f(z) = 1/z$ at the origin.) A function that is continuous at every point of its domain-set is called a *continuous function*. More generally, we say that a function f is *continuous on* (or *in*) *a set* S if S is contained in the domain-set of f and if the function obtained by restricting f to S is a continuous function.

2. Continuity and Limits of Functions

The definition of continuity at a point has a convenient reformulation in terms of sequences.

Theorem 2.1. *A function $f: A \to \mathbb{C}$ is continuous at a point z_0 of A if and only if for every sequence $\langle z_n \rangle$ in A converging to z_0 the image sequence $\langle f(z_n) \rangle$ converges to $f(z_0)$.*

Proof. Assume first that f is continuous at z_0 and consider a sequence $\langle z_n \rangle$ in A such that $z_n \to z_0$. We show that $f(z_n) \to f(z_0)$. Let $\epsilon > 0$ be given. We must exhibit an index N with the property that $f(z_n)$ lies in $\Delta(f(z_0), \epsilon)$ for all $n \geq N$. By the definition of continuity we can choose $\delta > 0$ so that (2.4) holds. Next, since $z_n \to z_0$, we can select an index N such that z_n is a point of $A \cap \Delta(z_0, \delta)$ whenever $n \geq N$. By (2.4), $f(z_n)$ belongs to $\Delta(f(z_0), \epsilon)$ for all $n \geq N$, as desired.

For the converse, we assume that f is discontinuous at z_0 and use this assumption to construct a sequence $\langle z_n \rangle$ in A such that $z_n \to z_0$, while $f(z_n) \not\to f(z_0)$. As observed, the discontinuity of f at z_0 means that there is some $\epsilon > 0$ such that (2.5) holds for every $\delta > 0$. Fix such an ϵ. Statement (2.5) is then true of $\delta = 1/n$, where n is a positive integer. This allows us to pick a point z_n of $A \cap \Delta(z_0, 1/n)$ for which $f(z_n)$ lies outside $\Delta(f(z_0), \epsilon)$. Doing so for $n = 1, 2, 3, \cdots$ we produce a sequence $\langle z_n \rangle$ in A that satisfies $|z_n - z_0| < 1/n$ for all n, forcing $z_n \to z_0$. However, $|f(z_n) - f(z_0)| \geq \epsilon$ for every n, so $f(z_n) \not\to f(z_0)$. ∎

In tandem, Theorems 1.5 and 2.1 make short work of the proofs of two noteworthy propositions.

Theorem 2.2. *Suppose that functions $f: A \to \mathbb{C}$ and $g: B \to \mathbb{C}$ are both continuous at a point z_0. Then the functions cf for a complex constant c, $\operatorname{Re} f$, $\operatorname{Im} f$, $\bar{f}$, $|f|$, $f + g$, fg, and, if $g(z_0) \neq 0$, f/g are all continuous at z_0. In particular, if f and g are continuous functions, then so is each of the functions just listed, provided its domain-set is not empty.*

Theorem 2.3. *Suppose that $f: A \to \mathbb{C}$ and $g: B \to \mathbb{C}$ are functions for which $f(A)$ is contained in B. If f is continuous at z_0 and g is continuous at $w_0 = f(z_0)$, then the composition $g \circ f$ is continuous at z_0. In particular, if f and g are continuous functions, then $g \circ f$ is a continuous function.*

In order to demonstrate a method of establishing such results, we prove Theorem 2.3. Consider an arbitrary sequence $\langle z_n \rangle$ in A satisfying $z_n \to z_0$. Since f is continuous at z_0, $w_n = f(z_n) \to f(z_0) = w_0$ (Theorem 2.1). Similarly, the continuity of g at w_0 implies that $g(w_n) \to g(w_0)$. Thus, $g \circ f(z_n) = g[f(z_n)] = g(w_n) \to g(w_0) = g[f(z_0)] = g \circ f(z_0)$. Theorem 2.1 vouches for the continuity of $g \circ f$ at z_0.

Since $f(z) = \operatorname{Re} z = x$ and $g(z) = \operatorname{Im} z = y$ quite obviously define continuous functions in $\mathbb{C}$, Theorems 2.2 and 2.3 can be applied to them repeatedly to generate a lot of other examples of continuous functions.

It follows from Theorem 2.2, to give one illustration, that any rational function of x and y is continuous. (See Section I.3.1.) As an important special case, we note that rational functions of z are continuous functions. If $\varphi: \mathbb{R} \to \mathbb{R}$ is a continuous function, then in view of Theorem 2.3 both $\varphi \circ f$ and $\varphi \circ g$ are continuous. Accordingly, $h(z) = e^x$, $k(z) = \cos y$, and $\ell(z) = \sin y$ represent continuous functions. So, therefore, does $E(z) = e^z = e^x \cos y + i e^x \sin y$. In a similar spirit, by appealing to the fact that the function φ defined on the interval $[-1, 1]$ by $\varphi(t) = \operatorname{Arcsin} t$ is continuous, we are able to conclude that

$$(2.6) \qquad \alpha(z) = \operatorname{Arcsin}\left(\frac{y}{|z|}\right)$$

defines a continuous function in $\mathbb{C} \sim \{0\}$. This particular function is significant due to the relationship $\alpha(z)$ bears to $\operatorname{Arg} z$. It plays a role in documenting a fact to which we shall refer more than once in the future.

Lemma 2.4. *The function $\theta: \mathbb{C} \sim \{0\} \to \mathbb{R}$ given by $\theta(z) = \operatorname{Arg} z$ is continuous in the set $D = \mathbb{C} \sim (-\infty, 0]$ and discontinuous at all points of the interval $(-\infty, 0)$.*

Proof. Let $\alpha: \mathbb{C} \sim \{0\} \to \mathbb{R}$ be the function in (2.6). Recalling our discussion of arguments in Section I.1.2, we express $\theta(z)$ for $z = x + iy$ as follows:

$$(2.7) \qquad \theta(z) = \begin{cases} \alpha(z) & \text{if } x \geq 0, \\ \pi - \alpha(z) & \text{if } x < 0 \text{ and } y \geq 0, \\ -\pi - \alpha(z) & \text{if } x < 0 \text{ and } y < 0. \end{cases}$$

Formula (2.7) and the continuity of α make it apparent that θ is continuous in the three open sets $U = \{z \in D : x > 0\}$, $V = \{z \in D : x < 0 \text{ and } y > 0\}$, and $W = \{z \in D : x < 0 \text{ and } y < 0\}$, so question marks concerning the continuity of θ in D occur only at points of the imaginary axis. Consider such a point z_0, say with $\operatorname{Im} z_0 > 0$. (The case $\operatorname{Im} z_0 < 0$ is treated similarly.) For $z = x + iy$ in D with $y > 0$ we observe that

$$|\theta(z) - \theta(z_0)| = |\alpha(z) - \alpha(z_0)|$$

if $x \geq 0$, while the fact that $\alpha(z_0) = \dfrac{\pi}{2}$ also gives

$$|\theta(z) - \theta(z_0)| = \left|\pi - \alpha(z) - \frac{\pi}{2}\right| = \left|\frac{\pi}{2} - \alpha(z)\right| = |\alpha(z) - \alpha(z_0)|$$

when $x < 0$. In other words, $|\theta(z) - \theta(z_0)| = |\alpha(z) - \alpha(z_0)|$ holds for every z in D with $\operatorname{Im} z > 0$. In conjunction with the continuity of α at z_0, this clearly implies the continuity of θ at z_0. The function θ is continuous, therefore, at all points of D.

2. Continuity and Limits of Functions

Finally, if x_0 is real and negative, look at the sequence $z_n = x_0 - n^{-1}i$. Then $z_n \to x_0$, but by (2.7)

$$\theta(z_n) = -\pi - \alpha(z_n) \to -\pi - \alpha(x_0) = -\pi \neq \pi = \theta(x_0),$$

showing that θ is discontinuous at x_0. ∎

Lemma 2.4 implies, among other things, that the principal logarithm function, $L(z) = \text{Log } z = \text{Log }|z| + i \text{ Arg } z$, has its only discontinuities at the points of the negative real axis. The last result of this section characterizes continuous functions with open domain-sets. Its proof is left as an exercise.

Theorem 2.5. *Suppose that U is an open set in the complex plane. A function $f: U \to \mathbb{C}$ is continuous if and only if for every open subset V of the complex plane the set $\{z \in U : f(z) \in V\}$ is also open.*

2.2 Limits of Functions

Closely associated with the notion of a function being continuous at a point is the concept of a function possessing a limit at a point. Roughly speaking, the limit of a function $f: A \to \mathbb{C}$ at a point z_0 (which need not belong to A) is the complex number w_0 (which may or may not exist) that renders the function $g: A \cup \{z_0\} \to \mathbb{C}$ defined by

$$g(z) = \begin{cases} f(z) & \text{if } z \in A \sim \{z_0\}, \\ w_0 & \text{if } z = z_0, \end{cases}$$

continuous at z_0. The single detail that prevents this from being a proper definition of limit is a technical one concerned with the relationship of the point z_0 to the set A. Specifically, we must insist that z_0 be a *limit point* of A, meaning that $A \cap \Delta^*(z_0, r) \neq \phi$ for every $r > 0$ or, equivalently, that there exists a sequence $\langle z_n \rangle$ in $A \sim \{z_0\}$ such that $z_n \to z_0$. For example, the origin is a limit point of the set $\{1, 1/2, 1/3, \cdots\}$, but not of the set $\{0, 1/2, 2/3, 3/4, \cdots\}$. With these preliminary remarks behind us, we make the definition: a function $f: A \to \mathbb{C}$ *has the complex number w_0 as its limit at a point z_0* if z_0 is a limit point of A and if corresponding to each $\epsilon > 0$ there exists a $\delta = \delta(\epsilon) > 0$ with the property that

(2.8) $$f[A \cap \Delta^*(z_0, \delta)] \subset \Delta(w_0, \epsilon).$$

(See Figure 3.) Assuming that a complex number w_0 fitting this description exists — as already suggested, its existence is by no means assured — there is only one such number. We write either $\lim_{z \to z_0} f(z) = w_0$ or $f(z) \to w_0$ as $z \to z_0$ to express symbolically the fact that the limit does exist and has value w_0. (N.B. As in the definition of continuity at a point, we have elected to use (2.8) in preference to the equivalent inequalities

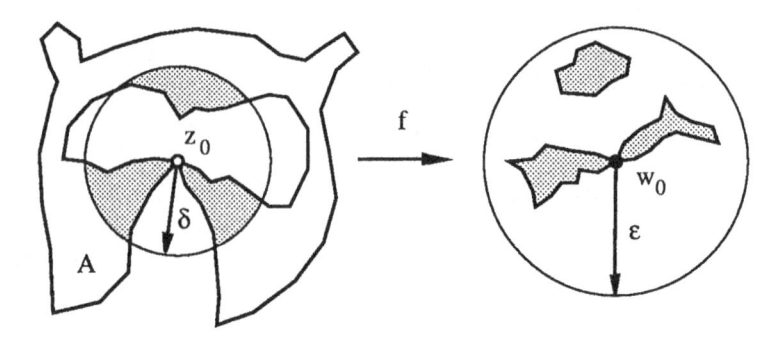

Figure 3.

— i.e., $|f(z) - w_0| < \epsilon$ for every z in A satisfying $0 < |z - z_0| < \delta$ — solely in the interest of fostering a smoother transition to limits in the extended complex plane when that topic is taken up in Chapter VIII.) It is a straightforward paraphrase of the definition of a limit to state

(2.9) $$\lim_{z \to z_0} f(z) = w_0 \Leftrightarrow \lim_{z \to z_0} |f(z) - w_0| = 0 \ .$$

Paralleling a remark made earlier in conjunction with convergent sequences is the comment that the right-hand side of (2.9) is convenient to work with because the real quantity $|f(z) - w_0|$ is subject to the "squeeze rule." To exemplify this point, let us verify that

$$\lim_{z \to 0} [z + 1 + z \operatorname{Log} z] = 1 \ .$$

We need only observe that for $z \neq 0$

$$|(z + 1 + z \operatorname{Log} z) - 1| = |z + z \operatorname{Log} z| = |z + z \operatorname{Log} |z| + iz \operatorname{Arg} z|$$

$$\leq |z| + ||z| \operatorname{Log} |z|| + |z||\operatorname{Arg} z| \leq |z| + ||z| \operatorname{Log} |z|| + \pi|z| \ .$$

It is not hard to see that the last expression tends to 0 as $z \to 0$, since with the aid of l'Hospital's rule we can evaluate the real limit $\lim_{t \to 0+} t \operatorname{Log} t$:

$$\lim_{t \to 0+} t \operatorname{Log} t = \lim_{t \to 0+} \frac{\operatorname{Log} t}{t^{-1}} = \lim_{t \to 0+} \frac{t^{-1}}{-t^{-2}} = -\lim_{t \to 0+} t = 0 \ .$$

We emphasize again that the question of the existence of $\lim_{z \to z_0} f(z)$ for a function $f: A \to \mathbb{C}$ is a meaningful one as soon as z_0 is a limit point of A. It is not required — or even relevant — that z_0 belong to A. The single most important type of limit with which we must come to terms in this book arises in just such a situation. Assume that we are presented with a function $f: A \to \mathbb{C}$ and an interior point z_0 of A. The function g whose rule of correspondence is

$$g(z) = \frac{f(z) - f(z_0)}{z - z_0}$$

2. Continuity and Limits of Functions   45

has domain-set $B = A \sim \{z_0\}$. Because z_0 is interior to A, the set B definitely has z_0 as a limit point. It therefore makes sense to ask: Does

$$\lim_{z \to z_0} g(z) = \lim_{z \to z_0} \frac{f(z) - f(z_0)}{z - z_0}$$

exist? It will not be lost on a student of calculus that what we are attempting to do here is to take a derivative of f at z_0, a "complex" derivative! This attempt represents the first step on the road to complex analysis, as we begin to learn in the next chapter.

We have already hinted at the connection between limits and continuity. Given the formal definitions of these two concepts, we can describe that connection precisely.

Theorem 2.6. *Suppose that z_0 belongs to a subset A of $\mathbb{C}$ and that, in addition, z_0 is a limit point of A. A function $f: A \to \mathbb{C}$ is continuous at z_0 if and only if $\lim_{z \to z_0} f(z) = f(z_0)$.*

A point of a plane set A that is not a limit point of A is called an *isolated point* of A. For such a point z_0 it is evident that $A \cap \Delta(z_0, r) = \{z_0\}$ once $r > 0$ is sufficiently small. A function $f: A \to \mathbb{C}$ is trivially continuous at all isolated points of A, the only points of A to which Theorem 2.6 does not address itself.

Two earlier results touching on continuity at a point have limit counterparts that are cited often enough to warrant explicit statement. The first corresponds to Theorem 2.1 and has a similar proof.

Theorem 2.7. *Let z_0 be a limit point of a subset A of the complex plane. A function $f: A \to \mathbb{C}$ has limit w_0 at z_0 if and only if for every sequence $\langle z_n \rangle$ in $A \sim \{z_0\}$ converging to z_0 the sequence $\langle f(z_n) \rangle$ converges to w_0.*

The second result certifies that the expected algebraic rules hold for limits of functions. It follows from Theorem 2.7 in the same way Theorem 2.2 does from Theorem 2.1.

Theorem 2.8. *Suppose that functions $f: A \to \mathbb{C}$ and $g: B \to \mathbb{C}$ satisfy $f(z) \to w_0$ and $g(z) \to w_0'$ as $z \to z_0$, where z_0 is a limit point of $A \cap B$. Then as $z \to z_0$: $cf(z) \to cw_0$ for any complex constant c; $\operatorname{Re} f(z) \to \operatorname{Re} w_0$ and $\operatorname{Im} f(z) \to \operatorname{Im} w_0$; $\overline{f(z)} \to \overline{w_0}$; $|f(z)| \to |w_0|$; $f(z) + g(z) \to w_0 + w_0'$; $f(z)g(z) \to w_0 w_0'$; if $w_0' \neq 0$, $f(z)/g(z) \to w_0/w_0'$.*

Here are some samples of limits evaluated with the help of the preceding material:

$$\lim_{z \to 1} [z^3 - 3\bar{z} + 2] = 0 \, ,$$

$$\lim_{z \to 0} \frac{2e^z + z^2 + i}{z + 1 + z \operatorname{Log} z} = \frac{\lim_{z \to 0} [2e^z + z^2 + i]}{\lim_{z \to 0} [z + 1 + z \operatorname{Log} z]} = 2 + i \, ,$$

$$\lim_{z\to 2i}\frac{z^3+8i}{z^2+4} = \lim_{z\to 2i}\frac{(z-2i)(z^2+2iz-4)}{(z-2i)(z+2i)} = \lim_{z\to 2i}\frac{z^2+2iz-4}{z+2i} = 3i\ .$$

In the first case the function involved is continuous at the point where the limit is being taken, so the problem of computing this limit reduces to the simple process of evaluating the function at that point. For the second limit we invoke Theorems 2.8 and 2.6, along with a limit computed earlier. The third example, which is of a standard type that one meets in beginning calculus, makes implicit use of the "local" character of limits: the existence and value of $\lim_{z\to z_0} f(z)$ are determined by the restriction of f to any pre-selected punctured disk centered at z_0. Thus, if two functions f and g agree in $\Delta^*(z_0,r)$ — meaning that their respective domain-sets have the same intersection with $\Delta^*(z_0,r)$ and $f(z) = g(z)$ for all z in that common intersection — then $\lim_{z\to z_0} f(z) = \lim_{z\to z_0} g(z)$, in the sense that both limits exist or neither does and that, when they exist, they are equal. In our example $f(z) = (z^3+8i)/(z^2+4)$ and $g(z) = (z^2+2iz-4)/(z+2i)$ are functions that agree in $\Delta^*(2i,r)$ for every $r > 0$. Also, $\lim_{z\to 2i} g(z) = g(2i) = 3i$, for g is continuous at $2i$. As a consequence, $\lim_{z\to 2i} f(z) = 3i$ as well.

A final remark about limits may help to ward off errors later on. It relates to the transition from limits that one encounters in the real-variable confines of calculus to their complex analogues. Consider, for a moment, the function $f(x) = e^{-1/x^2}$. We are aware from calculus that $\lim_{x\to 0} f(x) = \lim_{x\to 0} e^{-1/x^2} = 0$. There is a temptation — it must be resisted — to assume that one can merrily substitute the complex variable z for x here and write $\lim_{z\to 0} f(z) = \lim_{z\to 0} e^{-1/z^2} = 0$. However, if we set $z_n = (2n\pi)^{-1/2} e^{\pi i/4}$ for $n = 1, 2, \cdots$, we observe that $z_n \to 0$, whereas $f(z_n) = e^{-2n\pi i} = 1$ for every n, so $f(z_n) \to 1$. Thus, it is not true that $f(z) \to 0$ as $z \to 0$! The lesson: we cannot rely on standard limits from calculus to carry over automatically to the complex setting. Sometimes, of course, they do — as an example, we shall eventually explain how to make sense of $\sin z$ for complex z and verify that $\lim_{z\to 0} \sin z/z = 1$ — but, when a limit does generalize in this way, that fact usually requires new confirmation, often involving methods of proof quite different from the techniques employed to establish the original calculus limit. (This will be true, in particular, of the computation of complex derivatives.) The point is that the existence of $\lim_{z\to z_0} f(z)$ makes relatively severe "two-dimensional" demands on the function f near z_0: the quantity $f(z)$ is under a special kind of control for a totally random approach of z to z_0 through the domain-set of f. It is asking much less of a function to insist that its behavior be controlled only along a specific curve (e.g., the real axis) passing through z_0 or along sequences approaching z_0 from some specified direction.

3 Connected Sets

3.1 Disconnected Sets

An excellent "real life" example of a disconnected set is the state of Michigan. This "set" is the union of two "subsets", Upper and Lower Michigan, that would be separated from one another were it not for the Mackinac Bridge. Similarly, a subset A of the complex plane will be called "disconnected" if it can be realized as the union $A = B \cup C$ of non-empty sets B and C that are "separated" from each other in a way to be prescribed momentarily. (We shall allow for the possibility that the two "pieces" B and C of A might themselves be decomposable into still smaller pieces. This even happens in the Michigan example when various islands in the Great Lakes are taken into consideration.) By way of definition, a subset A of $\mathbb{C}$ is declared to be *disconnected* if there exist open sets U and V in $\mathbb{C}$ that obey the following three conditions: (i) U and V are disjoint; (ii) $A \cap U \neq \phi$ and $A \cap V \neq \phi$; (iii) $A \subset U \cup V$. Note especially that we can write $A = B \cup C$, where $B = A \cap U$ and $C = A \cap V$, and that these subsets B and C of A are separated by the open "buffer zones" created about them by U and V, respectively (Figure 4).

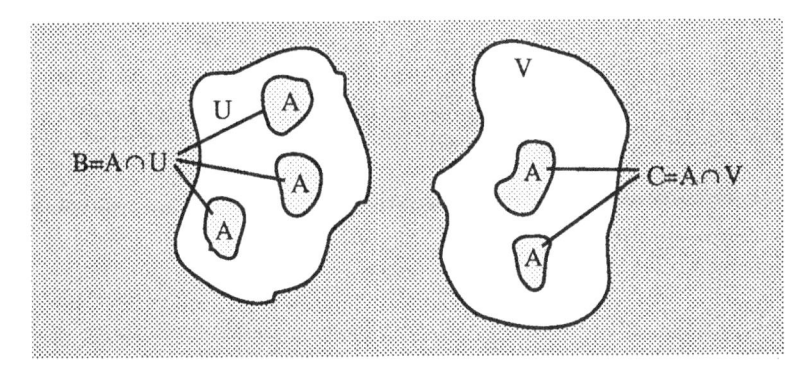

Figure 4.

The following lemma describes a property of sets that might initially appear different from disconnectedness, but is actually equivalent to it. In practice it is often easier to produce sets that meet condition (i) in this lemma than to find the disjoint sets required by condition (i) in the definition of a disconnected set. Such will be the case, for instance, in the proof of Theorem 3.8.

Lemma 3.1. *Let A be a set in the complex plane. Suppose that there exist a pair of open sets U^* and V^* in $\mathbb{C}$ satisfying the following three conditions: (i) $A \cap U^* \cap V^* = \phi$; (ii) $A \cap U^* \neq \phi$ and $A \cap V^* \neq \phi$; (iii) $A \subset U^* \cup V^*$. Then A is disconnected.*

Proof. The catch here is that (i) does not prevent the two sets U^* and V^* from intersecting at points away from A. The idea of the proof is to pare U^* and V^* down to disjoint open sets, call them U and V, that continue to meet requirements (ii) and (iii). This will confirm the disconnectedness of A. To carry out this plan, write $S = U^* \sim V^*$ and $T = V^* \sim U^*$. The sets S and T are non-empty — S contains $A \cap U^*$ and T contains $A \cap V^*$ — and disjoint. Because S lies in U^* and because U^* is open, corresponding to each point z of S we can fix a number $r_z > 0$ so that the disk $\Delta(z, 2r_z)$ is contained in U^*. We then set $\Delta_z = \Delta(z, r_z)$. In like fashion we choose for each w in T a disk $\Delta_w = \Delta(w, r_w)$ in such a way that $\Delta(w, 2r_w)$ is contained in V^*. Now take $U = \bigcup_{z \in S} \Delta_z$ and $V = \bigcup_{w \in T} \Delta_w$. As unions of open sets, U and V are themselves open. Also, $A \cap U = A \cap U^* \neq \phi$, $A \cap V = A \cap V^* \neq \phi$, and, since $A \cap U^* \cap V^* = \phi$,

$$A \subset S \cup T \subset U \cup V.$$

Finally, $U \cap V = \phi$. To prove this, it suffices to show that $\Delta_z \cap \Delta_w = \phi$ whenever z belongs to S and w to T. Suppose that z_0 were an element of $\Delta_z \cap \Delta_w$ for such z and w. Assume first that $r_z \leq r_w$. In this case we would have

$$|z - w| \leq |z - z_0| + |z_0 - w| < r_z + r_w \leq 2r_w ,$$

placing the element z of S in $\Delta(w, 2r_w)$, a subset of V^*. This would clearly violate the definition of $S = U^* \sim V^*$. Similar reasoning when $r_w \leq r_z$ would lead to an inconsistency with the definition of T. Therefore $U \cap V = \phi$ and A is seen to be disconnected. ∎

3.2 Connected Sets

A plane set A is said to be *connected* if it is not disconnected. Phrased in more positive terms, the statement that A is connected asserts that the only way to include A in the union of two disjoint open sets U and V is the trivial way, which is to have A contained in either U or V. Certainly a set consisting of a single point is connected. A more significant example, one whose connectedness is closely associated with the completeness of the real number system (see Appendix A.2.2), is a line segment in $\mathbb{C}$.

Theorem 3.2. *A line segment I in the complex plane is a connected set.*

Proof. Let I have endpoints z_0 and z_1. Then I is plainly described by $I = \{(1-t)z_0 + tz_1 : 0 \leq t \leq 1\}$. Suppose that I is contained in $U \cup V$, where U and V are disjoint open sets in $\mathbb{C}$. We must demonstrate that I lies in U or that it lies in V. We shall assume that z_0 belongs to U and verify the former. (If z_0 is a point of V, then an analogous argument shows that I is contained in V.) Using the fact that $f(t) = (1-t)z_0 + tz_1$ defines a

3. Connected Sets

continuous function on $[0,1]$ and that the sets U and V are open, we make the following observation: $(*)$ if t_0 is in $[0,1]$ and if $f(t_0)$ is an element of U (respectively, V), then $f\left([0,1]\cap [t_0-\delta, t_0+\delta]\right)$ lies in U (respectively, V) for some $\delta > 0$. Consider the set J of all t in $[0,1]$ with the property that $f([0,t])$ is a subset of U. By assumption $f(0) = z_0$ is a point of U, so $t=0$ belongs to J. In particular, $J \neq \phi$. As a non-empty subset of $[0,1]$, J has a supremum ($=$ least upper bound) t_0 that belongs to $[0,1]$. It is evident from the definition of t_0 that f must map the half-open interval $[0, t_0)$ into U. The observation $(*)$, coupled with the disjointness of U and V, then insures that $f(t_0)$ cannot be an element of V and so implies that $f([0, t_0])$ is a subset of U; i.e., t_0 is an element of J. If $t_0 < 1$, a second appeal to $(*)$ would produce a $\delta > 0$ with the feature that $f\left([t_0, t_0+\delta]\right)$ — and, hence, $f\left([0, t_0+\delta]\right)$ — is a subset of U. Thus $t_0+\delta$ would be a member of J larger than t_0, an impossibility. The conclusions: $t_0 = 1$, $J = [0,1]$, and $I = f(J)$ is contained in U. ∎

Taking the union of a collection of connected sets with a common point of intersection provides a mechanism for building up fairly complicated connected sets from simple ones. The principle underlying this procedure is:

Theorem 3.3. *Let $\mathcal{C}$ be a collection of connected sets in the complex plane, each of which contains a given point z_0. The union of the members of $\mathcal{C}$ is then a connected set.*

Proof. Write A for the union in question and assume that A is contained in $U\cup V$, where U and V are disjoint open sets. We must prove: A is either a subset of U or a subset of V. We suppose that z_0 belongs to U and check that A is contained in U. (The alternative case, where z_0 is an element of V, is dealt with similarly.) For this, we need only prove that each member C of $\mathcal{C}$ is a subset of U. The connectedness of any such C, together with the information that
$$z_0 \in C \subset A \subset U \cup V,$$
allows precisely that conclusion. ∎

In combination Theorems 3.2 and 3.3 confirm that a plane set A is connected if it contains a point z_0 with the following property: for each point z of $A \sim \{z_0\}$ the line segment with endpoints z_0 and z is contained in A. (We shall describe a set of this type as *starlike with respect to z_0*.) Included in this class of sets are a number of commonplace sets that we would intuitively think of as connected and that we can now officially certify as connected according to the technical understanding of the term. Prominent among them are the complex plane itself, the open disk $\Delta(z_0, r)$, and the closed disk $\overline{\Delta}(z_0, r)$. The same two theorems just cited and a straightforward induction argument verify the connectedness of any *polygonal arc*, the name we bestow on a set $A = \cup_{j=1}^n I_j$ formed by stringing together end-to-end a finite number of line segments $I_1, I_2 \ldots, I_n$, subject to the constraints

50   II. The Rudiments of Plane Topology

that I_{j+1} be disjoint from I_k when $k < j$ and that I_{j+1} intersect I_j only at a common endpoint of these two segments. Figure 5 shows an example with $n = 6$. The points z_0 and z_1 are the *endpoints* of this polygonal arc.

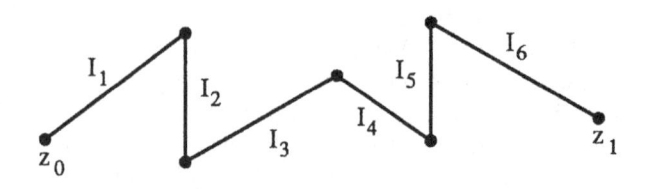

Figure 5.

3.3 Domains

A non-empty set D in the complex plane that is both open and connected is standardly referred to as a *domain* in $\mathbb{C}$. (This concept must not be confused with the "domain-set" of a function. The desire to avoid potential confusion between these two notions actually motivated our choice of the term "domain-set" — as opposed to just "domain" — for the set where a function is defined.) Throughout this book the letters D and G (for the German equivalent of "domain," *Gebiet*), when used to represent sets, will consistently stand for domains. The definition of connectedness has as a direct corollary a useful remark concerning domains.

Theorem 3.4. *If a plane domain D is expressed as the union $D = U \cup V$ of disjoint open sets U and V, then either $U = \phi$ or $V = \phi$.*

3.4 Components of Open Sets

Consider an open set U in $\mathbb{C}$ and a point z_0 of U. Let $D(z_0)$ designate the set consisting of z_0 and of all points z in U having the following property: there is a polygonal arc in U with one of its endpoints at z_0 and the other at z. Given a point z of $D(z_0)$, we set $A = \{z_0\}$ if $z = z_0$ and select a polygonal arc A in U with endpoints z_0 and z if $z \neq z_0$. We then take r small enough so that the open disk $\Delta = \Delta(z,r)$ is contained in U and, when $z \neq z_0$, intersects none of the line segments that make up A except the one with z as an endpoint. (See Figure 6.) If w is a point of Δ, then either w already lies on A — in this event w clearly belongs to $D(z_0)$ — or z_0 is linked to w by the polygonal arc in U formed when the segment from z to w is appended to A, which again places w in $D(z_0)$. In other words, Δ lies in $D(z_0)$. We have just demonstrated that $D(z_0)$ is an open set. Next,

3. Connected Sets

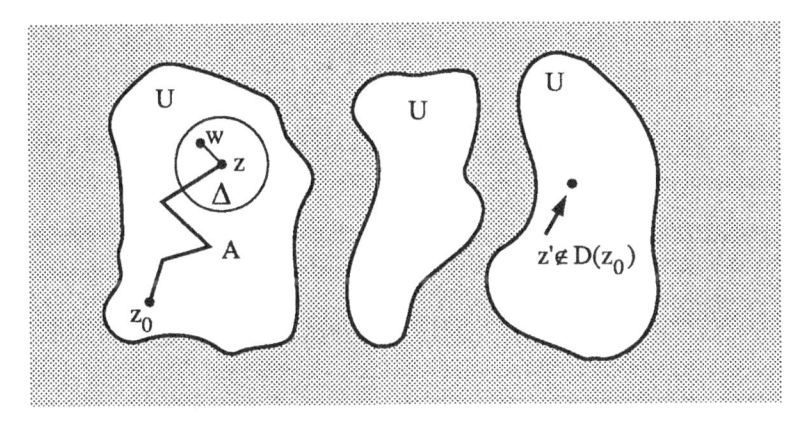

Figure 6.

if we choose for each z in $D(z_0)$, $z \neq z_0$, a polygonal arc A_z in U joining z_0 and z, it is evident that

$$D(z_0) = \bigcup_{z \neq z_0} A_z .$$

Theorem 3.3 lets us know that $D(z_0)$ is connected. Accordingly, this set is a domain. The domain $D(z_0)$ is called the *component of U containing z_0*. It is easy to see that, if z_0' is a second point of U, then $D(z_0) = D(z_0')$ when z_0' belongs to $D(z_0)$, whereas $D(z_0)$ and $D(z_0')$ are disjoint otherwise. Since every point of U belongs to some component of U, we can assert:

Theorem 3.5. *A non-empty open set U in the complex plane is the disjoint union of domains. To be specific, U is the union of its distinct components.*

For example, the open set $U = \{z : \operatorname{Re} z \neq 0\}$ has two components, the half-planes $D_1 = \{z : \operatorname{Re} z > 0\}$ and $D_2 = \{z : \operatorname{Re} z < 0\}$. The number of components of an open set may be infinite. This is true of $V = \{z : \operatorname{Re} z$ is not an integer$\}$, which has components $G_n = \{z : n < \operatorname{Re} z < n+1\}$ for $n = 0, \pm 1, \pm 2, \cdots$. It follows from Theorem 3.4 that a plane domain D has exactly one component, itself. Therefore, $D = D(z_0)$ for every z_0 in D, which proves the next proposition.

Theorem 3.6. *Any pair of distinct points z_0 and z_1 in a plane domain D can be made the endpoints of a polygonal arc lying in D.*

Another consequence of the definition of connectedness that will be needed later is:

Theorem 3.7. *Suppose that U is an open set in the complex plane and that A is a connected subset of U. Then A is contained in some component of U.*

One of the important characteristics of continuous functions is that they preserve connectedness.

Theorem 3.8. *Let $f: A \to \mathbb{C}$ be a continuous function, and let C be a connected subset of A. Then $f(C)$ is a connected set.*

Proof. The proof is indirect. We suppose that $f(C)$ is disconnected and derive a contradiction. Assume, therefore, that U and V are open sets in $\mathbb{C}$ satisfying (i) $U \cap V = \phi$, (ii) $f(C) \cap U \neq \phi$ and $f(C) \cap V \neq \phi$, and (iii) $f(C) \subset U \cup V$. Define $C_U = \{z \in C : f(z) \in U\}$ and $C_V = \{z \in C : f(z) \in V\}$. Then (i) implies that $C_U \cap C_V = \phi$, from (ii) it follows that C_U and C_V are non-empty, and (iii) shows that $C = C_U \cup C_V$. Taking advantage of the continuity of f and the openness of U, we choose for each z in C_U an open disk Δ_z centered at z with the property that $f(A \cap \Delta_z)$ is contained in U. Similarly, we pick for each w in C_V an open disk Δ_w such that $f(A \cap \Delta_w)$ lies in V. The sets $U^* = \cup_{z \in C_U} \Delta_z$ and $V^* = \cup_{w \in C_V} \Delta_w$ are open. Furthermore, by construction $C \cap U^* = C_U$ and $C \cap V^* = C_V$. This means that: (i) $C \cap U^* \cap V^* = C_U \cap C_V = \phi$; (ii) $C \cap U^* = C_U \neq \phi$ and $C \cap V^* = C_V \neq \phi$; (iii) $C = C_U \cup C_V \subset U^* \cup V^*$. Lemma 3.1 informs us that, contrary to hypothesis, C is disconnected. This is the contradiction we sought. Thus, $f(C)$ must be connected after all. ∎

4 Compact Sets

4.1 Bounded Sets and Sequences

A subset A of $\mathbb{C}$ is *bounded* if there is a constant $c > 0$ such that $|z| \leq c$ holds for every z in A; i.e., if A lies in some closed disk centered at the origin. A complex sequence $\langle z_n \rangle$ is bounded if there is a constant $c > 0$ such that $|z_n| \leq c$ holds for all n, or, stated differently, if $\{z_n : n = 1, 2, \cdots\}$ is a bounded set. A convergent sequence $\langle z_n \rangle$ is necessarily bounded. To see this, suppose that $z_n \to z_0$. Choose N such that $|z_n - z_0| < 1$ for all $n \geq N$ and set $c = \max\{1 + |z_0|, |z_1|, \ldots, |z_{N-1}|\}$. Then $|z_n| \leq c$ for every n. A fundamental theorem concerning the structure of the real number system states: *every bounded sequence of real numbers has a real accumulation point.* (See Theorem 2.2 in Appendix A.) This consequence of the completeness property of $\mathbb{R}$ is generally associated with the names of Bernard Bolzano (1781-1848) and Karl Weierstrass (1815-1897). It has a natural extension to complex sequences.

4. Compact Sets

Theorem 4.1. (Bolzano-Weierstrass Theorem) *Suppose that $\langle z_n \rangle$ is a bounded sequence in $\mathbb{C}$. Then $\langle z_n \rangle$ has at least one accumulation point. Furthermore, this sequence has exactly one accumulation point if and only if it is a convergent sequence having that unique accumulation point as its limit.*

Proof. Write $z_n = x_n + iy_n$. Then $\langle x_n \rangle$ and $\langle y_n \rangle$ are bounded sequences of real numbers. Appealing to the real version of the Bolzano-Weierstrass theorem, we choose a real accumulation point of $\langle x_n \rangle$ and label it x_0. Theorem 1.6 affirms the existence of a subsequence $\langle x_{n_k} \rangle$ of $\langle x_n \rangle$ such that $\lim_{k \to \infty} x_{n_k} = x_0$. We can then apply the same combination of theorems to the bounded real sequence $\langle y_{n_k} \rangle$ to produce a subsequence $\langle y_{n_{k_\ell}} \rangle$ of $\langle y_{n_k} \rangle$ having a limit y_0. According to Theorem 1.7, $x_{n_{k_\ell}} \to x_0$ as $\ell \to \infty$, so $z_{n_{k_\ell}} \to z_0 = x_0 + iy_0$ as $\ell \to \infty$. Again quoting Theorem 1.6, we conclude that z_0 is an accumulation point of $\langle z_n \rangle$.

We have earlier remarked that a convergent sequence has but one accumulation point, its limit. Assume, conversely, that our given bounded sequence $\langle z_n \rangle$ has a unique accumulation point z_0. We claim that $z_n \to z_0$. If not, there would exist an $\epsilon > 0$ with the property that z_n lies outside $\Delta(z_0, \epsilon)$ for infinitely many values of n. This would imply that $\langle z_n \rangle$ has a subsequence $\langle z_{n_k} \rangle$ in the set $\mathbb{C} \sim \Delta(z_0, \epsilon)$. As a bounded complex sequence $\langle z_{n_k} \rangle$ would, by the first part of the present theorem, have an accumulation point. Pick one and call it w_0. Then w_0 would certainly be an accumulation point of $\langle z_n \rangle$. However, $w_0 \neq z_0$, for by construction z_0 is not an accumulation point of $\langle z_{n_k} \rangle$. If $z_n \not\to z_0$, we would thus arrive at a contradiction. The conclusion that $z_n \to z_0$ is the alternative. ∎

4.2 Cauchy Sequences

A sequence $\langle z_n \rangle$ of complex numbers is christened a "Cauchy sequence," after Augustin-Louis Cauchy (1789-1857), if the distance between z_n and z_m tends to zero as n and m grow arbitrarily large. A precise definition proclaims the sequence $\langle z_n \rangle$ a *Cauchy sequence* if corresponding to each $\epsilon > 0$ there is an index $N = N(\epsilon)$ such that $|z_m - z_n| < \epsilon$ holds whenever m and n satisfy $m > n \geq N$. It is obvious that any convergent complex sequence is a Cauchy sequence. The converse is also true (Theorem 4.2), but not quite as obvious. The last fact is a significant one, for it equips us with a means to prove that the limit of a sequence exists without obliging us to identify that limit beforehand.

Theorem 4.2. (Cauchy Criterion for Convergence) *A sequence $\langle z_n \rangle$ in $\mathbb{C}$ is convergent if and only if it is a Cauchy sequence.*

Proof. Only the sufficiency is not evident. Suppose, therefore, that $\langle z_n \rangle$ is a Cauchy sequence. Then $\langle z_n \rangle$ is definitely bounded. Indeed, taking $\epsilon = 1$ in the definition of a Cauchy sequence, we can fix an index N such that $|z_m - z_N| < 1$ whenever $m > N$ and observe that

$$|z_m| \leq c = \max\{1 + |z_N|, |z_1|, \ldots, |z_N|\}$$

for all m. In light of Theorem 4.1, $\langle z_n \rangle$ must have an accumulation point in $\mathbb{C}$. In fact, it can have only one accumulation point. Assume, to the contrary, that z_0 and w_0 were distinct accumulation points of this sequence. Corresponding to $\epsilon = |z_0 - w_0|/3$ we could, according to the definition of a Cauchy sequence, choose an index N with the feature that $|z_m - z_n| < \epsilon$ whenever m and n satisfy $m > n \geq N$. Referring to the definition of an accumulation point, we would then be at liberty to select indices $n \geq N$ and $m > n$ for which

$$|z_n - z_0| < \epsilon \quad, \quad |z_m - w_0| < \epsilon\ .$$

With this selection of m and n we would be led to

$$|z_0 - w_0| \leq |z_0 - z_n| + |z_n - z_m| + |z_m - w_0| < \epsilon + \epsilon + \epsilon = |z_0 - w_0|,$$

a clear contradiction. We infer that $\langle z_n \rangle$ must have exactly one accumulation point. Theorem 4.1 attests to the fact that $\langle z_n \rangle$ is convergent. ∎

We mention in passing one important class of complex sequences that can be shown to be Cauchy sequences, the "contractive" sequences. (See Exercise 5.29.) A sequence $\langle z_n \rangle$ in $\mathbb{C}$ is *contractive* if there is a number λ, $0 < \lambda < 1$, with the property that

(2.10) $$|z_{n+2} - z_{n+1}| \leq \lambda |z_{n+1} - z_n|$$

for every $n \geq 1$. This definition simply demands that the distances between successive terms of the sequence decrease at an exponentially fast rate, a rate dictated by the number λ. As a specific example, let $\langle z_n \rangle$ be defined recursively by $z_1 = 0$, $z_2 = i$, and $z_n = (z_{n-1} + z_{n-2})/2$ for $n \geq 3$. We find here that

$$|z_{n+2} - z_{n+1}| = |z_{n+1} - z_n|/2$$

for all $n \geq 1$. Accordingly, $\langle z_n \rangle$ is contractive with $\lambda = 1/2$ — hence, a Cauchy sequence. We leave to the reader the problem of finding $\lim_{n \to \infty} z_n$ for this example (Exercise 5.9).

4.3 Compact Sets

A subset A of the complex plane is pronounced a *compact set* if each sequence in A possesses an accumulation point that belongs to A. (N.B. In a

4. Compact Sets

more general topological discussion a set obeying the preceding condition would quite probably be called "sequentially compact" rather than "compact." As it turns out, these generally different notions are equivalent in the present context, which fact justifies our use of the unqualified term "compact.") Two traits characterize the compact sets in $\mathbb{C}$. These are identified in the next theorem.

Theorem 4.3. *The subset A of $\mathbb{C}$ is compact if and only if A is both closed and bounded.*

Proof. Suppose first that A is compact. To show that A is closed we need only check that $\overline{A}$ is contained in A, for this makes it apparent that $A = \overline{A}$. Given z in $\overline{A}$, we can choose a sequence $\langle z_n \rangle$ in A such that $z_n \to z$ (Theorem 1.8). Then, on the one hand, z is the unique accumulation point of $\langle z_n \rangle$; on the other, the compactness of A guarantees the existence of an accumulation point of $\langle z_n \rangle$ in A. We conclude that z must belong to A. Therefore $A = \overline{A}$ and, as a result, A is a closed set. If A failed to be bounded, then A would not be contained in the disk $\overline{\Delta}(0, n)$ for any positive integer n. This would mean that we could select for each n a point w_n of A for which $|w_n| > n$. The sequence $\langle w_n \rangle$ in A so produced would have no accumulation points in $\mathbb{C}$, to say nothing of A, counter to the assumption that A is compact. It follows that a compact set in $\mathbb{C}$ is necessarily bounded.

For the converse, consider an arbitrary sequence $\langle z_n \rangle$ in a closed and bounded plane set A. Such a sequence is clearly bounded and, this being the case, $\langle z_n \rangle$ has at least one accumulation point (Theorem 4.1). As a closed set, however, A has to contain all accumulation points of $\langle z_n \rangle$ (Theorem 1.9). Consequently, $\langle z_n \rangle$ does have an accumulation point in A. This demonstrates the compactness of A. ∎

In several key situations later in this book we shall need to know that a compact set K sitting inside an open set U lies at a "positive distance" from the complement of U. We give this idea a precise formulation in a lemma whose conclusion we can interpret as saying that the distance from K to $\mathbb{C} \sim U$ is at least r.

Lemma 4.4. *Suppose that U is an open set in the complex plane and that K is a compact subset of U. There exists a radius $r > 0$ with the property that for each point z of K the disk $\Delta(z, r)$ is contained in U.*

Proof. Important here is that the same r should work for all points of K. If $U = \mathbb{C}$, there is nothing to prove: any $r > 0$ will do. We proceed under the assumption that $U \neq \mathbb{C}$, we suppose that no r with the described feature exists, and we derive a contradiction. Since $r = 1/n$ for n a positive integer does not have the desired property, it is possible to fix for each n a pair of points z_n in K and w_n in $F = \mathbb{C} \sim U$ satisfying $|z_n - w_n| < 1/n$. Because K is compact, the sequence $\langle z_n \rangle$ has an accumulation point in K. Let z_0

be such a point, and let $\langle z_{n_k}\rangle$ be a subsequence of $\langle z_n\rangle$ with the property that $\lim_{k\to\infty} z_{n_k} = z_0$. Then

$$|w_{n_k} - z_0| \leq |w_{n_k} - z_{n_k}| + |z_{n_k} - z_0| \leq \frac{1}{n_k} + |z_{n_k} - z_0| \to 0$$

as $k \to \infty$, showing that $\lim_{k\to\infty} w_{n_k} = z_0$ as well. This forces z_0 to be an accumulation point of $\langle w_n\rangle$. As $\langle w_n\rangle$ is a sequence in F and F is a closed set, Theorem 1.9 puts z_0 in F. In short, z_0 belongs to $K \cap F$, an impossible situation in view of the fact that K is a subset of $U = \mathbb{C} \sim F$. The contradiction proves that some $r > 0$ must have the stated property. ∎

Another basic fact concerning compact sets that will play a decisive role in certain future developments is usually attributed to Georg Cantor (1845-1918), one of the founders of modern set theory.

Theorem 4.5. (Cantor's Theorem) *Suppose that $\langle K_n\rangle$ is a sequence of non-empty compact sets in $\mathbb{C}$ satisfying $K_1 \supset K_2 \supset K_3 \supset \cdots$. Then $\bigcap_{n=1}^{\infty} K_n$ is not empty.*

Proof. Choose for each n a point z_n of K_n. Then $\langle z_n\rangle$ is a sequence in the compact set K_1, so we can be certain that $\langle z_n\rangle$ has at least one accumulation point. Let z_0 be such a point. We claim that z_0 belongs to $\bigcap_{n=1}^{\infty} K_n$. Fix $n \geq 1$. If z_0 happened to be in the set $\mathbb{C} \sim K_n$, then there would be an $\epsilon > 0$ such that the disk $\Delta(z_0, \epsilon)$ is contained in $\mathbb{C} \sim K_n$. (Recall that, by Theorem 4.3, $\mathbb{C} \sim K_n$ is an open set.) Because K_m is contained in K_n when $m \geq n$ and because z_m lies in K_m, we could infer that no term z_m with $m \geq n$ belongs to $\Delta(z_0, \epsilon)$, contrary to the definition of z_0 as an accumulation point of $\langle z_n\rangle$. As a result, z_0 must lie in K_n — this for each $n \geq 1$. ∎

Just as continuous functions transform connected sets to connected sets, so also do they preserve compactness.

Theorem 4.6. *Let $f: A \to \mathbb{C}$ be a continuous function, and let K be a compact subset of A. Then $f(K)$ is a compact set.*

Proof. Consider a sequence $\langle w_n\rangle$ in $f(K)$. We must show that this sequence has an accumulation point in $f(K)$. We begin by fixing z_n in K for which $w_n = f(z_n)$. Since K is given to be compact, we can choose a point z_0 of K at which the sequence $\langle z_n\rangle$ accumulates and then extract a subsequence $\langle z_{n_k}\rangle$ from $\langle z_n\rangle$ such that $\lim_{k\to\infty} z_{n_k} = z_0$. The point $w_0 = f(z_0)$ lies in $f(K)$, and the continuity of f at z_0 tells us that

$$\lim_{k\to\infty} w_{n_k} = \lim_{k\to\infty} f(z_{n_k}) = f(z_0) = w_0 \ .$$

This plainly marks w_0 as an accumulation point of $\langle w_n\rangle$ in $f(K)$. ∎

Theorems 4.3 and 4.6 inform us that the image of a closed and bounded subset of $\mathbb{C}$ under a continuous, complex-valued function is again both

4. Compact Sets

closed and bounded. It should be mentioned, however, that individually neither closedness nor boundedness is automatically preserved by a continuous function.

The least upper bound and greatest lower bound of a non-empty, bounded set S of real numbers are clearly points of $\overline{S}$. If such a set S is also closed, then S contains these points and so has both a largest element and a smallest one. This observation leads to an oft-quoted corollary of Theorem 4.6.

Corollary 4.7. *Suppose that A is a subset of $\mathbb{C}$, that $f: A \to \mathbb{R}$ is a continuous function, and that K is a compact subset of A. There exist points z_0 and w_0 in K such that $f(z_0) \leq f(z) \leq f(w_0)$ holds for every z in K; i.e., when restricted to K the function f attains both a maximum value and a minimum value.*

4.4 Uniform Continuity

A function $f: A \to \mathbb{C}$ is said to be *uniformly continuous on a subset S of A* if corresponding to each $\epsilon > 0$ there exists a $\delta = \delta(\epsilon) > 0$ such that

$$(2.11) \qquad f[S \cap \Delta(z, \delta)] \subset \Delta(f(z), \epsilon)$$

for every z in S; in other words, such that $|f(z) - f(w)| < \epsilon$ holds whenever z and w are points of S satisfying $|z - w| < \delta$. The thing that distinguishes this property from the mere continuity of f on S is the demand that one and the same δ serve in (2.11) for every point z of S. If, for example, f obeys an estimate of the type

$$|f(z) - f(w)| \leq c |z - w|$$

for all z and w in S, where $c > 0$ is a constant, then f is uniformly continuous on S: given $\epsilon > 0$ we can take $\delta = \epsilon/c$ and rest assured that $|f(z) - f(w)| < \epsilon$ for any points z and w of S with $|z - w| < \delta$. A connection between uniform continuity and compactness is established in the final result of this chapter.

Theorem 4.8. *Let $f: A \to \mathbb{C}$ be a continuous function, and let K be a compact subset of A. Then f is uniformly continuous on K.*

Proof. Assume that the conclusion is false. This means there is some $\epsilon > 0$ to which no corresponding δ exists, as required by the definition of uniform continuity for f on K. Fix such an ϵ. Since, in particular, taking $\delta = 1/n$ with n a positive integer does not "work" for this ϵ, we can choose for each n a pair of points z_n and w_n in K such that $|z_n - w_n| < 1/n$, but such that $|f(z_n) - f(w_n)| \geq \epsilon$. As K is compact, the sequence $\langle z_n \rangle$ has a subsequence

$\langle z_{n_k} \rangle$ with the property that $\lim_{k \to \infty} z_{n_k} = z_0$, where z_0 is a point of K. Indeed, any accumulation point z_0 of $\langle z_n \rangle$ belonging to K gives rise to such subsequences. For the reason that

$$|w_{n_k} - z_0| \leq |w_{n_k} - z_{n_k}| + |z_{n_k} - z_0| \leq \frac{1}{n_k} + |z_{n_k} - z_0| \to 0$$

as $k \to \infty$, we conclude that also $\lim_{k \to \infty} w_{n_k} = z_0$. The continuity of f at z_0 thus implies that

$$\lim_{k \to \infty} |f(z_{n_k}) - f(w_{n_k})| = |f(z_0) - f(z_0)| = 0 ,$$

in obvious conflict with the fact that $|f(z_{n_k}) - f(w_{n_k})| \geq \epsilon$ for every k. This contradiction resulted from assuming a lack of uniform continuity on the part of f on the set K. Therefore, f must be uniformly continuous on that set. ∎

5 Exercises for Chapter II

5.1 Exercises for Section II.1

5.1. Verify that $\Delta(z_0, r)$ is an open set and that $\overline{\Delta}(z_0, r)$ is a closed set.

5.2. Prove Theorems 1.1 and 1.2.

5.3. Classify each of the following sets as open, closed, or neither open nor closed: (i) $A = \{z : -\pi < \operatorname{Im} z \leq \pi\}$; (ii) $B = \{z : 1 < |z| < 2\}$; (iii) $C = \{z : |\operatorname{Re} z| + |\operatorname{Im} z| \leq 1\}$; (iv) $D = \{z : 0 < \max\{x, y\} \leq 1\}$; (v) $E = \{z : y > x^2\}$; (vi) $F = \{z : \operatorname{Re} z \text{ and } \operatorname{Im} z \text{ are rational}\}$. (*Hint.* At least for (i)-(v) it may help to view the set graphically.)

5.4. Determine the boundary, closure, and interior of each of the sets listed in Exercise 5.3.

5.5. Prove Theorem 1.3.

5.6. Compute the limit of $\langle z_n \rangle$ when: (i) $z_n = i^{n!} + 2^{-n}$; (ii) $z_n = (in^3 + 1)/(2n^3 + n^2)$; (iii) $z_n = 2^{-n+i\sqrt{n}}$; (iv) $z_n = \sqrt[n]{z}$ for z in $\mathbb{C}$; (v) $z_n = \sqrt[n]{n}$; (vi) $z_n = (1 + n^{-1})^n + i(1 - n^{-1})^n$; (vii) $z_n = z^n/n!$ for z in $\mathbb{C}$; (viii) $z_n = n\sin(1/n) + ine^{-n}$.

5.7. Assuming the usual algebraic rules for real sequential limits, derive Theorem 1.5 from Lemma 1.4.

5.8. Let $a \geq 0$ and $b \geq 0$. Show that $\lim_{n \to \infty} \sqrt[n]{a^n + b^n} = \max\{a, b\}$.

5.9. A complex sequence $\langle z_n \rangle$ is defined recursively by $z_1 = 0, z_2 = i$, and $z_n = (z_{n-1} + z_{n-2})/2$ for $n \geq 3$. Show that $z_n = (2i/3)[1 - (-1/2)^{n-1}]$ for $n \geq 2$. Conclude that $\langle z_n \rangle$ is convergent and determine $\lim_{n \to \infty} z_n$.

5.10. A sequence $\langle z_n \rangle$ is defined recursively by $z_1 = 1$ and $z_n = \sqrt{1 + z_{n-1}}$

5. Exercises for Chapter II

for $n \geq 2$. Demonstrate that $0 \leq z_n \leq 2$ for every n and that, in addition, $z_1 \leq z_2 \leq z_3 \leq \cdots$. Infer from Theorem 2.1 in Appendix A that $\langle z_n \rangle$ is convergent. Find $\lim_{n \to \infty} z_n$.

5.11. Identify all accumulation points of the following sequences; for each accumulation point exhibit a subsequence that converges to it: (i) $z_n = n^{-1} + (-1)^n$; (ii) $z_n = 2^{-n} + (-1)^n + i^n$; (iii) $z_n = n^{1/n} \sin(n\pi/2)$; (iv) $z_n = [2 + \cos(n\pi)] e^{n\pi i/5}$; (v) $z_n = \sin(n\pi/2) + i\cos(n\pi/6)$; (vi) $z_n = n\sin(n\pi/2) + i\cos(n\pi/6)$; (vii) $z_n = n[\sin(n\pi/2) + i\cos(n\pi/6)]$; (viii) $z_n = n[\sin(n\pi/3) + i\cos(n\pi/6)]$.

5.12. Prove Theorem 1.7.

5.2 Exercises for Section II.2

5.13. Prove Theorem 2.2.

5.14. Prove Theorem 2.5.

5.15. Find all points of discontinuity of the following functions: (i) $f(z) = \sqrt[5]{z}$; (ii) $g(z) = \text{Arg}(z^2)$; (iii) $h(z) = \text{Log}(z^3 + 1)$; (iv) $k(z) = \sqrt{1 - z^2}$; (v) $\ell(z) = \text{Arg}(1 - \sqrt{z})$.

5.16. Modify the proof of Theorem 2.1 to prove Theorem 2.7.

5.17. Compute the following limits:

(i) $\lim_{z \to i} [2z^2 - iz^3 + z \, \text{Arg} \, \bar{z}]$;

(ii) $\lim_{z \to i} [(z^4 + 1)/(z + i)]$;

(iii) $\lim_{z \to -i} [(z^4 - 1)/(z + i)]$;

(iv) $\lim_{z \to 2i} [(z^2 - iz + 2)/(z^2 + 4)]$;

(v) $\lim_{z \to 1} [(\sqrt{z} - 1)/(z - 1)]$;

(vi) $\lim_{z \to 0} [(e^z + z \, \text{Log} \, z)/(1 - z^2 \, \text{Arg} \, z)]$.

(*Warning.* The use of l'Hospital's rule to evaluate complex limits is prohibited here.)

5.3 Exercises for Section II.3

5.18. Prove that a straight line in the complex plane is a connected set.

5.19. Characterize the following sets as disconnected or connected: (i) $A = \{z : |z| > 1\}$; (ii) $B = \{z : |z - i| \neq 1\}$; (iii) $C = \{z : x^2 - y^2 = 0\}$;

(iv) $D = \{z: x^2 - y^2 < 1\}$; (v) $E = \{z: y^4 - x^2 > 0\}$; (vi) $F = \{z: \operatorname{Im} z$ is a rational multiple of $\operatorname{Re} z\}$.

5.20. Let $A = \mathbb{C} \sim \{z: \operatorname{Re} z$ and $\operatorname{Im} z$ are rational$\}$. Show that A is a connected set.

5.21. Identify all of the components of the following open sets: (i) $S = \{z: |x| + |y| \neq 1\}$; (ii) $T = \{z: x^2 y^2 \neq 1\}$; (iii) $U = \{z: z^3$ is not real$\}$; (iv) $V = \{z: e^z$ is not real$\}$; (v) $W = \{z: |z^2 - 1| \neq 1\}$. (*Hint.* Pictures help. For (v) recall Exercise I.4.20.)

5.22. Prove that the circle $K(z_0, r)$ is a connected set. (*Hint.* Make use of Theorems 3.2 and 3.8.)

5.23. If A is a connected set in the complex plane, then its closure $\overline{A}$ is also connected. Justify this statement.

5.24. Prove Theorem 3.7.

5.25. If G and D are plane domains such that $G \subset D$, prove that $\partial G \cap D \neq \phi$, except in the situation that $G = D$.

5.26. Suppose a continuous function $f: D \to \mathbb{C}$, where D is a plane domain, has the property that $|[f(z)]^2 - 1| < 1$ for every z in D. Demonstrate that either $|f(z) - 1| < 1$ for every z in D or $|f(z) + 1| < 1$ for every z in D. (*Hint.* Use Theorem 3.4.)

5.27. Let D be a plane domain, and let $f: D \to \mathbb{C}$ be a continuous function satisfying $e^{f(z)} = 1$ for every z in D. Show that f is a constant function whose sole value lies in the set $\{2k\pi i: k = 0, \pm 1, \pm 2, \cdots\}$.

5.28. A continuous function $f: D \to \mathbb{C}$, in which D is a plane domain, never takes a purely imaginary value and satisfies the condition $[f(z)]^2 = z$ for every z in D. Verify that either $f(z) = \sqrt{z}$ for every z in D or $f(z) = -\sqrt{z}$ for every z in D.

5.4 Exercises for Section II.4

5.29. Let $\langle z_n \rangle$ be a contractive sequence in $\mathbb{C}$ with contraction constant λ, $0 < \lambda < 1$. Confirm that $\langle z_n \rangle$ is a Cauchy sequence by first deriving the estimate
$$|z_m - z_n| \leq \frac{\lambda^{n-1}|z_2 - z_1|}{1 - \lambda},$$
valid when $m > n \geq 2$. (*Hint.* $|z_m - z_n| \leq |z_m - z_{m-1}| + |z_{m-1} - z_{m-2}| + \cdots + |z_{n+1} - z_n|$ if $m > n \geq 1$.)

5.30. Suppose that $\langle K_n \rangle$ is a sequence of non-empty, compact, and connected sets in the complex plane satisfying $K_1 \supset K_2 \supset K_3 \supset \cdots$. Show that $K = \bigcap_{n=1}^{\infty} K_n$ is also non-empty, compact, and connected. (*Hint.* For the connectedness part, let U and V be disjoint open sets whose union contains K. Argue that $K_n \subset U \cup V$ once n is sufficiently large. Do this by an indirect proof involving the compact sets $\widetilde{K}_n = K_n \sim (U \cup V)$.)

5.31. Assume that K is a compact set, that $f: K \to \mathbb{C}$ is a continuous function, and that $f(z) \neq 0$ holds for every z in K. Prove that $f(K)$ lies outside $\Delta(0, r)$ for some $r > 0$.

5.32. A continuous function $f: \mathbb{C} \to \mathbb{C}$ satisfies $\lim_{|z| \to \infty} f(z) = 0$. (This means that corresponding to each $\epsilon > 0$ there is an $r > 0$ with the property that $|f(z)| < \epsilon$ whenever $|z| \geq r$.) Prove that $|f|$ attains a maximum value at some point of the complex plane.

5.33. If a function f is uniformly continuous on a set S and if $\langle z_n \rangle$ is a Cauchy sequence in S, then $\langle f(z_n) \rangle$ is a Cauchy sequence. Confirm this fact. Deduce that f transforms any convergent sequence in S — even one whose limit does not belong to S — to a convergent sequence.

5.34. Given that $f: A \to \mathbb{C}$ is uniformly continuous on A and that z_0 is a limit point of A, establish the fact that $\lim_{z \to z_0} f(z)$ exists. (*Hint.* Use Theorem 2.7 and Exercise 5.33.)

5.35. Let $S = \{z : z = 0 \text{ or } |\operatorname{Arg} z| \leq \alpha\}$, where $0 < \alpha < \pi$. Verify that the function $f(z) = \sqrt{z}$ is uniformly continuous on S, but that it is not uniformly continuous on the set $\mathbb{C} \sim (-\infty, 0)$. (*Hint.* In the first half of the exercise begin by showing that $|f(z) - f(w)| \leq c|z - w|$ for z and w in S satisfying $|z| \geq 1$ and $|w| \geq 1$, where $c = 2^{-1} \sec(\alpha/2)$. Then direct attention to the compact set $S \cap \overline{\Delta}(0, 2)$.)

5.36. If a continuous function $f: \mathbb{C} \to \mathbb{C}$ satisfies $f(z) \to 0$ as $|z| \to \infty$, then f is uniformly continuous on $\mathbb{C}$. Support this assertion.

5.37. Let a continuous function $f: \mathbb{C} \to \mathbb{C}$ have the following two properties: (i) $|f(z)| \to \infty$ as $|z| \to \infty$ and (ii) $f(\mathbb{C})$ is an open set. Prove that $f(\mathbb{C}) = \mathbb{C}$. (*Hint.* Assume that $G = f(\mathbb{C}) \neq \mathbb{C}$ and use Exercise 5.25 to derive a contradiction.)

Chapter III

Analytic Functions

Introduction

The time has now come to begin the study of complex analysis proper by bringing into the picture a concept of differentiability for functions of a complex variable. Owing to certain superficial features — the notation, the basic terminology, and the derivative formulas for the standard elementary functions are cases in point — the derivative we are about to introduce may at first sight leave one with the impression of being nothing more than a clone of the usual derivative met in calculus. Such initial impressions notwithstanding, the differences between ordinary calculus and what might be called "complex calculus" are profound and far-reaching. Hopefully the extent of — and the reasons for — those differences will become abundantly clear as the fundamental principles of complex function theory are laid down in the present and subsequent chapters of this text. The modest first step in the process is explaining what differentiability in the complex setting is supposed to mean.

1 Complex Derivatives

1.1 Differentiability

A function $f: A \to \mathbb{C}$ is said to be *differentiable at a point* z_0 provided z_0 is an interior point of A at which the limit

$$\lim_{z \to z_0} \frac{f(z) - f(z_0)}{z - z_0}$$

exists. When this is the case we appropriate the familiar notation and terminology found in calculus, writing

(3.1) $$f'(z_0) = \lim_{z \to z_0} \frac{f(z) - f(z_0)}{z - z_0}$$

1. Complex Derivatives

and speaking of $f'(z_0)$ as the *derivative of f at z_0*. (Use of the Leibniz notation $\dfrac{df}{dz}(z_0)$ in place of $f'(z_0)$ is not uncommon.) It is sometimes preferable to express (3.1) in a slightly different fashion, that being

$$f'(z_0) = \lim_{h \to 0} \frac{f(z_0 + h) - f(z_0)}{h} \,.$$

There is an alternative (and geometrically more illuminating) way of formulating the definition of differentiability, one that involves the notion of a linear approximation to a function. In this formulation $f: A \to \mathbb{C}$ is pronounced differentiable at z_0 if z_0 is interior to A and if there exists a complex number c with the property that

(3.2) $$f(z) = f(z_0) + c(z - z_0) + E(z)$$

for every z in A, where $E: A \to \mathbb{C}$ is a function satisfying the condition

(3.3) $$\lim_{z \to z_0} \frac{|E(z)|}{|z - z_0|} = 0 \,.$$

Thus $L(z) = f(z_0) + c(z - z_0)$ defines a linear function of z that approximates $f(z)$ up to an "error" $E(z)$ which, for z close to z_0, has small magnitude in comparison with $|z - z_0|$. This alternate description of differentiability is easily seen to be equivalent to the original one. First, conditions (3.2) and (3.3) clearly imply that

$$c = \lim_{z \to z_0} \frac{f(z) - f(z_0)}{z - z_0} \,,$$

in other words, that $f'(z_0)$ exists and is equal to c. Conversely, under the assumption that $f'(z_0)$ exists, we arrive at (3.2) and (3.3) by taking $c = f'(z_0)$ and setting $E(z) = f(z) - f(z_0) - f'(z_0)(z - z_0)$ for z in A.

When $c = f'(z_0) \neq 0$ in (3.2), the term $c(z - z_0)$ strongly dominates $E(z)$ in the immediate vicinity of z_0, with the result that the contribution of the error term to $f(z)$ there becomes almost negligible. We might therefore reasonably expect the geometric behavior of the mapping $w = f(z)$ for z close to z_0 to mimic the behavior of the transformation $w = L(z) = f(z_0) + f'(z_0)(z - z_0)$. The geometry of L is readily understood: L is just a similarity transformation that rotates the complex plane about the point z_0 through the angle $\theta = \text{Arg}[f'(z_0)]$, then dilates the plane so that each ray emanating from z_0 is mapped to itself and all distances get scaled by the factor $|f'(z_0)|$, and finally translates the plane so as to move z_0 to $f(z_0)$ (cf., Exercise I.4.50). And indeed, at least when f is known to be differentiable at all points in some open set containing z_0, the description just given of L does convey a correct intuitive picture, accurate to a remarkable degree, of the nonlinear mapping $w = f(z)$ in the neighborhood of z_0. When $f'(z_0) = 0$, on the other hand, the analysis of the "approximate geometry" of $w = f(z)$ near z_0 becomes considerably more delicate and requires information of higher precision than that recorded in (3.2) and (3.3). This matter will be addressed in Chapter VIII.

With a function $f: A \to \mathbb{C}$ we can now associate a *derivative function* $f': B \to \mathbb{C}$ whose domain-set B consists of all points z in the interior of A at which f is differentiable and whose rule of correspondence is

$$f'(z) = \lim_{h \to 0} \frac{f(z+h) - f(z)}{h} \, .$$

We must allow for the eventuality that there are no such points z (i.e., that B is the empty set), in which case the function f' does not exist.

We remark in passing that differentiability on the part of a function f at a point z_0 entails the continuity of f at that point, for

$$\lim_{z \to z_0} f(z) = \lim_{z \to z_0} \left[f(z_0) + (z - z_0) \frac{f(z) - f(z_0)}{z - z_0} \right]$$

$$= f(z_0) + 0 \cdot f'(z_0) = f(z_0) \, .$$

1.2 Differentiation Rules

If each of two functions $f: A \to \mathbb{C}$ and $g: B \to \mathbb{C}$ is differentiable at a point z_0, so also are the functions cf for any complex constant c, $f + g$, fg, and, under the added constraint that $g(z_0) \neq 0$, f/g. Furthermore, the expected derivative rules are in force:

(3.4)
$$\begin{cases} (cf)'(z_0) = cf'(z_0) \, , \\ (f+g)'(z_0) = f'(z_0) + g'(z_0) \, , \\ (fg)'(z_0) = f'(z_0)g(z_0) + f(z_0)g'(z_0) \, , \\ \left(\frac{f}{g}\right)'(z_0) = \frac{f'(z_0)g(z_0) - f(z_0)g'(z_0)}{[g(z_0)]^2} \, . \end{cases}$$

The third formula, for instance, is verified as follows:

$$(fg)'(z_0) = \lim_{z \to z_0} \frac{f(z)g(z) - f(z_0)g(z_0)}{z - z_0}$$

$$= \lim_{z \to z_0} \left[\frac{f(z) - f(z_0)}{z - z_0} g(z) + f(z_0) \frac{g(z) - g(z_0)}{z - z_0} \right]$$

$$= \lim_{z \to z_0} \frac{f(z) - f(z_0)}{z - z_0} \lim_{z \to z_0} g(z) + f(z_0) \lim_{z \to z_0} \frac{g(z) - g(z_0)}{z - z_0}$$

1. Complex Derivatives

$$= f'(z_0)g(z_0) + f(z_0)g'(z_0) \ .$$

The verification of the remaining formulas in (3.4) is equally straightforward (Exercise 6.4). Notice that the continuity of g at z_0 was needed in the above computation. Observe, as well, that this continuity along with the condition $g(z_0) \neq 0$ accounts for the fact that z_0, plainly an interior point of $A \cap B$, is an interior point of the domain-set of f/g.

The simplest example of a function possessing a derivative is a function f that is constant on $\mathbb{C}$. In this case it is all but immediate that $f'(z) = 0$ for every z in $\mathbb{C}$. What follows next are further elementary examples illustrating the concept of the derivative and the use of the above differentiation rules.

EXAMPLE 1.1. Compute $f'(z)$ for $f(z) = z^n$, with n a positive integer.

For any z_0 in $\mathbb{C}$ we calculate

$$f'(z_0) = \lim_{z \to z_0} \frac{z^n - z_0^n}{z - z_0}$$

$$= \lim_{z \to z_0} \frac{(z - z_0)\left(z^{n-1} + z^{n-2}z_0 + \cdots + zz_0^{n-2} + z_0^{n-1}\right)}{z - z_0}$$

$$= \lim_{z \to z_0} \left(z^{n-1} + z^{n-2}z_0 + \cdots + zz_0^{n-2} + z_0^{n-1}\right) = nz_0^{n-1} \ .$$

In other words, the usual formula from calculus remains intact in the complex setting: $f'(z) = nz^{n-1}$ for arbitrary z in $\mathbb{C}$.

EXAMPLE 1.2. Determine $f'(z)$ if $f(z) = z^n$ for a negative integer n.

Here, of course, the origin is missing from the domain-set of f. Writing $f(z) = 1/(z^{-n})$, we employ the quotient rule for derivatives and the preceding example to arrive at

$$f'(z) = \frac{-(-n)z^{-n-1}}{z^{-2n}} = nz^{n-1} \ .$$

Again the customary formula holds.

EXAMPLE 1.3. Evaluate $f'(i)$ for $f(z) = 3z^4 + \pi z^3 + iz^{-1}$.

Using the foregoing examples along with (3.4), we quickly obtain $f'(z) = 12z^3 + 3\pi z^2 - iz^{-2}$, so that $f'(i) = 12i^3 + 3\pi i^2 - i(i)^{-2} = -3\pi - 11i$.

EXAMPLE 1.4. Determine the "critical points" of $f(z) = z/(z^2 - 1)$, meaning the points z for which $f'(z) = 0$.

Application of the quotient rule from (3.4) yields the formula $f'(z) = -(z^2 + 1)/(z^2 - 1)^2$. The critical points of f are seen to be i and $-i$.

EXAMPLE 1.5. Discuss the differentiability of $f(z) = \sqrt{z}$.

Write $\theta(z) = \operatorname{Arg} z$. Then $\sqrt{z} = \sqrt{|z|}\, e^{i\theta(z)/2}$. Recall that the function θ is continuous in the set $D = \mathbb{C} \sim (-\infty, 0]$ (Lemma II.2.4), a fact which makes it clear that f, too, is continuous in D. For z_0 in D we take advantage of this continuity in computing

$$f'(z_0) = \lim_{z \to z_0} \frac{\sqrt{z} - \sqrt{z_0}}{z - z_0} = \lim_{z \to z_0} \frac{\sqrt{z} - \sqrt{z_0}}{(\sqrt{z} - \sqrt{z_0})(\sqrt{z} + \sqrt{z_0})}$$

$$= \lim_{z \to z_0} \frac{1}{\sqrt{z} + \sqrt{z_0}} = \frac{1}{2\sqrt{z_0}}.$$

When $z_0 = 0$, we have

$$\frac{f(z) - f(z_0)}{z - z_0} = \frac{\sqrt{z}}{z} = \frac{1}{\sqrt{z}}.$$

Since $|1/\sqrt{z}| = 1/\sqrt{|z|} \to \infty$ as $z \to 0$, it is evident that f cannot be differentiable at the origin. Finally, if z_0 is real and negative, we observe that $\lim_{h \to 0+} \theta(z_0 - ih) = -\pi$, with the consequence that

$$\lim_{h \to 0+} f(z_0 - ih) = \lim_{h \to 0+} \sqrt{|z_0 - ih|}\, e^{i\theta(z_0 - ih)/2}$$

$$= \sqrt{|z_0|}\, e^{-\pi i/2} = -i\sqrt{|z_0|} = -\sqrt{z_0} \neq f(z_0).$$

This means that f is not even continuous at z_0, much less differentiable there. To summarize: $f(z) = \sqrt{z}$ is differentiable only at the points of the set $\mathbb{C} \sim (-\infty, 0]$, where $f'(z) = 1/(2\sqrt{z})$. Once more the formula is just what one would predict based on experience with calculus. The need to avoid the points of the negative real axis, however, might come as something of a surprise.

As in calculus, the most important differentiation rule in complex analysis is the "chain rule," which governs the differentiation of compositions: $(g \circ f)' = (g' \circ f) f'$.

Theorem 1.1. (Chain Rule) *Let functions $f: A \to \mathbb{C}$ and $g: B \to \mathbb{C}$ be such that $f(A)$ is contained in B. If f is differentiable at a point z_0 and g is differentiable at $w_0 = f(z_0)$, then the composition $g \circ f$ is differentiable at z_0 and $(g \circ f)'(z_0) = g'(w_0) f'(z_0)$.*

Proof. Consider the functions $F: A \to \mathbb{C}$ and $G: B \to \mathbb{C}$ defined by

$$F(z) = \begin{cases} [f(z) - f(z_0)]/[z - z_0] & \text{if } z \neq z_0, \\ f'(z_0) & \text{if } z = z_0, \end{cases}$$

and

$$G(w) = \begin{cases} [g(w) - g(w_0)]/[w - w_0] & \text{if } w \neq w_0, \\ g'(w_0) & \text{if } w = w_0. \end{cases}$$

1. Complex Derivatives

Since
$$\lim_{z \to z_0} F(z) = \lim_{z \to z_0} \frac{f(z) - f(z_0)}{z - z_0} = f'(z_0) = F(z_0),$$
F is continuous at z_0. Similarly, G is continuous at w_0, which implies that the composition $G \circ f$ is continuous at z_0. (Remember Theorem II.2.3. Being differentiable at z_0, f is certainly continuous there.) For z in A, $z \neq z_0$, we can write

$$\frac{g \circ f(z) - g \circ f(z_0)}{z - z_0} = \frac{g[f(z)] - g[f(z_0)]}{z - z_0}$$

$$= \begin{cases} \dfrac{g[f(z)] - g(w_0)}{f(z) - w_0} \dfrac{f(z) - f(z_0)}{z - z_0} & \text{if } f(z) \neq w_0, \\ 0 & \text{if } f(z) = w_0, \end{cases}$$

$$= G[f(z)]F(z).$$

It follows that

$$(g \circ f)'(z_0) = \lim_{z \to z_0} \frac{g \circ f(z) - g \circ f(z_0)}{z - z_0}$$

$$= \lim_{z \to z_0} G[f(z)]F(z)$$

$$= G[f(z_0)]F(z_0) = g'(w_0)f'(z_0),$$

as desired. ∎

EXAMPLE 1.6. Calculate $h'(z)$ for $h(z) = \left[(z^2 - 1)/(z^2 + 1)\right]^{10}$.

Noting that $h = g \circ f$, where $f(z) = (z^2 - 1)/(z^2 + 1)$ and $g(z) = z^{10}$, we compute $f'(z) = 4z/(z^2 + 1)^2$, $g'(z) = 10z^9$, and apply the chain rule to obtain

$$h'(z) = g'[f(z)]f'(z) = 10\left(\frac{z^2-1}{z^2+1}\right)^9 \frac{4z}{(z^2+1)^2} = \frac{40z(z^2-1)^9}{(z^2+1)^{11}}.$$

1.3 Analytic Functions

Suppose that U is a non-empty open subset of the complex plane, that f is a complex-valued function whose domain-set contains U, and that f is differentiable at every point of U. When these conditions are met we say that f is *analytic in U*. (Some books employ the terminology *holomorphic in U* in referring to a function of this type.) We repeat one point for emphasis: in this book use of the phrase "analytic in U" always presupposes that U is an open set in the complex plane, whether or not this is made explicit at the time. A function whose domain-set is an open set in which that function is analytic is known as an *analytic function*.

If a function f is analytic in U, then U is by definition contained in the domain-set of its derivative function f'. It is a remarkable fact, to be verified in Chapter V, that f' is itself analytic in U; i.e., the second derivative $f''(z) = (f')'(z)$ of f exists at each point of U! By induction it follows that, once a function f is determined to be analytic in U, the existence in U of its derivatives $f', f'', f''', \ldots, f^{(n)}, \cdots$ of all orders is assured! Accepting for a moment the truth of this statement, we recognize in it a radical departure from the situation confronted in calculus, where a function $f: \mathbb{R} \to \mathbb{R}$ can be constructed that has a continuous derivative $f': \mathbb{R} \to \mathbb{R}$ which, in turn, is differentiable at not a single point. This is just one of many phenomena we shall encounter that distinguish "complex calculus" from "real calculus."

The formulas in (3.4) confirm the fact that, if both f and g are analytic in U, then so too are the functions cf for constant c, $f+g$, and fg, while f/g is analytic in the open set $V = \{z \in U : g(z) \neq 0\}$, provided it is not empty. The composition of analytic functions is likewise analytic: if f is analytic in a set U, if g is analytic in a set V, and if $f(U)$ is contained in V, then $g \circ f$ is analytic in U. The chain rule justifies the last assertion.

At present the available stock of functions to which we are able to point as examples of analytic functions is decidedly meager. Certainly it includes all rational functions of z, meaning as we recall from Chapter I.3.1 polynomial functions $f(z) = a_0 + a_1 z + \cdots + a_n z^n$ and quotients thereof. Example 1.5 shows that $f(z) = \sqrt{z}$ defines a function which is analytic in any open set U not intersecting the negative real axis. These simple functions can be used to build more complicated examples. For instance, $h(z) = \sqrt{1+z^2}$ is analytic in $U = \mathbb{C} \sim \{z : \operatorname{Re} z = 0 \text{ and } |\operatorname{Im} z| \geq 1\}$, since $h = g \circ f$ with $f(z) = 1 + z^2$ and $g(z) = \sqrt{z}$, since f is analytic in U, and since $f(U) = \mathbb{C} \sim (-\infty, 0]$, a set in which g is analytic. (As a matter of fact, U is the largest open set where h is analytic: any larger open set contains points z satisfying $\operatorname{Re} z = 0$ and $|\operatorname{Im} z| > 1$, at which h is discontinuous.) Among our next orders of business is the expansion of this supply of examples to include exponential, trigonometric, logarithm, and general power functions. One method of achieving this goal appeals to the so-called "Cauchy-Riemann equations."

2 The Cauchy-Riemann Equations

2.1 The Cauchy-Riemann System of Equations

Suppose that a function $f = u + iv$ is differentiable at $z_0 = x_0 + iy_0$. By definition this requires that z_0 be interior to the domain-set of f and that $f'(z_0) = \lim_{z \to z_0}[f(z) - f(z_0)]/[z - z_0]$ exist. Important here is that no constraints are imposed or implied with regard to the manner in which z tends to z_0. Differentiability demands that the same limit $f'(z_0)$ be ob-

2. The Cauchy-Riemann Equations

tained no matter what possible sequence or curve or set z moves along in approaching z_0. Given that this limit exists, therefore, it can be used to shed light on certain other familiar limits which also involve the difference quotient $[f(z) - f(z_0)]/[z - z_0]$, but in which z is forced to approach z_0 in some specialized fashion. We can, for instance, elect to have z tend to z_0 along a line parallel to the real axis by taking $z = x + iy_0$ and letting $x \to x_0$. If we do so and if, at the same time, we think of f as a function of the two real variables x and y, we find that

$$f'(z_0) = \lim_{x \to x_0} \frac{f(x + iy_0) - f(x_0 + iy_0)}{x - x_0}$$

$$= \lim_{x \to x_0} \frac{f(x, y_0) - f(x_0, y_0)}{x - x_0}$$

$$= \lim_{x \to x_0} \frac{u(x, y_0) - u(x_0, y_0)}{x - x_0} + i \lim_{x \to x_0} \frac{v(x, y_0) - v(x_0, y_0)}{x - x_0}$$

$$= u_x(x_0, y_0) + iv_x(x_0, y_0) \;;$$

i.e., we conclude that the usual partial derivatives $u_x(z_0)$ and $v_x(z_0)$ — and, thus, the partial derivative $f_x(z_0) = u_x(z_0) + iv_x(z_0)$ — exist and satisfy

(3.5) $$f'(z_0) = u_x(z_0) + iv_x(z_0) = f_x(z_0) \;.$$

Similarly, by setting $z = x_0 + iy$ and allowing $y \to y_0$ we arrive at

$$f'(z_0) = \lim_{y \to y_0} \frac{f(x_0 + iy) - f(x_0 + iy_0)}{i(y - y_0)}$$

$$= \lim_{y \to y_0} \frac{v(x_0, y) - v(x_0, y_0)}{y - y_0} - i \lim_{y \to y_0} \frac{u(x_0, y) - u(x_0, y_0)}{y - y_0}$$

$$= v_y(z_0) - iu_y(z_0) \;.$$

This tells us that the partial derivative $f_y(z_0) = u_y(z_0) + iv_y(z_0)$ also exists and that

(3.6) $$f'(z_0) = v_y(z_0) - iu_y(z_0) = -if_y(z_0) \;.$$

Comparing (3.5) and (3.6) leads to the conclusion that

$$u_x(z_0) = v_y(z_0) \;, \quad u_y(z_0) = -v_x(z_0) \;.$$

The system of partial differential equations $u_x = v_y$, $u_y = -v_x$ is known as the *Cauchy-Riemann system* in honor of Augustin-Louis Cauchy (1789-1857) and Georg Bernhard Riemann (1826-1866), two of the architects of complex analysis. The foregoing discussion can be summarized in the statement: *a necessary condition for a function $f = u + iv$ to be differentiable at a point z_0 is that u and v satisfy the Cauchy-Riemann equations at z_0.*

We shall learn from Example 3.3 that the above necessary condition for differentiability at a point is not generally sufficient to imply differentiability at that point. Given proper supplementary information about a function, however, we can sometimes use the Cauchy-Riemann equations to establish its differentiability. For the purposes of the present book the next theorem will provide a satisfactory differentiability criterion of this type.

Theorem 2.1. *Suppose that a function $f = u + iv$ is defined in an open subset U of the complex plane and that the partial derivatives u_x, u_y, v_x, and v_y exist everywhere in U. If each of these partial derivatives is continuous at a point z_0 of U and if the Cauchy-Riemann equations are satisfied at z_0, then f is differentiable there and $f'(z_0) = f_x(z_0) = -if_y(z_0)$.*

Proof. In order to minimize notational complications, we shall write out the details of the proof only for the case $z_0 = 0$. Set $c = f_x(0) = a + ib$, so that
$$a = u_x(0) = v_y(0) \quad , \quad b = v_x(0) = -u_y(0) \ .$$
Our task is to verify that
$$\left| \frac{f(z) - f(0)}{z} - c \right| = \frac{|f(z) - f(0) - cz|}{|z|} \to 0$$
as $z \to 0$. To do this, we first fix an open disk $\Delta = \Delta(0, r)$ contained in U. Because u_x and u_y exist everywhere in Δ, we remark that, when u is restricted to a horizontal line segment in Δ, it reduces to a differentiable function of the real variable x with derivative u_x, while on vertical segments in this disk u is a function of y alone and has derivative u_y. We consider a point $z = x + iy$ satisfying $0 < |z| < r$ and write
$$u(z) - u(0) = u(x, y) - u(0, 0) = u(x, y) - u(0, y) + u(0, y) - u(0, 0) \ .$$
In view of the preceding comment we can invoke the mean value theorem from calculus and choose, corresponding to z, points ζ on the line segment between iy and $x + iy$ and η on the segment between 0 and iy for which
$$u(x, y) - u(0, y) = u_x(\zeta)x \quad , \quad u(0, y) - u(0, 0) = u_y(\eta)y \ .$$
(See Figure 1. To avoid any technical oversights here, we agree to set $\zeta = 0$ if $x = 0$ and $\eta = 0$ if $y = 0$.) Thus $|\zeta| \leq |z|$, $|\eta| \leq |z|$, and
$$u(z) - u(0) = u_x(\zeta)x + u_y(\eta)y \ .$$

2. The Cauchy-Riemann Equations

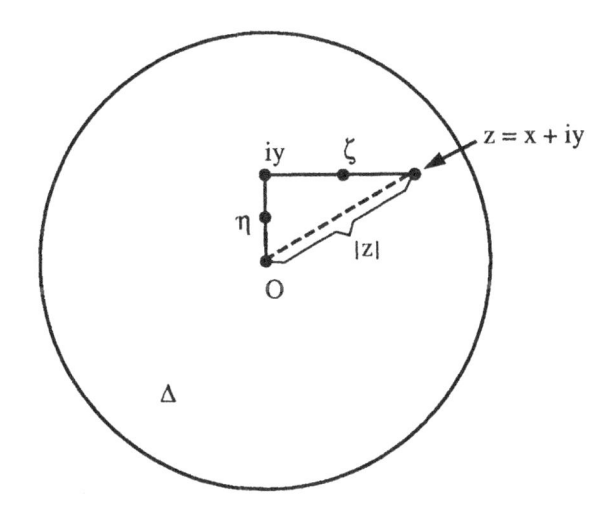

Figure 1.

In a like manner we can pick a pair of points θ and ξ, depending on z, such that $|\theta| \leq |z|$, $|\xi| \leq |z|$, and

$$v(z) - v(0) = v_x(\theta)x + v_y(\xi)y .$$

It is clear from their selection that ζ, η, θ, and ξ must all tend to 0 when z does.

Next, we compute

$$f(z) - f(0) - cz = u(z) - u(0) + i\{v(z) - v(0)\} - (a + ib)(x + iy)$$

$$= u(z) - u(0) - u_x(0)x - u_y(0)y + i\{v(z) - v(0) - v_x(0)x - v_y(0)y\}$$

$$= [u_x(\zeta) - u_x(0)]x + [u_y(\eta) - u_y(0)]y$$

$$+ i\{[v_x(\theta) - v_x(0)]x + [v_y(\xi) - v_y(0)]y\} .$$

Consequently, since $|x|/|z| \leq 1$ and $|y|/|z| \leq 1$,

$$\frac{|f(z) - f(0) - cz|}{|z|} \leq \frac{|u_x(\zeta) - u_x(0)||x|}{|z|} + \frac{|u_y(\eta) - u_y(0)||y|}{|z|}$$

$$+ \frac{|v_x(\theta) - v_x(0)||x|}{|z|} + \frac{|v_y(\xi) - v_y(0)||y|}{|z|}$$

$$\leq |u_x(\zeta) - u_x(0)| + |u_y(\eta) - u_y(0)| + |v_x(\theta) - v_x(0)| + |v_y(\xi) - v_y(0)| .$$

The last expression has limit 0 as $z \to 0$, for ζ, η, θ, and ξ tend to 0 along with z and, by assumption, u_x, u_y, v_x, and v_y are continuous at 0. We conclude that f is differentiable at 0 with $f'(0) = f_x(0)$. Owing to (3.6), we can rewrite this as $f'(0) = -if_y(0)$.

The proof when $z_0 \neq 0$ differs only cosmetically from the one presented. Alternatively, the result for $z_0 \neq 0$ can be retrieved from the case $z_0 = 0$ merely by applying the latter to the function $g(z) = f(z + z_0)$, which exhibits the same behavior at the origin that f does at z_0. ∎

The following examples demonstrate the use of the Cauchy-Riemann equations to test for differentiability.

EXAMPLE 2.1. At which points is the function $f(z) = \bar{z}$ differentiable?

Taking the real and imaginary parts of f, we see that $u(z) = x$ and $v(z) = -y$, functions whose partial derivatives are given by $u_x(z) = 1$, $v_y(z) = -1$, and $u_y(z) = v_x(z) = 0$ for every z. At no point are the Cauchy-Riemann conditions met. As a result, there are no points at which f is differentiable.

EXAMPLE 2.2. Identify all points of differentiability for the function $f(z) = 2xy + i(x^2 + y^2)$, and determine $f'(z)$ at such points.

In this example $u_x(z) = v_y(z) = 2y$ and $u_y(z) = v_x(z) = 2x$. These partial derivatives are continuous in $\mathbb{C}$. The only points where the Cauchy-Riemann equations are satisfied are the points with $x = 0$, i.e., the points of the imaginary axis. Theorem 2.1 implies that f is differentiable at these and only these points and that $f'(iy) = u_x(iy) + iv_x(iy) = 2y$.

Let U be an open set in the complex plane. We say that a function $f = u + iv$ is in the class $C^k(U)$ if U is contained in the domain-set of f and if all possible partial derivatives f_x, f_y, f_{xx}, f_{xy}, f_{yx}, f_{yy}, $\cdots$ up to and including those of order k exist throughout U and are continuous functions in U. Of course, by f_{xx} we mean $u_{xx} + iv_{xx}$, and we understand other higher order partials similarly. (N.B. In this book the notation $C^k(U)$ is used exclusively in conjunction with open sets U. Another author might well observe a different convention with regard to this notation.) The reader is reminded of the "equality of mixed partials" of functions in $C^k(U)$. For example, if f is a member of $C^3(U)$, it is true that $f_{xy} = f_{yx}$, $f_{xxy} = f_{xyx} = f_{yxx}$, and $f_{yyx} = f_{yxy} = f_{xyy}$ in U. To say that f belongs to $C^0(U)$, which is usually abbreviated to $C(U)$, merely asserts that f is continuous in U. A function that is in $C^k(U)$ for every non-negative integer k is said to be in the class $C^\infty(U)$. Any polynomial $f(z) = \sum_{j=0}^{n} \sum_{k=0}^{m} a_{jk} x^j y^k$ in the real variables x and y with complex coefficients a_{jk} gives an example of a function that is a member of $C^\infty(U)$ for every open subset U of $\mathbb{C}$.

Theorem 2.1 leads directly to a convenient criterion for deciding whether a function from the class $C^1(U)$ is analytic in U.

Theorem 2.2. *A function $f = u + iv$ belonging to the class $C^1(U)$ is analytic in U if and only if the pair of functions u and v provides a solution*

2. The Cauchy-Riemann Equations

to the Cauchy-Riemann system in that open set. When such is the case, $f' = f_x = -if_y$ in U.

2.2 Consequences of the Cauchy-Riemann Relations

The Cauchy-Riemann equations can be exploited to establish a number of the elementary properties of analytic functions by reducing their verifications to problems in two-variable calculus. The section at hand offers some demonstration of this technique. The principal calculus result needed here is Lemma 2.3, the proof of which we risk presenting in perhaps excessive detail because it provides an excellent model for the "connectedness arguments" that will be used in later chapters to pass from local statements about analytic functions to corresponding global statements. We remind the reader that in standard topological usage the term "domain" indicates a non-empty, open, connected set and that in this book the symbol D will, without exception, signify a domain. (See Section II.3.3.)

Lemma 2.3. *Suppose that u is a real-valued function which is defined in a plane domain D and that $u_x(z) = u_y(z) = 0$ for every z in D. Then u is constant in D.*

Proof. Consider an open disk Δ that is contained in D, say with center $c = a + ib$. If $z = x + iy$ is an arbitrary point of Δ, the mean value theorem can be quoted — recall the proof of Theorem 2.1 — to certify the existence of points ζ and η in Δ for which

$$u(z) - u(c) = u(x, y) - u(a, b) = u(x, y) - u(a, y) + u(a, y) - u(a, b)$$

$$= u_x(\zeta)(x - a) + u_y(\eta)(y - b) = 0 \cdot (x - a) + 0 \cdot (y - b) = 0 .$$

Accordingly, $u(z) = u(c)$ for all z in Δ. Expressed differently, u is constant in each open disk that lies entirely inside D.

Next, fix a point z_0 of D and define sets U and V by

$$U = \{z \in D : u(z) = u(z_0)\} \quad , \quad V = \{z \in D : u(z) \neq u(z_0)\} .$$

The foregoing discussion implies that U and V are open sets. We confirm this fact for the set U. The proof for V is similar. Suppose that z belongs to U. We must produce an open disk $\Delta = \Delta(z, r)$ such that Δ is a subset of U. We choose any open disk Δ centered at z and contained in D. Since z is an element of U and since u is constant in Δ, $u(z') = u(z) = u(z_0)$ for each z' in Δ, placing Δ inside U. Hence, U is open. Now quite clearly $D = U \cup V$ and $U \cap V = \phi$. Because D is connected, one of the sets U or V must be empty (Theorem II.3.4). That certainly can't be U, for by design z_0 belongs to U. The conclusion: $V = \phi$ or, what is the same, $U = D$.

Therefore $u(z) = u(z_0)$ for every z in D, which establishes the constancy of u there. ∎

From Lemma 2.3 we deduce immediately:

Theorem 2.4. *Suppose that a function f is analytic in a domain D and that $f'(z) = 0$ for every z in D. Then f is constant in D.*

Proof. Write $f = u + iv$. Recalling (3.5) and (3.6) we obtain $0 = f'(z) = u_x(z) + iv_x(z) = v_y(z) - iu_y(z)$ for each point z of D. It follows that $u_x = u_y = v_x = v_y = 0$ throughout the domain. Lemma 2.3 asserts that both u and v — and, therefore, f — are constant in D. ∎

The reader ought to recognize in the preceding theorem the analogue of a standard theorem in calculus concerning differentiable functions on open intervals. In Theorem 2.4 (and in many other results that occur later) a domain in the complex plane furnishes a natural two-dimensional substitute for an open interval.

The final result of this section is somewhat more surprising than Theorem 2.4. It allows us to catch our first glimpse of the special behavior we must learn to expect from analytic functions.

Theorem 2.5. *Let $f = u + iv$ be analytic in a domain D. If any one of the functions u, v, or $|f|$ is constant in this domain, then f itself is constant in D.*

Proof. Assume first that u is constant in D. Then $u_x(z) = u_y(z) = 0$ for all z in D. From the Cauchy-Riemann equations we infer that $f'(z) = u_x(z) + iv_x(z) = u_x(z) - iu_y(z) = 0$ for every z in D. Theorem 2.4 shows that f is constant in D. A similar argument leads to the identical conclusion under the assumption that v is constant in D.

Suppose now that $|f|$ is constant in D; i.e., $u^2 + v^2 = c$ throughout D for some constant $c \geq 0$. If $c = 0$, the conclusion that $f = 0$ everywhere in D is evident. We proceed assuming that $c > 0$. Differentiating both sides of the relation $u^2 + v^2 = c$ with respect to x and y produces the equations

$$2uu_x + 2vv_x = 0 \quad , \quad 2uu_y + 2vv_y = 0$$

in D. Appealing to the Cauchy-Riemann conditions, we see that this system of differential equations is equivalent to

$$uu_x - vu_y = 0 \quad , \quad uu_y + vu_x = 0 \; .$$

If we multiply the first of these new equations by u and the second by v, and if we then add the results, we arrive at

$$cu_x = (u^2 + v^2)u_x = 0 \; ,$$

which implies that $u_x = 0$ throughout D. It can likewise be shown that $u_y = 0$ in D. Lemma 2.3 states that u is constant in D — as then is f, in view of the first part of the present proof. ∎

In Chapter VIII we shall establish a very important theorem that has Theorem 2.5 and many results akin to it as instant corollaries. The theorem referred to is the "Open Mapping Theorem": *if a function f is analytic and non-constant in a domain D, then f(D) is a domain.* If, for example, $f = u + iv$ is analytic in D and if u is constant there, then $f(D)$ lies on a vertical line. Such being the case, $f(D)$ cannot be a domain, leaving only the option that f is constant in D.

3 Exponential and Trigonometric Functions

3.1 Entire Functions

The title *entire function* is traditionally bestowed on a function that is analytic in the whole complex plane. The most obvious example of an entire function is a polynomial in z, $f(z) = a_0 + a_1 z + \cdots + a_n z^n$. Not quite so evident a fact is the "entireness" of the exponential function $f(z) = e^z$. We include its confirmation in a short sequence of examples focussing on this function and on the general subject of entire functions.

EXAMPLE 3.1. Verify that $f(z) = e^z$ defines an entire function with $f'(z) = e^z$.

Breaking f down into its real and imaginary parts u and v, we have $u(z) = e^x \cos y$ and $v(z) = e^x \sin y$. It follows that

$$u_x(z) = e^x \cos y = v_y(z) \quad , \quad u_y(z) = -e^x \sin y = -v_x(z) .$$

From this it is apparent that f belongs to the class $C^1(\mathbb{C})$ and that the Cauchy-Riemann equations are satisfied by u and v everywhere in the complex plane. Theorem 2.2 insures that f is entire and that $f'(z) = f_x(z) = e^x \cos y + i e^x \sin y = e^z$.

EXAMPLE 3.2. What are the critical points of the entire function $f(z) = e^{z^4 + 2z^2}$?

Write $f = g \circ h$, where $g(z) = e^z$ and $h(z) = z^4 + 2z^2$. The chain rule can be invoked to give $f'(z) = g'[h(z)]h'(z) = (4z^3 + 4z)e^{z^4 + 2z^2}$. The zeros of f' are identifiable as $0, i,$ and $-i$. These are the only critical points of f.

EXAMPLE 3.3. Define a function $f \colon \mathbb{C} \to \mathbb{C}$ by

$$f(z) = \begin{cases} \exp(-z^{-4}) & \text{if } z \neq 0 , \\ 0 & \text{if } z = 0 . \end{cases}$$

Show that $u = \operatorname{Re} f$ and $v = \operatorname{Im} f$ obey the Cauchy-Riemann conditions at every point of the complex plane, but that, in spite of this, f is not an entire function.

The function f is clearly analytic in the set $U = \mathbb{C} \sim \{0\}$, where it has derivative $f'(z) = 4z^{-5} \exp(-z^{-4})$. This guarantees that u and v meet the Cauchy-Riemann requirements in U. What happens at the origin? To answer this question we first compute

$$\lim_{x \to 0^+} \frac{f(x) - f(0)}{x} = \lim_{x \to 0^+} \frac{e^{-1/x^4}}{x} = \lim_{t \to \infty} \frac{t^{1/4}}{e^t} = \lim_{t \to \infty} \frac{(1/4)t^{-3/4}}{e^t} = 0$$

by initially making the change of variable $t = x^{-4}$ and by then appealing to l'Hospital's rule. A similar calculation reveals that

$$\lim_{x \to 0^-} \frac{f(x) - f(0)}{x} = 0.$$

The existence and equality of these one-sided limits demonstrate that $f_x(0)$ exists and that $f_x(0) = u_x(0) + iv_x(0) = 0$. In an analogous fashion it can be established that $f_y(0) = u_y(0) + iv_y(0) = 0$. In particular, u and v satisfy the Cauchy-Riemann equations at the origin. The function f, however, is not differentiable there. Indeed, f is not even continuous at the origin. To see this, it is enough to observe that

$$\lim_{r \to 0^+} f\left(re^{\pi i/4}\right) = \lim_{r \to 0^+} e^{1/r^4} = \infty.$$

Therefore, f is not an entire function. The reason that the Cauchy-Riemann relations do not force the differentiability of f at the origin is that f_x and f_y are not continuous there, a fact which is easily verified directly.

EXAMPLE 3.4. Determine the entire function $f = u + iv$ satisfying $f(0) = i$ and $u(z) = 2x^3 y - 2xy^3 + x^2 - y^2$.

The functions u and v are required to satisfy the Cauchy-Riemann equations everywhere in $\mathbb{C}$. It is thus true of v that

$$v_x(z) = -u_y(z) = -2x^3 + 6xy^2 + 2y \quad , \quad v_y(z) = u_x(z) = 6x^2 y - 2y^3 + 2x.$$

The first equation can hold only if v has the form

$$v(z) = -\frac{x^4}{2} + 3x^2 y^2 + 2xy + g(y),$$

where $g(y)$ indicates a function depending on the single variable y. Consequently,

$$v_y(z) = 6xy + 2x + \frac{dg}{dy}(y).$$

Comparing this with the earlier expression for v_y, we deduce that $\frac{dg}{dy}(y) = -2y^3$ for all y. The implication is that $g(y) = (-1/2)y^4 + c$ for some constant c and so that

$$v(z) = -\frac{x^4}{2} + 3x^2 y^2 - \frac{y^4}{2} + 2xy + c.$$

3. Exponential and Trigonometric Functions

The demand that $f(0) = i$ requires us to take $c = 1$, which gives

$$f(z) = 2x^3y - 2xy^3 + x^2 - y^2 + i\left(-\frac{x^4}{2} + 3x^2y^2 - \frac{y^4}{2} + 2xy + 1\right).$$

Setting $x = (z + \bar{z})/2$, $y = (z - \bar{z})/2i$ and doing some laborious algebra — Exercise 6.20 suggests a quicker route to the same conclusion — we discover that

$$f(z) = -\frac{iz^4}{2} + z^2 + i.$$

3.2 Trigonometric Functions

In view of our definition of e^{it} for a real number t the verification of the identities

$$\cos t = \frac{e^{it} + e^{-it}}{2}, \quad \sin t = \frac{e^{it} - e^{-it}}{2i}$$

for real t is routine. Suppose that it is our wish to define the cosine and sine of a complex number z. Naturally, we would like the extended definitions to agree with the usual ones in case z is real. Prompted by this consideration and by the above identities we make the definitions in the only sensible fashion:

$$\cos z = \frac{e^{iz} + e^{-iz}}{2}, \quad \sin z = \frac{e^{iz} - e^{-iz}}{2i}.$$

The remaining elementary trigonometric functions of z — $\tan z$, $\cot z$, $\sec z$ and $\csc z$ — are defined, as they ordinarily are, in terms of $\cos z$ and $\sin z$. For example, we compute

$$\cos i = \frac{e^{i^2} + e^{-i^2}}{2} = \frac{e^{-1} + e}{2} = \frac{e^2 + 1}{2e},$$

$$\sin i = \frac{e^{i^2} - e^{-i^2}}{2i} = \frac{e^{-1} - e}{2i} = i\left(\frac{e^2 - 1}{2e}\right),$$

$$\tan i = \frac{\sin i}{\cos i} = i\left(\frac{e^2 - 1}{e^2 + 1}\right),$$

$$\sec i = \frac{1}{\cos i} = \frac{2e}{e^2 + 1}.$$

The standard trigonometric identities remain intact in the complex setting. Thus $\cos^2 z + \sin^2 z = 1$, $\cos(2z) = \cos^2 z - \sin^2 z$, $\cos[(\pi/2) - z] = \sin z$, $1 + \tan^2 z = \sec^2 z$, and $\sin(z + w) = \sin z \cos w + \cos z \sin w$, to cite just a few. We supply a justification for the first of these, leaving the others as exercises:

$$\cos^2 z + \sin^2 z = \left(\frac{e^{iz}+e^{-iz}}{2}\right)^2 + \left(\frac{e^{iz}-e^{-iz}}{2i}\right)^2$$

$$= \frac{e^{2iz}+2+e^{-2iz}}{4} - \frac{e^{2iz}-2+e^{-2iz}}{4} = \frac{4}{4} = 1 \, .$$

It will be advantageous for future considerations to have available expressions for the real and imaginary parts of $\cos z$ and $\sin z$. Setting $z = x + iy$, we calculate

$$\cos z = \frac{e^{iz}+e^{-iz}}{2} = \frac{e^{-y+ix}+e^{y-ix}}{2}$$

$$= \frac{e^{-y}(\cos x + i \sin x) + e^{y}(\cos x - i \sin x)}{2}$$

$$= \cos x \left(\frac{e^y + e^{-y}}{2}\right) - i \sin x \left(\frac{e^y - e^{-y}}{2}\right) \, .$$

The last line can be written more succinctly using hyperbolic functions. Recall that the *hyperbolic cosine* ($\cosh t$) and *hyperbolic sine* ($\sinh t$) of a real number t are defined by

$$\cosh t = \frac{e^t + e^{-t}}{2} \, , \quad \sinh t = \frac{e^t - e^{-t}}{2} \, .$$

We have thus demonstrated that

(3.7) $\qquad\qquad \cos z = \cos x \cosh y - i \sin x \sinh y \, .$

A similar calculation leads to the formula

(3.8) $\qquad\qquad \sin z = \sin x \cosh y + i \cos x \sinh y \, .$

As an application of (3.7) let us find all complex zeros of the cosine. A complex number $z = x + iy$ satisfies $\cos z = 0$ precisely when x and y solve the system of equations

$$\cos x \cosh y = 0 \, , \quad \sin x \sinh y = 0 \, .$$

Because $\cosh y$ is never zero, the first of these equations is satisfied if and only if $x = k\pi/2$ for an odd integer k. Since $\sin x = \pm 1$ for such values of x, the second equation then holds if and only if $\sinh y = 0$, which means exactly when $y = 0$. We conclude that the only zeros of the cosine are its familiar real zeros. A parallel analysis of the sine function would disclose that its only zeros are its real zeros.

3. Exponential and Trigonometric Functions

The functions $f(z) = \cos z$ and $g(z) = \sin z$ are quite obviously entire functions. As we might anticipate,

$$f'(z) = \frac{ie^{iz} - ie^{-iz}}{2} = i\left(\frac{e^{iz} - e^{-iz}}{2}\right) = -\left(\frac{e^{iz} - e^{-iz}}{2i}\right) = -\sin z$$

and, similarly, $g'(z) = \cos z$. The other four elementary trigonometric functions are analytic functions, but not entire functions. The quantity $\tan z$, for instance, is undefined at $z = k\pi/2$ when k is an odd integer. The expected derivative formulas hold for these functions as well; e.g., $h(z) = \tan z$ has $h'(z) = \sec^2 z$. Their derivatives are obtained exactly as they are in calculus, by expressing the functions in terms of the sine and cosine and then using the rule for differentiating quotients (Exercise 6.35).

Some examples dealing with trigonometric functions will prove helpful.

EXAMPLE 3.5. Determine all z for which $\sin z = (12/5)i$.

We seek all complex numbers z satisfying

$$\frac{e^{iz} - e^{-iz}}{2i} = \frac{12i}{5}.$$

A little algebraic manipulation transforms this equation into

$$5e^{2iz} + 24e^{iz} - 5 = 0,$$

a quadratic in e^{iz}. Straightforward factorization — or application of the quadratic formula — leads to $e^{iz} = 1/5$ or $e^{iz} = -5$. In other words, iz must be either a logarithm of $1/5$ or a logarithm of -5. The former case produces solutions $z = 2k\pi + i \operatorname{Log} 5$ for each integer k, while the solutions in the latter case take the form $z = (2k+1)\pi - i \operatorname{Log} 5$ for an arbitrary integer k. We conclude that the full set of solutions to $\sin z = (12/5)i$ consists of all complex numbers of the type $2k\pi + i \operatorname{Log} 5$ or of the type $(2k+1)\pi - i \operatorname{Log} 5$, where k is an integer.

EXAMPLE 3.6. Determine the largest open set U in which the function $f(z) = \tan(\pi z^2/2)$ is analytic, and compute $f'(z)$ in U.

We have $f = g \circ h$, where $g(z) = \tan z$ and $h(z) = \pi z^2/2$. The function h is entire, whereas g is an analytic function whose domain-set V is described by $V = \mathbb{C} \sim \{(2k+1)\pi/2 \colon k \text{ an integer}\}$. The largest open set in which f is analytic, therefore, is the set $U = \{z \colon \pi z^2/2 \in V\}$. To obtain U we must simply remove from $\mathbb{C}$ all z for which z^2 is an odd integer; i.e.,

$$U = \mathbb{C} \sim \left(\left\{\pm\sqrt{2k+1} \colon k = 0, 1, \cdots\right\} \cup \left\{\pm i\sqrt{2k+1} \colon k = 0, 1, \cdots\right\}\right).$$

Finally, the chain rule gives $f'(z) = \pi z \sec^2(\pi z^2/2)$ for z in U.

EXAMPLE 3.7. Discuss the geometry of the transformation defined in the strip $A = \{z \colon |\operatorname{Re} z| \leq \pi/2\}$ by $w = \sin z$.

Writing $w = u + iv$ and referring to (3.8), we note that this transformation can be described in real coordinates by

$$u = \sin x \cosh y \quad , \quad v = \cos x \sinh y$$

for $z = x + iy$ in A. We investigate the behavior of this transformation on horizontal and vertical lines in A.

Image of the horizontal segment $I = \{x + iy : y = y_0, |x| \leq \pi/2\}$.

Case 1: $y_0 = 0$. The image curve is parametrized by $u = \sin x$ and $v = 0$, with x varying over the interval $[-\pi/2, \pi/2]$. These equations obviously describe the line segment between -1 and 1 on the real axis of the w-plane.

Case 2: $y_0 \neq 0$. In this case the image of I has the parametric description $u = \sin x \cosh y_0$ and $v = \cos x \sinh y_0$ for $-\pi/2 \leq x \leq \pi/2$, which implies that the relation

$$\frac{u^2}{\cosh^2 y_0} + \frac{v^2}{\sinh^2 y_0} = 1$$

holds at each point w of the image curve. The last equation is the equation of an ellipse centered at the origin of the w-plane. Its major axis has length $2\cosh y_0$ and is situated on the u-axis; its minor axis has length $2\sinh y_0$ and is located on the v-axis. In point of fact, only half of this ellipse is obtained in transforming the given segment, the top half when $y_0 > 0$ — then $v \geq 0$ — and the bottom half when $y_0 < 0$. The images of the segments corresponding to $\pm y_0$ fit together to form a complete ellipse.

Image of the vertical line $J = \{x + iy : x = x_0, -\infty < y < \infty\}$.

Case 1: $x_0 = 0$. The image is determined by the conditions $u = 0$ and $v = \sinh y$ for $-\infty < y < \infty$. These are just parametric equations for the imaginary axis in the w-plane.

Case 2: $x_0 = \pm\pi/2$. When $x_0 = \pi/2$ the equations describing the image of J are $u = \cosh y$ and $v = 0$, with $-\infty < y < \infty$. The set parametrized by these equations is the interval $[1, \infty)$ on the w-plane's real axis. The image of J when $x_0 = -\pi/2$ is the interval $(-\infty, -1]$ on the same axis.

Case 3: $0 < |x_0| < \pi/2$. We now find the image curve to be given parametrically by $u = \sin x_0 \cosh y$ and $v = \cos x_0 \sinh y$ for $-\infty < y < \infty$ and check, using basic algebra, that

$$\frac{u^2}{\sin^2 x_0} - \frac{v^2}{\cos^2 x_0} = 1$$

is true of any point w on this curve. Accordingly, the image is seen to be a branch of a hyperbola, the right branch if $0 < x_0 < \pi/2$ — then $u \geq 0$ — and the left branch if $-\pi/2 < x_0 < 0$. The combined images of the lines corresponding to $\pm x_0$ constitute a full hyperbola.

When the various bits of information collected above are assembled, the picture in Figure 2 emerges.

3. Exponential and Trigonometric Functions   81

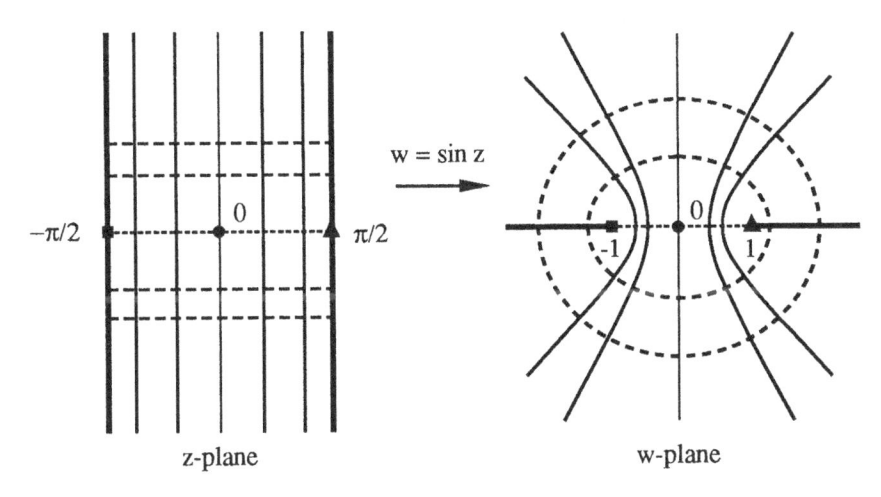

Figure 2.

EXAMPLE 3.8. Show that the function $f(z) = \tan z$ maps the set $T = \{z : -\pi/2 < \operatorname{Re} z \leq \pi/2, z \neq \pi/2\}$ in a univalent fashion to the twice-punctured plane $\mathbb{C} \sim \{\pm i\}$.

We first make the observation that

$$\tan z = \frac{\sin z}{\cos z} = \frac{1}{i} \frac{e^{iz} - e^{-iz}}{e^{iz} + e^{-iz}} = i\frac{1 - e^{2iz}}{1 + e^{2iz}},$$

and use this to decompose f as $f = f_4 \circ f_3 \circ f_2 \circ f_1$, in which

$$f_1(z) = 2iz \quad, \quad f_2(z) = e^z \quad, \quad f_3(z) = \frac{1-z}{1+z} \quad, \quad f_4(z) = iz .$$

It is then possible to track the image of T under the consecutive application of these simple mappings and arrive at the stated conclusion (Figure 3). We are putting to use here Examples I.3.3 and I.3.4, along with the comments following the latter example. We recall, especially, that f_3 is a function whose domain-set and range are $\mathbb{C} \sim \{-1\}$ and which is its own inverse.

3.3 The Principal Arcsine and Arctangent Functions

A perusal of Figure 2 ought to make the following fact geometrically plausible: *corresponding to each complex number w there is one and only one member z of the set S*,

$$S = \{z : |x| < \pi/2; \ x = \pi/2 \text{ and } y \leq 0; \ x = -\pi/2 \text{ and } y \geq 0\} ,$$

with the property that $\sin z = w$. (An algebraic demonstration of this fact is outlined below.) The complex number z determined in this way is called

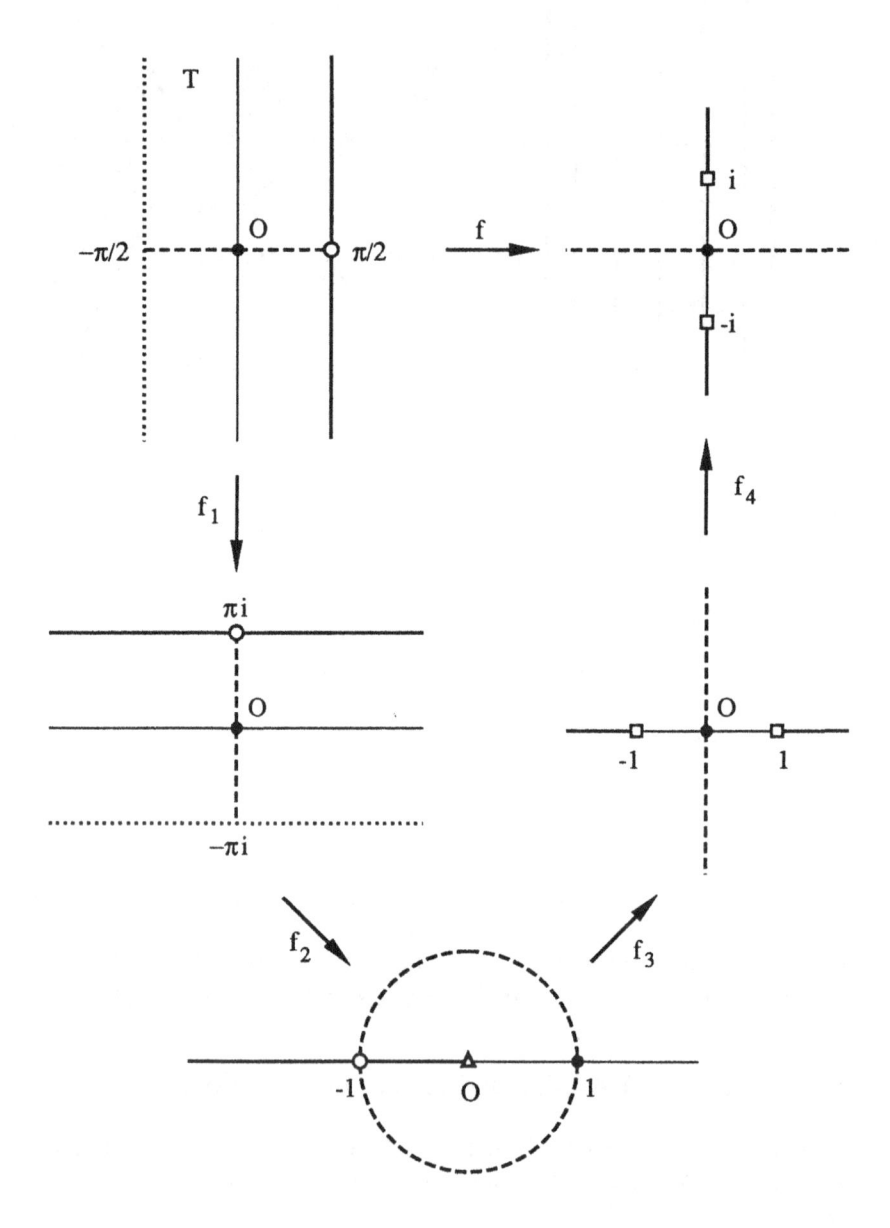

Figure 3.

3. Exponential and Trigonometric Functions

the *principal arcsine* of w. It is denoted by $\operatorname{Arcsin} w$. The function $f: \mathbb{C} \to S$ defined by $f(w) = \operatorname{Arcsin} w$ is termed the *principal arcsine function*. We shall now derive an explicit formula for the quantity $\operatorname{Arcsin} w$ and, in the process, validate its definition. Given w in $\mathbb{C}$, we wish to show that the equation

(3.9) $$\left(e^{iz} - e^{-iz}\right)/2i = w$$

has a unique solution z in S; above and beyond this, we want to exhibit that solution. Proceeding as in Example 3.5, we first make application of the quadratic formula to solve (3.9) for e^{iz},

$$e^{iz} = \pm(1 - w^2)^{1/2} + iw \ .$$

This allows us to identify the two solutions of (3.9) satisfying $-\pi < \operatorname{Re} z \leq \pi$; namely,

$$z_1 = -i \operatorname{Log}\left[(1 - w^2)^{1/2} + iw\right] \quad , \quad z_2 = -i \operatorname{Log}\left[-(1 - w^2)^{1/2} + iw\right] \ .$$

The fact that

$$-(1 - w^2)^{1/2} + iw = -\frac{1}{(1 - w^2)^{1/2} + iw}$$

comes in handy when verifying (Exercise 6.40) that

$$\operatorname{Re}\left[(1 - w^2)^{1/2} + iw\right] > 0 \quad , \quad \operatorname{Re}\left[-(1 - w^2)^{1/2} + iw\right] < 0$$

— unless w is real and $|w| \geq 1$. It follows that, with exceptions for such w, $|\operatorname{Re} z_1| < \pi/2$ and $|\operatorname{Re} z_2| > \pi/2$. In this way it becomes apparent that z_1 is the one and only solution of (3.9) belonging to S; i.e.,

(3.10) $$\operatorname{Arcsin} w = -i \operatorname{Log}\left[(1 - w^2)^{1/2} + iw\right] \ .$$

Finally, the reader can check directly (Exercise 6.41) that (3.10) is still true when w is real with $|w| \geq 1$. With the help of (3.10) we compute a few representative values of $\operatorname{Arcsin} w$:

$$\operatorname{Arcsin} 1 = \frac{\pi}{2} \quad , \quad \operatorname{Arcsin}\left(-\frac{\sqrt{2}}{2}\right) = -\frac{\pi}{4} \quad , \quad \operatorname{Arcsin} i = -i \operatorname{Log}(-1+\sqrt{2}) \ .$$

Figure 4 is worth bearing in mind in conjunction with the principal arcsine function.

Because the principal square root function has its only discontinuities at the points of the negative real axis, it is easy to see that the formula $h(w) = (1 - w^2)^{1/2} + iw$ defines a function which is continuous in the domain $D = \mathbb{C} \sim \{w : \operatorname{Im} w = 0 \text{ and } |w| \geq 1\}$. Also, we have earlier noted

84   III. Analytic Functions

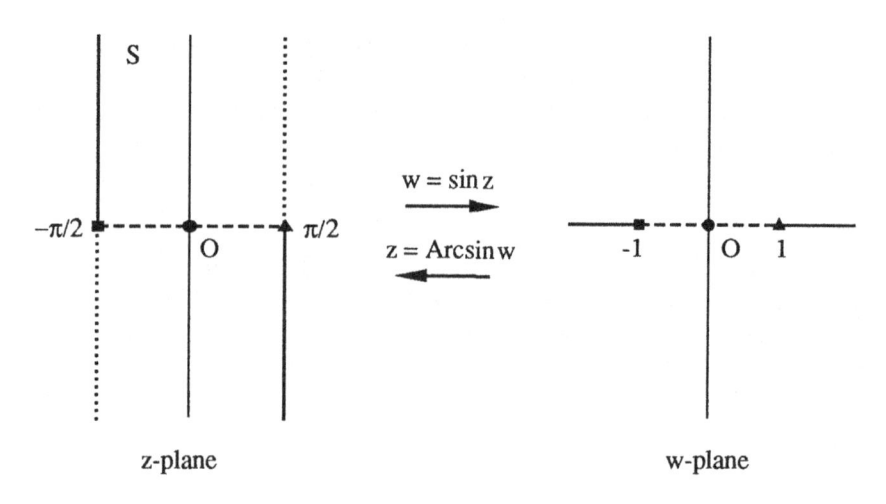

Figure 4.

that $\operatorname{Re} h(w) > 0$ for w in D. Since the principal logarithm function is continuous away from the real interval $(-\infty, 0]$, we infer from (3.10) that $f(w) = \operatorname{Arcsin} w$ defines a continuous function in D. In fact, we shall learn in the next section that f is analytic in D. The principal arcsine function is not, however, an entire function, for it has discontinuities at all points w of the real axis with $|w| > 1$.

In much the same manner, the *principal arctangent of w* ($\operatorname{Arctan} w$), where w is any complex number other than i or $-i$, is defined so: $\operatorname{Arctan} w$ *is the unique member of the set* $T = \{z : -\pi/2 < \operatorname{Re} z \leq \pi/2,\ z \neq \pi/2\}$ *with the property that* $\tan z = w$. Example 3.8 guarantees that this definition makes sense. This time a formula for $\operatorname{Arctan} w$ can be derived by referring to Figure 3. If we simply backtrack through this diagram, at each stage replacing a mapping with its obvious inverse mapping, and then compose these inverses, we are led to

$$(3.11) \qquad \operatorname{Arctan} w = -\frac{i}{2} \operatorname{Log}\left(\frac{1 + iw}{1 - iw}\right).$$

As sample values we record

$$\operatorname{Arctan} 1 = \frac{\pi}{4}\ ,\quad \operatorname{Arctan}(-1) = -\frac{\pi}{4}\ ,\quad \operatorname{Arctan}\left(\frac{i}{2}\right) = i \operatorname{Log} \sqrt{3}\ .$$

A graphic aid for dealing with the *principal arctangent function*, the function $g : \mathbb{C} \sim \{\pm i\} \to T$ defined by $g(w) = \operatorname{Arctan} w$, is provided in Figure 5.

The discontinuities of the principal logarithm function are at the points of the real interval $(-\infty, 0)$. In tandem with (3.11) this fact shows that $g(w) = \operatorname{Arctan} w$ is continuous at all points w, $w \neq \pm i$, with the property

4. Branches of Inverse Functions

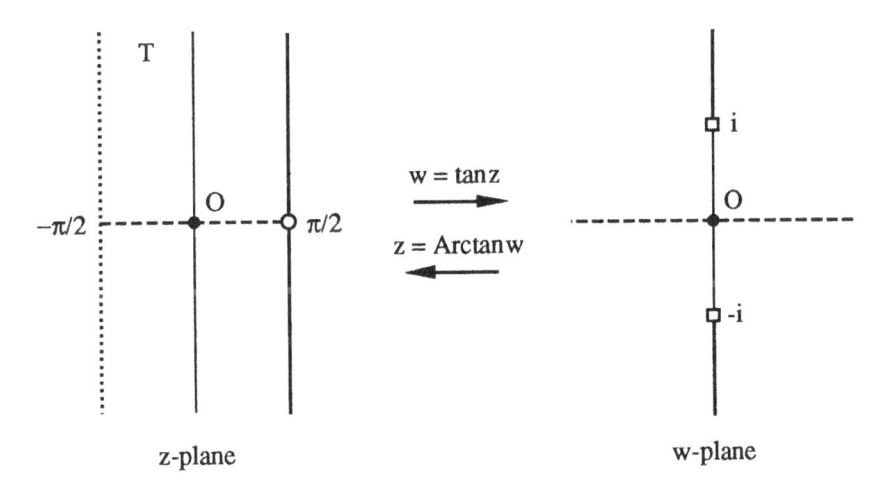

Figure 5.

that $(1+iw)/(1-iw)$ does not belong to $(-\infty, 0)$, which translates to all points w of the domain $D = \mathbb{C} \sim \{w : \operatorname{Re} w = 0 \text{ and } |w| \geq 1\}$. Indeed, g is analytic in D, as we shall presently see. The principal arctangent function is discontinuous at every point w of the imaginary axis satisfying $|w| > 1$.

4 Branches of Inverse Functions

4.1 Branches of Inverse Functions

Having enhanced our supply of elementary analytic functions through the addition of the exponential and trigonometric functions, we direct our attention next to logarithm functions and, following this, to power functions. There is a complication associated with these functions that will make it worth our while to spend a few preliminary words discussing the phenomenon of a "branch of a multi-valued function" or, to be more specific, the notion of a "branch of the inverse" of a given analytic function.

Consider a non-univalent analytic function $f: U \to \mathbb{C}$. Such a function does not have an inverse function, which would mean, as we recall, a function $g: f(U) \to U$ satisfying $g[f(z)] = z$ for every z in U and $f[g(z)] = z$ for all z in $f(U)$. What f does have — and has in abundance — are so-called *right-inverse functions*. This name is attached to any function $g: f(U) \to U$ enjoying only the latter property, $f[g(z)] = z$ for every z in $f(U)$. Functions g of this type are readily manufactured: to define g at z in $f(U)$, select any w in U satisfying $f(w) = z$ — since f is not univalent, there will usually be many choices for w — and set $g(z) = w$. So haphazard a method of assigning values to g at the points of $f(U)$ has, to say the least, little chance of producing a function with agreeable features. One might be optimistic, of course, and hope to find among the multitude of right-inverses of f at least a few that were analytic. Unfortunately, even the most carefully wrought right-inverses tend to have built-in discontinuities. Thus the most we can reasonably expect in a given right-inverse g of f is a function that is analytic in each non-empty open subset V of $f(U)$ which contains no point of discontinuity of g. Notice that, when testing g for analyticity, it suffices to concentrate the effort on the open sets V of this description that are also connected (i.e., that are domains), since any non-empty open set is the disjoint union of domains — namely, of its components (Theorem II.3.5). This observation explains in part why we restrict ourselves to a domain in defining a branch of an inverse function. As a concrete illustration of the preceding remarks, take the pair of functions $f(z) = z^2$ and $g(z) = \sqrt{z}$. Clearly g is a right-inverse of f, but is not analytic in $f(\mathbb{C}) = \mathbb{C}$, for it has discontinuities at all points of the negative real axis. However, g is analytic in the domain $D = \mathbb{C} \sim (-\infty, 0]$, which happens to be the largest open set that avoids those discontinuities.

Motivated by the foregoing comments, we make the following definition: if $f: U \to \mathbb{C}$ is an analytic function and if D is a domain contained in $f(U)$, then by a *branch of f^{-1} in D* is meant a continuous function $g: D \to U$ that satisfies the condition $f[g(z)] = z$ for every z in D. (We emphasize the need to specify the domain D in dealing with this concept. The phrase "let g be a branch of f^{-1}" has a hollow ring to it, for its raises the question: In what domain?) In light of the discussion above, we should not expect D to be the whole set $f(U)$, even when the latter is itself open and connected. As indicated, some "whittling down" of $f(U)$ is normally required to accomodate the continuity of g. Older books often speak of making "branch cuts" in $f(U)$ to excise the points of discontinuity of a given right-inverse of f, thereby producing branches of f^{-1} in the leftover domains. In particular, we do not intend to suggest that there will be a branch of f^{-1} in an arbitrary subdomain D of $f(U)$. On the other hand, when one branch of f^{-1} does exist in D, there will frequently be many other branches there. A benefit derived from the connectedness of D is that the collection of all such branches can sometimes be "catalogued," as they are in the case of branches of the logarithm function in a domain. (See Theorem 4.3.) We note in proceeding that a branch g of f^{-1} in D is necessarily a univalent function, since $g(z_1) = g(z_2)$ implies that $z_1 = f[g(z_1)] = f[g(z_2)] = z_2$.

A prime example of a branch of an inverse has already been cited: if $f(z) = z^2$, then a branch g of f^{-1} in the domain $D = \mathbb{C} \sim (-\infty, 0]$ is given by $g(z) = \sqrt{z}$. The function $h(z) = -\sqrt{z}$ represents a different branch of f^{-1} in the same domain. In fact, these are the only branches of f^{-1} in D (Theorem 4.2). A second example: the function $f(z) = \sin z$ has $g(z) = \text{Arcsin}\, z$ as a branch of its inverse in the domain $D = \mathbb{C} \sim \{z : \text{Im}\, z = 0 \text{ and } |z| \geq 1\}$. (Can the reader think of another branch there?)

4. Branches of Inverse Functions

Our premise in introducing branches of inverses at the present time was that such functions could enrich our pool of examples of analytic functions. Is the premise sound? Given that an analytic function $f: U \to \mathbb{C}$ has g as a branch of its inverse in a domain D, is it true that g is analytic in D? Theorem 4.1 gives an affirmative response to this question under the added assumption that f' has no zeros in the set $g(D)$. As far as the current chapter is concerned, this suits our needs perfectly. In the situations we examine here the extra hypothesis can be checked directly. The full truth of the matter is that the answer to the above question is always "yes," although we shall not be in a position to confirm this until Chapter VIII. (N.B. If we knew that g were analytic in D, then we could differentiate the relation $f[g(z)] = z$ and learn that $f'[g(z)]g'(z) = 1$ for every z in D. It would follow that $f'[g(z)] \neq 0$ for z in D. We infer that the absence of zeros of f' from $g(D)$ is certainly a necessary condition for g to be an analytic function.)

Theorem 4.1. *Suppose that $f: U \to \mathbb{C}$ is an analytic function and that g is a branch of f^{-1} in a domain D. Let z_0 be a point of D and let $w_0 = g(z_0)$. If $f'(w_0) \neq 0$, then g is differentiable at z_0 and $g'(z_0) = 1/f'(w_0)$. Consequently, if f' is free of zeros in $g(D)$, then g is analytic in D, where its derivative satisfies $g'(z) = 1/f'[g(z)]$.*

Proof. Recalling that g is univalent in D, we consider a point z of D different from z_0, we set $w = g(z)$ (then $w \neq w_0$) and we compute

$$\frac{g(z) - g(z_0)}{z - z_0} = \frac{g(z) - g(z_0)}{f[g(z)] - f[g(z_0)]} = \frac{w - w_0}{f(w) - f(w_0)} \to \frac{1}{f'(w_0)}$$

as $z \to z_0$. Here the continuity of g at z_0 is essential to make sure that $w = g(z) \to g(z_0) = w_0$ as $z \to z_0$. Therefore, g is seen to be differentiable at z_0 with $g'(z_0) = 1/f'(w_0)$. ∎

4.2 Branches of the pth-root Function

The simplest situation in which Theorem 4.1 gives rise to a "new" example of an analytic function occurs when we take for f the function $f(z) = z^p$, where $p \geq 2$ is an integer. Then $f'(z) = pz^{p-1}$, so the only zero of f' is at the origin. Suppose now that D is a domain not containing the origin and that g is a branch of f^{-1} in D. To be explicit here, $g: D \to \mathbb{C}$ is just a continuous function with the property that $[g(z)]^p = z$ everywhere in D. This identity and the fact that zero is not an element of D eliminate the possibility that $g(z) = 0$ for a point z of D. Stated in another way, $g(D)$ does not contain the sole zero of f'. Theorem 4.1 thus lets us know that g

is actually analytic in D. Its derivative can be determined as follows:

$$(3.12) \qquad g'(z) = \frac{1}{f'[g(z)]} = \frac{1}{p[g(z)]^{p-1}} = \frac{g(z)}{p[g(z)]^p} = \frac{g(z)}{pz}.$$

If $p \geq 2$ is an integer and if D is a plane domain, then by a *branch of the pth-root function in D* we shall understand an analytic function $g: D \to \mathbb{C}$ with the feature that $[g(z)]^p = z$ for all z in D. The chain rule implies that g obeys the condition $p[g(z)]^{p-1} g'(z) = 1$ throughout D. We infer that no such function g could exist if D contained the origin, for then we would clearly have $g(0) = 0$, a state of affairs incompatible with $p[g(0)]^{p-1} g'(0) = 1$. The message of the previous paragraph is this: once the origin is known to lie outside D, any continuous function $g: D \to \mathbb{C}$ satisfying $[g(z)]^p = z$ is automatically analytic in D and is, therefore, a branch of the pth-root function in that domain. Since continuity is a property that is relatively easy to check, this is a significant observation. As a specific example, we look at $g(z) = \sqrt[p]{z}$ in the domain $D = \mathbb{C} \sim (-\infty, 0]$.

EXAMPLE 4.1. Let $p \geq 2$ be an integer. Show that $g(z) = \sqrt[p]{z}$ defines an analytic function with derivative $g'(z) = (1/p) z^{(1/p)-1}$ in the domain $D = \mathbb{C} \sim (-\infty, 0]$. Conclude that g is a branch of the pth-root function in D.

We can write $g(z) = \sqrt[p]{|z|} \exp\left[(i \operatorname{Arg} z)/p\right]$. Since the function $\theta(z) = \operatorname{Arg} z$ is continuous in D, it is apparent that g is continuous there as well. Obviously $[g(z)]^p = z$ for every z in D. In light of the preceding discussion, g is analytic in D and (3.12) gives

$$g'(z) = \frac{g(z)}{pz} = \frac{\sqrt[p]{z}}{pz} = \frac{1}{p} z^{(1/p)-1},$$

the derivative formula we would anticipate. Accordingly, g is a branch of the pth-root function in D.

The absence of zero from a domain D offers no guarantee in itself that a branch of the pth-root function exists there, and, even if such a branch g does exist, it may not be possible to find a pleasant formula for it, such as $g(z) = \sqrt[p]{z}$ in Example 4.1. As a case in point, just contemplate the structure of a branch g of the square root function in a domain D. For fixed z in D either $g(z) = \sqrt{z}$ or $g(z) = -\sqrt{z}$, but the continuity of g will ordinarily dictate that the selection of sign vary from point to point of D. Depending on the domain D, the rule coordinating the choices of sign may be quite complicated. To get an inkling of how this can work in practice, consider $D = \mathbb{C} \sim [0, \infty)$.

EXAMPLE 4.2. Find the two branches of the square root function in the domain $D = \mathbb{C} \sim [0, \infty)$.

4. Branches of Inverse Functions

If we are able to produce one branch g of the square root function in D, the other branch will be $h = -g$. (See Theorem 4.2.) To find g, we observe that, for a point z_0 of the negative real axis, $\sqrt{z} \to \sqrt{z_0}$ as z approaches z_0 from above or from the side, whereas $\sqrt{z} \to -\sqrt{z_0}$ as z tends to z_0 from below. As a result, in defining a function $g: D \to \mathbb{C}$ by

$$g(z) = \begin{cases} \sqrt{z} & \text{if } z \in D \text{ and } \operatorname{Im} z \geq 0, \\ -\sqrt{z} & \text{if } z \in D \text{ and } \operatorname{Im} z < 0, \end{cases}$$

we prevent any discontinuities from cropping up in the interval $(-\infty, 0)$ and so obtain a continuous function in D satisfying $[g(z)]^2 = z$. By our earlier comments this function is analytic in D. In fact, (3.12) informs us that

$$g'(z) = \frac{g(z)}{2z} = \begin{cases} 1/(2\sqrt{z}) & \text{if } z \in D \text{ and } \operatorname{Im} z \geq 0, \\ -1/(2\sqrt{z}) & \text{if } z \in D \text{ and } \operatorname{Im} z < 0. \end{cases}$$

Therefore, g is one branch of the square root function in D and $h = -g$ is the other.

What is it that limits the number of branches in the preceding example to two? The answer is supplied by the following theorem, which demonstrates how connectedness serves to impose some order on the possible disarray of branches of the inverse of a given function in a given domain. The proof is a nice application of Theorem 2.4.

Theorem 4.2. *Suppose that a branch g of the pth-root function exists in a domain D. Then there are exactly p distinct branches of the pth-root function in D. Each has the form cg, where c is a pth-root of unity.*

Proof. Obviously each of the p functions cg, in which c is a pth-root of unity, is a branch of the pth-root function in D. Could there be others? Let h be an arbitrary branch of the pth-root function in D. As pointed out in the second paragraph of Section 4.2, the origin cannot lie in D. Also, g is zero-free in D, and the same is true of h. The function $f = h/g$, consequently, is analytic in D, has no zeros there, and satisfies

$$[f(z)]^p = \frac{[h(z)]^p}{[g(z)]^p} = \frac{z}{z} = 1$$

for all z in D. Differentiation yields $p[f(z)]^{p-1} f'(z) = 0$. Because $f(z) \neq 0$, this implies that $f'(z) = 0$ throughout D. By Theorem 2.4, f is constant in D — say $f(z) = c$. Then $c^p = [f(z)]^p = 1$, marking c as a pth-root of unity. The result: $h = cg$, which places h among the p branches we originally identified. ∎

This is neither the place nor the time to undertake a systematic analysis of which plane domains admit branches of the pth-root function and which do not. That matter is best left until Chapter V, when more of the machinery of complex analysis will be at our disposal. Our present goal is simply to lengthen our list of examples of elementary analytic functions. In a minor digression before moving on to logarithms, we remark that Theorem 4.1 allows us to put two additional functions on that list.

EXAMPLE 4.3. Verify that $g(z) = \text{Arcsin } z$ defines an analytic function in the domain $D = \mathbb{C} \sim \{z : \text{Im } z = 0 \text{ and } |z| \geq 1\}$ and that

$$g'(z) = \frac{1}{\sqrt{1-z^2}}$$

in D.

The function g is a branch of f^{-1} in D, where $f(z) = \sin z$. Furthermore, $f'(z) = \cos z \neq 0$ for z in the set $g(D) = \{z : |\text{Re } z| < \pi/2\}$. Theorem 4.1 asserts that g is analytic in D and that $g'(z) = 1/f'[g(z)] = 1/\cos(\text{Arcsin } z)$ for z in D. Invoking formula (3.10), we compute

$$\cos(\text{Arcsin } z) = \frac{1}{2}\left(e^{i\,\text{Arcsin } z} + e^{-i\,\text{Arcsin } z}\right)$$

$$= \frac{1}{2}\left[e^{\text{Log}(\sqrt{1-z^2}+iz)} + e^{-\text{Log}(\sqrt{1-z^2}+iz)}\right]$$

$$= \frac{1}{2}\left(\sqrt{1-z^2} + iz + \frac{1}{\sqrt{1-z^2}+iz}\right)$$

$$= \frac{1}{2}\left(\sqrt{1-z^2} + iz + \sqrt{1-z^2} - iz\right) = \sqrt{1-z^2},$$

which gives $g'(z) = 1/\sqrt{1-z^2}$, as asserted. (N.B. We have purposely avoided using $\cos(\text{Arcsin } z) = \pm\sqrt{1-\sin^2(\text{Arcsin } z)} = \pm\sqrt{1-z^2}$ here because of the extra work involved in justifying the choice of sign.)

EXAMPLE 4.4. Show that the function $g(z) = \text{Arctan } z$ is analytic in the domain $D = \mathbb{C} \sim \{z : \text{Re } z = 0 \text{ and } |z| \geq 1\}$, where it has derivative

$$g'(z) = \frac{1}{1+z^2}.$$

Taking $f(z) = \tan z$ we note that g is a branch of f^{-1} in D. Since $f'(z) = \sec^2 z$ is zero-free, an appeal to Theorem 4.1 confirms the analyticity of g in D. Furthermore,

$$g'(z) = \frac{1}{f'[g(z)]} = \frac{1}{\sec^2(\text{Arctan } z)} = \frac{1}{1+\tan^2(\text{Arctan } z)} = \frac{1}{1+z^2}.$$

We have, of course, used the identity $1 + \tan^2 w = \sec^2 w$.

4.3 Branches of the Logarithm Function

A branch of the logarithm function in a domain D (or, for short, a *branch of* $\log z$ *in* D) is an analytic function $L: D \to \mathbb{C}$ with the property that $e^{L(z)} = z$ for every point z of D. In other words, L is analytic in D and its value $L(z)$ at any point of D is some logarithm of z. Since $f(z) = e^z$ is an entire function with $f'(z) = e^z \neq 0$ for all z, Theorem 4.1 tells us that any continuous function $L: D \to \mathbb{C}$ satisfying $e^{L(z)} = z$ (i.e., any branch of f^{-1} in D) is necessarily analytic in D — and so is a branch of $\log z$ in that domain. (As was a similar comment concerning branches of the pth-root function, this is a noteworthy statement, for it is usually much easier to decide whether a function is continuous than to test that function directly for differentiability.) Moreover,

$$(3.13) \qquad L'(z) = \frac{1}{z},$$

since according to Theorem 4.1

$$L'(z) = \frac{1}{f'[L(z)]} = \frac{1}{e^{L(z)}} = \frac{1}{z}.$$

If a branch L of $\log z$ exists in a domain D, then D cannot contain the origin, for $z = e^{L(z)} \neq 0$ holds by definition at each point of D. In view of the general structure of a complex logarithm, we are able to represent L by a formula

$$(3.14) \qquad L(z) = \text{Log}\,|z| + i\,\theta(z),$$

where $\theta = \text{Im}\,L$ is a continuous real-valued function in D with the property that, for each z in D, $\theta(z)$ is an argument of z. Any continuous function $\theta: D \to \mathbb{R}$ fitting this description is termed a *branch of the argument function* (or just a *branch of* $\arg z$) in D. It is clear from (3.14) and the comments preceding it that each branch of $\log z$ in D gives rise to a corresponding branch of $\arg z$ there, and vice versa. The importance of continuity in all this cannot be overstated.

It may happen that in a given domain D — even one avoiding the origin — there is no branch of the logarithm function. (See Example 4.8.) Assuming that a branch L of $\log z$ does exist in D, we can add to L any constant of the form $2k\pi i$, where k is an integer, and produce another such branch. Conversely, if L_1 is an arbitrary branch of $\log z$ in D, then (3.13) gives $L'(z) = 1/z = L_1'(z)$ throughout D, implying that $L_1 - L$ is constant there (Theorem 2.4). Owing to the fact that $e^{L_1(z) - L(z)} = e^{L_1(z)}/e^{L(z)} = z/z = 1$, this constant is necessarily of the form $2k\pi i$ for a suitable integer k. Therefore, L_1 has the form $L_1 = L + 2k\pi i$. (Crucial here is that the integer k does not vary from point to point of D.) If we know one branch of $\log z$ in a domain, we know them all! For reference purposes we record this observation in a theorem.

Theorem 4.3. *Suppose that a branch L of the logarithm function exists in a domain D. Then the collection of all branches of the logarithm function in D consists of the functions $L + 2k\pi i$, where k is an integer.*

A corollary of Theorem 4.3 and (3.14) is this: if a branch θ of the argument function exists in a domain D, then the collection of all such branches is made up of the functions $\theta + 2k\pi$, with k an integer.

4.4 Branches of the λ-power Function

If L is a branch of the logarithm function in a domain D and if λ is a complex number, then the *branch of the λ-power function in D associated with L* is the function $h_\lambda: D \to \mathbb{C}$ defined by

$$h_\lambda(z) = e^{\lambda L(z)} \,.$$

(We should really write something like $h_{L,\lambda}$ here in order to emphasize the branch of $\log z$ in D being used, but the notation then gets too cumbersome. We shall leave the dependence on L implicit, with the understanding that, whenever we work with a branch of the λ-power function in a domain, the branch of the logarithm with which it is associated has been predetermined and is lurking somewhere in the background.) It is apparent that h_λ is analytic in D. We compute its derivative by using (3.13) and the chain rule:

$$h'_\lambda(z) = e^{\lambda L(z)} \cdot \frac{\lambda}{z} = e^{\lambda L(z)} \cdot \frac{\lambda}{e^{L(z)}} = \lambda e^{(\lambda-1)L(z)} \,,$$

revealing that

(3.15) $$h'_\lambda(z) = \lambda h_{\lambda-1}(z) \,.$$

Of course, when $\lambda = p$, an integer, the function h_λ is nothing more than the restriction to D of the rational function $f(z) = z^p$, regardless of which branch L of the logarithm function is employed in D. Observe, too, that if $\lambda = 1/p$ for an integer $p \geq 2$, then

$$\left[h_{1/p}(z)\right]^p = \left[e^{L(z)/p}\right]^p = e^{L(z)} = z \,,$$

making $h_{1/p}$ a branch of the pth-root function in D, the *branch of the pth-root function in D associated with L*.

Some examples should help instill in the reader a better feeling for branches of the logarithm and λ-power functions.

EXAMPLE 4.5. Confirm that $L(z) = \text{Log}\, z$ defines an analytic function in the domain $D = \mathbb{C} \sim (-\infty, 0]$ and, thus, that $L: D \to \mathbb{C}$ is a branch of the logarithm function in D. Identify the associated power function h_λ.

4. Branches of Inverse Functions

Since the only discontinuities of the principal logarithm function occur along the negative real axis, L is continuous in D. Certainly $e^{L(z)} = z$ for every z in D. By the comments at the beginning of Section 4.3, L is analytic in D and has derivative $L'(z) = 1/z$ there. The associated λ-power function is given in D by

$$h_\lambda(z) = e^{\lambda L(z)} = e^{\lambda \operatorname{Log} z} = z^\lambda ,$$

the principal λ-power of z. From (3.15) it follows that $h'_\lambda(z) = \lambda z^{\lambda-1}$ in the given domain. Observe that D is the largest open set in which the principal logarithm function is analytic.

EXAMPLE 4.6. Construct a branch L of $\log z$ in the domain $D = \mathbb{C} \sim \{z\colon \operatorname{Re} z = 0 \text{ and } \operatorname{Im} z \leq 0\}$ satisfying $L(i) = 5\pi i/2$. Compute $h'_\lambda(-1)$ for $\lambda = 3/2$, where h_λ is the branch of the λ-power function in D associated with L.

We shall first produce a branch θ of $\arg z$ in D. The principal argument function will not do, for it has discontinuities at all points z_0 of the negative real axis, a subset of D. Making the remark that $\operatorname{Arg} z \to \pi$ as z approaches such a point z_0 from above or from the side and that $\operatorname{Arg} z \to -\pi$ as z tends to z_0 from below, we eliminate these discontinuities if we define a function $\theta\colon D \to \mathbb{R}$ as follows:

$$\theta(z) = \begin{cases} \operatorname{Arg} z & \text{if } z \in D \text{ and } -\pi/2 < \operatorname{Arg} z \leq \pi , \\ 2\pi + \operatorname{Arg} z & \text{if } z \in D \text{ and } -\pi < \operatorname{Arg} z < -\pi/2 . \end{cases}$$

(Put differently, $\theta(z)$ is the unique argument of z that lies in the interval $(-\pi/2, 3\pi/2)$.) The function θ so defined is continuous, and its value $\theta(z)$ for any z in D is an argument of z; i.e., θ is a branch of $\arg z$ in D. We obtain a branch L_0 of $\log z$ in D by setting

$$L_0(z) = \operatorname{Log}|z| + i\,\theta(z) .$$

For L_0 we have

$$L_0(i) = \operatorname{Log}|i| + i\,\theta(i) = \operatorname{Log} 1 + i\operatorname{Arg} i = \pi i/2 .$$

To get the branch L desired, we need only increase L_0 by $2\pi i$:

$$L(z) = \operatorname{Log}|z| + i[\theta(z) + 2\pi] .$$

Finally, using (3.15) we compute

$$h'_{3/2}(-1) = (3/2)\,h_{1/2}(-1) = (3/2)\,e^{L(-1)/2} = (3/2)\,e^{3\pi i/2} = -3i/2 .$$

EXAMPLE 4.7. Show that there is a branch L of the logarithm function in the domain $D = \mathbb{C} \sim \{te^{it}\colon 0 \leq t < \infty\}$ having $L(1) = 0$. Find $L(e^\pi)$ and $h_\lambda(e^\pi)$ for $\lambda = i$.

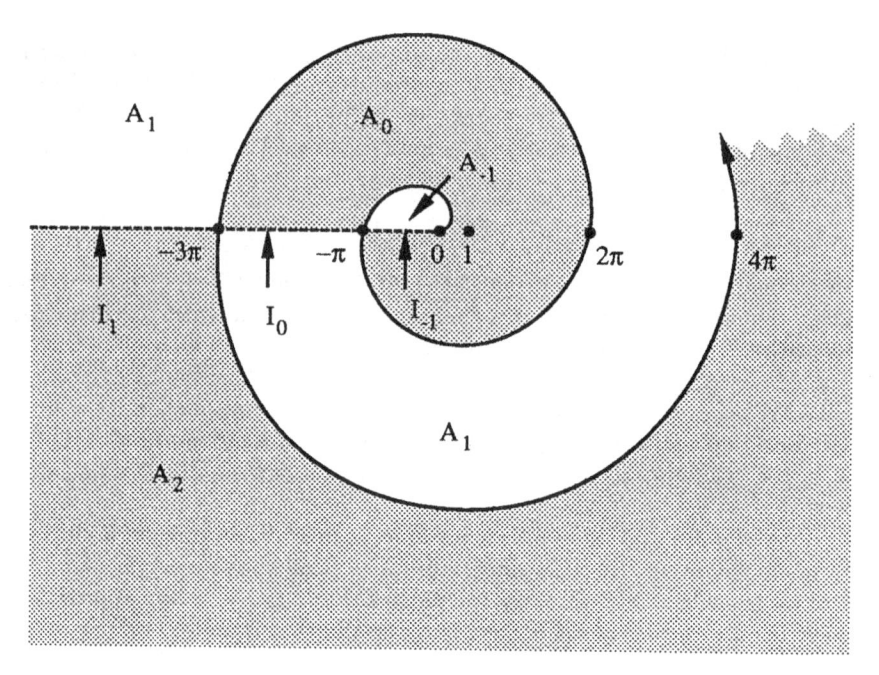

Figure 6.

Again the problem comes down to the construction of a branch θ of the argument function in D, one satisfying $\theta(1) = 0$. In order to accomplish this we first express D in the manner $D = A_{-1} \cup A_0 \cup A_1 \cup \cdots$ suggested by Figure 6. The open interval I_k on the negative real axis — $I_{-1} = (-\pi, 0)$, $I_0 = (-3\pi, -\pi)$, $I_1 = (-5\pi, -3\pi)$, $\cdots$ — is to be included in the set A_k, but not in A_{k+1}. We define $\theta\colon D \to \mathbb{C}$ by insisting that $\theta(z) = \operatorname{Arg} z + 2k\pi$ for z belonging to A_k, $k = -1, 0, 1, \cdots$. The only possible places discontinuities of θ could occur are the points of the intervals I_k. The definition of θ has been rigged, however, to prevent discontinuities at even these points! Suppose, namely, that z_0 belongs to I_k. Then $\theta(z) = \operatorname{Arg} z + 2k\pi \to (2k+1)\pi = \theta(z_0)$ as z approaches z_0 through A_k; it is also true that $\theta(z) = \operatorname{Arg} z + (2k+2)\pi \to (2k+1)\pi = \theta(z_0)$ when z tends to z_0 through A_{k+1}, for $\operatorname{Arg} z \to -\pi$ in this case. Thus $\lim_{z \to z_0} \theta(z) = \theta(z_0)$, so θ is continuous at z_0. Consequently, θ provides us with a branch of $\arg z$ in D. Since $z = 1$ lies in A_0, $\theta(1) = \operatorname{Arg} 1 = 0$. The corresponding branch L of the logarithm in D is defined by $L(z) = \operatorname{Log}|z| + i\,\theta(z)$. It has $L(1) = 0$. Furthermore, because $z = e^\pi = 23.14\cdots$ is a point of A_3,

$$L(e^\pi) = \operatorname{Log}(e^\pi) + i\,\theta(e^\pi) = \pi + 6\pi i$$

and

$$h_i(e^\pi) = e^{iL(e^\pi)} = e^{-6\pi + \pi i} = -e^{-6\pi}.$$

4. Branches of Inverse Functions

This example illustrates that a branch of $\log z$ in a domain can be quite a complicated function. In particular, it demonstrates that the range of a branch of $\arg z$ in a domain is not constrained to lie in an interval of length 2π, a common misconception among beginning complex analysis students.

EXAMPLE 4.8. Demonstrate that there can be no branch of the logarithm function in the domain $D = \mathbb{C} \sim \{0\}$.

Suppose, to the contrary, that $L: D \to \mathbb{C}$ is a branch of $\log z$ in D. The restriction of L to the domain $D^* = \mathbb{C} \sim (-\infty, 0]$ is then a branch of the logarithm function in D^*. On the other hand, $L^*(z) = \text{Log } z$ also defines a branch of the logarithm function in D^*. It follows from Theorem 4.3 that there must be an integer k such that $L(z) = \text{Log } z + 2k\pi i$ for every z in D^*. Being analytic in D, the function L is required to be continuous at $z_0 = -1$, from which fact we infer that

$$L(-1) = \lim_{y \to 0^+} L(-1 + iy) = \lim_{y \to 0^+} \text{Log}(-1 + iy) + 2k\pi i = (2k+1)\pi i$$

and, at the same time, that

$$L(-1) = \lim_{y \to 0^-} L(-1 + iy) = \lim_{y \to 0^-} \text{Log}(-1 + iy) + 2k\pi i = (2k-1)\pi i ,$$

a clear contradiction. Thus, no such L can exist.

The point of this example is that there exist plane domains which do not contain the origin yet still do not harbor branches of the logarithm function. We shall later identify the precise class of domains in which there do exist branches of $\log z$. (We refer the reader to Example V.4.1.)

EXAMPLE 4.9. Determine the largest open set U in which the function $f(z) = \text{Log}(z^3 + 1)$ is analytic, and compute $f'(z)$ for z in that set.

We have $f = g \circ h$, where $g(z) = \text{Log } z$ and $h(z) = z^3 + 1$. The function g is analytic in $D = \mathbb{C} \sim (-\infty, 0]$, but in no larger open set. It has derivative $g'(z) = 1/z$ in D. The function h is an entire function. It follows that f is definitely analytic in the open set $U = \{z : z^3 + 1 \in D\}$. Notice that a point z fails to belong to U if and only if z^3 is an element of the real interval $(-\infty, -1]$; i.e., if and only if z is a cube root of some number in $(-\infty, -1]$. To exhibit U we merely remove from the complex plane all such cube roots (Figure 7). Because f is either undefined or discontinuous at each of the deleted points, U is the largest open set in which the function f is analytic. Finally, the chain rule leads to

$$f'(z) = \frac{3z^2}{z^3 + 1}$$

whenever z is an element of U.

Now that we have in our grasp an ample working supply of analytic functions — our examples include complex facsimiles of virtually every

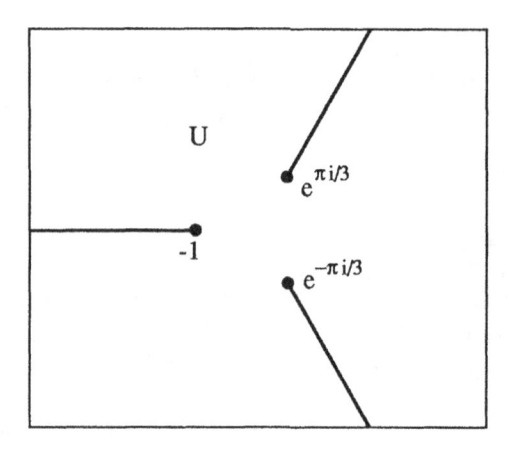

Figure 7.

elementary differentiable function of a real variable studied in beginning calculus and, in many cases, represent extensions of the latter functions from their real domain-sets to complex domain-sets — it is reasonable to start a serious inquiry into the general properties of analytic functions. That investigation, which will occupy us throughout the remainder of the book, will also entail the development of techniques for constructing new, interesting, and sometimes exotic-looking analytic functions, functions not easily expressed (if expressible at all) in terms of the standard elementary functions. In order to conduct such an inquiry a person must be equipped with the proper tools. In our case the main tool will be contour integration, which is introduced in the next chapter. In what remains of the present chapter we discuss the connection between the concept of differentiability that is the basis for complex analysis and the type of differentiability that is studied in ordinary multivariate calculus.

5   Differentiability in the Real Sense

5.1   Real Differentiability

It is not improbable that at some previous time the reader has been exposed to a notion of differentiability bearing a likeness to — but actually different from — the one under scrutiny in this book. In vector calculus one studies functions of the form $f: A \to \mathbb{R}^2$, where A is a subset of euclidean two-space $\mathbb{R}^2$, whose elements it will be momentarily convenient to represent as column vectors $z = \begin{bmatrix} x \\ y \end{bmatrix}$, rather than as ordered pairs $z = (x, y)$. The

5. Differentiability in the Real Sense

function $f: \mathbb{R}^2 \to \mathbb{R}^2$,

$$f\left(\begin{bmatrix} x \\ y \end{bmatrix}\right) = \begin{bmatrix} x^2 + e^y \\ x + \sin(xy) \end{bmatrix},$$

is the sort of thing we have in mind here. In this connection a definition of differentiability comes up that reads as follows: a function $f: A \to \mathbb{R}^2$ is differentiable at a point $z_0 = \begin{bmatrix} x_0 \\ y_0 \end{bmatrix}$ if z_0 is an interior point of A and if there exists a 2×2 matrix $\begin{bmatrix} \alpha & \beta \\ \gamma & \delta \end{bmatrix}$ with real entries such that

$$(3.16) \qquad f(z) = f(z_0) + \begin{bmatrix} \alpha & \beta \\ \gamma & \delta \end{bmatrix} \begin{bmatrix} x - x_0 \\ y - y_0 \end{bmatrix} + E(z)$$

for every $z = \begin{bmatrix} x \\ y \end{bmatrix}$ in A, where $E: A \to \mathbb{R}^2$ is a function that satisfies

$$(3.17) \qquad \lim_{z \to z_0} \frac{|E(z)|}{|z - z_0|} = 0.$$

Here $|z|$ has the expected meaning for $z = \begin{bmatrix} x \\ y \end{bmatrix}$: $|z| = \sqrt{x^2 + y^2}$. If we express f as $f(z) = \begin{bmatrix} u(z) \\ v(z) \end{bmatrix}$, then the entries of $\begin{bmatrix} \alpha & \beta \\ \gamma & \delta \end{bmatrix}$ are obtained by computing partial derivatives of the real-valued functions u and v:

$$(3.18) \quad \alpha = u_x(z_0) \;,\quad \beta = u_y(z_0) \;,\quad \gamma = v_x(z_0) \;,\quad \delta = v_y(z_0).$$

Until this point the term "complex number" has not entered the present discussion. There is, to be sure, a perfectly natural way to recast in terms of a complex variable what we have said above, merely by identifying the column vector $z = \begin{bmatrix} x \\ y \end{bmatrix}$ with the complex number $z = x + iy$. The functions under consideration can then once again be thought of as functions $f: A \to \mathbb{C}$, with A a subset of $\mathbb{C}$. If we carry out the matrix multiplication in (3.16), then transcribe all column vectors as complex numbers in the manner suggested, and finally make use of the identities $x = (z + \bar{z})/2$ and $y = (z - \bar{z})/2i$, we arrive at the translation into the language of complex analysis of this multivariate calculus notion of differentiability, a concept we henceforth characterize as "differentiability in the real sense": a function $f: A \to \mathbb{C}$ is *differentiable in the real sense at a point* z_0 if z_0 is an interior point of A and if there exist complex numbers c and d with the property that

$$(3.19) \qquad f(z) = f(z_0) + c(z - z_0) + d(\bar{z} - \bar{z}_0) + E(z)$$

for every z in A, where $E\colon A \to \mathbb{C}$ is a function satisfying condition (3.17). In fact, the computations that we outlined yield the values

(3.20) $\quad c = \dfrac{1}{2}[\alpha + \delta + i(\gamma - \beta)] \quad , \quad d = \dfrac{1}{2}[\alpha - \delta + i(\beta + \gamma)] \ .$

(N.B. In this book the unadorned term "differentiable" is reserved exclusively for reference to differentiability "in the complex sense," meaning in the sense of (3.1) or, equivalently, of (3.2) and (3.3). Whenever we intend to speak of real differentiability we shall in some way draw attention to the fact — generally by using the full phrase "differentiable in the real sense.")

A comparison of (3.2) with (3.19) reveals the discrepancy between the idea of differentiability that lies at the root of complex analysis and differentiability in the spirit of vector calculus, as just described. For the latter to reduce to the former at a point z_0 it is required that $d = 0$ in (3.19). (In view of (3.18) and (3.20) the condition $d = 0$ is nothing more than a restatement of the condition that the Cauchy-Riemann equations are satisfied at z_0. Thus, a function $f = u + iv$ is differentiable at a point z_0 if and only if f is differentiable in the real sense at z_0 and the functions u and v obey the Cauchy-Riemann relations at z_0.) While this difference may seem quite innocuous, it has major repercussions. Fundamentally, (3.19) ignores the fact that the complex numbers form a field in which the difference quotient $[f(z) - f(z_0)]/[z - z_0]$ and its potential limit as z tends to z_0 are meaningful. Instead, (3.19) ultimately exploits only the fact that $\mathbb{C}$ is a two-dimensional vector space over the real numbers.

5.2 The Functions f_z and $f_{\bar{z}}$

There is a formalism associated with equations (3.19) and (3.20) that is convenient for dealing in complex notation with questions concerning differentiability in the real sense. It merits a few words of discussion. Suppose that the partial derivatives $f_x(z_0)$ and $f_y(z_0)$ of a function $f = u + iv$ exist at a point z_0. The *formal z- and $\bar{z}$-partial derivatives of f at z_0*, denoted $f_z(z_0)$ and $f_{\bar{z}}(z_0)$ — or sometimes $\partial f(z_0)$ and $\bar{\partial} f(z_0)$ — are then defined by the formulas

(3.21)
$$f_z(z_0) = \dfrac{1}{2}[f_x(z_0) - if_y(z_0)]$$
$$= \dfrac{1}{2}[u_x(z_0) + v_y(z_0)] + \dfrac{i}{2}[v_x(z_0) - u_y(z_0)]$$

and

(3.22)
$$f_{\bar{z}}(z_0) = \dfrac{1}{2}[f_x(z_0) + if_y(z_0)]$$
$$= \dfrac{1}{2}[u_x(z_0) - v_y(z_0)] + \dfrac{i}{2}[v_x(z_0) + u_y(z_0)] \ .$$

5. Differentiability in the Real Sense

Notice that the statement "u and v satisfy the Cauchy-Riemann equations at z_0" condenses in this notation to the single assertion that $f_{\bar{z}}(z_0) = 0$. If f happens to be differentiable at z_0, it follows immediately that $f_{\bar{z}}(z_0) = 0$ and is then easily checked that $f_z(z_0) = f'(z_0)$. Lest there be the slightest chance of a misunderstanding, however, we stress that the existence of $f_z(z_0)$ places far weaker demands on a function f than does the existence of $f'(z_0)$. (Recall the function f in Example 3.3. There $f_z(0) = f_{\bar{z}}(0) = 0$, but $f'(0)$ does not exist.)

In determining the functions f_z and $f_{\bar{z}}$ we can, naturally, write f as a function of x and y, compute the partials f_x and f_y, and combine these in the manner prescribed by (3.21) and (3.22). There is also a different procedure for finding f_z and $f_{\bar{z}}$, one that often turns out in practice to be a great labor-saver. We indicate how that procedure works, but scrimp on the details of why it works, since a full explanation would stray further from the topic at hand than we wish to go. The secret is to treat z and $\bar{z}$ temporarily as if they were completely unrelated variables and to rewrite $f(z)$ in the form $f(z) = g(z, \bar{z})$, where g is the function of the complex variables z and $\bar{z}$ that is obtained by substituting $x = (z+\bar{z})/2$ and $y = (z-\bar{z})/2i$ in $f(x,y)$. Under mild restrictions on f — it is true in any open set U where f is of class C^1 — f_z and $f_{\bar{z}}$ reduce to the z- and $\bar{z}$-partial derivatives, respectively, of the function g. (Remember: as far as g is concerned, z and $\bar{z}$ — not x and y — are considered the independent variables. The chain rule for partial derivatives offers some insight into the method. Writing $w = g(z, \bar{z}) = f(x, y)$, a purely formal application of the chain rule would give

$$\frac{\partial w}{\partial z} = \frac{\partial w}{\partial x}\frac{\partial x}{\partial z} + \frac{\partial w}{\partial y}\frac{\partial y}{\partial z} = \frac{1}{2}\left(\frac{\partial w}{\partial x} - i\frac{\partial w}{\partial y}\right),$$

i.e., $\dfrac{\partial g}{\partial z} = f_z$. Similarly, $\dfrac{\partial g}{\partial \bar{z}} = f_{\bar{z}}$. Of course, in the absence of further justification this would represent a bogus use of the chain rule, for z and $\bar{z}$ are complex variables.) In the case of the function $f(z) = |z| = (x^2 + y^2)^{1/2}$, for example, the procedure gives $g(z, \bar{z}) = (z\bar{z})^{1/2}$ and leads when $z \neq 0$ to

$$f_z(z) = \frac{\partial g}{\partial z}(z, \bar{z}) = \frac{1}{2}(z\bar{z})^{-1/2}\bar{z} = \frac{\bar{z}}{2|z|},$$

$$f_{\bar{z}}(z) = \frac{\partial g}{\partial \bar{z}}(z, \bar{z}) = \frac{1}{2}(z\bar{z})^{-1/2}z = \frac{z}{2|z|}.$$

A second example follows.

EXAMPLE 5.1. Compute f_z and $f_{\bar{z}}$ for $f(z) = |z|^2 + (z/\bar{z})$. At which points is f differentiable?

We can write $z = x + iy$ and expand

$$f(z) = \left(x^2 + y^2 + \frac{x^2 - y^2}{x^2 + y^2}\right) + i\left(\frac{2xy}{x^2 + y^2}\right).$$

Then
$$f_x(z) = \left[2x + \frac{4xy^2}{(x^2+y^2)^2}\right] + i\left[\frac{2y^3 - 2x^2y}{(x^2+y^2)^2}\right]$$
and
$$f_y(z) = \left[2y - \frac{4x^2y}{(x^2+y^2)^2}\right] + i\left[\frac{2x^3 - 2xy^2}{(x^2+y^2)^2}\right].$$

According to the definitions (3.21) and (3.22),
$$f_z(z) = \left[x + \frac{x}{x^2+y^2}\right] + i\left[-y + \frac{y}{x^2+y^2}\right]$$
and
$$f_{\bar{z}}(z) = \left[x + \frac{3xy^2 - x^3}{(x^2+y^2)^2}\right] + i\left[y - \frac{3x^2y - y^3}{(x^2+y^2)^2}\right].$$

If we make the substitutions $x = (z+\bar{z})/2$ and $y = (z-\bar{z})/2i$, these expressions simplify to
$$f_z(z) = \bar{z} + (\bar{z})^{-1}, \quad f_{\bar{z}}(z) = z - z(\bar{z})^{-2}.$$

A much more efficient method to reach the same conclusion is to rewrite $f(z)$ in the form $f(z) = z\bar{z} + (z/\bar{z})$ and to take partial derivatives with respect to z and $\bar{z}$, treating them as independent variables. Note that f_z and $f_{\bar{z}}$ — hence, f_x and f_y — are continuous functions in the domain-set $U = \mathbb{C} \sim \{0\}$ of f, so f belongs to the class $C^1(U)$. Since $f_{\bar{z}}(z) = 0$ only when $z = \pm 1$, these are the only points at which the Cauchy-Riemann equations hold for f and, as a result, the only ones at which f is differentiable.

Among our chief reasons for introducing the z- and $\bar{z}$-partials of a function is the following companion to Theorem 2.1 dealing with differentiability in the real sense. We state the result in order to have it available for reference. Its proof, which closely parallels that of Theorem 2.1, is relegated to the exercises.

Theorem 5.1. *Suppose that a function $f = u + iv$ is defined in an open subset U of the complex plane and that the partial derivatives u_x, u_y, v_x, and v_y exist everywhere in U. If each of these partial derivatives is continuous at a point z_0 of U, then f is differentiable in the real sense at z_0. Furthermore,*
$$f(z) = f(z_0) + f_z(z_0)(z - z_0) + f_{\bar{z}}(z_0)(\bar{z} - \bar{z}_0) + E(z)$$
for z in U, where $E: U \to \mathbb{C}$ satisfies $\lim_{z \to z_0} |E(z)|/|z - z_0| = 0$.

6 Exercises for Chapter III

6.1 Exercises for Section III.1

6.1. Working directly from the definition of a derivative, show that $f(z) = 3ze^{iz} + |z|^2 + 4$ is differentiable at the origin and determine $f'(0)$.

6.2. Let $f(z) = z^2 + (z-1)(z^2-1)\operatorname{Log}(z^2-1)$ if $z \neq \pm 1$, while $f(1) = f(-1) = 1$. Show that f is differentiable at the point 1 with $f'(1) = 2$, but not differentiable at the point -1.

6.3. Armed with the information that $f(0) = 0$ and $f'(0) = 1$, evaluate:
(i) $\lim_{z \to 0} f(2z)/z$; (ii) $\lim_{z \to 0} f(z^2)/z$; (iii) $\lim_{z \to 0} f(z^2-z)/z$;
(iv) $\lim_{z \to i} f(z^2+1)/(z-i)$.

6.4. Verify the formulas for $(cf)'(z_0)$, $(f+g)'(z_0)$, and $(f/g)'(z_0)$ in (3.4).

6.5. Provide a discussion of the differentiability of $f(z) = \sqrt[3]{z}$ along the lines of Example 1.5.

6.6. Identify all points z at which the function $f(z) = z^{3/2}$ is differentiable and compute $f'(z)$ for such z.

6.7. For each of the following functions determine the largest open set in which it is analytic, and calculate its derivative there by putting to work the various differentiation rules: (i) $f(z) = z^4 - 2iz^3 + 1$; (ii) $g(z) = (z-i)/(z+i)$; (iii) $h(z) = z^4(1-z)^6$; (iv) $k(z) = [(z+1)/(z^3-8)]^4$; (v) $\ell(z) = \sqrt{4-z^2}$; (vi) $m(z) = (z - 2\sqrt{z})^{-1}$; (vii) $p(z) = \sqrt{1 - \sqrt{z}}$.

6.8. What are the critical points of (i) $f(z) = z^4(1-2z)^6$; (ii) $g(z) = 2z^3 - 9iz^2 - 12z + \pi i$; (iii) $h(z) = (z^4 - 8z)^{10}$?

6.9. A function f is analytic in an open set U. Define g by $g(z) = \overline{f(\bar{z})}$. Show that g is analytic in the open set $U^* = \{z : \bar{z} \in U\}$ and that $g'(z) = \overline{f'(\bar{z})}$ for z in U^*.

6.2 Exercises for Section III.2

6.10. Let $f(z) = (3/2)x^2 - xy + ixy^2$. Locate all points z at which f is differentiable, and determine $f'(z)$ for each such point.

6.11. If $f(z) = 1 - y^2 + i(2xy - y^2)$, identify the set of points z where f is differentiable, and compute $f'(z)$ for these points.

6.12. Find a polynomial function of x and y that is differentiable at every point of the parabola with equation $y = x^2$, but at no other point of the complex plane.

6.13. Find a polynomial function f of x and y that is differentiable at the origin, where $f'(0) = 1$, and at every point of the circle $K = K(0,1)$, but at no other point of the complex plane.

6.14. If functions f and g are both analytic in a domain D and if $f'(z) = g'(z)$ for every z in D, show that f and g differ by a constant in D.

6.15. A function f is analytic in a domain D and its derivative f' is constant in D. Prove that f is a linear function of z in D; i.e., $f(z) = az + b$ for z in D, where a and b are constants. (*Hint.* Exercise 6.14.)

6.16. If a function f is known to be n-times differentiable in a domain D and if $f^{(n)}(z) = 0$ for every z in D, then in this domain f is a polynomial function of z whose degree is at most $n - 1$. Confirm this. (*Hint.* Maybe the quickest argument proceeds by induction.)

6.17. Given that a real-valued function h is a solution of the "Laplace equation" in an open set U — this means that h belongs to the class $C^2(U)$ and satisfies $h_{xx}(z) + h_{yy}(z) = 0$ for every z in U — show that the function $f: U \to \mathbb{C}$ defined by $f(z) = h_x(z) - ih_y(z)$ is an analytic function. (N.B. A real-valued function that is a solution of Laplace's equation in an open set U is said to *harmonic in* U. We shall study harmonic functions in Chapter VI.)

6.18. Suppose that a function $f = u + iv$ is analytic in an open set U. Under the assumption that f is a member of $C^2(U)$, verify that both of the functions u and v are harmonic in U. (N.B. The analyticity of f in U actually implies that f is in $C^2(U)$ — indeed, in $C^\infty(U)$ — but this information will not be available to us until Chapter V.)

6.19. Suppose that f, a function analytic in an open set U, is a member of the class $C^2(U)$. Demonstrate that its derivative f' is then analytic in U. Conclude that a function f which is analytic in U and also belongs to the class $C^\infty(U)$ has derivatives $f', f'', f''', \cdots$ of all orders in U.

6.20. Let $f(z) = c_{00} + c_{10}x + c_{01}y + \cdots + c_{nm}x^n y^m$ be a polynomial function of x and y. If, in addition, f is an analytic function, show that f has to be a polynomial in z. Specifically, show that $f(z) = a_0 + a_1 z + \cdots + a_n z^n$, where $a_k = (k!)^{-1}[\partial^k f/\partial x^k](0)$ for $k = 0, 1, \ldots, n$. (*Hint.* Use Exercises 6.19 and 6.16.)

6.21. Assume that a function f is analytic in a domain D and satisfies $\text{Arg}[f(z)] = \alpha$, a constant, for every z in D where $f(z) \neq 0$. Without invoking the open mapping theorem, a result to which we alluded in the text, demonstrate that f is constant in D. (*Hint.* Look at $g(z) = e^{-i\alpha}f(z)$. What can you say about $\text{Im}[g(z)]$ for z in D?)

6.22. If a function $f = u + iv$ is analytic in a domain D and if $v(z) = [u(z)]^2$ for every z in D, then f is constant in D. Prove this. Do not quote the open mapping theorem. (*Hint.* Differentiate the relation $u^2 - v = 0$ with respect to x and y.)

6.23. Let a function $f = u + iv$ be analytic in a domain D. Assume the existence of an open set U containing $f(D)$ and a function Φ in the class $C^1(U)$ — one may view Φ either as a function of a complex variable $w = u + iv$ or as a function of two real variables u and v — such that the partial derivatives $\Phi_u(w)$ and $\Phi_v(w)$ do not vanish simultaneously for any point w of $f(D)$ and such that $\Phi[u(z), v(z)] = c$, a constant, for every z in D. Deduce that f must be constant in D. (N.B. This is a generalization of Theorem 2.5, whose conclusions can be inferred from it by taking $\Phi(u,v) = u$, $\Phi(u,v) = v$, and $\Phi(u,v) = u^2 + v^2$, respectively. It also includes, for example, Exercise 6.22, where $\Phi(u,v) = u^2 - v$.)

6.3 Exercises for Section III.3

6.24. Let $f = u + iv$, where $u(z) = x^3 + axy^2$ and $v(z) = bx^2y + cy^3 + 1$. Determine the values of the real numbers a, b, and c that make f an entire function. Then express f as a polynomial function of z. (For the last part refer to Exercise 6.20.)

6.25. Find the entire function $f = u + iv$ with $f(0) = 2$ such that $v(z) = (x \sin y + \sin y + y \cos y)e^x$. Try to express the final answer as a function of z — not of x and y — but be warned that this may not prove so easy without a stroke of factoring insight.

6.26. Find the critical points of $f(z) = \exp\left[(1/3)z^3 + z^2 - (2+4i)z + 1\right]$.

6.27. Determine the largest domain D containing the origin in which the function $f(z) = \sqrt{e^z}$ is analytic, and confirm that $f(z) = e^{z/2}$ for every z in D. (N.B. It is not true that $f(z) = e^{z/2}$ for every complex z. For which z is it the case that $f(z) = e^{z/2}$? What is true for other z?)

6.28. If a function f is analytic in a domain D and if f satisfies a differential equation $f' - \alpha f = 0$ in D — here α is a complex constant — demonstrate that in this domain f takes the form $f(z) = Ae^{\alpha z}$ for some constant A. (*Hint.* Consider the function g defined by $g(z) = f(z)e^{-\alpha z}$.)

6.29. Establish the following identities for complex z and w:
(i) $\cos z = \sin[(\pi/2) - z]$; (ii) $\sin(\bar{z}) = \overline{\sin z}$, $\cos(\bar{z}) = \overline{\cos z}$; (iii) $\cos(-z) = \cos z$, $\sin(-z) = -\sin z$; (iv) $\cos(z+w) = \cos z \cos w - \sin z \sin w$, $\sin(z+w) = \sin z \cos w + \cos z \sin w$; (v) $2\cos^2 z = 1 + \cos(2z)$, $2\sin^2 z = 1 - \cos(2z)$; (vi) $1 + \tan^2 z = \sec^2 z$; (vii) $\sin z/(1 + \cos z) = \tan(z/2)$.

6.30. (a) Find all complex numbers z for which $\cos z = 5/4$. (b) Find all z for which $\tan z = 2i$.

6.31. Given that $\sin w = \sin z$, prove that $w = z + 2k\pi$ for some integer k when $\cos w = \cos z$ and that $w = -z + (2k+1)\pi$ for some integer k when $\cos w = -\cos z$.

6.32. Let z and w be complex numbers, neither an odd multiple of $\pi/2$.

Show that $\tan w = \tan z$ if and only if $w = z + k\pi$ for some integer k.

6.33. Derive the bounds $|\sinh y| \leq |\sin z| \leq \cosh y$ for $z = x + iy$. (N.B. The same estimates are valid for $|\cos z|$.)

6.34. Define $h: \mathbb{C} \to \mathbb{C}$ by $h(z) = z^{-1} \sin z$ if $z \neq 0$ and $h(0) = 1$. Check that h is a continuous and even function. The latter property demands, of course, that $h(z) = h(-z)$.

6.35. Verify that $f(z) = \tan z$, $g(z) = \cot z$, $h(z) = \sec z$, and $k(z) = \csc z$ are analytic functions whose derivatives are given by $f'(z) = \sec^2 z$, $g'(z) = -\csc^2 z$, $h'(z) = \sec z \tan z$, and $k'(z) = -\csc z \cot z$.

6.36. Find the limits: (i) $\lim_{z \to 0} z^{-2}(e^{z^2} - 1)$; (ii) $\lim_{z \to 0} z^{-1} \sin(3z)$; (iii) $\lim_{z \to 0} z^{-1/2} \tan z$; (iv) $\lim_{z \to 1}(z^2 - 1)^{-1}[\exp(z^2 - z) - 1]$; (v) $\lim_{z \to 0}(z + 4\sqrt{z})^{-1} \sin(i\sqrt{z})$; (vi) $\lim_{z \to i}(z^4 - 1)\csc(z^2 + 1)$.

6.37. Verify that the formula $f(z) = \cos(\sqrt{z})$ defines an entire function, whereas $g(z) = \sin(\sqrt{z})$ does not. (*Hint.* Begin by proving that $f_x(z) = -if_y(z) = -\sin(\sqrt{z})/(2\sqrt{z})$ if $z \neq 0$, while $f_x(0) = -if_y(0) = -1/2$. Then argue that f_x and f_y are continuous in $\mathbb{C}$. A possibly helpful observation is this: if a function $h: \mathbb{C} \to \mathbb{C}$ is both continuous and even, then the function $k: \mathbb{C} \to \mathbb{C}$ given by $k(z) = h(\sqrt{z})$ is continuous.)

6.38. Generalize the preceding exercise as follows: if g is an even entire function, then the formula $f(z) = g(\sqrt{z})$ defines an entire function. Feel free to assume for the purposes of the present exercise that g' is also entire, although this added assumption will become superfluous once Theorem V.3.1 is established. (*Hint.* To start, show that $f_x(z) = -if_y(z) = g'(\sqrt{z})/(2\sqrt{z})$ if $z \neq 0$ and $f_x(0) = -if_y(0) = g''(0)/2$.)

6.39. For each of the following functions determine the largest open set in which it is analytic, and compute its derivative in that set: (i) $f(z) = (e^z - 1)/(e^z + 1)$; (ii) $g(z) = \csc(\sqrt{z})$; (iii) $h(z) = \sec(\sqrt{z})$; (iv) $k(z) = z^{-1/2} e^{\tan z}$; (v) $\ell(z) = \sin(\sqrt{iz + 1})/(z^2 + 1)$; (vi) $m(z) = \cos(\sqrt{e^z - 1})$.

6.40. Under the assumption that w is not a real number with $|w| \geq 1$ show that $\mathrm{Re}\left[(1 - w^2)^{1/2} + iw\right] > 0$ and $\mathrm{Re}\left[-(1 - w^2)^{1/2} + iw\right] < 0$. This information was utilized in the derivation of (3.10). (*Hint.* Treat the cases $\mathrm{Im}\, w > 0$, $\mathrm{Im}\, w < 0$, and $w \in (-1, 1)$ separately.)

6.41. Certify that the formula $\mathrm{Arcsin}\, w = -i\,\mathrm{Log}[(1 - w^2)^{1/2} + iw]$ remains valid when w is real and $|w| \geq 1$. Thus, complete the verification of (3.10).

6.42. Write in the form $x + iy$: (i) $\mathrm{Arcsin}(i\sqrt{3})$; (ii) $\mathrm{Arctan}[(2 - i)/5]$; (iii) $e^{i\,\mathrm{Arcsin}(\sqrt{2})}$; (iv) $\mathrm{Arcsin}(\sqrt{t^2 + 1})$ for $t \geq 0$.

6.4 Exercises for Section III.4

6.43. Exhibit all three branches of the cube root function in the domain $D = \mathbb{C} \sim (-\infty, 0]$.

6.44. Let $D = \mathbb{C} \sim [0, \infty)$. Define $f: D \to \mathbb{C}$ by $f(z) = \sqrt[3]{z}$ if $\operatorname{Im} z \geq 0$ and $f(z) = \omega\sqrt[3]{z}$ if $\operatorname{Im} z < 0$, where $\omega = (-1 + i\sqrt{3})/2$. Verify that f is a branch of the cube root function in D, and determine its derivative. Identify the two remaining branches of the cube root function in D.

6.45. Give a formula in the mold of the one in Example 4.2 for the branch f of the fourth root function in the domain $D = \mathbb{C} \sim [0, \infty)$ that satisfies $f(-1) = (\sqrt{2} + i\sqrt{2})/2$. Calculate $f'(-i)$. What are the other three branches of the fourth root function in D?

6.46. If $f(z) = \cos z$, then a branch g of f^{-1} in the domain $D = \mathbb{C} \sim \{z : \operatorname{Im} z = 0 \text{ and } |z| \geq 1\}$ is given by $g(z) = (\pi/2) - \operatorname{Arcsin} z$. Support this statement.

6.47. Let $f(z) = \sec z$. Show that $g(z) = (\pi/2) - \operatorname{Arcsin}(1/z)$ defines a branch of f^{-1} in the domain $D = \mathbb{C} \sim [-1, 1]$. (The function g is known as the *principal arcsecant function* and is normally written $g(z) = \operatorname{Arcsec} z$.) Demonstrate that g is an analytic function whose derivative at a point z of D is given as follows: $g'(z) = \left(z\sqrt{z^2 - 1}\right)^{-1}$ if $\operatorname{Re} z > 0$ or if $\operatorname{Re} z = 0$ and $\operatorname{Im} z > 0$; $g'(z) = -\left(z\sqrt{z^2 - 1}\right)^{-1}$ if $\operatorname{Re} z < 0$ or if $\operatorname{Re} z = 0$ and $\operatorname{Im} z < 0$. (N.B. Knowing that g' is continuous in D may prove helpful for the last part of the problem.)

6.48. Assume that g is a branch of the inverse tangent function in a domain D —i.e., $g: D \to \mathbb{C}$ is a continuous function with the property that $\tan[g(z)] = z$ for every z in D. Prove that g is necessarily analytic in D, with $g'(z) = (1 + z^2)^{-1}$ there. Prove, in addition, that any other branch h of the inverse tangent function in D differs from g by a constant integral multiple of π.

6.49. Let g be a branch of the inverse sine function in a domain D, meaning that $g: D \to \mathbb{C}$ is a continuous function and $\sin[g(z)] = z$ holds for every z in D. Assume that D contains neither the point 1 nor the point -1. (N.B. The existence of such a function g actually prevents either 1 or -1 from being in D, but we shall not have the means to prove this until Chapter V.) Demonstrate that g is an analytic function and that $g'(z) = \pm(1 - z^2)^{-1/2}$ for every z in D. (The sign may well vary from one point of D to another.) Use this information to show that any branch h of the inverse sine function in D has either the form $h = g + 2k\pi$ for some integer k or the form $h = -g + (2k+1)\pi$ for some integer k. (*Hint.* For the last part prove that g' and h' are continuous in D, and then consider the sets $U = \{z \in D : h'(z) = g'(z)\}$ and $V = \{z \in D : h'(z) = -g'(z)\}$.)

6.50. Prove that $\theta(z) = \text{Arg}(-z) + \pi$ defines a branch of $\arg z$ in the domain $D = \mathbb{C} \sim [0, \infty)$. What is the range of θ? Find formulas for the branch L of $\log z$ in D corresponding to θ and for the associated λ-power function h_λ.

6.51. In which of the following domains D is it possible to define a branch of the logarithm function: (i) $D = \{z : x+y < 0\}$; (ii) $D = \{z : 1 < |z| < e\}$; (iii) $D = \{z : 1 < |z| < e\} \sim \{ti : 1 < t < e\}$; (iv) $D = \{z : 0 < y-x < 1\}$; (v) $D = \{z : 0 < |x|+|y| < 1\}$? Justify the answer in each instance by constructing a branch of $\log z$ in D or by proving that none exists.

6.52. Let $D = \mathbb{C} \sim \{z : z = 0 \text{ or } z = e^{t+it}, -\infty < t < \infty\}$, and let L be the branch of $\log z$ in D that satisfies $L(e) = 1$. Determine: (i) $L(e^6)$; (ii) $L(-e^{-8})$; (iii) $L(ie^{\pi k})$ for any integer k. What is the range of L?

6.53. We define S_k for $k = 1, 2, \cdots$ as follows: if $k = 1, 3, 5, \cdots, S_k = \{z : |z-1| = k \text{ and } y \geq 0\}$; if $k = 2, 4, 6, \cdots, S_k = \{z : |z| = k \text{ and } y \leq 0\}$. Let $D = \mathbb{C} \sim \cup_{k=1}^\infty S_k$, let $L: D \to \mathbb{C}$ be the branch of $\log z$ in D for which $L(1) = 0$, and let h_λ indicate the branch of the λ-power function in D associated with L. Compute the following quantities: (i) $L(-1)$; (ii) $L(e)$; (iii) $h_\lambda(ei)$ for $\lambda = i$; (iv) $h'_\lambda(-\pi^2)$ for $\lambda = 1/2$; (v) $L(e^2)$; (vi) $L(2k+1)$ for any integer k. (*Hint.* Start by drawing a picture.)

6.54. For each of the following functions determine the largest open set in which it is analytic and compute its derivative: (i) $f(z) = \sqrt[3]{z^4-1}$; (ii) $g(z) = z \operatorname{Arcsin}(z^2)$; (iii) $h(z) = e^{\operatorname{Arctan}(i\sqrt{z})}$; (iv) $k(z) = (z^2+1)^z$; (v) $\ell(z) = \operatorname{Log}(\operatorname{Log} z)$; (vi) $m(z) = (1+e^z)\operatorname{Log}(1+e^z)$; (vii) $p(z) = e^{\sqrt{z}} + e^{-\sqrt{z}}$; (viii) $q(z) = \operatorname{Arcsin}(\sqrt{1-z^2})$.

6.5 Exercises for Section III.5

6.55. Compute the z- and $\bar{z}$-partials of the following functions: (i) $f(z) = 2x^3y^2 + i(x^2-y)$; (ii) $g(z) = 3z^2 + e^{\bar{z}} + z|z|^4$; (iii) $h(z) = \operatorname{Log}(1-|z|^2)$; (iv) $k(z) = z\operatorname{Arcsin}(|z|)$; (v) $\ell(z) = \sqrt{\bar{z}/z}$.

6.56. Let $\alpha > 0$. Define $f: \mathbb{C} \to \mathbb{C}$ by $f(z) = |z|^{\alpha-1}z$ for $z \neq 0$ and $f(0) = 0$. Compute $f_z(z)$ and $f_{\bar{z}}(z)$ at the points z where these quantities make sense.

6.57. Corresponding to each of the following sets S construct a function $f: \mathbb{C} \to \mathbb{C}$ in the class $C^1(\mathbb{C})$ with the property that S is precisely the set of points in the complex plane at which f is differentiable: (i) $S = \{z : |z|^4 - |z|^2 = 0\}$; (ii) $S = \{z : x^2 - y^2 = 1\}$; (iii) $S = \{z : \bar{z}e^z - |z|^2 e^{\bar{z}} = 0\}$; (iv) $S = \{z : |z| = 1 \text{ or } |z| = 2\}$.

6.58. Construct a function f in the class $C^1(\mathbb{C})$ that is differentiable with

$f'(z) = 1$ at every point z of the closed disk $S = \overline{\Delta}(0,1)$, but that is differentiable at no other point of the complex plane. (*Hint.* Let $\varphi: \mathbb{R} \to \mathbb{R}$ be the function given by $\varphi(t) = 0$ if $t \leq 1$ and $\varphi(t) = (t-1)^2$ if $t > 1$. Consider the function $\Phi: \mathbb{C} \to \mathbb{C}$ defined by $\Phi(z) = \varphi(|z|^2)$. Check first that Φ is a member of $C^1(\mathbb{C})$.)

6.59. Modify the construction in Exercise 6.58 so as to produce a function f in $C^1(\mathbb{C})$ that is differentiable with $f'(z) = z$ at every point z of the square $S = \{z : |x| \leq 1, |y| \leq 1\}$, but differentiable nowhere else. (*Hint:* Work with the function $\psi: \mathbb{R} \to \mathbb{R}$ given by $\psi(t) = 0$ if $|t| \leq 1$ and $\psi(t) = (1 - |t|)^2$ if $|t| > 1$ and with the function $\Psi: \mathbb{C} \to \mathbb{C}$ defined by $\Psi(z) = \psi(x^2) + \psi(y^2)$.)

6.60. For a function f in the class $C^1(U)$ verify that $(\overline{f})_z = \overline{f_{\overline{z}}}$ and $(\overline{f})_{\overline{z}} = \overline{f_z}$ in U.

6.61. Suppose that $f \in C^1(U)$, that $g \in C^1(V)$, and that $f(U)$ is contained in V. Derive the chain rules for the z- and $\overline{z}$-partials of $g \circ f$ in U:

$$(g \circ f)_z = (g_z \circ f)f_z + (g_{\overline{z}} \circ f)(\overline{f})_z$$

and

$$(g \circ f)_{\overline{z}} = (g_z \circ f)f_{\overline{z}} + (g_{\overline{z}} \circ f)(\overline{f})_{\overline{z}} \ .$$

Conclude that

$$(g \circ f)_z = (g_z \circ f)f' \quad , \quad (g \circ f)_{\overline{z}} = (g_{\overline{z}} \circ f)\overline{f'}$$

if f is analytic in U, while

$$(g \circ f)_z = (g' \circ f)f_z \quad , \quad (g \circ f)_{\overline{z}} = (g' \circ f)f_{\overline{z}}$$

if g is analytic in V.

6.62. Under the assumption that the partial derivatives $f_x(z_0)$ and $f_y(z_0)$ of a function $f = u + iv$ exist at a point z_0, the *Jacobian determinant of f* at z_0, denoted $J_f(z_0)$, is defined by the rule

$$J_f(z_0) = \begin{vmatrix} u_x(z_0) & u_y(z_0) \\ v_x(z_0) & v_y(z_0) \end{vmatrix} = u_x(z_0)v_y(z_0) - u_y(z_0)v_x(z_0) \ .$$

Show that $J_f(z_0)$ can be expressed with the aid of complex notation as $J_f(z_0) = |f_z(z_0)|^2 - |f_{\overline{z}}(z_0)|^2$.

6.63. Prove Theorem 5.1. As in the proof of Theorem 2.1 carry out details for the case $z_0 = 0$. Proceed by modifying that earlier proof.

6.64. The *directional derivative* $\partial_\theta f(z_0)$ of a function $f: A \to \mathbb{C}$ at an interior point z_0 of A in the direction $e^{i\theta}(-\pi < \theta \leq \pi)$ is given by

$$\partial_\theta f(z_0) = \lim_{r \to 0+} \frac{f(z_0 + re^{i\theta}) - f(z_0)}{r} \ ,$$

provided this limit exists. Assuming that f is differentiable in the real sense at z_0, show that $\partial_\theta f(z_0) = e^{i\theta} f_z(z_0) + e^{-i\theta} f_{\bar{z}}(z_0)$. Deduce that

$$\max_{-\pi < \theta \leq \pi} |\partial_\theta f(z_0)| = |f_z(z_0)| + |f_{\bar{z}}(z_0)|$$

and

$$\min_{-\pi < \theta \leq \pi} |\partial_\theta f(z_0)| = ||f_z(z_0)| - |f_{\bar{z}}(z_0)|| .$$

6.65. Suppose that a function f is differentiable in the real sense at a point z_0 and that, in addition, $L = \lim_{z \to z_0} |f(z) - f(z_0)|/|z - z_0|$ exists. Prove that at least one of the functions f or $\bar{f}$ is differentiable at z_0. (*Hint.* Look at $\lim_{r \to 0+} |f(z_0 + re^{i\theta}) - f(z_0)|/r$ for $0 \leq \theta \leq \pi$. Use the preceding exercise.)

Chapter IV

Complex Integration

Introduction

Our objective in the present chapter is to set up the main piece of technical machinery required to probe the structure of analytic functions, the complex line integral — or, as some people prefer to call it, the contour integral. There are several levels of generality at which it is possible to implement this mode of integration, levels dictated by the answers to certain questions. What class of functions is to serve as the class of integrands? What degree of regularity is imposed on paths of integration? Our response to the former question is a relatively standard one for beginning texts: the ability to form contour integrals of continuous functions is entirely adequate for our needs. As for the latter question, we choose to confine ourselves in this book to integration along "piecewise smooth" paths, for which the definition of the complex line integral is extremely simple. Certainly there exist circumstances in which it is desirable (or even necessary) to integrate a function along a so-called "rectifiable" path, but such circumstances can be avoided in developing the essential principles of basic complex analysis. It is only at more advanced stages in the subject that rectifiable paths begin to play a prominent role. Reasoning that our focus in this book should be on analytic functions, rather than on the refinements of integration theory, we opt for the more pragmatic approach.

1 Paths in the Complex Plane

1.1 Paths

By a *path* γ in the complex plane we understand a continuous function of the type $\gamma:[a,b] \to \mathbb{C}$, where $[a,b]$ is a non-degenerate ($a < b$) closed interval of real numbers. The range of a path γ is more commonly called

110   IV. Complex Integration

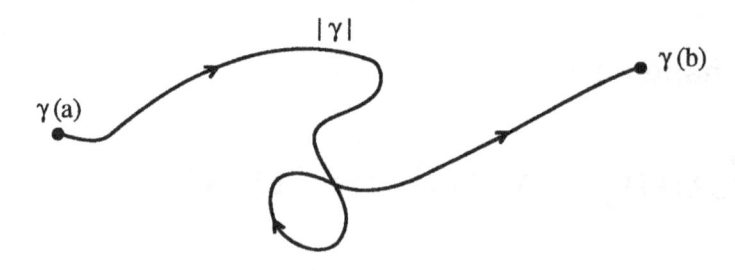

Figure 1.

its *trajectory*, a set that we denote by $|\gamma|$. (The context will always make clear whether the symbol | | stands for "modulus" or for "trajectory.") We speak of the function γ as providing a *parametrization* of $|\gamma|$. If A is a subset of $\mathbb{C}$, then γ is referred to as a *path in A* under the condition that its trajectory is contained in A. We sometimes employ the notation $\gamma : [a, b] \to A$ to signal that this is the case. Observe that the trajectory of a path, being the continuous image of a compact and connected set, is itself compact and connected.

In this book we shall often depict a path $\gamma : [a, b] \to \mathbb{C}$ by sketching its trajectory and by indicating with arrows the fashion in which that trajectory is traced out as t increases from a to b. (See Figure 1.) Naturally, this obscures the actual parametrization of $|\gamma|$ given by γ, but it does represent a useful mechanism for rendering a rather abstract object — a path is after all a function, not a set of points — more concrete and geometric.

The *initial* and *terminal points* of a path $\gamma : [a, b] \to \mathbb{C}$ are the points $\gamma(a)$ and $\gamma(b)$, respectively. When these values coincide we call γ a *closed path*. If $\gamma(t) \neq \gamma(s)$ for $t \neq s$, with the possible exception that $\gamma(a) = \gamma(b)$, γ is proclaimed a *simple path*. (We refer the reader to Figure 2.)

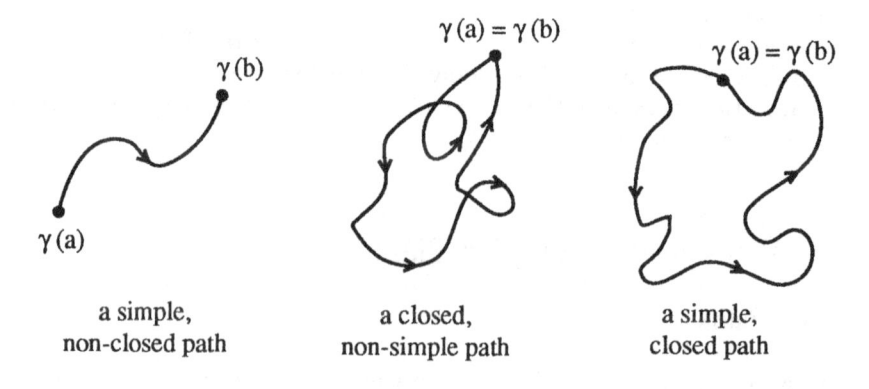

a simple,   a closed,   a simple,
non-closed path   non-simple path   closed path

Figure 2.

1. Paths in the Complex Plane

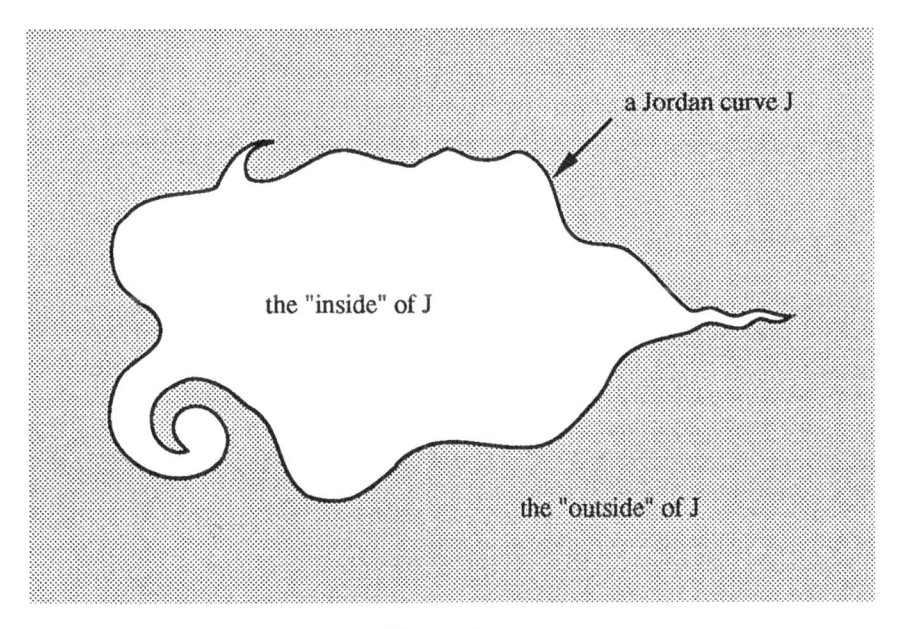

Figure 3.

A set J of points in the complex plane is known as a *Jordan curve* if J is the trajectory of some simple, closed path. Since J is compact — hence, is a closed set — its complement $\mathbb{C} \sim J$ is an open set. As such, $\mathbb{C} \sim J$ is a disjoint union of domains, its various components (Theorem II.3.5). The "Jordan Curve Theorem", a celebrated theorem in plane topology first asserted by Camille Jordan (1838–1892) in 1887 and finally given a rigorous proof by Oswald Veblen (1880–1960) in 1905, states that the complement of a plane Jordan curve J has exactly two components, each having J for its boundary. One of these components (the *inside* of J) is a bounded set, while the other (the *outside* of J) is unbounded (Figure 3).

A person must not be misled by diagrams like Figure 3 into thinking that the Jordan curve theorem is so "intuitively obvious" that it scarcely deserves the title "theorem." Jordan curves can be immensely more complicated than the typical sketch of one is apt to suggest. The proof of the Jordan curve theorem is surprisingly difficult, and its length alone would prohibit its inclusion in the present book. Our general policy in the development of complex analysis will be to avoid appeals to this theorem except in the most transparent of situations, situations where its validity is unlikely to be challenged. (The major departure from this policy occurs in Lemma V.2.1 and in subsequent results that depend on that lemma.) Nevertheless, we shall not hesitate on occasion to bring to the reader's attention topological facts associated with the Jordan curve theorem that have a direct bearing on complex analysis. Some of the fundamental results

1.2 Smooth and Piecewise Smooth Paths

A path given by $\gamma(t) = x(t) + iy(t)$ for $a \le t \le b$ is termed a *smooth path* if its derivative $\dot\gamma(t)$ with respect to the real parameter t, $\dot\gamma(t) = \dot x(t) + i\dot y(t)$, exists for each t in $[a,b]$ and if the function $\dot\gamma$ is continuous on the interval $[a,b]$. We have elected here to use Newton's dot notation $\dot\gamma(t)$ for the derivative, rather than $\gamma'(t)$, in order to head off any possible mix-ups with complex differentiation. Of course, by $\dot x(a)$, $\dot x(b)$, $\dot y(a)$, or $\dot y(b)$ we mean the appropriate one-sided derivative, e.g.,

$$\dot x(a) = \lim_{t \to a^+} \frac{x(t) - x(a)}{t - a}.$$

(There is, unfortunately, little consensus among mathematicians with regard to usage of the expression "smooth path." In defining this concept some authors would, for instance, append to our definition the requirement that $\dot\gamma(t) \ne 0$ for every t in $[a,b]$. The reader is advised to bear this inconsistency in mind when consulting other references in complex analysis.) For a value of t at which $\dot\gamma(t) \ne 0$ this derivative admits a simple geometric interpretation — namely, as a tangent vector to the trajectory of γ at the point $\gamma(t)$. (See Figure 4.)

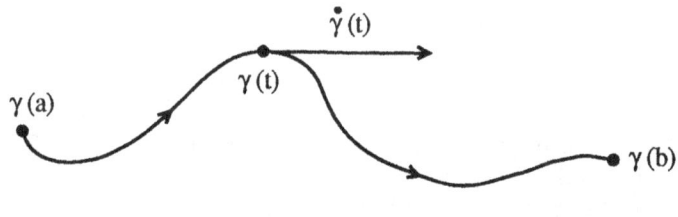

Figure 4.

A path $\gamma:[a,b] \to \mathbb{C}$ is said to be *piecewise smooth* provided there is a partition $P: a = t_0 < t_1 < \cdots < t_n = b$ of the interval $[a,b]$ with the property that the restriction of γ to each of the intervals $[t_{k-1}, t_k]$, $1 \le k \le n$, is a smooth path.

EXAMPLE 1.1. Describe the features of the path given by $\gamma(t) = t + it$ for $0 \le t \le 1$.

This path is simple and non-closed, with the line segment joining 0 and $1 + i$ for its trajectory. It is also smooth. Indeed, $\dot\gamma(t) = 1 + i$ for $0 \le t \le 1$.

1. Paths in the Complex Plane

EXAMPLE 1.2. Discuss the path γ defined on $[-1,1]$ by $\gamma(t) = t^2 + it^2$.

Here $\dot\gamma(t) = 2t + 2it$ for $-1 \leq t \leq 1$, so γ is smooth. It is closed, but distinctly non-simple, for $\gamma(t) = \gamma(-t)$. This path has the same trajectory as the path in the previous example. The two paths, however, parametrize that trajectory in vastly different ways.

EXAMPLE 1.3. Let $\gamma(t) = t + i|t|$ for $-1 \leq t \leq 1$. Discuss the properties of γ (Figure 5).

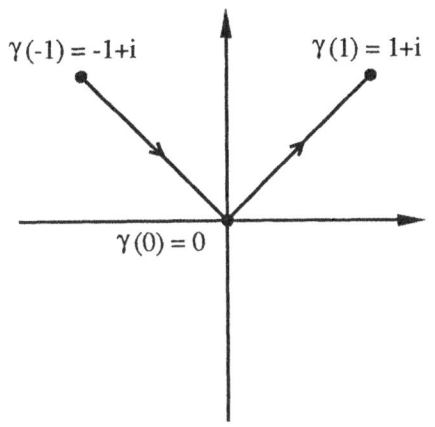

Figure 5.

Since $\dot\gamma(0)$ fails to exist, γ is not smooth. We have $\gamma(t) = t - it$ for $-1 \leq t \leq 0$ and $\gamma(t) = t + it$ for $0 \leq t \leq 1$. The restrictions of γ to $[-1,0]$ and to $[0,1]$ are thus easily seen to be smooth. Accordingly, γ is a piecewise smooth path. It is simple and non-closed. Incidentally, the appearance of a corner or cusp in the trajectory of a path, as occurs in this example, is not always indicative of non-smoothness. For instance, the path β defined on $[-1,1]$ by $\beta(t) = t^3 + i|t^3|$ is a smooth path that has the same trajectory as γ. When $t = 0$, the parameter value that corresponds to the corner, it is true that $\dot\beta(t) = 0$. This is typically what happens when a cusp or corner turns up in the trajectory of a smooth, simple path. A smooth, simple, closed path is also allowed by our definition of the term "smooth" to have a corner or cusp at its initial (= terminal) point, even if the relevant derivatives are non-zero.

EXAMPLE 1.4. What is the general character of the path $\gamma : [0,4] \to \mathbb{C}$ defined by

γ(3) = i   γ(2) = 1+i

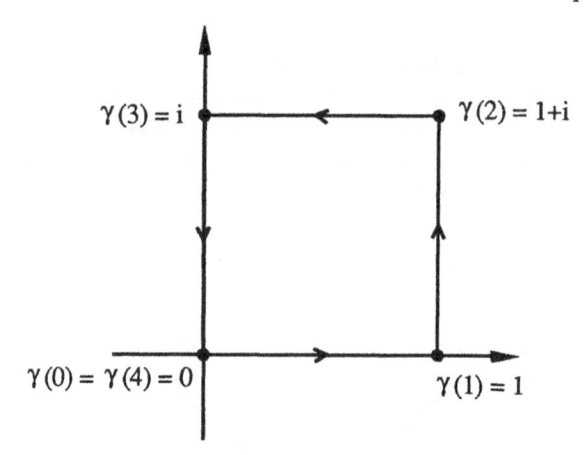

γ(0) = γ(4) = 0   γ(1) = 1

Figure 6.

$$\gamma(t) = \begin{cases} t & \text{if } 0 \le t \le 1, \\ 1 + i(t-1) & \text{if } 1 \le t \le 2, \\ (3-t) + i & \text{if } 2 \le t \le 3, \\ i(4-t) & \text{if } 3 \le t \le 4. \end{cases}$$

This path is piecewise smooth, simple, and closed. It has the boundary of a square as its trajectory (Figure 6).

EXAMPLE 1.5. Let $\gamma(t) = z_0 + re^{it}$ for $a \le t \le b$, where $r > 0$. Discuss this path.

The path γ is smooth, with $\dot{\gamma}(t) = ire^{it}$ for $a \le t \le b$. (Verify this!) It is simple when $b \le a + 2\pi$, and it is closed when $b = a + 2k\pi$ for some positive integer k. Its trajectory obviously lies on the circle $K(z_0, r)$ (Figure 7).

1.3 Parametrizing Line Segments

Given points z_1 and z_2 in $\mathbb{C}$, we consider the smooth path $\gamma \colon [0,1] \to \mathbb{C}$ defined by the formula $\gamma(t) = (1-t)z_1 + tz_2$. The trajectory of γ is the line segment with endpoints z_1 and z_2, which γ traces out at a uniform rate, starting at z_1 and terminating at z_2. Smooth paths of this type will be encountered frequently throughout the book. It will be convenient to reserve a special notation for them. One suggestive way of designating such a path is simply to use interval notation: $\gamma = [z_1, z_2]$. This is the notation we adopt. In so doing we allow for the possibility that $z_1 = z_2$, in which case $\gamma = [z_1, z_1]$ reduces to the constant path with domain-set $[0,1]$ whose trajectory is the single point z_1. We rely on the context to make evident

1. Paths in the Complex Plane   115

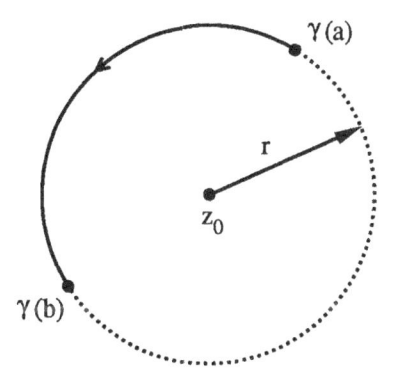

Figure 7.

whether the notation [,] merely signifies a closed interval of real numbers or, instead, indicates the specific parametrization under discussion here of a line segment in the complex plane.

1.4   Reverse Paths, Path Sums

Two elementary methods for modifying or combining paths, in order to produce new paths, will establish their merits as our treatment of complex integration unfolds. We introduce them here.

If $\gamma:[a,b] \to \mathbb{C}$ is a path, then the *reverse* of γ, which we denote by $-\gamma$, is the path defined for t in $[a,b]$ by

$$[-\gamma](t) = \gamma(b+a-t) .$$

A moment's thought reveals that γ and $-\gamma$ share the same trajectory, but that $\gamma(t)$ and $[-\gamma](t)$ traverse this trajectory in "opposite directions" as t grows from a to b. The path $-\gamma$ is readily seen to be piecewise smooth whenever γ has that feature (Exercise 4.2). As an illustration observe that, with the notation introduced in Section 1.3, $-[z_1, z_2] = [z_2, z_1]$ for any pair of complex numbers z_1 and z_2.

Next, suppose that paths $\gamma_1:[a_1,b_1] \to \mathbb{C}$ and $\gamma_2:[a_2,b_2] \to \mathbb{C}$ have the property that $\gamma_1(b_1) = \gamma_2(a_2)$. Under this condition we are at liberty to define a path $\gamma_1 + \gamma_2:[a_1, b_1 + b_2 - a_2] \to \mathbb{C}$ as follows:

$$[\gamma_1 + \gamma_2](t) = \begin{cases} \gamma_1(t) & \text{if } a_1 \leq t \leq b_1 , \\ \gamma_2(t - b_1 + a_2) & \text{if } b_1 \leq t \leq b_1 + b_2 - a_2 . \end{cases}$$

Consistent with this notation, the path $\gamma_1 + \gamma_2$ is called the *path sum* of γ_1 and γ_2. This path first traces out $|\gamma_1|$, then proceeds to describe $|\gamma_2|$. The somewhat strange-looking parameter interval of $\gamma_1 + \gamma_2$ is obtained by attaching an interval of the "correct" length, $b_2 - a_2$, to the right-hand end

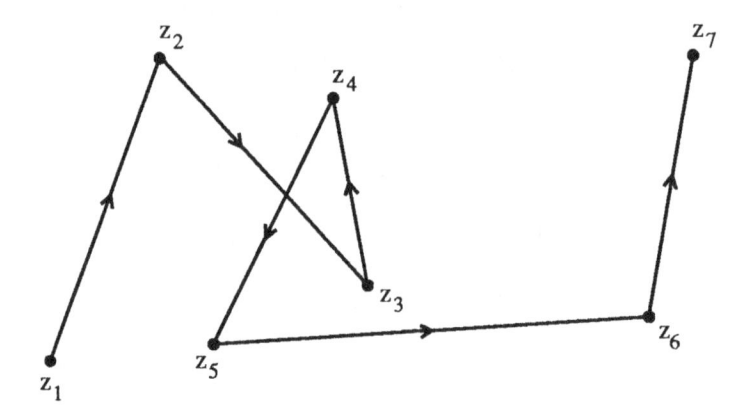

Figure 8.

of $[a_1, b_1]$. More generally, we define by induction the sum $\gamma_1 + \gamma_2 + \cdots + \gamma_n$ of paths $\gamma_k : [a_k, b_k] \to \mathbb{C}$, $1 \leq k \leq n$, provided $\gamma_k(b_k) = \gamma_{k+1}(a_{k+1})$ for $k = 1, 2, \ldots, n-1$: $\gamma_1 + \gamma_2 + \cdots + \gamma_n = (\gamma_1 + \cdots + \gamma_{n-1}) + \gamma_n$. It is verified without difficulty that the sum of piecewise smooth paths is again piecewise smooth (Exercise 4.2). We remark that any piecewise smooth path γ can be expressed in the form $\gamma = \gamma_1 + \gamma_2 + \cdots + \gamma_n$, where each γ_k is a smooth path (Exercise 4.6). Finally, the notation $\gamma_1 - \gamma_2$ is used as an abbreviation for $\gamma_1 + (-\gamma_2)$ whenever the latter path sum is well-defined.

To point out one instance of a path sum we recall the path γ that we ran into in Example 1.4. We can now represent this path in the manner $\gamma = [0, 1] + [1, 1+i] + [1+i, i] + [i, 0]$. Any path γ of the same general structure, meaning of the form $\gamma = [z_1, z_2] + [z_2, z_3] + \cdots + [z_{n-1}, z_n]$, is called for obvious reasons a *polygonal path*. (Question: What is the parameter interval of such a polygonal path?) Figure 8 displays an example in which $n = 7$. The trajectory of a simple and non-closed polygonal path is a polygonal arc (Section II.3.2).

1.5 Change of Parameter

A highly desirable feature in any theory of line integrals is the invariance of the integral under reasonable changes of parameter in the path of integration. Anticipating this point, we explain briefly what we shall view in this book as a "reasonable change of parameter." Suppose that $\gamma : [a, b] \to \mathbb{C}$ is a path. We say that a path $\beta : [c, d] \to \mathbb{C}$ can be *obtained from γ by the change of parameter h* if h is a non-decreasing continuous function with domain-set $[c, d]$, with range $[a, b]$, and with the property that $\beta(s) = \gamma[h(s)]$ for every s belonging to $[c, d]$. Thus γ and β have a common trajectory, which these two paths parametrize in ways specifically related by h. Figure 9 illus-

1. Paths in the Complex Plane

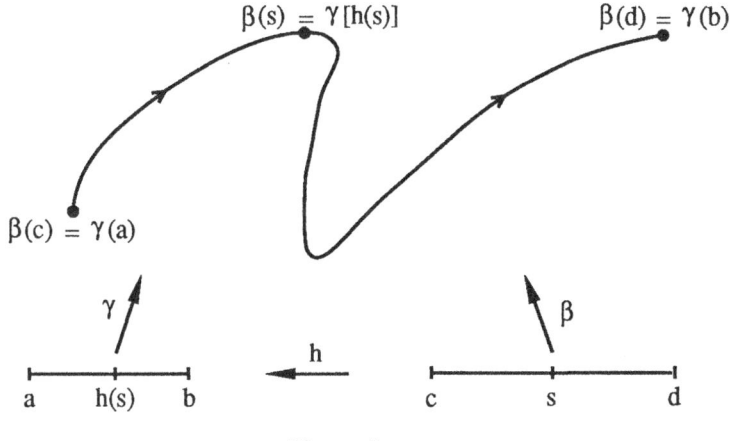

Figure 9.

trates this. If the function h is smooth (respectively, piecewise smooth), we speak of a *smooth* (respectively, *piecewise smooth*) *change of parameter*. It is piecewise smooth changes of parameter that turn out to be "reasonable" for our purposes.

We offer the following simple example of a smooth change of parameter. Let z_1 and z_2 be distinct points in the complex plane, and let $\gamma = [z_1, z_2]$. Recall what this means: $\gamma(t) = (1-t)z_1 + tz_2$ for $0 \leq t \leq 1$. An alternate parametrization of the directed line segment from z_1 to z_2 is furnished by β: $\beta(s) = z_1 + s[(z_2 - z_1)/|z_2 - z_1|]$ for $0 \leq s \leq |z_2 - z_1|$. The path β is obtained from γ by the smooth change of parameter $h(s) = s/|z_2 - z_1|$, where $0 \leq s \leq |z_2 - z_1|$.

Before moving on to the topic of integration we record for future reference the following observation concerning paths and plane domains.

Lemma 1.1. *Let D be a domain in the complex plane, and let z_0 and z_1 be points of D, not excluding the case $z_0 = z_1$. There exists a piecewise smooth path in D with initial point z_0 and terminal point z_1.*

Proof. If $z_0 = z_1$, then $\gamma = [z_0, z_0]$ is a path with the desired property. Assuming now that $z_0 \neq z_1$, we refer to Theorem II.3.6 and select a polygonal arc A in D having endpoints z_0 and z_1. Let $z_0 = w_1, w_2, \ldots, w_{n-1}, w_n = z_1$ list the endpoints of the line segments that make up A, arranged in the order that they are encountered when traversing A from z_0 to z_1. Then $\gamma = [w_1, w_2] + [w_2, w_3] + \cdots + [w_{n-1}, w_n]$ is a piecewise smooth path in D with initial point z_0 and terminal point z_1. ∎

2 Integrals Along Paths

2.1 Complex Line Integrals

As our point of departure in defining complex line integrals we presume that the reader has at least a rudimentary knowledge of Riemann integration theory, including an awareness of the following basic existence criterion: *if a real-valued function g is continuous on the interval $[a, b]$, then the Riemann integral $\int_a^b g(t)dt$ exists.* There is little difficulty in extending the definition of the Riemann integral to allow for the integration of a continuous function $g: [a, b] \to \mathbb{C}$. Writing $g = u + iv$ we make the obvious definition:

$$\int_a^b g(t)\, dt = \int_a^b u(t)\, dt + i \int_a^b v(t)\, dt \; .$$

As a matter of fact, the continuity of g at every point of $[a, b]$ is not a strict requirement here. Readers better versed in the ways of Riemann integration may know, for instance, that the above integral makes perfectly good sense if there exists a partition $P: a = t_0 < t_1 < \cdots < t_n = b$ of the interval $[a, b]$ such that the function g is continuous and bounded on each of the open intervals (t_{k-1}, t_k), $1 \leq k \leq n$. Moreover, in this situation

$$\int_a^b g(t)\, dt = \int_a^{t_1} g(t)\, dt + \int_{t_1}^{t_2} g(t)\, dt + \cdots + \int_{t_{n-1}}^b g(t)\, dt \; .$$

It is not even essential that g actually be defined at the points $t_0, \ldots, t_n$, because the values of g on this finite set of points can be assigned or changed arbitrarily with no effect on the integral.

The linearity of the Riemann integral carries over to the case of complex integrands. Thus, if $g, h: [a, b] \to \mathbb{C}$ are continuous functions and if c is a complex number, then it is true that

$$\int_a^b [g(t) + h(t)]\, dt = \int_a^b g(t)\, dt + \int_a^b h(t)\, dt$$

and

$$\int_a^b c g(t)\, dt = c \int_a^b g(t)\, dt \; .$$

The Second Fundamental Theorem of Calculus also remains valid in this setting: *if $g, G: [a, b] \to \mathbb{C}$ are continuous functions and if $\dot{G}(t) = g(t)$ for every t in (a, b), then*

$$\int_a^b g(t)\, dt = \Big[G(t) \Big]_a^b = G(b) - G(a) \; .$$

2. Integrals Along Paths

For example,

$$\int_0^2 2t + it^3 \, dt = \left[t^2 + i\left(\frac{t^4}{4}\right)\right]_0^2 = 4 + 4i$$

and

$$\int_0^\pi e^{it} \, dt = \left[\frac{e^{it}}{i}\right]_0^\pi = \frac{e^{\pi i} - e^0}{i} = 2i \; .$$

Suppose now that $\gamma: [a, b] \to \mathbb{C}$ is a smooth path and that f is a complex-valued function which is defined and continuous on the trajectory of γ. Under these conditions we define the *complex line integral* (or *contour integral*) *of f along γ*, denoted $\int_\gamma f(z)dz$, as follows:

(4.1) $$\int_\gamma f(z) \, dz = \int_a^b f[\gamma(t)] \dot\gamma(t) \, dt \; .$$

Since $g(t) = f[\gamma(t)]\dot\gamma(t)$ describes a function that is continuous on the interval $[a, b]$, there are no hidden pitfalls in this definition — the integral on the right is well-defined. A second type of integral to play a role in future proceedings is the *integral of f along γ with respect to arclength*, symbolized by $\int_\gamma f(z)|dz|$ and defined by

(4.2) $$\int_\gamma f(z) \, |dz| = \int_a^b f[\gamma(t)]|\dot\gamma(t)| \, dt \; .$$

(In this book the principal assignment of arclength integrals will be to serve as a means for estimating the magnitudes of complex line integrals, the latter being the integrals of primary concern to us. There are, however, areas of mathematics — and physics — where arclength integrals have great significance in their own right.) The preceding definitions extend with little fuss to the case in which the path γ is merely piecewise smooth. The integrands on the right-hand sides of (4.1) and (4.2) are then defined and continuous at all points of $[a, b]$ with the possible exception of a finite number of points in (a, b). They are also bounded on the open intervals into which any such exceptional points subdivide (a, b). We have remarked earlier that functions matching this description are Riemann integrable over $[a, b]$. The values, if any, assigned to the integrands at the exceptional points are immaterial. When it finally comes to evaluating these two integrals for a piecewise smooth path γ, often the most sensible approach is to choose a partition $P: a = t_0 < t_1 < \cdots < t_n = b$ of the interval $[a, b]$ with the property that γ_k, the restriction of γ to the interval $[t_{k-1}, t_k]$, is a smooth path for $1 \le k \le n$ and to make the observation that

(4.3) $$\int_\gamma f(z) \, dz = \int_{\gamma_1} f(z) \, dz + \int_{\gamma_2} f(z) \, dz + \cdots + \int_{\gamma_n} f(z) \, dz$$

and

(4.4) $\quad \int_\gamma f(z)\,|dz| = \int_{\gamma_1} f(z)\,|dz| + \int_{\gamma_2} f(z)\,|dz| \cdots + \int_{\gamma_n} f(z)\,|dz|$.

Readers uncomfortable about integrating discontinuous functions are free to regard formulas (4.3) and (4.4) as acceptable alternative definitions of $\int_\gamma f(z)\,dz$ and $\int_\gamma f(z)|dz|$, respectively, in the piecewise smooth situation.

If $\gamma(t) = x(t) + iy(t)$, $a \le t \le b$, is a smooth path, then

$$\int_\gamma |dz| = \int_a^b \sqrt{\dot{x}(t)^2 + \dot{y}(t)^2}\, dt \ .$$

The integral on the right ought to be familiar from calculus as the integral that gives the length of a plane parametric curve. More generally, the length $\ell(\gamma)$ of even a piecewise smooth path γ in $\mathbb{C}$ can be expressed by $\ell(\gamma) = \int_\gamma |dz|$. (This partly explains the description of $\int_\gamma f(z)|dz|$ as an "integral with respect to arclength.") We speak deliberately of the length of γ, as opposed to the length of $|\gamma|$. For instance, the path γ defined on $[0, 4\pi]$ by $\gamma(t) = e^{it}$ has for its trajectory a circle of unit radius. However, $\ell(\gamma) = 4\pi$ — not 2π — because $\gamma(t)$ traverses this trajectory twice as t increases from 0 to 4π. For further discussion of path length, the reader is referred to Section 3 of this chapter.

We compute some examples of the integrals we have just defined.

EXAMPLE 2.1. Evaluate the integrals $\int_\gamma z^{-1}\,dz$ and $\int_\gamma z^{-2}|dz|$, where γ is the parametrization of the circle $K(0,2)$ given by $\gamma(t) = 2e^{it}$ for $0 \le t \le 2\pi$.

Here γ is smooth, $\dot{\gamma}(t) = 2ie^{it}$, and $|\dot{\gamma}(t)| = 2$. In the first integral the integrand is $f(z) = 1/z$, so setting $z = \gamma(t) = 2e^{it}$ gives $f[\gamma(t)] = 1/(2e^{it})$. Consequently, by (4.1)

$$\int_\gamma \frac{dz}{z} = \int_0^{2\pi} \frac{2ie^{it}\,dt}{2e^{it}} = i\int_0^{2\pi} dt = 2\pi i \ .$$

For the second integral we obtain from (4.2) and the Second Fundamental Theorem of Calculus

$$\int_\gamma \frac{|dz|}{z^2} = \int_0^{2\pi} \frac{2\,dt}{4e^{2it}} = \frac{1}{2}\int_0^{2\pi} e^{-2it}\,dt = \left[-\frac{e^{-2it}}{4i}\right]_0^{2\pi} = 0 \ .$$

EXAMPLE 2.2. Evaluate $\int_\gamma |z|^2\,dz$ and $\int_\gamma x|dz|$, where $\gamma(t) = t + i(t^2/2)$ for $0 \le t \le 1$ (Figure 10).

In this example γ is smooth, with $\dot{\gamma}(t) = 1 + it$ and $|\dot{\gamma}(t)| = \sqrt{1+t^2}$. Therefore,

$$\int_\gamma |z|^2\,dz = \int_0^1 \left(t^2 + \frac{t^4}{4}\right)(1+it)\,dt$$

2. Integrals Along Paths

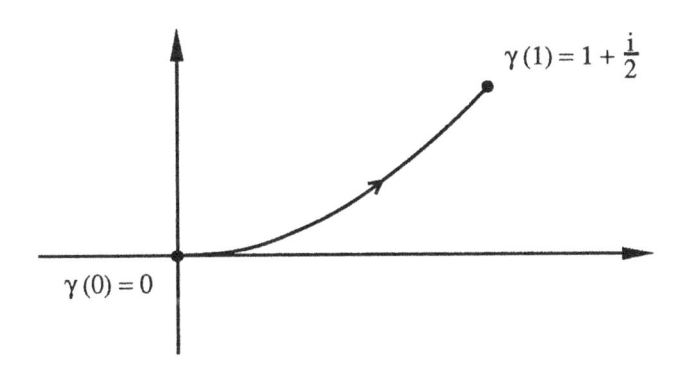

Figure 10.

$$= \int_0^1 t^2 + \frac{t^4}{4} \, dt + i \int_0^1 t^3 + \frac{t^5}{4} \, dt$$

$$= \left[\frac{t^3}{3} + \frac{t^5}{20} \right]_0^1 + i \left[\frac{t^4}{4} + \frac{t^6}{24} \right]_0^1 = \frac{23}{60} + \frac{7i}{24}$$

and

$$\int_\gamma x \, |dz| = \int_0^1 t\sqrt{1+t^2} \, dt = \left[\frac{(1+t^2)^{3/2}}{3} \right]_0^1 = \frac{2\sqrt{2}-1}{3}.$$

EXAMPLE 2.3. Evaluate $\int_\gamma (x+y) dz$ for $\gamma = [0, 1+i] + [1+i, i]$.

The path γ is piecewise smooth. Recalling the definitions of a path sum and of the path $[z_1, z_2]$, we observe that

$$\gamma(t) = \begin{cases} t(1+i) & \text{if } 0 \le t \le 1, \\ (2-t) + i & \text{if } 1 \le t \le 2. \end{cases}$$

The restrictions γ_1 and γ_2 of γ to the intervals [0,1] and [1,2], respectively, are smooth paths. We employ (4.3) and compute

$$\int_\gamma (x+y) \, dz = \int_{\gamma_1} (x+y) \, dz + \int_{\gamma_2} (x+y) \, dz$$

$$= \int_0^1 2t(1+i) \, dt + \int_1^2 (3-t)(-1) \, dt$$

$$= \left[t^2(1+i) \right]_0^1 - \left[3t - \frac{t^2}{2} \right]_1^2 = -\frac{1}{2} + i.$$

(This problem can also be done — and done more simply — by invoking property (iv) in Lemma 2.1.)

2.2 Properties of Contour Integrals

The following lemma summarizes the most important elementary properties of the complex line integral. The proofs presented are actually valid in the piecewise smooth case, although those squeamish about even mildly discontinuous integrands may prefer to regard them as legitimate in the smooth category only and to reduce the piecewise smooth case to the smooth one using (4.3). No proofs are given for assertions (i) and (ii), which follow easily from the definition of the contour integral and the linearity of Riemann integration.

Lemma 2.1. *Suppose that $f: A \to \mathbb{C}$ and $g: A \to \mathbb{C}$ are continuous functions and that γ and β are piecewise smooth paths in A.*

(i) $\displaystyle\int_\gamma [f(z) + g(z)]\, dz = \int_\gamma f(z)\, dz + \int_\gamma g(z)\, dz$;

(ii) $\displaystyle\int_\gamma cf(z)\, dz = c\int_\gamma f(z)\, dz$ *for any complex constant c* ;

(iii) $\displaystyle\int_{-\gamma} f(z)\, dz = -\int_\gamma f(z)\, dz$;

(iv) *if $\gamma + \beta$ is defined, then* $\displaystyle\int_{\gamma+\beta} f(z)\, dz = \int_\gamma f(z)\, dz + \int_\beta f(z)\, dz$;

(v) *if β is obtainable from γ by a piecewise smooth change of parameter, then* $\displaystyle\int_\gamma f(z)\, dz = \int_\beta f(z)\, dz$;

(vi) $\displaystyle\left|\int_\gamma f(z)\, dz\right| \leq \int_\gamma |f(z)|\, |dz|$.

Proof of (iii). Assume that $\gamma: [a, b] \to \mathbb{C}$. Since $[-\gamma](t) = \gamma(b + a - t)$, we see that the derivative of $-\gamma$ at t is $-\dot{\gamma}(b + a - t)$. We use the change of variable $u = b + a - t$, $du = -dt$ to calculate

$$\int_{-\gamma} f(z)\, dz = -\int_a^b f[\gamma(b+a-t)][\dot{\gamma}(b+a-t)]\, dt = \int_b^a f[\gamma(u)]\dot{\gamma}(u)\, du$$

$$= -\int_a^b f[\gamma(u)]\dot{\gamma}(u)\, du = -\int_\gamma f(z)\, dz .$$

Proof of (iv). Suppose that $\gamma:[a,b] \to \mathbb{C}$, $\beta:[c,d] \to \mathbb{C}$, and $\gamma(b) = \beta(c)$. Referring to the definition of a path sum, we note that $\gamma + \beta$ has derivative $\dot{\gamma}(t)$ in $[a, b]$ and $\dot{\beta}(t + c - b)$ in $[b, b + d - c]$. We reduce the integral along $\gamma + \beta$ into a sum of integrals over these two intervals and make the change of

2. Integrals Along Paths

variable $u = t+c-b$, $du = dt$ to expedite the computation over $[b, b+d-c]$:

$$\int_{\gamma+\beta} f(z)\,dz = \int_a^b f[\gamma(t)]\dot\gamma(t)\,dt + \int_b^{b+d-c} f[\beta(t+c-b)]\dot\beta(t+c-b)\,dt$$

$$= \int_a^b f[\gamma(t)]\dot\gamma(t)\,dt + \int_c^d f[\beta(u)]\dot\beta(u)\,du = \int_\gamma f(z)\,dz + \int_\beta f(z)\,dz \ .$$

Proof of (v). Assume, as above, that $\gamma\colon [a,b] \to \mathbb{C}$ and $\beta\colon [c,d] \to \mathbb{C}$. By hypothesis, there is a piecewise smooth change of parameter $h\colon [c,d] \to [a,b]$ such that $\beta(s) = \gamma[h(s)]$. Then $\dot\beta(s) = \dot\gamma[h(s)]\dot h(s)$. Performing the change of variable $t = h(s)$, $dt = \dot h(s)ds$ leads to

$$\int_\beta f(z)\,dz = \int_c^d f[\beta(s)]\dot\beta(s)\,ds = \int_c^d f\{\gamma[h(s)]\}\dot\gamma[h(s)]\dot h(s)\,ds$$

$$= \int_a^b f[\gamma(t)]\dot\gamma(t)\,dt = \int_\gamma f(z)\,dz \ .$$

Proof of (vi). We suppose that $\int_\gamma f(z)dz \neq 0$ — (vi) holds trivially otherwise — and set $u = e^{-i\theta}$, where θ is any argument of $\int_\gamma f(z)dz$. Thus $|u| = 1$ and

$$\left|\int_\gamma f(z)\,dz\right| = u\int_\gamma f(z)\,dz \ .$$

Then, assuming that $\gamma\colon [a,b] \to \mathbb{C}$,

$$\left|\int_\gamma f(z)\,dz\right| = \mathrm{Re}\left(\left|\int_\gamma f(z)\,dz\right|\right) = \mathrm{Re}\left(u\int_\gamma f(z)\,dz\right) = \mathrm{Re}\left(\int_\gamma uf(z)\,dz\right)$$

$$= \mathrm{Re}\left(\int_a^b uf[\gamma(t)]\dot\gamma(t)\,dt\right) = \int_a^b \mathrm{Re}\{uf[\gamma(t)]\dot\gamma(t)\}\,dt$$

$$\leq \int_a^b |uf[\gamma(t)]\dot\gamma(t)|\,dt = \int_a^b |f[\gamma(t)]||\dot\gamma(t)|\,dt = \int_\gamma |f(z)|\,|dz| \ ,$$

as desired. ∎

There is an obvious analogue of Lemma 2.1 in which the complex line integrals are replaced by integrals with respect to arclength. Only assertion (iii) requires modification. It becomes

$$\int_{-\gamma} f(z)\,|dz| = \int_\gamma f(z)\,|dz| \ .$$

(See Exercise 4.15.)

Statement (vi) of Lemma 2.1 is extremely important, for it provides us with a means of estimating the size of a contour integral. In a simple application, suppose it is known that $|f(z)| \leq m$, a constant, at every point z of $|\gamma|$. Then (vi) immediately yields the upper bound

(4.5) $$\left| \int_\gamma f(z)\, dz \right| \leq m\ell(\gamma) .$$

Indeed,

$$\int_\gamma |f(z)|\, |dz| = \int_a^b |f[\gamma(t)]||\dot\gamma(t)|\, dt \leq \int_a^b m|\dot\gamma(t)|\, dt = m \int_\gamma |dz| = m\ell(\gamma) .$$

We demonstrate the use of Lemma 2.1(vi) in two concrete problems.

EXAMPLE 2.4. Let $\gamma(t) = 2e^{it}$ for $-\pi/6 \leq t \leq \pi/6$ (Figure 11). Estimate $|\int_\gamma (z^3 + 1)^{-1} dz|$ from above.

If $z = \gamma(t)$, then $|z| = 2$. It follows that

$$|z^3 + 1| \geq |z|^3 - 1 = 8 - 1 = 7 .$$

As a result, we obtain for z on $|\gamma|$ the bound $|z^3 + 1|^{-1} \leq 1/7$. Since $\ell(\gamma) = 2\pi/3$, (4.5) produces the estimate

$$\left| \int_\gamma \frac{dz}{z^3 + 1} \right| \leq \frac{2\pi}{21} .$$

In fact, if we pay a little closer attention to basic geometry, we can achieve an estimate somewhat better than $2\pi/21$. Specifically, if $z = \gamma(t)$, then $w = z^3$ lies on the semi-circle $S = \{w = u + iv : |w| = 8 \text{ and } u \geq 0\}$. The

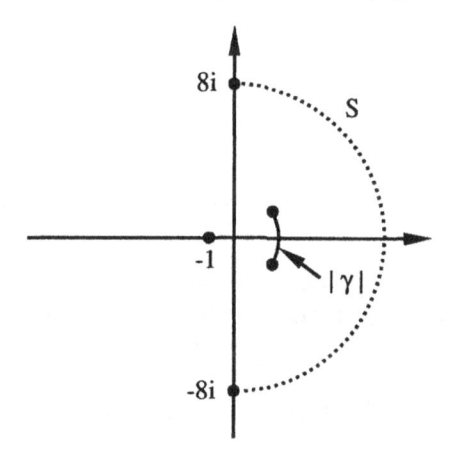

Figure 11.

2. Integrals Along Paths

closest points to -1 on S are the points $\pm 8i$, each at a distance $\sqrt{65}$ from -1. Accordingly, we conclude that $|z^3+1| \geq \sqrt{65}$ for z on $|\gamma|$, which allows us to sharpen our earlier bound to

$$\left| \int_\gamma \frac{dz}{z^3+1} \right| \leq \frac{2\pi}{3\sqrt{65}} .$$

EXAMPLE 2.5. Let $r > 0$. Verify that

$$\left| \int_\gamma e^{iz^2} dz \right| \leq \frac{\pi(1-e^{-r^2})}{4r} ,$$

where $\gamma(t) = re^{it}$ for $0 \leq t \leq \pi/4$.

We use the estimate provided by Lemma 2.1: $|\int_\gamma e^{iz^2} dz| \leq \int_\gamma |e^{iz^2}||dz|$. Now $|e^{iz^2}| = e^{\operatorname{Re}(iz^2)} = e^{-2xy}$. In particular, for $z = re^{it} = r\cos t + ir\sin t$ we have

$$|e^{iz^2}| = e^{-2r^2 \cos t \sin t} = e^{-r^2 \sin(2t)} .$$

Clearly $|\dot\gamma(t)| = r$. We set $u = 2t$, $du = 2dt$ in computing

$$\int_\gamma |e^{iz^2}||dz| = \int_0^{\pi/4} e^{-r^2 \sin(2t)} r\, dt = (r/2) \int_0^{\pi/2} e^{-r^2 \sin u}\, du .$$

But $\sin u \geq 2u/\pi$ whenever $0 \leq u \leq \pi/2$, with the consequence that

$$\left| \int_\gamma e^{iz^2} dz \right| \leq \int_\gamma |e^{iz^2}||dz| \leq (r/2) \int_0^{\pi/2} e^{-2ur^2/\pi}\, du$$

$$= \left[-\frac{\pi e^{-2ur^2/\pi}}{4r} \right]_0^{\pi/2} = \frac{\pi(1-e^{-r^2})}{4r} .$$

2.3 Primitives

Suppose that U is an open set in the complex plane and that f is a function whose domain-set includes U. A function $F: U \to \mathbb{C}$ is a *primitive* for f in U if F is analytic in U and has $F'(z) = f(z)$ for every z in that set. (The word "antiderivative" is synonymous with "primitive.") When F is a primitive for f in U, so also is the function $F+c$ for any complex constant c; if U is a domain, then every primitive of f in U can be obtained from F in this way. Some illustrations: $F(z) = z^2/2$ is a primitive for $f(z) = z$ in $\mathbb{C}$; the function $f(z) = e^z$ is its own primitive in the complex plane; the function $f(z) = z^{-1}$ has $F(z) = \operatorname{Log} z$ as a primitive in the domain $D = \mathbb{C} \sim (-\infty, 0]$. The reason for introducing primitives is supplied by the next theorem, which is the counterpart of the Second Fundamental Theorem of Calculus in the setting of contour integration.

Theorem 2.2. *Suppose that a function f is continuous in an open set U and that F is a primitive for f in U. If $\gamma:[a,b] \to U$ is a piecewise smooth path, then*

$$\int_\gamma f(z)\,dz = \Big[F(z)\Big]_{\gamma(a)}^{\gamma(b)}.$$

In particular, under the above hypotheses it is true that

$$\int_\gamma f(z)\,dz = 0$$

for every closed, piecewise smooth path γ in U.

Proof. Assume, initially, that γ is smooth. Write $\gamma(t) = x(t) + iy(t)$. Define a function G on $[a,b]$ by $G(t) = F[\gamma(t)] = F[x(t), y(t)]$. Then G is a continuous function. Moreover, the chain rule from vector calculus, in concert with the Cauchy-Riemann equations, permits us to compute

$$\dot{G}(t) = F_x[\gamma(t)]\dot{x}(t) + F_y[\gamma(t)]\dot{y}(t) = F_x[\gamma(t)]\dot{x}(t) + iF_x[\gamma(t)]\dot{y}(t)$$

$$= F_x[\gamma(t)][\dot{x}(t) + i\dot{y}(t)] = F'[\gamma(t)]\dot{\gamma}(t) = f[\gamma(t)]\dot{\gamma}(t)$$

for every t in (a, b). The use of the chain rule is justified here by the fact that F, being an analytic function in U, is certainly differentiable in the real sense at every point of U. By the Second Fundamental Theorem of Calculus

$$\int_\gamma f(z)\,dz = \int_a^b f[\gamma(t)]\dot{\gamma}(t)\,dt = G(b) - G(a)$$

$$= F[\gamma(b)] - F[\gamma(a)] = \Big[F(z)\Big]_{\gamma(a)}^{\gamma(b)}.$$

In case γ is only piecewise smooth, we can by definition choose a partition $P: a = t_0 < t_1 < \cdots < t_n = b$ of $[a, b]$ with the property that γ_k, the restriction of γ to $[t_{k-1}, t_k]$, is smooth for $1 \leq k \leq n$. In view of (4.3) and what has just been proved,

$$\int_\gamma f(z)\,dz = \int_{\gamma_1} f(z)\,dz + \int_{\gamma_2} f(z)\,dz + \cdots + \int_{\gamma_n} f(z)\,dz$$

$$= F[\gamma(t_1)] - F[\gamma(a)] + F[\gamma(t_2)] - F[\gamma(t_1)] + \cdots + F[\gamma(b)] - F[\gamma(t_{n-1})]$$

$$= F[\gamma(b)] - F[\gamma(a)] = \Big[F(z)\Big]_{\gamma(a)}^{\gamma(b)},$$

again as asserted. ∎

(N.B. The proof of Theorem 2.2 shows that its conclusion, $\int_\gamma f(z)dz = [F(z)]_{\gamma(a)}^{\gamma(b)}$, remains intact under slightly weaker assumptions than stated:

2. Integrals Along Paths

this formula is valid as long as f and F are continuous on $|\gamma|$, and F is differentiable with $F'(z) = f(z)$ for every z in $|\gamma| \sim \gamma(S)$, where S is a finite subset of $[a, b]$.) Theorem 2.2 insures that, if a continuous function f possesses a primitive in an open set U, then the value of the complex line integral of f along a piecewise smooth path in U depends entirely on the initial and terminal points of that path. Incidentally, the hypothesis in Theorem 2.2 that f be continuous in U turns out to be superfluous. Since $F' = f$ in U, it will later be seen — this fact has received mention earlier — that f is itself analytic in U. We do not yet, however, have the tools to prove such an assertion. Let it be stated emphatically that Theorem 2.2 applies *only* to contour integrals, *not* to arclength integrals. This theorem has no facsimile for integrals of the latter variety, which in practice are often quite amenable to estimation, but usually more resistant to exact evaluation than are contour integrals.

EXAMPLE 2.6. Evaluate $\int_\gamma z^2 dz$ and $\int_\gamma e^z dz$, where $\gamma(t) = t + i(t^2/\pi)$ for $0 \le t \le \pi$.

Since $F(z) = z^3/3$ is a primitive for $f(z) = z^2$, Theorem 2.2 gives:

$$\int_\gamma z^2\, dz = \left[\frac{z^3}{3}\right]_0^{\pi+\pi i} = \frac{(\pi + \pi i)^3}{3} = -\frac{2\pi^3}{3} + \left(\frac{2\pi^3}{3}\right)i \ .$$

The function $f(z) = e^z$ is its own primitive and thus

$$\int_\gamma e^z\, dz = \left[e^z\right]_0^{\pi+\pi i} = e^{\pi + \pi i} - e^0 = -e^\pi - 1 \ .$$

EXAMPLE 2.7. Evaluate $\int_\gamma z^{-1} dz$, if $\gamma(t) = e\cos t + i\sin t$ for $0 \le t \le \pi/2$.

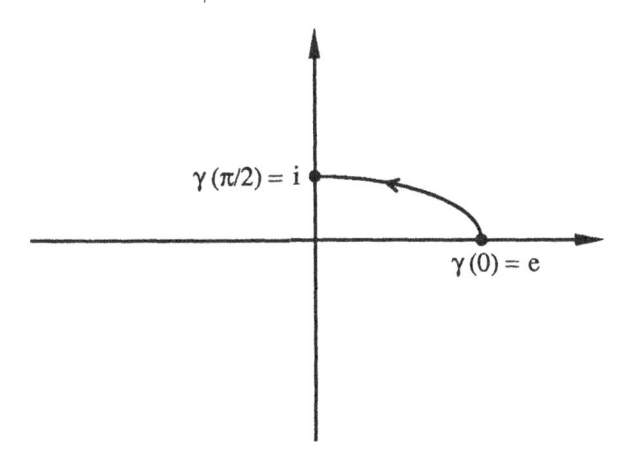

Figure 12.

As indicated in Figure 12, the trajectory of γ is an arc of the ellipse $(x^2/e^2)+y^2 = 1$ joining $\gamma(0) = e$ to $\gamma(\pi/2) = i$. The function $F(z) = \text{Log } z$ is a primitive for $f(z) = z^{-1}$ in an open set that contains this trajectory. It follows that

$$\int_\gamma \frac{dz}{z} = \Big[\text{Log } z\Big]_e^i = \text{Log } i - \text{Log } e = -1 + (\pi/2)i \ .$$

EXAMPLE 2.8. Show that $f(z) = z^{-1}$ has no primitive in the set $U = \mathbb{C} \sim \{0\}$.

It was observed in Example 2.1 that $\int_\gamma z^{-1} dz = 2\pi i$, where $\gamma(t) = 2e^{it}$ for $0 \leq t \leq 2\pi$. Were f to possess a primitive in U, the value of this integral would have to be 0. Consequently, no such primitive exists.

EXAMPLE 2.9. Evaluate $\int_\gamma z \sin z \, dz$, where $\gamma: [a, b] \to \mathbb{C}$ is the piecewise smooth path pictured in Figure 13.

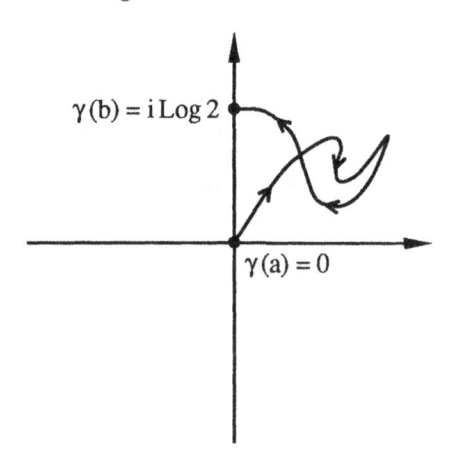

Figure 13.

We use the integration by parts formula, which for integration along a piecewise smooth path $\gamma: [a, b] \to \mathbb{C}$ takes the form

(4.6) $$\int_\gamma f(z)g'(z) \, dz = \Big[f(z)g(z)\Big]_{\gamma(a)}^{\gamma(b)} - \int_\gamma g(z)f'(z) \, dz$$

(Exercise 4.24). Applying this formula with $f(z) = z$, $g(z) = -\cos z$, and γ as depicted, we obtain

$$\int_\gamma z \sin z \, dz = \Big[-z \cos z\Big]_0^{i \text{Log } 2} + \int_\gamma \cos z \, dz$$

2. Integrals Along Paths

$$= \left[-z\cos z + \sin z \right]_0^{i\,\text{Log}\,2} = \left(\frac{3 - 5\,\text{Log}\,2}{4} \right) i \, .$$

2.4 Some Notation

We take the opportunity here to establish special notation for certain types of integrals that turn up frequently in later discussions. First, suppose that a function f is continuous on the line segment joining points z_1 and z_2 in the complex plane. (We do not rule out the degenerate case $z_1 = z_2$.) We write $\int_{z_1}^{z_2} f(z)\,dz$ and $\int_{z_1}^{z_2} f(z)|dz|$ in preference to the more cumbersome notations $\int_{[z_1,z_2]} f(z)\,dz$ and $\int_{[z_1,z_2]} f(z)|dz|$. Clearly

(4.7) $$\int_{z_1}^{z_1} f(z)\,dz = 0 \, ,$$

since $[z_1, z_1]$ is a constant path. Because $[z_2, z_1] = -[z_1, z_2]$, it follows from Lemma 2.1(iii) that

(4.8) $$\int_{z_2}^{z_1} f(z)\,dz = -\int_{z_1}^{z_2} f(z)\,dz \, .$$

Furthermore, if z_3 is a third point on the segment under consideration, it is the case that

(4.9) $$\int_{z_1}^{z_2} f(z)\,dz = \int_{z_1}^{z_3} f(z)\,dz + \int_{z_3}^{z_2} f(z)\,dz \, .$$

This assertion is not a direct consequence of Lemma 2.1(iv), for it is not true by our definition of a path sum that $[z_1, z_2] = [z_1, z_3] + [z_3, z_2]$. Assuming that z_3 is not equal to z_1 or z_2 — in view of (4.7), (4.9) holds trivially if $z_3 = z_1$ or $z_3 = z_2$ — we remark that $[z_1, z_2]$ can be obtained from $[z_1, z_3] + [z_3, z_2]$ by the change of parameter $h\colon [0,1] \to [0,2]$,

$$h(s) = \begin{cases} s/\lambda & \text{for } 0 \leq s \leq \lambda \, , \\ (s + 1 - 2\lambda)/(1 - \lambda) & \text{for } \lambda \leq s \leq 1 \, , \end{cases}$$

where $\lambda = |z_3 - z_1|/|z_2 - z_1|$ (Exercise 4.8). The combination of properties (iv) and (v) in Lemma 2.1 leads to (4.9).

Let it be noted that if $z_1 = x_1 + iy$ and $z_2 = x_2 + iy$ (in other words, if z_1 and z_2 lie on a horizontal line) then

(4.10) $$\int_{z_1}^{z_2} f(z)\,dz = \int_{x_1}^{x_2} f(x + iy)\,dx \, .$$

One obtains (4.10) from the definition of $[z_1, z_2]$ and (4.1) by writing

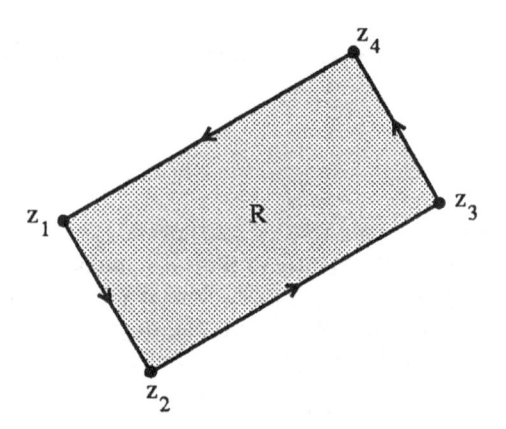

$$\int_{z_1}^{z_2} f(z)\,dz = \int_0^1 f[(1-t)x_1 + tx_2 + iy](x_2 - x_1)\,dt$$

and then making the change of variable $x = (1-t)x_1 + tx_2$, $dx = (x_2 - x_1)dt$. Similarly, for points $z_1 = x + iy_1$ and $z_2 = x + iy_2$ lying on a vertical line,

(4.11) $$\int_{z_1}^{z_2} f(z)\,dz = i\int_{y_1}^{y_2} f(x+iy)\,dy\ .$$

Suppose next that R is a closed rectangle in the complex plane. (The word "closed" is used here in the topological sense of "closed set," while "rectangle" is taken to mean the full two-dimensional figure, not just its boundary curve.) Let the vertices z_1, z_2, z_3, and z_4 of R be labeled in such a way that the polygonal path $\gamma = [z_1, z_2] + [z_2, z_3] + [z_3, z_4] + [z_4, z_1]$ parametrizes ∂R with "positive orientation" relative to R. This just requires that the point $\gamma(t)$ move along ∂R in the counterclockwise direction as t grows from 0 to 4, keeping R to its left (Figure 14). We employ the notation $\int_{\partial R} f(z)dz$ to signify the complex line integral of a function f along γ:

$$\int_{\partial R} f(z)\,dz = \int_{z_1}^{z_2} f(z)\,dz + \int_{z_2}^{z_3} f(z)\,dz + \int_{z_3}^{z_4} f(z)\,dz + \int_{z_4}^{z_1} f(z)\,dz\ .$$

The notation $\int_{\partial R} f(z)|dz|$ is interpreted in an analogous way. These definitions obviously do not depend on the particular choice of labeling for the vertices of R — provided the orientation convention is respected.

Finally, assume that a function f is continuous on the circle $K = K(z_0, r)$. We can realize K as the trajectory of a specific simple and closed path γ — namely, $\gamma(t) = z_0 + re^{it}$ for $t_0 \leq t \leq t_0 + 2\pi$, where t_0 is an arbitrarily chosen real number. Direct computation shows that $\int_\gamma f(z)dz$ and $\int_\gamma f(z)|dz|$ are independent of the value of t_0. The notations $\int_{|z-z_0|=r} f(z)dz$

3. Rectifiable Paths

and $\int_{|z-z_0|=r} f(z)|dz|$ are commonly utilized to represent these integrals. Thus, in Example 2.1 we could have written $\int_{|z|=2} z^{-1} dz = 2\pi i$ and, likewise, $\int_{|z|=2} z^{-2}|dz| = 0$.

3 Rectifiable Paths

3.1 Rectifiable Paths

As proclaimed in the introduction to this chapter, it is not our intention to get bogged down in a detailed technical discussion of contour integration along paths that fail to be piecewise smooth. Nevertheless, we feel it is quite pertinent to sketch at least one approach to the more general theory of complex line integrals that a reader investigating advanced topics in complex analysis will eventually run into. The emphasis is on "sketch," for no proofs are included in the discussion. None of the material presented in this section will be needed later in the text.

Crucial for what now follows is the notion of a "rectifiable" path. A path $\gamma\colon [a,b] \to \mathbb{C}$ is called *rectifiable* if the set of sums

$$\ell(\gamma, P) = \sum_{k=1}^{n} |\gamma(t_k) - \gamma(t_{k-1})|$$

obtained as $P\colon a = t_0 < t_1 < \cdots < t_n = b$ ranges over all possible partitions of the interval $[a,b]$ is a bounded set of numbers. (Geometrically, $\ell(\gamma, P)$ is nothing but the length of a certain polygonal path "inscribed" in $|\gamma|$. The case of a partition with $n = 4$ is illustrated in Figure 15.) When this is so,

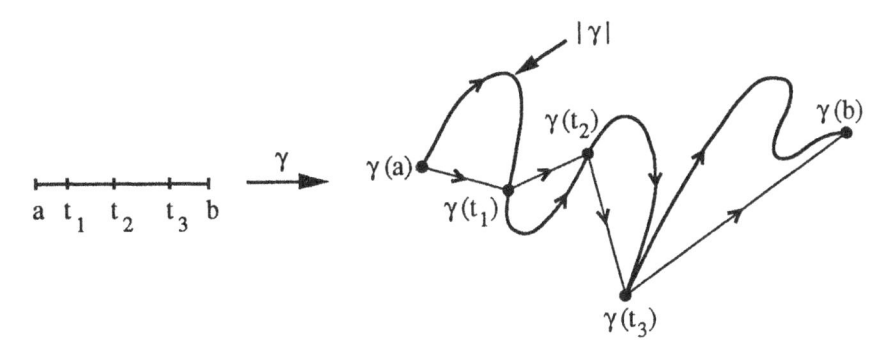

Figure 15.

the *length* $\ell(\gamma)$ *of* γ is defined to be the supremum of $\{\ell(\gamma, P) \colon P$ a partition of $[a, b]\}$. We have previously stated that the length of a piecewise smooth path γ is $\int_\gamma |dz|$. Lemma 3.1 will demonstrate that a piecewise smooth path is, in fact, rectifiable and that the definition of length just given does assign to such a path the length announced earlier. It should be stressed that a general rectifiable path is a far cry from being piecewise smooth.

We collect some relevant information about rectifiable paths. (The verifications of the statements that ensue involve nothing beyond the definitions of rectifiability and length. They are recommended to the ambitious reader as instructive exercises.) If a path γ is rectifiable, then its reverse path $-\gamma$ is also rectifiable and $\ell(-\gamma) = \ell(\gamma)$. Should a path γ have a decomposition $\gamma = \gamma_1 + \gamma_2 + \cdots + \gamma_n$, then γ is rectifiable if and only if each of the paths γ_k, $1 \leq k \leq n$, is rectifiable, in which event $\ell(\gamma) = \ell(\gamma_1) + \ell(\gamma_2) + \cdots + \ell(\gamma_n)$. If β is obtained from γ by a change of parameter, then one of these paths is rectifiable precisely when the other is; such being the case, $\ell(\gamma) = \ell(\beta)$. A path γ is constant if and only if $\ell(\gamma) = 0$.

With a rectifiable path $\gamma : [a, b] \to \mathbb{C}$ it is possible to associate its *length function*, the function $\sigma_\gamma : [a, b] \to \mathbb{R}$ defined as follows: $\sigma_\gamma(a) = 0$ and, for $a < t \leq b$, $\sigma_\gamma(t)$ is the length of the path obtained by restricting γ to the interval $[a, t]$. The function σ_γ is a non-decreasing, continuous function with range $[0, \ell(\gamma)]$. (Warning: the continuity of σ_γ is by no means obvious!) We now define a function $\gamma_0 : [0, \ell(\gamma)] \to \mathbb{C}$ by setting $\gamma_0(s) = \gamma(t)$ whenever $s = \sigma_\gamma(t)$. It is straightforward to check that γ_0 is well-defined and continuous. Assuming that $\ell(\gamma) > 0$, γ_0 is thus a path. Clearly γ is obtained from γ_0 by the change of parameter $h = \sigma_\gamma$. It follows from a comment in the preceding paragraph that γ_0 is rectifiable and has $\ell(\gamma_0) = \ell(\gamma)$. Furthermore, the length function σ_{γ_0} of γ_0 has the formula $\sigma_{\gamma_0}(s) = s$ for $0 \leq s \leq \ell(\gamma)$. Expressed in geometric terms, $\gamma_0(s)$ is the point reached by starting at $\gamma(a)$ and moving a length s along the trajectory of γ, always heeding the "directions" specified by γ for traversing this trajectory (Figure 16). This description of γ_0 accounts for the name given to it, *the reparametrization of γ by arclength*.

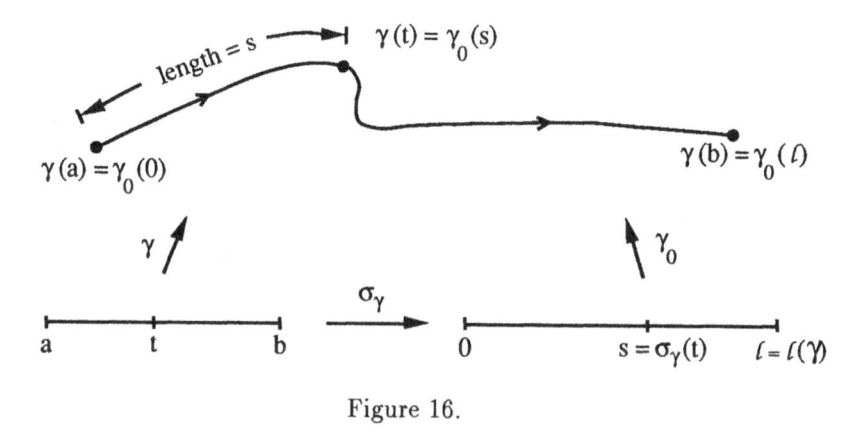

Figure 16.

If, for example, $\gamma = [z_1, z_2]$ with $z_1 \neq z_2$, we find that γ_0 is given by

3. Rectifiable Paths

$\gamma_0(s) = z_1 + s[(z_2 - z_1)/|z_2 - z_1|]$ for $0 \le s \le |z_2 - z_1|$. A second example: if $\gamma(t) = re^{it}$ for $0 \le t \le 2\pi$, where $r > 0$, then $\gamma_0(s) = re^{i(s/r)}$ for $0 \le s \le 2\pi r$.

Let γ and β be rectifiable paths in the complex plane. Elementary arguments confirm that $(-\gamma)_0 = -\gamma_0$, that $(\gamma + \beta)_0 = \gamma_0 + \beta_0$ when $\gamma + \beta$ is defined, and that $\beta_0 = \gamma_0$ if β is obtained from γ by a change of parameter. Finally, it can be demonstrated that at "almost every" point s of $[0, \ell(\gamma)]$ the derivative $\dot\gamma_0(s)$ exists and satisfies $|\dot\gamma_0(s)| = 1$. It is the last statement (in particular, the explanation of the expression "almost every") that transports the present discussion out of the territory of ordinary calculus and into the realm of its more sophisticated offspring, modern real analysis.

3.2 Integrals Along Rectifiable Paths

The machinery is now in place to give meaning to $\int_\gamma f(z)dz$ and $\int_\gamma f(z)|dz|$ in the situation where γ is a rectifiable path and f is a function continuous on $|\gamma|$. To do so, let γ_0 be the reparametrization of γ by arclength. (We shall assume that $\ell(\gamma) > 0$, since constant paths are already covered by the smooth case.) Because γ and γ_0 are related by a change of parameter, Lemma 2.1(v) makes it reasonable to insist that $\int_\gamma f(z)dz = \int_{\gamma_0} f(z)dz$ and that $\int_\gamma f(z)|dz| = \int_{\gamma_0} f(z)|dz|$. Prompted by this remark, by the fact that γ_0 is, all things considered, rather well-behaved, and by what happens in the piecewise smooth setting, we define

$$(4.12) \qquad \int_\gamma f(z)\, dz = \int_0^{\ell(\gamma)} f[\gamma_0(s)]\dot\gamma_0(s)\, ds\,,$$

and, remembering that $|\dot\gamma_0(s)| = 1$ for "most" values of s,

$$(4.13) \qquad \int_\gamma f(z)\, |dz| = \int_0^{\ell(\gamma)} f[\gamma_0(s)]\, ds\,.$$

The second definition causes no problems, for it involves a standard Riemann integral. The function $f[\gamma_0(s)]\dot\gamma_0(s)$, unfortunately, may have "too many discontinuities" to be Riemann integrable over $[0, \ell(\gamma)]$. For this reason the integral defining $\int_\gamma f(z)dz$ must be interpreted as a more general type of integral — namely, as a "Lebesgue integral." (It is the theory of integration invented by Henri Lebesgue (1875–1941) that forms the core of present-day real analysis.) Under definitions (4.12) and (4.13) the properties of path integrals catalogued in Lemma 2.1 and Theorem 2.2 are preserved. We emphasize that the formulas

$$(4.14) \qquad \int_\gamma f(z)\, dz = \int_a^b f[\gamma(t)]\dot\gamma(t)\, dt\,,\quad \int_\gamma f(z)\, |dz| = \int_a^b f[\gamma(t)]|\dot\gamma(t)|\, dt$$

need not hold for a general rectifiable path $\gamma\colon [a,b] \to \mathbb{C}$, even in the context of Lebesgue integration. In the presence of certain additional hypotheses on γ, they may become applicable. The formulas in (4.14) are valid, for instance, if γ satisfies a "Lipschitz condition": $|\gamma(t') - \gamma(t)| \le m|t' - t|$ for all t and t' in $[a,b]$, where m is a constant. Included among such Lipschitz paths are all piecewise smooth paths, for which the generalized definitions of path integrals are thus seen to reduce to the definitions given earlier.

There are other ways of treating integrals along rectifiable paths. One common approach, for example, defines these integrals as limits of appropriate Riemann sums. As we intend to make no use of integration in this generalized setting, however, we shall not pursue the matter further. Instead we conclude the chapter by showing, as promised, that a piecewise smooth path γ is rectifiable and has $\ell(\gamma) = \int_\gamma |dz|$.

Lemma 3.1. *Let γ be a piecewise smooth path in the complex plane. Then γ is rectifiable and $\ell(\gamma) = \int_\gamma |dz|$.*

Proof. Consider first the case when $\gamma\colon [a,b] \to \mathbb{C}$ is smooth. Take an arbitrary partition $P\colon a = t_0 < t_1 < \cdots < t_n = b$ of the interval $[a,b]$. Lemma 2.1(vi) applied to γ_k, the restriction of γ to $[t_{k-1}, t_k]$, gives

$$\left| \int_{t_{k-1}}^{t_k} \dot{\gamma}(t)\, dt \right| = \left| \int_{\gamma_k} dz \right| \le \int_{\gamma_k} |dz| = \int_{t_{k-1}}^{t_k} |\dot{\gamma}(t)|\, dt$$

for $1 \le k \le n$. We use this fact in the computation

$$\sum_{k=1}^n |\gamma(t_k) - \gamma(t_{k-1})| = \sum_{k=1}^n \left| \int_{t_{k-1}}^{t_k} \dot{\gamma}(t)\, dt \right| \le \sum_{k=1}^n \int_{t_{k-1}}^{t_k} |\dot{\gamma}(t)|\, dt$$

$$= \int_a^b |\dot{\gamma}(t)|\, dt = \int_\gamma |dz| \, ;$$

i.e., $\ell(\gamma, P) \le \int_\gamma |dz|$. From this it follows that γ is rectifiable and that

(4.15) $$\ell(\gamma) \le \int_\gamma |dz| \, .$$

Next, let $\epsilon > 0$ be given. Since $\dot{\gamma}$ is a continuous function on the interval $[a,b]$, it is uniformly continuous there (Theorem II.4.8). Moreover, by the elementary theory of the Riemann integral,

$$\int_a^b |\dot{\gamma}(t)|\, dt = \lim_{|P| \to 0} \sum_{k=1}^n |\dot{\gamma}(t_k)|(t_k - t_{k-1}) \, .$$

These facts permit us to choose a partition $P\colon a = t_0 < t_1 < \cdots < t_n = b$ of $[a,b]$ such that for $1 \le k \le n$ the inequality

$$|\dot{\gamma}(t) - \dot{\gamma}(t_k)| < \epsilon$$

3. Rectifiable Paths

is in force for every t in $[t_{k-1}, t_k]$ and, in addition, such that

$$\left| \int_a^b |\dot{\gamma}(t)| \, dt - \sum_{k=1}^n |\dot{\gamma}(t_k)|(t_k - t_{k-1}) \right| < \epsilon .$$

We infer that

$$\int_\gamma |dz| = \int_a^b |\dot{\gamma}(t)| \, dt < \epsilon + \sum_{k=1}^n |\dot{\gamma}(t_k)|(t_k - t_{k-1})$$

$$= \epsilon + \sum_{k=1}^n \left| \int_{t_{k-1}}^{t_k} \dot{\gamma}(t_k) \, dt \right|$$

$$= \epsilon + \sum_{k=1}^n \left| \gamma(t_k) - \gamma(t_{k-1}) + \int_{t_{k-1}}^{t_k} [\dot{\gamma}(t_k) - \dot{\gamma}(t)] \, dt \right|$$

$$\leq \epsilon + \sum_{k=1}^n |\gamma(t_k) - \gamma(t_{k-1})| + \sum_{k=1}^n \left| \int_{t_{k-1}}^{t_k} [\dot{\gamma}(t_k) - \dot{\gamma}(t)] \, dt \right|$$

$$\leq \epsilon + \ell(\gamma) + \sum_{k=1}^n \int_{t_{k-1}}^{t_k} |\dot{\gamma}(t_k) - \dot{\gamma}(t)| \, dt$$

$$\leq \epsilon + \ell(\gamma) + \sum_{k=1}^n \int_{t_{k-1}}^{t_k} \epsilon \, dt = \epsilon(1 + b - a) + \ell(\gamma) .$$

The implication is that

$$\int_\gamma |dz| \leq \epsilon(1 + b - a) + \ell(\gamma)$$

for every $\epsilon > 0$. Letting $\epsilon \to 0$ yields

$$\int_\gamma |dz| \leq \ell(\gamma) .$$

In conjuction with (4.15) this leads to

$$\ell(\gamma) = \int_\gamma |dz| ,$$

provided γ is smooth.

If $\gamma: [a, b] \to \mathbb{C}$ is merely piecewise smooth, we can choose a partition $P: a = t_0 < t_1 < \cdots < t_n = b$ of $[a, b]$ such that γ_k, once again the restriction of γ to $[t_{k-1}, t_k]$, is smooth for $1 \leq k \leq n$. Consequently, by what has just been proved and by comments in Section 3.1, $\gamma = \gamma_1 + \cdots + \gamma_n$ is rectifiable and

$$\int_\gamma |dz| = \int_{\gamma_1} |dz| + \cdots + \int_{\gamma_n} |dz| = \ell(\gamma_1) + \cdots + \ell(\gamma_n) = \ell(\gamma). \blacksquare$$

4 Exercises for Chapter IV

4.1 Exercises for Section IV.1

4.1. Mimicking the style of the discussions in Examples 1.1-1.5, describe the properties of the following paths; in each instance sketch a rough graph of the trajectory and indicate how the path in question traces it out: (i) $\gamma(t) = t^2 + it^4$ for $-1 \leq t \leq 1$; (ii) $\gamma(t) = e^{-it^2}$ for $0 \leq t \leq \sqrt{2\pi}$; (iii) $\gamma(t) = t^2 + it^3$ for $-1 \leq t \leq 1$; (iv) $\gamma(t) = 2\cos t + i\sin t$ for $0 \leq t \leq 2\pi$; (v) $\gamma(t) = t^3 + i|t^3|$ for $-1 \leq t \leq 1$; (vi) $\gamma(t) = e^t + ie^{-t}$ for $0 \leq t \leq 1$; (vii) $\gamma(t) = t^2 - t + i(t^3 - t^2)$ for $-1 \leq t \leq 2$; (viii) $\gamma(t) = \exp(\pi i \sqrt[3]{t})$ for $-1 \leq t \leq 1$.

4.2. If $\gamma:[a,b] \to \mathbb{C}$ and $\beta:[c,d] \to \mathbb{C}$ are piecewise smooth paths for which $\gamma(b) = \beta(c)$, confirm that $-\gamma$ and $\gamma + \beta$ are also piecewise smooth paths.

4.3. If $\alpha:[a_1,a_2] \to \mathbb{C}$, $\beta:[b_1,b_2] \to \mathbb{C}$, and $\gamma:[c_1,c_2] \to \mathbb{C}$ are paths for which $\alpha(a_2) = \beta(b_1)$ and $\beta(b_2) = \gamma(c_1)$, verify that $\alpha + (\beta + \gamma) = (\alpha + \beta) + \gamma$.

4.4. Define paths α, β, and γ on $[0,1]$ by $\alpha(t) = t + it$, $\beta(t) = t + it^2$, and $\gamma(t) = t^2 + it$. Write the rules of correspondence for the paths $-\alpha, -\beta, -\gamma$, $\alpha - \beta$, $\beta - \gamma$, $\alpha - \beta + \gamma$, and $-\alpha + \beta - \gamma$. Describe each of the preceding paths graphically.

4.5. If $\gamma:[a,b] \to \mathbb{C}$ is a path and if $a < c < b$, show that $\gamma = \alpha + \beta$, where α is the restriction of γ to $[a,c]$ and β is its restriction to $[c,b]$. How does this fact generalize for an arbitrary partition $P: a = t_0 < t_1 < \cdots < t_n = b$ of the interval $[a,b]$?

4.6. Certify that any piecewise smooth path γ can be expressed as a path sum of finitely many smooth paths. (*Hint.* Exercise 4.5.)

4.7. If $\gamma:[a,b] \to \mathbb{C}$ is a piecewise smooth path and if $\beta:[c,d] \to \mathbb{C}$ is a path that can be obtained from γ by making a piecewise smooth change of parameter, then β is also piecewise smooth. Prove this.

4.8. Let $\beta = [z_1, z_2]$ and $\gamma = [z_1, z_3] + [z_3, z_2]$, where z_1, z_2, and z_3 are distinct points in the complex plane. Assuming that z_3 lies on the line segment with endpoints z_1 and z_2, show that β can be obtained from γ by the piecewise smooth change of parameter $h:[0,1] \to [0,2]$ given as follows: with $\lambda = |z_3 - z_1|/|z_2 - z_1|$, $h(s) = s/\lambda$ when $0 \leq s \leq \lambda$ and

$h(s) = (s + 1 - 2\lambda)/(1 - \lambda)$ when $\lambda \leq s \leq 1$.

4.9. Let $\gamma(t) = e^{it}$ for $-\pi/2 \leq t \leq \pi/2$ and $\beta(s) = (1+s^2)^{-1}(1-s^2+2is)$ for $-1 \leq s \leq 1$. Show that β arises from γ by making a smooth change of parameter $h: [-1, 1] \to [-\pi/2, \pi/2]$ — namely, $h(s) = 2 \operatorname{Arctan} s$.

4.2 Exercises for Section IV.2

4.10. Given that $g, h: [a, b] \to \mathbb{C}$ are continuous functions and that c is a complex constant, verify that

$$\int_a^b [g(t) + h(t)]\, dt = \int_a^b g(t)\, dt + \int_a^b h(t)\, dt$$

and

$$\int_a^b c\, g(t)\, dt = c \int_a^b g(t)\, dt$$

by appealing to the corresponding facts for real-valued functions and real constants.

4.11. If $g, G: [a, b] \to \mathbb{C}$ are continuous functions and if $\dot{G}(t) = g(t)$ for every t in (a, b), check that $\int_a^b g(t)\, dt = G(b) - G(a)$. Again, the analogous result concerning real-valued functions is at one's disposal.

4.12. If $\gamma(t) = te^{it}$ for $0 \leq t \leq \pi$, evaluate: (i) $\int_\gamma \bar{z}\, dz$; (ii) $\int_\gamma |z||dz|$; (iii) $\int_\gamma z\, dz$; (iv) $\int |z|\, dz$; (v) $\int_\gamma z(1+|z|^2)^{-1/2}|dz|$.

4.13. If $\gamma(t) = 3t^2 + 2t^3 i$ for $0 \leq t \leq 1$, compute: (i) $\ell(\gamma)$; (ii) $\int_\gamma x\, dz$; (iii) $\int_\gamma x|dz|$; (iv) $\int_\gamma \sqrt{3x+9}\,|dz|$; (v) $\int_\gamma z\, dz$; (vi) $\int_\gamma |2x+3iy||dz|$.

4.14. Let $\gamma = \beta + [e^{2\pi}, 1]$, where β is given by $\beta(t) = e^{t+it}$ for $0 \leq t \leq 2\pi$. Evaluate: (i) $\int_\gamma z^{-1}\, dz$; (ii) $\int_\gamma z^{-1}|dz|$; (iii) $\int_\gamma |z|^{-1}\, dz$; (iv) $\int_\gamma |z|^{-1}|dz|$; (v) $\int_\beta e^z\, dz$; (vi) $\int_\beta e^{|z|}|dz|$.

4.15. Demonstrate that $\int_{-\gamma} f(z)\,|dz| = \int_\gamma f(z)\,|dz|$, where γ is a piecewise smooth path and f is a function that is continuous on $|\gamma|$.

4.16. If $\gamma(t) = e^{1+it}$ for $0 \leq t \leq \pi$, show that

$$\left| \int_\gamma (\operatorname{Log} z)^{-1} dz \right| \leq e \operatorname{Log}(\pi + \sqrt{\pi^2 + 1}) .$$

4.17. Assuming that $a > 0$ and $b > 0$, derive the estimate

$$\left| \int_0^{a+ib} \cos(z^2)\, dz \right| \leq (a^2 + b^2)^{1/2} \sinh(2ab)/(2ab) .$$

4.18. Let a and b be real numbers satisfying $a < b$, and let $I(c)$ be defined for any real number c by $I(c) = \int_{c+ia}^{c+ib} e^{-z^2} dz$. By deriving a suitable upper bound for $|I(c)|$ conclude that $I(c) \to 0$ as $c \to \pm\infty$.

4.19. For r obeying $r > 0$ and $r \neq 1$, set $I(r) = \int_{|z|=r} (z^2+1)^{-1} \operatorname{Log} z \, dz$. By obtaining estimates on $|I(r)|$ prove that $I(r) \to 0$ as $r \to 0$ and also as $r \to \infty$. (N.B. The integrand in $I(r)$ has a discontinuity at $z = -r$, but it is continuous and bounded on the rest of the circle $K(0,r)$, so the integral is still meaningful.)

4.20. For r with $0 < r < \infty$ let $I(r) = \int_{\gamma_r} z^{-1} e^{iz} dz$, where γ_r is the path defined on $[0, \pi]$ by $\gamma_r(t) = re^{it}$. Show that $I(r) \to 0$ as $r \to \infty$ and also that $I(r) \to \pi i$ as $r \to 0$. (*Hint.* For the second part, start by demonstrating that $\int_{\gamma_r} z^{-1}(e^{iz}-1) dz \to 0$ as $r \to 0$.)

4.21. Evaluate $\int_\beta x e^y |dz|$, where $\beta(s) = (1+s^2)^{-1}(1-s^2+2is)$ for s in the interval $[-1, 1]$. (*Hint.* One of the earlier exercises is pertinent here.)

4.22. If $\gamma(t) = 3 - 3t^2 + i(t^3 - 3t + 1)$ on the interval $[-1, 1]$, determine:
(i) $\int_\gamma (z^2 + iz) dz$; (ii) $\int_\gamma e^{\pi z} dz$; (iii) $\int_\gamma \sin^2(\pi z) dz$; (iv) $\int_\gamma z^{-2} dz$;
(v) $\int_\gamma z^{-1/2} dz$; (vi) $\int_\gamma z |dz|$.

4.23. For the path $\gamma = [e, 1] + [1, -1+i\sqrt{3}]$, evaluate: (i) $\int_\gamma z \, dz$;
(ii) $\int_\gamma z |dz|$; (iii) $\int_\gamma z^{-1} dz$; (iv) $\int_\gamma z^{-1} \operatorname{Log} z \, dz$; (v) $\int_\gamma (z^2+2z)^{-1} dz$;
(vi) $\int_\gamma \operatorname{Log} z \, dz$.

4.24. Under the assumption that γ is a piecewise smooth path in an open set U where functions f and g are analytic, verify the integration by parts formula (4.6). One may assume for the sake of this problem that f' and g' are already known to be continuous in U.

4.25. If γ is the path defined on the interval $[0, 1]$ by $\gamma(t) = t - t^2 + it^3$, calculate: (i) $\int_\gamma z \cos(\pi iz) dz$; (ii) $\int_\gamma z \operatorname{Log}(z+1) dz$; (iii) $\int_\gamma z^2 e^{\pi z} dz$;
(iv) $\int_\gamma z^3 (1-z^2)^{-1/2} dz$; (v) $\int_\gamma \operatorname{Arcsin} z \, dz$.

4.26. Evaluate the integrals: (i) $\int_{|z|=1} (z-2)^{-2} dz$; (ii) $\int_{|z|=1} (z^2-4)^{-1} dz$;
(iii) $\int_{|z|=1} (z^2+2z)^{-1} dz$; (iv) $\int_{|z|=1} (z^3+4z)^{-1} dz$; (v) $\int_{|z|=1} (4-z^2)^{-1/2} dz$;
(vi) $\int_{|z|=1} (z+z^{-1})^n dz$, with n a positive integer.

4.27. Let Q be the square with vertices $1, i, -1$, and $-i$. Evaluate:
(i) $\int_{\partial Q} z^3 dz$; (ii) $\int_{\partial Q} z^2 |dz|$; (iii) $\int_{\partial Q} z^{-2} dz$; (iv) $\int_{\partial Q} z^{-1} dz$;
(v) $\int_{\partial Q} z^{-1/2} dz$. (N.B. In (v) the integrand is discontinuous at -1, but the integral makes sense nonetheless, for $f(z) = z^{-1/2}$ is continuous and bounded on $\partial Q \sim \{-1\}$.)

4.28. Evaluate: (i) $\int_i^{-4} z^{-1/2} dz$; (ii) $\int_{-i}^i z^{1/2} dz$; (iii) $\int_{-1}^i (z \operatorname{Log} z)^{-1} dz$;

4. Exercises for Chapter IV

(iv) $\int_1^{i\sqrt{2}} \text{Log}(1+z^2)\,dz$; (v) $\int_1^{2i} \text{Arctan}\, z\, dz$. (N.B. The integrands in this problem are not analytic in open sets containing the paths of integration. Recall, however, the remark that immediately follows the proof of Theorem 2.2.)

Chapter V

Cauchy's Theorem and its Consequences

Introduction

The chapter at hand touches the very heart and soul of complex analysis. In Cauchy's theorem — this designation actually encompasses an entire class of theorems, owing to the variety of formulations that the result admits — can be traced the roots of virtually every advance in complex function theory from 1825, when Cauchy published his original version of the theorem, to the present day. Its central theme is sounded in the question: Under what conditions can it be inferred that the integral of an analytic function along a closed path vanishes? Our investigation to determine general situations in which this happens will also generate a number of lovely variations on the theme, Cauchy's integral formula prominent among them. At the same time it will divulge an intriguing relationship between complex analysis and plane topology. Most importantly, it will ultimately pave the way to an extremely clear and precise understanding of the local structure of analytic functions.

1 The Local Cauchy Theorem

1.1 Cauchy's Theorem For Rectangles

Lemma 1.1 serves as a prototype for results to which the title "Cauchy's theorem" can be applied. Even this simple case of the theorem has an unexpectedly complicated proof. The idea for the elegant subdivision procedure in the proof originated with Édouard Goursat (1858-1936), although the formal argument we present is modeled on one that was published in 1901 by Alfred Pringsheim (1850-1941).

1. The Local Cauchy Theorem

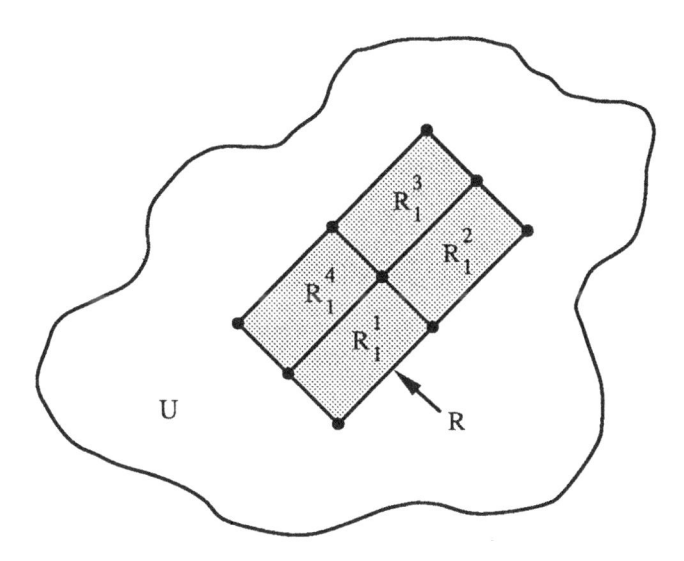

Figure 1.

Lemma 1.1. *If a function f is analytic in an open set U, then $\int_{\partial R} f(z)dz = 0$ for every closed rectangle R in U.*

Proof. Let R be a closed rectangle in U, and set $I = \int_{\partial R} f(z)dz$. We subdivide R into four congruent rectangles R_1^1, R_1^2, R_1^3, and R_1^4, as indicated in Figure 1, and write $I_1^j = \int_{\partial R_1^j} f(z)dz$ for $1 \leq j \leq 4$. It follows from properties of the complex line integral articulated in the previous chapter — recall especially (IV.4.8) and (IV.4.9) — that

$$I = I_1^1 + I_1^2 + I_1^3 + I_1^4.$$

The triangle inequality tells us that

$$|I| \leq |I_1^1| + |I_1^2| + |I_1^3| + |I_1^4|.$$

If every term in this sum were less than $|I|/4$, a contradiction would result. As a consequence,
$$|I_1^j| \geq 4^{-1}|I|$$
must hold for at least one of the indices j. It is therefore possible to assert the existence of a subrectangle R_1 of R whose dimensions are half those of R such that $I_1 = \int_{\partial R_1} f(z)dz$ satisfies

$$|I_1| \geq 4^{-1}|I|.$$

The same subdivision argument can, of course, be applied to the rectangle R_1 to produce an even smaller rectangle — call it R_2 — such that $I_2 =$

$\int_{\partial R_2} f(z)dz$ has
$$|I_2| \geq 4^{-1}|I_1| \geq 4^{-2}|I| .$$

By continuing this subdivision procedure inductively we establish the existence of a sequence of closed rectangles
$$R_1 \supset R_2 \supset R_3 \supset \cdots$$
in U such that $I_n = \int_{\partial R_n} f(z)dz$ satisfies
$$|I_n| \geq 4^{-1}|I_{n-1}| \geq \cdots \geq 4^{-n}|I| ;$$
i.e., such that the bound

(5.1) $$|I| \leq 4^n|I_n|$$

is available for $n = 1, 2, 3, \cdots$. Let d_n designate the length of a diagonal of the rectangle R_n, and let L_n be its perimeter. If d and L represent the corresponding quantities for the original rectangle R, then by construction
$$d_n = 2^{-n}d , \quad L_n = 2^{-n}L .$$

Invoking Cantor's theorem (Theorem II.4.5), we fix a point z_0 that belongs to R_n for all n. Then z_0 lies in U. By assumption the function f is differentiable at z_0, which fact entitles us to write

(5.2) $$f(z) = f(z_0) + f'(z_0)(z - z_0) + E(z)$$

for every z in U, where the function $E: U \to \mathbb{C}$ obeys the condition

(5.3) $$\lim_{z \to z_0} \frac{|E(z)|}{|z - z_0|} = 0 .$$

It is evident from (5.2) that E is a continuous function and that $E(z_0) = 0$. If γ is any closed and piecewise smooth path in U, we are allowed to integrate each side of (5.2) along γ. In so doing we discover that

$$\int_\gamma f(z)\,dz = f(z_0)\int_\gamma dz + f'(z_0)\int_\gamma (z - z_0)\,dz + \int_\gamma E(z)\,dz = \int_\gamma E(z)\,dz .$$

We have here used the fact that $\int_\gamma dz = \int_\gamma (z - z_0)dz = 0$, true because both integrands are plainly functions with primitives in the entire complex plane. Making an obvious special choice for γ, we can thus rewrite I_n in the form

(5.4) $$I_n = \int_{\partial R_n} E(z)\,dz .$$

Suppose now that $\epsilon > 0$ is given. We shall show that

(5.5) $$|I| \leq \epsilon dL .$$

1. The Local Cauchy Theorem

For this, we use (5.3) to choose an open disk $\Delta = \Delta(z_0, \delta)$ contained in U with the property that the estimate

(5.6) $$|E(z)| \leq \epsilon |z - z_0|$$

is valid for every z in Δ. Next, because $d_n \to 0$ as $n \to \infty$, we can fix an index n such that $d_n < \delta$. Since z_0 lies in R_n and since $|z - z_0| \leq d_n$ is certainly true of any z in R_n, it follows that R_n is a subset of Δ. Using (5.1), (5.4), and (5.6), we compute

$$|I| \leq 4^n |I_n| = 4^n \left| \int_{\partial R_n} E(z)\, dz \right| \leq 4^n \int_{\partial R_n} |E(z)|\, |dz|$$

$$\leq \epsilon 4^n \int_{\partial R_n} |z - z_0|\, |dz| \leq \epsilon 4^n d_n \int_{\partial R_n} |dz| = \epsilon 4^n d_n L_n = \epsilon dL \; ,$$

which establishes (5.5). Since ϵ in this inequality is an arbitrary positive number, we conclude that $I = 0$, precisely the assertion of the lemma. ∎

For certain future considerations — one of these is the derivation of Cauchy's integral formula — a minor extension of Lemma 1.1 will be needed. In it we see that the differentiability requirement on f in Lemma 1.1 can be relaxed a small amount without affecting the result.

Lemma 1.2. *If a function f is continuous in an open set U and analytic in $U \sim \{z_0\}$ for some point z_0 of U, then $\int_{\partial R} f(z) dz = 0$ for every closed rectangle R in U.*

Proof. Fix a closed rectangle R in U. We may assume that z_0 lies in R, for the conclusion of the lemma is a consequence of Lemma 1.1 otherwise. Given a positive integer n, we subdivide R into n^2 congruent rectangles $R_{k\ell}$. Figure 2 illustrates the subdivision for $n = 4$. From properties of

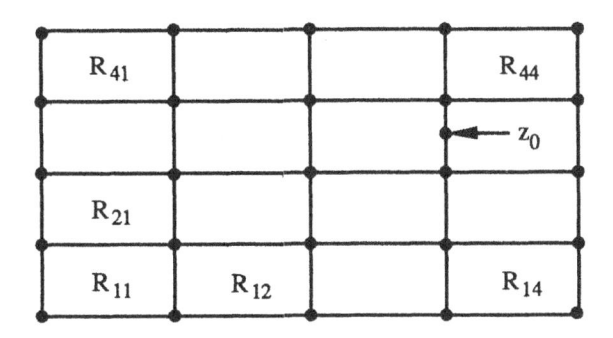

Figure 2.

complex line integrals it follows that

$$\int_{\partial R} f(z)\, dz = \sum_{k=1}^{n}\sum_{\ell=1}^{n} \int_{\partial R_{k\ell}} f(z)\, dz\ .$$

If z_0 is not a point of $R_{k\ell}$, then $\int_{\partial R_{k\ell}} f(z)dz = 0$ by Lemma 1.1. If, on the other hand, z_0 does belong to $R_{k\ell}$, then we have the estimate

$$\left|\int_{\partial R_{k\ell}} f(z)\, dz\right| \leq \int_{\partial R_{k\ell}} |f(z)|\, |dz| \leq m\ \mathrm{Perimeter}\,(R_{k\ell}) = \frac{mL}{n}\ ,$$

where L is the perimeter of R and m is the maximum value of the continuous function $|f|$ on the compact set R. The point z_0 can belong to no more than four of the rectangles $R_{k\ell}$. We infer, therefore, that

$$\left|\int_{\partial R} f(z)\, dz\right| = \left|\sum_{z_0 \in R_{k\ell}} \int_{\partial R_{k\ell}} f(z)\, dz\right| \leq \sum_{z_0 \in R_{k\ell}} \left|\int_{\partial R_{k\ell}} f(z)\, dz\right| \leq \frac{4mL}{n}\ .$$

Since the positive integer n was arbitrary, we can let $n \to \infty$ in the last estimate and deduce that $\int_{\partial R} f(z)dz = 0$. ∎

1.2 Integrals and Primitives

The next result shows that, at least on a local level, information concerning the vanishing of integrals around rectangles amounts to information about the existence of primitives.

Lemma 1.3. *Let Δ be an open disk in the complex plane, and let f be a continuous function in Δ with the property that $\int_{\partial R} f(z)dz = 0$ for every closed rectangle R in Δ whose sides are parallel to the coordinate axes. Then f has a primitive in Δ. In particular, $\int_\gamma f(z)dz = 0$ for every closed, piecewise smooth path γ in this disk.*

Proof. Let $z_0 = x_0 + iy_0$ be the center of Δ. For an arbitrary point $z = x + iy$ of this disk we set $z_1 = x + iy_0$ and $z_2 = x_0 + iy$, and we make the observation that

$$(5.7) \qquad \int_{z_0}^{z_1} f(\zeta)\, d\zeta + \int_{z_1}^{z} f(\zeta)\, d\zeta + \int_{z}^{z_2} f(\zeta)\, d\zeta + \int_{z_2}^{z_0} f(\zeta)\, d\zeta = 0\ .$$

(N.B. Since the letter z is being used here in a special role, a different symbol is needed for the "dummy variable" of integration. A traditional choice in this and similar situations is ζ, the Greek equivalent of z.) Relation (5.7) follows directly from the elementary properties of contour integrals in case $x = x_0$ or $y = y_0$; it is a consequence of the hypotheses of the lemma

1. The Local Cauchy Theorem

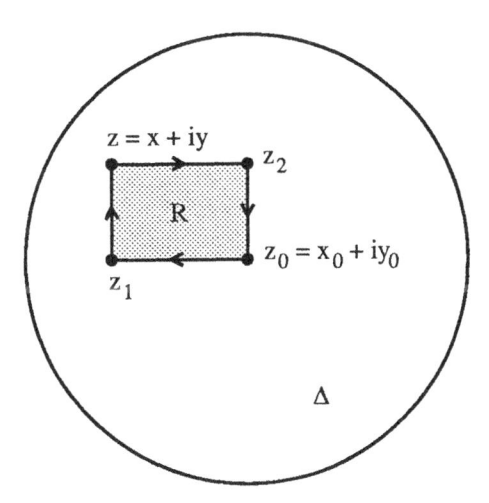

Figure 3.

otherwise. To be exact, when $x \neq x_0$ and $y \neq y_0$ the sum in (5.7) represents — depending on the location of z — either $\int_{\partial R} f(\zeta) d\zeta$ or $-\int_{\partial R} f(\zeta) d\zeta$, where R is the rectangle in Δ with vertices z_0, z_1, z, and z_2 (Figure 3).

Using the above notation, we define a function $F: \Delta \to \mathbb{C}$ as follows: for z in Δ,

$$F(z) = \int_{z_0}^{z_2} f(\zeta) d\zeta + \int_{z_2}^{z} f(\zeta) d\zeta$$

or, what in view of (5.7) is the same,

$$F(z) = \int_{z_0}^{z_1} f(\zeta) d\zeta + \int_{z_1}^{z} f(\zeta) d\zeta \ .$$

Again letting $z = x + iy$ and referring to (IV.4.10) and (IV.4.11), we rewrite these expressions for $F(z)$ in the forms

(5.8) $$F(z) = \int_{x_0}^{x} f(t + iy) \, dt + i \int_{y_0}^{y} f(x_0 + it) \, dt$$

and

(5.9) $$F(z) = \int_{x_0}^{x} f(t + iy_0) \, dt + i \int_{y_0}^{y} f(x + it) \, dt \ .$$

Formula (5.8) makes it easy to compute the partial derivative $F_x(z)$ at a point of Δ. Indeed, the second term is independent of x and the first integral is of the type to which the First Fundamental Theorem of Calculus applies:

$$\frac{d}{dx} \int_a^x g(t) \, dt = g(x)$$

on $(a-r, a+r)$ if $g:(a-r,a+r) \to \mathbb{C}$ is a continuous function. The implication in the present situation is that $F_x(z) = f(z)$ for each point z of Δ. In a like manner, we see using (5.9) that $F_y(z) = if(z)$ for such z. It follows that F is a member of the class $C^1(\Delta)$. Furthermore, $F_x = -iF_y$ throughout Δ, which in disguised form is the statement that the real and imaginary parts of F satisfy the Cauchy-Riemann equations in that disk. Theorem III.2.2 confirms that F is analytic in Δ and has derivative $F' = F_x = f$ there; i.e., F is a primitive for f in Δ. The final assertion in the lemma can then be inferred from Theorem IV.2.2. ∎

Before proceeding to our statement of the local version of Cauchy's theorem, we digress momentarily to record an important consequence of Lemma 1.3 that, although we put it to no immediate use, will prove to be quite handy in later deliberations. It is the converse of the last statement in Theorem IV.2.2.

Theorem 1.4. *Suppose that a function f is continuous in a plane domain D and that $\int_\gamma f(z)dz = 0$ for every closed, piecewise smooth path γ in D. Then f has a primitive in D.*

Proof. We construct a primitive F for f in D as follows. We first fix a point w_0 of D. Given z in D we choose a piecewise smooth path α in D with initial point w_0 and terminal point z (Lemma IV.1.1) and define: $F(z) = \int_\alpha f(\zeta)d\zeta$. (See Figure 4.) Can we be sure that $F(z)$ is well-defined,

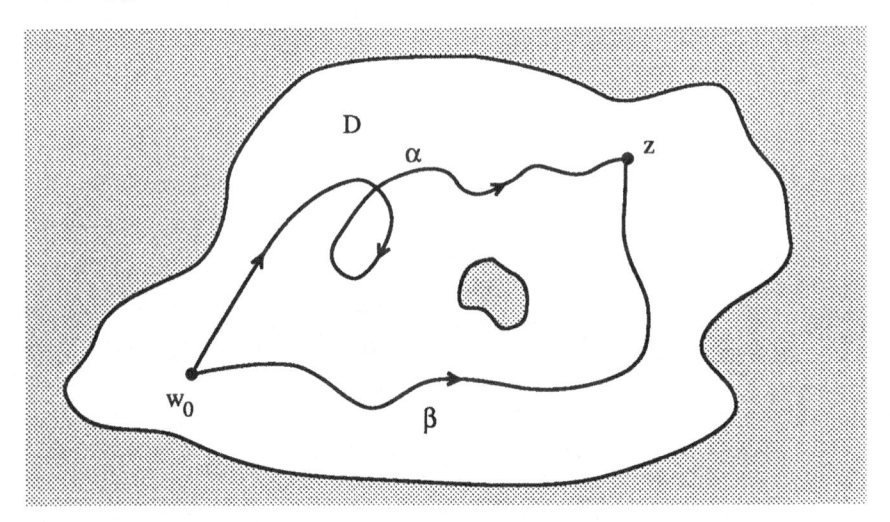

Figure 4.

meaning that this value does not depend on our selection of the path α? If β is a second piecewise smooth path in D from w_0 to z, then $\gamma = \alpha - \beta$ is a closed, piecewise smooth path in the domain. Lemma IV.2.1 and our

1. The Local Cauchy Theorem

assumptions about f certify that

$$\int_\alpha f(\zeta)\,d\zeta - \int_\beta f(\zeta)\,d\zeta = \int_\gamma f(\zeta)\,d\zeta = 0 ,$$

implying that $\int_\alpha f(\zeta)d\zeta = \int_\beta f(\zeta)d\zeta$. Therefore, the definition of $F(z)$ does not depend on the particular choice of α.

To prove that F is a primitive for f in D we consider an arbitrary point z_0 of D and verify that F is differentiable at z_0 with $F'(z_0) = f(z_0)$. For this we fix $\Delta = \Delta(z_0, r)$, an open disk centered at z_0 and contained in D, and choose a piecewise smooth path α_0 in D initiating at w_0 and terminating at z_0. In computing $F(z)$ for a point z of Δ we are at liberty by what has been said above to use any piecewise smooth path α of our liking that joints w_0 to z in D. We take $\alpha = \alpha_0 + [z_0, z]$ (Figure 5). Now

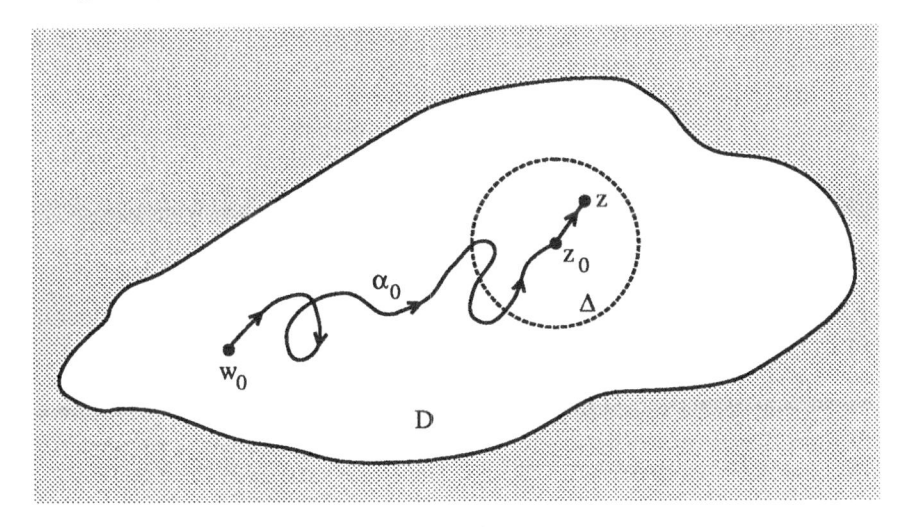

Figure 5.

the function f certainly satisfies the hypotheses of Lemma 1.3 in the disk Δ, a fact that enables us to select a primitive — call it G — for f in Δ. By Theorem IV.2.2 we obtain for any z in that disk

$$F(z) = \int_{\alpha_0} f(\zeta)\,d\zeta + \int_{z_0}^z f(\zeta)\,d\zeta = F(z_0) + G(z) - G(z_0) .$$

This shows that in Δ the function F differs from the analytic function G by the constant $F(z_0) - G(z_0)$. The result: F is indeed differentiable at z_0 and $F'(z_0) = G'(z_0) = f(z_0)$, as desired. ∎

1.3 The Local Cauchy Theorem

Returning to the main line of development, we combine Lemma 1.2 with Lemma 1.3 to arrive at:

Theorem 1.5. (Cauchy's Theorem – Local Form) *Suppose that Δ is an open disk in the complex plane and that f is a function which is analytic in Δ (or, more generally, is continuous in Δ and analytic in $\Delta \sim \{z_0\}$ for some point z_0 of Δ). Then $\int_\gamma f(z)dz = 0$ for every closed, piecewise smooth path γ in Δ.*

The word "local" refers here to the restriction of working in a disk. For a function f analytic in an arbitrary open set U the theorem insures that $\int_\gamma f(z)dz = 0$ if γ is a closed, piecewise smooth path whose trajectory can be enclosed in an open disk that is contained in U — i.e., the theorem can be applied "locally" in U. It makes no statement about integrals of f along paths not subject to such a constraint, as a "global" result might. In the following two examples we see that even this limited version of the Cauchy theorem already packs some punch. We use it in the evaluation of two integrals. The first is a contour integral that, though quite innocent in appearance, is actually somewhat unpleasant to evaluate by "brute force" methods. The second example shows how complex methods can come to one's aid in computing an ordinary calculus-style integral.

EXAMPLE 1.1. Show that $\int_{\partial R} (z - z_0)^{-1} dz = 2\pi i$, where R is a closed rectangle with center at z_0.

Let $K = K(z_0, r)$ be the circle that circumscribes R, and for $1 \le k \le 4$ let γ_k and β_k be the paths indicated in Figure 6, parametrized in such a fashion that $\gamma = \gamma_1 + \gamma_2 + \gamma_3 + \gamma_4$ and $\beta = \beta_1 + \beta_2 + \beta_3 + \beta_4$ parametrize ∂R and K, respectively, in the ways we have accepted as standard for rectangles and circles. We can choose for each k an open disk Δ_k which contains the trajectory of the closed path $\gamma_k - \beta_k$ and in which the integrand $f(z) = (z - z_0)^{-1}$ is analytic. Figure 6 displays Δ_1. According to Theorem 1.5,

$$\int_{\gamma_k} \frac{dz}{z - z_0} - \int_{\beta_k} \frac{dz}{z - z_0} = \int_{\gamma_k - \beta_k} \frac{dz}{z - z_0} = 0$$

for $1 \le k \le 4$. Therefore,

$$\int_{\partial R} \frac{dz}{z - z_0} = \sum_{k=1}^{4} \int_{\gamma_k} \frac{dz}{z - z_0} = \sum_{k=1}^{4} \int_{\beta_k} \frac{dz}{z - z_0}$$

$$= \int_{|z - z_0| = r} \frac{dz}{z - z_0} = \int_0^{2\pi} \frac{rie^{it} dt}{re^{it}} = 2\pi i \ .$$

1. The Local Cauchy Theorem   149

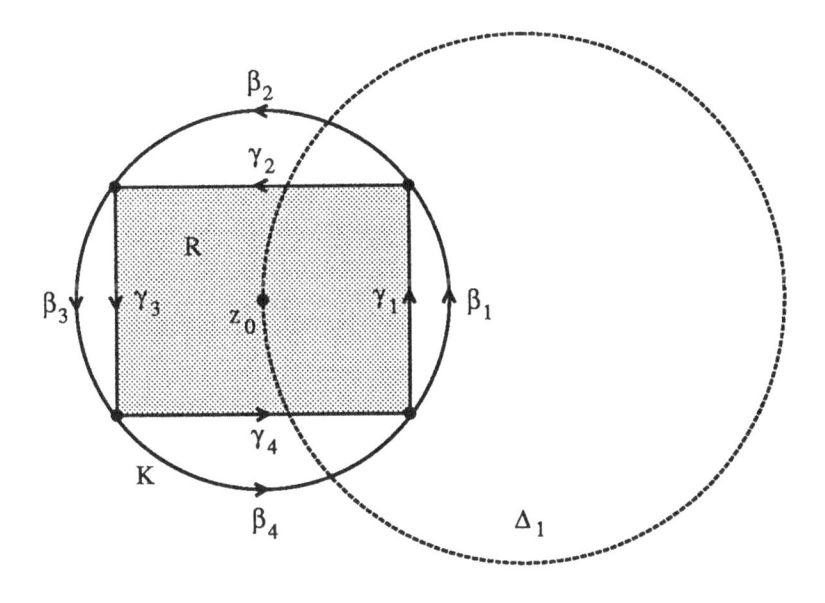

Figure 6.

EXAMPLE 1.2. Verify that

$$\int_0^\infty \cos(t^2)\,dt = \int_0^\infty \sin(t^2)\,dt = \frac{\sqrt{2\pi}}{4}.$$

The integrals in question are improper Riemann integrals. (They are often referred to as "Fresnel's integrals.") What we are required to show is that

$$\lim_{r\to\infty}\int_0^r \cos(t^2)\,dt = \lim_{r\to\infty}\int_0^r \sin(t^2)\,dt = \frac{\sqrt{2\pi}}{4}.$$

Since $\cos(t^2) + i\sin(t^2) = e^{it^2}$, it is enough to demonstrate that

$$\lim_{r\to\infty}\int_0^r e^{it^2}\,dt = \frac{(1+i)\sqrt{2\pi}}{4}$$

and then to take real and imaginary parts. We are thus led naturally to look at the function $f(z) = e^{iz^2}$. A much less obvious second step is to consider the integral of f along the path $\gamma = \gamma_1 + \gamma_2 - \gamma_3$ suggested by Figure 7. To be precise, we take $\gamma_1(t) = t$ and $\gamma_3(t) = te^{\pi i/4}$ for $0 \le t \le r$, while $\gamma_2(t) = re^{it}$ for $0 \le t \le \pi/4$.

V. Cauchy's Theorem and its Consequences

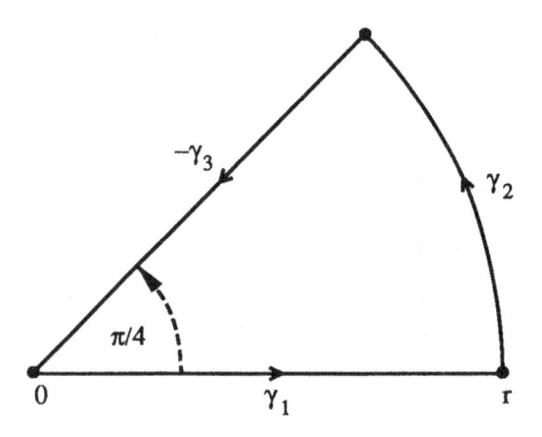

Figure 7.

Since f is entire — hence, certainly analytic in any open disk Δ containing the trajectory of the path γ — the local Cauchy theorem gives

$$\int_{\gamma_1} e^{iz^2}\,dz + \int_{\gamma_2} e^{iz^2}\,dz - \int_{\gamma_3} e^{iz^2}\,dz = \int_{\gamma} e^{iz^2}\,dz = 0.$$

Therefore,

$$\int_0^r e^{it^2}\,dt = \int_{\gamma_1} e^{iz^2}\,dz = -\int_{\gamma_2} e^{iz^2}\,dz + \int_{\gamma_3} e^{iz^2}\,dz$$

$$= -\int_{\gamma_2} e^{iz^2}\,dz + \frac{(1+i)\sqrt{2}}{2}\int_0^r e^{-t^2}\,dt \to \frac{(1+i)\sqrt{2}}{2}\int_0^\infty e^{-t^2}\,dt$$

as $r \to \infty$. In support of the last assertion we recall Example IV.2.5, which delivers the estimate

$$\left|\int_{\gamma_2} e^{iz^2}\,dz\right| \leq \frac{\pi\left(1 - e^{-r^2}\right)}{4r}.$$

As a result, it is apparent that

$$\int_{\gamma_2} e^{iz^2}\,dz \to 0$$

as $r \to \infty$.

It remains only to determine $I = \int_0^\infty e^{-t^2}\,dt$. This is a standard integral from advanced calculus, but we sketch its evaluation for the sake of completeness. The trick is to write

$$I^2 = \left(\int_0^\infty e^{-x^2}\,dx\right)\left(\int_0^\infty e^{-y^2}\,dy\right) = \iint_A e^{-(x^2+y^2)}\,dx\,dy,$$

1. The Local Cauchy Theorem

where $A = \{z : x \geq 0 \text{ and } y \geq 0\}$. After a change to polar coordinates we obtain
$$I^2 = \int_0^{\pi/2} \int_0^\infty e^{-r^2} r \, dr \, d\theta = \frac{\pi}{4},$$
so $I = \sqrt{\pi}/2$. We conclude that
$$\lim_{r \to \infty} \int_0^r e^{it^2} \, dt = \frac{(1+i)\sqrt{2\pi}}{4},$$
as we had hoped.

One is able to gain from the preceding example an intimation of the valuable techniques complex analysis often provides for evaluating integrals that arise in real-variable problems with no ostensible connection to complex numbers. The evolution of such complex methods culminates in Cauchy's "Residue Theorem," which will be discussed in Chapter VIII.

The hypotheses of the local Cauchy theorem are far too constrictive for this proposition to be of overriding significance in general applications of contour integration. Upgrading the theorem to a more flexible global result is one of our next major objectives. In the process, however, we shall discover that some of the most important properties of analytic functions can be derived using nothing more than Theorem 1.5. We conclude this section by establishing a lemma that will be invoked several times in implementing these plans.

Lemma 1.6. *Let γ be a piecewise smooth path in the complex plane, let h be a function that is continuous on $|\gamma|$, and let k be a positive integer. The function H defined in the open set $U = \mathbb{C} \sim |\gamma|$ by*
$$H(z) = \int_\gamma \frac{h(\zeta) \, d\zeta}{(\zeta - z)^k}$$
is an analytic function whose derivative is given by
$$H'(z) = k \int_\gamma \frac{h(\zeta) \, d\zeta}{(\zeta - z)^{k+1}}.$$

Proof. We intend to use this lemma only in the cases $k = 1$ and $k = 2$, so it is for these cases alone that the proof is given. A similar technique of proof will work in the general case, but the algebra involved becomes a little unwieldy. In fact, we provide full details when $k = 2$, and merely indicate the modifications that occur in the simpler case $k = 1$.

Assuming that $k = 2$, let z_0 be a point in U. It is our task to prove that
$$\lim_{z \to z_0} \frac{H(z) - H(z_0)}{z - z_0} = 2 \int_\gamma \frac{h(\zeta) \, d\zeta}{(\zeta - z_0)^3}$$

152   V. Cauchy's Theorem and its Consequences

or, what is equivalent to this, that

$$\left| \frac{H(z) - H(z_0)}{z - z_0} - 2 \int_\gamma \frac{h(\zeta)\, d\zeta}{(\zeta - z_0)^3} \right| \to 0$$

as $z \to z_0$. Fix $r > 0$ and $s > 0$ so that the disk $\Delta(z_0, 2r)$ is contained in U and so that $|\gamma|$ is contained in the disk $\Delta(0, s)$ (Figure 8). For z in U,

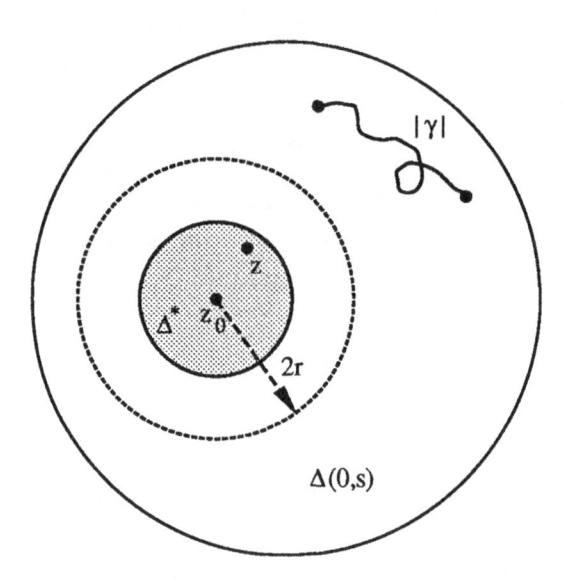

Figure 8.

$z \neq z_0$, we compute

$$\frac{H(z) - H(z_0)}{z - z_0} = \frac{1}{z - z_0} \int_\gamma h(\zeta) \left\{ \frac{1}{(\zeta - z)^2} - \frac{1}{(\zeta - z_0)^2} \right\} d\zeta$$

$$= \int_\gamma \frac{(2\zeta - z - z_0) h(\zeta)\, d\zeta}{(\zeta - z)^2 (\zeta - z_0)^2}$$

and

$$\frac{H(z) - H(z_0)}{z - z_0} - 2 \int_\gamma \frac{h(\zeta)\, d\zeta}{(\zeta - z_0)^3}$$

$$= \int_\gamma h(\zeta) \left\{ \frac{2\zeta - z - z_0}{(\zeta - z)^2 (\zeta - z_0)^2} - \frac{2}{(\zeta - z_0)^3} \right\} d\zeta$$

$$= (z - z_0) \int_\gamma \frac{(3\zeta - 2z - z_0) h(\zeta)\, d\zeta}{(\zeta - z)^2 (\zeta - z_0)^3} .$$

2. Winding Numbers and the Local Cauchy Integral Formula 153

If z lies in the punctured disk $\Delta^* = \Delta^*(z_0, r)$ and ζ on $|\gamma|$, then $|\zeta - z_0| \geq r$ and $|\zeta - z| \geq r$. Also,

$$|3\zeta - 2z - z_0| = |3\zeta - 2(z - z_0) - 3z_0| \leq 3|\zeta| + 2|z - z_0| + 3|z_0| \leq 3s + 2r + 3|z_0|.$$

Writing $c = 3s + 2r + 3|z_0|$, we learn that for such z

$$\left| \frac{H(z) - H(z_0)}{z - z_0} - 2 \int_\gamma \frac{h(\zeta)\, d\zeta}{(\zeta - z_0)^3} \right| \leq |z - z_0| \int_\gamma \frac{|3\zeta - 2z - z_0| |h(\zeta)| |d\zeta|}{|\zeta - z|^2 |\zeta - z_0|^3}$$

$$\leq \frac{c|z - z_0|}{r^5} \int_\gamma |h(\zeta)|\, |d\zeta|.$$

The last expression clearly tends to 0 as $z \to z_0$. (The parallel computations in the case $k = 1$ would result in

$$\left| \frac{H(z) - H(z_0)}{z - z_0} - \int_\gamma \frac{h(\zeta)\, d\zeta}{(\zeta - z_0)^2} \right| \leq \frac{|z - z_0|}{r^3} \int_\gamma |h(\zeta)|\, |d\zeta| \to 0$$

as $z \to z_0$.) Since z_0 was an arbitrary point of U, the conclusion of the lemma follows — at least when $k = 1$ or $k = 2$. ∎

2 Winding Numbers and the Local Cauchy Integral Formula

2.1 Winding Numbers

Suppose that γ is a closed and piecewise smooth path in the complex plane and that z is a point of $\mathbb{C} \sim |\gamma|$. The *winding number* $n(\gamma, z)$ *of* γ *about* z (also called the *index of* γ *with respect to* z) is defined by the formula

$$n(\gamma, z) = \frac{1}{2\pi i} \int_\gamma \frac{d\zeta}{\zeta - z}.$$

The importance of this concept will make itself clear as developments in this chapter unfold. In anticipation of such developments we now determine some of the main characteristics of winding numbers.

To begin, we note: $n = n(\gamma, z)$ *is an integer*. The easiest way to see this, assuming that $\gamma : [a, b] \to \mathbb{C}$, is to consider the function $g : [a, b] \to \mathbb{C}$,

$$g(t) = \int_a^t \frac{\dot\gamma(s)\, ds}{\gamma(s) - z}.$$

This function is continuous and satisfies $g(a) = 0$, $g(b) = 2\pi i n$. In fact, g is piecewise smooth and has derivative

(5.10) $$\dot g(t) = \frac{\dot\gamma(t)}{\gamma(t) - z}$$

at any point t where $\dot\gamma$ is continuous, as follows from the First Fundamental Theorem of Calculus. The function h defined on $[a,b]$ by

$$h(t) = [\gamma(t) - z]e^{-g(t)}$$

is likewise continuous and, owing to (5.10), satisfies

$$\dot h(t) = \{\dot\gamma(t) - [\gamma(t) - z]\dot g(t)\}\, e^{-g(t)} = [\dot\gamma(t) - \dot\gamma(t)]e^{-g(t)} = 0$$

at any point of continuity for $\dot\gamma$. Since γ is piecewise smooth, there are at most finitely many points t in $[a,b]$ for which $\dot h(t) = 0$ fails to be true. We deduce from this fact and from the continuity of h that h is constant on $[a,b]$. In particular, $h(a) = h(b)$ and, since $\gamma(a) = \gamma(b)$,

$$e^{-2\pi i n} = e^{-g(b)} = \frac{h(b)}{\gamma(b)-z} = \frac{h(a)}{\gamma(a)-z} = e^{-g(a)} = e^0 = 1\;,$$

which forces $n = n(\gamma, z)$ to be an integer.

Why the name "winding number"? We shall attempt to answer this question by offering a geometric interpretation of $n(\gamma, z)$. It will be instructive to look first at a special case. Preliminary even to this, we fix $r > 0$ and make some comments about the circular path α defined by $\alpha(t) = z + re^{it}$ for $a \leq t \leq a + 2\pi = b$. Easy calculations show that $n(\alpha, z) = 1$ and $n(-\alpha, z) = -1$. As t grows from a to b the radial vector $v(t)$ joining z to $\alpha(t)$ makes one complete ($= 360°$) revolution in the counterclockwise direction. In the case of $-\alpha$ the corresponding radial vector executes one complete clockwise revolution. The associated winding numbers suggest that we should regard the counterclockwise revolution as positive, the clockwise revolution as negative. This is a convention we adopt.

As to the "special case" hinted at earlier, we have in mind a path γ of the form $\gamma = \gamma_1 + \gamma_2 + \cdots + \gamma_m$, where for each k either $\gamma_k = \alpha$ or $\gamma_k = -\alpha$. Here α is the path just discussed. The trajectory of γ is still the circle of radius r centered at z, and we again use $v(t)$ to designate the radial vector from z to $\gamma(t)$. Then

$$n(\gamma, z) = \frac{1}{2\pi i}\int_\gamma \frac{d\zeta}{\zeta - z} = \frac{1}{2\pi i}\int_{\gamma_1}\frac{d\zeta}{\zeta - z} + \cdots + \frac{1}{2\pi i}\int_{\gamma_m}\frac{d\zeta}{\zeta - z}$$

$$= n(\gamma_1, z) + \cdots + n(\gamma_m, z) = P - N\;,$$

where P is the number of indices k for which $\gamma_k = \alpha$ and N the number for which $\gamma_k = -\alpha$. Expressed in terms of the behavior of $v(t)$, $n(\gamma, z)$ registers the "net number" of complete revolutions performed by $v(t)$ as $\gamma(t)$ traces out its trajectory; i.e., it gives the number of positive revolutions minus the number of negative revolutions.

We turn now to $\gamma\colon [a,b] \to \mathbb{C} \sim \{z\}$, an arbitrary closed and piecewise smooth path. While it is certainly possible to employ the vector from z

2. Winding Numbers and the Local Cauchy Integral Formula

to $\gamma(t)$ in interpreting $n(\gamma, z)$, we find it more convenient to work with vectors of uniform length. For this reason we introduce an auxiliary path $\beta: [a, b] \to \mathbb{C}$, defined as follows: we choose a radius $r > 0$ and set $\beta(t) = z + r\left([\gamma(t) - z]/|\gamma(t) - z|\right)$. Thus, the trajectory of β lies on the circle $K = K(z, r)$. We use $v(t)$ to denote the vector from z to $\beta(t)$. (See Figure 9. We have deliberately chosen r there small enough that $|\gamma|$ lies outside K. This makes it visually easier to keep track of $v(t)$ as t varies, but is otherwise of no consequence for what follows.) As $\gamma(t)$ traverses $|\gamma|$ the "pointer" $v(t)$ again rotates about the pivot z, although the nature of this motion may be considerably more complicated than that observed in our illustrative example. Nevertheless, the interpretation of $n(\gamma, z)$ stands: *the winding number $n(\gamma, z)$ records the net number of complete revolutions — meaning the number of positive revolutions less the number of negative revolutions — executed by the vector $v(t)$ when t increases from a to b.* Since the behavior of $v(t)$ is likely to be rather erratic, it is important to clarify what the expression "complete revolution" signifies in this setting and how, precisely, positive revolutions are distinguished from negative ones.

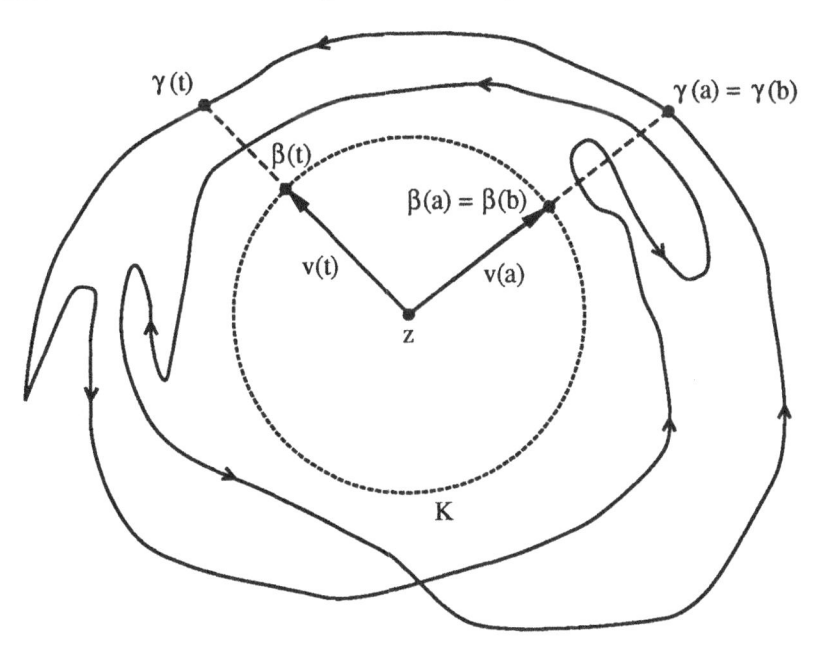

Figure 9.

In saying what it means for $v(t)$ to make "one complete revolution" we shall use the vector $v(a)$ as a reference vector and the angle described in the counterclockwise direction from $v(a)$ to $v(t)$ as a reference angle. Let $\theta(t)$ be the measurement of this angle in the interval $[0, 2\pi)$ (Figure 10). What properties should characterize a single complete revolution of

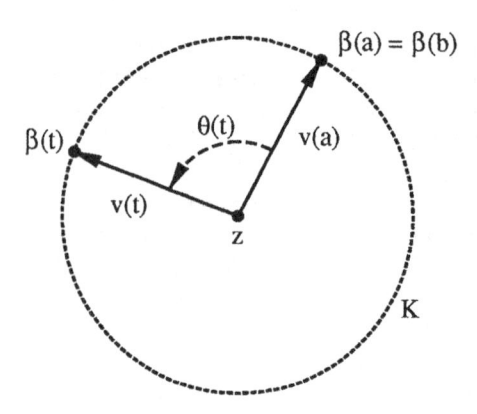

Figure 10.

$v(t)$? Certainly it ought to begin and end with $v(t)$ in reference position. Next, we would obviously like the tip of $v(t)$ to pass through each point of K at least once — to insist on exactly once would be overly restrictive — as $v(t)$ performs such a revolution. Finally, since a return of $v(t)$ to the reference position ought typically to mark the end of one revolution and the beginning of another, it is reasonable to demand that in the course of a revolution $v(t)$ not occupy the reference position except at the start and the finish. We accomodate these suggestions in a formal definition by declaring that $v(t)$ *makes one complete revolution over a subinterval $[c,d]$ of $[a,b]$* if the path β meets the following three conditions on $[c,d]$: $\beta(c) = \beta(d) = \beta(a)$; $\beta([c,d]) = K$; $\beta(t) \neq \beta(a)$ for $c < t < d$. We again stress that the restriction of β to $[c,d]$, which is a closed path, is not required here to be simple. A certain amount of "back-tracking" must be anticipated in the movement of $v(t)$ as t grows from c to d. Translated to requirements on the angle $\theta(t)$, the above conditions force the range of θ over the open interval (c,d) to be the entire interval $(0, 2\pi)$. Now the continuity of β and the fact that $\beta(t) \neq \beta(a)$ on $[c,d]$ except when $t = c$ and $t = d$ insure that θ is continuous on (c,d) and that both $\lim_{t \to d^-} \theta(t)$ and $\lim_{t \to c^+} \theta(t)$ exist, each being either 0 or 2π. The added demand that $\theta[(c,d)] = (0, 2\pi)$ implies that these limits must be different. (If both limits were the same, it would follow either that θ attains a maximum value strictly less than 2π on (c,d) or that θ has a strictly positive minimum on (c,d), neither of which is compatible with $\theta[(c,d)] = (0, 2\pi)$.) In short, two possibilities arise: either $\lim_{t \to c^+} \theta(t) = 0$ and $\lim_{t \to d^-} \theta(t) = 2\pi$ or else $\lim_{t \to c^+} \theta(t) = 2\pi$ and $\lim_{t \to d^-} \theta(t) = 0$. In the first case, where the overall change in $\theta(t)$ as t goes from c to d is an increase of 2π, the revolution is proclaimed *positive*; in the latter instance, where $\theta(t)$ experiences a net decrease from 2π to 0, we speak of a *negative revolution*. In other words, a positive revolution is in spirit (i.e., modulo back-tracking) just a counterclockwise revolution,

2. Winding Numbers and the Local Cauchy Integral Formula

whereas a negative revolution amounts essentially to a clockwise revolution.

Unless $|\beta| \neq K$, in which event $v(t)$ does not make any complete revolutions as t increases from a to b, the uniform continuity of β can be used to exhibit a subdivision $a \leq c_1 < d_1 \leq c_2 < \cdots \leq d_{p-1} \leq c_p < d_p \leq b$ of the interval $[a, b]$ about which the following is true: $v(t)$ performs one complete revolution, in the precise sense defined above, over each of the intervals $[c_k, d_k]$ — and no others. This guarantees, in particular, that the total number of such complete revolutions is finite.

Each of the terms in our statement describing the relationship between $n(\gamma, z)$ and the revolutions of $v(t)$ is now meaningful. (Incidentally, many books formulate this assertion somewhat differently, saying that $2\pi n(\gamma, z)$ represents the "net change in the argument of $\gamma(t) - z$ over $[a, b]$." We just find it easier to conceive of vectors turning than of arguments changing.) To be sure, the statement cries out for justification. Since that justification is quite technical — and would interrupt the flow of ideas in the present chapter, were it to be included here — we relegate it to Appendix B, along with certain other technical information concerning winding numbers. Maximum advantage will be gained here by using the geometric interpretation to determine winding numbers in some concrete examples. When finding $n(\gamma, z)$ geometrically it is always useful to draw the ray R issuing from z and passing through $\gamma(a)$, and then to mark the points where $|\gamma|$ intersects R. Only as $\gamma(t)$ traverses a "component" of $|\gamma| \sim R$ is it possible for $v(t)$ to undergo a complete revolution. Thus, to calculate $n(\gamma, z)$ for the path $\gamma : [a, b] \to \mathbb{C}$ pictured in Figure 11 we monitor the behavior of the vector $v(t)$ as $\gamma(t)$ moves from one point of $|\gamma| \cap R$ to the next and observe that $v(t)$ performs positive revolutions over $[c_1, d_1]$ and $[c_2, b]$, but executes no other complete revolutions. Therefore $n(\gamma, z) = 2$. Further examples are indicated in Figure 12.

The following lemma lists several key properties of winding numbers. We remind the reader that the trajectory of a path γ in the complex plane is a closed and bounded set. The set $U = \mathbb{C} \sim |\gamma|$ is a non-empty open set. As such, it is the disjoint union of domains, the components of U (Theorem II.3.5). Since $|\gamma|$ is bounded, exactly one of these components is unbounded. (This fact actually surfaces in the proof of Lemma 2.1(ii).) The assertion in Lemma 2.1(iii) assumes the validity of the Jordan curve theorem. The rather lengthy proof of this portion of the lemma is dispatched to Appendix B.

Lemma 2.1. *Let γ be a closed, piecewise smooth path in the complex plane and let $U = \mathbb{C} \sim |\gamma|$. Then:*

(i) $n(\gamma, z)$ remains constant as z varies over any component of U;

(ii) $n(\gamma, z) = 0$ for any z belonging to the unbounded component of U;

(iii) when γ is simple, either $n(\gamma, z) = 1$ for every z in the bounded component of U or $n(\gamma, z) = -1$ for all such z.

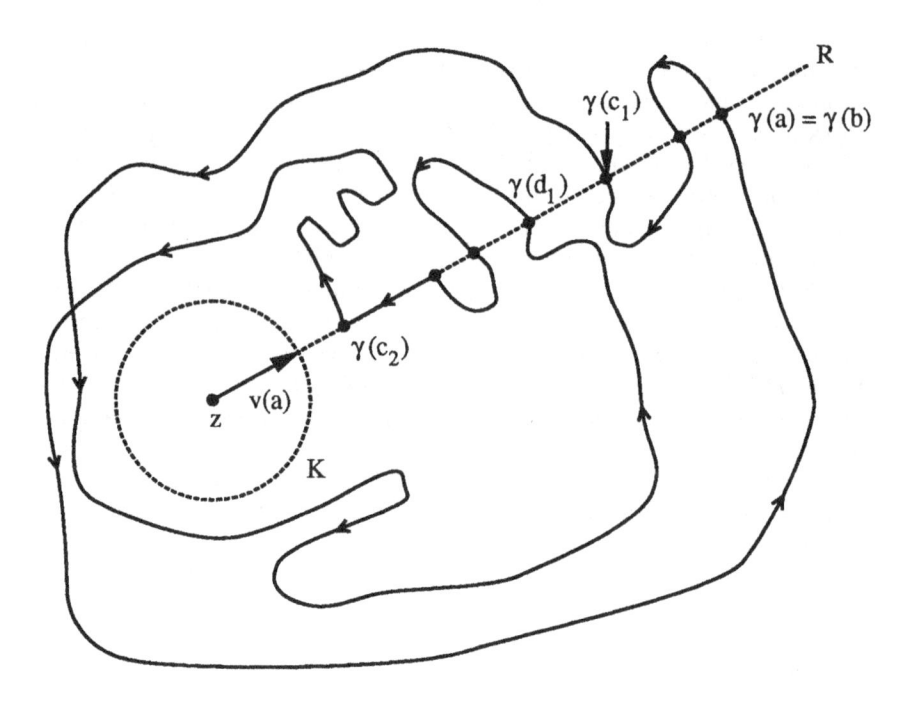

Figure 11.

Proof of (i). Applying Lemma 1.6 with $k = 1$ and with $h: |\gamma| \to \mathbb{C}$ given by $h(\zeta) = 1/(2\pi i)$, we remark that $n(\gamma, z)$ is an analytic function of the variable z in the set U, one whose derivative has the formula

$$n'(\gamma, z) = \frac{1}{2\pi i} \int_\gamma \frac{d\zeta}{(\zeta - z)^2} .$$

On the other hand, for fixed z in $\mathbb{C}$ the function $f(\zeta) = (\zeta - z)^{-2}$ is an analytic function of the variable ζ in $\mathbb{C} \sim \{z\}$ and has $F(\zeta) = -(\zeta - z)^{-1}$ as a primitive in that set. If z belongs to U, then γ is a closed path in $\mathbb{C} \sim \{z\}$. It follows from Theorem IV.2.2 that for z in U

$$\int_\gamma \frac{d\zeta}{(\zeta - z)^2} = \int_\gamma f(\zeta) \, d\zeta = 0 .$$

Thus, $n'(\gamma, z) = 0$ throughout U. Theorem III.2.4 implies that $n(\gamma, z)$ stays constant in each component of U.

Proof of (ii). We choose a number $r > 0$ so that the compact set $|\gamma|$ is contained in the disk $\Delta(0, r)$. The set $\mathbb{C} \sim \Delta(0, r)$ is a connected subset of U and, as a result, this set is contained in some component D of U (Theorem II.3.7). The specified component D is the unique unbounded

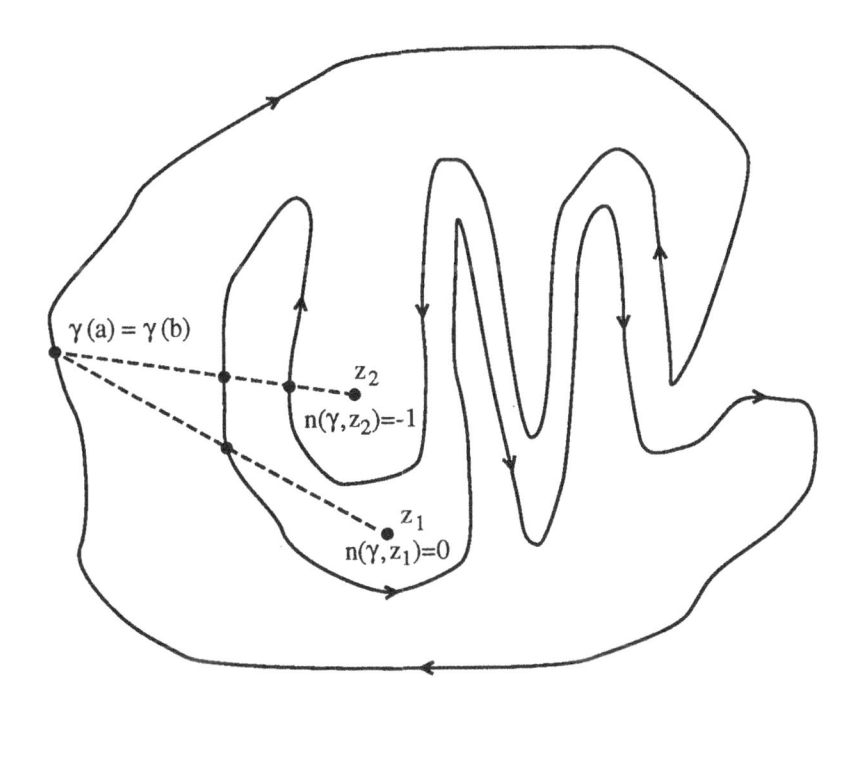

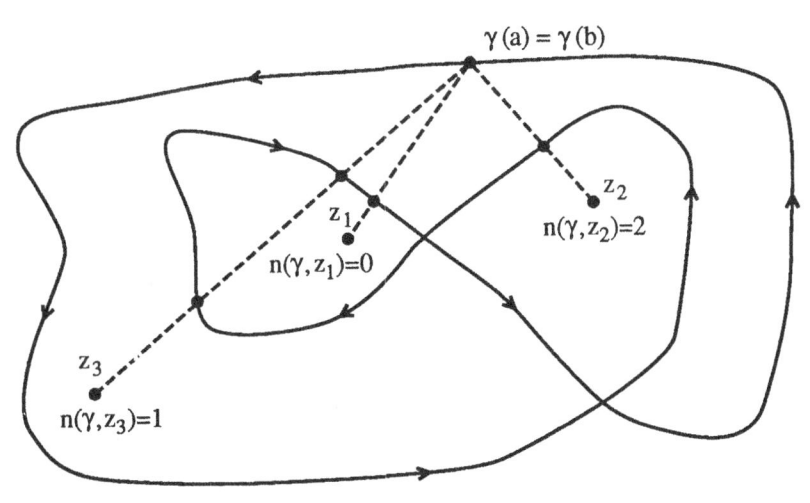

2. Winding Numbers and the Local Cauchy Integral Formula 159

Figure 12.

component of U, for all other components are clearly subsets of $\Delta(0,r)$. The point $z_0 = 2r$ lies in D and, since the function $f(\zeta) = (\zeta - z_0)^{-1}$ is analytic in $\Delta(0,r)$, we learn from the local Cauchy theorem that

$$n(\gamma, z_0) = \frac{1}{2\pi i} \int_\gamma \frac{d\zeta}{\zeta - z_0} = 0 \ .$$

Finally, (i) insures that $n(\gamma, z) = 0$ for every z in D. ∎

Here is a simple example that puts winding numbers to work.

EXAMPLE 2.1. Evaluate $\int_{\partial R} (\zeta - z)^{-1} d\zeta$, where R is a closed rectangle and z is a point not on ∂R.

By definition, $\int_{\partial R} (\zeta - z)^{-1} d\zeta = 2\pi i n(\gamma, z)$, where γ is a standard parameterization of ∂R. We abuse notation slightly and write $n(\partial R, z)$ for $n(\gamma, z)$ in this situation. There are several methods of determining $n(\partial R, z)$. For instance, we could obtain its value geometrically. Instead, we choose to use Lemma 2.1 and do the problem analytically. If the point z lies outside R, $\int_{\partial R} (\zeta - z)^{-1} d\zeta = 2\pi i n(\partial R, z) = 0$ by Lemma 2.1(ii). When z is an interior point of R, we recall Example 1.1 and appeal to Lemma 2.1(i) in computing

$$\int_{\partial R} \frac{d\zeta}{\zeta - z} = 2\pi i n(\partial R, z) = 2\pi i n(\partial R, z_0) = \int_{\partial R} \frac{d\zeta}{\zeta - z_0} = 2\pi i \ ,$$

where z_0 denotes the center of R. (N.B. We have just verified statement (iii) of Lemma 2.1 for a special case, one in which the conclusions of the Jordan curve theorem would probably not be disputed.)

2.2 Oriented Paths, Jordan Contours

Let $\gamma \colon [a, b] \to \mathbb{C}$ be a simple, closed, and piecewise smooth path. We refer to γ as *positively oriented* if $n(\gamma, z) = 1$ for every z in the inside D of the Jordan curve $J = |\gamma|$ and *negatively oriented* if $n(\gamma, z) = -1$ for all such z. Lemma 2.1(iii) justifies the terminology. The proof of Lemma 2.1(iii) found in Appendix B shows that γ is positively oriented if and only if, as J gets traced out, $\gamma(t)$ "keeps D to its left" in the following exact sense: for every t in (a, b) at which γ is differentiable with $\dot{\gamma}(t) \neq 0$ the "normal vector" $i\dot{\gamma}(t)$ to J at $\gamma(t)$ — the vector obtained by rotating the tangent vector $\dot{\gamma}(t)$ ninety degrees in a counterclockwise direction — points into D; i.e., the element $\gamma(t) + \epsilon i\dot{\gamma}(t)$ belongs to D for all sufficiently small $\epsilon > 0$. (See Figure 13.) If γ is positively oriented, it is easy to see that its reverse path is negatively oriented, and vice versa.

The term *Jordan contour* will serve in this book as a short expression for "positively oriented, simple, closed, piecewise smooth path." If γ and β are Jordan contours with the same trajectory J, it can be shown that

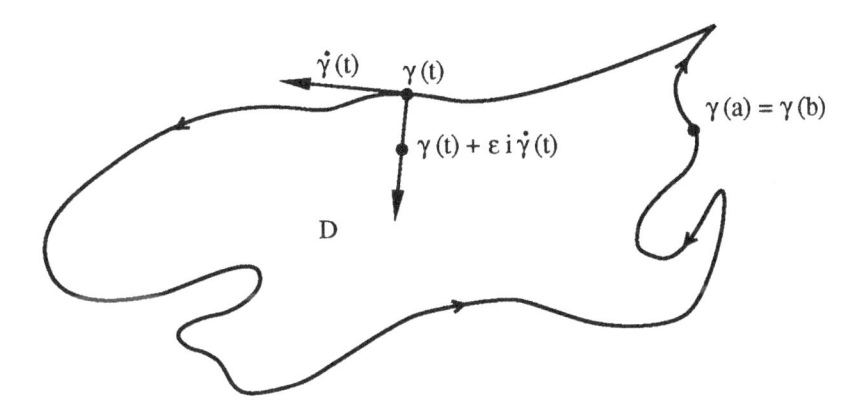

2. Winding Numbers and the Local Cauchy Integral Formula   161

Figure 13.

$\int_\gamma f(z)dz = \int_\beta f(z)dz$ for any continuous function $f: J \to \mathbb{C}$. The notation $\int_{\partial D} f(z)dz$, where D is the bounded component of $\mathbb{C} \sim J$, is used in many books to symbolize the common value of the integrals of f along γ and β in such a situation. In this book we employ it only when $\overline{D}$ is a rectangle.

2.3   The Local Integral Formula

All the pieces are now in place for the derivation of an absolutely fundamental representation formula for analytic functions, Cauchy's integral formula, from which a wealth of information about this class of functions will flow. We initially give a local version of the result, postponing the statement of its global analogue until Section 5 (Theorem 5.6).

Theorem 2.2. (Cauchy's Integral Formula – Local Form) *Suppose that a function f is analytic in an open disk Δ and that γ is a closed, piecewise smooth path in Δ. Then*

$$n(\gamma, z)f(z) = \frac{1}{2\pi i}\int_\gamma \frac{f(\zeta)\,d\zeta}{\zeta - z}$$

for every z in $\Delta \sim |\gamma|$.

Proof. Fix a point z of $\Delta \sim |\gamma|$ and consider the function $g: \Delta \to \mathbb{C}$ defined, using ζ to denote the independent variable, as follows:

$$g(\zeta) = \begin{cases} [f(\zeta) - f(z)]/[\zeta - z] & \text{if } \zeta \neq z, \\ f'(z) & \text{if } \zeta = z. \end{cases}$$

Then g is obviously analytic in $\Delta \sim \{z\}$. Moreover,

$$\lim_{\zeta \to z} g(\zeta) = \lim_{\zeta \to z} \frac{f(\zeta) - f(z)}{\zeta - z} = f'(z) = g(z),$$

so g is at least continuous at the point z. We thus find ourselves in a situation covered by Theorem 1.5. (At last a reason emerges to explain our allowing for an exceptional point in that result!) Because z lies off the trajectory of γ, we can compute

$$0 = \int_\gamma g(\zeta)\,d\zeta = \int_\gamma \frac{f(\zeta) - f(z)}{\zeta - z}\,d\zeta = \int_\gamma \frac{f(\zeta)\,d\zeta}{\zeta - z} - f(z)\int_\gamma \frac{d\zeta}{\zeta - z}$$

$$= \int_\gamma \frac{f(\zeta)\,d\zeta}{\zeta - z} - 2\pi i\,n(\gamma, z)f(z),$$

which gives

$$n(\gamma, z)f(z) = \frac{1}{2\pi i}\int_\gamma \frac{f(\zeta)\,d\zeta}{\zeta - z},$$

as asserted. ∎

An important aspect of Cauchy's integral formula is that in the case of a point z for which $n(\gamma, z) \neq 0$ it permits us to express the function value $f(z)$ in a very explicit fashion in terms of the values of f at points lying some distance away from z, points on the trajectory of γ. This fact has many valuable consequences for complex function theory, as we shall soon start to see. The local Cauchy integral formula can also serve as a direct aid in evaluating integrals. The following two examples document this statement.

EXAMPLE 2.2. Evaluate $\int_{|z|=2}(z^3 + z)^{-1}e^{\pi z}\,dz$.

We first decompose $(z^3 + z)^{-1}$ into partial fractions and multiply by $e^{\pi z}$ to arrive at

$$\frac{e^{\pi z}}{z^3 + z} = \frac{e^{\pi z}}{z} - \frac{e^{\pi z}}{2(z-i)} - \frac{e^{\pi z}}{2(z+i)}.$$

We then apply Cauchy's integral formula to the function $f(z) = e^{\pi z}$ in any open disk Δ that contains the circle $K(0,2)$ and to the path defined by $\gamma(t) = 2e^{it}$ for $0 \leq t \leq 2\pi$. We obtain

$$\int_{|z|=2}\frac{e^{\pi z}\,dz}{z^3 + z} = \int_{|z|=2}\frac{f(z)\,dz}{z} - \frac{1}{2}\int_{|z|=2}\frac{f(z)\,dz}{z-i} - \frac{1}{2}\int_{|z|=2}\frac{f(z)\,dz}{z+i}$$

$$= 2\pi i\,n(\gamma, 0)f(0) - \frac{2\pi i\,n(\gamma, i)f(i)}{2} - \frac{2\pi i\,n(\gamma, -i)f(-i)}{2}$$

$$= 2\pi i - \pi i e^{\pi i} - \pi i e^{-\pi i} = 4\pi i.$$

(N.B. Since we are concerned here with the values of f at the specific points $0, i$, and $-i$, no confusion results from retaining z as the variable of integration, rather than switching to the dummy variable ζ that appears in

2. Winding Numbers and the Local Cauchy Integral Formula

the general statement of the Cauchy integral formula. We have made use of the fact that $n(\gamma, z) = 1$ for every z in the disk $\Delta(0,2)$. This follows either geometrically or from Lemma 2.1(i) and the observation that $n(\gamma, 0) = 1$, as a simple calculation confirms.)

EXAMPLE 2.3. Evaluate $\int_0^\infty (t^2 + 1)^{-1} \cos t \, dt$.

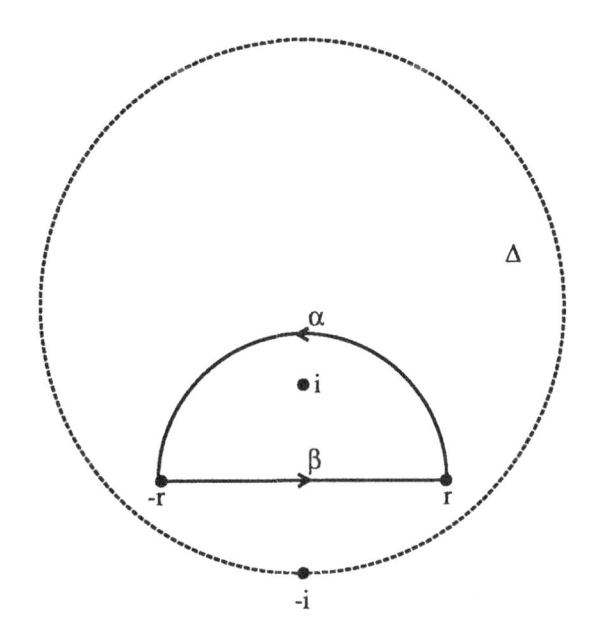

Figure 14.

Since the integrand is an even function and since the analogous function with $\sin t$ in place of $\cos t$ is odd, we remark that for $r > 0$

$$(5.11) \quad \int_0^r \frac{\cos t \, dt}{t^2 + 1} = \frac{1}{2} \int_{-r}^r \frac{\cos t \, dt}{t^2 + 1} + \frac{i}{2} \int_{-r}^r \frac{\sin t \, dt}{t^2 + 1} = \frac{1}{2} \int_{-r}^r \frac{e^{it} \, dt}{t^2 + 1}.$$

This prompts us to consider the analytic function $g(z) = (z^2 + 1)^{-1} e^{iz}$. We shall also want to think of $g(z)$ as $(z - i)^{-1} f(z)$, where $f(z) = (z + i)^{-1} e^{iz}$. The secret to working the problem is to integrate g along the path $\gamma = \beta + \alpha$, where β and α are depicted in Figure 14. Specifically, $\beta(t) = t$ for $-r \leq t \leq r$ and $\alpha(t) = re^{it}$ for $0 \leq t \leq \pi$. It is assumed here that $r > 1$. Using (5.11), we compute

$$\int_0^r \frac{\cos t \, dt}{t^2 + 1} = \frac{1}{2} \int_\beta \frac{e^{iz} \, dz}{z^2 + 1} = \frac{1}{2} \int_\gamma \frac{e^{iz} \, dz}{z^2 + 1} - \frac{1}{2} \int_\alpha \frac{e^{iz} \, dz}{z^2 + 1}$$

$$= \frac{1}{2}\int_\gamma \frac{f(z)dz}{z-i} - \frac{1}{2}\int_\alpha \frac{e^{iz}\,dz}{z^2+1} = \pi i\, n(\gamma,i)f(i) - \frac{1}{2}\int_\alpha \frac{e^{iz}\,dz}{z^2+1}$$

$$= \frac{\pi}{2e} - \frac{1}{2}\int_\alpha \frac{e^{iz}\,dz}{z^2+1}.$$

In this calculation we have applied the local Cauchy integral formula to the function f in a disk Δ that encompasses $|\gamma|$, but does not contain the point $-i$. The fact that $n(\gamma,i) = 1$ follows, for instance, from Lemma 2.1 (iii). We now call upon the information that

$$\left|\int_\alpha \frac{e^{iz}\,dz}{z^2+1}\right| \leq \int_\alpha \frac{|e^{iz}||dz|}{|z^2+1|} \leq \int_\alpha \frac{e^{-\operatorname{Im} z}|dz|}{|z|^2-1}$$

$$= \frac{r}{r^2-1}\int_0^\pi e^{-r\sin t}\,dt \leq \frac{\pi r}{r^2-1} \to 0$$

as $r \to \infty$. Here we've appealed to the inequality $|z^2 - 1| \geq |z|^2 - 1$ and taken advantage of the fact that $e^{-r\sin t} \leq 1$ when $0 \leq t \leq \pi$. Accordingly,

$$\int_0^\infty \frac{\cos t\,dt}{t^2+1} = \lim_{r\to\infty}\int_0^r \frac{\cos t\,dt}{t^2+1} = \frac{\pi}{2e}.$$

3 Consequences of the Local Cauchy Integral Formula

3.1 Analyticity of Derivatives

Already the first noteworthy theoretical conclusion we draw from Cauchy's integral formula is quite striking. It begins to expose the glaring differences between "complex calculus" and "real calculus."

Theorem 3.1. *If a function f is analytic in an open set U, then f' is also analytic in U. In particular, f belongs to the class $C^1(U)$.*

Proof. It plainly suffices to demonstrate that each point z_0 of U is the center of some open disk D in which f' is analytic, for then $f''(z_0) = (f')'(z_0)$ definitely exists. We fix z_0 and initially choose $r > 0$ so that the disk $\Delta = \Delta(z_0, r)$ lies in U. We next fix s satisfying $0 < s < r$ and set $D = \Delta(z_0, s)$. The local Cauchy integral formula, applied in the disk Δ to the function f and the path γ given by $\gamma(t) = z_0 + se^{it}$ for $0 \leq t \leq 2\pi$, permits us to represent $f(z)$ for z belonging to D in the form

$$f(z) = \frac{1}{2\pi i}\int_\gamma \frac{f(\zeta)\,d\zeta}{\zeta - z}.$$

3. Consequences of the Local Cauchy Integral Formula

(We appeal here to Lemma 2.1(i) to certify that

$$n(\gamma, z) = n(\gamma, z_0) = \frac{1}{2\pi i} \int_0^{2\pi} \frac{i r e^{it} \, dt}{r e^{it}} = 1$$

for every z in D.) The case $k = 1$ of Lemma 1.6 — in the present situation we take $h = f/(2\pi i)$ on $|\gamma| = \partial D$ and have $H = f$ in D — implies that

$$f'(z) = \frac{1}{2\pi i} \int_\gamma \frac{f(\zeta) \, d\zeta}{(\zeta - z)^2}$$

for all z in D. The case $k = 2$ of the same lemma informs us that the expression on the right-hand side of the last equation defines an analytic function of z in the set $\mathbb{C} \sim |\gamma|$. It thereby insures the analyticity of f' in D, an open subset of $\mathbb{C} \sim |\gamma|$.

The final assertion in the theorem follows from the fact that $f_x = f'$ and $f_y = if'$ in U, certainly implying that these partial derivatives are continuous there and thus placing f in the class $C^1(U)$. ∎

A simple induction argument leads from Theorem 3.1 to the following corollary (Exercise 8.15).

Corollary 3.2. *If a function f is analytic in an open set U, then it can be differentiated arbitrarily often in U and all its derivatives $f', f'', \ldots, f^{(k)}, \ldots$ are analytic there. In particular, f belongs to the class $C^\infty(U)$.*

A second consequence of Theorem 3.1 amounts to a converse of Lemma 1.1. It is attributed to Giacinto Morera (1856-1909).

Theorem 3.3. (Morera's Theorem) *Let a function f be continuous in an open set U. Assume that $\int_{\partial R} f(z) dz = 0$ for every closed rectangle R in U whose sides are parallel to the coordinate axes. Then f is analytic in U.*

Proof. It is enough to verify that f is analytic in each open disk contained in U. Fix such a disk Δ. In Δ the function f satisfies all the hypotheses of Lemma 1.3. According to that result, f possesses a primitive in Δ, say F. Being in Δ the derivative of the analytic function F, f must itself be analytic in this disk. ∎

For future reference we record the outcome when Morera's theorem is combined with with Lemma 1.2.

Theorem 3.4. *Suppose that a function f is continuous in an open set U and analytic in $U \sim \{z_0\}$ for some point z_0 of U. Then f is analytic in U.*

Theorem 3.4 anticipates a more general result, the "Riemann Extension Theorem" (Theorem VIII.2.1), which we shall encounter in the process of studying singularities of analytic functions.

The local Cauchy integral formula generalizes to provide representations not only for an analytic function, but for its derivatives as well.

Theorem 3.5. *Suppose that a function f is analytic in an open disk Δ, that k is a non-negative integer, and that γ is a closed, piecewise smooth path in Δ. Then*

$$n(\gamma,z)f^{(k)}(z) = \frac{k!}{2\pi i}\int_\gamma \frac{f(\zeta)\,d\zeta}{(\zeta-z)^{k+1}}$$

for every z in $\Delta \sim |\gamma|$.

Proof. The proof is by induction. When $k = 0$ the stated formula has already been established for all functions f that are analytic in Δ: since $f^{(0)}$ means f, this is just Cauchy's integral formula. We now assume that it has been verified for all such f in the case of an integer k, and we consider the situation for $k + 1$. Fix z in $\Delta \sim |\gamma|$. If f is analytic in Δ, so is its derivative f'. Accordingly, for any function f of this type we have

$$(5.12) \quad n(\gamma,z)f^{(k+1)}(z) = n(\gamma,z)(f')^{(k)}(z) = \frac{k!}{2\pi i}\int_\gamma \frac{f'(\zeta)\,d\zeta}{(\zeta-z)^{k+1}}$$

by the induction hypothesis applied to f'. Now the function g defined in the set $\Delta \sim \{z\}$ by

$$g(\zeta) = \frac{f(\zeta)}{(\zeta-z)^{k+1}}$$

is an analytic function and has derivative

$$(5.13) \quad g'(\zeta) = \frac{f'(\zeta)}{(\zeta-z)^{k+1}} - \frac{(k+1)f(\zeta)}{(\zeta-z)^{k+2}}\,.$$

Since g' is continuous in $\Delta \sim \{z\}$ and obviously has a primitive there, Theorem IV.2.2 lets us know that

$$\int_\gamma g'(\zeta)\,d\zeta = 0\,.$$

We infer using (5.12) and (5.13) that

$$n(\gamma,z)f^{(k+1)}(z) = \frac{k!}{2\pi i}\int_\gamma \frac{f'(\zeta)\,d\zeta}{(\zeta-z)^{k+1}}$$

$$= \frac{k!}{2\pi i}\int_\gamma g'(\zeta)\,d\zeta + \frac{(k+1)!}{2\pi i}\int_\gamma \frac{f(\zeta)\,d\zeta}{(\zeta-z)^{k+2}} = \frac{(k+1)!}{2\pi i}\int_\gamma \frac{f(\zeta)\,d\zeta}{(\zeta-z)^{k+2}}\,.$$

Because z was an arbitrary point of $\Delta \sim |\gamma|$ and f an arbitrary analytic function in Δ, we have verified the announced formula for $k + 1$ under the assumption that it is valid for k. An appeal to the Principle of Mathematical Induction confirms the formula for all k and so completes the proof. ∎

We use Theorem 3.5 to work an example.

3. Consequences of the Local Cauchy Integral Formula

EXAMPLE 3.1. Evaluate $\int_{\partial Q} z^{-3}(e^z + z^2 \sin z)dz$, where Q is the square bounded by the curve with equation $|x| + |y| = 1$.

We apply Theorem 3.5 for the case $k = 2$ to the entire function $f(z) = e^z + z^2 \sin z$ in any open disk Δ that contains Q. This gives — recall Example 2.1 —

$$\int_{\partial Q} \frac{(e^z + z^2 \sin z)\, dz}{z^3} = \int_{\partial Q} \frac{f(z)\, dz}{(z-0)^3} = \frac{2\pi i\, n(\partial Q, 0) f''(0)}{2!} = \pi i\,.$$

3.2 Derivative Estimates

Among the virtues of Theorem 3.5 is the power it gives one to pass from an estimate for an analytic function to estimates for its derivatives. The next result offers a standard illustration of this phenomenon.

Theorem 3.6. *Suppose that a function f is analytic in an open disk $\Delta = \Delta(z_0, r)$ and that $|f(z)| \leq m$ holds throughout Δ, where m is a constant. Then for each positive integer k the estimate*

$$\left|f^{(k)}(z)\right| \leq \frac{k!\, mr}{(r - |z - z_0|)^{k+1}}$$

is valid for every z in Δ. In particular, $|f^{(k)}(z_0)| \leq k!\, m r^{-k}$.

Proof. For a fixed z in Δ, let s satisfy $|z - z_0| < s < r$. If ζ is situated on the circle $K(z_0, s)$, then

$$|\zeta - z| = |(\zeta - z_0) + (z_0 - z)| \geq |\zeta - z_0| - |z - z_0| = s - |z - z_0|\,.$$

We apply Theorem 3.5 to estimate $\left|f^{(k)}(z)\right|$ as follows:

$$\left|f^{(k)}(z)\right| = \left|\frac{k!}{2\pi i} \int_{|\zeta - z_0|=s} \frac{f(\zeta)\, d\zeta}{(\zeta - z)^{k+1}}\right| \leq \frac{k!}{2\pi} \int_{|\zeta - z_0|=s} \frac{|f(\zeta)|\, |d\zeta|}{|\zeta - z|^{k+1}}$$

$$\leq \frac{k!}{2\pi} \int_{|\zeta - z_0|=s} \frac{m\, |d\zeta|}{(s - |z - z_0|)^{k+1}} = \frac{k!\, ms}{(s - |z - z_0|)^{k+1}}\,.$$

(N.B. As in the proof of Theorem 3.1, $n(\gamma, z) = 1$ for the path γ involved here.) We now let $s \to r$ to arrive at the desired inequality. ∎

The derivative bounds furnished by Theorem 3.6 are often referred to as "Cauchy's estimates." Even when $k = 1$ Theorem 3.6 has some surprising implications. A case in point is a well-known theorem of Joseph Liouville (1809-1882).

Theorem 3.7. (Liouville's Theorem) *The only bounded entire functions are the constant functions on $\mathbb{C}$.*

Proof. Suppose that f, an entire function, satisfies $|f(z)| \leq m$ for all z, where m is a constant. For a fixed complex number z and for any $r > |z|$ we can apply Theorem 3.6 to f in the disk $\Delta = \Delta(0,r)$ to conclude that $|f'(z)| \leq mr/(r-|z|)^2$. Letting $r \to \infty$, we deduce that $f'(z) = 0$. Since this is true for every z in $\mathbb{C}$, Theorem III.2.4 shows that f is constant in the complex plane. ∎

Liouville's theorem can be viewed as a statement about the range of a non-constant entire function: the range of such a function cannot be a bounded set. This fact admits many refinements and variations, most of them subsumed by a deep and beautiful theorem of Émile Picard (1856-1941): *if f is a non-constant entire function, then the set $\mathbb{C} \sim f(\mathbb{C})$ contains at most one point!* (For further discussion of this result, see Chapter VIII.) The succeeding example, for instance, can be modified to demonstrate that the range of a non-constant entire function is never contained in a half-plane. This piece of information constitutes an improvement of Liouville's theorem, but obviously falls far short of Picard's.

EXAMPLE 3.2. Let $f = u + iv$ be an entire function with the property that $u(z) \leq 0$ for every z. Show that f is a constant function.

We consider the new entire function $g = e^f$. By assumption

$$|g(z)| = e^{u(z)} \leq e^0 = 1$$

for every z. Liouville's theorem tells us that g must be a constant function, say $g(z) = c$ for all z. There are a number of ways to infer the constancy of f from that of g. (Hastily writing $f(z) = \log c$ is not, however, one of these ways! A correct solution must rule out the possibility that $f(z)$ runs through many different logarithms of c as z varies.) For instance, one argument notes that $0 = g'(z) = f'(z)e^{f(z)}$ throughout $\mathbb{C}$, which implies that $f'(z) = 0$ for every z and, therefore, that f is constant in $\mathbb{C}$.

The unpretentious appearance of Liouville's theorem belies its importance. For example, the theorem numbers among its corollaries a celebrated discovery of Carl Friedrich Gauss (1777-1855) — as a matter of fact, the subject of his doctoral dissertation — the so-called "Fundamental Theorem of Algebra." The reader is certain to recall this theorem from his or her high school advanced algebra text, where its statement was no doubt accompanied by a footnote to the effect that its proof was beyond the scope of the book. We now derive this result from Liouville's theorem. Underlying the argument is the easy-to-believe remark that if p, a polynomial function of z, has positive degree, then p is not a constant function. This fact is implicit in the observation, contained in the proof which follows, that $|p(z)| \to \infty$ as $|z| \to \infty$.

Theorem 3.8. (Fundamental Theorem of Algebra) *Any polynomial function $p(z) = a_0 + a_1 z + \cdots + a_n z^n$ of degree $n \geq 1$ has a root in $\mathbb{C}$.*

3. Consequences of the Local Cauchy Integral Formula

Proof. We wish to show that $p(z_0) = 0$ for some complex number z_0. Suppose that this is not so. Then $f(z) = 1/p(z)$ defines a non-constant entire function f. Since p is of degree n, $a_n \neq 0$. It follows with the aid of the triangle inequality that for $z \neq 0$

$$|p(z)| = |a_0 + a_1 z + \cdots + a_n z^n|$$

$$= |z|^n \left| \frac{a_0}{z^n} + \cdots + \frac{a_{n-1}}{z} + a_n \right| \geq |z|^n \left(|a_n| - \frac{|a_0|}{|z|^n} - \cdots - \frac{|a_{n-1}|}{|z|} \right).$$

From this inequality we are able to infer that $|p(z)| \to \infty$ as $|z| \to \infty$, which means that $|f(z)| \to 0$ as $|z| \to \infty$. We can thus fix $r > 0$ so that $|f(z)| \leq 1$ is true whenever $|z| \geq r$. As f is continuous — hence, bounded — on the compact set $\overline{\Delta}(0, r)$, there is a constant m, which we may take to be at least 1, such that $|f(z)| \leq m$ when $|z| \leq r$. In other words, $|f(z)| \leq m$ holds for every z in $\mathbb{C}$. Liouville's theorem asserts that f is constant after all! This contradiction forces us to conclude that p must, in fact, have a root in $\mathbb{C}$. ∎

The fundamental theorem of agebra is frequently quoted in the form of Theorem 3.9, which can be obtained from Theorem 3.8 by a straightforward induction argument based on the elementary "Factor Theorem": *if z_0 is a root of a polynomial $p(z)$, then $z - z_0$ is a factor of $p(z)$.* (Recall Exercise I.4.42.) We leave the proof of Theorem 3.9 as an exercise (Exercise 8.34).

Theorem 3.9. *A polynomial function $p(z) = a_0 + a_1 z + \cdots + a_n z^n$ of degree $n \geq 1$ has a factorization $p(z) = c(z - z_1)(z - z_2) \cdots (z - z_n)$, in which $z_1, z_2, \ldots, z_n$ are the roots of p and c is a constant.*

It is not assumed that the roots $z_1, z_2, \ldots, z_n$ of p listed in Theorem 3.9 are distinct. The factorization may well end up looking like $p(z) = iz(z-1)^2(z+i)^5$. In this example we would say that p has 0 as a *simple root*, 1 as a *double root* (or, *root of multiplicity* 2), and $-i$ as a *root of multiplicity* 5. In general, the factors provided by Theorem 3.9 can be grouped so as to represent p in the form

$$p(z) = c(z - z_1)^{m_1}(z - z_2)^{m_2} \cdots (z - z_r)^{m_r},$$

where now $z_1, z_2, \ldots, z_r$ are the different roots of p and $m_1, m_2, \ldots, m_r$ are positive integers that indicate the multiplicities of the respective roots. We also emphasize that Theorems 3.8 and 3.9 are pure existence theorems — neither gives us a clue about how to find the roots of p. The latter problem can be, to say the least, an extremely taxing one. More often than not one must settle for approximations of the roots, approximations arrived at by numerical methods.

3.3 The Maximum Principle

The final topic we take up in conjunction with the local version of Cauchy's integral formula is a result commonly known as the "Maximum Principle" (or "Maximum Modulus Theorem") for analytic functions.

Theorem 3.10. (Maximum Principle) *Let a function f be analytic in a domain D. Suppose that there exists a point z_0 of D with the property that $|f(z)| \leq |f(z_0)|$ for every z in D. Then f is constant in D.*

Proof. We consider the continuous function $w: D \to \mathbb{R}$ given by $w(z) = |f(z)|$. In view of Theorem III.2.5 it is enough to prove that w is constant in D. In doing this, we take advantage of the following property enjoyed by w: corresponding to each point z of D there exists a radius $\rho = \rho(z) > 0$ such that

$$(5.14) \qquad w(z) \leq \frac{1}{2\pi} \int_0^{2\pi} w(z + re^{it}) \, dt$$

holds whenever $0 < r < \rho$. Indeed, if $\rho > 0$ is chosen so that the open disk $\Delta(z, \rho)$ is contained in D, the Cauchy integral formula leads immediately to

$$w(z) = |f(z)| = \left| \frac{1}{2\pi i} \int_{|\zeta - z| = r} \frac{f(\zeta) \, d\zeta}{\zeta - z} \right|$$

$$\leq \frac{1}{2\pi} \int_{|\zeta - z| = r} \frac{|f(\zeta)| \, |d\zeta|}{|\zeta - z|} = \frac{1}{2\pi} \int_0^{2\pi} w(z + re^{it}) \, dt$$

for $0 < r < \rho$.

Proceeding now to the proof of the theorem, we write $m = w(z_0)$ and define sets U and V by

$$U = \{z \in D : w(z) = m\} \quad , \quad V = \{z \in D : w(z) < m\}.$$

We show that U and V are open sets.

Let z be a point of U, and let $\rho > 0$ be such that (5.14) is true whenever $0 < r < \rho$. For fixed r in this range we observe that, since z belongs to U and since by assumption $w \leq m$ everywhere in D,

$$m = w(z) \leq \frac{1}{2\pi} \int_0^{2\pi} w(z + re^{it}) \, dt \leq \frac{1}{2\pi} \int_0^{2\pi} m \, dt = m \, .$$

The presence of the same term at the extremes of this string of inequalities forces equality to hold throughout it. As a consequence, we see that

$$(5.15) \qquad \frac{1}{2\pi} \int_0^{2\pi} [m - w(z + re^{it})] \, dt = 0 \, .$$

3. Consequences of the Local Cauchy Integral Formula

The integrand in (5.15) is a non-negative continuous function of t on the interval $[0, 2\pi]$. It follows from (5.15) and from basic facts in calculus that this integrand must vanish identically on that interval. Thus $w(z + re^{it}) = m$ (i.e., $z + re^{it}$ lies in U) for $0 \leq t \leq 2\pi$. Put in a different way, the circle of radius r centered at z is a subset of U. Because this is the case for every r in the interval $(0, \rho)$, we conclude that the whole open disk $\Delta(z, \rho)$ is a subset of U. As z was an arbitrary point of U, the inference that U is an open set is clear.

Next, suppose that z is a member of V. Then $w(z) < m$. Owing to the continuity of w at z, we can choose corresponding to $\epsilon = m - w(z) > 0$ an open disk $\Delta = \Delta(z, r)$ in D with the property that $|w(z') - w(z)| < \epsilon$ whenever z' is a point of Δ. Accordingly,

$$w(z') = w(z') - w(z) + w(z) < \epsilon + w(z) = m$$

for each such z', which demonstrates that Δ is contained V. In this way V, too, is seen to be open.

It is evident that $D = U \cup V$ and that $U \cap V = \phi$. Also, $U \neq \phi$, for by the very definition of this set the point z_0 belongs to it. The connectedness of D — we refer to Theorem II.3.4 — guarantees that $V = \phi$, which means that $w(z) = m$ for every z in D. The appeal to Theorem III.2.5 finishes the proof. ∎

We can paraphrase Theorem 3.10 by saying that the modulus of a function which is analytic and non-constant in a domain D cannot assume a maximum value in D. The maximum principle has numerous other reformulations, of which the following is an especially useful sample.

Corollary 3.11. *Let D be a bounded domain in the complex plane, and let $f: \overline{D} \to \mathbb{C}$ be a continuous function that is analytic in D. Then $|f(z)|$ reaches its maximum at some point on the boundary of D.*

Proof. The continuous function $|f|$ certainly attains a maximum value at some point of the compact set $\overline{D}$ (Corollary II.4.7). If this maximum is achieved at a point of D, then, due to the maximum principle, the function f is constant in D — hence, by continuity, constant on $\overline{D}$. In this case $|f|$ assumes its maximum value, its only value, at each point of ∂D. In the alternative case where $|f|$ does not assume its maximum value at any point of D, this value can only be taken at a point or points of $\partial D = \overline{D} \sim D$. ∎

EXAMPLE 3.3. Let $f(z) = z^2 - 2z$. Compute the maximum value of $|f(z)|$ on the square $Q = \{z : 0 \leq x, y \leq 1\}$.

By Corollary 3.11 the maximum of $|f(z)|$ on Q is attained somewhere on ∂Q. We consider the sides of Q separately. On its bottom edge we have $|f(x + iy)| = |f(x)| = 2x - x^2$, which attains a maximum of 1 at $x = 1$. On the left side, $|f(x + iy)| = |f(iy)| = \sqrt{y^4 + 4y^2}$, yielding a maximum value $\sqrt{5}$ at i. As to the right side, $|f(x + iy)| = |f(1 + iy)| = 1 + y^2$, a quantity maximized by 2 at the upper right-hand corner of Q. Finally, along the top edge of Q, $|f(x+iy)| = |f(x+i)| = \sqrt{x^4 - 4x^3 + 6x^2 - 4x + 5}$, which achieves its maximum value $\sqrt{5}$ over the interval $[0,1]$ when $x = 0$. Conclusion: the maximum value of $|f(z)|$ on Q is $\sqrt{5}$.

Apart from the invocation of Theorem III.2.5 the proof given for Theorem 3.10 owed very little to the analyticity of f per se. It really hinged on two properties of the function $w = |f|$: (i) w is continuous in D and (ii) corresponding to each point z of D there exists a radius $\rho = \rho(z) > 0$ such that
$$w(z) \leq \frac{1}{2\pi} \int_0^{2\pi} w(z + re^{it})\, dt$$
holds whenever $0 < r < \rho$. (In the proof we were able to choose any ρ for which the open disk $\Delta(z,\rho)$ is contained in D, but this was in no way critical to the argument.) These two properties implied the openness of the sets U and V that entered into the proof. The connectedness of D did the rest. The upshot of this comment is that Theorem 3.10 and its corollary generalize naturally to results about a class of functions much broader than the moduli of analytic functions in D — namely, the class of real-valued functions w that are continuous in D and are endowed with property (ii). For reasons that will not become clear until the next chapter, which treats harmonic functions, any continuous function w that exhibits property (ii) is said to be *subharmonic in* D. (In the case of a function $w: D \to \mathbb{R}$ known to be in the class $C^2(D)$, the statement that w is *harmonic in* D means that $w_{xx} + w_{yy} = 0$ throughout D, whereas w turns out to be subharmonic in D if it has $w_{xx} + w_{yy} \geq 0$ everywhere in this domain.) Since we can use the "subharmonic maximum principle" and its corollary to great advantage in studying harmonic functions, we seize the opportunity here to state these by-products of the proof of Theorem 3.10 explicitly.

Theorem 3.12. *Let a continuous function w be subharmonic in a domain D. Suppose that there is a point z_0 of D with the property that $w(z) \leq w(z_0)$ for every z in D. Then w is constant in D.*

Corollary 3.13. *Let D be a bounded domain in the complex plane, and let $w: \overline{D} \to \mathbb{R}$ be a continuous function that is subharmonic in D. Then $w(z)$ attains its maximum at some point on the boundary of D.*

The maximum principle is a powerful tool for obtaining bounds on the size of analytic functions. We close this section by describing two instances in which it leads to quite handy estimates. The first of these is a result usually associated with the name of Hermann Schwarz (1843-1921), although the credit for its now classic formulation should rightfully go to Constantin Carathéodory (1873-1950).

Theorem 3.14. (Schwarz's Lemma) *Suppose that a function f is analytic in the disk $\Delta = \Delta(0,1)$ and that it obeys the conditions $f(0) = 0$ and*

3. Consequences of the Local Cauchy Integral Formula

$|f(z)| \leq 1$ for every z in Δ. Then $|f'(0)| \leq 1$ and $|f(z)| \leq |z|$ for every z in Δ. Furthermore, unless f happens to be a function of the type $f(z) = cz$ in Δ, where c is a complex constant of modulus one, it is actually true that $|f'(0)| < 1$ and that $|f(z)| < |z|$ when $0 < |z| < 1$.

Proof. The function $g: \Delta \to \mathbb{C}$ defined by $g(z) = f(z)/z$ when $0 < |z| < 1$ and $g(0) = f'(0)$ is continuous in Δ and analytic in the punctured disk $\Delta^*(0,1)$. Referring to Theorem 3.4, we conclude that g is analytic in Δ. We fix z in Δ and consider any r satisfying $|z| < r < 1$. Corollary 3.11, applied to g in the closed disk $\overline{\Delta}(0,r)$, gives

$$|g(z)| \leq \max_{|\zeta|=r} |g(\zeta)| = \max_{|\zeta|=r} \frac{|f(\zeta)|}{|\zeta|} \leq \frac{1}{r}.$$

Letting r tend to 1, we infer that $|g(z)| \leq 1$. Since z was an arbitrary point of Δ, that estimate is valid throughout this disk. In particular, we see that $|f'(0)| = |g(0)| \leq 1$ and that $|f(z)| = |g(z)||z| \leq |z|$ for every z in Δ. The maximum principle goes a step further and asserts that $|g(z)| < 1$ for every z in Δ, except in the circumstance that g is a constant function whose value has modulus one. This translates to the statement that $|f'(0)| < 1$ and that $|f(z)| < |z|$ when $0 < |z| < 1$, unless $f(z) = cz$ in Δ for some complex number c with $|c| = 1$. ∎

Our second application of the maximum principle delivers bounds for functions that are analytic in an infinite strip. The theorem is due to Jacques Hadamard (1865-1963).

Theorem 3.15. (Hadamard's Three-Lines Theorem) *Let S be the set $\{z: 0 < \operatorname{Re} z < 1\}$, and let $f: \overline{S} \to \mathbb{C}$ be a bounded, continuous function that is analytic in S. Suppose that $|f(iy)| \leq m_0$ and $|f(1+iy)| \leq m_1$ for all real y, where m_0 and m_1 are constants. Then*

$$|f(x+iy)| \leq m_0^{1-x} m_1^x$$

for all real y, whenever $0 < x < 1$.

Proof. There are a number of subtleties at play here, not the least of which is that the set S is unbounded. This means that Corollary 3.11 does not directly apply. The secret of the proof is to reduce the given situation to one covered by that corollary. This we accomplish in stages.

Case 1: $m_0 = m_1 = 1$. What the theorem asks us to prove in this case is that $|f(z)| \leq 1$ for every z in S. By assumption f is bounded on $\overline{S}$; i.e., there is some constant m such that $|f(z)| \leq m$ for all z in $\overline{S}$. We would like to be able to take $m = 1$. To show that this is possible, we fix z_0 in S and $t > 0$. We first demonstrate that $|1+tz_0|^{-1}|f(z_0)| \leq 1$. If we then let $t \to 0$ we arrive at $|f(z_0)| \leq 1$, the inequality desired. The auxiliary function $g: \overline{S} \to \mathbb{C}$ defined by $g(z) = (1+tz)^{-1} f(z)$ is continuous, and it is analytic in S. Now $|g(z)| = |1+tz|^{-1}|f(z)|$, which implies that $|g(z)| \leq m|1+tz|^{-1}$ for all z in $\overline{S}$. Also, $|g(z)| \leq |f(z)| \leq 1$ on the boundary of S, for obviously $|1+tz| \geq \text{Re}(1+tz) \geq 1$ when z belongs to ∂S. The big advantage that g has over f is the availability of an estimate for $|g(z)|$ in terms of $|y|$:

$$|g(x+iy)| \leq \frac{m}{|1+t(x+iy)|} \leq \frac{m}{t|y|}.$$

This inequality enables us to choose a number $b > 0$, which we are free to make larger than $|\text{Im } z_0|$, such that $|g(z)| \leq 1$ holds for every point $z = x+iy$ of $\overline{S}$ with $|y| = b$. For example, $b = \max\{m/t, 1+|\text{Im } z_0|\}$ will do. We now apply Corollary 3.11 to the function g in the domain $D = \{z: 0 < x < 1, |y| < b\}$ (Figure 15). By construction z_0 lies in D and

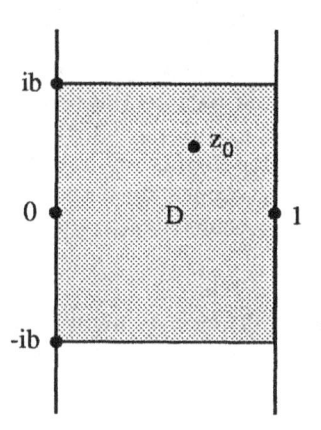

Figure 15.

$|g(z)| \leq 1$ on ∂D, so Corollary 3.11 gives

$$\frac{|f(z_0)|}{|1+tz_0|} = |g(z_0)| \leq 1.$$

This statement is true for every $t > 0$. As noted earlier, by letting $t \to 0$ it leads to $|f(z_0)| \leq 1$. Since z_0 was an arbitrary point of S, the proof of the theorem in the case $m_0 = m_1 = 1$ is achieved.

Case 2: $m_0 > 0$ and $m_1 > 0$ are arbitrary. We introduce another auxiliary function $h: \overline{S} \to \mathbb{C}$, this one being $h(z) = m_0^{z-1} m_1^{-z} f(z)$. Then h is continuous, it is analytic in S, and, because

$$|h(x+iy)| = m_0^{x-1} m_1^{-x} |f(x+iy)| \leq \max\{m_0^{-1}, m_1^{-1}\} |f(x+iy)|,$$

it is also seen to be bounded. Furthermore, $|h(iy)| \leq 1$ and $|h(1+iy)| \leq 1$ for all real y. The conclusion from Case 1 is that the estimate $|h(z)| \leq 1$ is valid throughout S or, what is the same, that $|f(x+iy)| \leq m_0^{1-x} m_1^x$ holds for $0 < x < 1$ and for all real y.

Case 3: either $m_0 = 0$ *or* $m_1 = 0$. We can apply the preceding case with m_0 and m_1 replaced by $m_0 + \epsilon$ and $m_1 + \epsilon$ for any $\epsilon > 0$ to infer that $|f(z)| \leq (m_0 + \epsilon)^{1-x}(m_1 + \epsilon)^x$ for every point z of S. Letting $\epsilon \to 0$ shows that $f(z) = 0$ throughout S, which is just what the theorem's conclusion comes down to in case $m_0 = 0$ or $m_1 = 0$. ∎

4 More About Logarithm and Power Functions

4.1 Branches of Logarithms of Functions

A windfall from Theorem 1.4 and some of the intervening ideas is a host of new analytic functions to add to our existing stock of examples. A word of caution is in order, however. It will not usually be possible to represent these new functions conveniently in closed form, which is to say by explicit formulas involving standard elementary functions. (We've already experienced a taste of the same phenomenon in dealing with branches of the logarithm and p^{th}-root functions.) This does not diminish the value of the functions. It is a simple fact that some of the most important and useful functions in mathematics are either defined implicitly or given to us in the form of an infinite series or, as is the case in the forthcoming discussion, expressed in terms of integrals that are impossible to evaluate in closed form. The key to understanding and working with such functions lies in taking advantage of any special properties they enjoy. This is definitely the recommended approach to the class of functions we are about to describe. In preparation we review a couple of relevant points from Chapter III.4.3.

Suppose that D is a plane domain and that $f : D \to \mathbb{C}$ is the identity function, the function defined by $f(z) = z$. Recall that a branch of the logarithm function in D is nothing but an analytic function $g : D \to \mathbb{C}$ that satisfies $e^{g(z)} = z = f(z)$ for every point z of D. A necessary (but not sufficient) condition for the existence of such a function g is that the origin not belong to D or, stated differently, that the function f not have its only root in D.

There is nothing sacred about the function f in the preceding comments. We can just as easily consider an arbitrary function f that is analytic in the domain D and ask the question: Does there exist an analytic function $g : D \to \mathbb{C}$ with the property that $e^g = f$ in D? Any such g is termed a *branch of the logarithm of* f (or, more succinctly, *a branch of* $\log f(z)$) *in* D. The answer to our question depends on both f and D. There can clearly be no function g of this type if f has a zero in D, but, as we have already seen with branches of $\log z$, the fact that f is zero-free in D does not by itself insure the existence of a branch of $\log f(z)$ there. A very serviceable criterion for settling the question is presented to us in the next theorem.

Theorem 4.1. *Suppose that a function f is analytic and free of zeros in a domain D. There exists a branch of $\log f(z)$ in D if and only if*

(5.16) $$\int_\gamma \frac{f'(z)\,dz}{f(z)} = 0$$

for every closed, piecewise smooth path γ in D. If g is a branch of $\log f(z)$ in D, then the collection of all such branches consists of the functions $g+2k\pi i$, where k is an integer.

Proof. Assume first that a branch g of the logarithm of f exists in D. Then $e^{g(z)} = f(z)$ throughout D. Taking derivatives on both sides of this relation leads to
$$g'(z)f(z) = g'(z)e^{g(z)} = f'(z)\,.$$
Since f never vanishes in D, we can divide by f there and conclude that $g' = f'/f$ in D. In other words, g is a primitive for the function f'/f in the given domain. Theorem IV.2.2 implies that (5.16) holds for every closed and piecewise smooth path γ in D.

Conversely, assume that the function f'/f satisfies the integral condition (5.16) in D. Theorem 1.4 permits us to find a primitive for f'/f in that domain. We choose one and call it F. We next fix a point z_0 of D and define $g: D \to \mathbb{C}$ by $g(z) = F(z) - F(z_0) + \text{Log}\, f(z_0)$. Then g is also a primitive for f'/f in D. The function $G = fe^{-g}$ is analytic in D, and its derivative satisfies
$$G'(z) = f'(z)e^{-g(z)} - f(z)e^{-g(z)}g'(z)$$
$$= f'(z)e^{-g(z)} - f(z)e^{-g(z)}[f'(z)/f(z)] = 0$$
for every z in D. As a result, G is constant in D. Moreover,
$$G(z_0) = f(z_0)e^{-g(z_0)} = f(z_0)e^{-\text{Log}\, f(z_0)} = \frac{f(z_0)}{f(z_0)} = 1\,,$$
implying that $G(z) = 1$ — or, equivalently, that $e^{g(z)} = f(z)$ — for all z in D. We have therefore produced a branch g of $\log f(z)$ in D.

Finally, if g and h are both branches of $\log f(z)$ in D, then the first part of this proof shows that $g' = h' = f'/f$ in D, so the derivative of the function $\ell = h - g$ is zero throughout the domain. This forces ℓ to be constant — say $\ell(z) = c$ — in D. Furthermore, for any z in D we have
$$e^c = e^{h(z)-g(z)} = \frac{e^{h(z)}}{e^{g(z)}} = \frac{f(z)}{f(z)} = 1\,,$$

4. More About Logarithm and Power Functions

so $c = 2k\pi i$ for some integer k. It follows that $h = g + 2k\pi i$. Since every function h of this type is plainly a branch of $\log f(z)$ in D, the last assertion of the theorem is confirmed. ∎

Under the assumption that a function f satisfies the hypotheses of Theorem 4.1 and obeys condition (5.16), the proofs of Theorems 1.4 and 4.1 allow us to exhibit in integral form a branch g of $\log f(z)$ in D: we arbitrarily select a point z_0 of D and define $g: D \to \mathbb{C}$ by

$$g(z) = \operatorname{Log} f(z_0) + \int_\alpha \frac{f'(\zeta)\,d\zeta}{f(\zeta)},$$

where α is any piecewise smooth path in D initiating at z_0 and terminating at z. (The proof of Theorem 1.4 shows that the value $g(z)$ is independent of the particular path α chosen.) As suggested in the introduction to this section, there is normally little chance that the integral representation of g can be exploited to recover anything approaching a satisfactory formula for this function. Fortunately, in many applications it is the explicit formula $g' = f'/f$ for the derivative of g, rather than a closed form expression for g itself, that supplies the crucial information.

Condition (5.16) has a nice geometric interpretation, about which we shall have more to say in Chapter VIII when we discuss the so-called "Argument Principle." For the present, we consider a zero-free function f that is analytic in a domain D and a closed, piecewise smooth path $\gamma:[a,b] \to D$. The closed path $\beta = f \circ \gamma$ — $\beta(t) = f[\gamma(t)]$ for $a \le t \le b$ — has $\dot\beta(t) = f'[\gamma(t)]\dot\gamma(t)$ for every point t of $[a,b]$ at which γ is differentiable. Since f' is continuous in D, β is thereby seen to be piecewise smooth. The relevance of β to (5.16) is shown by the calculation

$$\int_\gamma \frac{f'(z)\,dz}{f(z)} = \int_a^b \frac{f'[\gamma(t)]\dot\gamma(t)\,dt}{f[\gamma(t)]} = \int_a^b \frac{\dot\beta(t)\,dt}{\beta(t)} = \int_\beta \frac{dz}{z} = 2\pi i\, n(\beta, 0).$$

Thus, condition (5.16) merely imposes on f the requirement that it transform each closed and piecewise smooth path γ in D to a path $\beta = f \circ \gamma$ that "does not wind around the origin," i.e., to a path satisfying $n(\beta, 0) = 0$.

One potential source of confusion in dealing with logarithms of functions should be addressed. The phrase "a branch of the logarithm of f" is not to be interpreted as "a branch of the logarithm function composed with the function f." If a function f is analytic in a domain D and if L is a branch of $\log z$ in some domain that contains the set $f(D)$, then it is certainly true that $g = L \circ f$ provides a branch of $\log f(z)$ in D. What is not true, however, is that a branch of the logarithm of an arbitrary analytic and zero-free function will automatically have this simple composite structure. To realize this we need look no further than the pair of functions $f(z) = e^z$ and $g(z) = z$ in $D = \mathbb{C}$. It is a trivial observation that g is a branch of $\log f(z)$ in D. On the other hand, g cannot be written in the form $g = L \circ f$ with L a branch of $\log z$ in $f(D) = \mathbb{C} \sim \{0\}$, for we have already seen (Example III.4.8) that no such creature L exists.

4.2 Logarithms of Rational Functions

We turn at last to the examples spoken of at the beginning of this section. They arise by applying Theorem 4.1 to a rational function f. We assume that f is non-constant and in "lowest terms." The latter condition means that $f(z) = p(z)/q(z)$, where p and q are polynomial functions of z with no common factor of positive degree. By factoring p and q (Theorem 3.9) we can always arrange to express f in the manner

$$f(z) = c(z - z_1)^{m_1}(z - z_2)^{m_2} \cdots (z - z_r)^{m_r},$$

in which $c, z_1, z_2, \ldots, z_r$ are complex numbers satisfying $c \neq 0$ and $z_j \neq z_k$ for $j \neq k$, and $m_1, m_2, \ldots, m_r$ are non-zero integers. Given a domain D in the complex plane, we ask: Is there a branch of $\log f(z)$ in D? Here the answer appears in the form of a geometric condition.

Theorem 4.2. *Let* $f(z) = c(z - z_1)^{m_1}(z - z_2)^{m_2} \cdots (z - z_r)^{m_r}$, *where* $c, z_1, z_2, \ldots, z_r$ *are complex numbers satisfying* $c \neq 0$ *and* $z_j \neq z_k$ *for* $j \neq k$, *and where* $m_1, m_2, \ldots, m_r$ *are non-zero integers. There exists a branch of* $\log f(z)$ *in a plane domain* D *if and only if*

(5.17) $\qquad m_1\, n(\gamma, z_1) + m_2\, n(\gamma, z_2) + \cdots + m_r\, n(\gamma, z_r) = 0$

is true for every closed, piecewise smooth path γ *in* D.

Proof. We need only consider domains in which f is analytic and zero-free; i.e., domains that contain none of the points $z_1, z_2, \ldots, z_r$. We apply Theorem 4.1 to f in such a domain D. An easy calculation gives

$$\frac{f'(z)}{f(z)} = \frac{m_1}{z - z_1} + \frac{m_2}{z - z_2} + \cdots + \frac{m_r}{z - z_r},$$

so that

$$\int_\gamma \frac{f'(z)\,dz}{f(z)} = m_1 \int_\gamma \frac{dz}{z - z_1} + m_2 \int_\gamma \frac{dz}{z - z_2} + \cdots + m_r \int_\gamma \frac{dz}{z - z_r}$$

$$= 2\pi i\, [m_1\, n(\gamma, z_1) + m_2\, n(\gamma, z_2) + \cdots + m_r\, n(\gamma, z_r)]$$

for any closed and piecewise smooth path γ in D. It now becomes apparent that (5.16) reduces to (5.17) in the present setting. ∎

A sequence of examples in which the preceding ideas are applied to specific functions f may help to solidify them. The first of these redeems a promise made in Chapter III to characterize the plane domains that accomodate branches of $\log z$.

EXAMPLE 4.1. Show that there is a branch of $\log z$ in a plane domain D if and only if $n(\gamma, 0) = 0$ for every closed, piecewise smooth path γ in D.

4. More About Logarithm and Power Functions 179

We use Theorem 4.2 with $f(z) = z$. Thus $c = 1$, $r = 1$, $z_1 = 0$, and $m_1 = 1$. Condition (5.17) then collapses to the statement that $n(\gamma, 0) = 0$ whenever γ is a closed and piecewise smooth path in D. In Figure 16 the domain D_1 supports a branch of $\log z$: the origin clearly lies in the unbounded component of $\mathbb{C} \sim |\gamma|$ for every closed, piecewise smooth path γ in D_1, which implies that $n(\gamma, 0) = 0$ for all such paths (Lemma 2.1). There is no branch of $\log z$ in the domain D_2 in Figure 16. This follows from the observation that $n(\gamma, 0) = 1 \neq 0$ for the path γ pictured.

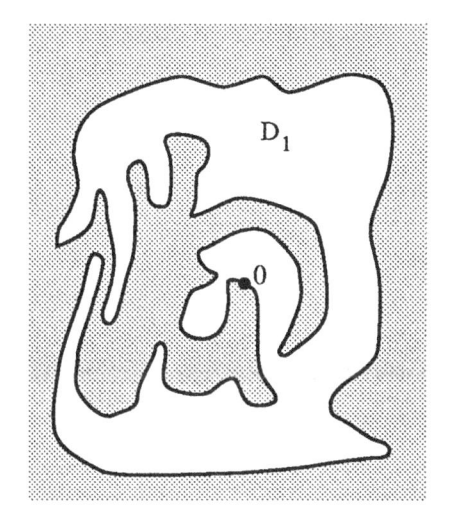

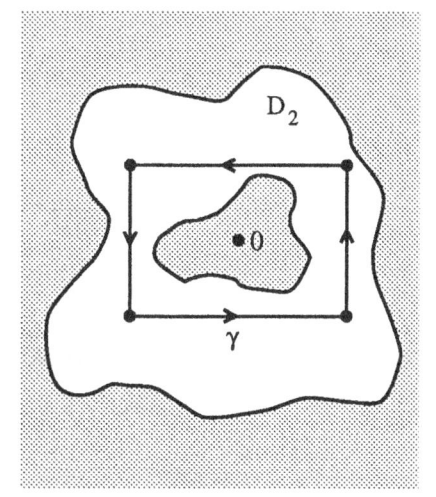

Figure 16.

EXAMPLE 4.2. Characterize the domains D for which there exists an analytic function $g: D \to \mathbb{C}$ satisfying $e^{g(z)} = (z+1)/(z-1)$.

We again appeal to Theorem 4.2. Here $f(z) = (z+1)(z-1)^{-1}$. The theorem informs us that a branch of $\log f(z)$ exists in a domain D if and only if $n(\gamma, 1) - n(\gamma, -1) = 0$ (or, equivalently, $n(\gamma, 1) = n(\gamma, -1)$) for every closed, piecewise smooth path γ in D. Two such domains are represented in Figure 17. In the case of D_1, $n(\gamma, 1) = n(\gamma, -1) = 0$ for all pertinent γ (Why?). Because the points 1 and -1 always lie in the same component of $\mathbb{C} \sim |\gamma|$ for any closed, piecewise smooth path γ in D_2, the requirement $n(\gamma, 1) = n(\gamma, -1)$ is also met by that domain (Lemma 2.1).

EXAMPLE 4.3. Let $D = \mathbb{C} \sim \{z : |z| = 1 \text{ and } \operatorname{Im} z \leq 0\}$ — see Figure 18 — and let g be the branch of $\log[(z+1)/(z-1)]$ in D with $g(0) = \pi i$. Compute $g'(z)$ for z in D, and use this to determine $g(i)$.

Set $f(z) = (z+1)(z-1)^{-1}$. Referring to the previous example we conclude that branches of $\log f(z)$ do exist in the domain D. Because $e^{g(0)} = f(0) = -1$ for any such branch g, it follows that $g(0) = (2k+1)\pi i$

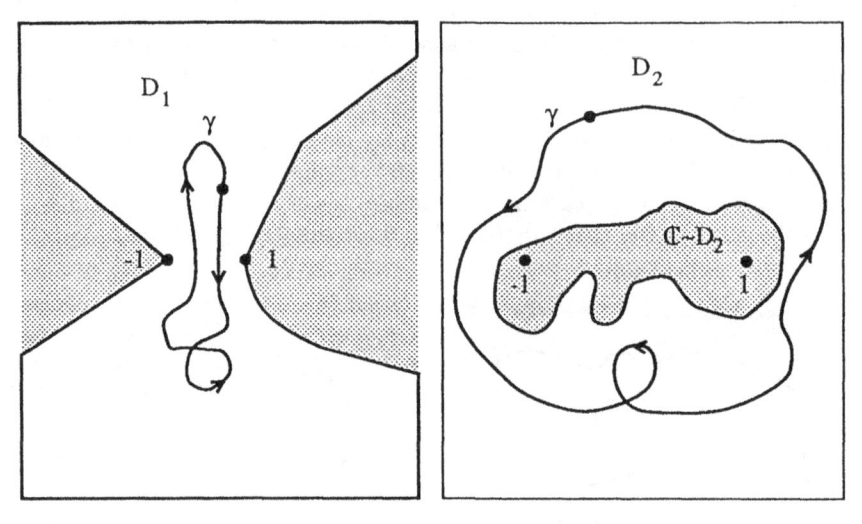

Figure 17.

for some integer k. In fact, Theorem 4.1 tells us that we can uniquely specify g by insisting that $k = 0$, so that $g(0) = \pi i$. This we do. Then

$$g'(z) = \frac{f'(z)}{f(z)} = -\frac{2}{z^2 - 1}$$

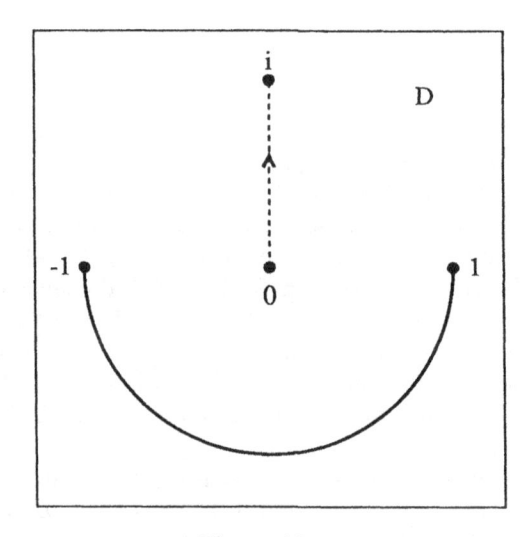

Figure 18.

4. More About Logarithm and Power Functions

for every z in D. Finally, since the line segment joining 0 and i lies in D, we obtain

$$g(i) = g(0) + \int_0^i g'(z)\, dz = \pi i + \int_0^1 \frac{2i\, dt}{t^2 + 1}$$

$$= \pi i + 2i \Big[\text{Arctan}\, t\Big]_0^1 = \frac{3\pi i}{2}.$$

EXAMPLE 4.4. In which of the domains pictured in Figure 19 is there defined an analytic function g that satisfies $e^{g(z)} = (z^2 + 1)/(z^2 - 1)$?

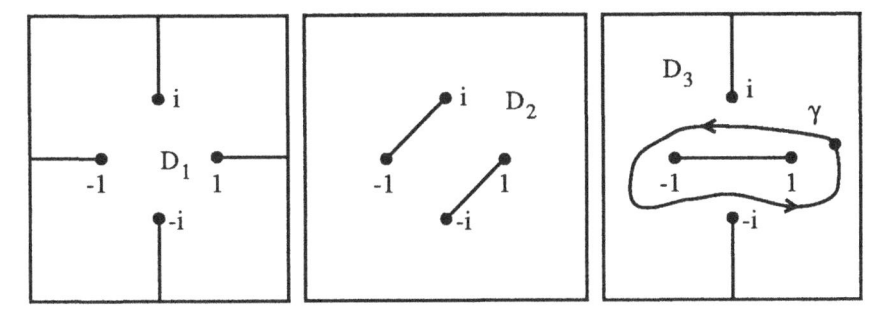

Figure 19.

We once more invoke Theorem 4.2, this time with

$$f(z) = (z+i)(z-i)(z+1)^{-1}(z-1)^{-1}.$$

In order for g as indicated to exist in a domain D the condition imposed on every closed, piecewise smooth path γ in D is:

$$n(\gamma, -i) + n(\gamma, i) - n(\gamma, -1) - n(\gamma, 1) = 0.$$

In D_1 we have $n(\gamma, -i) = n(\gamma, i) = n(\gamma, -1) = n(\gamma, 1) = 0$ for all relevant paths, while in D_2 it is clear that $n(\gamma, i) = n(\gamma, -1)$ and $n(\gamma, -i) = n(\gamma, 1)$ for the paths γ under scrutiny. In both D_1 and D_2, therefore, functions g of the stated type are seen to exist. In the domain D_3, by way of contrast, the path γ illustrated has

$$n(\gamma, -i) + n(\gamma, i) - n(\gamma, -1) - n(\gamma, 1) = -2,$$

so no function g with the properties demanded can exist there.

The methods under discussion here are not confined in their application solely to logarithms. In the next example these same methods are used to identify the plane domains in which there exist branches of the inverse tangent function. We remind the reader that, since the derivative of $f(z) = \tan z$ is free of zeros, a branch of the inverse tangent in a domain D (i.e., a continuous function $g\colon D \to \mathbb{C}$ with the property that $\tan g(z) = z$ for every z in D) is always an analytic function (Theorem III.4.1).

EXAMPLE 4.5. Let D be a plane domain. Show that a branch of the inverse tangent function exists in D if and only if $n(\gamma, i) = n(\gamma, -i)$ for every closed, piecewise smooth path γ in D.

Since the tangent function never assumes the values i and $-i$, only domains D avoiding these points need be considered. If g is a branch of the inverse tangent in D, then differentiation of the relation $z = \tan g(z)$ yields

$$1 = g'(z)\sec^2[g(z)] = g'(z)\left(1 + \tan^2[g(z)]\right) = g'(z)(1+z^2),$$

showing that $g'(z) = (1+z^2)^{-1}$ in D. Therefore, $h(z) = (1+z^2)^{-1}$ has a primitive in D. Theorem IV.2.2 informs us that

$$0 = \int_\gamma \frac{dz}{1+z^2} = \frac{1}{2i}\int_\gamma \frac{dz}{z-i} - \frac{1}{2i}\int_\gamma \frac{dz}{z+i} = \pi[n(\gamma,i) - n(\gamma,-i)]$$

for every closed and piecewise smooth path γ in D. It follows that $n(\gamma, i) = n(\gamma, -i)$ for any such γ.

For the converse we assume that the stated condition on winding numbers holds. Taking a cue from equation (3.11) in Chapter III, we choose a branch k of $\log[(1+iz)/(1-iz)]$ in D — since $(1+iz)/(1-iz) = -(z-i)(z+i)^{-1}$, Theorem 4.2 and the winding number assumption entitle us to do this — and set $g = k/(2i)$. Obviously g is analytic in D and

$$\tan g(z) = \frac{1}{i}\frac{e^{ig(z)} - e^{-ig(z)}}{e^{ig(z)} + e^{-ig(z)}} = \frac{1}{i}\frac{e^{2ig(z)} - 1}{e^{2ig(z)} + 1}$$

$$= \frac{1}{i}\frac{e^{k(z)} - 1}{e^{k(z)} + 1} = \frac{1}{i}\frac{[(1+iz)/(1-iz)] - 1}{[(1+iz)/(1-iz)] + 1} = z$$

for every z in D. The function g is thus a branch of the inverse tangent in D.

4.3 Branches of Powers of Functions

Suppose that f is a function known to be analytic in a domain D, that g is a branch of the logarithm of f in D, and that λ is an arbitrary complex number. In accordance with earlier usage (Chapter III.4.4) we declare the function $h\colon D \to \mathbb{C}$ defined by $h(z) = e^{\lambda g(z)}$ to be the the *branch of the λ-power of f in D associated with g*. The function h is analytic, and its

4. More About Logarithm and Power Functions

derivative is given by $h'(z) = \lambda f'(z) e^{(\lambda-1)g(z)}$, as an easy calculation shows. For $\lambda = p$, an integer, h reduces to the usual p^{th}-power of f — $h(z) = [f(z)]^p$ — regardless of which branch g of $\log f(z)$ is used. Attention is also directed to the case $\lambda = 1/p$ for an integer $p \geq 2$, in which $h = e^{g/p}$ constitutes a *branch of the pth-root of f in D*. This designation applies to any analytic function $h: D \to \mathbb{C}$ with the property that $[h(z)]^p = f(z)$ for every z in D. We remark that it is sometimes possible for a branch of the pth-root of f to exist in a domain despite the fact that no branch of the logarithm of f exists there. As an example, $h(z) = z$ gives a branch of the square root of $f(z) = z^2$ in $D = \mathbb{C}$, yet there can be no branch of $\log f(z)$ in D, for f is not free of zeros there. Our next example shows that this phenomenon is not in evidence in the case of the function $f(z) = z$ (cf., Example 4.1). In it we take for granted the following useful fact, whose proof the reader may treat as an exercise: *if D is a domain in $\mathbb{C}$, z_0 is a point of $\mathbb{C} \sim D$, and γ is a closed, piecewise smooth path in D for which $n(\gamma, z_0) \neq 0$, then there is a Jordan contour γ_1 in D for which $n(\gamma_1, z_0) = 1$.*

EXAMPLE 4.6. Let D be a plane domain, and let $p \geq 2$ be an integer. Show that there is a branch of the pth-root function in D if and only if $n(\gamma, 0) = 0$ for every closed, piecewise smooth path γ in D.

In light of Example 4.1 the stated winding number condition implies the existence of a branch of $\log z$ in D, which in turn gives rise to an associated branch of the pth-root function in D. As to the converse, let h be a branch of the pth-root function in D. By definition $h: D \to \mathbb{C}$ is an analytic function and satisfies $[h(z)]^p = z$ throughout D. The chain rule implies that $p[h(z)]^{p-1} h'(z) = 1$ for every z in D, which demonstrates that h has no zeros in — and, thus, that the origin is not a point of — the given domain. (This repeats an argument that we made once before, in Chapter III.4.3.) It also reveals that

(5.18) $$p \frac{h'(z)}{h(z)} = \frac{1}{z}$$

everywhere in D. We now consider an arbitrary closed and piecewise smooth path γ in D. We want to demonstrate that $n(\gamma, 0) = 0$. Recalling the remarks pursuant to the proof of Theorem 4.1, we use (5.18) to compute for the path $\beta = h \circ \gamma$:

$$p\, n(\beta, 0) = \frac{p}{2\pi i} \int_\gamma \frac{h'(z)\, dz}{h(z)} = \frac{1}{2\pi i} \int_\gamma \frac{dz}{z} = n(\gamma, 0)\ .$$

This shows that $n(\gamma, 0)$ is a multiple of p. Suppose that $n(\gamma, 0) \neq 0$. According to the fact cited above, there must be a Jordan contour γ_1 in D with $n(\gamma_1, 0) = 1$. The argument just used on γ applies equally well to γ_1 and proves that $n(\gamma_1, 0)$ must also be a multiple of p. Because $p \geq 2$ and $n(\gamma_1, 0) = 1$, this is plainly impossible. The contradiction leaves us no choice but to conclude that $n(\gamma, 0) = 0$.

184   V. Cauchy's Theorem and its Consequences

The final example of this section is concerned with branches of the square root of $f(z) = z^2 - 1$.

EXAMPLE 4.7. In which of the domains depicted in Figure 20 is there defined an analytic function h satisfying $[h(z)]^2 = z^2 - 1$?

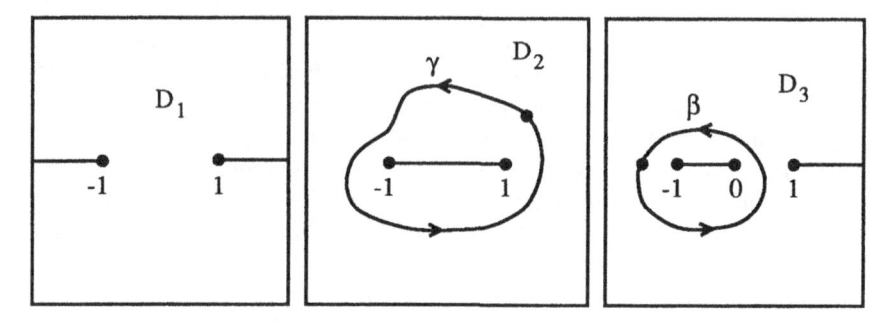

Figure 20.

Theorem 4.2 implies that there is a branch g of $\log(z^2 - 1)$ in D_1. With g is associated a branch h of the square root of $z^2 - 1$ in D_1. Unlike the situation in D_1, there is no branch of $\log(z^2 - 1)$ in D_2. (Condition (5.17) is violated by the path γ pictured.) There is, however, a branch g of $\log[(z+1)/(z-1)]$ in D_2 (Example 4.2). Let k be the branch of the square root of $(z+1)/(z-1)$ in D_2 associated with g. Then $h(z) = (z-1)k(z)$ defines an analytic function in D_2, one that does indeed have

$$[h(z)]^2 = (z-1)^2[k(z)]^2 = (z-1)^2 \left(\frac{z+1}{z-1}\right) = z^2 - 1$$

there. Lastly, suppose there were an analytic function $h: D_3 \to \mathbb{C}$ with the property that $[h(z)]^2 = z^2 - 1$. Differentiation would give $2h(z)h'(z) = 2z$ in D_3. Also, h would be zero-free in D_3, for the points 1 and -1 lie outside this domain. If β is the path indicated in Figure 20, then $\alpha = h \circ \beta$ would satisfy

$$2n(\alpha, 0) = \frac{1}{\pi i} \int_\beta \frac{h'(z)\,dz}{h(z)} = \frac{1}{\pi i} \int_\beta \frac{h(z)h'(z)\,dz}{[h(z)]^2} = \frac{1}{\pi i} \int_\beta \frac{z\,dz}{z^2 - 1}$$

$$= \frac{1}{2\pi i} \int_\beta \left(\frac{1}{z-1} + \frac{1}{z+1}\right) dz = n(\beta, 1) + n(\beta, -1) = 1,$$

which would stand in conflict with the fact that $n(\alpha, 0)$ is an integer. Accordingly, no such function h can exist.

5 The Global Cauchy Theorems

5.1 Iterated Line Integrals

We have reached a point in our development of complex function theory where further progress will be seriously impeded unless we make good our promise of global versions of Cauchy's theorem and integral formula. The ingenious proof we present for Cauchy's theorem was discovered by Alan Beardon, from whose book *Complex Analysis: the Argument Principle in Analysis and Topology* (John Wiley and Sons, New York, 1979) we have borrowed it. Beardon's argument assumes that the reader has a nodding acquaintance with continuous functions of two complex variables; i.e., continuous functions of the type $f: A \to \mathbb{C}$, where A is a subset of $\mathbb{C}^2 = \mathbb{C} \times \mathbb{C} = \{(z, \zeta): z, \zeta \in \mathbb{C}\}$. We ask the reader to accept the few topological statements we make about such functions for what they truly are, straightforward generalizations of ideas discussed at length in Chapter II for functions of a single complex variable. His proof presupposes, as well, awareness of a basic fact from two-variable calculus, a result we now recall.

Suppose that $R = \{z: a \leq x \leq b, c \leq y \leq d\}$ is a closed rectangle in the complex plane and that $h: R \to \mathbb{C}$ is a continuous function. The fundamental result to which we refer proclaims the equality of a pair of iterated integrals:

$$(5.19) \qquad \int_c^d \left\{ \int_a^b h(t,s)\,dt \right\} ds = \int_a^b \left\{ \int_c^d h(t,s)\,ds \right\} dt.$$

While this is typically stated for real-valued h, the passage to complex-valued functions is routine. Assume now that γ and β are piecewise smooth paths in $\mathbb{C}$ and that g is a function of two complex variables which is continuous on the set $|\gamma| \times |\beta| = \{(z, \zeta) \in \mathbb{C}^2 : z \in |\gamma|, \zeta \in |\beta|\}$. The generalization of (5.19) we shall need asserts:

$$(5.20) \qquad \int_\beta \left\{ \int_\gamma g(z,\zeta)\,dz \right\} d\zeta = \int_\gamma \left\{ \int_\beta g(z,\zeta)\,d\zeta \right\} dz.$$

Consider first the case in which $\gamma: [a,b] \to \mathbb{C}$ and $\beta: [c,d] \to \mathbb{C}$ are both smooth paths. Then $h(t,s) = g[\gamma(t), \beta(s)]\dot{\gamma}(t)\dot{\beta}(s)$ defines a continuous function on the rectangle R, and we are able to infer from (5.19) that

$$\int_\beta \left\{ \int_\gamma g(z,\zeta)\,dz \right\} d\zeta = \int_c^d \left\{ \int_a^b g[\gamma(t), \beta(s)]\dot{\gamma}(t)\,dt \right\} \dot{\beta}(s)\,ds$$

$$= \int_c^d \left\{ \int_a^b g[\gamma(t), \beta(s)]\dot{\gamma}(t)\dot{\beta}(s)\,dt \right\} ds$$

$$= \int_a^b \left\{ \int_c^d g[\gamma(t), \beta(s)] \dot{\gamma}(t) \dot{\beta}(s) \, ds \right\} dt$$

$$= \int_a^b \left\{ \int_c^d g[\gamma(t), \beta(s)] \dot{\beta}(s) \, ds \right\} \dot{\gamma}(t) dt = \int_\gamma \left\{ \int_\beta g(z, \zeta) \, d\zeta \right\} dz \, ,$$

as (5.20) claims. When γ and β are merely piecewise smooth, we can always write $\gamma = \gamma_1 + \gamma_2 + \cdots + \gamma_n$ and $\beta = \beta_1 + \beta_2 + \cdots + \beta_m$ with smooth "summands" γ_k and β_ℓ. In this event the smooth case of (5.20), in combination with elementary properties of contour integrals, yields

$$\int_\beta \left\{ \int_\gamma g(z, \zeta) \, dz \right\} d\zeta = \sum_{\ell=1}^m \sum_{k=1}^n \int_{\beta_\ell} \left\{ \int_{\gamma_k} g(z, \zeta) \, dz \right\} d\zeta$$

$$= \sum_{k=1}^n \sum_{\ell=1}^m \int_{\gamma_k} \left\{ \int_{\beta_\ell} g(z, \zeta) \, d\zeta \right\} dz = \int_\gamma \left\{ \int_\beta g(z, \zeta) \, d\zeta \right\} dz \, .$$

Once again (5.20) holds. In this derivation we have elected to gloss over one technical point necessary to make the above story complete: if we wish to remain true to the precise definition of a complex line integral, we cannot regard the iterated integrals in (5.20) as well-defined until it is shown that the functions sending z to $\int_\beta g(z, \zeta) d\zeta$ and ζ to $\int_\gamma g(z, \zeta) dz$ are continuous on the sets $|\gamma|$ and $|\beta|$, respectively. This is, indeed, the case. It is a consequence of the fact that g is uniformly continuous on the compact set $|\gamma| \times |\beta|$. We leave the details as an exercise.

5.2 Cycles

Before proceeding to the global Cauchy theorem we introduce some pertinent — and now quite standard — terminology. By a *cycle* — or, more accurately, a *piecewise smooth cycle* — we shall understand a finite sequence of closed, piecewise smooth paths in the complex plane. We write $\sigma = (\gamma_1, \gamma_2, \ldots, \gamma_p)$ if $\gamma_1, \gamma_2, \ldots, \gamma_p$ are the paths that make up a cycle σ. (The actual order in which the component paths of σ are listed is of no real consequence. Nevertheless, for technical reasons that will become clear shortly we prefer not to think of a cycle as just an unordered collection of closed paths.) Although the notion of a cycle has deep significance in the field of topology, we intend to use the concept in a rather superficial way; namely, as a convenient mechanism for helping to keep track of the results when an analytic function is integrated simultaneously along several different — but appropriately related — paths. A single closed and piecewise smooth path is, of course, the basic example of a cycle. For the applications we have in mind only cycles $\sigma = (\gamma_1, \gamma_2, \ldots, \gamma_p)$ in which the paths γ_k are distinct will play a role, but nothing in the definition is meant to bar the

5. The Global Cauchy Theorems

possibility that repetitions occur among the paths constituting σ. Quite to the contrary, in certain circumstances it is highly desirable to allow for this kind of redundancy. Thus, a cycle $\sigma = (\gamma, \beta, \gamma, -\alpha, -\alpha, \beta, \gamma)$, which might appear written in abbreviated form as $\sigma = (3\gamma, 2\beta, -2\alpha)$, is perfectly acceptable. (N.B. Many books would write $\sigma = 3\gamma + 2\beta - 2\alpha$ to indicate such a cycle and, in general, would employ additive notation instead of sequence notation in dealing with cycles; i.e., $\sigma = (\gamma_1, \gamma_2, \ldots, \gamma_p)$ would be written $\sigma = \gamma_1 + \gamma_2 + \cdots + \gamma_p$. We avoid additive notation for cycles, because we have already reserved that notation for another purpose, to signify a path sum. In this vein we might add that, even though they are formally different cycles, there is not a huge conceptual gap separating the n-component cycle $(n\gamma) = (\gamma, \gamma, \ldots, \gamma)$ from the n-fold path sum $\gamma + \gamma + \cdots + \gamma$. In situations such as those that come up in this book there is very little to be gained by stressing the distinction between the two.) If $\sigma = (\gamma_1, \gamma_2, \ldots, \gamma_p)$ is a cycle, we use $|\sigma|$ to denote the compact set $|\gamma_1| \cup |\gamma_2| \cup \cdots \cup |\gamma_p|$ and, mimicking the terminology established for paths, speak of σ as a *cycle in A*, where A is a subset of $\mathbb{C}$, provided $|\sigma|$ is contained in A.

Suppose that $\sigma = (\gamma_1, \gamma_2, \ldots, \gamma_p)$ is a cycle in a set A and that $f: A \to \mathbb{C}$ is a continuous function. We define $\int_\sigma f(z)dz$ in the natural fashion:

$$\int_\sigma f(z)\,dz = \int_{\gamma_1} f(z)\,dz + \int_{\gamma_2} f(z)\,dz + \cdots + \int_{\gamma_p} f(z)\,dz \ .$$

In particular, for z in $\mathbb{C} \sim |\sigma|$ we can define the winding number $n(\sigma, z)$ of σ about z by

$$n(\sigma, z) = \frac{1}{2\pi i} \int_\sigma \frac{d\zeta}{\zeta - z} \ .$$

Any z in $\mathbb{C} \sim |\sigma|$ obviously lies in $\mathbb{C} \sim |\gamma_k|$ for each of the paths γ_k that compose σ and, just as obviously,

$$n(\sigma, z) = n(\gamma_1, z) + n(\gamma_2, z) + \cdots + n(\gamma_p, z) \ .$$

If D is a component of the open set $\mathbb{C} \sim |\sigma|$, then $n(\sigma, z)$ remains constant as z varies over D. This is true because for each k the connected set D is contained in a component — call it D_k — of $\mathbb{C} \sim |\gamma_k|$ (Theorem II.3.7) and because by Lemma 2.1 the value of $n(\gamma_k, z)$ does not change in D_k. The set $\mathbb{C} \sim |\sigma|$ has a unique unbounded component D^*. (It is the component of $\mathbb{C} \sim |\sigma|$ that contains the complement of any disk which encloses $|\sigma|$.) Again by Lemma 2.1, $n(\sigma, z) = 0$ for every z in D^*.

Let U be an open set in the complex plane. A cycle σ in U is said to be *homologous to zero in* U if $n(\sigma, z) = 0$ for every z in $\mathbb{C} \sim U$. (This bit of terminology and others involving the expression "homologous" are drawn from topology, where the subject of "homology" is a fundamental concern.) Two cycles $\sigma_0 = (\gamma_1, \gamma_2, \ldots, \gamma_p)$ and $\sigma_1 = (\beta_1, \beta_2, \ldots, \beta_q)$ in U are pronounced *homologous in* U if the cycle $\sigma = (\gamma_1, \ldots, \gamma_p, -\beta_1, \ldots, -\beta_q)$

is homologous to zero in that set or, equivalently, if $n(\sigma_0, z) = n(\sigma_1, z)$ for every z in $\mathbb{C} \sim U$. Finally, two non-closed piecewise smooth paths λ_0 and λ_1 in U are declared to be *homologous in U* if λ_0 and λ_1 share both the same initial point and the same terminal point and if the closed path $\gamma = \lambda_0 - \lambda_1$ is homologous to zero in U. Note that in $U = \mathbb{C}$ any cycle is homologous to zero, for the defining condition is vacuous in this instance. In Figure 21 the cycle $\sigma = (\alpha, \beta, \gamma)$ is homologous to zero in U. However, α is not homologous to β in U, nor is λ_0 homologous to λ_1 there.

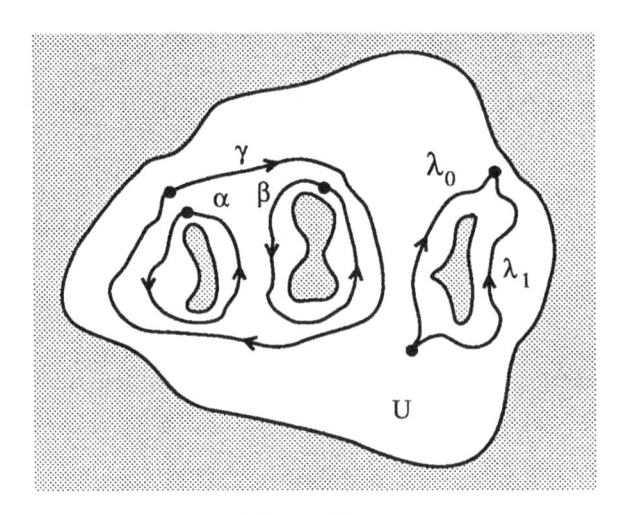

Figure 21.

5.3 Cauchy's Theorem and Integral Formula

The stage is set for a theorem that is without question one of the pivotal results in complex analysis.

Theorem 5.1. (Cauchy's Theorem) *Let σ be a cycle in an open set U. Then $\int_\sigma f(z)dz = 0$ for every function f that is analytic in U if and only if σ is homologous to zero in U.*

Proof. Assume first that σ is homologous to zero in U. Since $n(\sigma, z)$ remains constant as z varies over any component of $\mathbb{C} \sim |\sigma|$, the set V defined by $V = \{z \in \mathbb{C} \sim |\sigma| : n(\sigma, z) = 0\}$ must be a union of components of $\mathbb{C} \sim |\sigma|$. As such, V is open. The unbounded component of $\mathbb{C} \sim |\sigma|$ is contained in V and, because σ is homologous to zero in U, V also includes the complement

5. The Global Cauchy Theorems

of U. It follows that $K = \mathbb{C} \sim V$ is a closed and bounded set — in other words, a compact set — that lies in U and contains $|\sigma|$. We choose and fix for the duration of the proof a radius $\delta > 0$ with the following property: for each z in K the disk $\Delta(z, \delta)$ lies inside U (Lemma II.4.4). The systems of vertical and horizontal lines described, respectively, by the equations $x = n\delta/2$ and $y = n\delta/2$, with n ranging over all integers, partition the complex plane into a grid of non-overlapping closed squares, each of side-length $\delta/2$. The bounded set K can meet at most a finite number of these grid squares. Let $Q_1, Q_2, \ldots, Q_r$ be an enumeration of the set of grid squares that do have non-empty intersection with K. The preceding construction has been engineered to insure that, for $1 \leq j \leq r$, U contains the open disk Δ_j of radius $\delta/2$ concentric with Q_j. (Otherwise, the disk $\Delta(z, \delta)$ for any point z in $Q_j \cap K$ would intersect $\mathbb{C} \sim U$, contrary to our choice of δ.) Moreover, Δ_j contains Q_j. Figure 22 indicates an arrangement of "designated" squares that might emerge from this construction in the case where the cycle σ is composed of a single closed path. Recall that the notation Q_j^0 signifies the interior of Q_j.

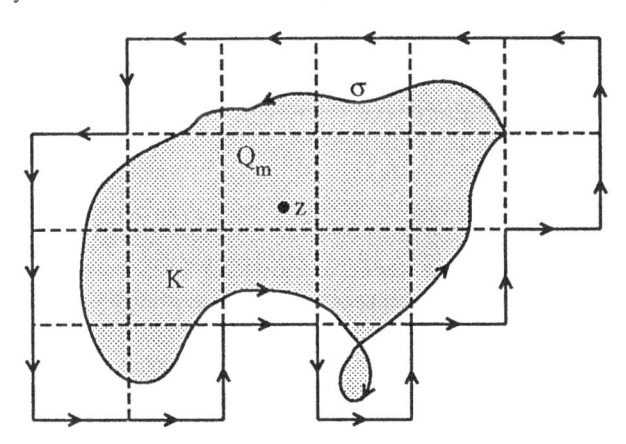

Figure 22.

Suppose now that f is a function known to be analytic in U. We desire to prove that $\int_\sigma f(z)dz = 0$. To begin, we consider a fixed square Q_m from our preferred list and a fixed point z of Q_m^0. By applying the local version of Cauchy's integral formula to the function f in the disk Δ_m we obtain

$$f(z) = n(\partial Q_m, z)f(z) = \frac{1}{2\pi i} \int_{\partial Q_m} \frac{f(\zeta)\,d\zeta}{\zeta - z} .$$

(Remember Example 2.1, which shows that $n(\partial Q_m, z) = 1$.) On the other hand, for $j \neq m$ Lemma 1.1 applied to $g(\zeta) = (\zeta - z)^{-1} f(\zeta)$ yields

$$0 = \frac{1}{2\pi i} \int_{\partial Q_j} \frac{f(\zeta)\,d\zeta}{\zeta - z} .$$

Summing the foregoing relations from $j = 1$ to $j = r$ produces the formula

$$(5.21) \qquad f(z) = \frac{1}{2\pi i} \sum_{j=1}^{r} \int_{\partial Q_j} \frac{f(\zeta) \, d\zeta}{\zeta - z} \, .$$

Equation (5.21) holds for each z in Q_m^0, this for $m = 1, 2, \ldots, r$; i.e., it holds for every z in the set $\cup_{j=1}^{r} Q_j^0$.

Next, let λ be one of the four parametrized line segments entering into a standard parametrization of ∂Q_j, so that $|\lambda|$ is a side of Q_j. It may happen that $|\lambda|$ is disjoint from K. If, however, $|\lambda|$ intersects K, then Q_j must share the side $|\lambda|$ with another square from our list, call it Q'_j. Obviously $-\lambda$ then occurs as a segment in the parametrization of $\partial Q'_j$. This being the case, the contributions that the integrals along λ and $-\lambda$ make to the right-hand side of (5.21) negate one another. The point of this observation is that for any z in $\cup_{j=1}^{r} Q_j^0$ formula (5.21) collapses to

$$(5.22) \qquad f(z) = \frac{1}{2\pi i} \sum_{k=1}^{q} \int_{\lambda_k} \frac{f(\zeta) \, d\zeta}{\zeta - z} \, ,$$

where now $\lambda_1, \lambda_2, \ldots, \lambda_q$ gives an enumeration of the collection of "side segments" λ which arise in the parametrizations of $\partial Q_1, \partial Q_2, \ldots, \partial Q_r$ and are blessed with the additional feature that $|\lambda|$ does not intersect K.

In light of Lemma 1.6 the function sending z to $\int_{\lambda_k} (\zeta - z)^{-1} f(\zeta) d\zeta$ is analytic — in particular, is continuous — in $\mathbb{C} \sim |\lambda_k|$. This means that the right-hand side of (5.22) defines a function of z which is continuous in $\mathbb{C} \sim \cup_{k=1}^{q} |\lambda_k|$, a set containing K. We use this fact to show that (5.22) holds for every z in K. Fix such a point z. If z is in $\cup_{j=1}^{r} Q_j^0$, then there is nothing left to prove. If not, then z lies on the boundary of two or more of the squares Q_j. We choose one of these — for simplicity call it Q — and select a sequence $\langle z_n \rangle$ in the interior of Q such that $z_n \to z$. Since both sides of (5.22) are continuous at z, we are led to

$$f(z) = \lim_{n \to \infty} f(z_n) = \lim_{n \to \infty} \frac{1}{2\pi i} \sum_{k=1}^{q} \int_{\lambda_k} \frac{f(\zeta) \, d\zeta}{\zeta - z_n} = \frac{1}{2\pi i} \sum_{k=1}^{q} \int_{\lambda_k} \frac{f(\zeta) \, d\zeta}{\zeta - z} \, ,$$

as desired. Most importantly, (5.22) is valid for any point z of $|\sigma|$.

To finish the proof, assume that $\sigma = (\gamma_1, \gamma_2, \ldots, \gamma_p)$. We refer to (5.22) and (5.20) in computing

$$\int_\sigma f(z) \, dz = \int_\sigma \left\{ \frac{1}{2\pi i} \sum_{k=1}^{q} \int_{\lambda_k} \frac{f(\zeta) \, d\zeta}{\zeta - z} \right\} dz$$

$$= \frac{1}{2\pi i} \sum_{\ell=1}^{p} \sum_{k=1}^{q} \int_{\gamma_\ell} \left\{ \int_{\lambda_k} \frac{f(\zeta) \, d\zeta}{\zeta - z} \right\} dz = \frac{1}{2\pi i} \sum_{k=1}^{q} \sum_{\ell=1}^{p} \int_{\lambda_k} \left\{ \int_{\gamma_\ell} \frac{f(\zeta) \, dz}{\zeta - z} \right\} d\zeta$$

5. The Global Cauchy Theorems

$$= \sum_{k=1}^{q} \int_{\lambda_k} f(\zeta) \left\{ \sum_{\ell=1}^{p} \frac{1}{2\pi i} \int_{\gamma_\ell} \frac{dz}{\zeta - z} \right\} d\zeta = -\sum_{k=1}^{q} \int_{\lambda_k} f(\zeta) \left\{ \frac{1}{2\pi i} \int_{\sigma} \frac{dz}{z - \zeta} \right\} d\zeta$$

$$= -\sum_{k=1}^{q} \int_{\lambda_k} f(\zeta) n(\sigma, \zeta) \, d\zeta = -\sum_{k=1}^{q} \int_{\lambda_k} f(\zeta) \cdot 0 \, d\zeta = 0 \, .$$

Here we have made use of the fact that $g(z, \zeta) = (\zeta - z)^{-1} f(\zeta)$ describes a function that is continuous on $|\gamma_\ell| \times |\lambda_k|$ for $1 \leq \ell \leq p$ and $1 \leq k \leq q$. (Since $|\gamma_\ell|$ and $|\lambda_k|$ are disjoint, the denominator in g, the only possible source of trouble, is never zero in $|\gamma_\ell| \times |\lambda_k|$.) A second critical observation is that $n(\sigma, \zeta) = 0$ whenever ζ belongs to $|\lambda_k|$, true because $|\lambda_k|$ does not intersect K — hence, is a subset of V. The proof of the sufficiency half of the theorem is now complete.

The proof in the converse direction is quite easy. If z is a point of $\mathbb{C} \sim U$, then the function $f: U \to \mathbb{C}$ given by $f(\zeta) = (\zeta - z)^{-1}$ is clearly analytic. By assumption,

$$0 = \int_{\sigma} f(\zeta) \, d\zeta = \int_{\sigma} \frac{d\zeta}{\zeta - z} = 2\pi i \, n(\sigma, z) \, .$$

Therefore $n(\sigma, z) = 0$ for every z in $\mathbb{C} \sim U$, which makes σ homologous to zero in U. ∎

Two immediate corollaries of Cauchy's theorem that are frequently invoked are:

Corollary 5.2. *If a function f is analytic in an open set U and if σ_0 and σ_1 are cycles in U that are homologous in this set, then $\int_{\sigma_0} f(z) dz = \int_{\sigma_1} f(z) dz$.*

Corollary 5.3. *If a function f is analytic in an open set U and if λ_0 and λ_1 are non-closed piecewise smooth paths in U that are homologous in this set, then $\int_{\lambda_0} f(z) dz = \int_{\lambda_1} f(z) dz$.*

Proof. For the first corollary, suppose that $\sigma_0 = (\gamma_1, \gamma_2, \ldots, \gamma_p)$ and $\sigma_1 = (\beta_1, \beta_2, \ldots, \beta_q)$. We can apply Cauchy's theorem to f and the cycle $\sigma = (\gamma_1, \ldots, \gamma_p, -\beta_1, \ldots, -\beta_q)$ to get

$$\int_{\sigma_0} f(z) \, dz - \int_{\sigma_1} f(z) \, dz = \int_{\sigma} f(z) \, dz = 0 \, .$$

As for the second corollary, we use Cauchy's theorem on the closed path $\sigma = \lambda_0 - \lambda_1$. This once again results in

$$\int_{\lambda_0} f(z) \, dz - \int_{\lambda_1} f(z) \, dz = \int_{\sigma} f(z) \, dz = 0 \, . \quad \blacksquare$$

The global Cauchy integral formula is also readily deduced from Cauchy's theorem.

Theorem 5.4. (Cauchy's Integral Formula) *Suppose that a function f is analytic in an open set U and that σ is a cycle in U which is homologous to zero in this set. Then*

$$n(\sigma, z)f(z) = \frac{1}{2\pi i} \int_\sigma \frac{f(\zeta)\, d\zeta}{\zeta - z}$$

for every z in $U \sim |\sigma|$.

Proof. We fix z in $U \sim |\sigma|$ and define $g: U \to \mathbb{C}$ as in the proof of the local version of Cauchy's integral formula,

$$g(\zeta) = \begin{cases} [f(\zeta) - f(z)]/(\zeta - z) & \text{if } \zeta \neq z, \\ f'(z) & \text{if } \zeta = z. \end{cases}$$

The function g is continuous in U and analytic in $U \sim \{z\}$. Corollary 3.4 informs us that g is actually analytic in U. Cauchy's theorem then leads to

$$0 = \int_\sigma g(\zeta)\, d\zeta = \int_\sigma \frac{f(\zeta) - f(z)}{\zeta - z}\, d\zeta = \int_\sigma \frac{f(\zeta)\, d\zeta}{\zeta - z} - 2\pi i\, n(\sigma, z)f(z)$$

and, therefore, to

$$n(\sigma, z)f(z) = \frac{1}{2\pi i} \int_\sigma \frac{f(\zeta)\, d\zeta}{\zeta - z}. \quad \blacksquare$$

Once in possession of Cauchy's integral formula, we can carry over the proof of Theorem 3.5 essentially verbatim to arrive at the global analogue of that result.

Theorem 5.5. *Suppose that a function f is analytic in an open set U, that k is a non-negative integer, and that σ is a cycle in U which is homologous to zero in this set. Then*

$$n(\sigma, z)f^{(k)}(z) = \frac{k!}{2\pi i} \int_\sigma \frac{f(\zeta)\, d\zeta}{(\zeta - z)^{k+1}}$$

for every z in $U \sim |\sigma|$.

To round out the present discussion we recapitulate the preceding results in a setting that closely approximates the one in which they were originally formulated by Cauchy. The validity of the Jordan curve theorem is taken for granted here.

Theorem 5.6. *Suppose that a function f is analytic in an open set U. Let γ be a Jordan contour in U with the property that the inside D of the Jordan curve $|\gamma|$ is contained in U. Then $\int_\gamma f(z)\, dz = 0$ and*

$$f^{(k)}(z) = \frac{k!}{2\pi i} \int_\gamma \frac{f(\zeta)\, d\zeta}{(\zeta - z)^{k+1}}$$

for every z in D and every non-negative integer k.

5. The Global Cauchy Theorems

Proof. Since $\overline{D} = D \cup |\gamma|$ is contained in U, $\mathbb{C} \sim U$ must lie in the unbounded component of $\mathbb{C} \sim |\gamma|$. Lemma 2.1(ii) shows that $n(\gamma, z) = 0$ for every z in $\mathbb{C} \sim U$; i.e., γ is homologous to zero in U. Also, $n(\gamma, z) = 1$ for every z in D, because γ is positively oriented (Section 2.4). The assertions now follow from Theorems 5.1 and 5.5. ∎

There is a refinement of Theorem 5.6 due to Goursat in which the requirement that f be analytic in an open set containing $\overline{D}$ is relaxed. Because it is of some historical interest, we state Goursat's theorem, but as we have no intention of using the result, we do not include a proof.

Theorem 5.7. (Goursat's Theorem) *Let γ be a Jordan contour in the complex plane, let D be the inside of the Jordan curve $|\gamma|$, and let $f: \overline{D} \to \mathbb{C}$ be a continuous function that is analytic in D. Then $\int_\gamma f(z)\, dz = 0$ and*

$$f^{(k)}(z) = \frac{k!}{2\pi i} \int_\gamma \frac{f(\zeta)\, d\zeta}{(\zeta - z)^{k+1}}$$

for every z in D and every non-negative integer k.

We conclude this section with a pair of examples in which Cauchy's theorems are utilized to evaluate integrals. These examples again anticipate the "Residue Theorem," an important refinement of Cauchy's theorem that will be discussed in Chapter VIII.

EXAMPLE 5.1. Evaluate $\int_\gamma (z^2 - 1)^{-1} \operatorname{Arctan} z\, dz$, where γ is the piecewise smooth path pictured in Figure 23.

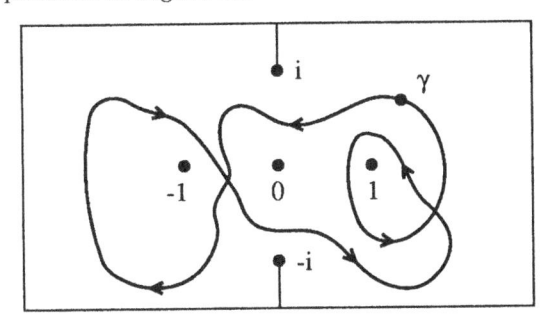

Figure 23.

We use the partial fractions decomposition of $(z^2 - 1)^{-1}$ in tandem with the Cauchy integral formula, here applied to the function $f(z) = \operatorname{Arctan} z$ in the set $U = \mathbb{C} \sim \{z : \operatorname{Re} z = 0 \text{ and } |z| \geq 1\}$. The outcome:

$$\int_\gamma \frac{\operatorname{Arctan} z\, dz}{z^2 - 1} = \frac{1}{2} \int_\gamma \frac{\operatorname{Arctan} z\, dz}{z - 1} - \frac{1}{2} \int_\gamma \frac{\operatorname{Arctan} z\, dz}{z + 1}$$

$$= \frac{2\pi i\, n(\gamma, 1) \operatorname{Arctan} 1}{2} - \frac{2\pi i\, n(\gamma, -1) \operatorname{Arctan}(-1)}{2} = \frac{\pi^2 i}{4}.$$

EXAMPLE 5.2. Evaluate $\int_\gamma (z^2+z+1)(z^3+z^2)^{-1}dz$, with γ as depicted in Figure 24.

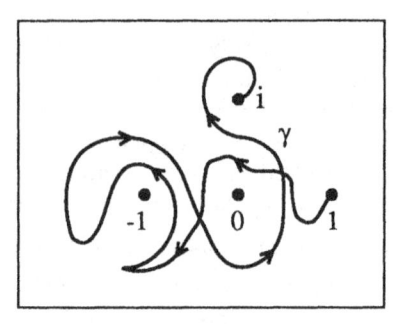

Figure 24.

We once more resort to partial fractions and write
$$\int_\gamma \frac{z^2+z+1}{z^3+z^2}\,dz = \int_\gamma \left(\frac{1}{z^2}+\frac{1}{z+1}\right)dz = \int_\gamma \frac{dz}{z^2} + \int_\gamma \frac{dz}{z+1}\;.$$

Since $F(z) = -z^{-1}$ is a primitive for $f(z) = z^{-2}$ in $\mathbb{C} \sim \{0\}$, the first integral is easily handled:
$$\int_\gamma \frac{dz}{z^2} = \left[-\frac{1}{z}\right]_1^i = 1+i\;.$$

As γ is homologous to the path $\beta = [1,i]$ in $\mathbb{C} \sim \{-1\}$, a set in which the function $g(z) = (z+1)^{-1}$ is analytic, Corollary 5.3 comes to our aid and gives
$$\int_\gamma \frac{dz}{z+1} = \int_\beta \frac{dz}{z+1} = \Big[\mathrm{Log}(z+1)\Big]_1^i = -\frac{\mathrm{Log}\,2}{2} + \frac{\pi i}{4}\;.$$

We have taken advantage of the fact that $G(z) = \mathrm{Log}(z+1)$ provides a primitive for g in the set $\mathbb{C} \sim (-\infty, -1]$, which contains the line segment with endpoints 1 and i. As a result,
$$\int_\gamma \frac{z^2+z+1}{z^3+z^2}\,dz = 1 - \mathrm{Log}\sqrt{2} + i\left(1+\frac{\pi}{4}\right)\;.$$

6 Simply Connected Domains

6.1 Simply Connected Domains

A class of plane domains that plays a conspicuous role in complex analysis is the class of "simply connected" domains. We shall refer to a domain D

6. Simply Connected Domains

in the complex plane as *simply connected* under the condition that every closed and piecewise smooth path in D — hence, every cycle in D — is homologous to zero in that domain. This is admittedly not the standard definition of the concept one is apt to come across in a topology textbook, where a more conventional definition might read: a plane domain D is simply connected provided every closed path in D is "contractible" (also called "null homotopic") in D. Intuitively speaking, a closed path γ in D is contractible there if $|\gamma|$ can be continuously "shrunk" to a point without leaving the confines of this domain. In the next section we shall give precise meaning to these words by introducing the notion of homotopy. Suffice it to say here that these competing descriptions of what is required for a plane domain to be simply connected are not really at odds, although the proof of their equivalence is by no means trivial. We shall ultimately confirm that equivalence by combining the homotopy invariance of winding numbers (Theorem 7.1) with the famous "Riemann Mapping Theorem." (See Theorem IX.3.6.) Until then the above homological definition of a simply connected domain will be the sole operative one.

The complex plane itself is clearly simply connected. If a domain D in $\mathbb{C}$ has the property that $\mathbb{C} \sim D$ is both unbounded and connected, then D is simply connected. Indeed, let γ be an arbitrary closed, piecewise smooth path in such a domain. As a connected subset of $\mathbb{C} \sim |\gamma|$, $\mathbb{C} \sim D$ lies in some component of $\mathbb{C} \sim |\gamma|$ (Theorem II.3.7); since $\mathbb{C} \sim D$ is unbounded, that component must be the unique unbounded one. It follows from Lemma 2.1(ii) that $n(\gamma, z) = 0$ for every z in $\mathbb{C} \sim D$, which stamps γ as homologous to zero in D. The strip $D = \{z \colon 0 < \operatorname{Re} z < 1\}$ is a simply connected domain whose complement is not connected: $\mathbb{C} \sim D$ is the disjoint union of the closed half-spaces $H_0 = \{z \colon \operatorname{Re} z \leq 0\}$ and $H_1 = \{z \colon \operatorname{Re} z \geq 1\}$. (See Exercise 8.71. In Chapter VIII we shall adjoin an ideal point ∞ to $\mathbb{C}$ to form the "extended complex plane" $\widehat{\mathbb{C}} = \mathbb{C} \cup \{\infty\}$. If we take the complement of the strip D in $\widehat{\mathbb{C}}$ we obtain $H_0 \cup H_1 \cup \{\infty\}$, which happens to be a connected subset of $\widehat{\mathbb{C}}$. This property actually characterizes simply connected plane domains, a second fact that we shall establish in Theorem IX.3.6: *a domain D in $\mathbb{C}$ is simply connected if and only if $\widehat{\mathbb{C}} \sim D$ is a connected subset of $\widehat{\mathbb{C}}$.*) Conceding the truth of the Jordan curve theorem, we remark that the inside of a plane Jordan curve is a simply connected domain.

6.2 Simple Connectivity, Primitives, and Logarithms

We refocus our attention on analytic functions by recording two theorems that typify the role simply connected domains play in the theory of these functions. A prominent feature of this class of domains is that statements about analytic functions which are valid only on a local scale in general domains tend to hold globally in simply connected ones. Both of the ensuing theorems illustrate this principle. In the first we learn that simply connected domains are the only "universal domains" when it comes to the existence of global primitives. (Lemma 1.3 implies that, when D is an arbitrary plane domain, any function which is analytic in D has "local primitives" there; i.e., it has a primitive in each open disk that is contained in D.)

Theorem 6.1. *Let D be a domain in the complex plane. Then D is simply connected if and only if every function that is analytic in D possesses a primitive in this domain.*

Proof. Assume first that D is simply connected, and let the function f be analytic in D. By Cauchy's theorem $\int_\gamma f(z)dz = 0$ whenever γ is a closed and piecewise smooth path in D, for each such path is homologous to zero in D. Theorem 1.4 certifies that f has a primitive in D.

Conversely, suppose that every function which is analytic in D has a primitive there. Let γ be an arbitrary closed, piecewise smooth path in D. We claim that γ is homologous to zero in the domain. To see this, let z be any point of $\mathbb{C} \sim D$. The function $f: D \to \mathbb{C}$ defined by $f(\zeta) = (\zeta - z)^{-1}$ is then analytic. Since f is assumed to have a primitive in D, we conclude using Theorem IV.2.2 that

$$n(\gamma, z) = \frac{1}{2\pi i} \int_\gamma \frac{d\zeta}{\zeta - z} = \frac{1}{2\pi i} \int_\gamma f(\zeta)\, d\zeta = 0\,,$$

as required to make γ homologous to zero in D. Thus, D is simply connected. ∎

Simply connected domains can also be characterized in terms of the global existence of branches of the logarithms of analytic functions. (At the local level, Cauchy's theorem and Theorem 4.1 enable us to construct branches of the logarithm of a function in any open disk where that function is both analytic and zero-free.)

Theorem 6.2. *Let D be a domain in the complex plane. Then D is simply connected if and only if for every function f that is both analytic and free of zeros in D there exists a branch of $\log f(z)$ in this domain.*

Proof. Suppose initially that D is simply connected and that f is a function which is analytic in D and has no zeros there. By Cauchy's theorem $\int_\gamma [f'(z)/f(z)]dz = 0$ for every closed, piecewise smooth path γ in D. Theorem 4.1 bears witness to the existence of a branch of $\log f(z)$ in D.

In the opposite direction we consider an arbitrary point z of $\mathbb{C} \sim D$ and associate with it the function $f: D \to \mathbb{C}$ defined by $f(\zeta) = \zeta - z$. Because f is analytic and zero-free in D, our hypothesis now tells us that branches of the logarithm of f exist in D. We choose one and call it g. It follows that

$g'(\zeta) = (\zeta - z)^{-1}$ in D, with the consequence that

$$n(\gamma, z) = \frac{1}{2\pi i} \int_\gamma \frac{d\zeta}{\zeta - z} = \frac{1}{2\pi i} \int_\gamma g'(\zeta) \, d\zeta = 0$$

whenever γ is a closed and piecewise smooth path in D. Since this is true for every z in $\mathbb{C} \sim D$, each such γ is homologous to zero in D; i.e., D is simply connected. ∎

As a corollary, Theorem 6.2 affords us the luxury in a simply connected domain D of being able to extract from any function f that is both analytic and zero-free in D and for any integer $p \geq 2$ a branch of the pth-root of f in D.

7 Homotopy and Winding Numbers

7.1 Homotopic Paths

The final section of this chapter is devoted to a proof of an important technical property of winding numbers, their invariance under homotopy. That invariance, though of interest in its own right, is needed in this book for one purpose only — to reconcile the definition we have adopted of a simply connected plane domain with the more traditional definitions of that concept found in topology books. Readers willing to accept on faith that the two definitions are equivalent — or at least uninterested in seeing a detailed proof of their equivalence — should feel free to skip over this section.

As suggested in the previous section, the concept of "homotopy" provides a natural framework in which to make precise the idea of deforming one path into another. Consider a pair of closed paths $\alpha: [a, b] \to A$ and $\beta: [a, b] \to A$, where A is a set in the complex plane. (N.B. It is significant here that α and β are mappings of the same parameter interval.) We speak of α and β being *freely homotopic in A* if there exists a continuous function $H: R = \{(t, s): a \leq t \leq b, 0 \leq s \leq 1\} \to A$ with the following two properties:

(5.23) $\begin{cases} \text{(i)} \ \ H(t, 0) = \alpha(t) \ , \ H(t, 1) = \beta(t) & \text{for } a \leq t \leq b \ ; \\ \text{(ii)} \ \ H(a, s) = H(b, s) & \text{for } 0 \leq s \leq 1 \ . \end{cases}$

Any such function H is known as a *free homotopy from α to β in A*. Thus, corresponding to each s in $[0, 1]$ the free homotopy H determines a closed path $\gamma_s: [a, b] \to A$ by means of the formula $\gamma_s(t) = H(t, s)$ (Figure 25). One way to think of the homotopy is this: as the parameter s increases from 0 to 1, the path $\alpha = \gamma_0$ "evolves" in a continuous manner into the path $\beta = \gamma_1$. The entire process unfolds within the confines of the set A. The path γ_s represents the "stage of development" at the instant s.

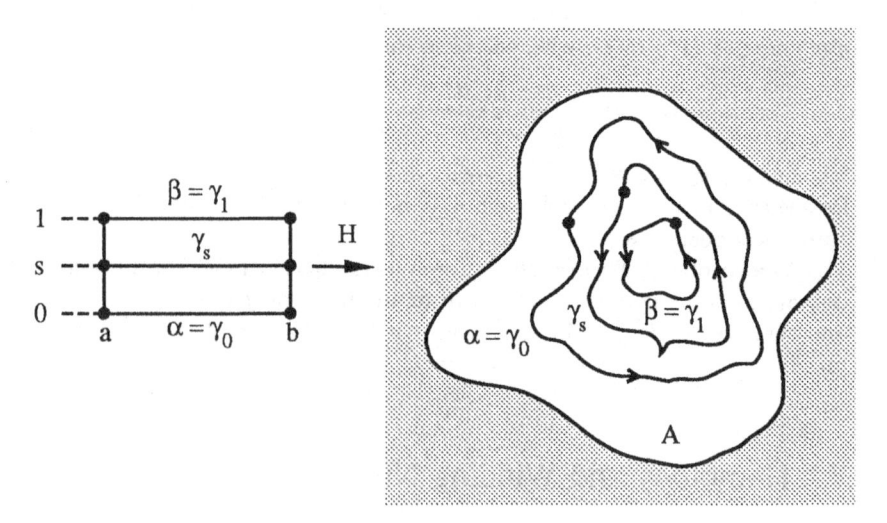

Figure 25.

Although our real concern at this point is with closed paths, it can do no harm to mention in passing a second type of homotopy, one that applies equally well to non-closed paths and that comes up in Chapter X in conjunction with the problem of "analytic continuation." Let $\alpha:[a,b] \to A$ and $\beta:[a,b] \to A$ be paths that have the same initial point and the same terminal point. We pronounce α and β *homotopic with fixed endpoints in A* if there is a continuous function $H: R = \{(t,s): a \leq t \leq b, 0 \leq s \leq 1\} \to A$ that satisfies in place of (5.23) the requirements

(5.24) $\begin{cases} \text{(i)} \ H(t,0) = \alpha(t) \ , \ H(t,1) = \beta(t) & \text{for } a \leq t \leq b \ ; \\ \text{(ii)} \ H(a,s) = \alpha(a) \ , \ H(b,s) = \alpha(b) & \text{for } 0 \leq s \leq 1 \ . \end{cases}$

Again, the homotopy H represents a set of instructions for continuously deforming α to β through a one-parameter family of paths γ_s in A, all having the same endpoints as α and β (Figure 26).

How does a person decide whether two given closed paths $\alpha:[a,b] \to A$ and $\beta:[a,b] \to A$ are freely homotopic in A? In general, this is likely to be a rather challenging exercise. For the present, we content ourselves with a simple observation: if for each t in $[a,b]$ the line segment joining $\alpha(t)$ and $\beta(t)$ lies completely in A, then α and β are freely homotopic in A. The function H defined for $a \leq t \leq b$ and $0 \leq s \leq 1$ by the formula $H(t,s) = (1-s)\alpha(t) + s\beta(t)$ furnishes a free homotopy from α to β in A. We illustrate this remark with a concrete example.

EXAMPLE 7.1. Let $\alpha(t) = e^{it}$ and $\beta(t) = (3/2) + 3e^{it}$ for t in the interval $[0, 2\pi]$. Show that α and β are freely homotopic in $A = \mathbb{C} \sim \{0\}$.

7. Homotopy and Winding Numbers

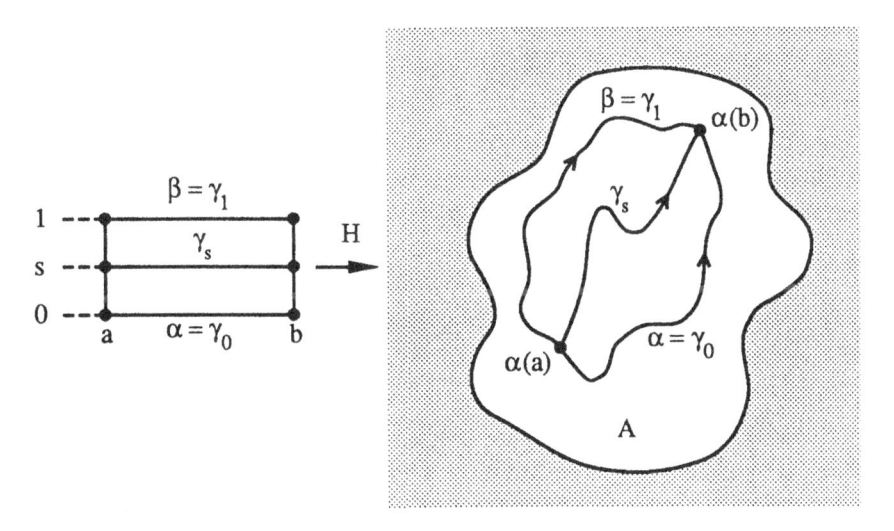

Figure 26.

In this example the function

$$H(t,s) = (1-s)e^{it} + s\left(\frac{3}{2} + 3e^{it}\right) = \frac{3s}{2} + (1+2s)e^{it},$$

where $0 \leq t \leq 2\pi$ and $0 \leq s \leq 1$, gives a free homotopy from α to β in A. The path γ_s is described here by $\gamma_s(t) = (3s/2) + (1+2s)e^{it}$ for

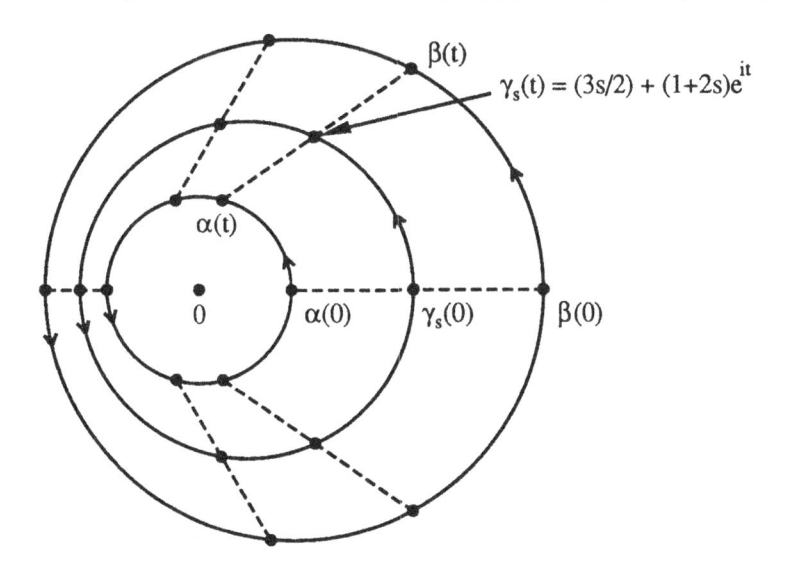

Figure 27.

$0 \leq t \leq 2\pi$. Its trajectory is a circle intermediate to the trajectories of α and β (Figure 27).

Our main goal in this section is to establish the following theorem.

Theorem 7.1. *Let z be a point of the complex plane. If α and β are closed, piecewise smooth paths in $\mathbb{C} \sim \{z\}$ that are freely homotopic in $\mathbb{C} \sim \{z\}$, then $n(\alpha, z) = n(\beta, z)$.*

Proof. Assuming that $\alpha, \beta : [a, b] \to \mathbb{C} \sim \{z\}$, we choose a free homotopy $H : R = \{(t, s) : a \leq t \leq b, 0 \leq s \leq 1\} \to \mathbb{C} \sim \{z\}$ from α to β. (*Warning*: It is not assumed here that the "intermediate paths" γ_s associated with H are piecewise smooth when $0 < s < 1$.) As the continuous image of a compact set, $K = H(R)$ is itself compact (Theorem II.4.6). Since K lies in the open set $\mathbb{C} \sim \{z\}$, we can appeal to Lemma II.4.4 and fix $\epsilon > 0$ with the property that, for each point w of K, the disk $\Delta(w, \epsilon)$ is contained in $\mathbb{C} \sim \{z\}$. Next, because the continuous function H is uniformly continuous on the compact set R (Theorem II.4.8), we can choose $\delta > 0$ to insure that

$$|H(t, s) - H(t', s')| < \epsilon$$

whenever (t, s) and (t', s') are points of R which satisfy $|t - t'| < \delta$ and $|s - s'| < \delta$. Finally, we can pick a positive integer $p \geq 2$ for which it is true that $\max\{1/p, (b - a)/p\} < \delta$ and subdivide R into p^2 congruent rectangles $R_{jk}(1 \leq j, k \leq p)$, labeled according to the pattern set for the case $p = 3$ by Figure 28. The sole purpose of the preceding maneuvers is this: corresponding to each pair of indices j and k we are able to select an open disk — call it Δ_{jk} — such that

$$H(R_{jk}) \subset \Delta_{jk} \subset \mathbb{C} \sim \{z\}.$$

For instance, we can just take Δ_{jk} to be the disk of radius ϵ whose center is the point to which the center of R_{jk} gets mapped by H.

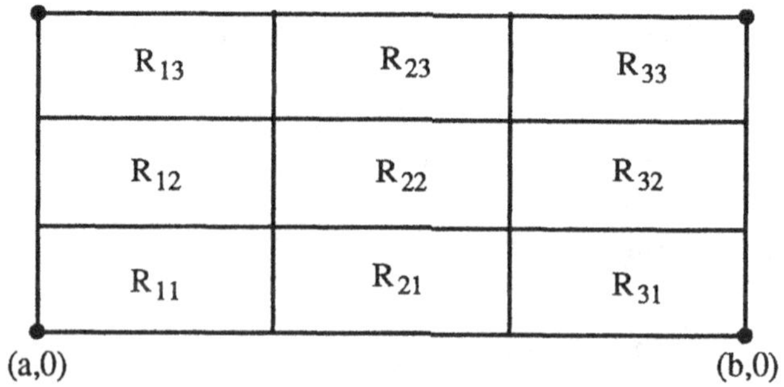

R_{13}	R_{23}	R_{33}
R_{12}	R_{22}	R_{32}
R_{11}	R_{21}	R_{31}

(a,0)   (b,0)

Figure 28.

7. Homotopy and Winding Numbers

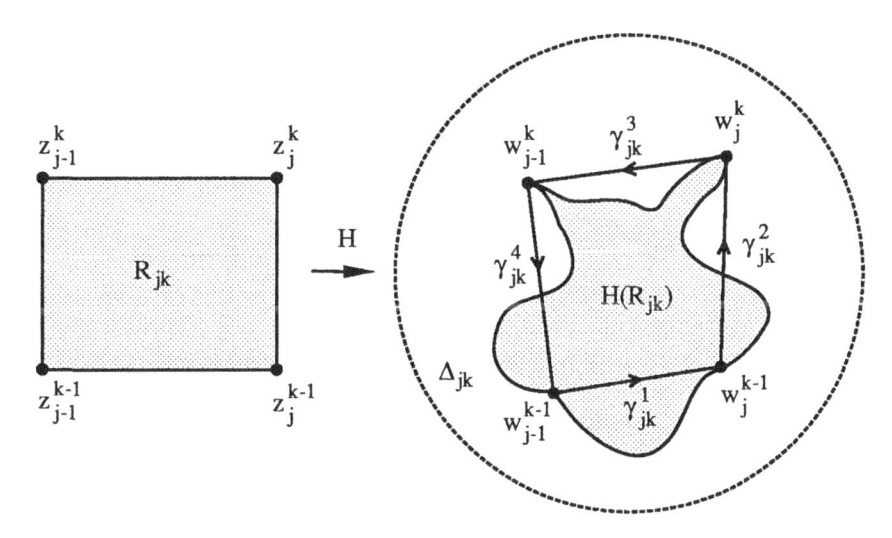

Figure 29.

For $0 \leq j, k \leq p$ let z_j^k be the point of R given by

$$z_j^k = \left(a + \frac{j(b-a)}{p}, \frac{k}{p}\right)$$

and set

$$w_j^k = H(z_j^k) \, .$$

Since $z_{j-1}^{k-1}, z_j^{k-1}, z_j^k$, and z_{j-1}^k are nothing but the vertices of R_{jk}, the points $w_{j-1}^{k-1}, w_j^{k-1}, w_j^k$, and w_{j-1}^k are all located in the disk Δ_{jk}. We consider the closed polygonal path $\gamma_{jk} = \gamma_{jk}^1 + \gamma_{jk}^2 + \gamma_{jk}^3 + \gamma_{jk}^4$ in Δ_{jk}, where now $1 \leq j, k \leq p$ and

$$\gamma_{jk}^1 = [w_{j-1}^{k-1}, w_j^{k-1}] \quad , \quad \gamma_{jk}^2 = [w_j^{k-1}, w_j^k] \, ,$$

$$\gamma_{jk}^3 = [w_j^k, w_{j-1}^k] \quad , \quad \gamma_{jk}^4 = [w_{j-1}^k, w_{j-1}^{k-1}] \, .$$

(See Figure 29.)

By Theorem 1.5,

$$0 = \int_{\gamma_{jk}} \frac{d\zeta}{\zeta - z} = \sum_{\ell=1}^{4} \int_{\gamma_{jk}^\ell} \frac{d\zeta}{\zeta - z} \, ,$$

for $f(\zeta) = (\zeta - z)^{-1}$ defines a function that is analytic in Δ_{jk}. Summing over j and k leads to

(5.25) $$\sum_{j=1}^{p} \sum_{k=1}^{p} \sum_{\ell=1}^{4} \int_{\gamma_{jk}^\ell} \frac{d\zeta}{\zeta - z} = 0 \, .$$

202   V. Cauchy's Theorem and its Consequences

This multiple sum is not nearly as fearsome as it might first appear. The definition of the paths γ_{jk}^ℓ guarantees that, in most instances, the integral of f along a path appearing in the sum can be paired with a corresponding integral along its reverse path, resulting in the cancellation of these integrals. To be precise, the integral along γ_{jk}^2 cancels the one along $\gamma_{(j+1)k}^4 = -\gamma_{jk}^2$ when $1 \leq j \leq p-2$ and $1 \leq k \leq p-1$, while the integral over γ_{jk}^3 negates the contribution of the integral along $\gamma_{j(k+1)}^1 = -\gamma_{jk}^3$ when $1 \leq j \leq p-1$ and $1 \leq k \leq p-2$. Furthermore, since we are dealing with a free homotopy of closed paths, it is also true that $w_0^k = w_p^k$ for $0 \leq k \leq p$, which implies that $\gamma_{1k}^4 = -\gamma_{pk}^2$ and, thus, that the terms in (5.25) corresponding to these paths can likewise be disregarded. The only integrals in the sum exempt from cancellation are those involving $\gamma_{11}^1, \gamma_{21}^1, \ldots, \gamma_{p1}^1$ and $\gamma_{1p}^3, \gamma_{2p}^3, \ldots, \gamma_{pp}^3$. In short, (5.25) reduces to the more manageable statement

$$(5.26) \qquad \sum_{j=1}^p \int_{\gamma_{j1}^1} \frac{d\zeta}{\zeta - z} + \sum_{j=1}^p \int_{\gamma_{jp}^3} \frac{d\zeta}{\zeta - z} = 0 \, .$$

For $1 \leq j \leq p$ let α_j designate the path obtained by restricting α to the interval with endpoints $a + [(j-1)(b-a)/p]$ and $a + [j(b-a)/p]$. Remembering that $H(t,0) = \alpha(t)$, we see that $\gamma_{j1}^1 - \alpha_j$ is a closed, piecewise smooth path in the disk Δ_{j1} (Figure 30). Appealing once again to Theorem 1.5, we obtain

$$\int_{\gamma_{j1}^1} \frac{d\zeta}{\zeta - z} - \int_{\alpha_j} \frac{d\zeta}{\zeta - z} = \int_{\gamma_{j1}^1 - \alpha_j} \frac{d\zeta}{\zeta - z} = 0$$

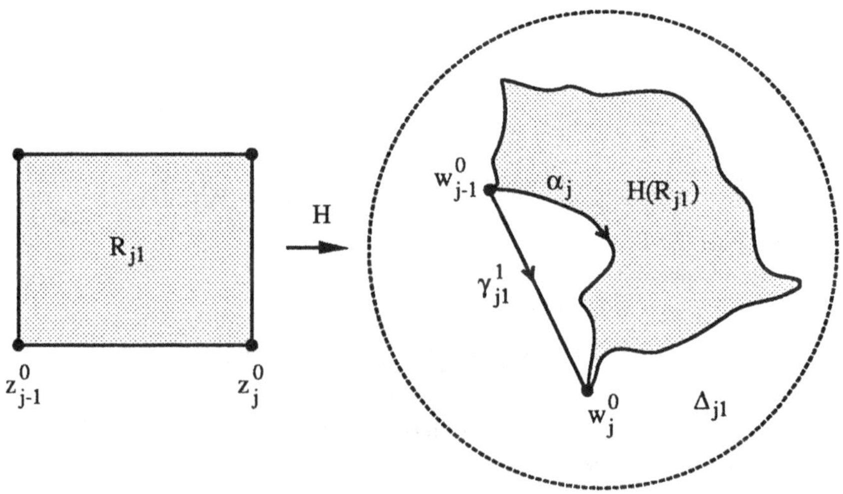

Figure 30.

7. Homotopy and Winding Numbers

and conclude that

$$\int_\alpha \frac{d\zeta}{\zeta - z} = \sum_{j=1}^p \int_{\alpha_j} \frac{d\zeta}{\zeta - z} = \sum_{j=1}^p \int_{\gamma_{j1}^1} \frac{d\zeta}{\zeta - z} .$$

Similar reasoning reveals that

$$\int_\beta \frac{d\zeta}{\zeta - z} = -\sum_{j=1}^p \int_{\gamma_{jp}^3} \frac{d\zeta}{\zeta - z} .$$

In view of (5.26) we are able to assert that

$$n(\alpha, z) = \frac{1}{2\pi i} \int_\alpha \frac{d\zeta}{\zeta - z} = \frac{1}{2\pi i} \int_\beta \frac{d\zeta}{\zeta - z} = n(\beta, z) ,$$

as desired. ∎

We obtain as a corollary of Theorem 7.1:

Corollary 7.2. *Let U be an open set in the complex plane, and let α and β be closed, piecewise smooth paths in U. If α and β are freely homotopic in U, then they are homologous in this set.*

Proof. If z is any point of $\mathbb{C} \sim U$, then α and β, paths given to be freely homotopic in U, are obviously freely homotopic in $\mathbb{C} \sim \{z\}$. According to Theorem 7.1, $n(\alpha, z) = n(\beta, z)$ for every such z, which is precisely what it takes for α and β to be homologous in U. ∎

7.2 Contractible Paths

A closed path γ in a set A is said to be *contractible* (or *null homotopic*) in A if γ is freely homotopic in that set to a constant path. To be more exact, assuming that $\gamma: [a, b] \to A$, this definition requires the existence of a point z_0 of A such that γ is freely homotopic in A to the path β described by $\beta(t) = z_0$ for $a \le t \le b$. Such is the case, for example, if there is a point z_0 in A with the property that for each z in $|\gamma|$ the line segment joining z_0 and z lies completely in A. Under this condition the function H defined by $H(t, s) = (1 - s)\gamma(t) + s z_0$ for $a \le t \le b$ and $0 \le s \le 1$ is a free homotopy in A from γ to the constant path β indicated above.

From Corollary 7.2 we derive:

Theorem 7.3. *Let U be an open set in the complex plane, and let γ be a closed, piecewise smooth path in U. If γ is contractible in U, then γ is homologous to zero in this set.*

Proof. Suppose that $\gamma: [a, b] \to U$. For γ to be contractible in U means that γ is freely homotopic in U to a constant path $\beta: [a, b] \to U$. By Theorem

7.1 we have for every z in $\mathbb{C} \sim U$

$$n(\gamma, z) = n(\beta, z) = \frac{1}{2\pi i} \int_\beta \frac{d\zeta}{\zeta - z} = \frac{1}{2\pi i} \int_a^b \frac{\dot{\beta}(t)\, dt}{\beta(t) - z} = \frac{1}{2\pi i} \int_a^b 0\, dt = 0 ,$$

which reveals that γ is homologous to zero in U. ∎

In a general open set U the converse of Theorem 7.1 is false. To see this, take $U = \mathbb{C} \sim \{1, -1\}$ and $\gamma = \alpha + \beta - \alpha - \beta$, where $\alpha(t) = 1 - e^{it}$ and $\beta(t) = -1 + e^{it}$ for $0 \le t \le 2\pi$. It is easy to check that γ is homologous to zero in U. It is true — but non-trivial to verify — that γ is not contractible in U.

We end this chapter with the observation that one direction in the equivalence between a topologist's definition of a simply connected plane domain and the definition we are using has now been established. For the converse we must wait until Chapter IX.

Theorem 7.4. *Let D be a domain in the complex plane. If every closed path in D is contractible in D, then D is simply connected.*

8 Exercises for Chapter V

8.1 Exercises for Section V.1

8.1. Modify the proof of Lemma 1.1 to establish the following: if a function f is analytic in an open set U, then $\int_{\partial T} f(z)\, dz = 0$ for every closed triangle T in U. (N.B. This approximates Pringsheim's actual formulation of Lemma 1.1 — he worked with triangles instead of rectangles. Of course, "triangle" means the two-dimensional figure, not just its boundary, and $\int_{\partial T} f(z)\, dz$ is defined in the obvious way to mimic the corresponding integral for a rectangle.)

8.2. Deduce Lemma 1.1 from Exercise 8.1. Does the statement in Exercise 8.1 follow immediately from Lemma 1.1? To what other geometric figures does Exercise 8.1 let one extend the conclusion of Lemma 1.1?

8.3. By means of the local Cauchy theorem find: (i) $\int_{|z|=1} \sqrt{9-z^2}\, dz$; (ii) $\int_{|z|=1} (z^2 + 2z)^{-1}\, dz$; (iii) $\int_{|z+i|=3/2} (z^4 + z^2)^{-1}\, dz$.

8.4. Let $\gamma(t) = 2\cos t + i\sin t$ for $0 \le t \le 2\pi$. Evaluate the integrals: (i) $\int_\gamma \sin(z^2)\, dz$; (ii) $\int_\gamma z^{-1}\, dz$; (iii) $\int_\gamma (z^2 + 2iz)^{-1}\, dz$. (*Hint for (ii).* How is this integral related to $\int_{|z|=1} z^{-1}\, dz$?)

8.5. Let $b > 0$. Show that $\int_{-\infty}^\infty e^{-t^2} \cos(2b\pi t)\, dt = \sqrt{\pi}\, e^{-b^2 \pi^2}$. (*Hint.* Consider $\int_{\partial R} e^{-z^2} dz$, where R is the rectangle with vertices $-c, c, c + b\pi i$, and $-c + b\pi i$ for $c > 0$. Recall Exercise IV.4.18.)

8.2 Exercises for Section V.2

8.6. Let γ and β be closed, piecewise smooth paths in $\mathbb{C}$ with the same initial point. Demonstrate that $n(-\gamma, z) = -n(\gamma, z)$ for every z in $\mathbb{C} \sim |\gamma|$ and that $n(\gamma + \beta, z) = n(\gamma, z) + n(\beta, z)$ for every z in $\mathbb{C} \sim (|\gamma| \cup |\beta|)$.

8.7. Suppose that γ is a closed and piecewise smooth path in the complex plane. Verify that $n(\overline{\gamma}, \overline{z}) = -n(\gamma, z)$ and that $n(a\gamma + b, az + b) = n(\gamma, z)$ for every z in $\mathbb{C} \sim |\gamma|$, where $a \neq 0$ and b are constants.

8.8. Let $\gamma = \gamma_1 + \gamma_2 + \gamma_3$, where $\gamma_1(t) = e^{it}$ for $0 \leq t \leq 2\pi$, $\gamma_2(t) = -1 + 2e^{-2it}$ for $0 \leq t \leq 2\pi$, and $\gamma_3(t) = 1 - i + e^{it}$ for $\pi/2 \leq t \leq 9\pi/2$. Determine all the values assumed by $n(\gamma, z)$ as z varies over $\mathbb{C} \sim |\gamma|$.

8.9. With the assistance of the local Cauchy integral formula find the values of: (i) $\int_{|z|=1} z^{-1} \operatorname{Log}(z+e)\, dz$; (ii) $\int_{|z-2|=2} (z^2-1)^{-1} \operatorname{Arctan} z\, dz$; (iii) $\int_{|z|=2} (z+1)^{-2} e^z\, dz$.

8.10. Let $\gamma = [-1, 1+i] + \beta + [-1+i, 1]$, where $\beta(t) = i + e^{it}$ for $0 \leq t \leq 3\pi$. Compute: $\int_\gamma (z^2 + 1)^{-1} dz$.

8.11. If $\gamma = \beta + [4\pi, 0]$, where $\beta(t) = te^{it}$ for $0 \leq t \leq 4\pi$, evaluate the integral $\int_\gamma (z^2+1)(z+1)^{-1}(z+4)^{-1} dz$.

8.12. Confirm that $\int_0^\infty t^{-1} \sin t\, dt = \pi/2$. (*Hint*. Consider $\int_\gamma z^{-1} e^{iz}\, dz$, where for $0 < s < r < \infty$ the contour of integration is described by $\gamma = [s,r] + \gamma_r + [-r,-s] - \gamma_s$, with $\gamma_r(t) = re^{it}$ and $\gamma_s(t) = se^{it}$ on $[0, \pi]$. Recall Exercise IV.4.20.)

8.13. For $k = 1, 2, 3, \cdots$, let $I_k = \int_{\gamma_k} z^{-1} \sin z\, dz$, where γ_k is the path defined on the interval $[-2k\pi, 2k\pi]$ by $\gamma_k(t) = e^{t+it}$. Determine $\lim_{k\to\infty} I_k$. (*Hint*. Make use of Exercise 8.12.)

8.14. Compute: $\int_{-\infty}^\infty t(t^2+4)^{-1} \sin(\pi t)\, dt$.

8.3 Exercises for Section V.3

8.15. Prove Corollary 3.2.

8.16. An entire function $f = u + iv$ has the feature that $u_x v_y - u_y v_x = 1$ throughout the complex plane. Demonstrate that f has the form $f(z) = az + b$, where a and b are constants and $|a| = 1$.

8.17. Suppose that an entire function $f = u + iv$ is blessed with the property that $u_x + v_y = 0$ everywhere in the complex plane. Verify that f must be a function of the type $f(z) = az + b$ for constants a and b, where $\operatorname{Re} a = 0$.

8.18. Let $D = \mathbb{C} \sim (-\infty, 0]$. Define $f: D \to \mathbb{C}$ by $f(z) = (z-1)/\operatorname{Log} z$ if

$z \neq 1$, while $f(1) = 1$. Verify that f is an analytic function.

8.19. Let g be a branch of the inverse sine function in a domain D. (Recall Exercise III.6.49.) Prove that g is an analytic function. Deduce from this that neither 1 nor -1 can belong to D. (*Hint*. The exercise just cited shows that g is analytic in the set $D \sim \{1, -1\}$.)

8.20. Apply Theorem 3.5 to find: (i) $\int_{|z-2i|=2} (z-i)^{-3} \operatorname{Arcsin} z \, dz$; (ii) $\int_{|z-2|=3/2} (z^2-1)^{-2} \operatorname{Log}(1+z) \, dz$; (iii) $\int_{|z|=2} z^k (z-1)^{-k-1} dz$ for k a positive integer ; (iv) $\int_{\partial Q} (z^3 + z^2)^{-1} \sin(\pi z) \, dz$, where Q is the square whose vertices are $2, 2i, -2,$ and $-2i$.

8.21. Let $\gamma = \gamma_1 + \gamma_2 + \gamma_3 + \gamma_4$, where $\gamma_1(t) = 2i + e^{it}$ for $\pi/2 \leq t \leq 7\pi/2$, $\gamma_2(t) = e^{-it}$ for $-\pi/2 \leq t \leq 9\pi/2$, $\gamma_3(t) = -2i + e^{it}$ for $\pi/2 \leq t \leq 3\pi/2$, and $\gamma_4(t) = 3e^{it}$ for $-\pi/2 \leq t \leq \pi/2$. Calculate $\int_\gamma (z^4 + 4z^2)^{-1} e^{\pi z} dz$.

8.22. Compute $\int_\gamma (2z-1)^{-3} \operatorname{Log}(1+z) \, dz$, where $\gamma(t) = (2\cos t - 1)e^{it}$ for $0 \leq t \leq 2\pi$.

8.23. Suppose that a function f is analytic in a convex domain D and that $|f'(z)| \leq m$ is satisfied for every z in D, where m is a constant. Prove that $|f(z_2) - f(z_1)| \leq m|z_2 - z_1|$ holds for every pair of points z_1 and z_2 in D. (N.B. A set A in the complex plane is called *convex* if for every pair of points z_1 and z_2 belonging to A the entire line segment with endpoints z_1 and z_2 is contained in A.)

8.24. If a function f is analytic in a convex domain D and if it is true that $\operatorname{Re} f'(z) \neq 0$ for every z in D, then f is univalent in D. Confirm this fact. (*Hint*. For distinct points z_1 and z_2 of D consider $\int_{z_1}^{z_2} f'(z) \, dz$.)

8.25. A function f is analytic in the disk $\Delta = \Delta(0,1)$ and satisfies $f(0) = 0$. Its derivative obeys the estimate $|f'(z)| \leq c(1-|z|)^{-n-1}$ throughout Δ, where n is a non-negative integer and $c > 0$ is a constant. Show that $|f(z)| \leq cn^{-1}(1-|z|)^{-n}$ for every z in Δ under the condition that $n \geq 1$, whereas $|f(z)| \leq c \operatorname{Log}[(1-|z|)^{-1}]$ for all such z when $n = 0$.

8.26. A function f is analytic in the disk $\Delta = \Delta(0,1)$ and admits the estimate $|f(z)| \leq c(1-|z|)^{-n}$ for every z in Δ, where n is a positive integer and $c > 0$ is a constant. Show that $|f'(z)| \leq 2^{n+1} c (1-|z|)^{-n-1}$ for every z in Δ. (*Hint*. For fixed z in Δ consider f in the disk $\Delta(z,r)$, where $r = (1-|z|)/2$.)

8.27. Let f be a function that is analytic in an open set U and satisfies a *Lipschitz condition of order* α ($0 < \alpha \leq 1$) there, by which is meant that $|f(z_2) - f(z_1)| \leq m|z_2 - z_1|^\alpha$ for all points z_1 and z_2 of U, where m is a constant. If the open disk $\Delta(z_0, r)$ is contained in U, then $|f'(z_0)| \leq mr^{\alpha-1}$. Prove this.

8.28. Assume that a function f is analytic in the disk $\Delta(0,r)$ and that $0 < s < r$. Define a path $\gamma: [0, 2\pi] \to \mathbb{C}$ by $\gamma(t) = f(se^{it})$. Demonstrate that $\ell(\gamma) \geq 2\pi s |f'(0)|$. (*Hint.* Apply Cauchy's integral formula, but not to f.)

8.29. Let f be an entire function with the property that $|f(z)| \leq c|z|^\lambda + d$ for all z, where λ, c, and d are positive constants. Prove that f is necessarily a polynomial function of z whose degree does not exceed λ. (*Hint.* Modify the proof of Liouville's theorem. Recall Exercise III.6.16.)

8.30. An entire function f with $f(0) = 0$ has the property that $\operatorname{Re} f(z) \to 0$ as $|z| \to \infty$. Demonstrate that $f(z) = 0$ for every z in $\mathbb{C}$. (The knowledge that $\operatorname{Im} f(z) \to 0$ as $|z| \to \infty$ would permit the same conclusion.)

8.31. Concerning an entire function f, one is presented with the information that $g(z) = \sqrt{f(z)}$ also defines an entire function. Show that f must be a constant function.

8.32. An entire function f obeys the estimate $|f(z)| \leq m e^{\alpha x}$ for all points $z = x + iy$, where m and α are positive constants. Verify that f has the form $f(z) = A e^{\alpha z}$ for some constant A. Could one draw the same conclusion if one knew only that $|f(z)| \leq m e^{\alpha |z|}$ held for every z?

8.33. If f is a non-constant entire function, prove that the range $f(\mathbb{C})$ of f must almost "fill up" the complex plane in the following precise sense: for every point w_0 in $\mathbb{C}$ and every $r > 0$ the range of f has non-empty intersection with the disk $\Delta(w_0, r)$. (N.B. The technical description of this situation is that $f(\mathbb{C})$ is *dense in* $\mathbb{C}$. *Hint.* Under the assumption that the assertion were false for some w_0 and r, derive a contradiction by considering the function $g(z) = [f(z) - w_0]^{-1}$.)

8.34. Prove Theorem 3.9.

8.35. Let $z_1, z_2, \ldots, z_n$ be the roots of $p(z) = a_0 + a_1 z + \cdots + a_n z^n$, a polynomial of degree $n \geq 1$. Show that any root ζ of p' can be expressed in the form $\zeta = \lambda_1 z_1 + \lambda_2 z_2 + \cdots + \lambda_n z_n$, where $\lambda_1, \lambda_2, \ldots, \lambda_n$ are non-negative real numbers satisfying $\lambda_1 + \lambda_2 + \cdots + \lambda_n = 1$. (N.B. The result described in this exercise is known as the "Gauss-Lucas Theorem." The set of all complex numbers that are expressible in the manner prescribed is called the "convex hull" of $\{z_1, z_2, \ldots, z_n\}$, being the smallest convex set that includes all points of this set. Thus, the zeros of p' lie in the convex hull of the zeros of p. *Hint.* Use Theorem 3.9. We may assume that $\zeta \neq z_1, z_2, \ldots, z_n$, for the assertion holds trivially otherwise. Compute $p'(z)/p(z)$ and evaluate the result at $z = \zeta$, remembering that $1/\overline{w} = w/|w|^2$.)

8.36. Let an analytic function f have domain-set the punctured disk $\Delta^*(z_0, r)$. Assume the existence of a complex number w_0 and a sequence $\langle r_n \rangle$ of positive real numbers satisfying $r > r_1 > r_2 > \cdots > r_n \to 0$ concerning which it is true that $M_n = \max\{|f(z) - w_0| : z \in K(z_0, r_n)\} \to 0$ as $n \to \infty$. Demonstrate that $\lim_{z \to z_0} f(z) = w_0$. Conclude that by defining $f(z_0) = w_0$ we can extend f to be a function that is analytic in the full disk $\Delta(z_0, r)$.

8.37. If a function f is analytic in a domain D and if there exists a point z_0 of D such that $|f(z)| \geq |f(z_0)| > 0$ is true for every z in D, then f is constant in D. Substantiate this claim.

8.38. Suppose that D is a bounded plane domain and that $f: \overline{D} \to \mathbb{C}$ is a non-constant continuous function which is analytic in D and satisfies $|f(z)| = 1$ for every z on ∂D. Verify that $f(z_0) = 0$ for some z_0 in D.

8.39. Generalize the previous exercise as follows: if D is a bounded domain and if $f: \overline{D} \to \mathbb{C}$ is a non-constant continuous function that is analytic in D and satisfies $|f(z)| = 1$ for every z on ∂D, then $f(D) = \Delta(0, 1)$. (*Hint.* To show that an arbitrary point c of $\Delta(0, 1)$ belongs to $f(D)$ look at the function $g(z) = [f(z) - c]/[1 - \bar{c} f(z)]$. Remember Exercise I.4.21.)

8.40. Let $p(z) = a_0 + a_1 z + \cdots + a_n z^n$ be a polynomial function of degree $n \geq 1$, and let $K = \{z: |p(z)| = 1\}$. (Sets K of this general type are known as "lemniscates.") It is easy to see that K is a closed set. Moreover, since $|p(z)| \to \infty$ as $|z| \to \infty$, K must also be bounded; i.e., K is a compact set. Verify that: (i) the open set $U = \mathbb{C} \sim K$ has at most $n + 1$ components; (ii) any bounded component of U has $\Delta(0, 1)$ as its image under p, whereas the unbounded component of U has for its image the set $\mathbb{C} \sim \overline{\Delta}(0, 1)$; (iii) when U has exactly $n + 1$ components, p is univalent in each bounded component of U.

8.41. A function f is continuous on the disk $\overline{\Delta}(0, 1)$, is analytic in $\Delta(0, 1)$, and satisfies the conditions $|f(z)| \leq a$ for every z on $K(0, 1)$ with $\operatorname{Im} z \geq 0$ and $|f(z)| \leq b$ for every z on $K(0, 1)$ with $\operatorname{Im} z \leq 0$. Here $a \geq 0$ and $b \geq 0$ are constants. Certify that $|f(0)| \leq \sqrt{ab}$. (*Hint.* Consider the function $g(z) = f(-z)$ along with f.)

8.42. Let Q be a closed square centered at the origin, and let S_1, S_2, S_3, and S_4 be an enumeration of its sides. If a function f is continuous on Q, is analytic in the interior of Q, and admits the bound $|f(z)| \leq m_j$, a constant, for every z on $S_j (1 \leq j \leq 4)$, then $|f(0)|^4 \leq m_1 m_2 m_3 m_4$. Justify this statement.

8.43. Prove the following generalization of the Schwarz lemma: if a function f is analytic in a disk $\Delta = \Delta(z_0, r)$ and if m is a constant such that $|f(z) - f(z_0)| \leq m$ holds for every z in Δ, then $|f'(z_0)| \leq m/r$ and the estimate $|f(z) - f(z_0)| \leq (m/r)|z - z_0|$ is true for every z in Δ.

8.44. Given that a function f is analytic in the disk $\Delta = \Delta(0, 1)$ and there satisfies the conditions $f(0) = 1$ and $\operatorname{Re} f(z) \geq 0$, derive the bounds $|f'(0)| \leq 2$ and $|f(z)| \leq (1 + |z|)/(1 - |z|)$ for every z in Δ. Determine the functions for which equality holds in the derivative estimate. (*Hint.* Recalling Example I.3.4, consider the function $g(z) = [1 - f(z)]/[1 + f(z)]$.)

8. Exercises for Chapter V

8.45. Suppose that a function f is analytic in the disk $\Delta = \Delta(0,1)$, that $f(0) = f'(0) = 0$, and that $|f'(z)| \leq 1$ for every z in Δ. Prove that $|f(z)| \leq |z|^2/2$ for every z in Δ. For which functions f can equality hold in this estimate at some $z \neq 0$?

8.46. Establish the following generalization of the preceding exercise: if f is analytic in $\Delta = \Delta(0,1)$, if $f(0) = f'(0) = \cdots = f^{(k)}(0) = 0$, and if $|f^{(k)}(z)| \leq 1$ for every z in Δ, then $|f(z)| \leq |z|^{k+1}/(k+1)!$ for every z in this disk.

8.47. Let $D = \{z : \alpha < \operatorname{Re} z < \beta\}$, where $\alpha < \beta$, and let $f : \overline{D} \to \mathbb{C}$ be a bounded, continuous function that is analytic in D. Assume that one has the bounds $|f(\alpha + iy)| \leq a$ and $|f(\beta + iy)| \leq b$ for all real y, where $a \geq 0$ and $b \geq 0$ are constants. Obtain the estimate

$$|f(x+iy)| \leq a^{(\beta-x)/(\beta-\alpha)} b^{(x-\alpha)/(\beta-\alpha)}$$

for $\alpha < x < \beta$ and all real y. (*Hint.* Look at the auxiliary function $g(z) = f[(\beta - \alpha)z + \alpha]$.)

8.48. Let $H = \{z : \operatorname{Im} z > 0\}$, and let $f : \overline{H} \to \mathbb{C}$ be a bounded, continuous function that is analytic in H. Assuming that $|f(z)| \leq 1$ for all real z, prove that $|f(z)| \leq 1$ for every z in H. Show by example that this conclusion need not follow if the boundedness assumption on f is dropped. (*Hint.* Consider the auxiliary function $g(z) = (i + tz)^{-1} f(z)$ for $t > 0$. Use an argument along the lines of the one in the first part of the proof of Theorem 3.15.)

8.49. A function f is analytic in a plane domain D and exhibits the following behavior: there is a constant m such that $\limsup_{n \to \infty} |f(z_n)| \leq m$ whenever $\langle z_n \rangle$ is a sequence in D which either converges to a point of ∂D or has the property that $|z_n| \to \infty$ as $n \to \infty$. (For the meaning of the symbol $\limsup_{n \to \infty}$ see Appendix A.) Prove that $|f(z)| < m$ for every z in D, unless f is a constant function of modulus m in D. (N.B. The result described in this exercise is sometimes called the "strong maximum principle." *Hint.* Start by proving that $\sup\{|f(z)| : z \in D\} \leq m$.)

8.4 Exercises for Section V.4

8.50. Characterize the plane domains D in which there exists a branch of the logarithm of the function $f(z) = 1 - z^{-2}$. For which of the following sets A does $D = \mathbb{C} \sim A$ have this property: (i) $A = (-\infty, 0] \cup [1, \infty)$; (ii) $A = [-1, 0] \cup [1, \infty)$; (iii) $A = \{te^{\pi i t} : -1 \leq t \leq 1\}$; (iv) $A = \{e^{it} : 0 \leq t \leq \pi\} \cup \{ti : -\infty \leq t \leq 0\}$?

8.51. For which of the following sets A does there exist an analytic function $f : \mathbb{C} \sim A \to \mathbb{C}$ that obeys the identity $e^{f(z)} = z^{-2}(z+1)^{-1}(z^2+1)$:

(i) $A = (-\infty, -1] \cup [0, \infty) \cup \{ti : t \in \mathbb{R} \text{ and } |t| \geq 1\}$; (ii) $A = (-\infty, -1] \cup \{ti : -1 \leq t \leq 1\}$; (iii) $A = (-\infty, 0] \cup \{e^{it} : -\pi/2 \leq t \leq \pi/2\}$; (iv) $A = \{e^{it} : \pi/2 \leq t \leq \pi\} \cup \{ti : -\infty < t \leq 0\}$?

8.52. There is a unique branch f of the inverse tangent function in the domain $D = \mathbb{C} \sim \{e^{it} : -\pi/2 \leq t \leq \pi/2\}$ meeting the condition $f(0) = 0$. (Recall Example 4.5.) Determine $f(-\sqrt{3})$ and $f(\sqrt{3})$.

8.53. Let f be a rational function of the type in Theorem 4.2. Assuming that a branch of the pth-root of f exists in a domain D which omits the points $z_1, z_2, \ldots, z_r$, show that $m_1 n(\gamma, z_1) + m_2 n(\gamma, z_2) + \cdots + m_r n(\gamma, z_r)$ must be a multiple of p for any closed, piecewise smooth path γ in D.

8.54. Let f be a quadratic polynomial function of z with two different roots z_1 and z_2. Given that a branch g of the square root of f exists in a domain D, demonstrate that neither z_1 nor z_2 can belong to D. If f had a double root, would the analogous statement be true?

8.55. Suppose that f is a quadratic polynomial function of z with two distinct roots z_1 and z_2, and that D is a plane domain. If $n(\gamma, z_1) = n(\gamma, z_2)$ for every closed, piecewise smooth path γ in D, then there exists an analytic function $g : D \to \mathbb{C}$ with the property that $g^2 = f$ in D. Confirm this.

8.56. Decide whether there is a branch of the pth-root of f in the domain $D = \mathbb{C} \sim A$: (i) $p = 3$, $f(z) = z(z-1)(z+1)$, $A = (-\infty, -1] \cup [0, 1]$; (ii) $p = 2$, $f(z) = z^2 - 2z$, $A = [0, 2]$; (iii) $p = 4$, $f(z) = z^3 + z$, $A = (-\infty, 0] \cup [1, \infty) \cup \{e^{it} : -\pi/2 \leq t \leq \pi/2\}$; (iv) $p = 2$, $f(z) = z^4 - 1$, $A = \{e^{it} : -\pi/2 \leq t \leq 0, \pi/2 \leq t \leq \pi\}$; (v) $p = 3$, $f(z) = z^3 + z^2 + z + 1$, $A = \{e^{it} : \pi/2 \leq t \leq 3\pi/2\}$.

8.57. Suppose that a function f is analytic and free of zeros in a domain D. Under the assumption that a branch g of the pth-root of f exists in D, show that there are exactly p distinct branches of the pth-root of f in D, each having the form cg for some p^{th}-root of unity c. (*Hint.* Modify the proof of Theorem III.4.2.)

8.58. Let $f(z) = (z-z_1)(z-z_2)\cdots(z-z_r)$, where $z_1, z_2, \ldots, z_r$ are distinct complex numbers, and let D be a domain with the following property: z_1 and z_2 belong to a connected subset C_1 of $\mathbb{C} \sim D$, z_3 and z_4 belong to a connected subset C_2 of $\mathbb{C} \sim D$, etc. (If r is odd, assume that z_r belongs to an unbounded connected subset of $\mathbb{C} \sim D$.) Verify that a branch of the square root of f exists in D.

8.59. Let g be the branch of the square root of $f(z) = (z-1)(z-2)(z+2)$ in the domain $D = \mathbb{C} \sim \{(-\infty, -2] \cup [1, 2]\}$ with $g(0) = 2$. Find a formula for g in terms of elementary functions. (*Warning.* The function g defined by the formula $g(z) = \sqrt{(z-1)(z-2)(z+2)}$ does not meet the requirement. This function has discontinuities in D.)

8.5 Exercises for Section V.5

8.60. Let $\gamma = [1,2] + \alpha + [2i, i] - \beta$, where $\alpha(t) = 2e^{it}$ and $\beta(t) = e^{it}$ for $0 \le t \le 5\pi/2$. In which of the following open sets is the path γ homologous to zero: (i) $\mathbb{C} \sim [0, 1/2]$; (ii) $\mathbb{C} \sim [3, \infty)$; (iii) $\{z: e^{-1} < |z| < e\}$; (iv) $\mathbb{C} \sim \{(3/2)e^{it} : 3\pi/4 \le t \le 5\pi/4\}$; (v) $\mathbb{C} \sim \{\sqrt[3]{2} + i\sqrt[3]{2}\}$?

8.61. Let $\alpha(t) = e^{it}$, $\beta(t) = (5/3) + e^{it}$, and $\gamma(t) = -1 + 2e^{it}$ for t in the interval $[0, 2\pi]$. In the following situations decide whether the cycle σ is homologous to zero in the open set U: (i) $\sigma = (\alpha, \beta, \gamma)$, $U = \mathbb{C} \sim \{2i, -2i\}$; (ii) $\sigma = (2\alpha, -\beta, -\gamma)$, $U = \{z: 1/5 < |z| < 5\}$; (iii) $\sigma = (\alpha, -\beta, \gamma)$, $U = \mathbb{C} \sim \{i\sqrt{2}\}$; (iv) $\sigma = (\alpha, -\beta)$, $U = \mathbb{C} \sim \{3/4, 2\}$; (v) $\sigma = (\alpha + \gamma, -2\beta)$, $U = \mathbb{C} \sim [-1/2, 1/2]$.

8.62. If $\alpha, \beta,$ and γ are as in the preceding exercise, determine whether the cycles σ and τ are homologous in the open set U: (i) $\sigma = \alpha$, $\tau = \beta$, $U = \mathbb{C} \sim \{2, -2\}$; (ii) $\sigma = \alpha - \gamma$, $\tau = 2\beta$, $U = \mathbb{C} \sim \overline{\Delta}(0, 1/2)$; (iii) $\sigma = (\alpha, \beta)$, $\tau = (\alpha, \gamma)$, $U = \mathbb{C} \sim [-5/2, -2]$; (iv) $\sigma = (\alpha, \beta, \gamma)$, $\tau = (2\gamma, \beta)$, $U = \mathbb{C} \sim \{0\}$.

8.63. Let $\gamma = [i, -1-i] + [-1-i, 1/2] + \beta + [-1/2, 1-i] + [1-i, i]$, where $\beta(t) = e^{it}/2$ for $0 \le t \le \pi$. Evaluate $\int_\gamma (z^4 + 2z^2)^{-1} \operatorname{Arcsin} z \, dz$.

8.64. If $\gamma = \beta + \alpha$, where $\beta(t) = 3\cos t + i\sin t$ for $-\pi/2 \le t \le \pi/2$ and $\alpha(t) = e^{-it}$ for $-\pi/2 \le t \le \pi/2$, compute $\int_\gamma z^{-1}(z-2)^{-2}(z-4)^{-1} \operatorname{Log} z \, dz$.

8.65. A function f is analytic in an annulus D centered at a point z_0 — say in $D = \{z: a < |z - z_0| < b\}$, where $0 \le a < b \le \infty$. Show that $\int_{|z-z_0|=r} f(z)\, dz = \int_{|z-z_0|=s} f(z)\, dz$ whenever $a < r < s < b$.

8.66. Demonstrate that $\lim_{r\to\infty} \int_{c-ir}^{c+ir} z^{-\lambda}(\operatorname{Log} z)^{-1} dz = 0$ when $c > 1$ and $\operatorname{Re}\lambda \ge 1$. (*Hint.* Consider $\int_\gamma z^{-\lambda}(\operatorname{Log} z)^{-1} dz$, where $\gamma(t) = c + re^{it}$ for $-\pi/2 \le t \le \pi/2$.)

8.67. Verify that $\lim_{r\to\infty} \int_{c-ir}^{c+ir} z^{-\lambda}(\operatorname{Log} z)^{-1} dz = -2\pi i$ when $0 < c < 1$ and $\operatorname{Re}\lambda \ge 1$. (*Hint.* Calling to mind Exercise 8.18, modify the approach in the preceding exercise.)

8.68. Assuming that a function f is continuous on the closed disk $\overline{\Delta}(0, r)$ and analytic in $\Delta(0, r)$, prove that $\int_{|z|=r} f(z)\, dz = 0$ and that $f^{(k)}(z) = (k!/2\pi i) \int_{|z|=r} (\zeta - z)^{-k-1} f(\zeta)\, d\zeta$ for every z in D and every non-negative integer k. (N.B. This establishes Theorem 5.7 in a special case. *Hint.* Consider the auxiliary function $g(z) = f(tz)$ for $0 < t < 1$. Recall Theorem II.4.8.)

8.69. Let R be a closed rectangle in the complex plane. If $f: R \to \mathbb{C}$ is a continuous function that is analytic in the interior R° of R, show that $\int_{\partial R} f(z)\, dz = 0$ and that $f^{(k)}(z) = (k!/2\pi i) \int_{\partial R} (\zeta-z)^{-k-1} f(\zeta)\, d\zeta$ for every

z in R° and every non-negative integer k. This will prove Goursat's theorem in another special case.

8.70. Let D be a plane domain with the property that $\mathbb{C} \sim D = \bigcup_{j=1}^{p+1} F_j$, where $F_1, F_2, \ldots, F_{p+1}$ are pairwise disjoint, closed, connected sets and only F_{p+1} is unbounded. For $j = 1, 2, \ldots, p$ let γ_j be a piecewise smooth, simple, closed path in D such that $n(\gamma_j, z) = 1$ for every z in F_j, whereas $n(\gamma_j, z) = 0$ whenever z belongs to F_k for $k \neq j$. Show that any cycle σ in D is homologous in D to a unique cycle τ of the type $\tau = (n_1\gamma_1, n_2\gamma_2, \ldots, n_p\gamma_p)$ for integers $n_1, n_2, \ldots, n_p$.

8.6 Exercises for Section V.6

8.71. A plane domain D has the property that each point z of $\mathbb{C} \sim D$ is contained in some unbounded, connected subset A_z of $\mathbb{C} \sim D$. Prove that D is simply connected.

8.72. Exhibit a simply connected domain D whose complement $\mathbb{C} \sim D$ is the disjoint union of an infinite number of closed, connected sets.

8.73. If D_1 and D_2 are simply connected domains in the complex plane whose intersection is non-empty and connected, then both of the domains $D_1 \cap D_2$ and $D_1 \cup D_2$ are simply connected. Prove this. Is this statement still true of $D_1 \cup D_2$ when $D_1 \cap D_2$ is non-empty, but disconnected? (*Hint.* Theorem 6.1 helps with the union.)

8.74. Suppose that a function f is analytic in a simply connected domain D. Let $\gamma : [a, b] \to D$ be a closed, piecewise smooth path, and let $\beta = f \circ \gamma$. Confirm that $n(\beta, w) = 0$ for every w in $\mathbb{C} \sim f(D)$. Given that $f(D)$ is a domain, does this imply that $f(D)$ is simply connected?

8.75. Demonstrate that a plane domain D is simply connected if and only if for every function f that is analytic and zero-free in D there is a branch of the square root of f in this domain. Can "square root" be replaced here by "p^{th}-root" for $p \geq 3$?

8.7 Exercises for Section V.7

8.76. If $\alpha : [a, b] \to A$ and $\beta : [a, b] \to A$ are closed paths in a set A that is starlike with respect to a point z_0 (see Chapter II.3.2) show that α and β are freely homotopic in A.

8.77. Given that closed paths α and β in a set A are freely homotopic there, prove that $-\alpha$ and $-\beta$ are also freely homotopic in A. (N.B. The same is true if "freely homotopic" is replaced by "homotopic with fixed

8. Exercises for Chapter V

endpoints.")

8.78. If $\alpha:[a,b] \to \mathbb{C}$ and $\beta:[a,b] \to \mathbb{C}$ are closed paths and if β arises from α by a change of parameter $h:[a,b] \to [a,b]$, then α and β are freely homotopic (and also homotopic with fixed endpoints) in any set A that contains $|\alpha|$. Confirm this fact.

8.79. Assume that paths α and β are freely homotopic (respectively, homotopic with fixed endpoints) in a subset A of the complex plane and that $f: A \to \mathbb{C}$ is a continuous function. Verify that the paths $f \circ \alpha$ and $f \circ \beta$ are freely homotopic (resp., homotopic with fixed endpoints) in the set $f(A)$.

8.80. If paths α_1 and β_1 are homotopic with fixed endpoints in a set A, if the same is true of paths α_2 and β_2, and if the path sum $\alpha_1 + \alpha_2$ — hence, also, $\beta_1 + \beta_2$ — is defined, then $\alpha_1 + \alpha_2$ and $\beta_1 + \beta_2$ are homotopic with fixed endpoints in A. Prove this.

8.81. If γ is a path in the complex plane, verify that the path $\gamma - \gamma$ is contractible in any set A that contains the trajectory of γ.

8.82. Decide whether α and β are freely homotopic in A:
  (i) $\alpha(t) = e^{it}$ and $\beta(t) = -1 + 2e^{it}$ for $0 \leq t \leq 2\pi$, $A = \mathbb{C} \sim \{0\}$;
  (ii) $\alpha(t) = e^{2it}$ and $\beta(t) = -1 + 2e^{it}$ for $0 \leq t \leq 2\pi$, $A = \mathbb{C} \sim \{0\}$;
  (iii) $\alpha(t) = 2e^{it}$ and $\beta(t) = 2\cos t + i\sin t$ for $0 \leq t \leq 2\pi$, $A = \mathbb{C} \sim [0,1]$;
  (iv) $\alpha(t) = e^{it}$ and $\beta(t) = i$ for $0 \leq t \leq 2\pi$, $A = \mathbb{C} \sim \{-2i\}$;
  (v) $\alpha(t) = e^{it}$ and $\beta(t) = i$ for $0 \leq t \leq 2\pi$, $A = \mathbb{C} \sim \{-i/2\}$;
  (vi) $\alpha(t) = 2ie^{it/2}$ for $0 \leq t \leq 4\pi$ and $\beta = \beta_1 + \beta_2$ with $\beta_1(t) = -1 + e^{it}$ and $\beta_2(t) = 1 - e^{it}$ for $0 \leq t \leq 2\pi$, $A = \mathbb{C} \sim \{1/2, -1/2\}$.

8.83. Show that a plane domain D which is starlike with respect to one of its points is simply connected.

Chapter VI

Harmonic Functions

Introduction

This chapter offers the reader a distinct change of pace. In it we apply some of the things we have learned about complex function theory to study a class of real-valued functions intimately connected with analytic functions, the class of plane harmonic functions. Even putting their close association with analytic functions aside, harmonic functions are tremendously important for the role they play in physics and engineering, where they turn up naturally in a variety of contexts, including electrostatics, fluid dynamics, acoustics, and heat transfer. Unfortunately, our limited treatment of the topic here can barely hope to scratch the surface of the rich theory of harmonic functions in two dimensions, to say nothing of the higher dimensional aspects of the subject. Readers eager to learn more about these interesting functions are encouraged to consult the classic monograph *Foundations of Modern Potential Theory* by O.D. Kellogg (Springer-Verlag, Berlin-Heidelberg-New York, 1967; reprint of 1929 edition) or another advanced reference, *Introduction to Potential Theory* by L. L. Helms (Krieger Publishing, Huntingdon, N.Y., 1975; reprint of Wiley-Interscience, Pure and Applied Mathematics, Vol. 22, 1969).

In the beginning section of this chapter we approach the theory of harmonic functions as a natural offshoot of the theory of analytic functions: chiefly through the judicious application of Cauchy's integral formula, we identify a few of the key properties of harmonic functions. But as the chapter develops the tables slowly begin to turn, until in the final pages of the chapter, building on knowledge gained from the solution of the Dirichlet problem for a disk, we extract a valuable piece of new information about analytic functions from our study of harmonic ones!

The discussions in Chapters VII through X do not depend in any critical way on the contents of Chapter VI. In reading this book, therefore, it is entirely possible to proceed directly from Chapter V to Chapter VII without fear of skipping over material indispensable to an understanding of the later chapters.

1 Harmonic Functions

1.1 Harmonic Conjugates

Given a plane domain D and a function $u: D \to \mathbb{R}$, we raise the question: Under what conditions is u the real part of an analytic function $f: D \to \mathbb{C}$? For such a function $f = u + iv$ to exist it is necessary (Corollary V.3.2) that u belong to the class $C^\infty(D)$. Furthermore, in light of the Cauchy-Riemann relations, the existence of f imposes another significant constraint on u: $u_{xx} + u_{yy} = 0$ throughout D. Indeed, we have

$$u_{xx} + u_{yy} = (u_x)_x + (u_y)_y = (v_y)_x + (-v_x)_y = v_{yx} - v_{xy} = 0 \ .$$

The partial differential equation $u_{xx} + u_{yy} = 0$ is known as *Laplace's equation* in honor of Pierre-Simon Laplace (1749-1827), one of the first persons to recognize its importance. The symbol Δu is traditionally used in the present setting as an abbreviation for $u_{xx} + u_{yy}$. Thus, the Laplace equation is frequently written in the form $\Delta u = 0$. The solutions of this equation are called "harmonic functions." To be a little more precise, a real-valued function u defined in an open subset U of $\mathbb{C}$ is said to be *harmonic in U* provided u is a member of the class $C^2(U)$ and $\Delta u = 0$ everywhere in U. A real-valued function with an open domain-set in which that function is harmonic is known as a *harmonic function*. The foregoing discussion merely points out that a function aspiring to be the real part of an analytic function in an open set had better be harmonic in this set. A similar argument shows that the imaginary part of an analytic function is likewise harmonic.

Taking into account what has just been said, we rephrase our opening question: Under the assumption that a function u is harmonic in a plane domain D, does there exist a harmonic function $v: D \to \mathbb{R}$ such that the function $f = u + iv$ is analytic in D? Any function v that fits this description is termed a *conjugate harmonic function* (or, more briefly, a *harmonic conjugate*) for u in D. Even so reformulated the question has the answer "not necessarily." For instance, we shall see in Example 1.3 that $u(z) = \text{Log}\,|z|$, a function readily checked to be harmonic in $D = \mathbb{C} \sim \{0\}$, has no harmonic conjugate in that domain. In general, the existence of a harmonic conjugate in a given domain D for a given harmonic function u depends on a number of factors, not the least of which may be the topological character of D. (See Theorem 1.1.) If u does admit a conjugate harmonic function in D, the latter function is certainly not unique — adding any real constant to one harmonic conjugate creates another. As a matter of fact, this procedure tells us how, starting with some harmonic conjugate v for u in D, to determine all others. Suppose, namely, that v_1 is a second harmonic conjugate for u in D. The functions $f = u + iv$ and $f_1 = u + iv_1$ are by definition analytic in D, as is $g = f - f_1 = i(v - v_1)$. Since the values of g are purely imaginary, Theorem III.2.5 implies that g is constant in D. This translates to: v_1 is obtained from v by adding to it a real constant.

There are certain domains in which every harmonic function has harmonic conjugates. The class of such domains is one we have met before.

Theorem 1.1. *Let D be a domain in the complex plane. Then it is true that every function which is harmonic in D possesses a harmonic conjugate in this domain if and only if D is simply connected.*

Proof. We assume first that D is simply connected and construct a harmonic conjugate in D for a function u known to be harmonic there. For this we consider the function $g = u_x - iu_y$. Since u belongs to $C^2(D)$, g is in the class $C^1(D)$. The fact that $u_{xx} + u_{yy} = 0$ in D permits us to verify that the real and imaginary parts of g satisfy the Cauchy-Riemann equations in the given domain:

$$(u_x)_x = u_{xx} = -u_{yy} = (-u_y)_y \quad , \quad (u_x)_y = u_{xy} = u_{yx} = -(-u_y)_x \ .$$

According to Theorem III.2.2, the function g is analytic in D. Theorem V.6.1 empowers us to choose a primitive $f = \tilde{u} + iv$ for g in this domain. Moreover, we are free to fix a point z_0 of D and to insist that $f(z_0) = u(z_0)$. (This determines f uniquely.) Thus $\tilde{u}(z_0) = u(z_0)$. Now

$$\tilde{u}_x - i\tilde{u}_y = f' = g = u_x - iu_y$$

in D, so $(\tilde{u} - u)_x = (\tilde{u} - u)_y = 0$ throughout D. Lemma III.2.3 asserts that $\tilde{u} - u$ is constant there. Because $\tilde{u}(z_0) = u(z_0)$, that constant is zero; i.e., $\tilde{u} = u$ in D. We conclude that $f = u + iv$ in D. Since f is by its very definition analytic in D, v furnishes a harmonic conjugate for u in said domain.

For the converse we begin with the assumption that every function which is harmonic in D possesses a harmonic conjugate there. We must deduce from this information that D is simply connected. Let γ be a closed, piecewise smooth path in D. We wish to show that γ is homologous to zero in D; i.e., given z_0 in $\mathbb{C} \sim D$, we want to verify that $n(\gamma, z_0) = 0$. To do this, we define a function u in $\mathbb{C} \sim \{z_0\}$ by $u(z) = \text{Log}\,|z - z_0|$. The function u is harmonic in $\mathbb{C} \sim \{z_0\}$ — hence, in D — as a simple calculation reveals (Exercise 4.1). By hypothesis we can choose a harmonic conjugate v for u in D. Thus, the function $f = u + iv$ is analytic in D. So, too, is the function $h: D \to \mathbb{C}$ given by $h(z) = (z - z_0)e^{-f(z)}$. We observe next that

$$|h(z)| = |z - z_0|e^{-u(z)} = |z - z_0|e^{-\text{Log}\,|z-z_0|} = \frac{|z - z_0|}{|z - z_0|} = 1$$

for every z in D. In view of Theorem III.2.5, h must be constant in D. This has the consequence that

$$0 = h'(z) = e^{-f(z)} - (z - z_0)e^{-f(z)}f'(z)$$

1. Harmonic Functions

throughout D, which in turn implies that $f'(z) = (z - z_0)^{-1}$ for every z in this domain. Appealing to Theorem IV.2.2 we can then assert that

$$n(\gamma, z_0) = \frac{1}{2\pi i} \int_\gamma \frac{d\zeta}{\zeta - z_0} = \frac{1}{2\pi i} \int_\gamma f'(\zeta)\, d\zeta = 0 \ ,$$

as we had hoped. Therefore, D is simply connected. ∎

Theorem 1.1 implies that every function which is harmonic in the whole complex plane is the real part of an entire function. In another special case the theorem certifies that a function u which is harmonic in an open disk Δ always has harmonic conjugates in Δ. A dividend paid by this comment and Corollary V.3.2 is that u, which is only required by the definition of harmonicity to be in the class $C^2(\Delta)$, is actually a member of the class $C^\infty(\Delta)$. The preceding remark applies locally to any harmonic function. It leads to a noteworthy corollary of Theorem 1.1.

Corollary 1.2. *If a function u is harmonic in an open set U, then u possesses a harmonic conjugate in each open disk that is contained in U. In particular, u belongs to the class $C^\infty(U)$.*

We explore the notion of a harmonic conjugate in some concrete examples.

EXAMPLE 1.1. Let $u(z) = 3xy^2 - x^3$. Show that u is harmonic in the complex plane, and find the harmonic conjugate v for u in $\mathbb{C}$ that obeys the condition $v(0) = 1$.

We have $u_{xx}(z) + u_{yy}(z) = -6x + 6x = 0$, demonstrating that u is harmonic in $\mathbb{C}$. Theorem 1.1 insures that u has harmonic conjugates v in $\mathbb{C}$, and the condition $v(0) = 1$ specifies one such v uniquely. In fact, from the Cauchy-Riemann equations we get

$$v_x(z) = -u_y(z) = -6xy \quad , \quad v_y(z) = u_x(z) = 3y^2 - 3x^2 \ .$$

In order for the first of these equations to hold, v must be of the form

$$v(z) = -3x^2 y + \varphi(y) \ ,$$

where φ is a function of y alone. Then

$$v_y(z) = -3x^2 + \frac{d\varphi}{dy}(y) \ .$$

Comparing this with our earlier expression for v_y, we find that

$$\frac{d\varphi}{dy}(y) = 3y^2 \ ,$$

which forces φ into the mold $\varphi(y) = y^3 + c$ for a real constant c and so gives v the form $v(z) = -3x^2y + y^3 + c$. Since $v(0) = 1$, we must take $c = 1$. We arrive, therefore, at

$$v(z) = -3x^2y + y^3 + 1 .$$

(N.B. Here $f = u + iv$ is the polynomial function $f(z) = -z^3 + i$.)

EXAMPLE 1.2. Let $u: \mathbb{C} \to \mathbb{R}$ be defined by

$$u(re^{i\theta}) = a_0 + \sum_{k=1}^{n} r^k \left[a_k \cos(k\theta) + b_k \sin(k\theta) \right]$$

for $r \geq 0$ and for real θ, where $a_0, a_1, \ldots, a_n$ and $b_1, b_2 \ldots, b_n$ are real constants. Demonstrate that u is a harmonic function in $\mathbb{C}$, and determine the harmonic conjugate v of u in the plane that satisfies $v(0) = 0$.

For $z = re^{i\theta}$ we compute

$$u(z) = a_0 + \sum_{k=1}^{n} r^k \left[a_k \cos(k\theta) + b_k \sin(k\theta) \right]$$

$$= a_0 + \sum_{k=1}^{n} r^k \left[a_k \operatorname{Re}(e^{ik\theta}) + b_k \operatorname{Im}(e^{ik\theta}) \right]$$

$$= a_0 + \sum_{k=1}^{n} r^k \left[a_k \operatorname{Re}(e^{ik\theta}) + b_k \operatorname{Re}(-ie^{ik\theta}) \right]$$

$$= a_0 + \sum_{k=1}^{n} \operatorname{Re}\left[(a_k - ib_k) r^k e^{ik\theta} \right]$$

$$= \operatorname{Re}\left[a_0 + \sum_{k=1}^{n} (a_k - ib_k) z^k \right] .$$

As the real part of the polynomial function $f(z) = a_0 + \sum_{k=1}^{n}(a_k - ib_k)z^k$, u is plainly harmonic in the complex plane. The function $v: \mathbb{C} \to \mathbb{R}$ given by $v(z) = \operatorname{Im}[f(z)]$ is a harmonic conjugate for u in the plane. Since $v(0) = \operatorname{Im}[f(0)] = \operatorname{Im} a_0 = 0$, it is this particular conjugate we seek. Thus, for $z = re^{i\theta}$ we obtain

$$v(z) = \operatorname{Im}\left[a_0 + \sum_{k=1}^{n} (a_k - ib_k) z^k \right]$$

$$= \sum_{k=1}^{n} \operatorname{Im}\left[(a_k - ib_k) r^k e^{ik\theta} \right]$$

2. The Mean Value Property

$$= \sum_{k=1}^{n} r^k \left[a_k \operatorname{Im}(e^{ik\theta}) - b_k \operatorname{Im}(ie^{ik\theta})\right]$$

$$= \sum_{k=1}^{n} r^k \left[-b_k \cos(k\theta) + a_k \sin(k\theta)\right] .$$

To summarize: the desired harmonic conjugate v is given by

$$v(re^{i\theta}) = \sum_{k=1}^{n} r^k \left[-b_k \cos(k\theta) + a_k \sin(k\theta)\right]$$

for $r \geq 0$ and real θ.

EXAMPLE 1.3. Verify that the function $u(z) = \operatorname{Log}|z|$ has no harmonic conjugate in the domain $D = \mathbb{C} \sim \{0\}$.

If v were such a conjugate for u in D, the analytic function $f = u + iv$ would have $f'(z) = z^{-1}$ for z in D, as the argument in the last part of the proof of Theorem 1.1 clearly demonstrates. We have already seen in Example IV.2.8 that there can be no function with that derivative in D — hence, no harmonic conjugate for u there.

In conjunction with the last example, we make the following remark. Suppose that $L : D \to \mathbb{C}$ is a branch of $\log z$ in a plane domain D. Then the associated branch θ of $\arg z$ in D — θ is just the imaginary part of L — is obviously a harmonic conjugate for the function $u(z) = \operatorname{Log}|z|$ in that domain. Conversely, if a harmonic conjugate v can be produced for this particular function u in a domain D and if v is adjusted (by the addition of a real constant) so that, at some specified point z_0 of D, the value $v(z_0)$ is an argument of z_0, then $L = u + iv$ will be a branch of $\log z$ in D (Exercise 4.8).

2 The Mean Value Property

2.1 The Mean Value Property

We begin this section with a short calculation. We suppose at the outset that u is a function which is harmonic in a plane open set U, that z is a point of U, and that $\rho > 0$ is small enough for the disk $D = \Delta(z, \rho)$ to be a subset of U. We then choose, as Theorem 1.1 entitles us to do, a harmonic conjugate v for u in D. Finally, for $0 < r < \rho$ we apply Cauchy's integral formula to $f = u + iv$ and compute

$$u(z) = \operatorname{Re}[f(z)] = \operatorname{Re}\left[\frac{1}{2\pi i} \int_{|\zeta - z| = r} \frac{f(\zeta)\, d\zeta}{\zeta - z}\right]$$

$$= \operatorname{Re}\left[\frac{1}{2\pi}\int_0^{2\pi} f(z+re^{i\theta})\,d\theta\right] = \frac{1}{2\pi}\int_0^{2\pi} u(z+re^{i\theta})\,d\theta\ .$$

The conclusion we draw from this is that

(6.1) $$u(z) = \frac{1}{2\pi}\int_0^{2\pi} u(z+re^{i\theta})\,d\theta$$

whenever $0 < r < \rho$. The right-hand side of (6.1) has a simple interpretation: it gives the mean (or average) value of u on the circle $K(z,r)$. Statement (6.1) expresses the fact that this mean value is the same for all r in the interval $(0,\rho)$ and coincides for such r with the value of u at the center of $K(z,r)$.

Motivated by (6.1) we make the following definition: a real-valued function w that is continuous in an open subset U of the complex plane has the *mean value property in* U if corresponding to each z in U there exists a radius $\rho = \rho(z) > 0$ such that

(6.2) $$w(z) = \frac{1}{2\pi}\int_0^{2\pi} w(z+re^{i\theta})\,d\theta$$

is true whenever $0 < r < \rho$. Implicit here is that the disk $\Delta(z,\rho)$ is contained in U. We do not demand, however, that (6.2) hold (as (6.1) would permit us to do in case w were known to be harmonic in U) for every radius ρ such that $\Delta(z,\rho)$ is contained in U, only that it hold when ρ is taken sufficiently small. We summarize the preceding remarks in a theorem.

Theorem 2.1. *If a function u is harmonic in an open set U, then u has the mean value property in U. Furthermore, any $\rho > 0$ such that the disk $\Delta(z,\rho)$ is contained in U can serve as the radius $\rho = \rho(z)$ corresponding to the point z in the aforementioned property.*

We shall later see (Theorem 3.3) that the mean value property distinguishes those functions that are harmonic in an open set U among all real-valued functions w that are continuous in U: if such a function w has the mean value property in U, then it is automatically harmonic in U. This fact begins to explain the comments made in Chapter V.3.3, where without much ceremony we declared a real-valued continuous function w to be *subharmonic in* U under the following condition: corresponding to each z in D there exists a radius $\rho = \rho(z) > 0$ with the property that the inequality

(6.3) $$w(z) \leq \frac{1}{2\pi}\int_0^{2\pi} w(z+re^{i\theta})\,d\theta$$

holds whenever $0 < r < \rho$. Consistent with this, the label *superharmonic in* U is attached to a continuous function that enjoys the analogous property with the direction of the inequality in (6.3) reversed. Thus, as it turns out, a function is harmonic in U if and only if it is simultaneously subharmonic

2. The Mean Value Property

and superharmonic there. It should be pointed out that, whereas a function known to be harmonic in U necessarily belongs to the class $C^\infty(U)$, a function that is only subharmonic or superharmonic in U need not be a member of the class $C^1(U)$. (The general definitions of subharmonic and superharmonic functions found in the literature do not even require that they be continuous, insisting only on "semi-continuity," a concept which will not be needed in this book, but the existence of which a reader pursuing the topic of harmonicity ought to be aware.) In view of Theorem 2.1 and the present discussion we are justified in stating for the record the versions of Theorem V.3.12 and its corollary that refer specifically to harmonic functions.

Theorem 2.2. *Let a function u be harmonic in a domain D. Suppose that there exists a point z_0 of D with the property that $u(z) \leq u(z_0)$ for every z in D. Then u is constant in D.*

Corollary 2.3. *Let D be a bounded domain in the complex plane, and let $u: \overline{D} \to \mathbb{R}$ be a continuous function that is harmonic in D. Then $u(z)$ attains its maximum at some point on the boundary of D.*

Here is a typical application of Corollary 2.3.

EXAMPLE 2.1. Let D be a bounded domain in the complex plane. Suppose that u and v are real-valued functions which are continuous on $\overline{D}$, are harmonic in D, and satisfy $u = v$ on ∂D. Prove that $u = v$ in D.

We consider the function $w: \overline{D} \to \mathbb{R}$ defined by $w(z) = u(z) - v(z)$. This function is continuous, it is harmonic in D, and it vanishes on ∂D. By Corollary 2.3, w attains its maximum value somewhere in ∂D, so $w \leq 0$ in D. The identical argument applies to $-w$ and leads to the conclusion that $w \geq 0$ in D. In combination these inequalities imply that $w = 0$ throughout D, whence $u = v$ there.

If we were to weaken the hypothesis on u in Example 2.1 from harmonicity in D to subharmonicity there, then the same argument would allow us to conclude that $w \leq 0$ in D — hence, that $u \leq v$ in the domain — for even under the weaker assumption on u the function w would remain subharmonic in D, meaning that Corollary V.3.13 could be invoked in place of Corollary 2.3. This inspires a perhaps more satisfactory explanation of the terminology "subharmonic" than the one given earlier: a subharmonic function always "lies below" any harmonic function with which it shares the same boundary values.

2.2 Functions Harmonic in Annuli

There is a second general situation in which harmonicity leads to some interesting information about mean values. We have in mind a function that is harmonic in an annulus and its mean values over circles concentric with the annulus.

Theorem 2.4. *Suppose that a function u is harmonic in the domain $D = \{z : a < |z - z_0| < b\}$, where $0 \leq a < b \leq \infty$. There exist constants c and d such that*

(6.4) $$\frac{1}{2\pi} \int_0^{2\pi} u(z_0 + re^{i\theta}) \, d\theta = c \operatorname{Log} r + d$$

for every r in (a, b).

Proof. The secret here is to work in polar coordinates. We look at the strip $V = \{(r, \theta) : a < r < b, -\infty < \theta < \infty\}$ and the function v defined in V by $v(r, \theta) = u(z_0 + re^{i\theta})$, a function that clearly belongs to the class $C^\infty(V)$ with respect to the variables r and θ. A direct calculation (Exercise 4.21), utilizing only the fact that u is a member of $C^2(D)$, discloses that

(6.5) $$v_{rr}(r, \theta) + r^{-1} v_r(r, \theta) + r^{-2} v_{\theta\theta}(r, \theta) = \Delta u(z_0 + re^{i\theta})$$

at every point (r, θ) of V. The requirement that u be a solution of Laplace's equation in D is thus equivalent to v satisfying

(6.6) $$r^2 v_{rr}(r, \theta) + r v_r(r, \theta) + v_{\theta\theta}(r, \theta) = 0$$

throughout the open set V.

Let $m: (a, b) \to \mathbb{R}$ be the function of the real variable r defined by

$$m(r) = \frac{1}{2\pi} \int_0^{2\pi} u(z_0 + re^{i\theta}) \, d\theta = \frac{1}{2\pi} \int_0^{2\pi} v(r, \theta) \, d\theta \ .$$

We claim that m is differentiable on the interval (a, b), where its derivative is given by

(6.7) $$m'(r) = \frac{1}{2\pi} \int_0^{2\pi} v_r(r, \theta) \, d\theta \ .$$

This follows from standard results in advanced calculus, but for the sake of completeness we sketch a proof.

Fixing r in (a, b), we set $t = \min\{(r-a)/2, (b-r)/2\}$. Given θ in $[0, 2\pi]$ and h satisfying $0 < |h| \leq t$, we use the mean value theorem to express

$$v(r + h, \theta) - v(r, \theta) = v_r(s, \theta) h$$

for some number $s = s(\theta, h)$ between r and $r + h$. This leads to the estimate

$$\left| \frac{v(r + h, \theta) - v(r, \theta)}{h} - v_r(r, \theta) \right| = |v_r(s, \theta) - v_r(r, \theta)| \leq M(h) \ ,$$

in which we set

$$M(h) = \max\{|v_r(s, \theta) - v_r(r, \theta)| : r - |h| \leq s \leq r + |h|, \ 0 \leq \theta \leq 2\pi\} \ .$$

2. The Mean Value Property

Accordingly, when $0 < |h| \leq t$ we discover that

$$\left| \frac{m(r+h) - m(r)}{h} - \frac{1}{2\pi} \int_0^{2\pi} v_r(r,\theta) \, d\theta \right|$$

$$\leq \frac{1}{2\pi} \int_0^{2\pi} \left| \frac{v(r+h,\theta) - v(r,\theta)}{h} - v_r(r,\theta) \right| d\theta \leq M(h) .$$

Because v_r is continuous — hence, uniformly continuous — on the rectangle $R = \{(s,\theta) : r-t \leq s \leq r+t , \ 0 \leq \theta \leq 2\pi\}$, it follows easily that $M(h) \to 0$ as $h \to 0$. We conclude in this way that

$$\lim_{h \to 0} \frac{m(r+h) - m(r)}{h} = \frac{1}{2\pi} \int_0^{2\pi} v_r(r,\theta) \, d\theta ,$$

so $m'(r)$ exists and satisfies (6.7). A similar argument demonstrates that

(6.8) $$m''(r) = \frac{1}{2\pi} \int_0^{2\pi} v_{rr}(r,\theta) \, d\theta$$

for r in (a,b).

As a consequence of (6.6), (6.7), and (6.8), we see that on (a,b)

$$\frac{d}{dr}[rm'(r)] = rm''(r) + m'(r) = \frac{1}{2\pi r} \int_0^{2\pi} \left[r^2 v_{rr}(r,\theta) + r v_r(r,\theta) \right] d\theta$$

$$= -\frac{1}{2\pi r} \int_0^{2\pi} v_{\theta\theta}(r,\theta) \, d\theta = -\frac{1}{2\pi r} [v_\theta(r,2\pi) - v_\theta(r,0)] = 0 .$$

(N.B. For fixed r the quantity $v_\theta(r,\theta)$ is a periodic function of the variable θ with period 2π.) This obviously forces m to be a function of the type $m(r) = c \operatorname{Log} r + d$, where c and d are constants — precisely the assertion of the theorem. ∎

Theorem 2.4 gives rise to a second proof that harmonic functions enjoy the mean value property, a proof that makes no mention of analytic functions. If u is harmonic in an open set U, if z is a point of U, and if $\Delta(z,\rho)$ is contained in U, then we can apply Theorem 2.4 to u in the punctured disk $\Delta^*(z,\rho)$ to conclude that

(6.9) $$\frac{1}{2\pi} \int_0^{2\pi} u(z + re^{i\theta}) \, d\theta = c \operatorname{Log} r + d$$

for all r in $(0,\rho)$, where c and d are constants. On the other hand, the continuity of u implies that

(6.10) $$u(z) = \lim_{r \to 0} \frac{1}{2\pi} \int_0^{2\pi} u(z + re^{i\theta}) \, d\theta .$$

To verify (6.10) observe that for r in $(0, \rho)$:

$$\left| \frac{1}{2\pi} \int_0^{2\pi} u(z + re^{i\theta}) \, d\theta - u(z) \right| = \left| \frac{1}{2\pi} \int_0^{2\pi} [u(z + re^{i\theta}) - u(z)] \, d\theta \right|$$

$$\leq \frac{1}{2\pi} \int_0^{2\pi} |u(z + re^{i\theta}) - u(z)| \, d\theta \leq \max\{|u(z + re^{i\theta}) - u(z)| : \theta \in [0, 2\pi]\}.$$

Owing to the continuity of u at z the last expression tends to 0 with r, and (6.10) follows. The only way to accomodate both (6.9) and (6.10) is to have $c = 0$ and $d = u(z)$ — i.e., to have

$$u(z) = \frac{1}{2\pi} \int_0^{2\pi} u(z + re^{i\theta}) \, d\theta$$

whenever $0 < r < \rho$.

A bonus gleaned from computations made in the proof of Theorem 2.4 (and, if the truth be told, an ulterior motive for bringing up the theorem in the first place) is a characterization of subharmonic functions that are twice continuously differentiable.

Theorem 2.5. *Let U be an open set in the complex plane, and let w be a real-valued function that belongs to the class $C^2(U)$. Then w is subharmonic in U if and only if $\Delta w \geq 0$ throughout this set.*

Proof. We assume first that $\Delta w \geq 0$ everywhere in U. Given a point z of U we wish to find a radius $\rho = \rho(z) > 0$ with the property that

(6.11) $$w(z) \leq \frac{1}{2\pi} \int_0^{2\pi} w(z + re^{i\theta}) \, d\theta$$

for every r in $(0, \rho)$. In fact, we fix any ρ for which $\Delta(z, \rho)$ is contained in U and verify that (6.11) is true for this ρ. To this end we consider the function $m : (0, \rho) \to \mathbb{R}$ with rule of correspondence

$$m(r) = \frac{1}{2\pi} \int_0^{2\pi} w(z + re^{i\theta}) \, d\theta = \frac{1}{2\pi} \int_0^{2\pi} v(r, \theta) \, d\theta \,,$$

where $v(r, \theta) = w(z + re^{i\theta})$ in $V = \{(r, \theta) : 0 < r < \rho \,, -\infty < \theta < \infty\}$. As in the proof of Theorem 2.4, we obtain

$$m'(r) = \frac{1}{2\pi} \int_0^{2\pi} v_r(r, \theta) \, d\theta \,, \quad m''(r) = \frac{1}{2\pi} \int_0^{2\pi} v_{rr}(r, \theta) \, d\theta \,,$$

for r belonging to the interval $(0, \rho)$. In view of our present assumption, relation (6.5) — u is now replaced by w — permits us to infer that

$$r^2 v_{rr}(r, \theta) + r v_r(r, \theta) + v_{\theta\theta}(r, \theta) = r^2 \Delta w(z + re^{i\theta}) \geq 0$$

2. The Mean Value Property

in V, which carries with it the implication that

$$\frac{d}{dr}[rm'(r)] = rm''(r) + m'(r) = \frac{1}{2\pi r}\int_0^{2\pi} [r^2 v_{rr}(r,\theta) + rv_r(r,\theta)]\, d\theta$$

$$\geq -\frac{1}{2\pi r}\int_0^{2\pi} v_{\theta\theta}(r,\theta)\, d\theta = 0$$

on the interval $(0,\rho)$. This means that $rm'(r)$ is a non-decreasing function of r. Also, since

$$|m'(r)| \leq \frac{1}{2\pi}\int_0^{2\pi} |v_r(r,\theta)|\, d\theta$$

$$= \frac{1}{2\pi}\int_0^{2\pi} |w_x(z+re^{i\theta})\cos\theta + w_y(z+re^{i\theta})\sin\theta|\, d\theta$$

$$\leq \frac{1}{2\pi}\int_0^{2\pi} \{|w_x(z+re^{i\theta})| + |w_y(z+re^{i\theta})|\}\, d\theta \to |w_x(z)| + |w_y(z)|$$

as $r \to 0$ — the last assertion has a proof similar to that of (6.10) — the derivative $m'(r)$ remains bounded as $r \to 0$, with the consequence that $rm'(r) \to 0$ as $r \to 0$. We conclude that $rm'(r) \geq 0$ — hence, that $m'(r) \geq 0$ — for every r in $(0,\rho)$. We have thus succeeded in demonstrating that m itself is a non-decreasing function. Referring once again to the proof of (6.10), we make the inference that

$$w(z) = \lim_{r \to 0}\frac{1}{2\pi}\int_0^{2\pi} w(z+re^{i\theta})\, d\theta$$

$$= \lim_{r \to 0} m(r) \leq m(r) = \frac{1}{2\pi}\int_0^{2\pi} w(z+re^{i\theta})\, d\theta$$

for all r in $(0,\rho)$, which confirms (6.11). As z was an arbitrary point of U, the subharmonicity of w in U has been established.

In the converse direction it suffices to show that for $\Delta w(z) < 0$ to hold at any point z of U is incompatible with inequality (6.11), the defining condition for subharmonicity. Assuming, therefore, that z is a point of U for which $\Delta w(z) < 0$, we exploit the continuity of Δw to choose $\rho > 0$ such that $\Delta(z,\rho)$ lies in U and such that $\Delta w < 0$ throughout this open disk. In these circumstances the function $m\colon (0,\rho) \to \mathbb{R}$ that was introduced earlier has $m'(r) < 0$ for every r in $(0,\rho)$, as reasoning similar to that in the first part of this proof readily shows. Since m is then a strictly decreasing function, we are forced to conclude that

$$w(z) = \lim_{r \to 0} m(r) > m(r) = \frac{1}{2\pi}\int_0^{2\pi} w(z+re^{i\theta})\, d\theta$$

whenever $0 < r < \rho$, in obvious conflict with (6.11). It follows that a function w which is subharmonic in U and is also a member of the class $C^2(U)$ must have $\Delta w \geq 0$ throughout U. ∎

As previously noted, there are plenty of subharmonic functions that are not of class C^2. For example, $w(z) = |z|$ is subharmonic in the whole complex plane — we observed in the proof of Theorem V.3.10 that taking the modulus of an analytic function generates a subharmonic function — even though both $w_x(0)$ and $w_y(0)$ fail to exist. The dual to Theorem 2.5 dealing with superharmonic functions in the class $C^2(U)$ finds them to be the real-valued functions w from that class which have $\Delta w \leq 0$ in U.

3 The Dirichlet Problem for a Disk

3.1 A Heat Flow Problem

We consider a thin plate (or lamina) of homogeneous composition, of uniform thickness, but of arbitrary shape. (See Figure 1.) We assume that at some initial time the temperature (on a convenient scale) is known at each point of the plate and that it varies continuously with the position of the point. Suppose now that by some mechanism or other we are able to maintain the temperature at every point on the edge of the plate at its originally measured reading and that we somehow manage to insulate the plate's faces so perfectly that no heat can escape across them. We then

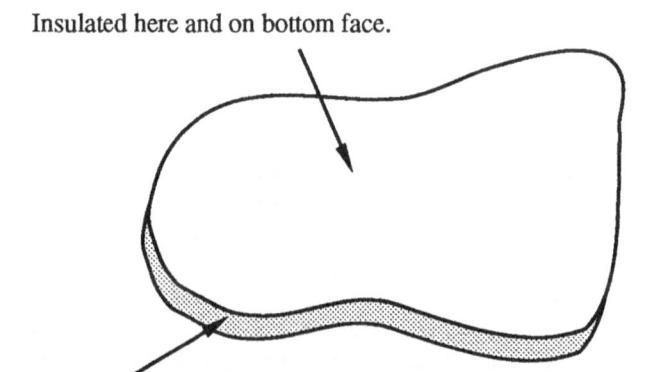

Figure 1.

pose the problem: Determine the temperature at any point of the lamina at any future time. As presented, of course, this is a three-dimensional

3. The Dirichlet Problem for a Disk

problem. Assuming that the lamina is extremely thin, however, and remarking that the insulation on the faces curtails the flow of heat in their directions, we might anticipate that only small variations of temperature would occur in the "vertical" direction; i.e., at any given time we might expect to see roughly the same temperature profile for every cross-section of the plate parallel to its faces. For a very thin plate, therefore, it is not physically outlandish to view the problem as essentially two-dimensional. A cross-section of the lamina can be thought of as the closure of a bounded domain D in the complex plane. The initial temperature data provides us with a continuous function $h: \overline{D} \to \mathbb{R}$. The problem comes down to finding the temperature $u(z,t)$ at each point z of $\overline{D}$ and at each time $t \geq 0$. Under the conditions described, the change of temperature in D is governed to good approximation by the *heat diffusion equation*,

(6.12) $$\alpha^2 u_t = u_{xx} + u_{yy},$$

subject to the boundary constraints $u(z,0) = h(z)$ for every z in D and $u(z,t) = h(z)$ for every z on ∂D and every $t \geq 0$. Here α is a constant determined by such factors as the substance of which the plate is composed, the units of measurement chosen, etc.

As time passes, we would expect the situation in the above system to stabilize and to approach some equilibrium configuration — or *steady state*, as it is usually styled. In this limiting state the temperature $u(z) = \lim_{t \to \infty} u(z,t)$ in our cross-section D would no longer fluctuate with time, but would be solely a function of position. Another way to express such time-independence would be to say that $u_t = 0$ in the steady state. It is thus suggested by (6.12) that the temperature distribution in D when the system reaches steady state might be modeled by the Laplace equation, $u_{xx} + u_{yy} = 0$, along with the only remaining boundary condition of relevance, $u(z) = h(z)$ for z on ∂D. This suggestion turns out in practice to be a tolerably good one.

If D is a bounded plane domain and if $h: \partial D \to \mathbb{R}$ is a continuous function, then the problem of solving the Laplace equation in D subject to the boundary condition $u = h$ on ∂D is known as the *Dirichlet problem for D with boundary data h*. (Much of the interest in this problem was stimulated by the work of P.G. Lejeune Dirichlet (1805-1859), whose name adorns it.) Posed more carefully the problem is to find a continuous function $u: \overline{D} \to \mathbb{R}$ that is harmonic in D and agrees with h on ∂D. It actually suffices to produce a harmonic function $u: D \to \mathbb{R}$ with the property that $\lim_{z \to \zeta} u(z) = h(\zeta)$ at each point ζ of ∂D, for in that case it is easy to see that the function which coincides with u in D and with h on ∂D is, in fact, continuous on $\overline{D}$. The steady-state heat conduction problem discussed above is among the many physical problems that reduce to the Dirichlet problem (or to its analogues in more than two variables). There is no assurance that in the generality stated the Dirichlet problem need have a solution. (See Example 3.2.) When a solution does exist, however, it is unique, as Example 2.1 makes clear. A domain D blessed with the property that the Dirichlet problem has a solution for every continuous boundary function h is said to be *regular for the Dirichlet problem*. We show in this chapter that an open disk is such a domain. In Chapter IX we shall employ conformal mapping techniques to prove that the inside of any plane Jordan curve is regular for the Dirichlet problem. A treatment of the problem for domains more arbitrary than these would involve ideas and methods not entirely appropriate in a book of this kind, so we leave the matter there.

In the following example we indicate a method for solving the Dirichlet problem in a disk centered at the origin when the boundary function is a polynomial in x and y.

EXAMPLE 3.1. Solve the Dirichlet problem for the disk $D = \Delta(0,2)$ with boundary data $h(z) = x^2 + 2xy^2$.

We set $z = 2e^{i\theta}$ and use trigonometric identities to rewrite $h: \partial D \to \mathbb{R}$ in the form

$$h(z) = 4\cos^2\theta + 16\cos\theta\sin^2\theta = 2[1 + \cos(2\theta)] + 8\cos\theta[1 - \cos(2\theta)]$$

$$= 2 + 8\cos\theta + 2\cos(2\theta) - 8\cos\theta\cos(2\theta)$$

$$= 2 + 8\cos\theta + 2\cos(2\theta) - 4\cos(3\theta) - 4\cos\theta$$

$$= 2 + 4\cos\theta + 2\cos(2\theta) - 4\cos(3\theta) .$$

Recalling Example 1.2, we note that for any choice of real constants a_1, a_2, and a_3 the function $u: \mathbb{C} \to \mathbb{R}$ defined by

$$u(re^{i\theta}) = 2 + 4a_1 r\cos\theta + 2a_2 r^2 \cos(2\theta) - 4a_3 r^3 \cos(3\theta)$$

is harmonic in the complex plane We simply pick a_1, a_2, and a_3 so as to make u agree with h when $r = 2$; i.e., we take $a_1 = 1/2$, $a_2 = 1/4$, and $a_3 = 1/8$. The solution of the given Dirichlet problem is thus provided by

$$u(z) = 2 + 2r\cos\theta + \frac{r^2 \cos(2\theta)}{2} - \frac{r^3 \cos(3\theta)}{2}$$

$$= \operatorname{Re}\left(2 + 2z + \frac{z^2}{2} - \frac{z^3}{2}\right)$$

$$= 2 + 2x + \frac{x^2}{2} - \frac{y^2}{2} - \frac{x^3}{2} + \frac{3xy^2}{2} .$$

3.2 Poisson Integrals

In preparation for the general solution of the Dirichlet problem in a disk we introduce the *Poisson kernel*. This is the name we give to the function

3. The Dirichlet Problem for a Disk

P of two complex variables that is defined at every point (z, ζ) of $\mathbb{C}^2$ for which $z \neq \zeta$ by the formula

$$P(z, \zeta) = \frac{|\zeta|^2 - |z|^2}{|\zeta - z|^2} = \operatorname{Re}\left(\frac{\zeta + z}{\zeta - z}\right).$$

Observe that, when $\zeta \neq 0$ is held fixed, $P(z, \zeta)$ becomes a positive harmonic function of the variable z in the disk $\Delta(0, |\zeta|)$. A property of P that will be crucial for the approaching discussion is this: if $r > 0$, then

(6.13) $$\frac{1}{2\pi} \int_0^{2\pi} P(z, re^{i\theta}) \, d\theta = 1$$

whenever $|z| < r$. To verify it, we calculate

$$\frac{1}{2\pi} \int_0^{2\pi} P(z, re^{i\theta}) \, d\theta = \frac{1}{2\pi} \int_0^{2\pi} \operatorname{Re}\left(\frac{re^{i\theta} + z}{re^{i\theta} - z}\right) d\theta$$

$$= \operatorname{Re}\left(\frac{1}{2\pi} \int_0^{2\pi} \frac{re^{i\theta} + z}{re^{i\theta} - z} \, d\theta\right) = \operatorname{Re}\left(\frac{1}{2\pi i} \int_{|\zeta|=r} \frac{\zeta + z}{\zeta - z} \frac{d\zeta}{\zeta}\right)$$

$$= \operatorname{Re}\left[\frac{1}{2\pi i} \int_{|\zeta|=r} \left(\frac{2}{\zeta - z} - \frac{1}{\zeta}\right) d\zeta\right] = 2 - 1 = 1.$$

Here we have made use of the fact that for any z in $\Delta(0, r)$

$$\frac{1}{2\pi i} \int_{|\zeta|=r} \frac{d\zeta}{\zeta - z} = n(\gamma, z) = 1,$$

where $\gamma(\theta) = re^{i\theta}$ for $0 \leq \theta \leq 2\pi$.

The "Poisson" in "Poisson kernel" is Siméon Poisson (1781-1840), one of the early investigators of harmonic functions. In the literature it is not uncommon to find the expression

$$P_r(\theta) = \frac{1 - r^2}{1 - 2r \cos \theta - r^2},$$

where $0 \leq r < 1$ and $-\infty < \theta < \infty$, identified as the Poisson kernel. (Notice that $P_r(\theta)$ is nothing but $P(re^{i\theta}, 1)$.) There are good reasons for this variance in terminology, as well as for the alternate notation. As none of these reasons is especially germane to our objectives in this book, however, we stick with the interpretation of "Poisson kernel" indicated earlier.

Suppose that $D = \Delta(z_0, r)$ and that $h: \partial D \to \mathbb{R}$ is a continuous function. The function $u: D \to \mathbb{R}$ given by

$$u(z) = \frac{1}{2\pi} \int_0^{2\pi} P(z - z_0, re^{i\theta}) h(z_0 + re^{i\theta}) \, d\theta$$

is termed the *Poisson integral of h in D*. A theorem of Hermann Schwarz shows: u solves the Dirichlet problem for the disk D with the boundary data h.

Theorem 3.1. (Schwarz's Theorem) *Let D be an open disk in the complex plane, let $h: \partial D \to \mathbb{R}$ be a continuous function, and let u be the Poisson integral of h in D. Then u is harmonic in D and $\lim_{z \to \zeta} u(z) = h(\zeta)$ for every point ζ of ∂D. In particular, D is regular for the Dirichlet problem.*

Proof. In order to conserve on notation, we shall assume that D has its center at the origin, say $D = \Delta(0, r)$. Our first job is to certify that u is harmonic in D. For this we appeal to Lemma V.1.6 and point out that the function $H: D \to \mathbb{C}$ defined by

$$H(z) = \frac{1}{2\pi i} \int_{|\zeta|=r} \frac{h(\zeta)\, d\zeta}{\zeta - z}$$

is analytic. Now for z in D

$$u(z) = \frac{1}{2\pi} \int_0^{2\pi} P(z, re^{i\theta}) h(re^{i\theta})\, d\theta$$

$$= \mathrm{Re}\left[\frac{1}{2\pi} \int_0^{2\pi} \left(\frac{re^{i\theta} + z}{re^{i\theta} - z}\right) h(re^{i\theta})\, d\theta\right] = \mathrm{Re}\left[\frac{1}{2\pi i} \int_{|\zeta|=r} \left(\frac{\zeta + z}{\zeta - z}\right) \frac{h(\zeta)\, d\zeta}{\zeta}\right]$$

$$= \mathrm{Re}\left[\frac{1}{2\pi i} \int_{|\zeta|=r} \left(\frac{2}{\zeta - z} - \frac{1}{\zeta}\right) h(\zeta)\, d\zeta\right] = \mathrm{Re}[2H(z) - H(0)] \ .$$

As the real part of a function that is analytic in D, u is harmonic there.

It remains to prove that u has the asserted boundary behavior: for each ζ in ∂D, $u(z) \to h(\zeta)$ as $z \to \zeta$. We fix such a point ζ and write $\zeta = re^{i\psi}$ with $0 \le \psi < 2\pi$. Let $\epsilon > 0$. The definition of a limit requires us to find a $\delta > 0$ with the property that

(6.14) $$|u(z) - h(\zeta)| < \epsilon$$

for every z in D satisfying $|z - \zeta| < \delta$. The key to doing this lies in employing (6.13) and profiting by the periodicity of an integrand to derive a convenient expression for $u(z) - h(\zeta)$:

$$u(z) - h(\zeta) = \frac{1}{2\pi} \int_0^{2\pi} P(z, re^{i\theta}) h(re^{i\theta})\, d\theta - \frac{h(re^{i\psi})}{2\pi} \int_0^{2\pi} P(z, re^{i\theta})\, d\theta$$

$$= \frac{1}{2\pi} \int_0^{2\pi} P(z, re^{i\theta}) [h(re^{i\theta}) - h(re^{i\psi})]\, d\theta$$

$$= \frac{1}{2\pi} \int_{\psi - \pi}^{\psi + \pi} P(z, re^{i\theta}) [h(re^{i\theta}) - h(re^{i\psi})]\, d\theta \ .$$

For $0 < t \le \pi$, we write

$$M(t) = \max\{|h(re^{i\theta}) - h(re^{i\psi})| : \theta \in [\psi - t, \psi + t]\} \ .$$

3. The Dirichlet Problem for a Disk 231

Since the function that sends θ to $|h(re^{i\theta}) - h(re^{i\psi})|$ is continuous on the real line, its restriction to the interval $[\psi - t, \psi + t]$ definitely does attain a maximum value, so $M(t)$ is well-defined. Clearly, $M(t) \leq M(\pi)$. Furthermore, for the reason that $h(re^{i\theta}) \to h(re^{i\psi})$ as $\theta \to \psi$, we see that $M(t) \to 0$ as $t \to 0$.

Let $0 < t < \pi$. For any z in D we estimate $|u(z) - h(\zeta)|$ in the ensuing fashion:

$$|u(z) - h(\zeta)| = \left| \frac{1}{2\pi} \int_{\psi-\pi}^{\psi+\pi} P(z, re^{i\theta}) \left[h(re^{i\theta}) - h(re^{i\psi}) \right] d\theta \right|$$

$$\leq \frac{1}{2\pi} \int_{\psi-\pi}^{\psi+\pi} P(z, re^{i\theta}) |h(re^{i\theta}) - h(re^{i\psi})| d\theta$$

$$\leq \frac{M(t)}{2\pi} \int_{|\theta-\psi|\leq t} P(z, re^{i\theta}) d\theta + \frac{M(\pi)}{2\pi} \int_{t\leq|\theta-\psi|\leq\pi} P(z, re^{i\theta}) d\theta$$

$$\leq \frac{M(t)}{2\pi} \int_0^{2\pi} P(z, re^{i\theta}) d\theta + \frac{M(\pi)}{2\pi} \int_{t\leq|\theta-\psi|\leq\pi} P(z, re^{i\theta}) d\theta$$

$$= M(t) + \frac{M(\pi)(r^2 - |z|^2)}{2\pi} \int_{t\leq|\theta-\psi|\leq\pi} \frac{d\theta}{|re^{i\theta} - z|^2} ;$$

i.e., for any t in $(0, \pi)$ and any z in D we obtain the estimate

(6.15) $\quad |u(z) - h(\zeta)| \leq M(t) + \dfrac{M(\pi)(r^2 - |z|^2)}{2\pi} \displaystyle\int_{t\leq|\theta-\psi|\leq\pi} \dfrac{d\theta}{|re^{i\theta} - z|^2}.$

In order to bound the final integral in (6.15) for points z of D that are near ζ, we call upon the inequality

(6.16) $\quad |e^{i\theta} - e^{i\psi}| \geq \dfrac{2t^2}{\pi^2},$

valid when $t \leq |\theta - \psi| \leq \pi$. Remembering that $\sin x \geq 2x/\pi$ is true whenever $0 \leq x \leq \pi/2$, we arrive at (6.16) as follows:

$$|e^{i\theta} - e^{i\psi}| = \left| e^{i\psi} \left(e^{i(\theta-\psi)} - 1 \right) \right| = |e^{i(\theta-\psi)} - 1|$$

$$\geq 1 - \cos(\theta - \psi) \geq 1 - \cos t = 2\sin^2\left(\frac{t}{2}\right) \geq \frac{2t^2}{\pi^2}.$$

For every z in D that satisfies $|z - \zeta| < rt^2/\pi^2$ we can thus assert:

$$|re^{i\theta} - z| \geq |re^{i\theta} - \zeta| - |\zeta - z| = r|e^{i\theta} - e^{i\psi}| - |\zeta - z| \geq \frac{2rt^2}{\pi^2} - \frac{rt^2}{\pi^2} = \frac{rt^2}{\pi^2},$$

provided $t \leq |\theta - \psi| \leq \pi$. For such z, therefore,

$$\int_{t \leq |\theta - \psi| \leq \pi} \frac{d\theta}{|re^{i\theta} - z|^2} \leq \frac{\pi^4}{r^2 t^4} \int_{t \leq |\theta - \psi| \leq \pi} d\theta \leq \frac{\pi^4}{r^2 t^4} \int_{-\pi}^{\pi} d\theta = \frac{2\pi^5}{r^2 t^4}.$$

In summary, (6.15) and the intervening computations reveal that for any t in $(0, \pi)$ we have the estimate

(6.17) $$|u(z) - h(\zeta)| \leq M(t) + \frac{\pi^4 M(\pi)(r^2 - |z|^2)}{r^2 t^4}$$

for any z in D which obeys the condition $|z - \zeta| < rt^2/\pi^2$.

Having done the necessary spadework, we return to the task of determining $\delta > 0$ so that (6.14) holds for every z in D satisfying $|z - \zeta| < \delta$. We first take advantage of the fact that $M(t) \to 0$ as $t \to 0$ to choose t in $(0, \pi)$ for which $M(t) < \epsilon/2$. With this t now fixed, we see that the last term in (6.17) tends to zero as $|z| \to r$. In particular, this happens as $z \to \zeta$. We can, as a result, select δ in $(0, rt^2/\pi^2)$ so that the last term in (6.17) is smaller than $\epsilon/2$ for every z in D that satisfies $|z - \zeta| < \delta$. The restriction $0 < \delta < rt^2/\pi^2$ insures that (6.17) holds — hence, owing to the selections of t and δ, that $|u(z) - h(\zeta)| < \epsilon$ holds — for all such z. This completes the proof that $u(z) \to h(\zeta)$ as $z \to \zeta$, so finishes the proof of the theorem. ∎

The beauty of Theorem 3.1 is that it not only establishes the existence of a solution for the Dirichlet problem in a disk $D = \Delta(z_0, r)$, but also equips us with an explicit integral representation for the solution,

$$u(z) = \frac{(r^2 - |z - z_0|^2)}{2\pi} \int_0^{2\pi} \frac{h(z_0 + re^{i\theta}) \, d\theta}{|re^{i\theta} - (z - z_0)|^2},$$

a formula from which a great deal of useful information about harmonic functions can be extracted. To exemplify this point we employ Theorem 3.1 in deriving upper and lower bounds for a harmonic function in a disk where that function is non-negative. These extremely handy estimates are usually associated with the name of Axel Harnack (1851-1888).

Theorem 3.2. (Harnack's Inequalities) *Suppose that a function u is both harmonic and non-negative in a disk $D = \Delta(z_0, r)$. Then*

(6.18) $$u(z_0) \frac{r - |z - z_0|}{r + |z - z_0|} \leq u(z) \leq u(z_0) \frac{r + |z - z_0|}{r - |z - z_0|}$$

for every z in D.

Proof. Fix z in D. For s satisfying $|z - z_0| < s < r$ we write $D_0 = \Delta(z_0, s)$, and we let h denote the restriction of u to the circle ∂D_0. The unique solution of the Dirichlet problem for the disk D_0 with the boundary data specified by h is obviously the restriction of u to $\overline{D}_0$. On the other hand,

3. The Dirichlet Problem for a Disk

Theorem 3.1 states that the solution is represented in D_0 by the Poisson integral of h. This means that we are authorized to express $u(z)$ in the form

$$(6.19) \qquad u(z) = \frac{(s^2 - |z - z_0|^2)}{2\pi} \int_0^{2\pi} \frac{u(z_0 + se^{i\theta})\,d\theta}{|se^{i\theta} - (z - z_0)|^2}.$$

Since $s - |z - z_0| \leq |se^{i\theta} - (z - z_0)| \leq s + |z - z_0|$ for any real number θ, we deduce that

$$(6.20) \qquad \frac{s - |z - z_0|}{s + |z - z_0|} \leq \frac{s^2 - |z - z_0|^2}{|se^{i\theta} - (z - z_0)|^2} \leq \frac{s + |z - z_0|}{s - |z - z_0|}$$

for such θ. Multiplying (6.20) by the non-negative quantity $u(z_0 + se^{i\theta})/2\pi$, integrating from 0 to 2π, and referring to (6.19), we obtain

$$\left(\frac{1}{2\pi} \int_0^{2\pi} u(z_0 + se^{i\theta})\,d\theta\right) \frac{s - |z - z_0|}{s + |z - z_0|} \leq u(z)$$

and also

$$u(z) \leq \left(\frac{1}{2\pi} \int_0^{2\pi} u(z_0 + se^{i\theta})\,d\theta\right) \frac{s + |z - z_0|}{s - |z - z_0|}.$$

On account of Theorem 2.1, this all simplifies to

$$u(z_0) \frac{s - |z - z_0|}{s + |z - z_0|} \leq u(z) \leq u(z_0) \frac{s + |z - z_0|}{s - |z - z_0|}.$$

Since the last inequalities are valid whenever $|z - z_0| < s < r$, statement (6.18) is the end-product when s tends to r. ∎

Another gain from Theorem 3.1 is a characterization of harmonic functions in terms of the mean value property, a result alluded to earlier. We anticipate one step in the upcoming proof by remarking that a function which has the mean value property in an open set U automatically retains that property in any open subset of U.

Theorem 3.3. *Suppose that a real-valued function w is continuous in a plane open set U. Then w is harmonic in U if and only if it has the mean value property in this set.*

Proof. Theorem 2.1 asserts that a function which is harmonic in U has the mean value property in that set. For the converse, assume that w has the mean value property in U. To prove that w is harmonic there it plainly suffices to verify that each point of U is the center of some open disk in which w is harmonic. We thus fix z_0 in U, choose $D = \Delta(z_0, r)$ so that $\overline{D}$ is contained in U, and claim: w is harmonic in D. Theorem 3.1 guarantees the existence of a continuous function $u: \overline{D} \to \mathbb{R}$ that is harmonic in D and agrees with w on ∂D. We need only demonstrate that $w = u$ in D. For this we consider the continuous function $v \colon \overline{D} \to \mathbb{R}$ defined by $v(z) = w(z) - u(z)$. Both w and u have the mean value property in D (w by the comment just preceding the statement of the theorem and u by Theorem 2.1), from which it follows without difficulty that v is endowed with the same property in that disk. This means, in particular, that v is a subharmonic function in D. Corollary V.3.13 informs us that $v(z)$ attains its maximum at some point of ∂D. As $v = 0$ on ∂D, we conclude that $v \leq 0$ throughout D. The same argument applies to $-v$ and shows that $v \geq 0$ in D. Accordingly, $v = 0$ in D; i.e., $w = u$ in D, as desired. ∎

Theorem 3.3 furnishes us with a means to establish the harmonicity of functions in situations where it is awkward to compute partial derivatives. A wonderful application of this result is the following "reflection principle" for harmonic functions.

Theorem 3.4. *Let D be a domain that is symmetric about the real axis, let G and G^* be the components of $D \sim \mathbb{R}$, and let $I = D \cap \mathbb{R}$. If $u \colon G \to \mathbb{R}$ is a harmonic function that obeys the condition $\lim_{\zeta \to z} u(\zeta) = 0$ for every point z of I, then the function $w \colon D \to \mathbb{R}$ defined by*

$$w(z) = \begin{cases} u(z) & \text{if } z \in G, \\ 0 & \text{if } z \in I, \\ -u(\bar{z}) & \text{if } z \in G^*, \end{cases}$$

is harmonic.

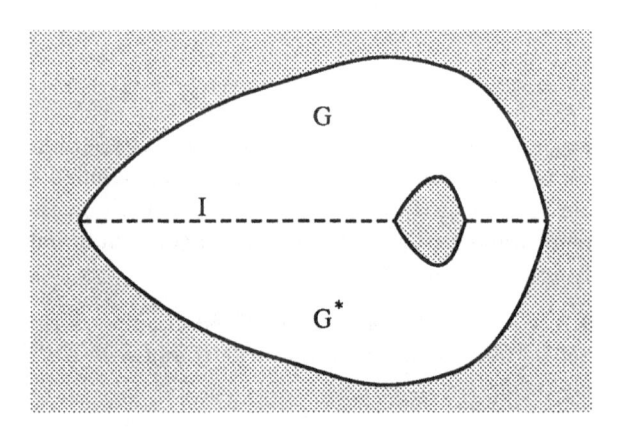

Figure 2.

Proof. By assumption, the function w is harmonic in G. In G^* we have $w(z) = -u(\bar{z}) = -u(x, -y)$, making it evident that w belongs to the class $C^2(G^*)$. Furthermore, for z in G^* we obtain

$$w_x(z) = -u_x(\bar{z}) \quad , \quad w_{xx}(z) = -u_{xx}(\bar{z}) ,$$

3. The Dirichlet Problem for a Disk

$$w_y(z) = u_y(\bar{z}) \quad , \quad w_{yy}(z) = -u_{yy}(\bar{z}) .$$

Thus, $\Delta w(z) = -\Delta u(\bar{z}) = 0$ for every such z, which shows that w is also harmonic in G^*. The troubles arise at points z of I, where even the existence of the partial derivative $w_y(z)$ is in question, to say nothing of the existence and continuity of second order partials. What our hypotheses do afford us, however, is the knowledge that w is at least continuous at the points of I — hence, throughout D. We now show that w has the mean value property in D. Given z in D, we choose $\rho = \rho(z) > 0$ so small that $\Delta(z,\rho)$ is contained in D. We shall insist, in addition, that $\Delta(z,\rho)$ be contained in G (respectively, G^*) if z happens to be an element of G (resp., G^*). Let r satisfy $0 < r < \rho$. We assert that

$$(6.21) \qquad w(z) = \frac{1}{2\pi} \int_0^{2\pi} w(z + re^{i\theta}) \, d\theta .$$

Since w is harmonic in both G and G^*, Theorem 2.1 and our restrictions on ρ dispel any doubts about (6.21) when z lies in one of the sets G or G^*. If z belongs to I, then $w(z) = 0$ and we calculate

$$\int_0^{2\pi} w(z + re^{i\theta}) \, d\theta = \int_{-\pi}^{\pi} w(z + re^{i\theta}) \, d\theta$$

$$= \int_{-\pi}^{0} w(z + re^{i\theta}) \, d\theta + \int_0^{\pi} w(z + re^{i\theta}) \, d\theta$$

$$= -\int_{-\pi}^{0} u(z + re^{-i\theta}) \, d\theta + \int_0^{\pi} u(z + re^{i\theta}) \, d\theta$$

$$= -\int_0^{\pi} u(z + re^{i\theta}) \, d\theta + \int_0^{\pi} u(z + re^{i\theta}) \, d\theta = 0 ,$$

which confirms (6.21) in this case also. Theorem 3.3 vouches for the harmonicity of w in D. ∎

We have earlier made reference to the fact that not every bounded plane domain is regular for the Dirichlet problem. We conclude this chapter with an example of a domain for which the Dirichlet problem does not always admit a solution. This example is also the pretext for highlighting one last important feature of harmonic functions, the "removability of isolated singularities" for bounded functions in this class.

Theorem 3.5. *If a function u is both harmonic and bounded in a punctured disk $\Delta^*(z_0, r)$, then $\lim_{z \to z_0} u(z)$ exists and the function $\tilde{u}$ defined in the full disk $\Delta(z_0, r)$ by*

$$\tilde{u}(z) = \begin{cases} u(z) & \text{if } z \in \Delta^*(z_0, r) , \\ \lim_{z \to z_0} u(z) & \text{if } z = z_0 , \end{cases}$$

is a harmonic function.

Proof. It is our assumption that $-m \leq u(z) \leq m$ holds for every z in $\Delta^*(z_0, r)$, where $m > 0$ is some constant. We fix a number d satisfying the condition $0 < d < \min\{1, r\}$, and we let v denote the solution of the Dirichlet problem for the disk $D = \Delta(z_0, d)$ that agrees with u on the circle ∂D. Since v is just the Poisson integral of the restriction of u to ∂D and since $-m \leq u(z) \leq m$ is assumed to be true for every point z of ∂D, it is easy to see with the help of (6.13) that $-m \leq v(z) \leq m$ holds for all z in $\overline{D}$. We contend that $u = v$ in $D^* = \Delta^*(z_0, d)$. Once this is shown the proof of the theorem will be complete, for then

$$\lim_{z \to z_0} u(z) = \lim_{z \to z_0} v(z) = v(z_0)$$

will exist, and the function $\widetilde{u}$, already known to be harmonic in $\Delta^*(z_0, r)$, will agree with v in the smaller disk D, making it harmonic there, too. Accordingly, $\widetilde{u}$ will be harmonic in $\Delta(z_0, r)$.

We consider the function $w = u - v$. We claim that $w = 0$ throughout D^*. In point of fact, we shall only verify that $w \leq 0$ in D^*. The same argument is applicable to $-w$, however, and shows that $w \geq 0$ in D^* as well. In tandem these two inequalities prove that the function w vanishes identically in D^*. Observe that w is harmonic in D^* and continuous in the set $A = \{z : 0 < |z - z_0| \leq d\}$, satisfies $-2m \leq w \leq 2m$ everywhere in A, and has $w(z) = 0$ when $|z - z_0| = d$, since by construction $u(z) = v(z)$ for such z. We would love to be able to apply Corollary 2.3 to w in D^*. Unfortunately, we are prevented from doing so by the annoying fact that there is a point of ∂D^*, namely z_0, at which we have virtually no information about w. We get around this detail by resorting to a bit of trickery — to wit, we introduce for each $t > 0$ the auxiliary function w_t,

$$w_t(z) = w(z) + t \operatorname{Log} |z - z_0| \, .$$

We shall check that $w_t(z) \leq 0$ for every z in D^* and every $t > 0$. This being the case, we can let $t \to 0$ and deduce that $w(z) \leq 0$ for all z in D^*, the desired conclusion. The entire proof thus boils down to confirming that $w_t \leq 0$ in D^*.

Fix $t > 0$ and let z be a point of D^*. We can choose a number c in the interval $(0, |z - z_0|)$ with the property that

$$2m + t \operatorname{Log} c \leq 0 \, .$$

The function w_t is harmonic in the annulus $G = \{\zeta : c < |\zeta - z_0| < d\}$ and continuous on $\overline{G}$. When $|\zeta - z_0| = d$, we have

$$w_t(\zeta) = w(\zeta) + t \operatorname{Log} d = t \operatorname{Log} d < 0 \, ,$$

since $w(\zeta) = 0$ and $0 < d < 1$. By the choice of c it is also true that

$$w_t(\zeta) = w(\zeta) + t \operatorname{Log} c \leq 2m + t \operatorname{Log} c \leq 0$$

3. The Dirichlet Problem for a Disk

if $|\zeta - z_0| = c$. We can apply Corollary 2.3 to w_t in the domain G to certify that $w_t \leq 0$ throughout G. In particular, $w_t(z) \leq 0$. Because $t > 0$ and z in D^* were arbitrary, this finishes the proof. ∎

Theorem 3.5 permits us to do a complete turnabout from our approach at the beginning of this chapter. There we used facts about analytic functions to obtain information concerning harmonic functions. Now we are in a position to benefit from our newly acquired knowledge of harmonic functions by improving an earlier theorem (Theorem V.3.4) that dealt with the class of analytic functions. The upgraded result is known as the "Riemann Extension Theorem."

Theorem 3.6. (Riemann Extension Theorem) *If a function f is both analytic and bounded in a punctured disk $\Delta^*(z_0, r)$, then $\lim_{z \to z_0} f(z)$ exists and the function $\tilde{f}$ defined in the full disk $\Delta(z_0, r)$ by*

$$\tilde{f}(z) = \begin{cases} f(z) & \text{if } z \in \Delta^*(z_0, r) \text{,} \\ \lim_{z \to z_0} f(z) & \text{if } z = z_0 \text{,} \end{cases}$$

is an analytic function.

Proof. We write $f = u + iv$. The functions u and v are harmonic and bounded in $\Delta^*(z_0, r)$. Invoking Theorem 3.5 we conclude that

$$\lim_{z \to z_0} f(z) = \lim_{z \to z_0} u(z) + i \lim_{z \to z_0} v(z)$$

exists. The function $\tilde{f}$ is thus seen to be well-defined and continuous in $\Delta(z_0, r)$. As $\tilde{f}$ is analytic in $\Delta^*(z_0, r)$, Theorem V.3.4 informs us that $\tilde{f}$ is analytic in $\Delta(z_0, r)$. ∎

We shall return to the Riemann extension theorem in Chapter VIII, at which time we shall present a second proof for it, one that does not depend on knowledge of harmonic functions. We close the present chapter with the promised example of a bounded domain that is not regular for the Dirichlet problem.

EXAMPLE 3.2. Let $D = \{z : 0 < |z| < 1\}$ and let $h: \partial D \to \mathbb{R}$ be given by $h(z) = 1 - |z|$. Show that D is not regular for the Dirichlet problem by demonstrating that the problem has no solution in D for the boundary data h.

Suppose that there did exist a continuous function $u: \overline{D} \to \mathbb{R}$ which is harmonic in D and equal to h on ∂D. Theorem 3.5 would imply that u is actually harmonic in the disk $\Delta = \Delta(0, 1)$. In view of Corollary 2.3, u would attain its maximum value at some point of $\partial \Delta$, a set on which this function, being equal to h, vanishes. Therefore, $u \leq 0$ would hold throughout Δ. However, $u(0) = h(0) = 1$. This contradiction shows that no such function u can exist.

4 Exercises for Chapter VI

4.1 Exercises for Section VI.1

4.1. Show that the formula $u(z) = \text{Log}\,|z-z_0|$ defines a harmonic function.

4.2. Generalize the preceding exercise as follows: if a function f is both analytic and zero-free in an open set U, then the function u defined by $u(z) = \text{Log}\,|f(z)|$ is harmonic in U.

4.3. If functions u and v are both harmonic in an open set U, check that the functions cu for any real number c and $u+v$ are also harmonic there. Is the product uv necessarily harmonic in U?

4.4. Let u be a real-valued function that is a member of the class $C^2(U)$, where U is an open set in the complex plane. Show that Δu can be expressed in U using z- and $\bar{z}$-derivatives as follows: $\Delta u = 4u_{z\bar{z}} = 4u_{\bar{z}z}$.

4.5. Assuming that $f: U \to \mathbb{C}$ is an analytic function and that a real-valued function v belongs to the class $C^2(V)$ in some open set V which contains the range of f, verify that the composite function $u = v \circ f$ has $\Delta u(z) = \Delta v[f(z)]|f'(z)|^2$ for every z in U. If v is harmonic in V, conclude that u is a harmonic function. (*Hint.* This problem can be done quite efficiently with the aid of Exercise 4.4 and Exercise III.6.61.)

4.6. In each of the following cases show that the function u is harmonic in the domain D and find a harmonic conjugate for u there, if any exist: (i) $u(z) = xy(x^2+y^2)^{-2}$ and $D = \mathbb{C} \sim \{0\}$; (ii) $u(z) = \text{Arg}(\sqrt{z})$ and $D = \mathbb{C} \sim (-\infty,0]$; (iii) $u(z) = \text{Log}\,|z^2-1|$ and $D = \mathbb{C} \sim \{1,-1\}$; (iv) $u(re^{i\theta}) = 1+r^2\sin\theta\cos\theta$ and $D = \mathbb{C}$; (v) $u(z) = \text{Log}\,|z^2-1|$ and $D = \{z : \text{Im}\, z > 0\}$.

4.7. Define u in $\mathbb{C} \sim \{0\}$ by $u(re^{i\theta}) = \sum_{k=0}^{n} r^{-k}[a_k\cos(k\theta) + b_k\sin(k\theta)]$ for $r > 0$ and for real θ, where $a_0, a_1, \ldots, a_n$ and $b_1, b_2, \ldots, b_n$ are real constants. Verify that u is harmonic in $\mathbb{C} \sim \{0\}$ and find a harmonic conjugate for u in this domain.

4.8. Let D be a plane domain that does not contain the origin, and let $u: D \to \mathbb{R}$ be defined by $u(z) = \text{Log}\,|z|$. Suppose that v is a harmonic conjugate for u in D and that, for some point z_0 of D, $v(z_0)$ is an argument of z_0. Demonstrate that the function $L = u+iv$ satisfies $e^{L(z)} = z$ for every z in D — hence, that L is a branch of $\log z$ in the domain D.

4.9. If a function u is harmonic in the whole complex plane and is either bounded above or bounded below, then u is constant in $\mathbb{C}$. Justify this assertion. (*Hint.* Recall Example V.3.2.)

4.10. A function u is harmonic in $\mathbb{C}$ and has the property that $u(z) \to 0$ as $|z| \to \infty$. Prove that $u(z) = 0$ for every z in $\mathbb{C}$.

4.2 Exercises for Section VI.2

4.11. If a function u is continuous on the closed disk $\overline{\Delta}(z_0, r)$ and harmonic in the open disk $\Delta(z_0, r)$, then $u(z_0) = (2\pi)^{-1} \int_0^{2\pi} u(z_0 + re^{i\theta}) \, d\theta$. Establish this minor extension of the mean value property for harmonic functions.

4.12. If a function u is harmonic in a plane domain D and if there is a point z_0 of D such that $u(z) \geq u(z_0)$ for every z in D, then u is constant in D. Prove this.

4.13. Let D be a bounded domain in the complex plane, and let $u: \overline{D} \to \mathbb{R}$ be a continuous function that is harmonic in D. Verify that $u(z)$ attains its minimum at some point of ∂D.

4.14. Let D be a convex domain in the complex plane, $D \neq \mathbb{C}$, and let $u: D \to \mathbb{R}$ be the function defined by $u(z) = \min\{|z - \zeta| : \zeta \in \partial D\}$ — i.e., $u(z)$ is the distance from z to the boundary of D. (i) Given that D is an open half-plane, prove that u is harmonic in D. (ii) Assuming, conversely, that u is harmonic in D, prove that D is necessarily an open half-plane. (*Hint for (ii)*. Fix z_0 in D and a point ζ_0 of ∂D for which $u(z_0) = |z_0 - \zeta_0|$. Start by showing that D lies in one of the two half-planes into which the complex plane is divided by the line through ζ_0 perpendicular to the segment with endpoints z_0 and ζ_0. Call that half-plane H. Compare u with the function $v: D \to \mathbb{R}$ given by $v(z) = \min\{|z - \zeta| : \zeta \in \partial H\}$.)

4.15. Confirm that subharmonicity — like harmonicity — is a local property of a function, in the following sense: a continuous function w is subharmonic in an open set U if and only if each point of U is the center of some open disk in which w is subharmonic.

4.16. Assume that continuous functions u and v are subharmonic in an open set U. Certify that the functions cu for a non-negative real number c, $u+v$, and $\max\{u, v\}$ are all subharmonic in U. What are the counterparts of these facts in the category of superharmonic functions?

4.17. Given that a function u is harmonic in an open set U, verify that $w = |u|$ is subharmonic in that set.

4.18. Prove the following generalization of the last exercise: if a function u is harmonic in an open set U, if I is an open interval that contains the range of u, and if $\varphi: I \to \mathbb{R}$ is a convex function, then the function $w = \varphi \circ u$ is subharmonic in U. (N.B. To say that φ is convex means that $\varphi[(1 - \lambda)x + \lambda y] \leq (1 - \lambda)\varphi(x) + \lambda \varphi(y)$ whenever x and y belong to I and $0 \leq \lambda \leq 1$. When φ is differentiable on I, this is equivalent to the requirement that φ' be a non-decreasing function on I. The definition of convexity implies that φ is a continuous function and that the inequality $\varphi[(x_1 + x_2 + \cdots + x_n)/n] \leq [\varphi(x_1) + \varphi(x_2) + \cdots + \varphi(x_n)]/n$ holds whenever n is a positive integer and $x_1, x_2, \ldots, x_n$ are points of I. Feel free to use these facts without proof. *Hint*. If $\psi: [0, 2\pi] \to \mathbb{R}$ is a continuous function, then $\int_0^{2\pi} \psi(\theta)d\theta = \lim_{n\to\infty}(2\pi/n)\sum_{k=1}^n \psi(2\pi k/n)$.)

4.19. Assuming that a continuous function u is subharmonic in an open set U and that $\varphi: I \to \mathbb{R}$ is a non-decreasing convex function, where I is an open interval that contains the range of u, show that $w = \varphi \circ u$ is subharmonic in U.

4.20. Let D be a bounded plane domain, and let u and v be real-valued functions that are continuous on $\overline{D}$. If it is known that u is harmonic in D, that v is subharmonic in D, and that $v \leq u$ on ∂D, deduce that either $v = u$ throughout D or $v < u$ everywhere in D.

4.21. Let $D = \{z: a < |z - z_0| < b\}$, where $0 \leq a < b \leq \infty$, and let u be a function in the class $C^2(D)$. Show that the function v defined in the strip $V = \{(r,\theta): a < r < b, -\infty < \theta < \infty\}$ by $v(r,\theta) = u(z_0 + re^{i\theta})$ satisfies

$$v_{rr}(r,\theta) + r^{-1}v_r(r,\theta) + r^{-2}v_{\theta\theta}(r,\theta) = \Delta u(z_0 + re^{i\theta})$$

for every point (r,θ) of V.

4.22. Let $w(z) = |z|^\lambda$, where $\lambda > 0$. Show that w is subharmonic in the complex plane. (*Hint.* Start by showing that for any real λ the function w so defined is subharmonic in the set $\mathbb{C} \sim \{0\}$. For this look at Δw.)

4.23. Establish the following extension of Exercise 4.22: if a function f is analytic in an open set U and if $\lambda > 0$, then the function w defined by $w(z) = |f(z)|^\lambda$ is subharmonic in U. (*Hint.* Begin by showing that w is subharmonic in the open set $V = \{z \in U: f(z) \neq 0\}$. The latter is also the case when $\lambda < 0$.)

4.24. Under the assumptions that a function f is analytic in an open set U and that a function v is of class C^2 and subharmonic in an open set V which contains the range of f, demonstrate that the composition $w = v \circ f$ is subharmonic in U. (N.B. This conclusion would still be true if the requirement that v belong to the class $C^2(V)$ were dropped, but the standard proofs of this fact make use of non-trivial information about analytic functions that will not be available to us until Chapter VIII.)

4.25. Prove the "strong maximum principle" for subharmonic functions: if a continuous function w is subharmonic in a plane domain D and if m is a constant with the property that $\limsup_{n\to\infty} w(z_n) \leq m$ whenever $\langle z_n \rangle$ is a sequence in D which converges to a point of ∂D or satisfies $|z_n| \to \infty$ as $n \to \infty$, then either $w(z) < m$ for every z in D or $w(z) = m$ for every z in D. (*Hint.* Show first that $\sup w(D) \leq m$.)

4.3 Exercises for Section VI.3

4.26. Find the solution of the Dirichlet problem for D with the specified boundary data h: (i) $D = \Delta(0,1)$ and $h(z) = x + x^2 y^2$; (ii) $D = \Delta(0,2)$

and $h(z) = x^2 - 2y^2$; (iii) $D = \Delta(0, 1/2)$ and $h(z) = 4x^2y - 8xy^2$.

4.27. Evaluate the integral $\int_0^{2\pi} |2e^{i\theta} - z|^{-2} \cos^2\theta \, d\theta$ when $|z| \neq 2$.

4.28. Let D and D_1 be bounded domains in the complex plane, and let $f: \overline{D} \to \mathbb{C}$ be a univalent continuous function that is analytic in D and has range $\overline{D}_1$. Given the knowledge that D is regular for the Dirichlet problem, deduce that D_1 also has this property. Indeed, certify that the solution u_1 of the Dirichlet problem for D_1 with boundary data h_1 is furnished by $u_1 = u \circ f^{-1}$, where u is the solution of the Dirichlet problem for D whose boundary values are set by $h(z) = h_1[f(z)]$. You may use without proof the following information: $f(D) = D_1$ — hence, $f(\partial D) = \partial D_1$ — and $f^{-1}: \overline{D}_1 \to \overline{D}$ is a continuous function that is analytic in D_1. These things will all be established later in the text.

4.29. Find the solution of the Dirichlet problem for D_1 with boundary data h_1: (i) $D_1 = \Delta(1, 1)$ and $h_1(z) = |z|^2$; (ii) $D_1 = \Delta(i, 2)$ and $h_1(z) = x^2y^2$; (iii) $D_1 = \Delta(2, 1)$ and $h_1(z) = x^3$.

4.30. Let D_1 be the domain inside the cardioid whose polar equation is $r = 2(1+\cos\theta)$. Determine the solution of the Dirichlet problem for D_1 with boundary values given by $h_1(z) = |z|$. (*Hint*. The function $f(z) = (z+1)^2$ maps $\overline{\Delta}(0, 1)$ in a univalent fashion to $\overline{D}_1$.)

4.31. Let $G = \{z : |z| < 1 \text{ and } \operatorname{Im} z > 0\}$. Find the solution of the Dirichlet problem for G with boundary data h specified as follows: $h(z) = 0$ if z belongs to $[-1, 1]$ and $h(z) = y - y^3$ if $|z| = 1$. (*Hint*. Use Theorem 3.4.)

4.32. A continuous function $h: K(0, 1) \to \mathbb{R}$ is known to have the property that $h(z) + h(\overline{z}) = 0$ for every z in $K(0, 1)$. Show that u, the solution of the Dirichlet problem for $\Delta(0, 1)$ with boundary data h, has $u(z) = 0$ for every z in the interval $[-1, 1]$.

4.33. Suppose that a function w is continuous on a closed disk $\overline{\Delta}(z_0, r)$ and subharmonic in $\Delta(z_0, r)$. For $0 < s \leq r$ define the number $m(s)$ by $m(s) = (2\pi)^{-1} \int_0^{2\pi} w(z_0 + se^{i\theta}) d\theta$. Show that $w(z_0) \leq m(s) \leq m(r)$. (*Hint*. Compare w with u, the solution of the Dirichlet problem for $\Delta(z_0, r)$ that coincides with w on the circle $K(z_0, r)$. Recall Exercise 4.11.)

4.34. Suppose that a continuous function w is subharmonic in an open set U and that the open disk $\Delta(z, \rho)$ is contained in U. Prove that (6.3) actually holds for all r in the interval $(0, \rho)$. (*Hint*. Exercise 4.33.)

4.35. A real-valued function w is continuous in a plane open set U, where it has the so-called "harmonic majorant property": for every bounded domain D whose closure lies in U and every continuous function $u: \overline{D} \to \mathbb{R}$ that is harmonic in D and satisfies $u \geq w$ on ∂D, it is true that $u \geq w$ throughout D. Demonstrate that any such function w is necessarily subharmonic in U. (*Hint*. If $\Delta(z, \rho)$ is contained in U, check that (6.3) holds for any r

satisfying $0 < r < \rho$ by looking at w in $D = \Delta(z,r)$ and coming up with an appropriate comparison function $u \colon \overline{D} \to \mathbb{R}$.)

4.36. Let $f \colon U \to \mathbb{C}$ be a univalent analytic function. Assume it to be known, as we shall prove in Chapter VIII, that $V = f(U)$ is always an open set in this situation and that $f^{-1} \colon V \to U$ is an analytic function. If a continuous function v is subharmonic in V, prove that the composition $w = v \circ f$ is subharmonic in U. (*Hint.* Show that w has the harmonic majorant property in U. Exercise 4.5 might come in handy.)

4.37. Let w be a continuous function that is subharmonic in an open set U, and let Δ be an open disk whose closure is a subset of U. Define a function v in U as follows: $v(z) = w(z)$ if z belongs to $U \sim \Delta$ and $v(z) = u(z)$ if z is an element of Δ, where $u \colon \overline{\Delta} \to \mathbb{R}$ is the solution of the Dirichlet problem for Δ whose boundary values are assigned by w on $\partial\Delta$. Verify that $w(z) \leq v(z)$ for every z in U and that v is subharmonic in U. (N.B. Under the conditions detailed in this problem we say that v is obtained from w by means of a "Poisson modification." Notice that v is "more harmonic" than w — v is harmonic in Δ, where w may fail to have this property — but looks exactly like w near the boundary of U.)

4.38. Let $0 < d \leq 1$, let u be a function that is continuous in the closed disk $\overline{\Delta}(z_0, d)$ and subharmonic in the punctured disk $\Delta^*(z_0, d)$, and let v be the function that is harmonic in $\Delta(z_0, d)$ and agrees with u on the circle $K(z_0, d)$. Verify that $u \leq v$ throughout $\overline{\Delta}(z_0, d)$. (*Hint.* First show that $u(z) \leq v(z)$ for every z in $\Delta^*(z_0, d)$. For this adapt the argument used to show that $w \leq 0$ in the proof of Theorem 3.5.)

4.39. If a function w is continuous in an open set U and is subharmonic in $U \sim \{z_0\}$ for some point z_0 of U, then w is subharmonic in U. Prove this analogue of Theorem V.3.4. (*Hint.* Put Exercises 4.38 and 4.11 to work.)

4.40. A function u is harmonic in $\Delta^*(z_0, r)$ and satisfies the condition that $(\operatorname{Log}|z - z_0|)^{-1} u(z) \to 0$ as $z \to z_0$. Modify the proof of Theorem 3.5 to show that $\lim_{z \to z_0} u(z)$ exists. Deduce that u admits a harmonic extension to the full disk $\Delta(z_0, r)$.

4.41. Prove: if a function w is harmonic in $D = \{z : a < |z| < b\}$, then there is a unique real number c for which the function $u(z) = w(z) - c \operatorname{Log} |z|$ has a harmonic conjugate in D. (*Hint.* Let $f = w_x - i w_y$ and $c = (2\pi i)^{-1} \int_{|z|=r} f(z)\,dz$, where $a < r < b$. Show that $g(z) = f(z) - cz^{-1}$ has a primitive in D.)

4.42. Assume that a function u is both harmonic and non-negative in a disk $\Delta(z_0, r)$. If $K = \overline{\Delta}(z_0, s)$ with $0 < s < r$, show that the inequality $c^{-1} u(z_0) \leq u(z) \leq c u(z_0)$ is satisfied for every z in K, in which $c = (r+s)(r-s)^{-1}$. Conclude, in particular, that $M = \max\{u(z) : z \in K\}$ and $m = \min\{u(z) : z \in K\}$ obey the relation $M \leq c^2 m$.

Chapter VII

Sequences and Series of Analytic Functions

Introduction

Some of the most significant properties of analytic functions only come to light because of the special means available for representing these functions. We have already experienced one occurrence of this phenomenon in the Cauchy integral formula and the many conclusions we were able to draw with that formula in hand. A second important method of representing analytic functions — as sums of certain kinds of infinite series — constitutes the subject matter of the present chapter. In particular, the Taylor series expansion of a function in a disk where it is analytic and the Laurent series development of a function that is analytic in an annulus will be discussed in detail. The former provides the key to elucidating many of the delicate local features of analytic functions; the latter plays a decisive role in the classification of isolated singularities of such functions and in our treatment of the residue theorem.

We begin the chapter with a review of some basic facts and terminology relating to the convergence of sequences and series of functions.

1 Sequences of Functions

1.1 Uniform Convergence

Suppose that $\langle f_n \rangle$ is a sequence of complex-valued functions, each of whose domain-sets contains a subset A of the complex plane. There is a perfectly obvious and natural way to speak of this sequence being convergent in A and having the function $f: A \to \mathbb{C}$ as its limit when $n \to \infty$: it can simply be demanded that for each point z of A the sequence of values $\langle f_n(z) \rangle$, which is just an ordinary sequence of complex numbers, be convergent and have the value $f(z)$ for its limit; i.e., we can insist that

$$f(z) = \lim_{n \to \infty} f_n(z)$$

hold at every z in A. When this happens we say that the sequence $\langle f_n \rangle$ *converges pointwise in A* to the limit function f and express the fact symbolically by writing $f_n \to f$ in A or $f = \lim_{n \to \infty} f_n$ in A. (N.B. In the absence of any pronouncement to the contrary, it will be our convention when talking of pointwise convergence in A to regard A as the complete domain-set of the limit function.) Although it provides an appropriate starting point for a discussion of the convergence of function sequences, pointwise convergence has a number of serious drawbacks that render it less than satisfactory for the ends we have in mind. Consider, for example, the sequence of functions $f_n : [0, 1] \to \mathbb{C}$ given by $f_n(t) = t^n$. Clearly $\langle f_n \rangle$ converges pointwise in $[0, 1]$ to the function f defined by $f(t) = 0$ when $0 \leq t < 1$ and $f(1) = 1$. Thus, whereas each of the functions f_n is continuous, the limit function f is not. This example exposes one of the major shortcomings of pointwise convergence: the property of continuity is not necessarily preserved under pointwise passage to a limit.

There is a stronger mode of convergence that is often used in conjunction with sequences of functions, one from which many of the negative features inherent in pointwise convergence are absent. It is called "uniform convergence." To say that a function sequence $\langle f_n \rangle$ converges uniformly to f on a set A means not only that $f_n(z)$ tends to $f(z)$ for each z in A, but also that this convergence takes place, roughly speaking, at the same rate everywhere in A. Formulated precisely the definition states: a sequence $\langle f_n \rangle$ of functions defined on a plane set A *converges uniformly on A* to the limit function f if corresponding to each $\epsilon > 0$ there is an index $N = N(\epsilon)$ such that $|f_n(z) - f(z)| < \epsilon$ holds for every z in A once $n \geq N$. When n is large, therefore, $f_n(z)$ is required to be "uniformly close" to $f(z)$ throughout A. We emphasize that the uniform convergence of a sequence $\langle f_n \rangle$ on A carries with it the pointwise convergence of this sequence in A. The next theorem describes two of the principal virtues of uniform convergence as they touch on matters of interest in this book.

Theorem 1.1. *Suppose that each function in a sequence $\langle f_n \rangle$ is continuous on a set A and that this sequence converges uniformly on A to the limit function f. Then f is also continuous on A. Furthermore,*

$$\int_\gamma f(z)\,dz = \lim_{n \to \infty} \int_\gamma f_n(z)\,dz$$

for every piecewise smooth path γ in A.

Proof. We fix a point z_0 of A and verify that $f : A \to \mathbb{C}$ is continuous at

1. Sequences of Functions

z_0. Let $\epsilon > 0$ be given. We must find a $\delta > 0$ with the property that

$$|f(z) - f(z_0)| < \epsilon$$

whenever z belongs to A and $|z - z_0| < \delta$. We observe using the triangle inequality that for any index n and any point z of A

(7.1)
$$|f(z) - f(z_0)| \leq |f(z) - f_n(z)| + |f_n(z) - f_n(z_0)|$$
$$+ |f_n(z_0) - f(z_0)|.$$

We now exploit the uniform convergence of $\langle f_n \rangle$ on A to select and fix an index n for which it is true that

$$|f_n(z) - f(z)| < \frac{\epsilon}{3}$$

for every z in A. Next, since the function f_n is by hypothesis continuous on A, it is possible to choose $\delta > 0$ with the property that

$$|f_n(z) - f_n(z_0)| < \frac{\epsilon}{3}$$

whenever z lies in A and $|z - z_0| < \delta$. Referring to (7.1) we conclude that

$$|f(z) - f(z_0)| < \frac{\epsilon}{3} + \frac{\epsilon}{3} + \frac{\epsilon}{3} = \epsilon$$

for every z belonging to A and satisfying $|z - z_0| < \delta$. This establishes the continuity of f at z_0, an arbitrary point of A — hence, its continuity on A.

Let γ be a piecewise smooth path in A. Given $\epsilon > 0$, we can again take advantage of the uniform convergence of $\langle f_n \rangle$ to pick N so that

$$|f_n(z) - f(z)| < \frac{\epsilon}{1 + \ell(\gamma)}$$

holds for every z in A as soon as $n \geq N$. It follows that

$$\left| \int_\gamma f_n(z)\, dz - \int_\gamma f(z)\, dz \right| = \left| \int_\gamma [f_n(z) - f(z)]\, dz \right|$$
$$\leq \int_\gamma |f_n(z) - f(z)|\, |dz| \leq \int_\gamma \frac{\epsilon}{\ell(\gamma) + 1}\, |dz| = \frac{\epsilon \ell(\gamma)}{1 + \ell(\gamma)} < \epsilon$$

whenever $n \geq N$. This proves that $\int_\gamma f_n(z)\, dz \to \int_\gamma f(z)\, dz$, the second assertion of the theorem. ∎

How is it possible to detect whether a sequence of functions converges uniformly on a set? In general, demonstrating uniform convergence can be a tricky business. There is, however, a criterion for the uniform convergence of function sequences that is analogous to the Cauchy criterion for the convergence of complex numerical sequences. A sequence of functions $\langle f_n \rangle$

is called a *uniform Cauchy sequence on a set* A if corresponding to each $\epsilon > 0$ there exists an index $N = N(\epsilon)$ such that $|f_m(z) - f_n(z)| < \epsilon$ is satisfied for all z in A whenever $m > n \geq N$. If $\langle f_n \rangle$ converges uniformly on A to the limit function f, then the inequaltiy

$$|f_m(z) - f_n(z)| \leq |f_m(z) - f(z)| + |f_n(z) - f(z)|$$

makes it clear that $\langle f_n \rangle$ is a uniform Cauchy sequence on A. It is the other direction in the following proposition that is frequently helpful in substantiating the uniform convergence of a sequence of functions.

Theorem 1.2. (Cauchy Criterion For Uniform Convergence) *Suppose that each function in a sequence $\langle f_n \rangle$ is defined on a set A. The sequence converges uniformly on A if and only if it is a uniform Cauchy sequence on A.*

Proof. As noted, only the sufficiency is a question mark. We assume, therefore, that $\langle f_n \rangle$ is a uniform Cauchy sequence on A and establish the existence of a function $f \colon A \to \mathbb{C}$ to which $\langle f_n \rangle$ converges uniformly. It is clear from the definition of a uniform Cauchy sequence that for each fixed z in A the sequence of values $\langle f_n(z) \rangle$ is a Cauchy sequence of complex numbers. Theorem II.4.2 thus certifies the existence of the pointwise limit of $\langle f_n \rangle$ in A; i.e., a function $f \colon A \to \mathbb{C}$ is properly defined by the rule of correspondence $f(z) = \lim_{n \to \infty} f_n(z)$. It remains to check that the convergence of $\langle f_n \rangle$ to f is actually uniform on A. Let $\epsilon > 0$ be given. We wish to exhibit an index N with the property that

$$(7.2) \qquad |f_n(z) - f(z)| < \epsilon$$

for every z in A, provided only that $n \geq N$. The uniform Cauchy condition enables us to choose N so that

$$(7.3) \qquad |f_n(z) - f_m(z)| < \frac{\epsilon}{2}$$

is satisfied for all z in A whenever $m > n \geq N$. If we now fix $n \geq N$ and let $m \to \infty$ in (7.3), we find that

$$|f_n(z) - f(z)| \leq \frac{\epsilon}{2} < \epsilon$$

for every z in A. As $n \geq N$ was arbitrary, we have verified (7.2) and so demonstrated that $f_n \to f$ uniformly on the set A. ∎

1.2 Normal Convergence

Our primary concern within the realm of function sequences is the convergence of sequences of analytic functions. We now indicate the framework in

1. Sequences of Functions

which the convergence of such sequences is usually discussed. Suppose that each function in a sequence $\langle f_n \rangle$ is defined — but not necessarily analytic — in an open subset U of $\mathbb{C}$. There is notion of convergence for $\langle f_n \rangle$ that lies between pointwise convergence in U and uniform convergence on U. We refer to it as "normal convergence" in U: $\langle f_n \rangle$ *converges normally in U to the limit function f* if $\langle f_n \rangle$ is pointwise convergent to f in U and if, in addition, the convergence is uniform on each compact set in U. The literature contains an assortment of other names for this type of convergence, two of the most popular being "locally uniform convergence in U" and "uniform convergence on compacta in U." We favor "normal convergence" for its conciseness. To test $\langle f_n \rangle$ for normal convergence in U it is not really necessary to check uniform convergence on every compact set in U — checking it on the closed disks in U is enough, as the next result points out.

Lemma 1.3. *Let $\langle f_n \rangle$ be a sequence of functions defined in an open set U. The sequence converges normally in U if and only if it converges uniformly on each closed disk that is contained in U.*

Proof. The "only if" implication is obvious. We must prove the converse. Suppose then that $\langle f_n \rangle$ converges uniformly on each closed disk in U. This, of course, entails the pointwise convergence of the sequence in U. We denote its limit by f. Let K be an arbitrary non-empty compact set in U. We claim that $f_n \to f$ uniformly on K. To see this, we first invoke Lemma II.4.4 and fix $r > 0$ with the following property: for each z in K the closed disk $\overline{\Delta}(z, r)$ is contained in U. We now pick a point z_1 of K. Either K is contained in $\Delta_1 = \overline{\Delta}(z_1, r)$, or we can select a point z_2 from $K \sim \Delta_1$. In the latter case, set $\Delta_2 = \overline{\Delta}(z_2, r)$. Then either K is contained in $\Delta_1 \cup \Delta_2$, or we can choose a point z_3 from the set $K \sim (\Delta_1 \cup \Delta_2)$ — and so forth. After a finite number of steps — call that number p — we must arrive at a collection $\Delta_1, \Delta_2, \ldots, \Delta_p$ of closed disks in U whose union covers K. (The alternative would be the construction of a sequence $\langle z_n \rangle$ in K having the feature that $|z_n - z_m| \geq r$ when $n \neq m$. Such a sequence could hardly have an accumulation point in K — or anywhere else in $\mathbb{C}$, for that matter — contrary to the definition of compactness.) Because $f_n \to f$ uniformly on each of the disks $\Delta_1, \Delta_2, \ldots, \Delta_p$, it is easy to see that convergence is uniform on $\Delta_1 \cup \Delta_2 \cup \cdots \cup \Delta_p$ — hence, on K. It follows that $f_n \to f$ normally in U. ∎

With a little extra effort, we could improve Lemma 1.3 somewhat: $\langle f_n \rangle$ *converges normally in U if and only if each point z of U is the center of some closed disk on which this sequence converges uniformly.* We shall make no use of this fact and, for that reason, do not include its proof. For reference purposes we record the counterparts of Theorems 1.1 and 1.2 in the setting of normal convergence. The straightforward proofs are left as exercises.

Theorem 1.4. *Suppose that each function in a sequence $\langle f_n \rangle$ is continuous in an open set U and that the sequence converges normally in U to the limit function f. Then f is continuous in U. Furthermore,*

$$\int_\gamma f(z)\,dz = \lim_{n\to\infty} \int_\gamma f_n(z)\,dz$$

for every piecewise smooth path γ in U.

Theorem 1.5. *A sequence $\langle f_n \rangle$ of functions defined in an open set U converges normally in U if and only if it is a uniform Cauchy sequence on each closed disk that is contained in U.*

2 Infinite Series

2.1 Complex Series

We are about ready to take up the topic of paramount interest in this chapter, convergent infinite series of functions. Before embarking on a treatment of such series, however, it will be wise to recall some elementary facts about series of complex numbers.

Presented with a sequence $\langle z_n \rangle$ of complex numbers, we can use it to generate a sequence $\langle s_n \rangle$, the *sequence of partial sums associated with $\langle z_n \rangle$*, by defining $s_1 = z_1$, $s_2 = z_1 + z_2$, $s_3 = z_1 + z_2 + z_3, \cdots$. Thus

$$s_n = z_1 + z_2 + \cdots + z_n\ .$$

For the sequence $\langle s_n \rangle$ there are, naturally, two possibilities. First, it may happen that $\langle s_n \rangle$ is convergent, say with the limit s. It is customary to express this fact by writing $s = \sum_{n=1}^\infty z_n$ and, in these circumstances, to speak of $\sum_{n=1}^\infty z_n$ as a *convergent infinite series with sum s*. Alternatively, $\langle s_n \rangle$ may fail to have a limit in the complex plane, in which case we declare the infinite series $\sum_{n=1}^\infty z_n$ to be *divergent*. The term z_n from the original sequence is also referred to as the nth *term* of the series $\sum_{n=1}^\infty z_n$.

Although it can sometimes be a far from trivial task to decide whether a series $\sum_{n=1}^n z_n$ converges or diverges, we point out one condition that must be met if it is to have the slightest chance of converging: *the nth term of a convergent infinite series $\sum_{n=1}^\infty z_n$ must tend to zero as $n \to \infty$.* The reason for this is that when $n \geq 2$ we can rewrite z_n as $s_n - s_{n-1}$, with the result that $z_n \to s - s = 0$ as $n \to \infty$ in case $s = \sum_{n=1}^\infty z_n$ exists. When trying to determine if the partial sum sequence $\langle s_n \rangle$ associated with a sequence $\langle z_n \rangle$ has a limit, one always has recourse to the Cauchy criterion. Since $s_m - s_{n-1} = \sum_{k=n}^m z_k$ for $m \geq n$, we can formulate the "Cauchy criterion for infinite series" as follows: *a series of complex numbers $\sum_{n=1}^\infty z_n$ converges*

2. Infinite Series

if and only if corresponding to each $\epsilon > 0$ there is an index $N = N(\epsilon)$ such that $|\sum_{k=n}^{m} z_k| < \epsilon$ holds whenever $m \geq n \geq N$.

In attempting to ascertain whether a complex series $\sum_{n=1}^{\infty} z_n$ converges or diverges, it is often instructive to consider along with the given series the corresponding series of moduli, $\sum_{n=1}^{\infty} |z_n|$. If the latter series converges, the original series $\sum_{n=1}^{\infty} z_n$ is said to be *absolutely convergent*. The inequality $|\sum_{k=n}^{m} z_k| \leq \sum_{k=n}^{m} |z_k|$ and the Cauchy criterion for infinite series insure that an absolutely convergent series does, in fact, converge. Furthermore, it is then an easy matter to check that

$$\left|\sum_{n=1}^{\infty} z_n\right| \leq \sum_{n=1}^{\infty} |z_n|.$$

The advantage gained by passing to the absolute value series $\sum_{n=1}^{\infty} |z_n|$ is that, being a series of non-negative real numbers, this series is subject to the standard convergence tests studied in calculus — in particular, to the comparison, root, ratio, and integral tests.

An important characteristic of an absolutely convergent series $\sum_{n=1}^{\infty} z_n$ is that the terms of such a series can be permuted in an arbitrary fashion without influencing either the fact of convergence or the value of the sum. We mean by this statement that

$$\sum_{n=1}^{\infty} z_n = \sum_{n=1}^{\infty} z_{\sigma(n)}$$

for every one-to-one mapping σ of the set of positive integers onto itself (Exercise 5.17). This rearrangement invariance (it is a property definitely not enjoyed by a series that converges but fails to converge absolutely, a so-called *conditionally convergent series*) can sometimes be instrumental in actually computing the sum of an absolutely convergent series.

If each of two series $\sum_{n=1}^{\infty} z_n$ and $\sum_{n=1}^{\infty} w_n$ is convergent (respectively, absolutely convergent) and if c is a complex number, then the series $\sum_{n=1}^{\infty} cz_n$ and $\sum_{n=1}^{\infty} (z_n + w_n)$ also converge (resp., converge absolutely). Moreover,

$$\sum_{n=1}^{\infty} cz_n = c \sum_{n=1}^{\infty} z_n \quad , \quad \sum_{n=1}^{\infty} (z_n + w_n) = \sum_{n=1}^{\infty} z_n + \sum_{n=1}^{\infty} w_n.$$

Notice especially that, if $z_n = x_n + iy_n$, the complex series $\sum_{n=1}^{\infty} z_n$ converges (resp., converges absolutely) if and only if both real series $\sum_{n=1}^{\infty} x_n$ and $\sum_{n=1}^{\infty} y_n$ converge (resp., converge absolutely) — in which case, of course,

$$\sum_{n=1}^{\infty} z_n = \sum_{n=1}^{\infty} x_n + i \sum_{n=1}^{\infty} y_n.$$

We shall frequently be forced to deal with infinite series of a more general type than those we've considered up to this point, series such as $\sum_{n=0}^{\infty} z_n$ or $\sum_{n=-3}^{\infty} z_n$ or even doubly infinite series $\sum_{n=-\infty}^{\infty} z_n$. Examples like the first two cause no problems: one simply makes the obvious adjustment in the definition of the partial sum s_n — in the first case use $s_n = z_0 + z_1 + \cdots + z_n$, in the second $s_n = z_{-3} + z_{-2} + \cdots + z_n$ — and proceeds as earlier. The situation with a series of the form $\sum_{n=-\infty}^{\infty} z_n$ is more complicated, but only mildly so. For each pair of positive integers m and n we form the partial sum $s_{m,n} = z_{-m} + z_{-m+1} + \cdots + z_{n-1} + z_n$. The series $\sum_{n=-\infty}^{\infty} z_n$ is defined to be convergent and to have the complex number s as its sum — again indicated by writing $s = \sum_{n=-\infty}^{\infty} z_n$ — if $s_{m,n} \to s$ as m and n tend independently to ∞. (The technical definition of the last condition reads: corresponding to each $\epsilon > 0$ there is a positive integer $N = N(\epsilon)$ such that $|s_{m,n} - s| < \epsilon$ is true whenever both $n \geq N$ and $m \geq N$.) When no complex number s with this property exists, $\sum_{n=-\infty}^{\infty} z_n$ is pronounced divergent. Often the most efficient method of handling a doubly infinite series $\sum_{n=-\infty}^{\infty} z_n$ is to deal separately with the series $\sum_{n=0}^{\infty} z_n$ and $\sum_{n=-\infty}^{-1} z_n = \sum_{n=1}^{\infty} z_{-n}$. This approach is the subject of a short lemma.

Lemma 2.1. *A doubly infinite series of complex numbers $\sum_{n=-\infty}^{\infty} z_n$ converges if and only if both of the series $\sum_{n=0}^{\infty} z_n$ and $\sum_{n=1}^{\infty} z_{-n}$ converge, in which event*

(7.4) $$\sum_{n=-\infty}^{\infty} z_n = \sum_{n=1}^{\infty} z_{-n} + \sum_{n=0}^{\infty} z_n .$$

Proof. For $m \geq 1$ and $n \geq 1$, we write

$$s_{-m} = z_{-m} + z_{-m+1} + \cdots + z_{-1} \quad , \quad s_n = z_0 + z_1 + \cdots + z_n ,$$

$$s_{m,n} = z_{-m} + z_{-m+1} + \cdots + z_{n-1} + z_n .$$

Thus $s_{m,n} = s_{-m} + s_n$. Suppose first that $s^- = \sum_{n=1}^{\infty} z_{-n}$ and $s^+ = \sum_{n=0}^{\infty} z_n$ both exist. Then clearly $s_{m,n} = s_{-m} + s_n \to s^- + s^+$ as $m \to \infty$ and $n \to \infty$. Accordingly, $\sum_{n=-\infty}^{\infty} z_n$ converges and (7.4) holds. For the converse, assume that $\sum_{n=-\infty}^{\infty} z_n$ converges and has the complex number s for its sum. We employ the Cauchy criterion for series to establish the convergence of $\sum_{n=0}^{\infty} z_n$. Let $\epsilon > 0$ be given. We want to produce an index N with the property that $|\sum_{k=n}^{m} z_k| < \epsilon$, provided $m \geq n \geq N$. Since $s_{p,q} \to s$ as p and q tend to ∞, we are able to fix a positive integer M such that $|s_{p,q} - s| < \epsilon/2$ holds whenever both $p \geq M$ and $q \geq M$. In particular, $|s_{M,q} - s| < \epsilon/2$ for $q \geq M$. Set $N = M + 1$. When $m \geq n \geq N$, we see that

$$\left| \sum_{k=n}^{m} z_k \right| = |s_{M,m} - s_{M,n-1}| \leq |s_{M,m} - s| + |s - s_{M,n-1}| < \frac{\epsilon}{2} + \frac{\epsilon}{2} = \epsilon .$$

2. Infinite Series   251

The Cauchy criterion for series informs us that $\sum_{n=0}^{\infty} z_n$ is convergent. A similar argument demonstrates the convergence of $\sum_{n=1}^{\infty} z_{-n}$. By the first part of this proof formula (7.4) is then valid. ∎

The notion of absolute convergence carries over to doubly infinite series, with the expected results: if $\sum_{n=-\infty}^{\infty} |z_n|$ converges, then $\sum_{n=-\infty}^{\infty} z_n$ converges; moreover, the terms of the latter series can be arbitrarily rearranged without affecting either the convergence or the value of the sum. Lemma 2.1 implies that $\sum_{n=-\infty}^{\infty} z_n$ is absolutely convergent precisely when both $\sum_{n=0}^{\infty} z_n$ and $\sum_{n=1}^{\infty} z_{-n}$ have that feature. The algebraic properties of convergent infinite series are, of course, inherited by convergent series of the doubly infinite variety.

Before moving on to series of functions, we pause to look at a few examples involving complex numerical series.

EXAMPLE 2.1. Discuss the convergence of the geometric series $\sum_{n=0}^{\infty} z^n$.

The geometric series is one series for which it is possible to express the partial sum $s_n = 1 + z + \cdots + z^n$ in an explicit and convenient form. If $z = 1$, then $s_n = n+1$; if $z \neq 1$, then the formula for the sum of a geometric progression gives

$$s_n = \frac{1 - z^{n+1}}{1 - z} = \frac{1}{1-z} - \frac{z^{n+1}}{1-z}.$$

When $z = 1$, $\langle s_n \rangle$ is unbounded and so has no limit. The geometric series thus diverges for $z = 1$. Assuming that $z \neq 1$, we recall from Section II.1.7 that $\lim_{n \to \infty} z^n = 0$ when $|z| < 1$ and that this limit fails to exist when $|z| \geq 1$. In the former case the geometric series is convergent, and its sum is found to be $\lim_{n \to \infty} s_n = (1-z)^{-1}$; in the latter case the series diverges. To summarize:

$$\sum_{n=0}^{\infty} z^n = \frac{1}{1-z}$$

when $|z| < 1$ and $\sum_{n=0}^{\infty} z^n$ diverges for all other complex numbers z. The convergence of $\sum_{n=0}^{\infty} z^n$ is plainly absolute when $|z| < 1$.

EXAMPLE 2.2. Test each of the series $\sum_{n=1}^{\infty} n(1+i)^n (2i)^{-n}$ and $\sum_{n=1}^{\infty} n^2 2^n (1+i)^{-n}$ for convergence.

For the first series we apply the root test to the associated series of moduli; i.e., we compute

$$L = \lim_{n \to \infty} \sqrt[n]{\frac{n|1+i|^n}{|2i|^n}} = \frac{|1+i|}{2} \lim_{n \to \infty} \sqrt[n]{n} = \frac{|1+i|}{2} = \frac{\sqrt{2}}{2}.$$

Since $L < 1$, the series $\sum_{n=1}^{\infty} n(1+i)^n (2i)^{-n}$ converges absolutely. We test the second series for convergence by using the ratio test on the absolute value series. This time we calculate

$$L = \lim_{n\to\infty} \frac{\frac{(n+1)^2 2^{n+1}}{|1+i|^{n+1}}}{\frac{n^2 2^n}{|1+i|^n}} = \frac{2}{|1+i|} \lim_{n\to\infty} \left(\frac{n+1}{n}\right)^2 = \frac{2}{|1+i|} = \sqrt{2}.$$

Because $L > 1$, the series $\sum_{n=1}^{\infty} n^2 2^n (1+i)^{-n}$ definitely does not converge absolutely. In fact, more can be said. If, in applying the root test or ratio test to the absolute value series $\sum_{n=1}^{\infty} |z_n|$ corresponding to a complex series $\sum_{n=1}^{\infty} z_n$, it turns out that $L = \lim_{n\to\infty} \sqrt[n]{|z_n|} > 1$ or that $L = \lim_{n\to\infty} |z_{n+1}|/|z_n| > 1$, then it is necessarily true that $z_n \not\to 0$ as $n \to \infty$ — hence, that $\sum_{n=1}^{\infty} z_n$ diverges. In particular, we conclude that the series $\sum_{n=1}^{\infty} n^2 2^n (1+i)^{-n}$ diverges.

EXAMPLE 2.3. Test the series $\sum_{n=1}^{\infty} n^{-1} i^n$ for convergence.

The absolute value series associated with this series is the harmonic series $\sum_{n=1}^{\infty} n^{-1}$. It diverges, as an application of the integral test demonstrates. Consequently, $\sum_{n=1}^{\infty} n^{-1} i^n$ is not absolutely convergent. If we now write $n^{-1} i^n$ in the form $x_n + i y_n$, we find that $x_n = 0$ for odd n and $x_{2n} = (-1)^n/(2n)$ for $n = 1, 2, 3, \cdots$, while $y_n = 0$ for even n and $y_{2n-1} = (-1)^{n-1}/(2n-1)$ for $n = 1, 2, 3, \cdots$. In other words, $\sum_{n=1}^{\infty} x_n$ reduces to $\sum_{n=1}^{\infty} (-1)^n/(2n)$ and $\sum_{n=1}^{\infty} y_n$ to $\sum_{n=1}^{\infty} (-1)^{n-1}/(2n-1)$. That each of these real series converges follows from the alternating series test. We deduce that $\sum_{n=1}^{\infty} n^{-1} i^n$ is conditionally convergent.

EXAMPLE 2.4. Discuss the convergence of $\sum_{n=-\infty}^{\infty} 2^{-|n|} z^n$.

According to Lemma 2.1 we are allowed to split the given series into the parts corresponding to $n \geq 0$ and $n < 0$, respectively, and to consider each of these subseries on its own. For the portion whose terms go with non-negative indices, Example 2.2 shows that

$$\sum_{n=0}^{\infty} 2^{-|n|} z^n = \sum_{n=0}^{\infty} \left(\frac{z}{2}\right)^n = \frac{1}{1-(z/2)} = \frac{2}{2-z}$$

when $|z/2| < 1$ — i.e., when $|z| < 2$ — and that this series diverges for all remaining z. For the negatively indexed half of the series we obtain

$$\sum_{n=-\infty}^{-1} 2^{-|n|} z^n = \sum_{n=1}^{\infty} 2^{-|-n|} z^{-n} = \sum_{n=1}^{\infty} \left(\frac{1}{2z}\right)^n = \frac{1}{2z} \sum_{n=1}^{\infty} \left(\frac{1}{2z}\right)^{n-1}$$

$$= \frac{1}{2z} \sum_{n=0}^{\infty} \left(\frac{1}{2z}\right)^n = \frac{1}{2z} \frac{1}{1-(1/2z)} = \frac{1}{2z-1}$$

when $|1/(2z)| < 1$ — i.e., when $|z| > 1/2$ — but we observe divergence

2. Infinite Series

otherwise. On the strength of Lemma 2.1 we can say that

$$\sum_{n=-\infty}^{\infty} 2^{-|n|} z^n = \frac{2}{2-z} + \frac{1}{2z-1} = \frac{3z}{(2-z)(2z-1)}$$

when $1/2 < |z| < 2$, whereas this series is divergent for every other z.

2.2 Series of Functions

We return attention once more to a sequence $\langle f_n \rangle$ of complex-valued functions, each of which we assume to be defined in an open subset U of the complex plane. It is natural to mimic the process carried out for complex numerical series and to associate with $\langle f_n \rangle$ its sequence of partial sums s_n:

$$s_n = f_1 + f_2 + \cdots + f_n \,.$$

If the sequence of functions $\langle s_n \rangle$ converges pointwise in the set U to the limit function f, then we write $f = \sum_{n=1}^{\infty} f_n$ and say that the infinite series $\sum_{n=1}^{\infty} f_n$ is *pointwise convergent in U with sum f*. (Another way to phrase this definition is to state that for each z in U the series of complex numbers $\sum_{n=1}^{\infty} f_n(z)$ is convergent and has sum $f(z)$.) If $\langle s_n \rangle$ converges uniformly on a subset A of U, we refer to $\sum_{n=1}^{\infty} f_n$ as *uniformly convergent on A*. Finally, if $\langle s_n \rangle$ converges uniformly on each compact set in U, $\sum_{n=1}^{\infty} f_n$ is termed *normally convergent in U*. Owing to Lemma 1.3, in order to certify that a series $\sum_{n=1}^{\infty} f_n$ is normally convergent in U we need only check that it is uniformly convergent on each closed disk in U. We speak of $\sum_{n=1}^{\infty} f_n$ as *absolutely convergent in U* if the absolute value series $\sum_{n=1}^{\infty} |f_n|$ is pointwise convergent there. When this is so, the series $\sum_{n=1}^{\infty} f_n$ is itself pointwise convergent in U, and its sum remains unchanged under an arbitrary reordering of its terms.

A necessary condition for a series of functions $\sum_{n=1}^{\infty} f_n$ to converge uniformly on a set A is that $f_n \to 0$ uniformly on that set. This is evident from the fact that $f_n = s_n - s_{n-1}$ when $n \geq 2$. Theorem 1.2 leads directly to the counterpart in the context of uniform convergence of function series to the Cauchy convergence criterion for series of complex numbers: *the series $\sum_{n=1}^{\infty} f_n$ converges uniformly on A if and only if corresponding to each $\epsilon > 0$ there is an index $N = N(\epsilon)$ such that $|\sum_{k=n}^{m} f_k(z)| < \epsilon$ is satisfied for every z in A whenever $m \geq n \geq N$.* In practice this criterion is usually implemented via comparison with an appropriate numerical series, as detailed in the next theorem.

Theorem 2.2. (Weierstrass M-test) *Suppose that each term in a function series $\sum_{n=1}^{\infty} f_n$ is defined on a set A. If there exists a sequence $\langle M_n \rangle$ of real numbers such that the estimate $|f_n(z)| \leq M_n$ holds for every z in A and such that the series $\sum_{n=1}^{\infty} M_n$ converges, then $\sum_{n=1}^{\infty} f_n$ converges absolutely and uniformly on A.*

Proof. That $\sum_{n=1}^{\infty} |f_n(z)|$ converges for each z in A follows immediately from the comparison test for real series. To verify that the convergence of $\sum_{n=1}^{\infty} f_n$ is uniform on A we invoke the Cauchy criterion cited above. Given $\epsilon > 0$, we first appeal to the Cauchy criterion for numerical series to select an index N with the property that $\sum_{k=n}^{m} M_k < \epsilon$ whenever $m \geq n \geq N$. This is possible because $\sum_{n=1}^{\infty} M_n$ is, by assumption, convergent. It follows that for every z in A

$$\left| \sum_{k=n}^{m} f_k(z) \right| \leq \sum_{k=n}^{m} |f_k(z)| \leq \sum_{k=n}^{m} M_k < \epsilon,$$

as soon as $m \geq n \geq N$. By the Cauchy criterion for uniform convergence $\sum_{n=1}^{\infty} f_n$ is seen to converge uniformly on A. ∎

We illustrate the application of the last theorem with two examples.

EXAMPLE 2.5. Verify that the geometric series $\sum_{n=0}^{\infty} z^n$ converges uniformly on $A_r = \overline{\Delta}(0, r)$ when $0 < r < 1$. Conclude that this series is normally convergent in the open disk $\Delta = \Delta(0, 1)$.

We are dealing here with the series of functions $\sum_{n=0}^{\infty} f_n$ in Δ, where $f_n(z) = z^n$. Since $|f_n(z)| = |z^n| = |z|^n \leq r^n$ for every z in A_r and since $\sum_{n=0}^{\infty} r^n$ converges when $0 < r < 1$, we can appeal to the Weierstrass M-test with $M_n = r^n$ to establish the uniform convergence of the geometric series on A_r for each fixed r in $(0, 1)$. Because any compact subset K of Δ lies in A_r for r suitably close to 1, we conclude that $\sum_{n=0}^{\infty} z^n$ converges uniformly on each such K — and, as a result, normally in Δ. Note, however, that $\sum_{n=0}^{\infty} z^n$ does not converge uniformly on Δ. If it did, it would follow that $f_n \to 0$ uniformly there. This is simply not the case. For instance, given any positive integer n, one can easily exhibit points z in Δ for which $|z^n| \geq 1/2$; e.g., one can take $z = \exp(-n^{-1} \operatorname{Log} 2)$.

EXAMPLE 2.6. Show that the series $\sum_{n=1}^{\infty} n^{-z}$ converges absolutely and uniformly on the set $A_\sigma = \{z : \operatorname{Re} z \geq \sigma\}$ for $\sigma > 1$. Deduce that this series is normally convergent in the open half-plane $U = \{z : \operatorname{Re} z > 1\}$.

This example is concerned with $\sum_{n=1}^{\infty} f_n$, where $f_n(z) = n^{-z}$. For $z = x + iy$ in A_σ we have

$$|f_n(z)| = |n^{-z}| = \left| e^{-z \operatorname{Log} n} \right| = e^{-x \operatorname{Log} n} \leq e^{-\sigma \operatorname{Log} n} = n^{-\sigma}.$$

The integral test shows that $\sum_{n=1}^{\infty} n^{-\sigma}$ converges when $\sigma > 1$. Using $M_n = n^{-\sigma}$ in the Weierstrass M-test we infer the absolute and uniform convergence of $\sum_{n=1}^{\infty} n^{-z}$ on A_σ, provided $\sigma > 1$. As each compact subset of U is contained in A_σ for some $\sigma > 1$, the normal convergence of the series in U then becomes evident.

Doubly infinite series of functions will also be commonplace throughout the rest of this book. The definitions of pointwise, uniform, and normal

2. Infinite Series

convergence for a series $\sum_{n=-\infty}^{\infty} f_n$ merely demand the respective type of convergence for the partial sums $s_{m,n} = f_{-m} + f_{-m+1} + \cdots + f_{n-1} + f_n$ as m and n tend independently to ∞. To test such a series for uniform convergence we can take a hint from Lemma 2.1 and break the series into its "positive" and "negative" parts, meaning $\sum_{n=0}^{\infty} f_n$ and $\sum_{n=1}^{\infty} f_{-n}$, for separate treatment. This method is justified by the analogue of Lemma 2.1 in the framework of uniform convergence of function series. Its proof, which essentially retraces the steps in the proof of Lemma 2.1, is relegated to the exercises (Exercise 5.28).

Lemma 2.3. *A doubly infinite series of functions $\sum_{n=-\infty}^{\infty} f_n$ converges uniformly on a set A if and only if both of the series $\sum_{n=0}^{\infty} f_n$ and $\sum_{n=1}^{\infty} f_{-n}$ converge uniformly on A, in which event it is true that*

$$\sum_{n=-\infty}^{\infty} f_n = \sum_{n=1}^{\infty} f_{-n} + \sum_{n=0}^{\infty} f_n$$

in this set.

We close this section by establishing the series version of Theorem 1.4.

Theorem 2.4. *Suppose that each term in a function series $\sum_{n=1}^{\infty} f_n$ is continuous in an open set U and that the series converges normally in U, with the function f as its sum. Then f is continuous in U. Furthermore,*

(7.5) $$\int_\gamma f(z)\,dz = \sum_{n=1}^{\infty} \int_\gamma f_n(z)\,dz$$

for every piecewise smooth path γ in U.

Proof. If $s_n = f_1 + f_2 + \cdots + f_n$, then s_n is continuous in U and $s_n \to f$ normally in this open set. Theorem 1.4 informs us that f is a continuous function in U and that

$$\int_\gamma f(z)\,dz = \lim_{N\to\infty} \int_\gamma s_N(z)\,dz = \lim_{N\to\infty} \sum_{n=1}^{N} \int_\gamma f_n(z)\,dz = \sum_{n=1}^{\infty} \int_\gamma f_n(z)\,dz$$

for every piecewise smooth path γ in U. ∎

We remark that it is only the uniform convergence of $\sum_{n=1}^{\infty} f_n$ on the specific compact set $|\gamma|$ that was required for the proof of (7.5). Theorem 2.4 has an obvious counterpart for doubly infinite series. We leave its formulation and proof to the reader.

3 Sequences and Series of Analytic Functions

3.1 General Results

The discussions in the preceding sections lay the groundwork for two theorems of the utmost importance in the theory of analytic functions.

Theorem 3.1. *Suppose that each function in a sequence $\langle f_n \rangle$ is analytic in an open set U and that the sequence converges normally in U to the limit function f. Then f is analytic in U. Moreover, $f_n^{(k)} \to f^{(k)}$ normally in U for each positive integer k.*

Proof. Theorem 1.4 certifies the continuity of the limit f in U. The same theorem, in combination with Lemma V.1.1, tells us that

$$\int_{\partial R} f(z)\, dz = \lim_{n \to \infty} \int_{\partial R} f_n(z)\, dz = \lim_{n \to \infty} 0 = 0$$

for every closed rectangle R in U. Morera's theorem then bears witness to the analyticity of f in that open set.

We next show that $f_n' \to f'$ normally in U. By Lemma 1.3 it is enough to check that $f_n' \to f'$ uniformly on each closed disk in U. Fix such a disk Δ, say $\Delta = \overline{\Delta}(z_0, r)$. Given $\epsilon > 0$ we wish to determine an index N with the property that $|f_n'(z) - f'(z)| < \epsilon$ for every z in Δ and every $n \geq N$. We begin by fixing $s > r$ such that the disk $\overline{\Delta}(z_0, s)$ is still contained in U. Let z belong to Δ. We appeal to Theorem V.3.4 in making the estimate

$$|f_n'(z) - f'(z)| = \left| \frac{1}{2\pi i} \int_{|\zeta - z_0| = s} \frac{f_n(\zeta)\, d\zeta}{(\zeta - z)^2} - \frac{1}{2\pi i} \int_{|\zeta - z_0| = s} \frac{f(\zeta)\, d\zeta}{(\zeta - z)^2} \right|$$

$$= \frac{1}{2\pi} \left| \int_{|\zeta - z_0| = s} \frac{[f_n(\zeta) - f(\zeta)]\, d\zeta}{(\zeta - z)^2} \right| \leq \frac{1}{2\pi} \int_{|\zeta - z_0| = s} \frac{|f_n(\zeta) - f(\zeta)|\, |d\zeta|}{|\zeta - z|^2}$$

$$\leq \frac{s}{(s-r)^2} \max\{|f_n(\zeta) - f(\zeta)| : \zeta \in K\},$$

where $K = K(z_0, s)$. (N.B. $|\zeta - z| \geq s - r$ when z belongs to Δ and ζ lies on the circle K.) By hypothesis, $f_n \to f$ uniformly on K. This fact enables us to select N so that, once $n \geq N$, we can count on the inequality

$$|f_n(\zeta) - f(\zeta)| < \frac{(s-r)^2 \epsilon}{s}$$

being in force for every ζ on K. In particular, it is applicable at any point ζ of K where the quantity $|f_n(\zeta) - f(\zeta)|$ is maximized. If $n \geq N$, therefore, the previous estimate insures that $|f_n'(z) - f'(z)| < \epsilon$ holds for every z in Δ, as desired. Accordingly, $f_n' \to f'$ uniformly on each closed disk in U

3. Sequences and Series of Analytic Functions

— hence, normally in U. Applying this fact to the sequence of derivatives $\langle f_n' \rangle$, we find that $f_n'' = (f_n')' \to (f')' = f''$ normally in U — and so forth for higher derivatives. ∎

The conclusion in Theorem 3.1 that the limit function f is analytic in U would not follow, in general, if the convergence assumption in the theorem were weakened from normal convergence in U to pointwise convergence there, although it is not an easy matter to write down an explicit counterexample. For more information on this topic we refer the reader to the article "Pointwise limits of analytic functions" by K.R. Davidson in *The American Mathematical Monthly*, Vol. 90, No. 6, 1983.

The series companion to Theorem 3.1 is:

Theorem 3.2. *Suppose that each term in a function series $\sum_{n=1}^{\infty} f_n$ is analytic in an open set U and that the series converges normally in U, with the function f as its sum. Then f is analytic in U. Moreover, $f^{(k)} = \sum_{n=1}^{\infty} f_n^{(k)}$ in U for each positive integer k. The convergence of these derived series is also normal in U.*

Proof. The partial sum $s_n = f_1 + f_2 + \cdots + f_n$ is analytic in U, and $s_n \to f$ normally in U. On the basis of Theorem 3.1 we can assert that f is analytic in U and that $s_n^{(k)} \to f^{(k)}$ normally in this set for each positive integer k. As a result, we discover that

$$f^{(k)} = \lim_{N \to \infty} s_N^{(k)} = \lim_{N \to \infty} \sum_{n=1}^{N} f_n^{(k)} = \sum_{n=1}^{\infty} f_n^{(k)}$$

in U and learn, too, that the convergence of the series of kth-derivatives is normal there. ∎

Theorem 3.2 can also be stated for a doubly infinite series of analytic functions. As all the necessary changes are obvious and purely cosmetic in nature, we omit the details. Theorems 3.1 and 3.2 equip us with a powerful new set of tools for constructing analytic functions. The extent of their power will not be fully appreciated until Chapters IX and X, where these theorems will be put to effective use in establishing basic results on the existence of analytic functions subject to various kinds of constraints. For the time being we must be content with a couple of examples that give the flavor of those later constructions.

EXAMPLE 3.1. Show that the formula $f(z) = \sum_{n=1}^{\infty} n^{-z}$ defines an analytic function in the open set $U = \{z : \operatorname{Re} z > 1\}$. Find a series expansion for f' that is valid in U.

It was observed in Example 2.6 that the series $\sum_{n=1}^{\infty} n^{-z}$ converges normally in U. Because $f_n(z) = n^{-z}$ describes an entire function, Theorem

3.2 implies that f is analytic in U and that

$$f'(z) = \sum_{n=1}^{\infty} f'_n(z) = -\sum_{n=1}^{\infty} n^{-z} \operatorname{Log} n = -\sum_{n=2}^{\infty} n^{-z} \operatorname{Log} n$$

for each z in U. More generally, for any positive integer k we obtain a series expansion of $f^{(k)}$ in U:

$$f^{(k)}(z) = (-1)^k \sum_{n=2}^{\infty} n^{-z} (\operatorname{Log} n)^k \ .$$

The function f in Example 3.1 has a multitude of interesting properties. For starters, it is known that f can be extended beyond the confines of U to a function that is analytic in the set $\mathbb{C} \sim \{1\}$. Since the series $\sum_{n=1}^{\infty} n^{-z}$ diverges when $\operatorname{Re} z < 1$, it is not at all apparent from the definition of f that such an extension would be possible, to say nothing of how it might be achieved. We give no details of the extension process. The only point we wish to make here is that through this extension one can construct from our example f a function that ranks among the most fascinating in all of mathematics. It is called the *Riemann zeta-function*. (In his studies of this function Riemann used the Greek letter ζ to designate it.) With the zeta-function is associated one of the truly monumental unsolved problems of mathematics, to prove or disprove the veracity of the "Riemann hypothesis." Riemann was able to show that $\zeta(-2k) = 0$ for every positive integer k. (These zeros are now referred to as the "trivial zeros" of the zeta-function.) Every other zero of the zeta-function that Riemann managed to identify had real part $1/2$. This led to the *Riemann hypothesis*: every non-trivial zero of the Riemann zeta-function has real part $1/2$. Whether this statement is true or false remains an unanswered question despite sustained efforts over the past century by some of the world's finest mathematicians to settle it. The Riemann hypothesis continues to be a great stimulus to mathematical research — especially in the field of analytic number theory, where progress on this problem concerned with a specific, granted somewhat exotic, analytic function translates in the long run into better knowledge of something as down to earth as the distribution of primes among the natural numbers. From simple examples mighty theories sometimes spring!

EXAMPLE 3.2. Verify that the formula $f(z) = \sum_{n=-\infty}^{\infty} (z-n)^{-2}$ defines an analytic function in the open set $U = \mathbb{C} \sim \{0, \pm 1, \pm 2, \cdots\}$.

For each integer n the function f_n given by $f_n(z) = (z-n)^{-2}$ is analytic in U. If we can demonstrate that the series $\sum_{n=-\infty}^{\infty} f_n$ converges normally in U, then the analyticity of f in U will follow from the analogue of Theorem 3.2 for doubly infinite series. In view of Lemma 2.3 it is enough to check

3. Sequences and Series of Analytic Functions   259

that the series $\sum_{n=0}^{\infty} f_n$ and $\sum_{n=0}^{\infty} f_{-n}$ both converge normally in U. Let p be a positive integer. We shall prove that the truncated series $\sum_{n=p}^{\infty} f_n$ and $\sum_{n=p}^{\infty} f_{-n}$ converge uniformly on the set $A_p = \overline{\Delta}(0, p/2)$. As any given compact subset K of U lies in A_p for p suitably large, we conclude that the full series $\sum_{n=0}^{\infty} f_n$ and $\sum_{n=1}^{\infty} f_{-n}$ converge uniformly on K — the extra terms $\sum_{n=0}^{p-1} f_n$ and $\sum_{n=1}^{p-1} f_{-n}$ clearly do not affect the convergence — and, thus, normally in U. Assuming that z belongs to A_p and that $|n| \geq p$, we remark that

$$|z - n| \geq |n| - |z| \geq |n| - \frac{p}{2} \geq |n| - \frac{|n|}{2} = \frac{|n|}{2}.$$

This gives rise to the estimate

$$|f_n(z)| = \frac{1}{|z-n|^2} \leq \frac{4}{n^2}$$

for all z in A_p, provided $|n| \geq p$. Because $\sum_{n=p}^{\infty} (4/n^2)$ converges, the Weierstrass M-test stamps $\sum_{n=p}^{\infty} f_n$ and $\sum_{n=p}^{\infty} f_{-n}$ as absolutely and uniformly convergent on A_p. We can now safely proclaim the function f analytic in U, where according to Theorem 3.2 its derivative is given by

$$f'(z) = -2 \sum_{n=-\infty}^{\infty} (z-n)^{-3}.$$

Notice that the series defining f is absolutely convergent in U, which means that we are free to rearrange its terms without changing the sum. For instance, we can place the terms $(z-n)^{-2}$ and $(z+n)^{-2}$ for $n \geq 1$ side by side and then combine them to arrive at a different representation for f; namely,

$$f(z) = \frac{1}{z^2} + \sum_{n=1}^{\infty} \left\{ \frac{1}{(z-n)^2} + \frac{1}{(z+n)^2} \right\} = \frac{1}{z^2} + 2 \sum_{n=1}^{\infty} \frac{z^2 + n^2}{(z^2 - n^2)^2}.$$

Observe, also, that the function f is periodic, with period 1: $f(z+1) = f(z)$ for every z in U. This fact is confirmed by the calculation

$$f(z+1) = \sum_{n=-\infty}^{\infty} \frac{1}{[(z+1)-n]^2} = \sum_{n=-\infty}^{\infty} \frac{1}{[z-(n-1)]^2}$$

$$= \sum_{n=-\infty}^{\infty} \frac{1}{(z-n)^2} = f(z).$$

3.2 Limit Superior of a Sequence

To insure a smooth treatment of Taylor and Laurent series we must say a few words about the so-called "limit superior" of a sequence $\langle r_n \rangle$ of non-negative real numbers. (For additional discussion of this concept consult Appendix A.) Suppose first that the sequence $\langle r_n \rangle$ is bounded. Its set of accumulation points is then easily seen to be non-empty, closed, and bounded. As a non-empty compact set of real numbers, the set of accumulation points of $\langle r_n \rangle$ has a largest element. That element is called the *limit superior of* $\langle r_n \rangle$. We denote it by $\limsup_{n\to\infty} r_n$. If the given sequence $\langle r_n \rangle$ is unbounded, on the other hand, we express that fact by writing $\limsup_{n\to\infty} r_n = \infty$. In the bounded case the limit superior of $\langle r_n \rangle$ is the unique real number r that meets the following conditions: for each $\epsilon > 0$ it is the case that (i) $r_n \geq r + \epsilon$ holds for at most finitely many values of n and (ii) there are infinitely many indices n for which $r_n > r - \epsilon$. The first requirement makes certain that $\langle r_n \rangle$ has no accumulation points larger than r; the two conditions together imply that r is itself an accumulation point of $\langle r_n \rangle$. Of course, if $\langle r_n \rangle$ happens to be a convergent sequence, then we note that $\limsup_{n\to\infty} r_n = \lim_{n\to\infty} r_n$. Some examples:

$$\limsup_{n\to\infty} [1 + (-1)^n] = 2 ,$$

since 0 and 2 are the only accumulation points of the sequence $0, 2, 0, 2, \cdots$;

$$\limsup_{n\to\infty} [n + (-1)^n n] = \infty ,$$

because the sequence $0, 4, 0, 8, \cdots$ is unbounded;

$$\limsup_{n\to\infty} \sqrt[n]{n} = \lim_{n\to\infty} \sqrt[n]{n} = 1 ;$$

$$\limsup_{n\to\infty} \left(1 + \frac{1}{n}\right)^n = \lim_{n\to\infty} \left(1 + \frac{1}{n}\right)^n = e .$$

3.3 Taylor Series

Suppose that z_0 is a point of the complex plane. We refer to a function series of the type

$$\sum_{n=0}^{\infty} a_n (z - z_0)^n = a_0 + a_1 (z - z_0) + a_2 (z - z_0)^2 + \cdots ,$$

where $a_0, a_1, a_2, \cdots$ is a sequence of complex numbers, as a *Taylor series* (or, alternatively, as a *power series*) *centered at* z_0. (The Taylor in the name is Brook Taylor (1685-1731), who along with Colin Maclauren (1698-1746) pioneered the study of these series.) The numbers a_n in this polynomial-like expression are known as its *coefficients*. With any such Taylor series we associate an extended real number ρ by the rule

(7.6) $$\rho = \left(\limsup_{n\to\infty} \sqrt[n]{|a_n|} \right)^{-1} .$$

3. Sequences and Series of Analytic Functions

Here we observe the conventions $1/0 = \infty$ and $1/\infty = 0$. The quantity ρ is known as the *radius of convergence* of the given series. When $\rho > 0$ the open disk $\Delta(z_0, \rho)$ is called its *disk of convergence*. (For $\rho = \infty$ this "disk" is actually the whole complex plane.) An explanation for the terminology is supplied by the following theorem.

Theorem 3.3. *Suppose that ρ is the radius of convergence of a Taylor series $\sum_{n=0}^{\infty} a_n(z-z_0)^n$ centered at z_0. The series diverges for any z satisfying $|z-z_0| > \rho$. If $\rho > 0$, the series converges absolutely and normally in the disk $\Delta = \Delta(z_0, \rho)$, so the function f defined by $f(z) = \sum_{n=0}^{\infty} a_n(z-z_0)^n$ is analytic in Δ. The coefficient a_n is then related to f through the formula*

$$(7.7) \qquad a_n = \frac{f^{(n)}(z_0)}{n!}.$$

Proof. Assume first that z satisfies $|z - z_0| = r > \rho$. Thus $r^{-1} < \rho^{-1}$. It follows from (7.6) and the definition of $\limsup_{n \to \infty} \sqrt[n]{|a_n|}$ that the statement $\sqrt[n]{|a_n|} > r^{-1}$ — hence, $|a_n| > r^{-n}$ — must be true for infinitely many values of n. Now

$$|a_n(z-z_0)^n| = |a_n||z-z_0|^n > r^{-n}r^n = 1$$

for any such n. This implies that $a_n(z-z_0)^n \not\to 0$ as $n \to \infty$, a fact which marks the series $\sum_{n=0}^{\infty} a_n(z-z_0)^n$ as divergent.

We assume next that $\rho > 0$ and write $\Delta = \Delta(z_0, \rho)$. In order to establish that $\sum_{n=0}^{\infty} a_n(z-z_0)^n$ converges absolutely and normally in Δ it suffices to demonstrate that this series converges absolutely and uniformly on the closed disk $\Delta_r = \overline{\Delta}(z_0, r)$ for each r in the interval $(0, \rho)$. (Any compact set K in Δ can be enclosed in Δ_r by taking r suitably close to ρ.) We fix r in $(0, \rho)$ and then fix s satisfying $r < s < \rho$. Because $\rho^{-1} < s^{-1}$, (7.6) implies that $\sqrt[n]{|a_n|} < s^{-1}$ once n is sufficiently large. Assume this to be so for all n larger than N. Setting $c = \max\{1, |a_0|, |a_1|s, \ldots, |a_N|s^N\}$, we observe that $|a_n| \le cs^{-n}$ then holds for every n. This gives rise to the estimate $|a_n(z-z_0)^n| \le c(r/s)^n = M_n$ for any z belonging to the disk Δ_r. As $r/s < 1$, $\sum_{n=1}^{\infty} M_n$ converges. The Weierstrass M-test thus guarantees that $\sum_{n=0}^{\infty} a_n(z-z_0)^n$ is absolutely and uniformly convergent on Δ_r, as we had set out to prove.

If $\rho > 0$, Theorem 3.2 assures us that the rule of correspondence $f(z) = \sum_{n=0}^{\infty} a_n(z-z_0)^n$ defines a function which is analytic in the disk Δ. Furthermore, the theorem tells us how to compute the derivatives of f:

$$f'(z) = \sum_{n=1}^{\infty} na_n(z-z_0)^{n-1} = a_1 + 2a_2(z-z_0) + \cdots,$$

$$f''(z) = \sum_{n=2}^{\infty} n(n-1)a_n(z-z_0)^{n-2} = 2a_2 + 6a_3(z-z_0) + \cdots,$$

$$f^{(k)}(z) = \sum_{n=k}^{\infty} n(n-1)\cdots(n-k+1)a_n(z-z_0)^{n-k}$$

$$= k!a_k + (k+1)!a_{k+1}(z-z_0) + \cdots .$$

Inserting $z = z_0$ into the formula for $f^{(k)}$ for any non-negative integer k leads to $f^{(k)}(z_0) = k!a_k$, which corroborates (7.7), the final assertion of the theorem. ∎

We emphasize that, when $0 < \rho < \infty$, Theorem 3.3 makes no statement whatsoever about the convergence or divergence of $\sum_{n=0}^{\infty} a_n(z-z_0)^n$ for z satisfying $|z-z_0| = \rho$. Depending on the coefficients, convergence may occur for all, for some, or for no such z. The use of formula (7.6) is by no means the exclusive way (or, for that matter, invariably the preferred way) to determine the radius of convergence of $\sum_{n=0}^{\infty} a_n(z-z_0)^n$. For instance, if $a_n \neq 0$ for all n — or, more generally, for all sufficiently large n — and if it happens that either $\lim_{n\to\infty}|a_n|/|a_{n+1}|$ exists in the strict sense or $\lim_{n\to\infty}|a_n|/|a_{n+1}| = \infty$, then this limit is equal to ρ (Exercise 5.38). In some situations — see Example 3.4 — a quite effective method of finding ρ is simply to apply the root test or ratio test to $\sum_{n=0}^{\infty}|a_n||z-z_0|^n$ on a pointwise basis.

EXAMPLE 3.3. Determine the radii of convergence of the Taylor series $\sum_{n=1}^{\infty} n^{-1}i^n z^n$ and $\sum_{n=0}^{\infty}\left[(n!)^2/(2n)!\right](z+i)^n$.

For the first series we calculate

$$\frac{1}{\rho} = \limsup_{n\to\infty} \sqrt[n]{|a_n|} = \limsup_{n\to\infty}\frac{1}{\sqrt[n]{n}} = \lim_{n\to\infty}\frac{1}{\sqrt[n]{n}} = 1,$$

which gives $\rho = 1$ as its radius of convergence. Due to the presence of the factorials in the latter series it becomes more convenient to evaluate

$$\lim_{n\to\infty}\frac{|a_n|}{|a_{n+1}|} = \lim_{n\to\infty}\frac{(n!)^2(2n+2)!}{(2n)![(n+1)!]^2} = \lim_{n\to\infty}\frac{4n+2}{n+1} = 4$$

and to conclude that $\rho = 4$ for this series. Its disk of convergence is $\Delta(-i,4)$.

EXAMPLE 3.4. Discuss the convergence of the series $\sum_{n=0}^{\infty} 2^n z^{n^2}$.

The given series $\sum_{n=0}^{\infty} 2^n z^{n^2} = 1 + 2z + 4z^4 + 8z^9 + \cdots$ is a Taylor series centered at the origin in which the coefficient a_n is zero for any non-square value of n. Rather than using (7.6) we test convergence by resorting to the root test, applied pointwise to the absolute value series $\sum_{n=0}^{\infty} 2^n |z|^{n^2}$. From the computation

$$\lim_{n\to\infty}\sqrt[n]{2^n|z|^{n^2}} = \lim_{n\to\infty} 2|z|^n = \begin{cases} 0 & \text{if } |z| < 1, \\ 2 & \text{if } |z| = 1, \\ \infty & \text{if } |z| > 1, \end{cases}$$

3. Sequences and Series of Analytic Functions

we learn that the series $\sum_{n=0}^{\infty} 2^n z^{n^2}$ converges absolutely when $|z| < 1$ and diverges otherwise. This state of affairs is only compatible with $\rho = 1$ being its radius of convergence. The approach we've opted to take here has the extra advantage of making clear in this example the status of convergence at all points on the circle bounding the disk of convergence, not always an easy thing to do.

The next theorem declares that the collection of functions which are representable in an open disk $\Delta = \Delta(z_0, r)$ as sums of convergent Taylor series centered at z_0 embraces all of the functions that are analytic in Δ. This important fact opens the door to a systematic examination of the local structural properties of analytic functions, a line of inquiry we shall pursue in the succeeding chapters of this book.

Theorem 3.4. *Suppose that a function f is analytic in an open set U, that z_0 is a point of U, and that the open disk $\Delta = \Delta(z_0, r)$ is contained in U. Then f can be represented in Δ as the sum of a Taylor series centered at z_0. This expansion is uniquely determined by f: if $f(z) = \sum_{n=0}^{\infty} a_n (z - z_0)^n$ in Δ, then the coefficient a_n is given by $a_n = f^{(n)}(z_0)/n!$.*

Proof. Define a sequence $\langle a_n \rangle_{n=0}^{\infty}$ by $a_n = f^{(n)}(z_0)/n!$. It is our contention that $f(z) = \sum_{n=0}^{\infty} a_n (z - z_0)^n$ for every z in Δ. Fix such a point z, and fix along with it a number s satisfying $|z - z_0| < s < r$. For any ζ on the circle $K = K(z_0, s)$ it is true that $|(z - z_0)/(\zeta - z_0)| = |z - z_0|/s < 1$. On the basis of Example 2.1 the following expansion of $f(\zeta)/(\zeta - z)$ is permitted:

$$\frac{f(\zeta)}{\zeta - z} = \frac{f(\zeta)}{\zeta - z_0} \frac{1}{1 - \left(\frac{z - z_0}{\zeta - z_0}\right)} = \frac{f(\zeta)}{\zeta - z_0} \sum_{n=0}^{\infty} \left(\frac{z - z_0}{\zeta - z_0}\right)^n ;$$

i.e., for ζ on K we can write

(7.8) $$\frac{f(\zeta)}{\zeta - z} = \sum_{n=0}^{\infty} \frac{f(\zeta)(z - z_0)^n}{(\zeta - z_0)^{n+1}} .$$

Viewed as a series of functions of the variable ζ the series in (7.8) converges uniformly on K. (This assertion is supported by the Weierstrass M-test — take $M_n = ct^n/s$, where $t = |z - z_0|/s < 1$ and $c = \max\{|f(\zeta)| : \zeta \in K\}$.) Appealing to the Cauchy integral formula, then to Theorem 2.4 for authorization to interchange summation and integration, and lastly to Cauchy's integral formula for derivatives, we find that

$$f(z) = \frac{1}{2\pi i} \int_{|\zeta - z_0| = s} \frac{f(\zeta) \, d\zeta}{\zeta - z}$$

$$= \frac{1}{2\pi i} \int_{|\zeta - z_0| = s} \left[\sum_{n=0}^{\infty} \frac{f(\zeta)(z - z_0)^n}{(\zeta - z_0)^{n+1}} \right] d\zeta$$

$$= \frac{1}{2\pi i} \sum_{n=0}^{\infty} \int_{|\zeta-z_0|=s} \frac{f(\zeta)(z-z_0)^n \, d\zeta}{(\zeta-z_0)^{n+1}}$$

$$= \sum_{n=0}^{\infty} \left[\frac{1}{2\pi i} \int_{|\zeta-z_0|=s} \frac{f(\zeta) \, d\zeta}{(\zeta-z_0)^{n+1}} \right] (z-z_0)^n$$

$$= \sum_{n=0}^{\infty} \left[\frac{f^{(n)}(z_0)}{n!} \right] (z-z_0)^n = \sum_{n=0}^{\infty} a_n (z-z_0)^n ,$$

the result desired.

The existence of at least one Taylor series development for f in Δ has now been substantiated. If $f(z) = \sum_{n=0}^{\infty} b_n (z-z_0)^n$ were a potentially different Taylor series representation of f in Δ, then the radius of convergence ρ of $\sum_{n=0}^{\infty} b_n (z-z_0)^n$ could clearly be no smaller than r, and the function $g(z) = \sum_{n=0}^{\infty} b_n (z-z_0)^n$, known by Theorem 3.3 to be analytic in $\Delta(z_0, \rho)$, would agree with f in Δ. According to the last statement in Theorem 3.3 this would mean that

$$b_n = \frac{g^{(n)}(z_0)}{n!} = \frac{f^{(n)}(z_0)}{n!} = a_n ,$$

so the second expansion would not really be different after all. The uniqueness assertion in the present theorem follows. ∎

Given that a function f is analytic in an open set U, we are now at liberty to expand f in a Taylor series about an arbitrary point z_0 of U, say $f(z) = \sum_{n=0}^{\infty} a_n (z-z_0)^n$. We are assured by Theorem 3.4 that the radius of convergence ρ of this Taylor series is not less than d, the radius of the largest open disk centered at z_0 that is contained in U. It may, in fact, happen that $\rho > d$. (See Figure 1.) Assuming this to be the situation, we know that $g(z) = \sum_{n=0}^{\infty} a_n (z-z_0)^n$ defines a function which is analytic in the disk $\Delta(z_0, \rho)$. *We wish to maintain that $f(z) = g(z)$, however, only for those z belonging to $\Delta(z_0, d)$.* For z outside $\Delta(z_0, d)$, as in Figure 1, it is frequently the case that $f(z) \neq g(z)$. Example 3.11 will provide a concrete illustration of this phenomenon. We remark that the condition $\rho > d$ sometimes opens the possibility of extending the function f, originally assumed to be analytic just in U, to a function that is analytic in an open set larger than U. For instance, we shall learn in the next chapter that, if $\rho > d$ and if the set $U \cap \Delta(z_0, \rho)$ is connected, then the rule of correspondence

$$h(z) = \begin{cases} f(z) & \text{if } z \in U, \\ g(z) & \text{if } z \in \Delta(z_0, \rho), \end{cases}$$

describes a function h that is analytic in $V = U \cup \Delta(z_0, \rho)$, an open set which properly includes U. This method of extending an analytic function is the basic step in a process called "analytic continuation," a topic that will be taken up in Chapter X.

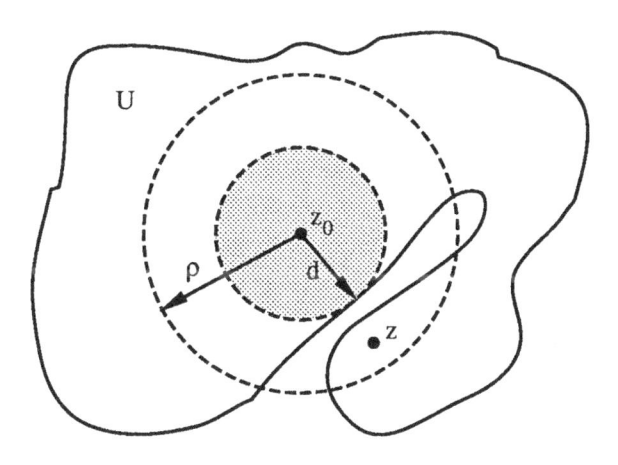

3. Sequences and Series of Analytic Functions   265

Figure 1.

We look now at a string of examples.

EXAMPLE 3.5. Find the Taylor series expansion of $f(z) = e^z$ that is centered at the origin.

Since $f^{(n)}(z) = e^z$ for every non-negative integer n, we obtain $f^{(n)}(0) = 1$ for all n. This produces the Taylor series representation

$$e^z = \sum_{n=0}^{\infty} \frac{z^n}{n!} = 1 + z + \frac{z^2}{2!} + \frac{z^3}{3!} + \cdots$$

for any z in the complex plane.

EXAMPLE 3.6. Determine the power series expansions of $f(z) = \sin z$ and $g(z) = \cos z$ about the origin.

Here we have $f^{(n)}(0) = 0$ for even n and $f^{(2n+1)}(0) = (-1)^n$ for $n = 0, 1, 2, \cdots$. The resulting Taylor series expansion,

$$\sin z = \sum_{n=0}^{\infty} \frac{(-1)^n z^{2n+1}}{(2n+1)!} = z - \frac{z^3}{3!} + \frac{z^5}{5!} - \frac{z^7}{7!} + \cdots ,$$

is valid throughout the complex plane. To arrive at the expansion of g we need only differentiate the series representing f term by term,

$$\cos z = \sum_{n=0}^{\infty} \frac{(-1)^n z^{2n}}{(2n)!} = 1 - \frac{z^2}{2!} + \frac{z^4}{4!} - \frac{z^6}{6!} + \cdots .$$

EXAMPLE 3.7. Expand $f(z) = \text{Log } z$ in a Taylor series about the point $z_0 = 1$.

The function f is analytic in $U = \mathbb{C} \sim (-\infty, 0]$, where its derivatives are given by $f^{(n)}(z) = (-1)^{n-1}(n-1)! \, z^{-n}$ for $n \geq 1$. Hence $f(1) = 0$ and $f^{(n)}(1) = (-1)^{n-1}(n-1)!$ for $n \geq 1$, which yields the expansion

$$\text{Log } z = \sum_{n=1}^{\infty} \frac{(-1)^{n-1}(z-1)^n}{n}$$

$$= (z-1) - \frac{(z-1)^2}{2} + \frac{(z-1)^3}{3} - \frac{(z-1)^4}{4} + \cdots .$$

This representation is applicable in $\Delta(1,1)$, the largest open disk centered at 1 that is contained in U.

EXAMPLE 3.8. Expand $f(z) = 2z(z^2 - 1)^{-1}$ in a power series centered at $z_0 = i$.

In doing this example one could, of course, attempt to find $f^{(n)}(z)$ and evaluate $f^{(n)}(i)/n!$ straight away. Taking just a couple of derivatives should quickly disabuse the reader of the wisdom of that approach. A slicker method uses a partial fractions decomposition of f and puts the geometric series to work:

$$\frac{2z}{z^2-1} = \frac{1}{z-1} + \frac{1}{z+1} = \frac{1}{(i-1)+(z-i)} + \frac{1}{(i+1)+(z-i)}$$

$$= \frac{1}{i-1} \frac{1}{1-\left(-\frac{z-i}{i-1}\right)} + \frac{1}{i+1} \frac{1}{1-\left(-\frac{z-i}{i+1}\right)}$$

$$= \frac{1}{i-1} \sum_{n=0}^{\infty} (-1)^n \left(\frac{z-i}{i-1}\right)^n + \frac{1}{i+1} \sum_{n=0}^{\infty} (-1)^n \left(\frac{z-i}{i+1}\right)^n$$

$$= \sum_{n=0}^{\infty} (-1)^n \left[(i-1)^{-n-1} + (i+1)^{-n-1}\right] (z-i)^n .$$

We have already accomplished what we intended to in this example : we have determined the Taylor expansion of f about i. With a small amount of extra effort, however, we can express its coefficients in a more readily computed form. First, by rationalizing denominators we see that

$$(i-1)^{-n-1} + (i+1)^{-n-1} = \frac{(-i-1)^{n+1} + (-i+1)^{n+1}}{2^{n+1}}$$

$$= \frac{(-1)^{n+1} \left[(i+1)^{n+1} + (i-1)^{n+1}\right]}{2^{n+1}} .$$

Next, for any positive integer k we have

$$(i+1)^k + (i-1)^k = 2^{k/2} e^{k\pi i/4} + 2^{k/2} e^{3k\pi i/4}$$

3. Sequences and Series of Analytic Functions 267

$$= 2^{k/2} e^{k\pi i/2} \left(e^{k\pi i/4} + e^{-k\pi i/4} \right) = 2^{(k+2)/2} i^k \cos(k\pi/4) \ .$$

These observations lead to the revised expansion

$$\frac{2z}{z^2 - 1} = \sum_{n=0}^{\infty} 2^{-(n-1)/2} i^{n+3} \cos[(n+1)\pi/4](z - i)^n$$

$$= -i - \frac{i(z-i)^2}{2} + \frac{(z-i)^3}{2} + \cdots .$$

It is valid for z in $\Delta(i, \sqrt{2})$, the largest open disk with center at i that lies in the set $U = \mathbb{C} \sim \{\pm 1\}$ where f is analytic.

EXAMPLE 3.9. Find the Taylor series expansion of $f(z) = e^{z^2}$ about the origin. Use this to determine $f^{(n)}(0)$ for every non-negative integer n.

Again in this example direct computation of $f^{(n)}(z)$ would be an unpleasant task at best. Instead, we start with the power series expansion

$$e^z = \sum_{n=0}^{\infty} \frac{z^n}{n!} = 1 + z + \frac{z^2}{2!} + \cdots$$

derived in Example 3.5 and substitute z^2 for z. This procedure results in

$$e^{z^2} = \sum_{n=0}^{\infty} \frac{z^{2n}}{n!} = 1 + z^2 + \frac{z^4}{2!} + \cdots$$

for every z in $\mathbb{C}$. Since the series on the right definitely is a Taylor series centered at the origin, the uniqueness statement in Theorem 3.4 certifies it as the expansion of f we were after. Formula (7.7) then makes it clear that $f^{(n)}(0) = 0$ when n is odd, while $f^{(2n)}(0) = (2n)!/n!$ for $n = 0, 1, 2, \cdots$.

EXAMPLE 3.10. Determine the Taylor series expansion of $f(z) = \cos^2 z$ that is centered at the origin.

We take advantage of the identity $\cos^2 z = [1 + \cos(2z)]/2$ and of the power series representation for $g(z) = \cos z$ described in Example 3.6 to calculate

$$\cos^2 z = \frac{1 + \cos(2z)}{2} = \frac{1}{2} + \frac{1}{2} \sum_{n=0}^{\infty} \frac{(-1)^n (2z)^{2n}}{(2n)!}$$

$$= 1 + \sum_{n=1}^{\infty} \frac{(-1)^n 2^{2n-1} z^{2n}}{(2n)!} = 1 - z^2 + \frac{z^4}{3} - \cdots ,$$

an expansion valid for all complex z.

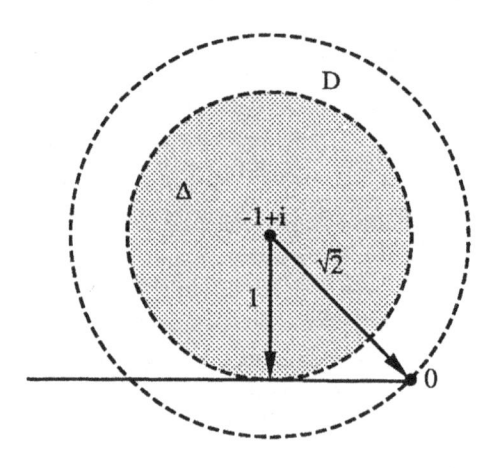

Figure 2.

EXAMPLE 3.11. Discuss the Taylor series expansion of the function $f(z) = \text{Log } z$ about the point $z_0 = -1 + i$.

As observed in Example 3.7 this function is analytic in the set $U = \mathbb{C} \sim (-\infty, 0]$, where $f^{(n)}(z) = (-1)^{n-1}(n-1)! \, z^{-n}$ for $n \geq 1$. It follows that
$$\text{Log } z = \text{Log } \sqrt{2} + \frac{3\pi i}{4} + \sum_{n=1}^{\infty} \frac{(-1)^{n-1}(z+1-i)^n}{n(-1+i)^n}$$
for z in $\Delta = \Delta(-1+i, 1)$, the largest open disk centered at $z_0 = -1 + i$ that is contained in U. On the other hand, the radius of convergence ρ of this Taylor series is quickly computed: $\rho = \sqrt{2}$. By Theorem 3.3
$$g(z) = \text{Log } \sqrt{2} + \frac{3\pi i}{4} + \sum_{n=1}^{\infty} \frac{(-1)^{n-1}(z+1-i)^n}{n(-1+i)^n}$$
actually defines a function that is analytic in the disk in $D = \Delta(-1+i, \sqrt{2})$. (See Figure 2.) What function could it be? To answer this question we consider the function L given by
$$L(z) = \begin{cases} \text{Log } z & \text{if } z \in D \text{ and } \text{Im } z \geq 0, \\ \text{Log } z + 2\pi i & \text{if } z \in D \text{ and } \text{Im } z < 0. \end{cases}$$

Then L is continuous and satisfies $e^{L(z)} = z$ for z in D — i.e., L is a branch of the logarithm in D. In particular, L is an analytic function. As such, L admits a Taylor series expansion about the point $-1+i$, an expansion which is definitely valid throughout the disk D. Moreover, since $L(z) = f(z)$ for every z in Δ, we see that $L^{(n)}(-1+i) = f^{(n)}(-1+i)$ for $n = 0, 1, 2, \cdots$, which means that the functions f and L generate exactly the same Taylor series at $-1+i$. It follows that $g = L$. We note especially that, for z in D with $\operatorname{Im} z < 0$, $g(z) = \operatorname{Log} z + 2\pi i \neq f(z)$. In the present example we see exhibited in a concrete situation the behavior alluded to in the comments following Theorem 3.4.

3.4 Laurent Series

Let z_0 again be a point in the complex plane. By a *Laurent series centered at z_0* is meant a doubly infinite function series of the form

$$\sum_{n=-\infty}^{\infty} a_n(z-z_0)^n$$

$$= \cdots + \frac{a_{-2}}{(z-z_0)^2} + \frac{a_{-1}}{z-z_0} + a_0 + a_1(z-z_0) + a_2(z-z_0)^2 + \cdots,$$

where for $n = 0, \pm 1, \pm 2, \cdots$ the coefficient a_n is a complex constant. Included among these series, which are named in honor of Pierre Alphonse Laurent (1813-1854), are all the Taylor series centered at z_0, the latter being the Laurent series in which $a_n = 0$ for every negative integer n. In line with our policy for Taylor series we assign to any such Laurent series two non-negative extended real numbers ρ_0 and ρ_I, its *outer* and *inner radii of convergence*, via the formulas

(7.9) $$\rho_0 = \left(\limsup_{n\to\infty} \sqrt[n]{|a_n|}\right)^{-1}, \quad \rho_I = \limsup_{n\to\infty} \sqrt[n]{|a_{-n}|}.$$

When $\rho_I < \rho_0$, $D = \{z : \rho_I < |z-z_0| < \rho_0\}$ is called the *ring* (or *annulus*) *of convergence* of the given series. In the extreme case where $\rho_I = 0$ and $\rho_0 = \infty$ this set is just the punctured plane $\mathbb{C} \sim \{z_0\}$. Figure 3 indicates the other possibilities for D. The terminology is suggested by a Laurent series companion to Theorem 3.3.

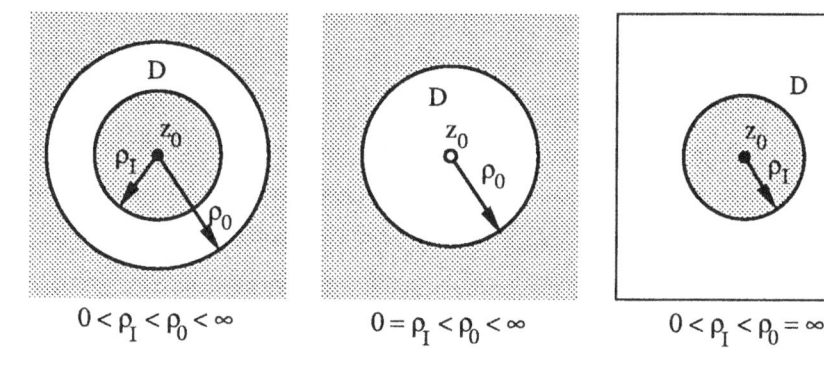

Figure 3.

Theorem 3.5. *Suppose that ρ_0 and ρ_I are the outer and inner radii of convergence of a Laurent series $\sum_{n=-\infty}^{\infty} a_n(z-z_0)^n$ centered at z_0. The series diverges for any z satisfying $|z-z_0| > \rho_0$ or $|z-z_0| < \rho_I$. If $\rho_0 > 0$, the series $\sum_{n=0}^{\infty} a_n(z-z_0)^n$ converges absolutely and normally in the disk $D_0 = \Delta(z_0, \rho_0)$, so $f_0(z) = \sum_{n=0}^{\infty} a_n(z-z_0)^n$ defines a function f_0 that is analytic in D_0. If $\rho_I < \infty$, the series $\sum_{n=1}^{\infty} a_{-n}(z-z_0)^{-n}$ converges absolutely and normally in the open set $D_I = \{z : |z-z_0| > \rho_I\}$, so $f_I(z) = \sum_{n=1}^{\infty} a_{-n}(z-z_0)^{-n}$ defines a function f_I that is analytic in D_I. If $\rho_I < \rho_0$, the full Laurent series $\sum_{n=-\infty}^{\infty} a_n(z-z_0)^n$ converges absolutely and normally in the set $D = \{z : \rho_I < |z-z_0| < \rho_0\}$, so the function f defined by $f(z) = \sum_{n=-\infty}^{\infty} a_n(z-z_0)^n = f_I(z) + f_0(z)$ is analytic in D. The coefficient a_n is then related to f through the formula*

$$(7.10) \qquad a_n = \frac{1}{2\pi i} \int_{|z-z_0|=r} \frac{f(z)\,dz}{(z-z_0)^{n+1}}$$

for any number r in the interval (ρ_I, ρ_0).

Proof. First, the series $\sum_{n=0}^{\infty} a_n(z-z_0)^n$ is nothing but a Taylor series with radius of convergence ρ_0. It thus diverges for any z satisfying $|z-z_0| > \rho_0$. Under the condition that $\rho_0 > 0$ it converges absolutely and normally in $D_0 = \Delta(z_0, \rho_0)$, and the function f_0 given by $f_0(z) = \sum_{n=0}^{\infty} a_n(z-z_0)^n$ is analytic in D_0 (Theorem 3.3). We next consider $\sum_{n=1}^{\infty} a_{-n}\zeta^n$, a Taylor series in the variable ζ whose radius of convergence is clearly $1/\rho_I$. This series diverges for any ζ with $|\zeta| > 1/\rho_I$. Also, assuming that $\rho_I < \infty$, it converges absolutely and normally in the disk $\Delta = \Delta(0, 1/\rho_I)$. If we make the substitution $\zeta = (z-z_0)^{-1}$ in $\sum_{n=1}^{\infty} a_{-n}\zeta^n$, we obtain the series $\sum_{n=1}^{\infty} a_{-n}(z-z_0)^{-n}$. This process plainly results in a divergent series when $|z-z_0| < \rho_I$ and an absolutely convergent one when $|z-z_0| > \rho_I$. Suppose now that $\rho_I < \infty$. The function $g(z) = (z-z_0)^{-1}$ is continuous and maps the set $D_I = \{z: |z-z_0| > \rho_I\}$ into Δ. In particular, g carries any compact set in D_I to a compact set in Δ. Given a compact set K in D_I and given $\epsilon > 0$, we can exploit the uniform convergence of $\sum_{n=1}^{\infty} a_{-n}\zeta^n$ on the compact set $g(K)$ to choose an index N such that the inequality $\left|\sum_{k=n}^{m} a_{-k}\zeta^k\right| < \epsilon$ is satisfied for every ζ in $g(K)$, provided $m \geq n \geq N$. It follows immediately that $\left|\sum_{k=n}^{m} a_{-k}(z-z_0)^{-k}\right| < \epsilon$ holds for every z in K whenever $m \geq n \geq N$. The Cauchy criterion for uniform convergence then attests to the uniform convergence of $\sum_{n=1}^{\infty} a_{-n}(z-z_0)^{-n}$ on K. We conclude that this series is normally convergent in D_I. By Theorem 3.2 the function f_I with rule of correspondence $f_I(z) = \sum_{n=1}^{\infty} a_{-n}(z-z_0)^{-n}$ is analytic in D_I.

We turn to the doubly infinite series $\sum_{n=-\infty}^{\infty} a_n(z-z_0)^n$ and draw some conclusions about it. In light of Lemma 2.1 this series diverges when z satisfies either $|z-z_0| > \rho_0$ or $|z-z_0| < \rho_I$. If $\rho_I < \rho_0$ — which implies that $\rho_0 > 0$ and $\rho_I < \infty$ — both the series $\sum_{n=0}^{\infty} a_n(z-z_0)^n$ and

3. Sequences and Series of Analytic Functions

its mate $\sum_{n=1}^{\infty} a_{-n}(z - z_0)^{-n}$ converge absolutely and normally in the set $D = D_0 \cap D_I = \{z : \rho_I < |z - z_0| < \rho_0\}$, which fact insures the absolute and normal convergence of $\sum_{n=-\infty}^{\infty} a_n(z - z_0)^n$ in D. (Lemma 2.3 comes into play here.) The function f defined by $f(z) = \sum_{n=-\infty}^{\infty} a_n(z - z_0)^n$ is analytic in D. Indeed, $f(z) = f_I(z) + f_0(z)$ for z in D.

Finally, let r satisfy $\rho_I < r < \rho_0$. If k is an arbitrary integer, we can write

$$\frac{f(z)}{(z - z_0)^{k+1}} = \sum_{n=-\infty}^{\infty} a_n (z - z_0)^{n-k-1}$$

for every z belonging to D, a set in which the convergence of the series on the right is normal. (See Exercise 5.21.) Invoking the analogue of (7.5) for doubly infinite series to justify an interchange of summation and integration, we discover that

$$\frac{1}{2\pi i} \int_{|z-z_0|=r} \frac{f(z)\,dz}{(z - z_0)^{k+1}} = \frac{1}{2\pi i} \int_{|z-z_0|=r} \left\{ \sum_{n=-\infty}^{\infty} a_n(z - z_0)^{n-k-1} \right\} dz$$

$$= \sum_{n=-\infty}^{\infty} \frac{a_n}{2\pi i} \int_{|z-z_0|=r} (z - z_0)^{n-k-1}\,dz = a_k \; ,$$

because

$$\int_{|z-z_0|=r} (z - z_0)^{n-k-1}\,dz = \begin{cases} 2\pi i & \text{if } n = k \; , \\ 0 & \text{if } n \neq k \; . \end{cases}$$

This verifies (7.10). ∎

Theorem 3.4, too, has its dual in the setting of Laurent series. Theorem 3.6 will provide us with a jumping-off point for our discussion of isolated singularities of analytic functions, the topic of the next chapter.

Theorem 3.6. *Suppose that a function f is analytic in an annulus $D = \{z : a < |z - z_0| < b\}$, where $0 \leq a < b \leq \infty$. Then f can be represented in D as the sum of a Laurent series centered at z_0. This expansion is uniquely determined by f: if $f(z) = \sum_{n=-\infty}^{\infty} a_n(z - z_0)^n$ in D, then the coefficient a_n is given by*

$$a_n = \frac{1}{2\pi i} \int_{|z-z_0|=r} \frac{f(z)\,dz}{(z - z_0)^{n+1}}$$

for any number r satisfying $a < r < b$.

Proof. For each integer n we define a complex number a_n in the manner prescribed by the statement of the theorem: we select r in the interval (a, b) and set $a_n = (2\pi i)^{-1} \int_{|z-z_0|=r} (z - z_0)^{-n-1} f(z)\,dz$. The specific choice of r is of no consequence, for the number a_n is independent of that choice. Indeed, since the paths γ and β defined on $[0, 2\pi]$ by $\gamma(t) = z_0 + re^{it}$ and $\beta(t) = z_0 + se^{it}$ (Figure 4) are homologous in D whenever $a < r < s < b$

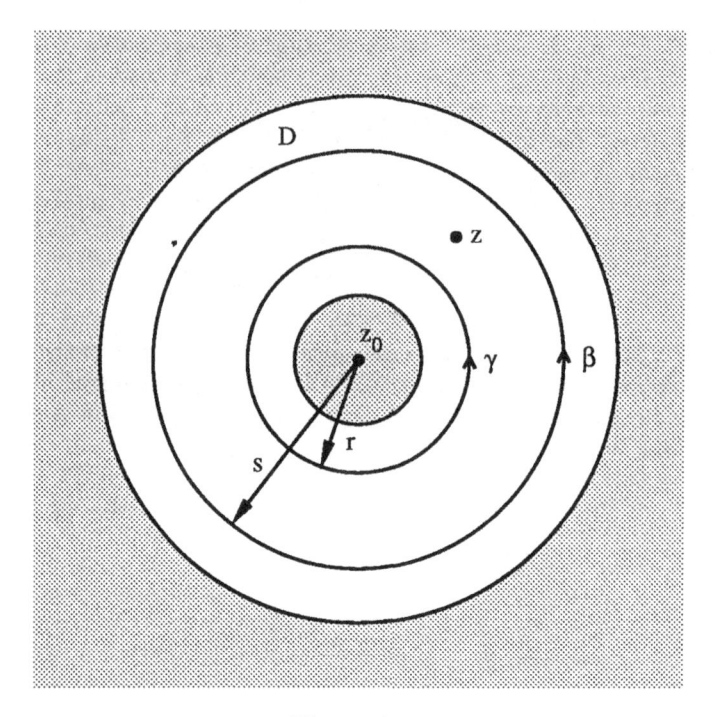

Figure 4.

and since the function g defined by $g(z) = (z - z_0)^{-n-1} f(z)$ is analytic in D, Corollary V.5.2 gives

(7.11) $\quad a_n = \dfrac{1}{2\pi i} \displaystyle\int_{|z-z_0|=r} \dfrac{f(z)\,dz}{(z-z_0)^{n+1}} = \dfrac{1}{2\pi i} \displaystyle\int_{|z-z_0|=s} \dfrac{f(z)\,dz}{(z-z_0)^{n+1}}$.

With this determination of coefficients a_n we make the assertion that $f(z) = \sum_{n=-\infty}^{\infty} a_n (z-z_0)^n$ for every z in D.

Fix z in D. Fix, also, numbers r and s obeying $a < r < |z-z_0| < s < b$. Let paths γ and β be defined as they were above. We apply the Cauchy integral formula in D to the function f and the cycle $\sigma = (\beta, -\gamma)$, which is obviously homologous to zero in D. Because $n(\sigma, z) = 1$, this step yields

$$f(z) = \dfrac{1}{2\pi i} \int_\sigma \dfrac{f(\zeta)\,d\zeta}{\zeta - z},$$

which in expanded form becomes

(7.12) $\quad f(z) = \dfrac{1}{2\pi i} \displaystyle\int_{|\zeta-z_0|=s} \dfrac{f(\zeta)\,d\zeta}{\zeta - z} - \dfrac{1}{2\pi i} \displaystyle\int_{|\zeta-z_0|=r} \dfrac{f(\zeta)\,d\zeta}{\zeta - z}$.

The argument now proceeds along lines established in the proof of Theorem 3.4. For ζ belonging to the circle $K = K(z_0, s)$ we capitalize on

3. Sequences and Series of Analytic Functions

the fact that $|(z - z_0)/(\zeta - z_0)| = |z - z_0|/s < 1$ to write

$$\frac{f(\zeta)}{\zeta - z} = \frac{f(\zeta)}{\zeta - z_0} \frac{1}{1 - \left(\dfrac{z - z_0}{\zeta - z_0}\right)} = \sum_{n=0}^{\infty} \frac{f(\zeta)(z - z_0)^n}{(\zeta - z_0)^{n+1}}.$$

We check, as we did in the proof of Theorem 3.4, that convergence is uniform on K and then integrate the series termwise to get

$$\frac{1}{2\pi i} \int_{|\zeta - z_0| = s} \frac{f(\zeta)\, d\zeta}{\zeta - z} = \sum_{n=0}^{\infty} \left[\frac{1}{2\pi i} \int_{|\zeta - z_0| = s} \frac{f(\zeta)\, d\zeta}{(\zeta - z_0)^{n+1}} \right] (z - z_0)^n.$$

In accordance with (7.11) this reduces to

(7.13) $$\frac{1}{2\pi i} \int_{|\zeta - z_0| = s} \frac{f(\zeta)\, d\zeta}{\zeta - z} = \sum_{n=0}^{\infty} a_n (z - z_0)^n.$$

In a like manner, since $|(\zeta - z_0)/(z - z_0)| = r/|z - z_0| < 1$ for ζ lying on $K^* = K(z_0, r)$, we obtain for such ζ the expansion

$$\frac{f(\zeta)}{\zeta - z} = -\frac{f(\zeta)}{z - z_0} \frac{1}{1 - \left(\dfrac{\zeta - z_0}{z - z_0}\right)} = -\sum_{n=0}^{\infty} \frac{f(\zeta)(\zeta - z_0)^n}{(z - z_0)^{n+1}}$$

$$= -\sum_{n=1}^{\infty} \frac{f(\zeta)(\zeta - z_0)^{n-1}}{(z - z_0)^n}.$$

Convergence here is uniform on K^*. In this instance, integration leads to

$$\frac{1}{2\pi i} \int_{|\zeta - z_0| = r} \frac{f(\zeta)\, d\zeta}{\zeta - z} = -\sum_{n=1}^{\infty} \left[\frac{1}{2\pi i} \int_{|\zeta - z_0| = r} \frac{f(\zeta)\, d\zeta}{(\zeta - z_0)^{-n+1}} \right] (z - z_0)^{-n}$$

and so by way of (7.11) to

(7.14) $$\frac{1}{2\pi i} \int_{|\zeta - z_0| = r} \frac{f(\zeta)\, d\zeta}{\zeta - z} = -\sum_{n=1}^{\infty} a_{-n}(z - z_0)^{-n}.$$

The effect of (7.13), (7.14), and (7.12) is to confirm that both of the series $\sum_{n=0}^{\infty} a_n(z - z_0)^n$ and $\sum_{n=1}^{\infty} a_{-n}(z - z_0)^{-n}$ are convergent and that

$$f(z) = \sum_{n=1}^{\infty} a_{-n}(z - z_0)^{-n} + \sum_{n=0}^{\infty} a_n(z - z_0)^n.$$

Owing to Lemma 2.1 we can conclude that $f(z) = \sum_{n=-\infty}^{\infty} a_n(z - z_0)^n$ — this for arbitrary z in D.

Suppose that $f(z) = \sum_{n=-\infty}^{\infty} b_n(z-z_0)^n$ were a competing Laurent series expansion of f in D. The convergence of this series in D would insure that its radii of convergence ρ_0 and ρ_I satisfy $\rho_I \leq a$ and $\rho_0 \geq b$. If r belongs to the interval (a,b), then Theorem 3.5 and (7.11) would imply that

$$b_n = \frac{1}{2\pi i} \int_{|z-z_0|=r} \frac{f(z)\,dz}{(z-z_0)^{n+1}} = a_n ,$$

so the presumed alternative expansion would coincide with our original one. The uniqueness assertion in the theorem is the consequence. ∎

Although the formula $a_n = (2\pi i)^{-1} \int_{|z-z_0|=r} (z-z_0)^{-n-1} f(z)dz$ for the nth Laurent coefficient of a function is of theoretical value, it is safe to say that this formula is all but useless when it ultimately gets down to producing the Laurent series representations of functions in concrete situations. As a matter of fact the integral involved here is, more often than not, sufficiently horrendous in its own right to dampen any enthusiasm for using the formula. Several of the more practical techniques employed to determine Laurent series expansions are indicated in the succeeding examples.

EXAMPLE 3.12. Expand the function $f(z) = e^{1/z}$ in a Laurent series in the ring $D = \{z : 0 < |z| < \infty\}$.

From Example 3.5 we recall that the Taylor series development

$$e^z = \sum_{n=0}^{\infty} \frac{z^n}{n!} = 1 + z + \frac{z^2}{2!} + \cdots$$

is valid for all complex numbers z. Assuming that $0 < |z| < \infty$, therefore, we are permitted to substitute $1/z$ for z in this series to obtain the expansion

$$e^{1/z} = \sum_{n=0}^{\infty} \frac{1}{n!z^n} = 1 + \frac{1}{z} + \frac{1}{2!z^2} + \cdots .$$

When expressed in standard Laurent series form this turns into

$$e^{1/z} = \sum_{n=-\infty}^{0} \frac{z^n}{(-n)!} .$$

EXAMPLE 3.13. Determine the Laurent series representation of $f(z) = (z-1)^{-3} \sin \pi z$ in the ring $D = \{z : 0 < |z-1| < \infty\}$.

We start by finding the Taylor series expansion of $g(z) = \sin \pi z$ about $z_0 = 1$. Computing derivatives we learn that $g^{(n)}(1) = 0$ for even n and $g^{(2n+1)}(1) = (-1)^{n+1}\pi^{2n+1}$ for $n = 0, 1, 2, \cdots$. We conclude that

$$\sin \pi z = \sum_{n=0}^{\infty} \frac{(-1)^{n+1}\pi^{2n+1}(z-1)^{2n+1}}{(2n+1)!}$$

3. Sequences and Series of Analytic Functions 275

$$= -\pi(z-1) + \frac{\pi^3(z-1)^3}{3!} - \frac{\pi^5(z-1)^5}{5!} + \cdots,$$

an expansion which applies everywhere in $\mathbb{C}$. Division by $(z-1)^3$ gives

$$\frac{\sin \pi z}{(z-1)^3} = \sum_{n=0}^{\infty} \frac{(-1)^{n+1}\pi^{2n+1}(z-1)^{2n-2}}{(2n+1)!}$$

$$= -\frac{\pi}{(z-1)^2} + \frac{\pi^3}{3} - \frac{\pi^5(z-1)^2}{5!} + \cdots,$$

as the Laurent series that represents f in D.

EXAMPLE 3.14. Develop the functions $f(z) = z^{-1}$ and $g(z) = z^{-2}$ in Laurent series in the annulus $D = \{z : 1 < |z - i| < \infty\}$.

The secret here is to make shrewd use of the geometric series. The initial temptation might be to write

$$\frac{1}{z} = \frac{1}{i + (z-i)} = \frac{1}{i} \frac{1}{1 - i(z-i)}.$$

The last expression does appear primed for application of the geometric series. Unfortunately, however, $|i(z-i)| > 1$ when z belongs to D. A little algebraic reorganization is sufficient to remedy the situation; namely, we manipulate the above expression into

$$\frac{1}{z} = \frac{1}{z-i} \frac{1}{1 - \frac{1}{i(z-i)}}.$$

For z in D the geometric series can then be utilized to arrive at

$$\frac{1}{z} = \frac{1}{z-i} \sum_{n=0}^{\infty} \frac{1}{i^n(z-i)^n} = \frac{1}{z-i} \sum_{n=-\infty}^{0} i^n(z-i)^n$$

$$= \sum_{n=-\infty}^{0} i^n(z-i)^{n-1} = \sum_{n=-\infty}^{-1} i^{n+1}(z-i)^n ;$$

i.e., the desired Laurent expansion of f in D takes the form

$$\frac{1}{z} = \sum_{n=-\infty}^{-1} i^{n+1}(z-i)^n = \frac{1}{z-i} - \frac{i}{(z-i)^2} - \frac{1}{(z-i)^3} + \cdots .$$

Concerning the function g, we note that $g(z) = -f'(z)$. To get the Laurent series for f' in D we have only to differentiate termwise the Laurent series obtained for f there. This produces the representation

$$-\frac{1}{z^2} = \sum_{n=-\infty}^{-1} ni^{n+1}(z-i)^{n-1} = \sum_{n=-\infty}^{-2} (n+1)i^{n+2}(z-i)^n$$

$$= -\sum_{n=-\infty}^{-2} (n+1)i^n(z-i)^n$$

for z in D. Accordingly,

$$\frac{1}{z^2} = \sum_{n=-\infty}^{-2} (n+1)i^n(z-i)^n = \frac{1}{(z-i)^2} - \frac{2i}{(z-i)^3} - \frac{3}{(z-i)^4} + \cdots$$

delivers the Laurent expansion of g in D.

EXAMPLE 3.15. Find the Laurent series expansion of the function $f(z) = (2 - 3z + z^2)^{-1}$ in each of the annuli $D = \{z : 1 < |z| < 2\}$ and $D^* = \{z : \sqrt{2} < |z+i| < \sqrt{5}\}$ (Figure 5).

We first express f in its partial fraction decomposition,

$$f(z) = \frac{1}{2 - 3z + z^2} = \frac{1}{1-z} - \frac{1}{2-z}.$$

For z lying in D we appeal to the geometric series in computing

$$\frac{1}{1-z} = -\frac{1}{z}\frac{1}{1-(1/z)} = -\frac{1}{z}\sum_{n=0}^{\infty}\left(\frac{1}{z}\right)^n = -\sum_{n=0}^{\infty}\frac{1}{z^{n+1}} = -\sum_{n=-\infty}^{-1} z^n$$

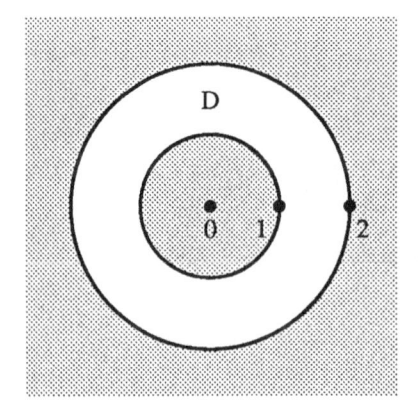

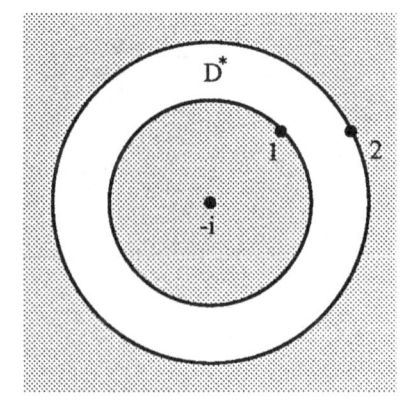

Figure 5.

3. Sequences and Series of Analytic Functions

and
$$\frac{1}{2-z} = \frac{1}{2}\frac{1}{1-(z/2)} = \frac{1}{2}\sum_{n=0}^{\infty}\left(\frac{z}{2}\right)^n = \sum_{n=0}^{\infty} 2^{-n-1}z^n.$$

Since $|1/z| < 1$ and $|z/2| < 1$ when z belongs to D, the use of the geometric series is legitimate in each instance. Consequently,

$$\frac{1}{2-3z+z^2} = -\sum_{n=-\infty}^{-1} z^n - \sum_{n=0}^{\infty} 2^{-n-1}z^n$$

$$= \cdots - \frac{1}{z^2} - \frac{1}{z} - \frac{1}{2} - \frac{z}{4} - \frac{z^2}{8} - \cdots$$

for every point z of D.

When z is in D^* the corresponding calculations run as follows:

$$\frac{1}{1-z} = \frac{1}{(1+i)-(z+i)} = -\frac{1}{z+i}\frac{1}{1-[(1+i)/(z+i)]}$$

$$= -\frac{1}{z+i}\sum_{n=0}^{\infty}\left(\frac{1+i}{z+i}\right)^n = -\frac{1}{z+i}\sum_{n=-\infty}^{0}\left(\frac{z+i}{1+i}\right)^n$$

$$= -\sum_{n=-\infty}^{0}(1+i)^{-n}(z+i)^{n-1} = -\sum_{n=-\infty}^{-1}(1+i)^{-n-1}(z+i)^n,$$

$$\frac{1}{2-z} = \frac{1}{(2+i)-(z+i)} = \frac{1}{2+i}\frac{1}{1-[(z+i)/(2+i)]}$$

$$= \frac{1}{2+i}\sum_{n=0}^{\infty}\left(\frac{z+i}{2+i}\right)^n = \sum_{n=0}^{\infty}(2+i)^{-n-1}(z+i)^n.$$

Here the fact that $|(1+i)/(z+i)| = \sqrt{2}/|z+i| < 1$ and $|(z+i)/(2+i)| = |z+i|/\sqrt{5} < 1$ for z belonging to D^* justifies the applications made of the geometric series. The Laurent series development of f in D^* thus reads

$$\frac{1}{2-3z+z^2} = -\sum_{n=-\infty}^{-1}(1+i)^{-n-1}(z+i)^n - \sum_{n=0}^{\infty}(2+i)^{-n-1}(z+i)^n$$

$$= \cdots - \frac{1+i}{(z+i)^2} - \frac{1}{z+i} - \frac{1}{2+i} - \frac{(z+i)}{(2+i)^2} - \frac{(z+i)^2}{(2+i)^3} - \cdots.$$

The combined methods of the last two examples — i.e., partial fraction decomposition, use of the geometric series, and termwise differentiation — can, in principle at least, be adapted to determine the Laurent series development of an arbitrary rational function in any ring where it is analytic.

4 Normal Families

4.1 Normal Subfamilies of $C(U)$

In Chapter II we found out that a bounded sequence of complex numbers $\langle z_n \rangle$ always possesses at least one accumulation point, which fact implies that $\langle z_n \rangle$ has convergent subsequences. In this section we consider the analogous situation for sequences of functions. Specifically, given a sequence of functions $\langle f_n \rangle$ each of whose terms is analytic in an open set U, we explore the possibility of extracting from $\langle f_n \rangle$ a subsequence $\langle f_{n_k} \rangle$ that converges normally in U. The principal result of the section, Montel's theorem, asserts that, if the sequence $\langle f_n \rangle$ is "bounded" in a sense soon to be explained, then the existence of such a subsequence is assured. Thus, the situation for sequences of analytic functions exactly parallels that for sequences of complex numbers. The material presented here will resurface just once later in this book — namely, in the proof of the "Riemann Mapping Theorem" (Theorem IX.3.4). Readers prepared to gloss over the single step in that proof where Montel's theorem gets invoked should not feel hesitant about proceeding directly to Chapter VIII.

In order to render less cumbersome the statements of propositions in this section, we adopt some convenient terminology. Recall that the notation $C(U)$ signifies the collection of all complex-valued functions which are continuous in an open subset U of the complex plane. We define a subfamily $\mathcal{F}$ of $C(U)$ to be *normal in U* — *pre-compact in U* is a different name for the same concept — provided each sequence $\langle f_n \rangle$ from $\mathcal{F}$ has at least one subsequence $\langle f_{n_k} \rangle$ that converges normally in U. (N.B. The reader is advised that this interpretation of "normal family" does not totally jibe with the general usage of the term in the literature of complex analysis, where allowances are usually made for sequences that "tend to infinity uniformly on compact sets in U." See, for example, Theorem 4.6. It would be wise for the reader to keep the discrepancy in mind when consulting other references.) We stress: the normality of $\mathcal{F}$ does not require that every function obtained as the limit of a normally convergent sequence from $\mathcal{F}$ be itself a member of $\mathcal{F}$. A normal subfamily of $C(U)$ endowed with this extra property is called a *compact subfamily* of $C(U)$, a definition that accords nicely with the concept of a compact subset of $\mathbb{C}$. Our ultimate objective in this section is the derivation of a reasonably simple criterion for detecting normality in those subfamilies of $C(U)$ whose members are analytic in U. Along the way, however, we shall also obtain a useful characterization of an arbitrary normal subfamily of $C(U)$.

4.2 Equicontinuity

How does one recognize that a given subfamily $\mathcal{F}$ of $C(U)$ is normal in U? An important element in the eventual answer to this question is the idea of "equicontinuity." The family $\mathcal{F}$ is said to be *equicontinuous at a point z_0 of U* if corresponding to each $\epsilon > 0$ there exists a number $\delta > 0$ with the property that $|f(z) - f(z_0)| < \epsilon$ holds for every f in $\mathcal{F}$ whenever $|z - z_0| < \delta$. Crucial here is that δ does not depend on the function f (although it will typically depend on the point z_0): the same δ must work for all members of $\mathcal{F}$. When the family $\mathcal{F}$ is equicontinuous at every point of U, we pronounce it *equicontinuous in U*. (In slightly less precise language we sometimes just call $\mathcal{F}$ an *equicontinuous subfamily of $C(U)$*.) For example, the family $\mathcal{F}$ that consists of all functions of the form $f(z) = 2z + c$, where c is an arbitrary complex constant, is an equicontinuous subfamily of the continuous functions on $\mathbb{C}$. In fact, since $|f(z) - f(z_0)| = 2|z - z_0|$ for all f in this family and for all complex numbers z and z_0, the choice $\delta = \epsilon/2$ corresponding to a given $\epsilon > 0$ meets the stated requirement for any z_0. This is more than the definition of equicontinuity for $\mathcal{F}$ in $\mathbb{C}$ strictly demands — δ would, in general, be permitted to vary from one point z_0 to another.

For present purposes the most significant feature of equicontinuity is that it bridges the gap between pointwise convergence and normal convergence.

Theorem 4.1. *Let $\langle f_n \rangle$ be a sequence from an equicontinuous subfamily $\mathcal{F}$ of $C(U)$. Suppose that this sequence converges pointwise in U. Then it converges normally in U.*

Proof. Denote by f the pointwise limit of $\langle f_n \rangle$ in U. Let K be an arbitrary compact set in U. We must demonstrate that $f_n \to f$ uniformly on K. On the basis of Theorem 1.2, we need only establish that $\langle f_n \rangle$ is a uniform Cauchy sequence on K. We do this indirectly; i.e., we assume the contrary and argue to a contradiction. Under the assumption that $\langle f_n \rangle$ fails to be a uniform Cauchy sequence on K, there must be a number $\epsilon > 0$ (we choose one and keep it fixed for the duration of the proof) about which the following statement is true: there is no integer N with the property that the inequality $|f_m(z) - f_n(z)| < \epsilon$ is valid for every z in K and every pair of indices m and n satisfying $m > n \geq N$. In particular, for any positive integer k the choice $N = k$ does not do the job, meaning that there have to be indices n_k and m_k with $m_k > n_k \geq k$ for which the inequality $|f_{m_k}(z) - f_{n_k}(z)| < \epsilon$ breaks down somewhere in K. Accordingly, corresponding to each positive integer k we can assert the existence of integers n_k and m_k with $m_k > n_k \geq k$ and a point z_k of K such that

$$(7.15) \qquad |f_{m_k}(z_k) - f_{n_k}(z_k)| \geq \epsilon \,.$$

This reasoning gives rise to a sequence $\langle z_k \rangle$ in K. As K is compact, $\langle z_k \rangle$ has at least one accumulaton point in K — say that z_0 is such a point. Utilizing the equicontinuity of the family $\mathcal{F}$ at z_0, we select $\delta > 0$ in a manner which insures that, for any z obeying the condition $|z - z_0| < \delta$, the inequality

(7.16) $$|f_n(z) - f_n(z_0)| < \frac{\epsilon}{3}$$

holds for every n. By construction $n_k \to \infty$ and $m_k \to \infty$ as $k \to \infty$, from which it follows that

$$|f_{m_k}(z_0) - f_{n_k}(z_0)| \to |f(z_0) - f(z_0)| = 0$$

as $k \to \infty$. This allows us to fix an index k_0 with the property that

(7.17) $$|f_{m_k}(z_0) - f_{n_k}(z_0)| < \frac{\epsilon}{3}$$

once $k \geq k_0$. The fact that $\langle z_k \rangle$ accumulates at z_0 permits us to pick an index $k \geq k_0$ for which $|z_k - z_0| < \delta$. The triangle inequality, in combination with (7.15), (7.16), and (7.17), leads for this selection of k to

$$\epsilon \leq |f_{m_k}(z_k) - f_{n_k}(z_k)|$$
$$\leq |f_{m_k}(z_k) - f_{m_k}(z_0)| + |f_{m_k}(z_0) - f_{n_k}(z_0)| + |f_{n_k}(z_0) - f_{n_k}(z_k)|$$
$$< \frac{\epsilon}{3} + \frac{\epsilon}{3} + \frac{\epsilon}{3} = \epsilon \, ,$$

the contradiction we were seeking. Our conclusion: $\langle f_n \rangle$ is, in the end, a uniform Cauchy sequence on K — hence, $f_n \to f$ uniformly on K. ∎

In actuality it is a mildly stronger version of Theorem 4.1 that we require for the application toward which we are heading. Let U be a nonempty open set in $\mathbb{C}$. We say that a subset S of U is *dense in* U if U is contained in the closure of S or, to put it differently, if $S \cap \Delta(z,r) \neq \phi$ for every z in U and every $r > 0$. An example of such a set (the only example we shall really need) is $S_0 = \{z \in U : \operatorname{Re} z \text{ and } \operatorname{Im} z \text{ are rational}\}$. This particular dense subset of U has another valuable property: its elements can be arranged as the terms of a sequence! To see this, first recall that the set of all rational numbers can be listed in a sequence. Figure 6 suggests a way in which to create such a listing, albeit one with duplicatons. By passing to a subsequence one can exhibit the rationals as the terms of a sequence $\langle q_n \rangle$ in which $q_n \neq q_m$ for $n \neq m$. If we now construct a sequence $\langle w_n \rangle$ by continuing the pattern established in

$$w_1 = q_1 + iq_1 \, ,$$
$$w_2 = q_1 + iq_2 \quad , \quad w_3 = q_2 + iq_1 \, ,$$
$$w_4 = q_1 + iq_3 \quad , \quad w_5 = q_2 + iq_2 \quad , \quad w_6 = q_3 + iq_1 \, , \cdots$$

4. Normal Families

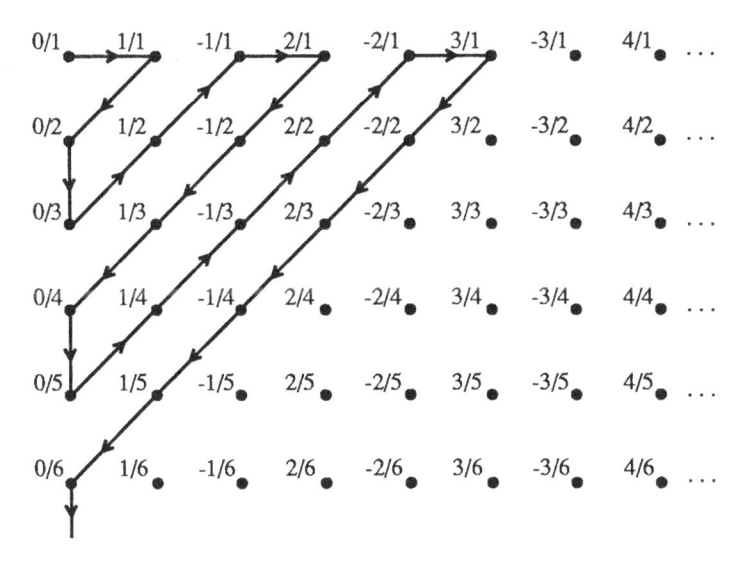

Figure 6.

we obtain a sequential listing of all complex numbers with rational real and imaginary parts. Also, $w_n \neq w_m$ for $n \neq m$. Finally, if we strike from this sequence all terms not belonging to the set U, what remains is a subsequence of $\langle w_n \rangle$ whose terms run through all the elements of S_0.

The refinement of Theorem 4.1 we desire is:

Lemma 4.2. *Let $\langle f_n \rangle$ be a sequence from an equicontinuous subfamily $\mathcal{F}$ of $C(U)$. Suppose that the sequence $\langle f_n(\zeta) \rangle$ is convergent for every ζ belonging to some dense subset S of U. Then $\langle f_n \rangle$ converges normally in U.*

Proof. Owing to Theorem 4.1, it suffices to prove that $\langle f_n \rangle$ converges pointwise in U. To achieve this, we have only to verify that for each element z of U the sequence $\langle f_n(z) \rangle$ is a Cauchy sequence. Fix such a point z, and let $\epsilon > 0$ be given. The challenge is to produce an integer N so that $|f_m(z) - f_n(z)| < \epsilon$ is satisfied whenever $m > n \geq N$. We first make use of the equicontinuity of $\mathcal{F}$ at z to pick a $\delta > 0$ with the property that $|f_n(w) - f_n(z)| < \epsilon/3$ for all indices n, provided only that $|w - z| < \delta$. Next, the fact that the set S is dense in U enables us to choose a point ζ of S satisfying $|\zeta - z| < \delta$. By hypothesis the sequence $\langle f_n(\zeta) \rangle$ is convergent. As a result, this sequence is a Cauchy sequence. There is, consequently, an integer N such that $|f_m(\zeta) - f_n(\zeta)| < \epsilon/3$ is true whenever $m > n \geq N$. We conclude that

$$|f_m(z) - f_n(z)| \leq |f_m(z) - f_m(\zeta)| + |f_m(\zeta) - f_n(\zeta)| + |f_n(\zeta) - f_n(z)|$$

$$< \frac{\epsilon}{3} + \frac{\epsilon}{3} + \frac{\epsilon}{3} = \epsilon \, ,$$

as long as $m > n \geq N$. Thus $\langle f_n(z) \rangle$ is a Cauchy sequence. Since this is the case for every z in U, the pointwise convergence of $\langle f_n \rangle$ in U is established. The quoting of Theorem 4.1 finishes the proof. ∎

4.3 The Arzelà-Ascoli and Montel Theorems

We still lack the final ingredient present in most elementary normal family criteria. It is a fitting notion of boundedness for a subfamily $\mathcal{F}$ of $C(U)$. As far as matters in this book are concerned, there are two relevant concepts of boundedness for such a family. First, we say that $\mathcal{F}$ is *pointwise bounded in U* if for each fixed z in U the set of values $\{f(z) : f \in \mathcal{F}\}$ is a bounded set of complex numbers. Secondly, we speak of the family $\mathcal{F}$ as *locally bounded in U* if its members are uniformly bounded on each compact set in U, which means that there exists for each compact subset K of U a constant $m = m(K)$ with the property that $|f(z)| \leq m$ for every point z in K and every function f in $\mathcal{F}$. Following the precedent set by normal convergence, to verify the local boundedness of $\mathcal{F}$ in U it is enough to check that its members are uniformly bounded on each closed disk in U (or, even less, that each point z of U is the center of some closed disk on which the functions in $\mathcal{F}$ are uniformly bounded). The reason for this is that an arbitrary compact set in U can be covered by a finite number of such disks. (Recall the proof of Lemma 1.3.) If $\mathcal{F}$ is locally bounded in U, then it is plainly pointwise bounded there. The converse is, in general, false.

Our first "normal family theorem" characterizes the normal subfamilies of $C(U)$. The names attached to it are those of Cesare Arzelà (1847-1912) and Giulio Ascoli (1843-1896), who share credit for its discovery. Their theorem and its generalizations have applications in many areas of mathematics.

Theorem 4.3. (Arzelà-Ascoli Theorem) *A subfamily $\mathcal{F}$ of $C(U)$ is normal in U if and only if it is both equicontinuous and pointwise bounded in this open set.*

Proof. Assume initially that $\mathcal{F}$ is both equicontinuous and pointwise bounded in U. Let $\langle f_n \rangle$ be a sequence from $\mathcal{F}$. It is our job to demonstrate the existence of a normally convergent subsequence of $\langle f_n \rangle$. For this we recall that the set $S_0 = \{z \in U : \operatorname{Re} z \text{ and } \operatorname{Im} z \text{ are rational}\}$ is dense in U and that it is possible to list the elements of S_0 in a sequence. Let $\langle z_n \rangle$ be such a listing. We begin a construction by considering the sequence of complex numbers $\langle f_n(z_1) \rangle$. Owing to the pointwise boundedness of $\mathcal{F}$ this sequence is bounded. In light of the Bolzano-Weierstrass theorem it has at least one accumulation point in $\mathbb{C}$. We choose such a point and label it w_1. Then $\langle f_n(z_1) \rangle$ has a subsequence converging to w_1. In other words, it is possible

4. Normal Families

to select a sequence of indices $m_1^{(1)} < m_2^{(1)} < m_3^{(1)} \cdots$ such that

$$\lim_{k \to \infty} f_{m_k^{(1)}}(z_1) = w_1 \, .$$

The superscript in $m_k^{(1)}$ is there to serve notice that this sequence of integers is associated with the point z_1 of S_0. In the interest of preserving some notational sanity we shall abbreviate $f_{m_k^{(1)}}$ to $f_{1,k}$ — and do likewise in similar situations throughout the course of this argument. Now $\langle f_{1,k}(z_2) \rangle_{k=1}^{\infty}$ is another bounded sequence of complex numbers. We can, therefore, single out one of its accumulation points — call it w_2 — and extract a subsequence $m_1^{(2)} < m_2^{(2)} < m_3^{(2)} < \cdots$ from $\langle m_k^{(1)} \rangle$ for which

$$\lim_{k \to \infty} f_{2,k}(z_2) = w_2 \, .$$

Continuing this process inductively we construct corresponding to each positive integer ℓ a complex number w_ℓ and a strictly increasing sequence of positive integers $\langle m_k^{(\ell)} \rangle$ such that

$$\lim_{k \to \infty} f_{\ell,k}(z_\ell) = w_\ell$$

and such that $\langle m_k^{(\ell+1)} \rangle$ is a subsequence of $\langle m_k^{(\ell)} \rangle$. For $k \geq 1$ we set $n_k = m_k^{(k)}$. By construction $n_1 < n_2 < n_3 < \cdots$. As a result, $\langle f_{n_k} \rangle$ is a legitimate subsequence of $\langle f_n \rangle$. Moreover, for fixed $\ell \geq 1$ the sequence $\langle f_{n_k} \rangle$ is, with the possible exception of its first $\ell - 1$ terms, also a subsequence of $\langle f_{\ell,k} \rangle$. This observation has the consequence that $\lim_{k \to \infty} f_{n_k}(z_\ell) = \lim_{k \to \infty} f_{\ell,k}(z_\ell) = w_\ell$ for each ℓ, which is to say that $\langle f_{n_k}(\zeta) \rangle$ possesses a limit for every point ζ of the set S_0. Lemma 4.2 sees to it that the subsequence $\langle f_{n_k} \rangle$ of $\langle f_n \rangle$ converges normally in U. We have thus established the normality of $\mathcal{F}$ in U.

Turning to the converse, assume that $\mathcal{F}$ is normal in U. Let z_0 be a point of U. If $\mathcal{F}$ fails to be equicontinuous at z_0, then there must exist an $\epsilon > 0$ — we choose such an ϵ and fix it for the rest of the proof — to which there corresponds no $\delta > 0$ fulfilling the condition imposed by the definition of equicontinuity at z_0: the inequality $|f(z) - f(z_0)| < \epsilon$ is supposed to hold for every f in $\mathcal{F}$ and for every z in U satisfying $|z - z_0| < \delta$. In particular, this condition is not met by making the choice $\delta = n^{-1}$, where n is a positive integer. We can therefore select for each n a function f_n in $\mathcal{F}$ and a point z_n of U such that $|z_n - z_0| < n^{-1}$, but such that $|f_n(z_n) - f_n(z_0)| \geq \epsilon$. By hypothesis, the sequence $\langle f_n \rangle$ has a subsequence $\langle f_{n_k} \rangle$ that converges normally in U, say to the limit function f. Then f belongs to $C(U)$ (Theorem 1.4). The continuity of f at z_0 entitles us to choose $\delta > 0$ so that the closed disk $K = \overline{\Delta}(z_0, \delta)$ is contained in U and so that $|f(z) - f(z_0)| < \epsilon/3$ holds whenever z lies in K. Because $f_{n_k} \to f$ uniformly on K and because $z_{n_k} \to z_0$, we can pick an index k with the property that $|f_{n_k}(z) - f(z)| < \epsilon/3$ is satisfied for every z in K and also with the property that z_{n_k} is an element of K. For this choice of k we arrive at

$$\epsilon \leq |f_{n_k}(z_{n_k}) - f_{n_k}(z_0)|$$
$$\leq |f_{n_k}(z_{n_k}) - f(z_{n_k})| + |f(z_{n_k}) - f(z_0)| + |f(z_0) - f_{n_k}(z_0)|$$
$$< \frac{\epsilon}{3} + \frac{\epsilon}{3} + \frac{\epsilon}{3} = \epsilon \, .$$

This contradiction forces us to reject the suggestion that $\mathcal{F}$ is not equicontinuous at the point z_0 and so to conclude that $\mathcal{F}$ must be an equicontinuous subfamily of $C(U)$.

Finally, if $\mathcal{F}$ were not pointwise bounded in U, there would be a point z_0 of U and a sequence $\langle f_n \rangle$ from $\mathcal{F}$ with the property that $|f_n(z_0)| \to \infty$ as $n \to \infty$. Such a sequence could scarcely have a subsequence that converges normally in U, so its existence would conflict with the assumed normality of $\mathcal{F}$ in U. It follows that a normal subfamily of $C(U)$ is of necessity both equicontinuous and pointwise bounded in U. ∎

The Arzelà-Ascoli theorem makes it clear that the property of normality in a subfamily $\mathcal{F}$ of $C(U)$ is a local one: $\mathcal{F}$ is normal in U if and only if corresponding to each point z of U there is an open disk $\Delta = \Delta(z, r)$ contained in U such that the restrictions of the members of $\mathcal{F}$ to Δ constitute a normal subfamily of $C(\Delta)$.

The statement that a normal subfamily of $C(U)$ is pointwise bounded in U can be strengthened. We express the stronger conclusion in the form of a theorem.

Theorem 4.4. *A normal subfamily $\mathcal{F}$ of $C(U)$ is locally bounded in U.*

Proof. Suppose the assertion to be false; i.e., suppose that there is a compact set K in U on which the members of $\mathcal{F}$ are not uniformly bounded. Then for each positive integer n we are at liberty to choose a function f_n in $\mathcal{F}$ and a point z_n of K for which $|f_n(z_n)| \geq n$. (Otherwise n would be a uniform bound for the members of $\mathcal{F}$ on K.) The normality of $\mathcal{F}$ enables us to extract from the sequence $\langle f_n \rangle$ a subsequence $\langle f_{n_k} \rangle$ that converges normally in U. Call its limit f, a function that belongs to $C(U)$. The continuous function $|f|$ attains a maximum value — let m be that value — on K (Corollary II.4.7). Because $f_{n_k} \to f$ uniformly on K there is an index k_0 with the property that $|f_{n_k}(z) - f(z)| < 1$ holds for all z in K once $k \geq k_0$. When $k \geq k_0$, therefore, we can use the triangle inequality to infer that

$$n_k \leq |f_{n_k}(z_{n_k})| \leq |f_{n_k}(z_{n_k}) - f(z_{n_k})| + |f(z_{n_k})| \leq 1 + m \, .$$

This is a contradiction, for $n_k \to \infty$ as $k \to \infty$. We conclude that $\mathcal{F}$ must be locally bounded in U, as claimed. ∎

4. Normal Families

We return at last to the sphere of analytic functions and derive from the Arzelà-Ascoli theorem the prototype of normal family theorems in complex analysis. It is due to Paul Montel (1876-1975).

Theorem 4.5. (Montel's Theorem) *Let $\mathcal{F}$ be a family of functions that are analytic in an open set U. Suppose that $\mathcal{F}$ is locally bounded in U. Then $\mathcal{F}$ is a normal family in this set.*

Proof. The family $\mathcal{F}$ is obviously pointwise bounded in U. In view of the Arzelà-Ascoli theorem, to prove that $\mathcal{F}$ is a normal subfamily of $C(U)$ we need only to check that $\mathcal{F}$ is equicontinuous in U. We fix a point z_0 in U and establish the equicontinuity of $\mathcal{F}$ at z_0. For this we choose $r > 0$ with the property that the closed disk $K = \overline{\Delta}(z_0, 2r)$ lies in U. By hypothesis there exists a constant $m = m(K) > 0$ such that $|f(\zeta)| \leq m$ holds whenever f belongs to $\mathcal{F}$ and ζ to K. For z lying in the disk $\Delta = \Delta(z_0, r)$ we call upon Cauchy's integral formula to provide an estimate valid for each member f of $\mathcal{F}$:

$$|f(z) - f(z_0)| = \left| \frac{1}{2\pi i} \int_{|\zeta - z_0| = 2r} \frac{f(\zeta)\, d\zeta}{\zeta - z} - \frac{1}{2\pi i} \int_{|\zeta - z_0| = 2r} \frac{f(\zeta)\, d\zeta}{\zeta - z_0} \right|$$

$$= \frac{|z - z_0|}{2\pi} \left| \int_{|\zeta - z_0| = 2r} \frac{f(\zeta)\, d\zeta}{(\zeta - z)(\zeta - z_0)} \right|$$

$$\leq \frac{|z - z_0|}{2\pi} \int_{|\zeta - z_0| = 2r} \frac{|f(\zeta)||d\zeta|}{|\zeta - z||\zeta - z_0|} \leq \frac{m|z - z_0|}{r}.$$

(N.B. $|\zeta - z_0| = 2r$ implies that $|\zeta - z| \geq r$ for z in Δ.) Given $\epsilon > 0$, we now set $\delta = \min\{r, r\epsilon/m\}$. The above estimate implies that the inequality $|f(z) - f(z_0)| < \epsilon$ is valid for every f from the family $\mathcal{F}$, as long as z satisfies $|z - z_0| < \delta$. This confirms the equicontinuity of $\mathcal{F}$ at z_0, an arbitrary point of U. The normality of $\mathcal{F}$ in U follows. ∎

Here is a simple example in which Montel's theorem is used to certify the normality of a specific family of analytic functions.

EXAMPLE 4.1. Let $c > 0$ be a constant, and let $\mathcal{F}$ consist of all the functions f which are analytic in the disk $D = \Delta(0, 1)$ and enjoy the property that $\int_0^{2\pi} |f(re^{i\theta})|\, d\theta \leq c$ for every r in $(0, 1)$. Show that $\mathcal{F}$ is a normal family in D.

Given a compact set K in D, we shall exhibit a constant $m = m(K)$ such that $|f(z)| \leq m$ holds for every z in K whenever f belongs to $\mathcal{F}$. We first fix s in $(0, 1)$ such that K lies in $\Delta(0, s)$, and we then fix r in $(s, 1)$. For f in $\mathcal{F}$ and z in K we obtain from Cauchy's integral formula and from the fact that $|\zeta - z| \geq r - s$ when $|\zeta| = r$ the estimate

$$|f(z)| = \left|\frac{1}{2\pi i}\int_{|\zeta|=r}\frac{f(\zeta)\,d\zeta}{\zeta - z}\right| \leq \frac{1}{2\pi}\int_{|\zeta|=r}\frac{|f(\zeta)||d\zeta|}{|\zeta - z|}$$

$$\leq \frac{1}{2\pi(r-s)}\int_{|\zeta|=r}|f(\zeta)||d\zeta| = \frac{r}{2\pi(r-s)}\int_0^{2\pi}|f(re^{i\theta})|\,d\theta \leq \frac{c}{2\pi(r-s)}.$$

Thus, $m = c/[2\pi(r-s)]$ is a uniform bound for the members of $\mathcal{F}$ on K. We have just verified that $\mathcal{F}$ is locally bounded in D. Montel's theorem states that $\mathcal{F}$ is a normal family in D.

It is not automatically the case that a pointwise bounded subfamily of $C(U)$ whose members are analytic in U is normal in this set. Montel, however, discovered a number of far-reaching generalizations of Theorem 4.5 that sometimes permit one to infer normality from pointwise boundedness. We state a typical example of these results as an apt finishing touch to this section. No proof is included, for all existing proofs rely on mathematical machinery that we do not wish to introduce in the present text.

Theorem 4.6. *Let $\mathcal{F}$ be a family of functions that are analytic in a domain D. Assume the existence of distinct complex numbers a and b such that the conditions $f(z) \neq a$ and $f(z) \neq b$ are met for every z in D by every f in $\mathcal{F}$. If $\langle f_n \rangle$ is a sequence from $\mathcal{F}$, then either $|f_n(z)| \to \infty$ as $n \to \infty$ for every z in D or $\langle f_n \rangle$ has a subsequence $\langle f_{n_k} \rangle$ that converges normally in D. In particular, if such a family $\mathcal{F}$ is bounded at some point of D, then it is a normal family in this domain.*

5 Exercises for Chapter VII

5.1 Exercises for Section VII.1

5.1. Let $\langle f_n \rangle$ be the sequence of functions defined on the interval $[0,1]$ as follows: $f_n(t) = 4n^2 t$ when $0 \leq t \leq (2n)^{-1}$, $f_n(t) = 4n - 4n^2 t$ for t satisfying $(2n)^{-1} \leq t \leq n^{-1}$, and $f_n(t) = 0$ if $n^{-1} \leq t \leq 1$. Check that $f_n \to 0$ in $[0,1]$, but that $\int_0^1 f_n(t)\,dt \not\to 0$. This example exposes another of the disadvantages of pointwise convergence: under this type of convergence the integral of a limit function is not always the limit of the integrals of the functions that converge to it.

5.2. If the functions in a sequence $\langle f_n \rangle$ are all continuous on a set A and if $f_n \to f$ uniformly on A, then $\int_\gamma f(z)|dz| = \lim_{n\to\infty}\int_\gamma f_n(z)|dz|$ for every piecewise smooth path γ in A. Establish this variation on the final conclusion of Theorem 1.1.

5.3. If $\langle f_n \rangle$ is a sequence of functions that converges uniformly on a set A and if $g: B \to \mathbb{C}$ is a function whose range $g(B)$ is contained in A, then

5. Exercises for Chapter VII

the sequence of compositions $\langle f_n \circ g \rangle$ is uniformly convergent on B. Prove this.

5.4. Suppose that a function sequence $\langle f_n \rangle$ converges uniformly on a set A and that S is a set which contains $f_n(A)$ for every n. Given that a function g is uniformly continuous on S, show that the sequence $\langle g \circ f_n \rangle$ converges uniformly on A.

5.5. Let $\langle f_n \rangle$ be a function sequence that is uniformly convergent on a set A, and let $g: A \to \mathbb{C}$ be a bounded function. Demonstrate that the sequence of products $\langle g f_n \rangle$ converges uniformly on A.

5.6. Function sequences $\langle f_n \rangle$ and $\langle g_n \rangle$ are known to converge uniformly on a set A. Show that the sequence $\langle f_n + g_n \rangle$ does the same. With the added hypothesis that $\langle f_n \rangle$ and $\langle g_n \rangle$ are uniformly bounded on A (i.e., there is a constant c such that $|f_n(z)| \leq c$ and $|g_n(z)| \leq c$ hold for every index n and every z in A), prove that the product sequence $\langle f_n g_n \rangle$ also converges uniformly on A. Confirm that the latter is true, in particular, when A is a compact set and all the terms in the given sequences $\langle f_n \rangle$ and $\langle g_n \rangle$ are continuous on A.

5.7. Each function in $\langle f_n \rangle$ is continuous on a compact set K and $f_n \to 0$ in K. Under the assumption that $|f_1(z)| \geq |f_2(z)| \geq |f_3(z)| \geq \cdots$ for every z in K, show that $f_n \to 0$ uniformly on K. (*Hint.* Given $\epsilon > 0$, consider the sequence of sets $\langle K_n \rangle - K_n = \{z \in K : |f_n(z)| \geq \epsilon\}$. Prove that there is an index N with the property that $K_n = \phi$ for every $n \geq N$. Do this by assuming the contrary to be true and deriving a contradiction to Cantor's theorem.)

5.8. Let D be a bounded domain in the complex plane. Suppose that every function in a sequence $\langle f_n \rangle$ is continuous on $\overline{D}$ and analytic in D. Given that this sequence converges uniformly on ∂D, prove that it converges uniformly on D.

5.9. Let $\langle f_n \rangle$ be a sequence of functions that are continuous on the closed disk $\overline{\Delta}(z_0, r)$ and analytic in its interior D. Assume that $\langle f_n \rangle$ converges pointwise on the circle ∂D to a continuous function φ. Assume, additionally, that $\int_{|\zeta - z_0| = r} |f_n(\zeta) - \varphi(\zeta)||d\zeta| \to 0$ as $n \to \infty$. (The latter condition will certainly be met if $f_n \to \varphi$ uniformly on ∂D.) Show that $f_n \to f$ normally in D, where f is the function defined in D by the rule of correspondence $f(z) = (2\pi i)^{-1} \int_{|\zeta - z_0| = r} (\zeta - z)^{-1} \varphi(\zeta) d\zeta$.

5.10. Let the terms of a function sequence $\langle f_n \rangle$ be continuous in an open set U. If $\langle f_n \rangle$ converges normally in U to the limit function f and if $\langle z_n \rangle$ is a sequence in U that converges to a point z_0 of U, then $f_n(z_n) \to f(z_0)$ as $n \to \infty$. Corroborate this assertion.

5.11. Each function in a sequence $\langle u_n \rangle$ is harmonic in an open set U and $u_n \to u$ normally in U. Confirm that the limit function u is harmonic in this set.

5.12. Assume that each function in a sequence $\langle u_n \rangle$ is harmonic in a plane domain D and that $0 \leq u_1 \leq u_2 \leq u_3 \leq \cdots$ in D. Then for each fixed z in D one of two things happens: either $\lim_{n \to \infty} u_n(z)$ exists (in the strict sense) or else $u_n(z) \to \infty$ as $n \to \infty$. Let $U = \{z \in D : \lim_{n \to \infty} u_n(z) \text{ exists}\}$ and $V = \{z \in D : u_n(z) \to \infty\}$. Show that either $D = U$ — in which event $\langle u_n \rangle$ is pointwise convergent in D — or $D = V$. (*Hint.* Let z_0 belong to D and let $\Delta = \Delta(z_0, r)$ be an open disk whose closure lies in D. Use Exercise VI.4.41 to establish the existence of a constant $c > 0$, independent of n, with the property that $c^{-1} u_n(z_0) \leq u_n(z) \leq c u_n(z_0)$ for every index n and every z in Δ. Conclude from this that Δ lies in either U or V.)

5.13. With the aid of Exercises 5.11 and 5.12 prove the following theorem of Harnack: if $\langle u_n \rangle$ is a non-decreasing sequence of functions that are harmonic in a domain D, then either $\langle u_n \rangle$ converges normally in D to a limit function u that is itself harmonic in D or $u_n(z) \to \infty$ as $n \to \infty$ for every z in D. (*Hint.* We may suppose that $u_n \geq 0$ in D for every n — if not, just consider in place of $\langle u_n \rangle$ the sequence $\langle v_n \rangle$, where $v_n = u_n - u_1$. Thus, we may assume that we find ourselves in the context of Exercise 5.12. If $\langle u_n \rangle$ is known to converge pointwise in D and if $\Delta = \overline{\Delta}(z_0, r)$ is an arbitrary closed disk in D, prove that $\langle u_n \rangle$ converges uniformly on Δ. Do this by applying Exercise VI.4.41 to $u_m - u_n$ for $m > n$ in a disk slightly larger than Δ.)

5.2 Exercises for Section VII.2

5.14. Decide whether the following series converge or diverge; in the case of convergence, indicate whether it is absolute: (i) $\sum_{n=1}^{\infty} n^{-1/2} i^n$; (ii) $\sum_{n=2}^{\infty} n^{-1} [\text{Log}(n + iy)]^{-2}$; (iii) $\sum_{n=0}^{\infty} (n!)^2 [(2n)!]^{-1} (3 + 4i)^n$; (iv) $\sum_{n=-\infty}^{\infty} 2^n \sec(ni)$; (v) $\sum_{n=1}^{\infty} (1 + n^{-1})^{-n^2} (1 + 2i)^n e^{ni}$.

5.15. Let $\langle r_n \rangle$ be a sequence of non-negative real numbers, let $\langle s_n \rangle$ be the sequence of partial sums associated with $\langle r_n \rangle$, and let $c \geq 0$. Show that the series $\sum_{n=1}^{\infty} r_n$ is convergent and satisfies $\sum_{n=1}^{\infty} r_n \leq c$ if and only if $s_n \leq c$ for every n.

5.16. Let $\langle z_n \rangle$ be a sequence of complex numbers whose associated sequence of partial sums $\langle s_n \rangle$ is bounded — but not necessarily convergent — and let $\langle r_n \rangle$ be a non-increasing sequence of real numbers with the property that $\lim_{n \to \infty} r_n = 0$. Use the Cauchy criterion to prove that the series $\sum_{n=1}^{\infty} r_n z_n$ converges. (N.B. The classic example here is to take $z_n = (-1)^{n-1}$ — then $s_n = 1$ for odd n and $s_n = 0$ for even n — and to conclude that $\sum_{n=1}^{\infty} (-1)^{n-1} r_n$ is convergent, the result known as the "alternating series test." (*Hint.* Start by demonstrating that $\sum_{k=n}^{m} r_k z_k =$

$r_m s_m - r_n s_{n-1} + \sum_{k=n}^{m-1}(r_k - r_{k+1})s_k$ when $m > n \geq 2$. Don't forget: $z_k = s_k - s_{k-1}$ for $k \geq 2$.)

5.17. Let $\sum_{n=1}^{\infty} z_n$ be an absolutely convergent series of complex numbers with sum s. Show that $s = \sum_{n \in \mathbb{N}} z_n$ — this notation is intended to suggest that the sum is independent of the actual manner in which its terms are put into sequence — interpreted as follows: corresponding to each $\epsilon > 0$ there is a finite subset F_0 of $\mathbb{N}$ such that $|s - \sum_{n \in F} z_n| < \epsilon$ holds for every finite subset F of $\mathbb{N}$ which contains F_0. Confirm, in particular, that a series $\sum_{n=1}^{\infty} w_n$ which is derived from $\sum_{n=1}^{\infty} z_n$ by permuting its terms — i.e., $w_n = z_{\sigma(n)}$, where $\sigma: \mathbb{N} \to \mathbb{N}$ is a one-to-one function with range $\mathbb{N}$ — is also convergent and has sum s. (*Hint.* Given $\epsilon > 0$, one can choose N so that $\sum_{n=N+1}^{\infty} |z_n| < \epsilon$. (Why?) Consider $F_0 = \{1, 2, \ldots, N\}$.)

5.18. Establish the following "grouping principle" for absolutely convergent series: if $\sum_{n=1}^{\infty} z_n$ is an absolutely convergent series of complex numbers with sum s and if $A_1, A_2, A_3, \cdots$ is a (finite or infinite) collection of nonempty, disjoint subsets of $\mathbb{N}$ whose union is $\mathbb{N}$, then $s = \sum_{j=1}^{\infty}(\sum_{n \in A_j} z_n)$.

5.19. Let $\langle z_n \rangle$ and $\langle w_n \rangle$ be complex sequences such that both $\sum_{n=1}^{\infty} |z_n|^2$ and $\sum_{n=1}^{\infty} |w_n|^2$ are convergent. Prove that the series $\sum_{n=1}^{\infty} z_n w_n$ is absolutely convergent and that $|\sum_{n=1}^{\infty} z_n w_n|^2 \leq (\sum_{n=1}^{\infty} |z_n|^2)(\sum_{n=1}^{\infty} |w_n|^2)$. (*Hint.* Recall Exercise I.4.22 and Exercise 5.15.)

5.20. Suppose that a series of functions $\sum_{n=1}^{\infty} f_n$ is uniformly convergent on a set A and that $g: B \to \mathbb{C}$ is a function whose range lies in A. Show that the series $\sum_{n=1}^{\infty} f_n \circ g$ converges uniformly on B.

5.21. A function series $\sum_{n=1}^{\infty} f_n$ converges uniformly on a set A. Assuming that $g: A \to \mathbb{C}$ is a bounded function, prove that the series $\sum_{n=1}^{\infty} g f_n$ is also uniformly convergent on A.

5.22. Assume that each term in a function series $\sum_{n=1}^{\infty} f_n$ is continuous in an open set U, that the series $\sum_{n=1}^{\infty} |f_n|$ converges pointwise in U, and that the function $f = \sum_{n=1}^{\infty} |f_n|$ is continuous in U. Prove that $\sum_{n=1}^{\infty} f_n$ is normally convergent in U. (*Hint.* Apply Exercise 5.7 on any compact subset K of U to the sequence $\langle g_n \rangle$, where $g_n = f - \sum_{k=1}^{n} |f_k|$.)

5.23. Show that the representation $\sec \pi z = 2 \sum_{n=0}^{\infty}(-1)^n e^{(2n+1)\pi i z}$ is valid for every z in the half-plane $H = \{z : \operatorname{Im} z > 0\}$ and that the convergence of this series of functions is absolute and normal there.

5.24. Verify that the series $\sum_{n=0}^{\infty}(1-z)^n(1+z)^{-n}$ converges absolutely and uniformly on the set $A_t = \{z : x \geq t, |z| \leq t^{-1}\}$ for every $t > 0$ — hence, absolutely and normally in the half-plane $U = \{z : x > 0\}$ — and find its sum. Show that the series diverges for every other z. Disregard the point $z = -1$, where the terms of the series are undefined.

5.25. A function sequence $\langle f_n \rangle$ converges normally in an open set U to the limit function f. Show that the series $\sum_{n=1}^{\infty}(f_n - f_{n+1})$ converges normally in U, and identify its sum.

5.26. Verify that the series $\sum_{n=1}^{\infty}[z^2 + (2n+1)z + n^2 + n]^{-1}$ converges normally in the open set $U = \mathbb{C} \sim \{-1, -2, -3, \cdots\}$. Determine its sum.

5.27. Show that the series $\sum_{n=1}^{\infty}(z^2 + n^2)^{-1}$ converges absolutely and normally in the set $U = \mathbb{C} \sim \{ni : n = \pm 1, \pm 2, \cdots\}$. If f denotes its sum, verify that $\int_{|z|=r} f(z)dz = 0$ for every non-integral radius $r > 0$ and that

$$\int_{|z-ki|=r} f(z)\,dz = \pi \sum_{|k-r|<n<k+r} n^{-1}$$

for every such r and every positive integer k.

5.28. Prove Lemma 2.3.

5.29. Show that the function series $\sum_{n=-\infty}^{\infty} a^n e^{-|n|z}$, where a is a non-zero constant, converges absolutely and normally in the open half-plane $U = \{z : \operatorname{Re} z > |\operatorname{Log}|a||\}$. Compute its sum.

5.30. Verify that the series $\sum_{n=-\infty}^{\infty}(n^2+1)^n e^{n^2 z}$ converges absolutely and normally in the half-plane $U = \{z : \operatorname{Re} z < 0\}$ and diverges in $\mathbb{C} \sim U$.

5.3 Exercises for Section VII.3

5.31. Assume that each function in a sequence $\langle f_n \rangle$ is continuous in a domain D, where the sequence is known to converge normally to the limit function f. Assume, beyond this, that a primitive F_n for f_n in D has been found and normalized so as to insure the convergence of the sequence $\langle F_n(z_0) \rangle$ for some predetermined point z_0 of D. Show that $\langle F_n \rangle$ converges normally in D and that its limit F is a primitive for f in this domain. (*Hint.* First demonstrate that $\langle F_n \rangle$ converges pointwise in D.)

5.32. Show that the series $\sum_{n=1}^{\infty} \operatorname{Arcsin}(n^{-2}z)$ converges normally in the whole complex plane and that the function $f : \mathbb{C} \to \mathbb{C}$ defined by the rule of correspondence $f(z) = \sum_{n=1}^{\infty} \operatorname{Arcsin}(n^{-2}z)$ is analytic in the domain $D = \mathbb{C} \sim \{z : \operatorname{Im} z = 0 \text{ and } |z| \geq 1\}$. Deduce from this that $g(z) = \sum_{n=1}^{\infty}(n^4 - z^2)^{-1/2}$ also defines a function that is analytic in D.

5.33. Confirm that the formula $F(z) = \sum_{n=1}^{\infty} n^{-1} \operatorname{Arctan}(n^{-1}z)$ describes a primitive for the function introduced in Exercise 5.27 in the domain $D = \mathbb{C} \sim \{z : \operatorname{Re} z = 0 \text{ and } |z| \geq 1\}$.

5.34. Demonstrate that the formula $f(z) = \sum_{n=1}^{\infty}[1 - \cos(n^{-1}z)]$ defines an entire function.

5.35. Show that $g(z) = (1/2)z^{-2} + \sum_{n=1}^{\infty}(z^2 - n^2)^{-1}$ defines an analytic function g in the open set $U = \mathbb{C} \sim \{0, \pm 1, \pm 2, \cdots\}$. Show, in addition, that $F(z) = -2zg(z)$ furnishes a primitive in U for the function f presented in

Example 3.2.

5.36. Verify that the function f defined by $f(z) = \sum_{n=0}^{\infty} z^n(1+z^{2n})^{-1}$ is analytic in the open set $U = \{z: |z| \neq 1\}$.

5.37. Suppose that a function f is analytic in the open disk $D = \Delta(0,1)$, that $f(0) = 0$, and that $|f(z)| \leq 1$ for every z in D. Document the fact that $g(z) = \sum_{n=1}^{\infty} f(z^n)$ defines a function g which is analytic in D and which has $|g'(0)| \leq 1$. For which initial functions f will $|g'(0)| = 1$ hold?

5.38. Let $a_0, a_1, a_2, \cdots$ be a sequence of non-zero complex numbers, and let $\lambda = \limsup_{n \to \infty} |a_n|/|a_{n+1}|$. Show that the radius of convergence ρ of the Taylor series $\sum_{n=0}^{\infty} a_n(z-z_0)^n$ satisfies $\rho \leq \lambda$ and that equality holds when $\lim_{n \to \infty} |a_n|/|a_{n+1}|$ exists or when $|a_n|/|a_{n+1}| \to \infty$ as $n \to \infty$. Illustrate, by example, that $\rho < \lambda$ is a possibility otherwise.

5.39. Determine the disks of convergence of: (i) $\sum_{n=1}^{\infty} \sqrt[n]{n}(z-1)^n$; (ii) $\sum_{n=1}^{\infty} 3^n n^3 (n!)^3 [(3n)!]^{-1} z^{3n}$; (iii) $\sum_{n=1}^{\infty} n^2 (z+2)^{2^n}$; (iv) $\sum_{n=1}^{\infty} n!(z-i)^{n!}$; (v) $\sum_{n=2}^{\infty} (\text{Log } n)^n (z+1)^{n^2}$.

5.40. Show that the Taylor series $\sum_{n=1}^{\infty} n^{-1} z^n$ has radius of convergence $\rho = 1$ and that it converges conditionally for any z with $|z| = 1$ save for $z = 1$, at which point it diverges. (*Hint*. Exercise 5.16 is useful in dealing with z for which $|z| = 1$.)

5.41. Demonstrate that the radius of convergence of the Taylor series $\sum_{n=0}^{\infty} (-1)^n (2n+1)^{-1} z^{2n+1}$ is one and that the series is conditionally convergent at any point z of the circle $K(0,1)$ with the exception of i and $-i$, where it is divergent.

5.42. Verify that the Taylor series $\sum_{n=1}^{\infty} n! n^{-n} z^n$ has radius of convergence $\rho = e$ and that it diverges for every z satisfying $|z| = e$.

5.43. The "Fibonacci sequence" $\langle a_n \rangle$ is defined recursively by $a_0 = 0$, $a_1 = 1$, and $a_n = a_{n-1} + a_{n-2}$ for $n \geq 2$. Verify that the radius of convergence ρ of the associated Taylor series $\sum_{n=1}^{\infty} a_n z^n$ is positive. Then, through the process of actually finding the sum of this series, determine ρ exactly. (*Hint.* For the first part check that $0 \leq a_n \leq 2^n$ for every n.)

5.44. Let $\sum_{n=0}^{\infty} a_n(z-z_0)^n$ be a Taylor series with radius of convergence $\rho > 0$, and let f be its sum in the disk $\Delta = \Delta(z_0, \rho)$. Support the statement that the series $\sum_{n=0}^{\infty} (n+1)^{-1} a_n (z-z_0)^{n+1}$ also converges in Δ, where its sum F furnishes a primitive for f.

5.45. Confirm the observation that a Taylor series $\sum_{n=0}^{\infty} a_n(z-z_0)^n$, its corresponding derived series $\sum_{n=0}^{\infty} na_{n-1}(z-z_0)^{n-1}$, and its associated primitive series $\sum_{n=0}^{\infty} (n+1)^{-1} a_n (z-z_0)^n$ all share the same radius of convergence ρ.

5.46. Find the Taylor series expansion of f about the origin, and identify the largest open disk $\Delta(0,d)$ in which the expansion is valid: (i) $f(z) =$

$(1-z)^{-2}$; (ii) $f(z) = (1+z^2)^{-3}$; (iii) $f(z) = \text{Arctan } z$; (iv) $f(z) = \text{Log}(1+z^2)$; (v) $f(z) = \text{Log}[(1-z)(1+z)^{-1}]$; (vi) $f(z) = (z^2+z-2)^{-2}$; (vii) $f(z) = \sin(3z)\cos(2z)$.

5.47. Obtain the Taylor series development of the function f about the specified point z_0, and determine the largest d for which the expansion is valid in $\Delta(z_0, d)$: (i) $f(z) = z^{-1}$, $z_0 = i$; (ii) $f(z) = (z^2 - z - 2)^{-1}$, $z_0 = 1$; (iii) $f(z) = (z^2+1)^{-1}$, $z_0 = 1$; (iv) $f(z) = \sin z$, $z_0 = \pi/2$; (v) $f(z) = z \sin z$, $z_0 = \pi/2$; (vi) $f(z) = \cos^2 z$, $z_0 = \pi$; (vii) $f(z) = z \cos^2 z$, $z_0 = \pi$.

5.48. (i) Derive the expansion: $e^z \cos z = \sum_{n=0}^{\infty} (n!)^{-1} 2^{n/2} \cos(n\pi/4) z^n$. (ii) Use the fact that $f^{(n)}(0)/n! = (2\pi i)^{-1} \int_{|z|=1} z^{-n-1} f(z) dz$ for any entire function f to arrive at an alternate description of the expansion in part (i) — namely,

$$e^z \cos z = \sum_{n=0}^{\infty} \left\{ \sum_{0 \le k \le n/2} (-1)^k [(2k)!(n-2k)!]^{-1} \right\} z^n .$$

5.49. Recall that for any non-zero complex number λ and any non-negative integer n the *binomial coefficient* $\binom{\lambda}{n}$ is defined by $\binom{\lambda}{0} = 1$ and $\binom{\lambda}{n} = \lambda(\lambda-1)\cdots(\lambda-n+1)/n!$ for $n \ge 1$. Thus $\binom{\lambda}{1} = \lambda$, $\binom{\lambda}{2} = \lambda(\lambda-1)/2$, etc. Certify that the expansion $(1+z)^\lambda = \sum_{n=0}^{\infty} \binom{\lambda}{n} z^n$ is valid when $|z| < 1$.

5.50. Expand $f(z) = \text{Arcsin } z$ in a Taylor series centered at the origin. In what open disk about the origin does this series represent f? (*Hint.* To get started, consider f'.)

5.51. Let $z_t = e^{it}$, where $-\pi < t < \pi$. Demonstrate that the Taylor series expansion of $f(z) = \sqrt{z}$ centered at z_t takes the form $f(z) = \sum_{n=0}^{\infty} \binom{1/2}{n} z_t^{(1-2n)/2} (z-z_t)^n$, and determine the radius d_t of the largest open disk centered at z_t in which this representation is valid. Show that the above series has radius of convergence $\rho = 1$. For those t with $d_t < 1$ identify the function to which this series sums in the disk $\Delta(z_t, 1)$.

5.52. Find the disk of convergence Δ of the series $\sum_{n=1}^{\infty} (-1)^{n-1} n^2 z^n$; compute its sum in this disk.

5.53. Confirm that the function series $\sum_{n=0}^{\infty} (n!)^{-1} \cos(nz)$ converges normally in the complex plane, and find the entire function that is its sum.

5.54. Let f and g be functions that are analytic in an open disk $\Delta = \Delta(z_0, r)$, where their Taylor expansions read $f(z) = \sum_{n=0}^{\infty} a_n (z-z_0)^n$ and $g(z) = \sum_{n=0}^{\infty} b_n (z-z_0)^n$. Verify that Taylor expansion of the product

5. Exercises for Chapter VII

$h = fg$ about z_0 has the structure $h(z) = \sum_{n=0}^{\infty} c_n(z - z_0)^n$, in which $c_n = \sum_{k=0}^{n} a_k b_{n-k}$. Deduce from this that the radius of convergence of the Taylor series $\sum_{n=0}^{\infty} c_n(z - z_0)^n$ is at least the minimum of the radii of convergence of $\sum_{n=0}^{\infty} a_n(z - z_0)^n$ and $\sum_{n=0}^{\infty} b_n(z - z_0)^n$.

5.55. Show that $(1-z)^{-1}e^z = \sum_{n=0}^{\infty} \left[\sum_{k=0}^{n}(k!)^{-1}\right] z^n$ when $|z| < 1$.

5.56. Verify that $z^{-1}\operatorname{Log} z = \sum_{n=1}^{\infty}(-1)^{n-1}\left[\sum_{k=1}^{n} k^{-1}\right](z-1)^n$ and that $\operatorname{Log}^2 z = \sum_{n=2}^{\infty}(-1)^n \left[\sum_{k=1}^{n-1} k^{-1}(n-k)^{-1}\right](z-1)^n$ when $|z-1| < 1$.

5.57. If a function f has the Taylor expansion $f(z) = \sum_{n=0}^{\infty} a_n(z-z_0)^n$ in the disk $\Delta(z_0, r)$ and if $f(z_0) \neq 0$, then the coefficients in the Taylor expansion $1/f(z) = \sum_{n=0}^{\infty} b_n(z-z_0)^n$ of its reciprocal are given recursively by $b_0 = a_0^{-1}$ and $b_n = -a_0^{-1}\sum_{k=1}^{n} a_k b_{n-k}$ for $n \geq 1$. Prove this.

5.58. Use Exercises 5.54 and 5.57 to find the terms up to and including the one involving z^5 in the Taylor expansions about the origin of the functions:
(i) $f(z) = \sec z$; (ii) $g(z) = \tan z$; (iii) $h(z) = (1 + e^z)^{-1}$; (iv) $k(z) = (1 - e^z)(1 + e^z)^{-1}$; (v) $\ell(z) = [\operatorname{Log}(1+z)]^{-1}$.

5.59. Let a function f be analytic in an open disk Δ centered at the origin, where it thus admits a Taylor series expansion $f(z) = \sum_{n=0}^{\infty} a_n z^n$. Demonstrate that f is an even function in Δ if and only if $a_n = 0$ for all odd n; that f is an odd function in Δ if and only if $a_n = 0$ for all even n; that f is symmetric about the real axis in Δ — i.e., $f(\bar{z}) = \overline{f(z)}$ for every z in Δ — if and only if a_n is real for every n.

5.60. If a function f is both analytic and even in a disk $\Delta = \Delta(0, r)$, then the function $g: \Delta \to \mathbb{C}$ defined by $g(z) = f(\sqrt{z})$ is an analytic function and $g^{(n)}(0) = [(2n)(2n-1)\cdots(n+1)]^{-1}f^{(2n)}(0)$. Justify these claims. (This exercise redoes, only much more simply, Exercise III.6.38.)

5.61. A non-constant entire function f has the feature that $f(\lambda z) = f(z)$ for every z, where $\lambda \neq 1$ is a constant. Prove that there must be a positive integer m for which $\lambda^m = 1$. Moreover, if m is the smallest such integer, prove that f has the structure $f(z) = g(z^m)$ for some entire function g.

5.62. The function f defined in the disk $\Delta(0, 2\pi)$ by $f(z) = z(e^z - 1)^{-1}$ if $z \neq 0$ and $f(0) = 1$ is an analytic function. (Why is f differentiable at the orgin?) Show that the Taylor expansion of f about the origin can be written in the form $f(z) = 1 - (1/2)z + \sum_{n=1}^{\infty}(-1)^{n-1}[(2n)!]^{-1}B_n z^{2n}$. The numbers $B_1, B_2, B_3, \cdots$ that turn up here — i.e., $B_n = (-1)^{n-1}f^{(2n)}(0)$ — are called the *Bernoulli numbers*. Verify that

$$\sum_{k=1}^{n}(-1)^{k-1}\binom{2n+1}{2k}B_k = \frac{2n-1}{2}$$

for $n = 1, 2, 3, \cdots$ and use this combinatorial identity to compute B_1 through B_7. (*Hint*. For the first part of the problem confirm the observation that the function g given by $g(z) = (1/2)z + f(z)$ is an even function.)

5.63. Derive the expansion

$$z \cot z = 1 - \sum_{n=1}^{\infty} \frac{B_n 4^n z^{2n}}{(2n)!}$$

for z in $\Delta^*(0, \pi)$, where $B_1, B_2, B_3, \cdots$ are the Bernoulli numbers. Use this to conclude that

$$\tan z = \sum_{n=1}^{\infty} \frac{B_n 4^n (4^n - 1) z^{2n-1}}{(2n)!}$$

when $|z| < \pi/2$. (*Hint.* Exploit the identities $z \cot z = iz + 2iz(e^{2iz} - 1)^{-1}$ and $\tan z = \cot z - 2\cot(2z)$.)

5.64. Demonstrate that the Taylor series $\sum_{n=1}^{\infty} z^{n!} = z + z^2 + z^6 + \cdots$ has radius of convergence $\rho = 1$ and that the analytic function g defined in $\Delta = \Delta(0,1)$ by $g(z) = \sum_{n=1}^{\infty} z^{n!}$ has the following behavior: for every rational number θ, $|g(re^{2\pi i\theta})| \to \infty$ as $r \to 1$ from the left.

5.65. Show that $f(z) = \sum_{n=1}^{\infty} (n!)^{-1} z^{n!}$ describes a function that is continuous on the closed disk $\overline{\Delta}(0,1)$, is analytic in the open disk $\Delta = \Delta(0,1)$, and is endowed with the following property: if D is any domain that properly includes Δ, then there exists no analytic function $F: D \to \mathbb{C}$ whose restriction to Δ coincides with f there. (*Hint.* Assume such a function F were to exist. Obtain a contradiction by looking at F'.)

5.66. Let $\sum_{n=1}^{\infty} a_n z^n$ be a Taylor series with non-negative coefficients and with radius of convergence $\rho = 1$. If U is an open set that contains both the disk $\Delta(0,1)$ and the point $z = 1$, demonstrate that there can be no function f which is analytic in U and satisfies $f(z) = \sum_{n=0}^{\infty} a_n z^n$ for every z in $\Delta(0,1)$. (*Hint.* Argue by contradiction. Assume that a function f of this description exists, expand f in a Taylor series about the point $z_0 = 1/2$, and, by regrouping the terms of this series, arrive at the conclusion that the series $\sum_{n=0}^{\infty} a_n (1+r)^n$ converges for some $r > 0$. Why is this unacceptable?)

5.67. Let $\sum_{n=0}^{\infty} a_n$ be a convergent series of complex numbers, and let s be its sum. Making note that the radius of convergence of the Taylor series $\sum_{n=0}^{\infty} a_n z^n$ is at least one, prove that the analytic function f defined in $\Delta(0,1)$ by $f(z) = \sum_{n=0}^{\infty} a_n z^n$ satisfies $\lim_{r \to 1^-} f(r) = s$. (N.B. This result is called "Abel's Theorem" after its discoverer, Henrik Abel (1802-1829). *Hint.* Let $s_n = \sum_{k=0}^{n} a_k$ for $n = 0, 1, 2, \cdots$ and set $s_{-1} = 0$. Use the fact that $a_n = s_n - s_{n-1}$ to get the formula $f(z) = (1-z) \sum_{n=0}^{\infty} s_n z^n$ for f in $\Delta(0,1)$. Then observe that $s = (1-z) \sum_{n=0}^{\infty} sz^n$. This information enables one to get a handle on $|f(r) - s|$ when $0 \leq r < 1$.)

5.68. Compute the sum of the series $\sum_{n=1}^{\infty} n^{-1} e^{in\theta}$ for θ in the interval $(0, 2\pi)$. Use it to find $\sum_{n=1}^{\infty} n^{-1} \cos(n\theta)$ and $\sum_{n=1}^{\infty} n^{-1} \sin(n\theta)$ for such θ.

(*Hint.* For fixed θ consider the function $f(z) = -\operatorname{Log}(1 - e^{i\theta}z)$ in the disk $\Delta(0,1)$. Recall Exercise 5.40.)

5.69. Calculate the sum of the series $\sum_{n=0}^{\infty}(-1)^n(2n+1)^{-1}e^{(2n+1)i\theta}$ for θ that is not an odd multiple of $\pi/2$. With the aid of the result sum the series $\sum_{n=0}^{\infty}(-1)^n(2n+1)^{-1}\cos[(2n+1)\theta]$ and $\sum_{n=0}^{\infty}(-1)^n(2n+1)^{-1}\sin[(2n+1)\theta]$ for such θ. (*Hint.* Consider the function $f(z) = \operatorname{Arctan}(e^{i\theta}z)$.)

5.70. A function f is analytic in the disk $\Delta = \Delta(0,1)$, where it has the Taylor series representation $f(z) = \sum_{n=0}^{\infty} a_n z^n$. If $0 < r < 1$, verify that $\int_0^{2\pi} |f(re^{i\theta})|^2 d\theta = 2\pi \sum_{n=0}^{\infty} |a_n|^2 r^{2n}$. Conclude, in particular, that the latter series is convergent. (*Hint.* $|f(z)|^2 = f(z)\overline{f(z)} = \sum_{n=0}^{\infty} \bar{a}_n \bar{z}^n f(z)$ in Δ, and the convergence of this series is normal there.)

5.71. If a function f is continuous on the closed disk $\overline{\Delta}(0,1)$ and analytic in the open disk $\Delta(0,1)$, and if it has the Taylor series expansion $f(z) = \sum_{n=0}^{\infty} a_n z^n$ about the origin, demonstrate (i) that the series $\sum_{n=0}^{\infty} |a_n|^2$ is convergent and (ii) that $\int_0^{2\pi} |f(e^{i\theta})|^2 d\theta = 2\pi \sum_{n=0}^{\infty} |a_n|^2$. (*Hint.* First show that $\sum_{n=0}^{N} |a_n|^2 \leq (2\pi)^{-1} \int_0^{2\pi} |f(e^{i\theta})|^2 d\theta$ by considering $\sum_{n=0}^{N} |a_n|^2 r^{2n}$ for r in $(0,1)$ and proving that $\int_0^{2\pi} |f(re^{i\theta})|^2 d\theta \to \int_0^{2\pi} |f(e^{i\theta})|^2 d\theta$ as $r \to 1^-$.)

5.72. Let $\Delta = \Delta(0,1)$ and let $g\colon \Delta \to \mathbb{C}$ be an analytic function satisfying $g(0) = 0$ and $|g'(0)| = 1$ whose derivative is a bounded function in Δ. Show that $|w| \geq (4m)^{-1}$ for every point w of $\mathbb{C} \sim g(\Delta)$, where $m = \sup\{|g'(z)| : z \in \Delta\}$; i.e., show that the range of g contains the disk $\Delta(0, (4m)^{-1})$. (*Hint.* Fix w belonging to $\mathbb{C} \sim g(\Delta)$. Then $w \neq 0$. The function h defined by $h(z) = 1 - w^{-1}g(z)$ is analytic and zero-free in Δ. Let f denote the branch of the square root of h in Δ for which $f(0) = 1$, and let $f(z) = \sum_{n=0}^{\infty} a_n z^n$ be the Taylor expansion of this function about the origin. Make use of the knowledge, a corollary of Exercise 5.70, that $2\pi(1 + |a_1|^2 r^2) \leq \int_0^{2\pi} |f(re^{i\theta})|^2 d\theta$ whenever $0 < r < 1$. Notice, too, that $|g(z)| \leq m$ for any z in Δ. Why?)

5.73. Under the assumptions that a function f is analytic in the open disk $\Delta = \Delta(0,1)$, that f' is bounded in Δ, and that $|f'(0)| = 1$, prove that the set $f(\Delta)$ must contain some open disk of radius $1/16$. (N.B. The disk in question need not have center $f(0)$. The result stated here is a watered-down version of a famous theorem of André Bloch (1893-1948). The argument we outline is ascribed to Edmund Landau (1877-1938). *Hint.* Define $\varphi\colon (0,1] \to \mathbb{R}$ by $\varphi(s) = \max\{|f'(z)| : |z| \leq 1 - s\}$ and set $r = \min\{s : s\varphi(s) = 1\}$. Since φ is a continuous function with $\varphi(1) = 1$ and since $s\varphi(s) \to 0$ as $s \to 0$ — φ is a bounded function — r is well-defined. Then $0 < r \leq 1$, $r\varphi(r) = 1$ by continuity, and $s\varphi(s) < 1$ when $0 < s < r$. The maximum principle allows us to choose a point z_0 with $|z_0| = 1 - r$ for which $|f'(z_0)| = \varphi(r) = r^{-1}$. Apply Exercise 5.72 in Δ to the function

296   VII. Sequences and Series of Analytic Functions

g defined in $\Delta(0,2)$ by $g(z) = 2\left[f(z_0 + 2^{-1}rz) - f(z_0)\right]$.)

5.74. Let $D = \Delta(0,r)$ and let f be a function that is continuous on $\overline{D}$ and analytic in D. Confirm the existence of a sequence $\langle p_n \rangle$ whose terms are polynomial functions of z such that $p_n \to f$ uniformly on $\overline{D}$. (N.B. This is a special case of a celebrated approximation theorem of S.N. Mergelyan: if K is a compact set in the complex plane and if $\mathbb{C} \sim K$ is a domain, then every function that is continuous on K and analytic in the interior of K is the uniform limit on K of some sequence of polynomials in z.)

5.75. Let $f(z) = z^{-2}\sin z$. Demonstrate that there is no sequence of polynomial functions of z that converges to f uniformly on the circle $K(0,1)$. Can the same statement be made about $f(z) = z^{-1}\sin z$? What about $f(z) = z^{-3}\sin z$?

5.76. Let D be a bounded domain in the complex plane such that $\partial\overline{D} = \partial D$. (This condition would not allow D to be a punctured disk, for example.) Assume that D is blessed with the following property: for every function f that is continuous on $\overline{D}$ and analytic in D there exists a sequence $\langle p_n \rangle$ of polynomials in z such that $p_n \to f$ uniformly on $\overline{D}$. Prove that D has to be simply connected. Find an example of a bounded simply connected domain D that satisfies $\partial\overline{D} = \partial D$, yet fails to have the approximation property just described.

5.77. For which complex numbers λ does the Laurent series $\sum_{n=-\infty}^{\infty} \lambda^{|n|} z^n$ have a non-empty ring of convergence D? For such λ identify the function represented by this series in D.

5.78. Determine the annulus of convergence of the Laurent series $\sum_{n=-\infty}^{\infty} n(-1)^n 2^{-|n|} z^n$, and compute its sum there.

5.79. Determine the Laurent series expansion of the function f in the specified ring D: (i) $f(z) = (z^2+1)^{-1}$, $D = \{z : 1 < |z - 2i| < 3\}$; (ii) $f(z) = z^{-4}(e^{z^2} - 1)$, $D = \Delta^*(0, \infty)$; (iii) $f(z) = (z-1)^{-2}\operatorname{Log} z$, $D = \Delta^*(1,1)$; (iv) $f(z) = (z^2 - 4z)^{-1}$, $D = \{z : 3 < |z - 3i| < 5\}$; (v) $f(z) = z(z-2)^{-4}\cos\pi z$, $D = \Delta^*(2, \infty)$; (vi) $f(z) = (z-1)\sin(z^{-1})$, $D = \Delta^*(0,\infty)$; (vii) $f(z) = z^6 \cos^2(z^{-2})$, $D = \Delta^*(0,\infty)$.

5.80. Obtain the Laurent expansions of the functions $f(z) = (z^2 - z)^{-1}$ and $g(z) = (1 - 2z)(z^2 - z)^{-2}$ in the annulus D: (i) $D = \Delta^*(0,1)$; (ii) $D = \Delta^*(1,1)$; (iii) $D = \{z : |z| > 1\}$; (iv) $D = \{z : |z-1| > 1\}$; (v) $D = \{z : 1 < |z+1| < 2\}$; (vi) $D = \{z : 1 < |z+i| < \sqrt{2}\}$.

5.81. Functions f and g are analytic in $D = \{z : a < |z - z_0| < b\}$, with $0 \leq a < b \leq \infty$, where they exhibit the Laurent expansions $f(z) = \sum_{n=-\infty}^{\infty} a_n (z - z_0)^n$ and $g(z) = \sum_{n=-\infty}^{\infty} b_n (z - z_0)^n$. Demonstrate that the Laurent expansion of the product $h = fg$ in D takes the form $h(z) = \sum_{n=-\infty}^{\infty} c_n (z - z_0)^n$, in which $c_n = \sum_{k=-\infty}^{\infty} a_k b_{n-k}$. Show, additionally, that the series which gives c_n is absolutely convergent. (*Hint.* Observe that

5. Exercises for Chapter VII

$h(z) = \sum_{n=-\infty}^{\infty} a_n(z-z_0)^n g(z)$ in D, the convergence being uniform on $K(z_0, r)$ for r in (a,b).)

5.82. Obtain the expansions

$$\exp(z + z^{-1}) = \sum_{n=-\infty}^{\infty} \left\{ \sum_{k=\max\{0,n\}}^{\infty} \frac{1}{k!(k-n)!} \right\} z^n$$

and

$$\exp(z - z^{-1}) = \sum_{n=-\infty}^{\infty} (-1)^n \left\{ \sum_{k=\max\{0,n\}}^{\infty} \frac{(-1)^k}{k!(k-n)!} \right\} z^n$$

for z satisfying $0 < |z| < \infty$.

5.83. A function f that is analytic and free of zeros in the domain $D = \mathbb{C} \sim \{0\}$ can be written in the form $f(z) = z^p g(z) h(z^{-1})$, where p is an integer and where both g and h are zero-free entire functions. Justify this statement. (*Hint.* Set $p = (2\pi i)^{-1} \int_{|z|=1} [f'(z)/f(z)] dz$; i.e., $p = n(\beta, 0)$, where $\beta(t) = f(e^{it})$ for $0 \le t \le 2\pi$. Show that in D there exists a branch of $\log k(z)$, where $k: D \to \mathbb{C}$ is the function given by $k(z) = z^{-p} f(z)$. For this use Theorem V.4.1.)

5.84. A function series of the type $\sum_{n=1}^{\infty} a_n n^{-z}$, where $\langle a_n \rangle$ is a sequence of complex constants, is called a *Dirichlet series*. (The series defining the Riemann zeta-function when $\operatorname{Re} z > 1$ is one such series.) Assuming that $\langle s_n \rangle$, the sequence of partial sums associated with $\langle a_n \rangle$, is bounded — say $|s_n| \le c$ for all n — show that $\sum_{n=1}^{\infty} a_n n^{-z}$ converges uniformly on the set $A_\theta = \{z: |\operatorname{Arg} z| < \theta, \operatorname{Re} z \ge (\pi/2) - \theta\}$ for any θ in the interval $(0, \pi/2)$ — hence, normally in the half-plane $D = \{z: \operatorname{Re} z > 0\}$. (*Hint.* Check that $\sum_{k=n}^{m} a_k k^{-z} = s_m m^{-z} - s_{n-1} n^{-z} + \sum_{k=n}^{m-1} s_k [k^{-z} - (k+1)^{-z}]$ when $m > n \ge 2$. Make use of this information, along with the identity $z \int_k^{k+1} t^{-z-1} dt = k^{-z} - (k+1)^{-z}$, to derive the estimate $|\sum_{k=n}^{m} a_k k^{-z}| \le 2c(1 + x^{-1}|z|)n^{-x}$ for $z = x + iy$ in D and for $m \ge n \ge 2$. Apply the Cauchy criterion in A_θ.)

5.85. Given that a Dirichlet series $\sum_{n=1}^{\infty} a_n n^{-z}$ converges at the point $z_0 = x_0 + iy_0$, certify that it converges normally in the half-plane $\{z: \operatorname{Re} z > x_0\}$ and that the convergence is absolute when $\operatorname{Re} z > 1 + x_0$. Infer the existence of an extended real number σ_0, known as the *abscissa of convergence* of $\sum_{n=1}^{\infty} a_n n^{-z}$, such that this series diverges for any z with $\operatorname{Re} z < \sigma_0$ and converges normally in the half-plane $D = \{z: \operatorname{Re} z > \sigma_0\}$, the convergence being absolute if $\operatorname{Re} z > 1 + \sigma_0$. Conclude that $f(z) = \sum_{n=1}^{\infty} a_n n^{-z}$ defines an analytic function in D, provided $\sigma_0 < \infty$. (*Hint.* Apply Exercise 5.84 to the series $\sum_{n=1}^{\infty} (a_n n^{-z_0}) n^{-\zeta}$ and then substitute $\zeta = z - z_0$.)

5.86. Verify that the formula $g(z) = \sum_{n=1}^{\infty} (-1)^{n-1} n^{-z}$ defines an analytic function in the half-plane $D = \{z: \operatorname{Re} z > 0\}$ and that $g(z) = (1 - 2^{1-z}) f(z)$

when $\text{Re}\, z > 1$, where f is the function in Example 3.1. Deduce from this fact that f admits an analytic extension to the set $\{z: \text{Re}\, z > 0, z \neq 1\}$.

5.4 Exercises for Section VII.4

5.87. Let $\langle f_n \rangle$ be a sequence from a normal subfamily $\mathcal{F}$ of $C(U)$. Under the assumption that $\langle f_n \rangle$ is not itself normally convergent in U, prove that $\langle f_n \rangle$ must have subsequences $\langle f_{n_k} \rangle$ and $\langle f_{m_k} \rangle$ which converge normally in U to different limit functions.

5.88. Let $\mathcal{F}$ be a normal subfamily of $C(U)$ each of whose members is analytic in U. Demonstrate that the family $\mathcal{F}' = \{f' : f \in \mathcal{F}\}$ is also normal in U.

5.89. Let $\mathcal{F}$ be a normal subfamily of $C(D)$, where D is a domain in the complex plane. Suppose that $\widetilde{\mathcal{F}}$ is a family of functions which are analytic in D and that $\widetilde{\mathcal{F}}$ has these two properties: (i) F' belongs to $\mathcal{F}$ whenever F is a member of $\widetilde{\mathcal{F}}$ and (ii) for some point z_0 of D the set $\{F(z_0) : F \in \widetilde{\mathcal{F}}\}$ is a bounded set of complex numbers. Prove that $\widetilde{\mathcal{F}}$ is a normal family in D.

5.90. Let $c > 0$ and $\lambda \geq 0$ be constants. Confirm that the family $\mathcal{F}$ of functions f which (i) are analytic in the disk $D = \Delta(0,1)$, (ii) satisfy $f(0) = 0$, and (iii) admit the derivative estimate $|f'(z)| \leq c(1 - |z|)^{-\lambda}$ for every z in D is a normal family in that disk.

5.91. Let $c > 0$ and let $\mathcal{F}$ be the family of all analytic functions f in the disk $D = \Delta(0,1)$ whose Taylor coefficients $a_n = f^{(n)}(0)/n!$ obey the condition $\sum_{n=0}^{\infty} |a_n|^2 \leq c$. Prove that $\mathcal{F}$ is a normal family in D. (*Hint.* Derive the estimate $|f(z)| \leq c^{1/2}(1-r^2)^{-1/2}$ for f in $\mathcal{F}$ and z in the closed disk $\overline{\Delta}(0,r)$, where $0 < r < 1$. Exercise 5.19 might be of some use.)

5.92. Let $c > 0$ be a constant. If $\mathcal{F}$ is the family of all analytic functions f in the disk $D = \Delta(0,1)$ for which $\iint_{\overline{\Delta}(0,r)} |f(z)|^2 dx dy \leq c$ when $0 < r < 1$, then $\mathcal{F}$ is a normal family in D. Prove this. (*Hint:* One solution begins by verifying that $\iint_{\overline{\Delta}(0,r)} |f(z)|^2 dx dy = \pi \sum_{n=0}^{\infty} (n+1)^{-1} |a_n|^2 r^{2n+2}$, where $f(z) = \sum_{n=0}^{\infty} a_n z^n$ is the Taylor expansion of f about the origin.)

5.93. Let D be a plane domain, and let $\mathcal{G}$ be a pointwise bounded family of analytic functions $g: D \to \mathbb{C}$. Assume the existence of a constant $c \geq 1$ about which the following is true: for each member g of $\mathcal{G}$ the set $\mathbb{C} \sim g(D)$ contains a pair of points a_g and b_g satisfying $|a_g| \leq c, |b_g| \leq c$, and $|a_g - b_g| \geq c^{-1}$. Use Theorem 4.6 to prove that $\mathcal{G}$ is a normal family in D. (*Hint.* Consider the family $\mathcal{F}$ whose member functions f have the form $f(z) = [g(z) - a_g]/[b_g - a_g]$, where g belongs to $\mathcal{G}$.)

5.94. Let $c \geq 1$ be a constant. With the help of Theorem 4.6 confirm

5. Exercises for Chapter VII

the following fact: corresponding to each pair of distinct complex numbers a and b there exists a radius $r > 0$ such that any analytic function $g: \Delta(0,r) \to \mathbb{C}$ for which $|g(0)| \leq c$ and $|g'(0)| \geq c^{-1}$ must have at least one of the points a or b in its range. (*Hint.* Suppose that no such r existed. Then for each positive integer n there would have to be an analytic function $g_n: \Delta(0,n) \to \mathbb{C}$ satisfying $|g_n(0)| \leq c$ and $|g'_n(0)| \geq c^{-1}$, but having its range in $\mathbb{C} \sim \{a,b\}$. Consider the sequence $\langle f_n \rangle$ in the disk $D = \Delta(0,1)$, where $f_n(z) = g_n(nz)$.)

5.95. From Exercise 5.94 deduce the following theorem of Picard: if f is a non-constant entire function, then $\mathbb{C} \sim f(\mathbb{C})$ contains at most one point. (*Hint.* Suppose that $\mathbb{C} \sim f(\mathbb{C})$ contains points a and b, $a \neq b$. Choose a point z_0 for which both $f(z_0) \neq 0$ and $f'(z_0) \neq 0$. Why must such a point exist? Look at the function $g(z) = f(z + z_0)$ to get a contradiction.)

5.96. Use Theorem 4.6 to generalize Schwarz's lemma as follows: if D is a plane domain with at least two boundary points, if $f: D \to D$ is an analytic function, and if f fixes a point z_0 of D, then $|f'(z_0)| \leq 1$. Show by example that the same needn't be true if $D = \mathbb{C}$ or if ∂D has only one point. (*Hint.* For the first part, assume that $|f'(z_0)| > 1$ and obtain a contradiction by examining the sequence $\langle f_n \rangle$, where $f_1 = f, f_2 = f \circ f, f_3 = f \circ f \circ f, \cdots$.)

Chapter VIII

Isolated Singularities of Analytic Functions

Introduction

The scene in the present chapter is dominated by the residue theorem and a number of its many applications. The residue theorem is a general form of Cauchy's theorem. It is concerned with integrals of the type $\int_\gamma f(z)dz$, in which γ is a closed path in a set where the function f is analytic, but in which this contour of integration is permitted to wind about isolated points where the integrand is either undefined or non-differentiable, so-called "isolated singularities" of f. One must not count on the integral of f along such a path being zero. Quite to the contrary, the integral typically picks up a contribution from each of the singularities in question. The residue theorem details the exact character of those contributions, as well as the manner in which they must be combined to evaluate the given integral. Our treatment of the residue theorem presupposes some understanding of the behavior of an analytic function in the vicinity of an isolated singularity, so it is this matter we address first. To make that important topic accessible, however, we must upgrade our knowledge of the local structure of analytic functions. This we do in the initial section of the chapter.

1 Zeros of Analytic Functions

1.1 The Factor Theorem for Analytic Functions

We are well aware that an analytic function f whose first derivative has value zero at each point of a domain D must be constant in the domain. The next theorem allows us to draw the same conclusion if it is known that at some particular point of D every derivative of f vanishes. In this

1. Zeros of Analytic Functions

instance extensive information about a function at one point is a fair trade for a small amount of information about it at every point.

Theorem 1.1. *If a function f is analytic in a domain D and if there exists a point ζ_0 of D with the property that $f^{(n)}(\zeta_0) = 0$ for every positive integer n, then f is constant in D.*

Proof. We consider the sets $U = \{z \in D : f^{(n)}(z) = 0 \text{ for every } n \geq 1\}$ and $V = D \sim U$. Quite obviously $D = U \cup V$ and $U \cap V = \phi$. By assumption the point ζ_0 lies in U, rendering this set non-empty. We shall prove that $U = D$. To accomplish this we need only verify that both U and V are open sets, for then we can invoke Theorem II.3.4.

We look first at a point z_0 of U. We choose an open disk Δ centered at z_0 that is contained in the domain D. From Theorem VII.3.4 and from the definition of the set U it follows that

$$f(z) = \sum_{n=0}^{\infty} \frac{f^{(n)}(z_0)}{n!}(z - z_0)^n = f(z_0)$$

for every z in Δ, so f is constant in that disk. This implies that $f^{(n)}(z) = 0$ for each $n \geq 1$ and for each z in Δ, which makes Δ a subset of U. The implication that the set U is open becomes clear.

Next, we focus on an element z_0 of V. By the definition of V we can find and fix a positive integer n for which $f^{(n)}(z_0) \neq 0$. Since $f^{(n)}$ is a continuous function in D, there exists an open disk $\Delta = \Delta(z_0, r)$ lying in D with the property that $f^{(n)}(z) \neq 0$ for every z belonging to Δ. This places the disk Δ inside the set V. It, too, is thus seen to be an open set.

Theorem II.3.4 informs us that $V = \phi$ or, equivalently, that $U = D$. In particular, $f'(z) = 0$ for every z in D, a fact already known to entail the constancy of f in D. ∎

An important consequence of the result just established is a generalization for analytic functions of the "Factor Theorem" for polynomial functions. It asserts, albeit in more precise terms, that if z_0 is a root of an analytic function f, then f is the product of the linear function ℓ, $\ell(z) = z - z_0$, and another analytic function. In other words, $z - z_0$ is a factor of $f(z)$ in the class of analytic functions.

Theorem 1.2. *Suppose that a function f is analytic and non-constant in a domain D and that z_0 is a point of D for which $f(z_0) = 0$. Then f can be uniquely represented in D in the fashion*

$$f(z) = (z - z_0)^m g(z) ,$$

where m is a positive integer and $g : D \to \mathbb{C}$ is an analytic function that obeys the condition $g(z_0) \neq 0$.

Proof. Since f is non-constant in D, Theorem 1.1 implies that there is at least one positive integer n for which $f^{(n)}(z_0) \neq 0$. Let m be the smallest such integer. We select and fix an open disk $\Delta = \Delta(z_0, r)$ that lies in D. In Δ we can represent f by its Taylor series expansion about the point z_0: $f(z) = \sum_{n=0}^{\infty} a_n(z-z_0)^n$, where $a_n = f^{(n)}(z_0)/n!$. Owing to the definition of m and to the assumption that $a_0 = f(z_0) = 0$, we note that $a_n = 0$ when $0 \leq n \leq m-1$, whereas $a_m \neq 0$. Consequently, we are entitled to write

$$f(z) = \sum_{n=m}^{\infty} a_n(z-z_0)^n = (z-z_0)^m \sum_{n=m}^{\infty} a_n(z-z_0)^{n-m}$$

for z belonging to Δ. Now the function $g: D \to \mathbb{C}$ defined by

$$g(z) = \begin{cases} \dfrac{f(z)}{(z-z_0)^m} & \text{if } z \neq z_0, \\ a_m & \text{if } z = z_0, \end{cases}$$

is plainly analytic in $D \sim \{z_0\}$. Moreover, inside Δ it admits the Taylor series development $g(z) = \sum_{n=m}^{\infty} a_n(z-z_0)^{n-m} = a_m + a_{m+1}(z-z_0) + \cdots$, which according to Theorem VII.3.3 marks g as analytic in Δ. In particular, g is differentiable at z_0. Therefore, g is an analytic function, $g(z_0) = a_m \neq 0$, and $f(z) = (z-z_0)^m g(z)$ holds throughout D.

Could f enjoy a second representation in D of the type described — say $f(z) = (z-z_0)^\ell h(z)$, where ℓ is a positive integer and $h: D \to \mathbb{C}$ is an analytic function with $h(z_0) \neq 0$? Assume so. For any z in D we have $(z-z_0)^m g(z) = (z-z_0)^\ell h(z)$. Then $m \leq \ell$, for the inequality $m > \ell$ would lead to

$$0 = \lim_{z \to z_0} (z-z_0)^{m-\ell} g(z) = \lim_{z \to z_0} h(z) = h(z_0) \neq 0,$$

a contradiction. Similarly, $\ell > m$ can be ruled out. It must then be the case that $\ell = m$. This, in turn, implies that $g(z) = h(z)$ for any z belonging to $D \sim \{z_0\}$. By continuity, $g(z_0) = h(z_0)$ as well. We conclude that $m = \ell$ and $g = h$ — hence, that the alleged alternate representation of f is nonexistent. The uniqueness assertion of the theorem follows. ∎

As a direct corollary of Theorem 1.2, we record:

Corollary 1.3. *Suppose that a function f is analytic and non-constant in a domain D and that z_0 is a point of D. Then f can be uniquely represented in D in the fashion*

$$f(z) = f(z_0) + (z-z_0)^m g(z),$$

where m is a positive integer and $g: D \to \mathbb{C}$ is an analytic function that obeys the condition $g(z_0) \neq 0$.

Proof. Apply Theorem 1.2 to the function f_1 defined in D by $f_1(z) = f(z) - f(z_0)$. ∎

1. Zeros of Analytic Functions

1.2 Multiplicity

Let f be a function that is analytic and non-constant in some open disk Δ centered at the point z_0. Corollary 1.3 states that f admits a unique representation in Δ of the form

$$f(z) = w_0 + (z - z_0)^m g(z) ,$$

where $w_0 = f(z_0)$, m is a positive integer, and g is an analytic function in Δ satisfying $g(z_0) \neq 0$. The integer m — which, incidentally, does not depend in any way on the choice of disk Δ — is called the *multiplicity* (or, synonymously, the *order*) *of f at z_0*. To be a little more precise, we say that *f takes the value w_0 with multiplicity* (or *order*) *m at z_0*. For instance, when $w_0 = 0$ we speak of f having a zero of multiplicity (or order) m at z_0. (The expressions "simple zero," "double zero," "triple zero," etc. are optional terms of reference to zeros of order one, two, three, etc.) The proof of Theorem 1.2 shows that the multiplicity of f at z_0 is the smallest positive integer m for which $f^{(m)}(z_0) \neq 0$, although, as the third in the following series of examples points out, evaluating derivatives is not always the most efficient method to ascertain multiplicity.

EXAMPLE 1.1. Determine the multiplicity of the zero of $f(z) = \cos z$ at the point $z_0 = \pi/2$.
Since $f'(\pi/2) = -\sin(\pi/2) = -1 \neq 0$, we see that the given function has a simple zero at $\pi/2$.

EXAMPLE 1.2. Find the order of the zero of $f(z) = e^z - z - 1$ at the origin.
Again we simply compute: $f'(0) = 0$, $f''(0) = 1$. It follows that $f(z) = e^z - z - 1$ has a double zero at the origin.

EXAMPLE 1.3. Determine the multiplicity with which $f(z) = \cos(z^3)$ assumes its value at the origin.
In this case we use to advantage the Taylor expansion of $\cos z$ about the origin by writing

$$\cos(z^3) = \sum_{n=0}^{\infty} \frac{(-1)^n (z^3)^{2n}}{(2n)!} = 1 - \frac{z^6}{2!} + \frac{z^{12}}{4!} - \cdots$$

$$= 1 + z^6 \left(-\frac{1}{2} + \frac{z^6}{4!} - \cdots \right) = 1 + z^6 g(z) ,$$

where

$$g(z) = \sum_{n=1}^{\infty} \frac{(-1)^n z^{6n-6}}{(2n)!} .$$

The function g is obviously an entire function that has $g(0) = -1/2 \neq 0$. We conclude that $f(z) = \cos(z^3)$ takes the value $w_0 = 1$ with multiplicity $m = 6$ at the origin. Observe that in the present example it would have involved considerably more work to arrive at the multiplicity by computing derivatives.

As an application of Theorem 1.2 and the concept of multiplicity we establish an extension to the complex setting of one of L'Hospital's rules for evaluating limits.

Theorem 1.4. (L'Hospital's Rule) *Let f and g be functions that are analytic and non-constant in a disk $\Delta = \Delta(z_0, r)$. Assume that each of these functions has a zero at the point z_0. Then*

$$(8.1) \qquad \lim_{z \to z_0} \frac{f(z)}{g(z)} = \lim_{z \to z_0} \frac{f'(z)}{g'(z)},$$

understood to mean that either both limits exist and are the same, or else neither limit exists.

Proof. Let m and ℓ be the respective orders of the zeros of f and g at z_0. In the disk Δ we can write $f(z) = (z - z_0)^m f_1(z)$ and $g(z) = (z - z_0)^\ell g_1(z)$, where f_1 and g_1 are analytic functions in Δ, neither of which has a zero at z_0. Furthermore, the proof of Theorem 1.2 gives $f_1(z_0) = f^{(m)}(z_0)/m!$ and $g_1(z_0) = g^{(\ell)}(z_0)/\ell!$. We infer that

$$\frac{f(z)}{g(z)} = (z - z_0)^{m-\ell} \frac{f_1(z)}{g_1(z)}$$

at any z in the punctured disk $\Delta^* = \Delta^*(z_0, r)$ for which $g_1(z) \neq 0$. Since g_1 is continuous at z_0 and $g_1(z_0) \neq 0$, this includes all z in some smaller punctured disk centered at z_0. Due to the fact that

$$\lim_{z \to z_0} \frac{f_1(z)}{g_1(z)} = \frac{f_1(z_0)}{g_1(z_0)} = \frac{\ell! \, f^{(m)}(z_0)}{m! \, g^{(\ell)}(z_0)},$$

it then becomes apparent that

$$(8.2) \qquad \lim_{z \to z_0} \frac{f(z)}{g(z)} = \begin{cases} 0 & \text{if } m > \ell, \\ f^{(m)}(z_0)/g^{(m)}(z_0) & \text{if } m = \ell, \\ \text{fails to exist} & \text{if } m < \ell. \end{cases}$$

What about the limit of $f'(z)/g'(z)$ as $z \to z_0$? Assume initially that both $m > 1$ and $\ell > 1$. Because the first $m - 1$ derivatives of f vanish at z_0 and $f^{(m)}(z_0) \neq 0$, we see that f' and its first $m - 2$ derivatives are zero at z_0, whereas $(f')^{(m-1)}(z_0) \neq 0$ — i.e., we see that f' has a zero of order $m - 1$ at z_0. Similarly, g' has a zero of order $\ell - 1$ at z_0. The reasoning used

1. Zeros of Analytic Functions

to arrive at (8.2) then applies equally well to the pair of functions f' and g', yielding

$$(8.3) \quad \lim_{z \to z_0} \frac{f'(z)}{g'(z)} = \begin{cases} 0 & \text{if } m-1 > \ell-1, \\ (f')^{(m-1)}(z_0)/(g')^{(m-1)}(z_0) & \text{if } m-1 = \ell-1, \\ \text{fails to exist} & \text{if } m-1 < \ell-1, \end{cases}$$

at least when $m > 1$ and $\ell > 1$. Next, suppose that $\ell = 1$. Then $g'(z) \to g'(z_0) \neq 0$ as $z \to z_0$, so $f'(z)/g'(z) \to f'(z_0)/g'(z_0)$. Since $m \geq 1$ and since $f'(z_0) = 0$ if $m > 1$, (8.3) continues to hold when $\ell = 1$. Finally, in the case $m = 1$ we have $f'(z) \to f'(z_0) \neq 0$ as $z \to z_0$, which implies that $f'(z)/g'(z) \to f'(z_0)/g'(z_0)$ if $g'(z_0) \neq 0$ — this happens only for $\ell = 1$ — and that $\lim_{z \to z_0}[f'(z)/g'(z)]$ fails to exist if $g'(z_0) = 0$, i.e., if $\ell > 1$. Again when $m = 1$, (8.3) is valid. Thus, (8.3) summarizes the behavior of $f'(z)/g'(z)$ as $z \to z_0$ for all admissible values of m and ℓ. Because the right-hand side of (8.3) is just a paraphrase of the right-hand side of (8.2), formula (8.1), with the stated interpretation, follows. ∎

The proof of Theorem 1.4 actually demonstrates that

$$\lim_{z \to z_0} \frac{f(z)}{g(z)} = \lim_{z \to z_0} \frac{f'(z)}{g'(z)} = \cdots = \lim_{z \to z_0} \frac{f^{(m)}(z)}{g^{(m)}(z)} = \frac{f^{(m)}(z_0)}{g^{(m)}(z_0)} \neq 0$$

when $m = \ell$, and that

$$\lim_{z \to z_0} \frac{f(z)}{g(z)} = 0$$

when $m > \ell$. This implies, naturally, that

$$\lim_{z \to z_0} \frac{g(z)}{f(z)} = 0$$

when $m < \ell$. Accordingly, $|f(z)|/|g(z)| \to \infty$ as $z \to z_0$ in case $m < \ell$.

EXAMPLE 1.4. Evaluate $\lim_{z \to 0} z^{-2}(e^z - z - 1)$.

We apply Theorem 1.4 twice in doing the calculation

$$\lim_{z \to 0} \frac{e^z - z - 1}{z^2} = \lim_{z \to 0} \frac{e^z - 1}{2z} = \lim_{z \to 0} \frac{e^z}{2} = \frac{1}{2}.$$

Since both $f(z) = e^z - z - 1$ and $g(z) = z^2$ have double zeros at the origin, we were assured of a finite limit from the outset.

EXAMPLE 1.5. Evaluate $\lim_{z \to 0}(z^3 \cos z - z^3 - 2z^4)/(e^{z^4} - 1)$.

We once again appeal to Theorem 1.4 and compute

$$\lim_{z \to 0} \frac{z^3 \cos z - z^3 - 2z^4}{e^{z^4} - 1} = \lim_{z \to 0} \frac{3z^2 \cos z - z^3 \sin z - 3z^2 - 8z^3}{4z^3 e^{z^4}}$$

$$= \lim_{z \to 0} \frac{3\cos z - z\sin z - 3 - 8z}{4z} \lim_{z \to 0} e^{-z^4}$$

$$= \lim_{z \to 0} \frac{-3\sin z - \sin z - z\cos z - 8}{4} = -2 .$$

Notice that we did not mechanically continue to take derivatives here until the limit popped out, as l'Hospital's rule would really permit us to do. Instead, we combined the use of l'Hospital's rule with some simple algebraic manipulation in order to shorten the calculation.

1.3 Discrete Sets, Discrete Mappings

Let U be an open set in the complex plane. A subset E of U is termed a *discrete subset of U* if E has no limit point that belongs to U. (Recall: for z_0 to be a limit point of E means that $E \cap \Delta^*(z_0, r) \neq \phi$ for every $r > 0$ or, equivalently, that there is a sequence $\langle z_n \rangle$ in $E \sim \{z_0\}$ such that $z_n \to z_0$.) Examples of discrete subsets of U include the empty set ϕ, any finite subset of U (a finite set has no limit points, period), and a set of the type $E = \{z_n : n = 1, 2, 3, \cdots\}$, where $\langle z_n \rangle$ is a sequence in U without any accumulation points in U. We emphasize that a discrete subset of U is allowed to have — indeed, is likely to have — limit points in $\mathbb{C} \sim U$.

We record some general observations about a discrete subset E of a plane open set U. First of all, E must be a set of isolated points: if z_0 belongs to E, then there exists an $r > 0$ for which $E \cap \Delta(z_0, r) = \{z_0\}$. (Otherwise, z_0 would itself be a limit point of E in U.) Secondly, the set $U \sim E$ is open and, if U is domain, then $U \sim E$ is also a domain (Exercise 5.4). Finally, if K is an arbitrary compact set in U, then $K \cap E$ has at most finitely many elements. (If $K \cap E$ were an infinite set, then we could construct a sequence $\langle z_n \rangle$ in $K \cap E$ with the property that $z_n \neq z_m$ for $n \neq m$. This sequence would have an accumulation point in K — hence, in U. Any such accumulation point would clearly be a limit point of E in U, contrary to the definition of a discrete subset of U.)

Consider again an open set U in $\mathbb{C}$, and let f be a complex-valued function whose domain-set includes U. We speak of f as a *discrete mapping of U* if for each fixed complex number w the set $E_w = \{z \in U : f(z) = w\}$ is a discrete subset of U. (This does, of course, allow for the possibility of E_w being empty.) In particular, for each fixed w the set of solutions z in U to the equation $f(z) = w$ — assuming any such exist — is a set of isolated points of U.

The reason for introducing the above terminology is supplied by the following important theorem.

Theorem 1.5. (Discrete Mapping Theorem) *If a function f is analytic and non-constant in a plane domain D, then f is a discrete mapping of D.*

Proof. Fix w in $\mathbb{C}$ and set $E = E_w = \{z \in D : f(z) = w\}$. We must prove that E is a discrete subset of D. For this, assume that z_0 is a limit point of E. We are required to demonstrate that z_0 lies in $\mathbb{C} \sim D$. We shall suppose the contrary to be true (i.e., that z_0 belongs to D) and derive a contradiction. Let $\langle z_n \rangle$ be a sequence in $E \sim \{z_0\}$ such that $z_n \to z_0$. The continuity of f at z_0 insures that

$$f(z_0) = \lim_{n \to \infty} f(z_n) = \lim_{n \to \infty} w = w$$

and thus places z_0 in the set E. Invoking Corollary 1.3, we express $f(z)$ for z in D as

$$f(z) = w + (z - z_0)^m g(z) ,$$

where m is a positive integer and where $g : D \to \mathbb{C}$ is an analytic function for which $g(z_0) \neq 0$. Putting to use the fact that g, too, is continuous at z_0, we choose a disk $\Delta = \Delta(z_0, r)$ in D such that $g(z) \neq 0$ holds for every z in Δ. This implies that $f(z) \neq w$ whenever z is a point of the punctured disk $\Delta^*(z_0, r)$. The result: $E \cap \Delta^*(z_0, r) = \phi$, in contradiction with the definition of z_0 as a limit point of E. The contradiction is traceable directly to the assumption that z_0 lies in D. Therefore, any and all limit points of E must belong to $\mathbb{C} \sim D$. ∎

We take special note of the following corollary of Theorem 1.5: *the set of zeros of a function in a domain D where that function is analytic and non-constant, if non-empty, consists entirely of isolated points.* Two other corollaries of the discrete mapping theorem are frequently cited. The first of these is known as the "Principle of Analytic Continuation." It comes into play in conjunction with the problem of extending (or "continuing") a function known to be analytic in some set to a function that is analytic in a larger set. (For more on this subject see Chapter X.)

Corollary 1.6. (Principle of Analytic Continuation) *If functions f and g are analytic in a domain D and if $f(z) = g(z)$ for all z belonging to some subset A of D that has a limit point in D, then $f(z) = g(z)$ for every z in D.*

Proof. The function $h = g - f$ is analytic in D, and its set of zeros in that domain is not a discrete subset of D, for h has a zero at each point of A, a set that by hypothesis has a limit point in D. In view of Theorem 1.5, h must vanish identically in D or, what is the same, $f(z) = g(z)$ must hold everywhere in D. ∎

The second corollary alluded to above states that the collection of functions which are analytic in a domain D is free of so-called "zero divisors." This term refers to a non-zero element in a multiplicative algebraic system whose product with some other non-zero element is zero. Such elements exist, for instance, in the system of 2×2 real matrices: if $A = \begin{bmatrix} 1 & 0 \\ 0 & 0 \end{bmatrix}$ and

$B = \begin{bmatrix} 0 & 0 \\ 0 & 1 \end{bmatrix}$, then $AB = \begin{bmatrix} 0 & 0 \\ 0 & 0 \end{bmatrix}$, the zero in that system, even though both A and B are non-zero.

Corollary 1.7. *If functions f and g are analytic in a domain D and if $f(z)g(z) = 0$ for every z in D, then either $f(z) = 0$ for every z in D or $g(z) = 0$ for every z in D.*

Proof. Suppose that there is a point z_0 of D for which $f(z_0) \neq 0$. We prove that in this situation $g(z) = 0$ throughout the domain. By the continuity of f we can choose an open disk $\Delta = \Delta(z_0, r)$ contained in D such that the condition $f(z) \neq 0$ persists for all z in Δ. But then $g(z) = 0$ is true whenever z lies in Δ, showing that the zeros of g in D are not isolated points. By Theorem 1.5, g must be constant in D and, thus, must vanish identically there. ∎

In a third application of the discrete mapping theorem, we tie up a loose end that we left hanging in Section III.4.1. There it was asserted that a branch of the inverse of an analytic function is automatically analytic. We now have the tools to prove this.

Theorem 1.8. *Suppose that U is an open set in the complex plane, that $f: U \to \mathbb{C}$ is an analytic function, and that g is a branch of f^{-1} in a domain D. Then g is an analytic function.*

Proof. By definition, $g: D \to U$ is nothing but a continuous function endowed with the property that $f[g(z)] = z$ for every z in D. We must show that g is differentiable at every point of D. Let G denote the component of U that contains the connected set $g(D)$. (Recall Theorems II.3.8 and II.3.7.) If $w_1 = g(z_1)$ and $w_2 = g(z_2)$, where z_1 and z_2 are distinct points of D, then $f(w_1) = z_1 \neq z_2 = f(w_2)$, which shows that that f is not constant in G. As a consequence, f' does not vanish identically there. It follows from the discrete mapping theorem, applied to f', that $E = \{w \in G : f'(w) = 0\}$ is a discrete subset of G. Now consider an arbitrary point z_0 of D and write $w_0 = g(z_0)$. If w_0 is not an element of E, then the differentiability of g at z_0 is assured by Theorem III.4.1. If the theorem we are now trying to establish were already known to be true, the chain rule would imply that $f'(w_0)g'(z_0) = 1$, so it could not really happen that w_0 is a point of E! In the present proof, however, we do not have the benefit of such hindsight. Here we must treat the possibility of w_0 being in E as genuine — and somehow get around it. Assuming then that w_0 does belong to E, we profit by the discreteness of that set to choose $r > 0$ for which $E \cap \Delta(w_0, r) = \{w_0\}$. Next, we use the continuity of g at z_0 to select an open disk $\Delta = \Delta(z_0, s)$ in D such that $g(\Delta)$ is a subset of $\Delta(w_0, r)$. Because the function g is univalent and $g(z_0) = w_0$, the image $g(z)$ of any point z in the punctured disk $\Delta^* = \Delta^*(z_0, s)$ lies in $\Delta^*(w_0, r)$ — hence, is not a point of E. As above, Theorem III.4.1 certifies that g is differentiable at every point of

Δ^*; i.e., g is analytic in Δ^*. Since g is continuous in Δ, we can appeal to Theorem V.3.4 and conclude that g is differentiable at z_0 as well. The differentiability of g throughout D is thus demonstrated. ∎

2 Isolated Singularities

2.1 Definition and Classification of Isolated Singularities

We say that a function f has an *isolated singularity* at a point z_0 of the complex plane provided there exists an $r > 0$ with the property that f is analytic in the punctured disk $\Delta^*(z_0, r)$, yet not analytic in the full open disk $\Delta(z_0, r)$. Given that f is analytic in $\Delta^*(z_0, r)$, this situation can come about for one of two reasons: either z_0 does not belong to the domain-set of f from the start or, alternatively, z_0 is a member of that domain-set, but is a point at which f is discontinuous. (Recall: if f is analytic in $\Delta^*(z_0, r)$ and continuous in $\Delta(z_0, r)$, then by Theorem V.3.4 it is actually analytic in $\Delta(z_0, r)$.) The first case is exemplified by the function $f(z) = z^{-1}$ at $z_0 = 0$; at $z_0 = 1$ the function defined by $f(z) = z$ for $z \neq 1$ and $f(1) = 2$ fits the second description. We call a function f *analytic modulo isolated singularities in an open set* U under the following conditions: there is a discrete subset E of U, the *singular set of f in U*, with the feature that f is analytic in the open set $U \sim E$, but has a singularity, in the sense just defined, at each point of E. This usage is not intended to bar the possibility that the singular set E is empty, which would simply mean that f is analytic in U.

Assume now that f is a function with an isolated singularity at z_0. Let $\Delta^* = \Delta^*(z_0, r)$ be a punctured disk in which f is analytic. It follows from Theorem VII.3.6 that f can be represented in Δ^* as the sum of a uniquely determined Laurent series centered at z_0: $f(z) = \sum_{n=-\infty}^{\infty} a_n (z - z_0)^n$. The singularity of f at z_0 falls into one of three categories, depending on the character of this Laurent expansion. We declare f to have a *removable singularity* at z_0 if $a_n = 0$ for every negative index n; to have a *pole* at z_0 if $a_n \neq 0$ holds for at least one, but for at most finitely many negative values of n; to have an *essential singularity* at z_0 if $a_n \neq 0$ is true for an infinite number of negative integers n. Since the given Laurent series $\sum_{n=-\infty}^{\infty} a_n(z - z_0)^n$ obviously has inner radius of convergence $\rho_I = 0$, we can appeal to Theorem VII.3.5 and affirm that $S(z) = \sum_{n=1}^{\infty} a_{-n}(z - z_0)^{-n}$ defines an analytic function with domain-set $\mathbb{C} \sim \{z_0\}$. The function S is called the *singular part* (or *principal part*) *of f at z_0*. The difference $f - S$ also has an isolated singularity at z_0, and its Laurent series in Δ^* is easily written down, being nothing more than the Taylor series $\sum_{n=0}^{\infty} a_n(z - z_0)^n$, the non-singular part of f at z_0. As a consequence, the function $f - S$ has a removable singularity at z_0. The coefficient a_{-1} that appears in the singular function S has special significance, as we shall soon learn. This number is known as the *residue of f at z_0*. The notation $\text{Res}(z_0, f)$ (or, as an option, $\text{Res}[z_0, f(z)]$) represents this quantity.

We proceed next to subject each kind of isolated singularity to closer scrutiny.

2.2 Removable Singularities

Consider a function f that is analytic in a punctured disk $\Delta^* = \Delta^*(z_0, r)$ and has an isolated singularity at z_0. Suppose initially that the singularity is removable. According to our definition this means that the Laurent series expansion of f in Δ^* reduces to the form $f(z) = \sum_{n=0}^{\infty} a_n(z - z_0)^n = a_0 + a_1(z - z_0) + \cdots$. If we define — or redefine, as the case may be — the value of f at z_0 by setting $f(z_0) = a_0$, we arrive at a function that, as the sum of a Taylor series in the disk $\Delta = \Delta(z_0, r)$, is certainly analytic there. In particular, we obtain a function that is differentiable at z_0. Conversely, if we make no presupposition about the nature of its singularity at z_0, but if we are somehow able to assign to the function f a value at z_0 that causes the resulting function to become differentiable at that point, then the extended function becomes analytic in Δ and so can be expanded there in a Taylor series centered at z_0. The restriction of this Taylor series to Δ^* furnishes the Laurent series development of the original function f in Δ^*, from which fact we infer that f has a removable singularity at z_0. Conclusion: *an isolated singularity of a function f at a point z_0 is removable if and only if $f(z_0)$ can be defined — or redefined — so as to render f differentiable* at z_0. Take, for example, the function $f(z) = z^{-1} \sin z$. It is analytic in its domain-set $D = \{z : 0 < |z| < \infty\}$ — hence, exhibits an isolated singularity at $z_0 = 0$. For z in D we have the expansion

$$\frac{\sin z}{z} = \sum_{n=0}^{\infty} \frac{(-1)^n z^{2n}}{(2n+1)!} = 1 - \frac{z^2}{3!} + \frac{z^4}{5!} - \cdots,$$

which shows that the singularity of f at the origin is removable. By defining $f(0) = 1$ we effectively "remove" the singularity and, in the process, create out of f an entire function.

It is possible to characterize removable singularities without reference to Laurent series. One theorem that does so is known as "Riemann's Extension Theorem." This result has already been discussed in Chapter VI (Theorem VI.3.6), but no harm is done by presenting a second version of it here.

Theorem 2.1. (Riemann Extension Theorem) *Let a function f have an isolated singularity at a point z_0. The singularity is removable if and only if f is bounded in some punctured disk centered at z_0.*

2. Isolated Singularities

Proof. Suppose at first that f is bounded in the punctured disk $\Delta^* = \Delta^*(z_0, r)$, say $|f(z)| \leq m$ for every z in Δ^*. By taking r sufficiently small we may, of course, assume that f is analytic in Δ^*. The coefficient a_n in the Laurent series expansion of f in Δ^* is given by

$$a_n = \frac{1}{2\pi i} \int_{|z-z_0|=s} \frac{f(z)\, dz}{(z-z_0)^{n+1}}$$

for any s satisfying $0 < s < r$. Standard estimation procedures lead to

$$|a_n| \leq \frac{1}{2\pi} \int_{|z-z_0|=s} \frac{|f(z)|\,|dz|}{|z-z_0|^{n+1}} \leq \frac{m}{s^n}.$$

When $n < 0$, we can let $s \to 0$ in this inequality and conclude that $|a_n| = 0$. Therefore, $a_n = 0$ for every negative value of n, which is exactly what the definition requires in order for the singularity of f at z_0 to be classified as removable.

As to the converse, assume that the singularity of f at z_0 is given to be removable. We are free to suppose that the value $f(z_0)$ has been defined — or adjusted — to make f differentiable at z_0. Naturally, $|f(z_0)| = \lim_{z \to z_0} |f(z)|$ must then be true. We can thus choose $r > 0$ in a way to insure, for instance, that $|f(z)| < |f(z_0)| + 1$ holds for every z in the punctured disk $\Delta^* = \Delta^*(z_0, r)$. This shows that f is bounded in Δ^*. ∎

A corollary of the Riemann extension theorem (or, rather, a corollary of this theorem and the deliberations that preceded it) is the following:

Theorem 2.2. *Let a function f have an isolated singularity at a point z_0. The singularity is removable if and only if $\lim_{z \to z_0} |f(z)|$ exists.*

Needless to say, the condition that $\lim_{z \to z_0} f(z)$ exists could be substituted for the existence of $\lim_{z \to z_0} |f(z)|$ in Theorem 2.2 to produce another characterization of a removable singularity. A key point in Theorems 2.1 and 2.2, however, is that the nature of the singularity is already discernible in the behavior of the magnitude of f, a real quantity, as z approaches z_0.

2.3 Poles

We turn next to a function f that has a pole at a point z_0. Assuming that f is analytic in the punctured disk $\Delta^* = \Delta^*(z_0, r)$, we expand it in a Laurent series there: $f(z) = \sum_{n=-\infty}^{\infty} a_n (z - z_0)^n$. The definition of a pole demands the existence of a positive integer m such that $a_{-m} \neq 0$, while $a_{-n} = 0$ whenever $n > m$. In other words, the expansion takes the form

$$(8.4) \qquad f(z) = \frac{a_{-m}}{(z-z_0)^m} + \cdots + \frac{a_{-1}}{z-z_0} + \sum_{n=0}^{\infty} a_n (z-z_0)^n,$$

with $a_{-m} \neq 0$. Under these conditions we say that f has a *pole of order m* at z_0. (Paralleling the situation with zeros, we often speak of a pole of order one as a *simple pole*, a pole of order two as a *double pole*, etc.) If we multiply both sides of (8.4) by $(z-z_0)^m$, we obtain for all z in Δ^*

$$(z-z_0)^m f(z) = a_{-m} + a_{-m+1}(z-z_0) + \cdots = \sum_{n=0}^{\infty} a_{n-m}(z-z_0)^n \ .$$

The last series is a Taylor series that converges at every point of the disk $\Delta = \Delta(z_0, r)$. It follows that the function g defined by the formula $g(z) = \sum_{n=0}^{\infty} a_{n-m}(z-z_0)^n$ is analytic in Δ, that $g(z_0) = a_{-m} \neq 0$, and that $f(z) = (z-z_0)^{-m} g(z)$ for every z in Δ^*. This establishes one direction in the following characterization of a pole of order m. The other direction is equally straightforward.

Theorem 2.3. *Let m be a positive integer. A function f that is analytic in a punctured disk $\Delta^* = \Delta^*(z_0, r)$ has a pole of order m at z_0 if and only if f can be represented in Δ^* in the fashion*

$$(8.5) \qquad f(z) = \frac{g(z)}{(z-z_0)^m} ,$$

where g is a function that is analytic in the disk $\Delta(z_0, r)$ and obeys the condition $g(z_0) \neq 0$.

The residue at z_0 of a function f matching the description in (8.5) is just the coefficient of $(z-z_0)^{m-1}$ in the Taylor expansion of g about z_0. That coefficient is given by $g^{(m-1)}(z_0)/(m-1)!$. Since f itself is undefined at z_0, we prefer to express this as $\lim_{z \to z_0} g^{(m-1)}(z)/(m-1)!$. The result of this observation is a frequently applied residue formula: *under the assumption that f has a pole of order m at z_0,*

$$(8.6) \qquad \text{Res}(z_0, f) = \frac{1}{(m-1)!} \lim_{z \to z_0} \frac{d^{m-1}}{dz^{m-1}} [(z-z_0)^m f(z)] \ .$$

Formula (8.6) is especially valuable in the case of a single or double pole, for which the task of evaluating the expression on the right-hand side usually remains quite manageable.

An operation that often causes poles to appear is forming quotients of analytic functions. To be specific, let g and h be functions that are analytic in an open disk $\Delta = \Delta(z_0, r)$, neither of them being identically zero there, and let $h(z_0) = 0$. We consider their quotient, $f = g/h$. We assume, as we certainly may by taking r suitably small, that both g and h are free of zeros in the punctured disk $\Delta^* = \Delta^*(z_0, r)$. This insures that the function f is analytic in Δ^*. Since f remains undefined at z_0, it has an isolated singularity at that point. In analyzing the character of this singularity we denote by m and ℓ the orders of the zeros that g and h, respectively, have

2. Isolated Singularities

at z_0. (N.B. We have deliberately not made the assumption that $g(z_0) = 0$. If $g(z_0) \neq 0$, we agree to set $m = 0$ and $g_1 = g$ in what follows.) In Δ we can write $g(z) = (z - z_0)^m g_1(z)$ and $h(z) = (z - z_0)^\ell h_1(z)$, where g_1 and h_1 are zero-free analytic functions defined in that disk. This leads to a representation $f(z) = (z - z_0)^{m-\ell} f_1(z)$ for the function f in Δ^*. Here $f_1 = g_1/h_1$ is both analytic and free of zeros in Δ. With the aid of Theorems 2.2 and 2.3 we conclude that the quotient f has a removable singularity at z_0 when $m \geq \ell$ and a pole of order $\ell - m$ at z_0 when $m < \ell$. Furthermore, when $m > \ell$ the process of removing this singularity endows f with a zero of order $m - \ell$ at z_0. One special instance of the preceding discussion deserves to be highlighted: *if an analytic function f has a zero of order m at z_0, then $1/f$ has a pole of order m at z_0.* Theorem 2.3 implies that something tantamount to a converse is likewise true: *if a function f has a pole of order m at z_0, then $1/f$ has a zero of order m there, in the sense that $1/f$ has a removable singularity at z_0 and that, upon its removal, $1/f$ acquires a zero of order m at that point.*

A look at some concrete examples will shed further light on the subject of poles.

EXAMPLE 2.1. Describe the character of the singularity that $f(z) = (e^z - 1)^{-1}$ has at the origin, and calculate $\operatorname{Res}(0, f)$.

The entire function $g(z) = e^z - 1$ has a simple zero at the origin, so $f = 1/g$ has a simple pole there. We use (8.6) for the case $m = 1$, along with L'Hospital's rule, in computing

$$\operatorname{Res}(0, f) = \lim_{z \to 0} z\, f(z) = \lim_{z \to 0} \frac{z}{e^z - 1} = \lim_{z \to 0} \frac{1}{e^z} = 1 \,.$$

EXAMPLE 2.2. Determine the nature of the singularity of $f(z) = (z - 1)^{-3} \sin \pi z$ at $z_0 = 1$. Compute $\operatorname{Res}(1, f)$.

Since $g(z) = \sin \pi z$ has a simple zero at z_0, while $h(z) = (z - 1)^3$ has a zero of order three there, $f = g/h$ has a pole of order two at z_0. Referring to Example VII.3.13, we write down the Laurent expansion of f in $D = \{z : 0 < |z - 1| < \infty\}$:

$$\frac{\sin \pi z}{(z-1)^3} = \sum_{n=0}^{\infty} \frac{(-1)^{n+1} \pi^{2n+1} (z-1)^{2n-2}}{(2n+1)!}$$

$$= -\frac{\pi}{(z-1)^2} + \frac{\pi^3}{3!} - \frac{\pi^5 (z-1)^2}{5!} + \cdots \,.$$

From this representation we read off directly that $\operatorname{Res}(1, f) = 0$. Alternatively, we can apply (8.6) with $m = 2$ to arrive at the same conclusion:

$$\operatorname{Res}(1, f) = \lim_{z \to 1} \frac{d}{dz} \left(\frac{\sin \pi z}{z - 1} \right) = \lim_{z \to 1} \frac{\pi (z-1) \cos \pi z - \sin \pi z}{(z-1)^2}$$

$$= \lim_{z \to 1} \frac{(-\pi^2)(z-1)\sin \pi z}{2(z-1)} = -\frac{\pi^2}{2} \lim_{z \to 1} \sin \pi z = 0 \ .$$

L'Hospital's rule justifies the third step in this computation.

EXAMPLE 2.3. Classify the singularity that $f(z) = [\sin z][\cos(z^3) - 1]^{-1}$ exhibits at the origin. Compute $\text{Res}(0, f)$.

The function $g(z) = \sin z$ shows a simple zero at the origin, and $h(z) = \cos(z^3) - 1$ has a zero of multiplicity six there. The latter assertion is clear from the Taylor expansion

$$\cos(z^3) - 1 = -\frac{z^6}{2!} + \frac{z^{12}}{4!} - \cdots = z^6 \left(-\frac{1}{2} + \frac{z^6}{4!} - \cdots \right) \ .$$

As a result, $f = g/h$ is seen to have a pole of order five at the origin.

We want now to evaluate $\text{Res}(0, f)$. We could certainly find this residue by applying formula (8.6) with $m = 5$. The amount of time and effort required to carry out the calculation, however, would be, if not prohibitive, at least discouraging. Fortunately, a more efficient method is available here. It involves doing some simple algebraic manipulation of power series in order to display explicitly the singular part of f at the origin, from which the desired residue is instantly retrievable. In the ensuing discussion — and in others later — the symbol $O[(z - z_0)^N]$ will serve as an abbreviation for "an expression containing only terms of degree N and above." More precisely, $O[(z - z_0)^N]$ will indicate a well-determined, but unspecified Taylor series centered at z_0 whose radius of convergence is positive and in which the coefficient of $(z - z_0)^n$ is zero for $n < N$. The notation $O(1) = O[(z - z_0)^0]$ will signify a "generic" Taylor series centered at z_0 with a positive radius of convergence. Just to demonstrate the use of this notation, we might elect to write

(8.7) $$\sin z = z + O(z^3) \ .$$

The right-hand side is read "z plus a Taylor series that contains only terms of order three and above." In this instance $O(z^3)$ stands for the series $-(z^3/3!) + (z^5/5!) - \cdots$. We operate algebraically with expressions of type (8.7) in the manner suggested by the following illustration:

$$(z+1)\sin z = (z+1)[z + O(z^3)] = z^2 + z\,O(z^3) + z + O(z^3)$$

$$= z + z^2 + O(z^3) \ .$$

We remark that $O(z^3)$ represents a different power series in the final expression than it did in the penultimate one. It is this flexibility that makes the notation so ideal for our present purposes.

Returning to the example at hand, we express $f(z)$ in the form

2. Isolated Singularities

$$f(z) = \frac{z - \frac{z^3}{3!} + \frac{z^5}{5!} - \cdots}{-\frac{z^6}{2!} + \frac{z^{12}}{4!} - \frac{z^{18}}{6!} + \cdots} = \frac{1}{z^5}\left(\frac{1 - \frac{z^2}{3!} + \frac{z^4}{5!} - \cdots}{-\frac{1}{2} + \frac{z^6}{4!} - \frac{z^{12}}{6!} + \cdots}\right).$$

The idea is to find a_0, a_1, a_2, a_3, and a_4 so that

$$\frac{1 - \frac{z^2}{3!} + \frac{z^4}{5!} - \cdots}{-\frac{1}{2} + \frac{z^6}{4!} - \frac{z^{12}}{6!} + \cdots} = a_0 + a_1 z + a_2 z^2 + a_3 z^3 + a_4 z^4 + O(z^5),$$

for then dividing by z^5 will expose the singular part of f at the origin. In the computations that follow we systematically disregard terms of degree five or more, terms that do not influence the singular part of f, by absorbing them into the $O(z^5)$ term. Thus, we write

$$1 - \frac{z^2}{3!} + \frac{z^4}{5!} - \cdots = 1 - \frac{z^2}{6} + \frac{z^4}{120} + O(z^5)$$

and

$$-\frac{1}{2} + \frac{z^6}{4!} - \frac{z^{12}}{6!} + \cdots = -\frac{1}{2} + O(z^5) = -\frac{1}{2}\left[1 - O(z^5)\right].$$

Now $|O(z^5)| < 1$ when z is near the origin, so for z of small magnitude we can use the geometric series to obtain

$$\left(-\frac{1}{2} + \frac{z^6}{4!} - \frac{z^{12}}{6!}\right)^{-1} = -2\left[1 - O(z^5)\right]^{-1}$$

$$= -2\left\{1 + O(z^5) + \left[O(z^5)\right]^2 + \cdots\right\} = -2\left[1 + O(z^5)\right] = -2 + O(z^5).$$

This leads to

$$f(z) = \frac{1}{z^5}\left[1 - \frac{z^2}{6} + \frac{z^4}{120} + O(z^5)\right]\left[-2 + O(z^5)\right]$$

$$= \frac{1}{z^5}\left[-2 + \frac{z^2}{3} - \frac{z^4}{60} + O(z^5)\right]$$

$$= -\frac{2}{z^5} + \frac{1}{3z^3} - \frac{1}{60z} + O(1).$$

The singular part S of f at the origin is immediately recognized to be

$$S(z) = -\frac{2}{z^5} + \frac{1}{3z^3} - \frac{1}{60z},$$

implying that $\operatorname{Res}(0, f) = -1/60$.

EXAMPLE 2.4. Find the singular part of $f(z) = (e^z - 1)^{-3}$ at the origin, and use it to determine $\operatorname{Res}(0, f)$.

The function we are dealing with in this example has a pole of order 3 at the origin. To exhibit its singular part there we employ the same method as in the preceding example. We first calculate

$$(e^z - 1)^3 = \left(z + \frac{z^2}{2!} + \frac{z^3}{3!} + \frac{z^4}{4!} + \cdots \right)^3$$

$$= z^3 \left[1 + \frac{z}{2} + \frac{z^2}{6} + O(z^3) \right]^3$$

$$= z^3 \left[1 + \frac{z}{2} + \frac{z^2}{6} + O(z^3) \right]^2 \left[1 + \frac{z}{2} + \frac{z^2}{6} + O(z^3) \right]$$

$$= z^3 \left[1 + z + \frac{7z^2}{12} + O(z^3) \right] \left[1 + \frac{z}{2} + \frac{z^2}{6} + O(z^3) \right]$$

$$= z^3 \left[1 + \frac{3z}{2} + \frac{5z^2}{4} + O(z^3) \right] .$$

An appeal to the geometric series gives

$$\left[1 + \frac{3z}{2} + \frac{5z^2}{4} + O(z^3) \right]^{-1}$$

$$= 1 - \left[\frac{3z}{2} + \frac{5z^2}{4} + O(z^3) \right] + \left[\frac{3z}{2} + \frac{5z^2}{4} + O(z^3) \right]^2 + O(z^3)$$

$$= 1 - \frac{3z}{2} - \frac{5z^2}{4} + \frac{9z^2}{4} + O(z^3)$$

$$= 1 - \frac{3z}{2} + z^2 + O(z^3) .$$

Consequently, we obtain the expansion

$$f(z) = \frac{1}{(e^z - 1)^3} = \frac{1}{z^3} - \frac{3}{2z^2} + \frac{1}{z} + O(1) .$$

The requested singular part is now seen to be

$$S(z) = \frac{1}{z^3} - \frac{3}{2z^2} + \frac{1}{z} ,$$

2. Isolated Singularities

which reveals that $\text{Res}(0, f) = 1$.

Theorem 2.2 has an analogue characterizing poles.

Theorem 2.4. *Let a function f have an isolated singularity at a point z_0. The singularity is a pole if and only if $\lim_{z \to z_0} |f(z)| = \infty$. Moreover, the singularity is a pole of order m if and only if m is the unique positive exponent for which $\lim_{z \to z_0} |z - z_0|^m |f(z)|$ is a positive real number.*

Proof. Assume first that f has a pole at z_0, say of order m. In some punctured disk $\Delta^* = \Delta^*(z_0, r)$ we can write $f(z) = (z - z_0)^{-m} g(z)$, where g is a function that is analytic in $\Delta = \Delta(z_0, r)$ and meets the condition $g(z_0) \neq 0$. It follows that

$$\lim_{z \to z_0} |f(z)| = \lim_{z \to z_0} \frac{|g(z)|}{|z - z_0|^m} = \infty$$

and that, for any positive real number ℓ,

$$(8.8) \qquad \lim_{z \to z_0} |z - z_0|^\ell |f(z)| = \begin{cases} \infty & \text{if } \ell < m, \\ |g(z_0)| & \text{if } \ell = m, \\ 0 & \text{if } \ell > m. \end{cases}$$

Only when $\ell = m$ is the limit in (8.8) a positive real number.

Turning to the converses, let it be assumed that $|f(z)| \to \infty$ as $z \to z_0$. We can then fix a punctured disk $\Delta^* = \Delta^*(z_0, r)$ such that f is analytic in Δ^* and satisfies $|f(z)| \geq 1$ throughout Δ^*. From this it is evident that $h = 1/f$ has an isolated singularity at z_0. Since $|h(z)| \leq 1$ for every z in Δ^*, the Riemann extension theorem informs us that the singularity of h at z_0 is removable. We remove it by defining $h(z_0) = \lim_{z \to z_0} h(z) = 0$. As h is obviously not identically zero in Δ^*, it acquires through this extension process a zero of some positive order at z_0, implying that $f = 1/h$ has a pole of that same order there.

Finally, if it is known that for some $m > 0$ the quantity $|z - z_0|^m |f(z)|$ tends to a positive real limit as $z \to z_0$, then certainly

$$\lim_{z \to z_0} |f(z)| = \lim_{z \to z_0} \frac{|z - z_0|^m |f(z)|}{|z - z_0|^m} = \infty.$$

We infer that f has a pole at z_0. A glance at (8.8) discloses that this pole is necessarily of order m. ∎

We introduce the expression "a function f has no worse than a pole at a point z_0" as a catch-all phrase to describe the following state of affairs: there exists a radius $r > 0$ such that the given function f is either analytic in the full open disk $\Delta = \Delta(z_0, r)$ or else analytic in the punctured disk $\Delta^* = \Delta^*(z_0, r)$ with a pole or removable singularity at its center. Except when f vanishes identically near z_0, it is always possible by choosing the radius r appropriately small to arrange that such a function be represented in Δ^* as $f(z) = (z-z_0)^m f_1(z)$, where m is an integer and f_1 is a function that is both analytic and zero-free in Δ. Partly by taking advantage of this sort of representation, partly by direct examination of Laurent expansions, one deduces the contents of the next theorem. The detailed proof is left as an exercise (Exercise 5.30).

Theorem 2.5. *If neither of two functions f and g has worse than a pole at a point z_0, then none of the functions f', $f+g$, fg, and, unless g vanishes identically in some punctured disk centered at z_0, f/g has worse than a pole at z_0.*

2.4 Meromorphic Functions

Let U be an open set in the complex plane. We characterize a function f as *meromorphic in U* provided f has at no point of U worse than a pole. Put differently, this definition requires f to be analytic modulo isolated singularities in U and insists that any non-removable singularities of f in that set be poles. It does, however, permit f to have removable singularities in U. Many authors prefer to define a function as meromorphic in U if it is analytic in U except for possible poles, with no heed paid to removable singularities. Of course, under the latter definition it is not technically true that the sum of functions meromorphic in U is automatically meromorphic there, for the sum may well have removable singularities — e.g., if $f(z) = z^{-1}$ and $g(z) = z - z^{-1}$, then f and g have poles at the origin, but $f+g$ has a removable singularity there. Whatever definition of a meromorphic function one opts to use, it is necessary sooner or later to deal with the niggling annoyances caused by removable singularities. We handle this problem by adopting the following "removable singularity policy" in conjunction with meromorphic functions: *whenever it is established that a function f is meromorphic (according to our definition of the term) in an open set U, it will always be tacitly assumed that f undergoes immediate modification to rid it of any removable singularities in U, this by defining or redefining the function at each such singular point so as to make it continuous there.*

Any function that is analytic in an open set U certainly qualifies as meromorphic in U. The most obvious example of a function that is meromorphic in the entire plane $\mathbb{C}$ is a rational function of z, for the only singularities such a function has in $\mathbb{C}$ are removable singularities or poles at the zeros of its denominator. The function $f(z) = \tan z$, which has a simple pole at every odd multiple of $\pi/2$ but is otherwise free of singularities in $\mathbb{C}$, is another function that is meromorphic in the whole plane. The function defined by $f(z) = (z^2+1)^{-2} e^{1/z}$ is meromorphic in $U = \mathbb{C} \sim \{0\}$, where its only singularities are double poles at the points i and $-i$. It is

2. Isolated Singularities

not, however, meromorphic in $\mathbb{C}$, for its singularity at the origin turns out to be an essential singularity.

Suppose that functions f and g are both meromorphic in an open set U. Theorem 2.5 implies directly that the functions f', $f+g$, fg, and, unless g is identically zero in some component of U, f/g are likewise meromorphic in this open set. Notice especially that the quotient f/g, where f and g are analytic functions in U and where g does not vanish identically in any component of U, is meromorphic in U. It is true (but not so easy to prove) that every function which is meromorphic in U admits a representation as the quotient of two functions which are analytic in U. We shall confirm this fact in Chapter X.

2.5 Essential Singularities

The last — and by far most interesting — type of isolated singularity that a function f can exhibit at a point z_0 is an essential singularity, one for which the singular part of f at z_0 contains infinitely many non-zero terms. The function $f(z) = e^{1/z} = 1 + z^{-1} + (z^{-2}/2!) + \cdots$ at $z_0 = 0$ provides a simple example of this kind of singularity. Dealing with essential singularities is, in general, a much more delicate matter than working with removable singularities or poles. There is, for instance, no foolproof formula along the lines of (8.6) for computing the residue of a function at an essential singularity. Frequently one is forced to extract such information directly from a Laurent expansion. Here are two examples.

EXAMPLE 2.5. Determine the character of the singularity of $f(z) = z\cos[(z-1)^{-1}]$ at the point $z_0 = 1$, and calculate $\text{Res}(1, f)$.

We do the problem by deriving the Laurent expansion of f in the set $D = \mathbb{C} \sim \{1\}$. To start, we substitute $(z-1)^{-1}$ for z in the Taylor expansion of $\cos z$ about the origin and obtain

$$\cos\left(\frac{1}{z-1}\right) = \sum_{n=0}^{\infty} \frac{(-1)^n}{(2n)!(z-1)^{2n}} = 1 - \frac{1}{2!(z-1)^2} + \frac{1}{4!(z-1)^4} - \cdots .$$

We are thus led to

$$f(z) = z\cos\left(\frac{1}{z-1}\right) = (z-1)\cos\left(\frac{1}{z-1}\right) + \cos\left(\frac{1}{z-1}\right)$$

$$= \sum_{n=0}^{\infty} \frac{(-1)^n}{(2n)!(z-1)^{2n-1}} + \sum_{n=0}^{\infty} \frac{(-1)^n}{(2n)!(z-1)^{2n}}$$

$$= (z-1) + 1 - \frac{1}{2!(z-1)} - \frac{1}{2!(z-1)^2} + \frac{1}{4!(z-1)^3} + \frac{1}{4!(z-1)^4} - \cdots ,$$

which makes it evident that f has an essential singularity at 1 and that $\operatorname{Res}(1,f) = -1/2$.

EXAMPLE 2.6. Compute the Laurent coefficients of $f(z) = \exp(z + z^{-1})$ in $D = \{z : 0 < |z| < \infty\}$. Use the information to classify the singularity of f at the origin and to determine $\operatorname{Res}(0,f)$.

The Laurent coefficient a_n of f in D is given by

(8.9) $$a_n = \frac{1}{2\pi i} \int_{|z|=1} z^{-n-1} f(z)\, dz .$$

Fixing n, we write

$$z^{-n-1} f(z) = z^{-n-1} e^{1/z} e^z = z^{-n-1} e^{1/z} \sum_{k=0}^{\infty} \frac{z^k}{k!} = \sum_{k=0}^{\infty} \frac{z^{k-n-1} e^{1/z}}{k!} .$$

As the convergence of the last series is uniform on the circle $K(0,1)$, we are justified in integrating it term by term to get

(8.10) $$\int_{|z|=1} z^{-n-1} f(z)\, dz = \sum_{k=0}^{\infty} \frac{1}{k!} \int_{|z|=1} z^{k-n-1} e^{1/z}\, dz .$$

Since the Laurent expansion of $g(z) = e^{1/z}$ in D is already known to us — namely,

$$e^{1/z} = \sum_{k=0}^{\infty} \frac{1}{k! z^k}$$

— we can appeal to the analogue of (8.9) for the function g and deduce that

$$\int_{|z|=1} z^{k-n-1} e^{1/z}\, dz = \begin{cases} \dfrac{2\pi i}{(k-n)!} & \text{if } k - n \geq 0 , \\ 0 & \text{if } k - n < 0 . \end{cases}$$

Coupled with (8.9) and (8.10) this shows that

$$a_n = \sum_{k=\max\{0,n\}}^{\infty} \frac{1}{k!(k-n)!} .$$

Noting that $a_n > 0$ for all negative n, we infer that the singularity of f at the origin is an essential singularity. Taking $n = -1$ yields

$$\operatorname{Res}(0,f) = \sum_{k=0}^{\infty} \frac{1}{k!(k+1)!} .$$

Although the representation of a_n — and, in particular, of $\operatorname{Res}(1,f)$ — as the sum of an infinite series may not seem altogether satisfactory, it is

2. Isolated Singularities

about the best we can hope for in the present example. This phenomenon is not untypical of residue calculations involving essential singularities.

In view of Theorems 2.2 and 2.4 we arrive by a process of elimination at a characterization of isolated essential singularities.

Theorem 2.6. *Let a function f have an isolated singularity at a point z_0. The singularity is essential if and only if $\lim_{z \to z_0} |f(z)|$ fails to exist either in the strict sense or as an infinite limit.*

Theorem 2.6 does not even begin to tell the tale of how truly ill-behaved a function can be in the vicinity of an isolated essential singularity. A better sense of that behavior is conveyed by a result of Felice Casorati (1835-1890) and Weierstrass.

Theorem 2.7. (Casorati-Weierstrass Theorem) *If a function f is analytic in a punctured disk $\Delta^* = \Delta^*(z_0, r)$ and has an essential singularity at its center, then $f(\Delta^*)$ is dense in the complex plane — i.e., the set $\mathbb{C} \sim f(\Delta^*)$ has no interior points.*

Proof. We argue by contradiction: we assume that $\mathbb{C} \sim f(\Delta^*)$ does have interior points and deduce that f can have no worse than a pole at z_0, contrary to hypothesis. Let w_0 be an interior point of $\mathbb{C} \sim f(\Delta^*)$, and let $s > 0$ be such that the disk $\Delta(w_0, s)$ is contained in $\mathbb{C} \sim f(\Delta^*)$. Then $|f(z) - w_0| \geq s$ plainly holds for every z in Δ^*. It follows that the function $g: \Delta^* \to \mathbb{C}$ defined by $g(z) = [f(z) - w_0]^{-1}$ is analytic and satisfies $|g(z)| \leq s^{-1}$ throughout Δ^*. Riemann's extension theorem tells us that the singularity of g at z_0 is removable. Moreover, since g is clearly zero-free in Δ^*, its reciprocal $1/g$ also has an isolated singularity at z_0, a singularity that must be either a pole or a removable singularity, depending on whether $\lim_{z \to z_0} |g(z)|$ is zero or not. This, in turn, insures that the singularity of $f = w_0 + (1/g)$ at z_0 can be no worse than a pole, the contradiction anticipated. Accordingly, the set $\mathbb{C} \sim f(\Delta^*)$ must be devoid of interior points. ∎

The Casorati-Weierstrass theorem states that the image of a punctured disk $\Delta^*(z_0, r)$, no matter how small, under a function f with an isolated essential singularity at z_0 effectively fills up the whole complex plane. Among its consequences is this: corresponding to any given complex number w_0 there exists a sequence $\langle z_n \rangle$ in $\mathbb{C} \sim \{z_0\}$ such that $z_n \to z_0$ and $f(z_n) \to w_0$. (Apply the Casorati-Weierstrass theorem to choose for $n = 1, 2, 3, \cdots$ a point z_n in $\Delta^*(z_0, 1/n)$ with $f(z_n)$ in $\Delta(w_0, 1/n)$.) An even more remarkable fact was announced in 1879 by Émile Picard:

Theorem 2.8. (Picard's Theorem) *If a function f is analytic in a punctured disk $\Delta^* = \Delta^*(z_0, r)$ and has an essential singularity at its center, then the set $\mathbb{C} \sim f(\Delta^*)$ contains at most one point.*

We shall not prove this deep and beautiful theorem, for the methods required to do so go beyond the scope of this book. (One standard proof is based on Theorem VII.4.6. Exercise 5.35 outlines the derivation of Picard's theorem from that result of Montel.) Picard's theorem implies the following fact concerning a function f that has an essential singularity at z_0: with allowance for one possible exceptional value of w_0, there exists for any complex number w_0 a sequence $\langle z_n \rangle$ in $\mathbb{C} \sim \{z_0\}$ such that $z_n \to z_0$ and such that $f(z_n) = w_0$ for every n. The function $f(z) = e^{1/z}$, which has an essential singularity at the origin and which maps each punctured disk centered there to $\mathbb{C} \sim \{0\}$, demonstrates that the "exceptional value" permitted by Picard's theorem may, in fact, exist. The existence of such a value is by no means, however, a foregone conclusion. For instance, the singularity of $f(z) = \sin(1/z)$ at the origin is also essential and, in this case, $f(\Delta^*) = \mathbb{C}$ for every punctured disk $\Delta^* = \Delta^*(0, r)$.

2.6 Isolated Singularities at Infinity

Having treated in some detail the types of isolated singularities a function might experience in the complex plane, we now briefly discuss the character of a function that is analytic in the complement of an arbitrarily large disk centered at the origin. To streamline the discussion we introduce the notation $\Delta^*(\infty, r)$ to represent the set $\{z : |z| > r^{-1}\}$ for $r > 0$. (N.B. This definition is designed to make sure that $\Delta^*(\infty, r)$ will be contained in $\Delta(\infty, s)$ when $r < s$. For further insight into the notation we refer the reader to Section 4 of this chapter.) We speak of a function that is analytic in $\Delta^*(\infty, r)$ for some $r > 0$ as having an *isolated singularity at* ∞. The example which immediately leaps to mind is that of an entire function, which is analytic in $\Delta^*(\infty, r)$ for every $r > 0$. The function $f(z) = \tan z$, by contrast, does not have an isolated singularity at ∞, because for each $r > 0$ the set $\Delta^*(\infty, r)$ includes poles of f.

Assuming that a function f is analytic in $\Delta^*(\infty, r)$, we can expand it there in a Laurent series centered at the origin, $f(z) = \sum_{n=-\infty}^{\infty} a_n z^n$. The singularity of f at ∞ is subject to classification based on this expansion: f has a *removable singularity at* ∞ if $a_n = 0$ for every positive integer n; f has a *pole at* ∞ if there exists a positive integer m — the *order* of the pole — such that $a_m \neq 0$, whereas $a_n = 0$ for every n greater than m; f has an *essential singularity at* ∞ if $a_n \neq 0$ holds for infinitely many positive integers n. The entire function S defined by $S(z) = \sum_{n=1}^{\infty} a_n z^n$ is called the *singular* (or *principal*) *part of* f at ∞. Just as was true for singularities in the complex plane, the function $f - S$ has a removable singularity at ∞. For technical reasons that we shall not go into here the *residue of* f *at* ∞ is taken to be $-a_{-1}$ — not a_{-1} itself or some quantity associated with the singular part of f at ∞. In the way of simple illustrations we remark that $f(z) = z^{-1}$ has a removable singularity at ∞, that a polynomial function

$f(z) = a_0 + a_1 z + \cdots + a_m z^m$ of positive degree m has a pole of order m at ∞, and that any non-polynomial entire function ($f(z) = e^z$, to name one) has an essential singularity at ∞.

It is evident from the definition of the concept that a function f has an isolated singularity at ∞ if and only if the function g given by $g(z) = f(z^{-1})$ has an isolated singularity at the origin, in which event the type of singularity f exhibits at ∞ is precisely the same as the type of singularity g displays at the origin. For example, $f(z) = (z^4 + 2z^2 + 1)(z^2 - z - 3)^{-1}$ has a double pole at ∞ by reason of the fact that the associated function $g(z) = f(z^{-1}) = z^{-2}(1 + 2z^2 + z^4)(1 - z^3 - 3z^4)^{-1}$ quite obviously shows a double pole at 0. Through this mechanism it is possible to convert virtually any question about an isolated singularity at ∞ to one concerned with an isolated singularity at the origin, where results established earlier in this chapter can be brought to bear on it. To see this principle in action, consider a function f that is known to be analytic and bounded in $\Delta^*(\infty, r)$. The Riemann extension theorem for a singularity at a finite point would suggest that the singularity of f at ∞ ought to be removable. We actually prove this by passing to the function $g(z) = f(z^{-1})$, which is clearly analytic and bounded in $\Delta^*(0, r)$ — hence, which has a removable singularity at the origin. It follows that the singularity of f at ∞ is indeed removable, as expected. Similar reasoning establishes the Casorati-Weierstrass theorem for essential singularities at ∞: if a function f is analytic in $\Delta^* = \Delta^*(\infty, r)$ and has an essential singularity at ∞, then $f(\Delta^*)$ is dense in $\mathbb{C}$. Finally, we remark that the analogue of Theorem 2.5 in which the finite point z_0 gets replaced by ∞ remains a valid theorem.

3 The Residue Theorem and its Consequences

3.1 The Residue Theorem

We have laid sufficient groundwork for the proof of — as well as for many applications of — the result that is the centerpiece of this chapter.

Theorem 3.1. (Residue Theorem) *Suppose that a function f is analytic modulo isolated singularities in an open set U, that $E \neq \phi$ is the singular set of f in U, and that σ is a cycle in $U \sim E$ which is homologous to zero in U (Figure 1). Then*

(8.11) $$\int_\sigma f(z)\, dz = 2\pi i \sum_{z \in E} n(\sigma, z) \operatorname{Res}(z, f) \ .$$

Proof. We first demonstrate that, even though the set E may have infinitely many elements, the condition $n(\sigma, z) \neq 0$ can hold for at most a

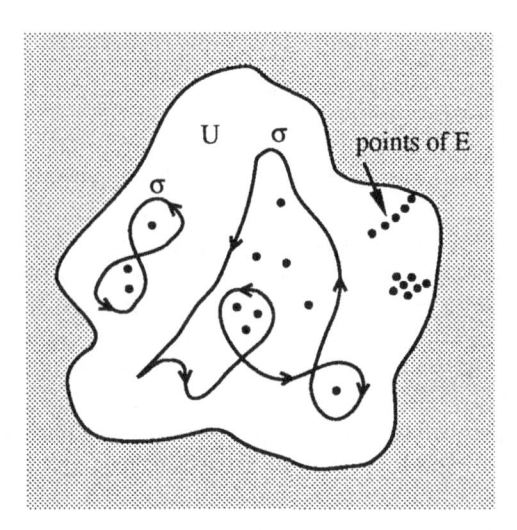

Figure 1.

finite number of those elements, with the result that the sum on the right-hand side of (8.11) reduces to a finite sum. Suppose, to the contrary, that $n(\sigma, z) \neq 0$ happens to be true for infinitely many points z belonging to E. This assumption enables us to choose a sequence $\langle z_k \rangle$ of distinct points of E such that $n(\sigma, z_k) \neq 0$ holds for every k. None of the points z_k can then lie in the unbounded component of $\mathbb{C} \sim |\sigma|$, a fact which implies that $\langle z_k \rangle$ is a bounded sequence. Therefore $\langle z_k \rangle$ must have at least one accumulation point in the complex plane. Choose such a point and label it z_0. The point z_0 is, among other things, a limit point of E. Because E, a discrete subset of U, has no limit points in U, z_0 has to be be located in $\mathbb{C} \sim U$. As σ is homologous to zero in U, we infer that $n(\sigma, z_0) = 0$. Now fix a disk $\Delta = \Delta(z_0, r)$ that does not intersect the set $|\sigma|$. Then $n(\sigma, z)$ remains constant as z varies over Δ; i.e., $n(\sigma, z) = n(\sigma, z_0) = 0$ whenever z is a point of Δ (Lemma V.2.1). On the other hand, since the sequence $\langle z_k \rangle$ accumulates at z_0, we can pick an N for which z_N belongs to Δ and, by definition, $n(\sigma, z_N) \neq 0$. This contradiction leaves us no choice but to conclude that there can exist only finitely many elements z of E for which $n(\sigma, z) \neq 0$. Let $\zeta_1, \zeta_2, \ldots, \zeta_p$ be an enumeration of the points of E enjoying this property, and let V be the open set obtained by starting with U and removing from it every member of E other than $\zeta_1, \zeta_2, \ldots, \zeta_p$. Thus σ is a cycle in $V \sim \{\zeta_1, \zeta_2, \ldots, \zeta_p\}$ and, since $\mathbb{C} \sim V = (\mathbb{C} \sim U) \cup \{z \in E : z \neq \zeta_1, \zeta_2, \ldots, \zeta_p\}$, it is clear that $n(\sigma, z) = 0$ for every z in $\mathbb{C} \sim V$. In other words, σ is homologous to zero in V. Incidentally, it is not out of the question that $n(\sigma, z) = 0$ could hold from the outset for every z in E. In that situation (8.11) would convey no information not already available to us, for the right-hand side of (8.11) would then be zero, the set V would be $U \sim E$, and (8.11) would just re-

3. The Residue Theorem and its Consequences

state the conclusions of Cauchy's theorem, as applied to f and σ in V. We shall proceed, therefore, under the premise that points $\zeta_1, \zeta_2, \ldots, \zeta_p$ really do exist.

Let S_k designate the singular part of f at the point ζ_k. The function S_k is analytic in $\mathbb{C} \sim \{\zeta_k\}$, and $f - S_k$ has a removable singularity at ζ_k. It follows that the function $g = f - S_1 - S_2 \cdots - S_p$ is analytic in V except for removable singularities at $\zeta_1, \zeta_2, \ldots, \zeta_p$. We take the liberty of removing these singularities, so are free to assume that g is analytic in V. Appealing to Cauchy's theorem, we can thus assert that

$$0 = \int_\sigma g(z)\,dz = \int_\sigma f(z)\,dz - \sum_{k=1}^p \int_\sigma S_k(z)\,dz$$

or, equivalently, that

$$(8.12) \qquad \int_\sigma f(z)\,dz = \sum_{k=1}^p \int_\sigma S_k(z)\,dz \ .$$

If $S(z) = \sum_{n=1}^\infty a_{-n}(z - \zeta_0)^{-n}$ is the singular part of f at an arbitrary point ζ_0 of E, then the series defining S converges normally in $\mathbb{C} \sim \{\zeta_0\}$. Most importantly, it converges uniformly on $|\sigma|$, which justifies the interchange of integration and summation in the computation

$$(8.13) \qquad \begin{aligned} \int_\sigma S(z)\,dz &= \int_\sigma \left(\sum_{n=1}^\infty \frac{a_{-n}}{(z-\zeta_0)^n} \right) dz = \sum_{n=1}^\infty a_{-n} \int_\sigma \frac{dz}{(z-\zeta_0)^n} \\ &= a_{-1} \int_\sigma \frac{dz}{z - \zeta_0} = 2\pi i\, n(\sigma, \zeta_0) \operatorname{Res}(\zeta_0, f) \ . \end{aligned}$$

(N.B. $\int_\sigma (z - \zeta_0)^{-n}\,dz = 0$ when $n > 1$, for in that case the integrand is blessed with a primitive in $\mathbb{C} \sim \{\zeta_0\}$.) In tandem (8.12) and (8.13) lead to

$$\int_\sigma f(z)\,dz = 2\pi i \sum_{k=1}^p n(\sigma, \zeta_k)\operatorname{Res}(\zeta_k, f) = 2\pi i \sum_{z \in E} n(\sigma, z)\operatorname{Res}(z, f) \ ,$$

as stated in (8.11). ∎

One special case of the residue theorem is invoked sufficiently often that it warrants a separate statement.

Corollary 3.2. *Suppose that a function f is analytic modulo isolated singularities in an open set U, that E is the singular set of f in U, and that γ is a Jordan contour in $U \sim E$ with the property that the inside D of the Jordan curve $|\gamma|$ is contained in U. If $D \cap E$ is non-empty, then*

$$\int_\gamma f(z)\,dz = 2\pi i \sum_{k=1}^p \operatorname{Res}(z_k, f) \ ,$$

where $z_1, z_2, \ldots, z_p$ lists the elements of E that belong to D.

326   VIII. Isolated Singularities of Analytic Functions

Corollary 3.2 takes for granted the validity of the Jordan curve theorem. That the points of E in D (i.e., those z in E for which $n(\gamma, z) = 1$) are finite in number would follow here directly from the discreteness of E and compactness of $\overline{D}$. Corollary 3.2 can be strengthened slightly if one assumes knowledge of Goursat's theorem (Theorem V.5.7). This is described in Exercise 5.42.

3.2 Evaluating Integrals with the Residue Theorem

The residue theorem is of tremendous importance in complex analysis both on a concrete level as a practical device for evaluating integrals, including many obstinate integrals from ordinary calculus, and on a more abstract plane as a powerful theoretical tool. The following series of examples is aimed at documenting its value in the former capacity. Interspersed with the examples are a couple of technical theorems pertinent to the discussion. The theoretical applications of the residue theorem are held in reserve until the next section.

EXAMPLE 3.1. Evaluate $\int_{|z|=1}(z^2 + 2z)\csc^2 z\, dz$.

The integrand, $f(z) = (z^2+2z)\csc^2 z$, has its singularities at the integral multiples of π. It has but one singularity in the closed disk $\overline{\Delta}(0, 1)$, that being a simple pole at the origin. With the aid of l'Hospital's rule we obtain from formula (8.6)

$$\operatorname{Res}(0, f) = \lim_{z \to 0} z f(z) = \lim_{z \to 0} \frac{z^3 + 2z^2}{\sin^2 z} = \lim_{z \to 0} \frac{3z^2 + 4z}{2 \sin z \cos z}$$

$$= \lim_{z \to 0} \frac{6z + 4}{2 \cos^2 z - 2 \sin^2 z} = 2 .$$

Corollary 3.2 informs us that

$$\int_{|z|=1} (z^2 + 2z)\csc^2 z\, dz = 4\pi i .$$

EXAMPLE 3.2. Evaluate $\int_{\partial Q}[\sin z][\cos(z^3) - 1]^{-1} dz$, where Q is the square bounded by the curve with equation $|x| + |y| = 1$.

The singularities of $f(z) = [\sin z][\cos(z^3) - 1]^{-1}$ are located at the cube roots of integral multiples of 2π. Each of these points, save one, has magnitude greater than 1 — hence, exerts no influence on the given integral. The sole exception occurs at the origin, where f has a fifth order pole. In Example 2.3 we computed $\operatorname{Res}(0, f)$ and found it to be $-1/60$. We deduce from Corollary 3.2 that

$$\int_{\partial Q} \frac{\sin z\, dz}{\cos(z^3) - 1} = -\frac{\pi i}{30} .$$

3. The Residue Theorem and its Consequences

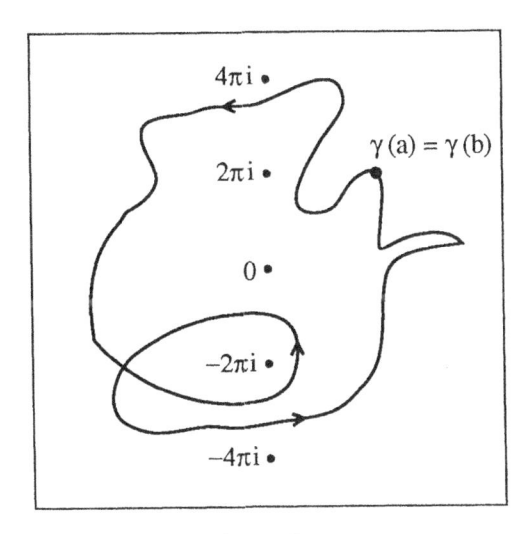

Figure 2.

EXAMPLE 3.3. Evaluate $\int_\gamma z(e^z - 1)^{-1} dz$, where $\gamma: [a, b] \to \mathbb{C}$ is the path pictured in Figure 2.

The function $f(z) = z(e^z - 1)^{-1}$ has isolated singularities at $z_k = 2k\pi i$ for $k = 0, \pm 1, \pm 2, \cdots$. Since $n(\gamma, z_k) = 0$ for $k = \pm 2, \pm 3, \cdots$, only the singularities at the points $0, 2\pi i$, and $-2\pi i$ have an effect on the integral. The singularity of f at the origin is removable, so $\text{Res}(0, f) = 0$. At each of the points $2\pi i$ and $-2\pi i$ this function has a simple pole. By (8.6)

$$\text{Res}(2\pi i, f) = \lim_{z \to 2\pi i} (z - 2\pi i) f(z) = \lim_{z \to 2\pi i} \frac{z(z - 2\pi i)}{e^z - 1}$$

$$= \lim_{z \to 2\pi i} \frac{2z - 2\pi i}{e^z} = 2\pi i$$

and, similarly, $\text{Res}(-2\pi i, f) = -2\pi i$. The residue theorem yields

$$\int_\gamma \frac{z\, dz}{e^z - 1} = 2\pi i \sum_{k=-\infty}^{\infty} n(\gamma, z_k) \text{Res}(z_k, f)$$

$$= 2\pi i \left[(2)(-2\pi i) + 0 + (1)(2\pi i) \right] = 4\pi^2 \ .$$

EXAMPLE 3.4. Evaluate $\int_\sigma (1 - z)^{-1} e^{1/z} dz$ for the cycle $\sigma = (\gamma, \beta)$, where $\gamma(t) = 2e^{it}$ and $\beta(t) = (-1/2) + e^{it}$ for $0 \leq t \leq 2\pi$ (Figure 3).

The integrand $f(z) = (1-z)^{-1} e^{1/z}$ has two singularities, a simple pole at the point 1, for which $n(\sigma, 1) = 1$, and an essential singularity at the origin, for which $n(\sigma, 0) = 2$. The residue at the first of these is easy to

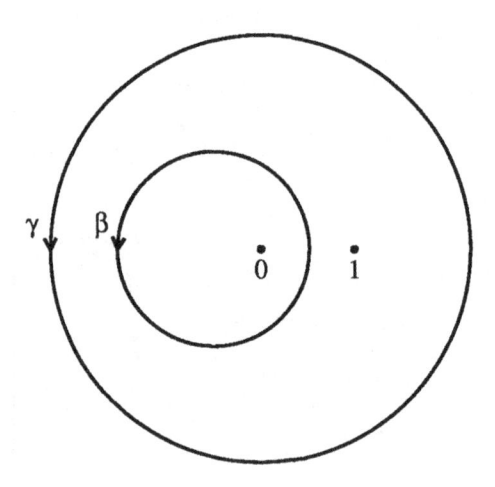

Figure 3.

handle with the assistance of (8.6),
$$\operatorname{Res}(1, f) = \lim_{z \to 1}(z - 1)f(z) = -\lim_{z \to 1} e^{1/z} = -e \ .$$
As might be expected, the residue at the essential singularity poses somewhat more of a problem. We take advantage of two familiar series — namely, the geometric series,
$$\frac{1}{1-z} = \sum_{n=0}^{\infty} z^n \ ,$$
and the Laurent expansion of $e^{1/z}$ in $D = \mathbb{C} \sim \{0\}$,
$$e^{1/z} = \sum_{n=0}^{\infty} \frac{1}{n! z^n}$$
— and revert to the definition of a Laurent coefficient in order to ascertain $\operatorname{Res}(0, f)$:
$$\operatorname{Res}(0, f) = \frac{1}{2\pi i} \int_{|z|=1/2} f(z)\, dz = \frac{1}{2\pi i} \int_{|z|=1/2} (1-z)^{-1} e^{1/z}\, dz$$
$$= \frac{1}{2\pi i} \int_{|z|=1/2} \left(\sum_{n=0}^{\infty} z^n e^{1/z} \right) dz = \sum_{n=0}^{\infty} \frac{1}{2\pi i} \int_{|z|=1/2} z^n e^{1/z}\, dz$$
$$= \sum_{n=0}^{\infty} \frac{1}{(n+1)!} = \sum_{n=1}^{\infty} \frac{1}{n!} = e - 1 \ .$$
The exchange of integration and summation is allowable because the series $\sum_{n=0}^{\infty} z^n e^{1/z}$ is uniformly convergent on the circle $K(0, 1/2)$. That
$$\frac{1}{2\pi i} \int_{|z|=1/2} z^n e^{1/z}\, dz = \frac{1}{(n+1)!}$$

3. The Residue Theorem and its Consequences

follows from the fact that this integral expression represents the coefficient of z^{-n-1} in the above Laurent series expansion for $e^{1/z}$. Having calculated the necessary residues, we learn from the residue theorem that

$$\int_\sigma \frac{e^{1/z}\, dz}{1-z} = 2\pi i\, [2\mathrm{Res}(0,f) + \mathrm{Res}(1,f)] = 2\pi i(2e - 2 - e) = (2e-4)\pi i\ .$$

We were, of course, quite fortunate here in being able to determine explicitly the residue at the essential singularity.

Preparations for the next two examples are met in the following theorem. One piece of notation in the theorem deserves a comment. Given a rational function $R(x,y)$ of the real variables x and y we generate a rational function of z by substituting $(z+z^{-1})/2$ for x and $(z-z^{-1})/2i$ for y in the formula describing R. We indicate these substitutions in the obvious way, by writing $R\left[(z+z^{-1})/2, (z-z^{-1})/2i\right]$. Notice, however, that the substituted quantities are only real when $|z| = 1$. (Remember that, in general, $x = (z+\bar{z})/2$ — not $x = (z+z^{-1})/2$. A similar remark applies to y.) If $R(x,y) = y/x$, for example, we obtain

$$R\left[(z+z^{-1})/2, (z-z^{-1})/2i\right] = \frac{(z-z^{-1})/2i}{(z+z^{-1})/2} = \frac{1}{i}\frac{z^2-1}{z^2+1}\ .$$

Theorem 3.3. *Let R be a rational function of x and y whose domain-set includes the circle $K(0,1)$. Then*

(8.14) $$\int_0^{2\pi} R(\cos\theta, \sin\theta)\, d\theta = 2\pi \sum_{k=1}^{p} \mathrm{Res}(z_k, f)\ ,$$

where $f(z) = z^{-1} R\left[(z+z^{-1})/2, (z-z^{-1})/2i\right]$ and $z_1, z_2, \ldots, z_p$ are the poles of f in the disk $\Delta(0,1)$.

Proof. The function f is a rational function of z. Each of its finitely many singularities is at worst a pole, and by hypothesis it has no singularities on the circle $K(0,1)$. Let $z_1, z_2, \ldots, z_p$ be the poles of f in $\Delta(0,1)$. Because the residue of a function is zero at any removable singularity, we infer from Corollary 3.2 that

$$\int_{|z|=1} f(z)\, dz = 2\pi i \sum_{k=1}^{p} \mathrm{Res}(z_k, f)\ .$$

On the other hand, by the very definition of this complex line integral

$$\int_{|z|=1} f(z)\, dz = \int_0^{2\pi} e^{-i\theta} R\left[(e^{i\theta}+e^{-i\theta})/2, (e^{i\theta}-e^{-i\theta})/2i\right] i e^{i\theta}\, d\theta$$

$$= i \int_0^{2\pi} R(\cos\theta, \sin\theta)\, d\theta\ .$$

Formula (8.14) now follows easily. ∎

EXAMPLE 3.5. Given that $a > 1$, evaluate $\int_0^{2\pi} (a + \cos\theta)^{-1} d\theta$.

This integral is of the type covered by Theorem 3.3. Here $R(x,y) = (a + x)^{-1}$. We consider the function f given by

$$f(z) = \frac{1}{z} R\left(\frac{z + z^{-1}}{2}, \frac{z - z^{-1}}{2i}\right) = \frac{1}{z} \frac{1}{a + [(z + z^{-1})/2]} = \frac{2}{z^2 + 2az + 1},$$

a rational function whose only poles are simple ones located at the points $-a + \sqrt{a^2 - 1}$ and $-a - \sqrt{a^2 - 1}$. Since $a > 1$, just the first of these, $z_1 = -a + \sqrt{a^2 - 1}$, finds itself in $\Delta(0,1)$. Formula (8.6) gives

$$\text{Res}(z_1, f) = \lim_{z \to z_1} (z - z_1) f(z) = \lim_{z \to z_1} \frac{2(z - z_1)}{z^2 + 2az + 1}$$

$$= \lim_{z \to z_1} \frac{2}{2z + 2a} = \frac{1}{\sqrt{a^2 - 1}}.$$

Referring to Theorem 3.3 we conclude that

$$\int_0^{2\pi} \frac{d\theta}{a + \cos\theta} = \frac{2\pi}{\sqrt{a^2 - 1}}$$

for $a > 1$.

EXAMPLE 3.6. Evaluate $\int_0^{\pi} \sin^{2n} \theta\, d\theta$ for n a positive integer.

Since $\sin^{2n} \theta$ has period π, $\int_0^{\pi} \sin^{2n} \theta\, d\theta = (1/2) \int_0^{2\pi} \sin^{2n} \theta\, d\theta$. The latter integral falls within the scope of Theorem 3.3, this time with $R(x,y) = y^{2n}$. We utilize the binomial theorem in computing

$$f(z) = \frac{1}{z} R\left(\frac{z + z^{-1}}{2}, \frac{z - z^{-1}}{2i}\right) = \frac{1}{z}\left(\frac{z - z^{-1}}{2i}\right)^{2n}$$

$$= \frac{1}{z} \sum_{k=0}^{2n} \binom{2n}{k} (2i)^{-2n} z^k (-z^{-1})^{2n-k}$$

$$= \sum_{k=0}^{2n} (-1)^{n-k} 4^{-n} \binom{2n}{k} z^{2k - 2n - 1}.$$

The only singularity of f in $\Delta(0,1)$ is a pole at the origin, where the residue can be extracted directly from the expression derived for f:

$$\text{Res}(0, f) = 4^{-n} \binom{2n}{n}.$$

3. The Residue Theorem and its Consequences

By Theorem 3.3,
$$\int_0^\pi \sin^{2n}\theta\, d\theta = \frac{1}{2}\int_0^{2\pi} \sin^{2n}\theta\, d\theta = 4^{-n}\binom{2n}{n}\pi$$
for any positive integer n.

It goes without saying that there are means other than the residue theorem for evaluating integrals of the type $\int_0^{2\pi} R(\cos\theta, \sin\theta)\, d\theta$, where R is a rational function of x and y. To name but one of these, the substitution $t = \tan(\theta/2)$ transforms such an integral into an integral to which the method of partial fractions is applicable. What the residue theorem can provide in this kind of problem is a labor-saving device of sometimes major proportions.

The next theorem addresses itself to a class of integrals more resistant to standard elementary techniques of integration than the integrals in the two preceding examples.

Theorem 3.4. *If $f(z) = (a_0 + a_1 z + \cdots + a_n z^n)/(b_0 + b_1 z + \cdots + b_m z^m)$ is a rational function in which $m \geq n+2$ and in which the denominator has no real roots, then for $c \geq 0$*

(8.15) $$\int_{-\infty}^{\infty} f(x)e^{icx}\, dx = 2\pi i \sum_{k=1}^{p} \text{Res}\left[z_k, f(z)e^{icz}\right],$$

where $z_1, z_2, \ldots, z_p$ are the poles of f in the half-plane $H = \{z : \text{Im}\, z > 0\}$. Furthermore, if all the coefficients in f are real numbers, then

(8.16) $$\int_{-\infty}^{\infty} f(x)\cos(cx)\, dx = \text{Re}\left\{2\pi i \sum_{k=1}^{p} \text{Res}\left[z_k, f(z)e^{icz}\right]\right\}$$

and

(8.17) $$\int_{-\infty}^{\infty} f(x)\sin(cx)\, dx = \text{Im}\left\{2\pi i \sum_{k=1}^{p} \text{Res}\left[z_k, f(z)e^{icz}\right]\right\}.$$

Proof. Because $m \geq n+2$ it is clear that $|z^2 f(z)|$ tends to a finite limit L as $|z| \to \infty$. This fact permits us to fix a number $M > |L|$ and then to choose $r_0 > 0$ with the property that

(8.18) $$|f(z)| \leq \frac{M}{|z|^2}$$

for every z having $|z| \geq r_0$. Together with the assumption that f is free of poles on the real axis, (8.18) makes certain that the improper integral $\int_{-\infty}^{\infty} f(x)e^{icx}\, dx$ is convergent; i.e.,

$$\int_{-\infty}^{\infty} f(x)e^{icx}\, dx = \lim_{a\to-\infty}\int_a^0 f(x)e^{icx}\, dx + \lim_{b\to\infty}\int_0^b f(x)e^{icx}\, dx$$

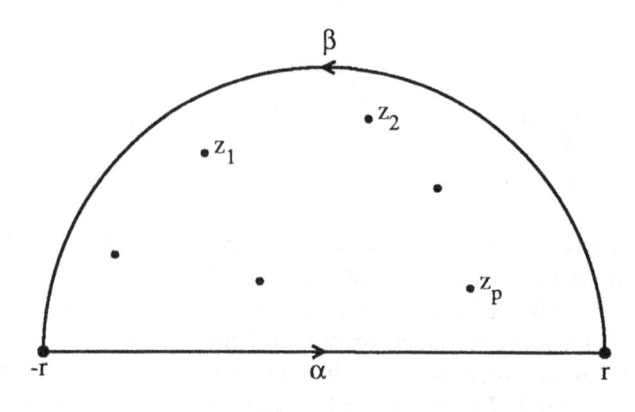

VIII. Isolated Singularities of Analytic Functions

Figure 4.

exists.

Let $r \geq r_0$. Inequality (8.18) implies that the disk $\Delta(0, r)$ contains all the poles of f. We consider the integral of f along the contour $\gamma = \alpha + \beta$, where $\alpha(t) = t$ for $-r \leq t < r$ and $\beta(t) = re^{it}$ for $0 \leq t \leq \pi$ (Figure 4). By Corollary 3.2,

$$(8.19) \qquad \int_\gamma f(z)e^{icz}\, dz = 2\pi i \sum_{k=1}^p \operatorname{Res}\left[z_k, f(z)e^{icz}\right] .$$

Here $z_1, z_2, \ldots, z_p$ is a list of the poles of f in $H = \{z : \operatorname{Im} z > 0\}$. The left-hand side of (8.19) can be expressed in the form

$$\int_\gamma f(z)e^{icz}\, dz = \int_{-r}^r f(t)e^{ict}\, dt + \int_\beta f(z)e^{icz}\, dz .$$

We now let $r \to \infty$. The right-hand side of (8.19) is unaffected by the passage to the limit, whereas on the left we have

$$\int_{-r}^r f(t)e^{ict}\, dt \to \int_{-\infty}^\infty f(t)e^{ict}\, dt ,$$

since the latter integral is known to be convergent, and

$$\int_\beta f(z)e^{icz}\, dz \to 0 .$$

Justification for the last assertion is found in (8.18) and the by now familiar method of estimation:

$$\left|\int_\beta f(z)e^{icz}\, dz\right| \leq \int_\beta |f(z)||e^{icz}||dz| \leq \frac{M}{r^2} \int_\beta e^{\operatorname{Re}(icz)}|dz|$$

3. The Residue Theorem and its Consequences

$$= \frac{M}{r} \int_0^\pi e^{-cr\sin t}\, dt \le \frac{M\pi}{r} \to 0$$

as $r \to \infty$. Putting these observations together, we obtain

$$\int_{-\infty}^\infty f(t)e^{ict}\, dt = 2\pi i \sum_{k=1}^p \operatorname{Res}\left[z_k, f(z)e^{icz}\right],$$

as asserted in (8.15). Formulas (8.16) and (8.17) follow by taking real and imaginary parts in (8.15) — provided the function f has real coefficients and is, as a result, real-valued on the real axis. ∎

If $c > 0$, then the conclusions of Theorem 3.4 are valid even in the case $m = n + 1$. However, a different argument is needed to confirm them, for inequality (8.18) becomes $|f(z)| \le M/|z|$ when $m = n + 1$, an estimate which is insufficient on its own to force the convergence of $\int_{-\infty}^\infty f(x)e^{icx}\, dx$. Exercise 5.51 indicates a method of circumventing this difficulty.

EXAMPLE 3.7. Evaluate $\int_{-\infty}^\infty (x^2 + 1)^{-1}(x^2 + 4)^{-1}\, dx$.

We apply Theorem 3.4 to $f(z) = (z^2 + 1)^{-1}(z^2 + 4)^{-1}$ with $c = 0$. The poles of f in H are simple poles at the points i and $2i$. The corresponding residues are easily calculated with the help of (8.6): $\operatorname{Res}(i, f) = -i/6$, $\operatorname{Res}(2i, f) = i/12$. Consequently,

$$\int_{-\infty}^\infty \frac{dx}{(x^2+1)(x^2+4)} = 2\pi i\left(-\frac{i}{6} + \frac{i}{12}\right) = \frac{\pi}{6}.$$

We would be quick to agree that the integral in Example 3.7 could have been worked out using partial fractions, but only at a much greater expenditure of effort than was involved in the solution presented. The integral in the next example would be even more difficult to evaluate without recourse to residue theorem methods.

EXAMPLE 3.8. Evaluate $\int_0^\infty x(x^2 + 1)^{-2} \sin x\, dx$.

In this instance we employ Theorem 3.4 with $f(z) = z(z^2 + 1)^{-2}$ and $c = 1$. The only pole of f in H is a double pole at the point i. We take $m = 2$ in (8.6) and compute

$$\operatorname{Res}\left[i, f(z)e^{iz}\right] = \lim_{z \to i} \frac{d}{dz}\left[(z-i)^2 f(z)e^{iz}\right] = \lim_{z \to i} \frac{d}{dz}\left[\frac{(z-i)^2 z e^{iz}}{(z^2+1)^2}\right]$$

$$= \lim_{z \to i} \frac{d}{dz}\left[\frac{z e^{iz}}{(z+i)^2}\right] = \lim_{z \to i} \frac{(z+i)^2(1+iz)e^{iz} - 2z(z+i)e^{iz}}{(z+i)^4} = \frac{1}{4e}.$$

A glance at (8.17) reveals that
$$\int_{-\infty}^{\infty} \frac{x \sin x \, dx}{(x^2+1)^2} = \frac{\pi}{2e}.$$
Since the integrand is an even function we deduce that
$$\int_0^{\infty} \frac{x \sin x \, dx}{(x^2+1)^2} = \frac{1}{2}\int_{-\infty}^{\infty} \frac{x \sin x \, dx}{(x^2+1)^2} = \frac{\pi}{4e}.$$

Only a small fraction of the integrals to which the residue theorem is relevant fit convenient standard patterns like the ones exemplified in Theorems 3.3 and 3.4. More typically the use of this theorem to evaluate a definite integral calls for a bit of ingenuity, especially in finding a contour appropriate to the integral under consideration and in establishing necessary estimates. (Even experienced practitioners of complex analysis can find their resourcefulness put to the test in coming up with a "correct" path of integration!) Although it may at first seem that the contours which appear in the next two examples are plucked from thin air, these examples are quite representative of the kind of inventiveness that enters into making the residue theorem a truly effective tool.

An important concept in harmonic analysis is the so-called *Hilbert transform* Hf of a function $f: \mathbb{R} \to \mathbb{R}$, which for the sake of the present discussion we assume to be continuous. The value of Hf at the real number x is given by

(8.20)
$$Hf(x) = -\frac{1}{\pi} \lim_{\substack{s \to 0^+ \\ r \to \infty}} \int_{s \le |t| \le r} \frac{f(x+t)\,dt}{t}$$
$$= -\frac{1}{\pi} \lim_{\substack{s \to 0^+ \\ r \to \infty}} \left\{ \int_{-r}^{-s} \frac{f(x+t)\,dt}{t} + \int_s^r \frac{f(x+t)\,dt}{t} \right\},$$

provided the limit exists. (The interpretation of the double limit becomes clearer when it is rewritten as the sum of
$$\lim_{s \to 0^+} \left\{ \int_{-1}^{-s} \frac{f(x+t)\,dt}{t} + \int_s^1 \frac{f(x+t)\,dt}{t} \right\}$$
and
$$\lim_{r \to \infty} \left\{ \int_{-r}^{-1} \frac{f(x+t)\,dt}{t} + \int_1^r \frac{f(x+t)\,dt}{t} \right\}.$$
The requirement is that both limits should exist.) We have intentionally avoided writing
$$Hf(x) = -\frac{1}{\pi} \int_{-\infty}^{\infty} \frac{f(x+t)\,dt}{t}$$

3. The Residue Theorem and its Consequences

here — and for good reason. The normal improper integral,

$$\int_{-\infty}^{\infty} \frac{f(x+t)\,dt}{t} = \lim_{a \to -\infty} \int_{a}^{-1} \frac{f(x+t)\,dt}{t} + \lim_{b \to 0^{-}} \int_{-1}^{b} \frac{f(x+t)\,dt}{t}$$

$$+ \lim_{c \to 0^{+}} \int_{c}^{1} \frac{f(x+t)\,dt}{t} + \lim_{d \to \infty} \int_{1}^{d} \frac{f(x+t)\,dt}{t} ,$$

is frequently divergent in situations where the extremely symmetric limits involved in (8.20) exist. One often finds the Hilbert transform expressed as

$$Hf(x) = -\frac{1}{\pi} \text{P.V.} \int_{-\infty}^{\infty} \frac{f(x+t)\,dt}{t} .$$

The symbol P.V. $\int$ stands for "principal value integral" and distinguishes this object from a standard improper integral. In our next example we use the residue theorem to compute the Hilbert transforms of two elementary functions.

EXAMPLE 3.9. Compute the Hilbert transforms of the functions $f(x) = \cos x$ and $g(x) = \sin x$.

We shall demonstrate that

$$\lim_{\substack{s \to 0^+ \\ r \to \infty}} \int_{s \le |t| \le r} \frac{e^{i(x+t)}\,dt}{t} = e^{ix} \lim_{\substack{s \to 0^+ \\ r \to \infty}} \int_{s \le |t| \le r} \frac{e^{it}\,dt}{t} = \pi i e^{ix}$$

for any real x. Multiplying by $-1/\pi$ and taking real and imaginary parts, we obtain $Hf(x) = \sin x$ and $Hg(x) = -\cos x$, respectively. We are thus led naturally to look at the meromorphic function $h(z) = z^{-1}e^{iz}$. The only singularity of h is a simple pole at the origin, and $\text{Res}(0, h) = 1$. What we need is an expression for the integral $\int_{s \le |t| \le r} h(t)\,dt$, say with $0 < s < 1$ and $r > 1$, that makes transparent its behavior when $s \to 0^+$ and $r \to \infty$. It is here that, by making a shrewd choice of contour, one can bring the residue theorem into the picture. The trick is to integrate h along the path $\gamma = \alpha_1 + \beta_1 + \alpha_2 + \beta_2$ depicted in Figure 5. The paths that compose γ are: $\alpha_1(t) = t$ for $s \le t \le r$, $\beta_1(t) = re^{it}$ for $0 \le t \le \pi$, $\alpha_2(t) = t$ for $-r \le t \le -s$, and $\beta_2(t) = se^{it}$ for $\pi \le t \le 2\pi$. Owing to the residue theorem, integrating h along γ gives

$$\int_\gamma h(z)\,dz = 2\pi i\, \text{Res}(0, h) = 2\pi i .$$

It follows that

$$\int_{s \le |t| \le r} \frac{e^{it}\,dt}{t} = \int_{-r}^{-s} h(t)\,dt + \int_{s}^{r} h(t)\,dt = \int_{\alpha_2} h(z)\,dz + \int_{\alpha_1} h(z)\,dz$$

$$= 2\pi i - \int_{\beta_1} h(z)\,dz - \int_{\beta_2} h(z)\,dz .$$

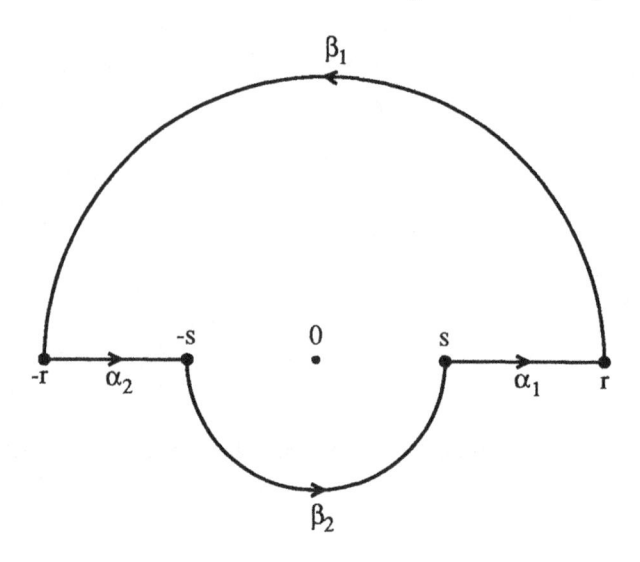

VIII. Isolated Singularities of Analytic Functions

Figure 5.

The behavior of $\int_{\beta_1} h(z)\,dz$ as $r \to \infty$ and $\int_{\beta_2} h(z)\,dz$ as $s \to 0^+$ is not hard to decipher. First,

$$\left| \int_{\beta_1} \frac{e^{iz}\,dz}{z} \right| \leq \frac{\pi(1 - e^{-r})}{r} \to 0$$

as $r \to \infty$. (A similar estimate was derived in the proof of Theorem 3.4.) Secondly,

$$\int_{\beta_2} \frac{e^{iz}\,dz}{z} = \int_{\beta_2} \frac{(e^{iz} - 1)\,dz}{z} + \int_{\beta_2} \frac{dz}{z} = \int_{\beta_2} \frac{(e^{iz} - 1)}{z} + \pi i \to \pi i$$

as $s \to 0^+$, since

$$\left| \int_{\beta_2} \frac{(e^{iz} - 1)\,dz}{z} \right| \leq \int_{\beta_2} \frac{|e^{iz} - 1||dz|}{|z|} \leq \pi \max\left\{ |e^{iz} - 1| : z \in K(0, s) \right\} \to 0$$

as $s \to 0^+$. We conclude that

$$\lim_{\substack{s \to 0^+ \\ r \to \infty}} \int_{s \leq |t| \leq r} \frac{e^{it}\,dt}{t} = \pi i \,,$$

as desired.

The final example in this section is another improper Riemann integral, an integral whose innocent appearance is misleading. Despite being armed with the residue theorem, we shall still not have a terribly easy time carrying out its evaluation.

3. The Residue Theorem and its Consequences   337

EXAMPLE 3.10. Assuming that $0 < \lambda < 1$ and $b > 0$, show that

(8.21) $$\int_0^\infty \frac{dt}{t^\lambda(t+b)} = \frac{\pi}{b^\lambda \sin(\lambda\pi)}.$$

Our job is to verify that

$$\lim_{\substack{s \to 0^+ \\ r \to \infty}} \int_s^r \frac{dt}{t^\lambda(t+b)} = \frac{\pi}{b^\lambda \sin(\lambda\pi)}.$$

We shall do this by using the residue theorem to derive the estimate

(8.22) $$\left| \pi b^{-\lambda} - \sin(\lambda\pi) \int_s^r \frac{dt}{t^\lambda(t+b)} \right| \leq \frac{\pi r^{1-\lambda}}{r-b} + \frac{\pi s^{1-\lambda}}{b-s},$$

valid when $0 < s < b$ and $r > b$. Once (8.22) is established, (8.21) follows almost immediately: since $0 < \lambda < 1$, the right-hand side of (8.21) tends to zero as $s \to 0^+$ and $r \to \infty$, implying that

$$\sin(\lambda\pi) \lim_{\substack{s \to 0^+ \\ r \to \infty}} \int_s^r \frac{dt}{t^\lambda(t+b)} = \frac{\pi}{b^\lambda}.$$

We now fix s in $(0, b)$ and r in (b, ∞). Even the clue that the residue theorem plays a role in the verification of (8.22) does not quickly suggest a way to proceed. To what function are we supposed to apply the theorem? The initial guess is very likely to be $f(z) = z^{-\lambda}(z+b)^{-1}$, but there is a problem with it: there are no isolated singularities of f in sight! This function is discontinuous at every point of the negative real axis, which means, in particular, that an apparent pole at $-b$ is not an isolated singularity at all. It is just the point $-b$, as it so happens, that is the key source of information in the derivation of (8.22). The secret to gaining access to that information lies in passing to a different branch of the λ-power function, one that is analytic near $-b$. Specifically, we set $D = \mathbb{C} \sim [0, \infty)$ and let h denote the branch of the λ-power function in D associated with the branch of $\arg z$ in D whose range is the interval $(0, 2\pi)$; i.e., if $z = te^{i\theta}$ with $t > 0$ and $0 < \theta < 2\pi$, then $h(z) = t^\lambda e^{i\lambda\theta}$. The function g defined by $g(z) = 1/[(z+b)h(z)]$ is then meromorphic in D, its only singularity in this domain is a simple pole at $-b$, and, since $h(-b) = h(be^{\pi i}) = b^\lambda e^{\lambda\pi i}$,

$$\operatorname{Res}(-b, g) = \lim_{z \to -b} (z+b)g(z) = \frac{1}{h(-b)} = b^{-\lambda} e^{-\lambda\pi i}.$$

Having settled on the function g as a reasonable candidate for the integrand in our application of the residue theorem, we are now faced with the problem of selecting a suitable contour in D along which to integrate it. We want to pick a path that manages to link the singularity of g at $-b$

to the integral $\int_s^r t^{-\lambda}(t+b)^{-1} dt$. The critical observation involved in our eventual choice is this: for each fixed $t > 0$,

$$\lim_{\theta \to 0+} [g(te^{i\theta})e^{i\theta}] = \lim_{\theta \to 0+} \frac{e^{i(1-\lambda)\theta}}{t^\lambda(te^{i\theta}+b)} = \frac{1}{t^\lambda(t+b)}$$

and

$$\lim_{\theta \to 2\pi^-} [g(te^{i\theta})e^{i\theta}] = \lim_{\theta \to 2\pi^-} \frac{e^{i(1-\lambda)\theta}}{t^\lambda(te^{i\theta}+b)} = \frac{e^{-2\lambda\pi i}}{t^\lambda(t+b)}.$$

Furthermore, in both cases the convergence is seen without difficulty to be uniform on the interval $[s,r]$, which permits us to conclude that

(8.23)
$$\lim_{\theta \to 0+} \int_s^r g(te^{i\theta})e^{i\theta} dt = \int_s^r \frac{dt}{t^\lambda(t+b)},$$
$$\lim_{\theta \to 2\pi^-} \int_s^r g(te^{i\theta})e^{i\theta} dt = e^{-2\lambda\pi i} \int_s^r \frac{dt}{t^\lambda(t+b)}.$$

It is now important to recognize that

(8.24)
$$\int_s^r g(te^{i\theta})e^{i\theta} dt = \int_\alpha g(z) dz,$$

where $\alpha(t) = te^{i\theta}$ for $s \leq t \leq r$. Coupled with (8.23), (8.24) and the need to encircle the point $-b$ in order to benefit from the residue information there make it not totally farfetched to consider, for any fixed θ in $(0, \pi/2)$, the path $\gamma = \alpha_1 + \beta_1 - \alpha_2 - \beta_2$ represented in Figure 6. To be precise: $\alpha_1(t) = te^{i\theta}$ and $\alpha_2(t) = te^{i(2\pi-\theta)}$ for $s \leq t \leq r$; $\beta_1(t) = re^{it}$ and $\beta_2(t) = se^{it}$ for $\theta \leq t \leq 2\pi - \theta$.

The residue theorem tells us that

$$\int_{\alpha_1} g(z) dz + \int_{\beta_1} g(z) dz - \int_{\alpha_2} g(z) dz - \int_{\beta_2} g(z) dz$$
$$= \int_\gamma g(z) dz = 2\pi i \text{Res}(-b, g) = 2\pi i b^{-\lambda} e^{-\lambda \pi i}.$$

Therefore, for θ in $(0, \pi/2)$ we can appeal to (8.24) and assert that

$$\left| 2\pi i b^{-\lambda} e^{-\lambda \pi i} - \int_s^r g(te^{i\theta})e^{i\theta} dt + \int_s^r g(te^{i\theta})e^{i(2\pi-\theta)} dt \right|$$
$$= \left| \int_{\beta_1} g(z) dz - \int_{\beta_2} g(z) dz \right| \leq \int_{\beta_1} |g(z)| |dz| + \int_{\beta_2} |g(z)| |dz|$$
$$\leq \frac{(2\pi - 2\theta)r}{r^\lambda(r-b)} + \frac{(2\pi - 2\theta)s}{s^\lambda(b-s)},$$

3. The Residue Theorem and its Consequences

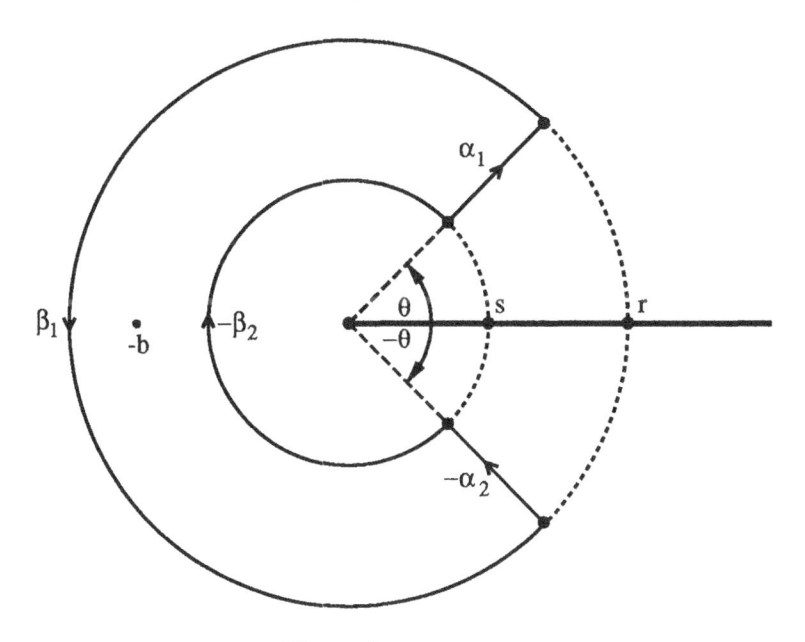

Figure 6.

since $|g(z)| \leq r^{-\lambda}(r-b)^{-1}$ for z in $|\beta_1|$ and $\ell(\beta_1) = (2\pi - 2\theta)r$, while $|g(z)| \leq s^{-\lambda}(b-s)^{-1}$ on $|\beta_2|$ and $\ell(\beta_2) = (2\pi - 2\theta)s$. In view of (8.23), letting $\theta \to 0^+$ results in the inequality

$$\left| 2\pi i b^{-\lambda} e^{-\lambda \pi i} - (1 - e^{-2\lambda \pi i}) \int_s^r \frac{dt}{t^\lambda(t-b)} \right| \leq \frac{2\pi r^{1-\lambda}}{r-b} + \frac{2\pi s^{1-\lambda}}{b-s}.$$

Dividing through by $|2ie^{-\lambda \pi i}|$, which is 2 in disguise, we arrive at (8.22).

3.3 Consequences of the Residue Theorem

Having seen the residue theorem at work in the evaluation of integrals, we move on to consider some of its theoretical implications for analytic functions. To say that the influence of this theorem on complex analysis extends far beyond the few ideas that will be treated in the forthcoming pages would be a vast understatement.

The information recoverable from the residue theorem is typified by a result known as the "Argument Principle." This proposition admits many different formulations. The version we present is based on Corollary 3.2 and, as such, involves a temporary departure from our usual policy of avoiding appeal to results that depend on the Jordan curve theorem. It is in the setting of Corollary 3.2 that the argument principle has its simplest — and, some might argue, most elegant — formulation. Variants of this basic

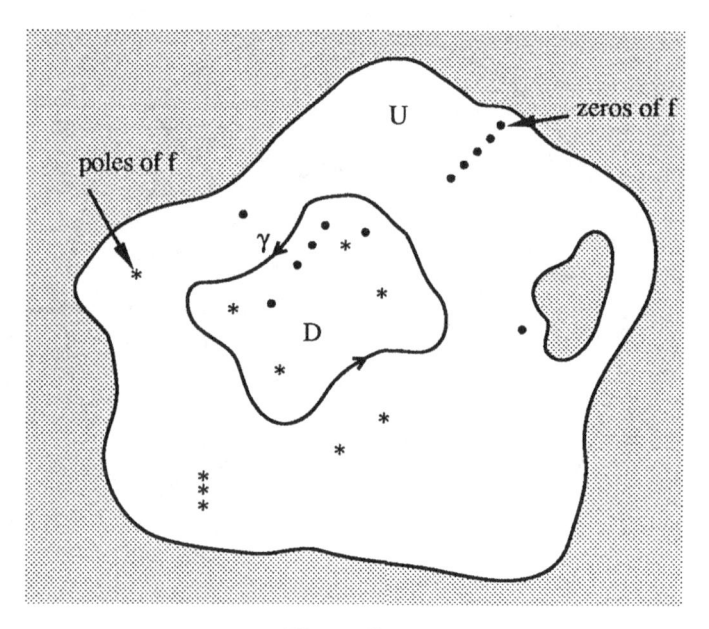

Figure 7.

form of the principle can be found in the exercises. Phrases like "taking multiplicity into account" and "counted with due regard for multiplicity" appear regularly in the discussion that follows. They refer to the way in which zeros and poles of a function are tallied: a zero of order m counts as m zeros, a pole of order m as m poles.

Theorem 3.5. (Argument Principle) *Assume that a function f is meromorphic in an open set U. Let γ be a Jordan contour in U such that the Jordan curve $|\gamma|$ does not pass through any zero or pole of f and such that the inside D of $|\gamma|$ is contained in U (Figure 7). Then*

$$\frac{1}{2\pi i}\int_\gamma \frac{f'(z)\,dz}{f(z)} = Z - P\,,$$

where Z and P indicate the number of zeros and the number of poles, respectively, that f has in D, multiplicity being taken into account.

Proof. The hypotheses imply that f does not vanish identically in G, the component of U that contains $\overline{D}$. This means that the function f'/f is meromorphic in G, its only singularities appearing at zeros and poles of f. Consider, first, a point z_0 of G at which f has a zero of order m. In some open disk $\Delta = \Delta(z_0, r)$ we can express f in the form $f(z) = (z - z_0)^m g(z)$, where $g: \Delta \to \mathbb{C}$ is a function that is both analytic and zero-free. This gives rise to the representation $f'(z) = m(z - z_0)^{m-1} g(z) + (z - z_0)^m g'(z)$ in Δ

3. The Residue Theorem and its Consequences

and has the consequence that

$$\frac{f'(z)}{f(z)} = \frac{m}{z - z_0} + \frac{g'(z)}{g(z)}$$

for z belonging to the punctured disk $\Delta^*(z_0, r)$. Because g'/g is analytic in Δ, it becomes evident that f'/f has a simple pole at z_0 and that $\text{Res}(z_0, f'/f) = m$. A completely analogous computation reveals that, at a point z_0 of G where f has a pole of order m, f'/f has a simple pole with $\text{Res}(z_0, f'/f) = -m$. On the authority of Corollary 3.2 we can assert that

$$\frac{1}{2\pi i} \int_\gamma \frac{f'(z)\, dz}{f(z)} = \sum_{k=1}^{p} \text{Res}(z_k, f'/f) \, ,$$

where $z_1, z_2, \ldots, z_p$ are the (distinct) poles of f'/f in D. (Intrepret an "empty" sum to mean 0.) Our analysis of the singularities of f'/f makes clear that the residue sum reduces to the number of zeros of f in D less the number of poles of f in this set, assuming that both are computed with multiplicity taken into consideration. Thus, the residue sum is $Z - P$. ∎

Suppose that the path γ in Theorem 3.5 is parametrized on the interval $[a, b]$. Then, as we noted in the comments after Theorem V.4.1,

$$\frac{1}{2\pi i} \int_\gamma \frac{f'(z)\, dz}{f(z)} = n(\beta, 0) \, ,$$

where β is the image of γ under f, meaning the path defined on $[a, b]$ by $\beta(t) = f[\gamma(t)]$. In supplying a geometric interpretation of winding numbers we observed that $2\pi n(\beta, 0)$ can be thought of as the net change in the argument of $\beta(t)$ when t increases from a to b; i.e., the net change in the argument of $w = f(z)$ as $z = \gamma(t)$ moves around $J = |\gamma|$. This explains the name "argument principle": the theorem establishes a precise relationship between the number of zeros and poles of f inside the Jordan curve J and the change in the argument of $f(z)$ as z traverses J once in the positive direction.

The argument principle has consequences galore. Prominent among these is a theorem of Eugène Rouché (1832-1910). We state Rouché's theorem in the context of analytic functions, the setting where the result is most frequently applied. Its straightforward generalization for meromorphic functions is relegated to the exercises (Exercise 5.62).

Theorem 3.6. (Rouché's Theorem) *If D is the domain inside the trajectory of a Jordan contour in the complex plane, if f and g are functions that are analytic in some open set which contains $\overline{D}$, and if the inequality*

(8.25) $$|f(z) - g(z)| < |f(z)| + |g(z)|$$

holds at every point z of ∂D, then f and g have the same number of zeros in D, provided that zero-counts are made with due regard for multiplicity.

Proof. Inequality (8.25) prevents either f or g from having a zero on the Jordan curve $J = \partial D$. In particular, J passes through no zero or pole of $h = f/g$, a function that is meromorphic in some open set containing $\overline{D}$. Furthermore, dividing both sides of (8.25) by $|g(z)|$ leads to

$$|1 - h(z)| < 1 + |h(z)| \tag{8.26}$$

for every z on J. Since $|1 - h(z)| = 1 + |h(z)|$ would plainly be true of any z for which $h(z)$ is real and non-positive, (8.26) implies that $h(J)$ is disjoint from the real interval $(-\infty, 0]$. It follows that the origin lies in the unbounded component of $\mathbb{C} \sim h(J)$ and, thus, that the path $\beta = h \circ \gamma$, where γ is any Jordan contour with trajectory J, has $n(\beta, 0) = 0$. Recalling the remarks pursuant to the proof of the argument principle and noting, as a simple calculation verifies, that $h'/h = (f'/f) - (g'/g)$, we conclude that

$$\frac{1}{2\pi i}\int_\gamma \frac{f'(z)\,dz}{f(z)} - \frac{1}{2\pi i}\int_\gamma \frac{g'(z)\,dz}{g(z)} = \frac{1}{2\pi i}\int_\gamma \frac{h'(z)\,dz}{h(z)} = n(\beta, 0) = 0. \tag{8.27}$$

In light of the argument principle and the assumption that neither f nor g has poles in D, the first term in (8.27) translates to $Z_f - Z_g$, where Z_f and Z_g designate the zero-totals for f and g in D, computed with allowance for multiplicity. Therefore, $Z_f = Z_g$. ∎

Condition (8.25) is definitely met if the inequality $|f(z) - g(z)| < |g(z)|$ is satisfied at every point of ∂D. It is this inequality, rather than (8.25), that appears in the classical statement of Rouché's theorem. The theorem in the form we have presented is of more recent vintage. (For extra commentary on this subject we refer the reader to the article "A remark on Rouché's theorem" by Irving Glicksberg in *The American Mathematical Monthly*, Vol. 83, No. 3, 1976.) The great value of Rouché's theorem stems from its capacity to provide information on the existence, number, and location of the zeros of an analytic function f solely on the basis of comparisons between $|f - g|$ and $|f| + |g|$ for suitably chosen "test functions" g, whose zeros are known from the start, along properly selected Jordan curves. Two concrete examples will demonstrate how this can work.

EXAMPLE 3.11. Let $f(z) = z^5 + 5z^3 + z - 2$. Show that f has three of its roots in the disk $\Delta(0, 1)$ and all of its roots in $\Delta(0, 5/2)$.

There is faint hope of actually finding the zeros of f, but Rouché's theorem will enable us to get some fix on their locations. When $|z|$ is near 1 the dominant term in $f(z)$ is $5z^3$. The suggestion is that this term might govern the number of zeros of f in the disk $\Delta(0, 1)$. If we take $g(z) = 5z^3$ we find that

$$|f(z) - g(z)| = |z^5 + z - 2| \le |z|^5 + |z| + 2 = 4 < 5 = |g(z)|$$

when $|z| = 1$, so by Rouché's theorem f and g have an equal number of zeros in $\Delta(0, 1)$. The only zero of g in that disk is a zero of order three at

3. The Residue Theorem and its Consequences

the origin. Therefore, making allowances for multiple roots, f has three of its roots in $\Delta(0,1)$. Once $|z|$ exceeds 2 the term z^5 begins to take over as the dominant one in $f(z)$. If we switch to $h(z) = z^5$ as our "comparison function" in Rouché's theorem and notice that

$$|f(z) - h(z)| = |5z^3 + z - 2| \le 2644/32 < 3125/32 = |h(z)|$$

when $|z| = 5/2$, we learn that f must have all five of its zeros in $\Delta(0, 5/2)$. (N.B. Other sources of information can, of course, be tapped in order to pinpoint further the zeros of f. For instance, since $f(0) = -2 < 0$ and $f(1) = 5 > 0$, the intermediate value theorem guarantees the presence of a root of f in the interval $(0, 1)$. As $f'(z) = 5z^4 + 15z^2 + 1 > 0$ for every real z, f has no other real roots. Because the non-real roots of f occur in conjugate pairs, we conclude that exactly two of these roots have positive imaginary parts. It follows that f has no multiple roots. Finally, since the sum of the roots of f — i.e., the negative of the coefficient of z^4 in $f(z)$ — is zero, since three of these roots lie in $\Delta(0,1)$, and since $z + \bar{z} = 2\operatorname{Re} z$, we discover that the two roots of f outside $\Delta(0,1)$ have their shared real part in the interval $(-3/2, 3/2)$.)

EXAMPLE 3.12. Assuming that c is a complex number with $|c| > e$, show that the equation $e^z = cz$ has exactly one solution in the half-plane $H = \{z \colon \operatorname{Re} z < 1\}$.

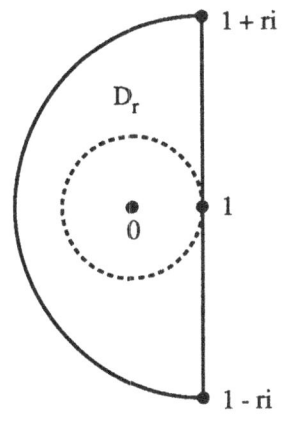

Figure 8.

It is enough to verify that for every $r \ge 2$ the given equation has one and only one solution located in the domain $D_r = H \cap \Delta(1, r)$. (See Figure 8.) We fix $r \ge 2$ and compare the function $f(z) = cz - e^z$ with $g(z) = cz$ on the Jordan curve $J = \partial D_r$. Since $r \ge 2$, D_r encompasses the disk $\Delta(0, 1)$, which forces $|z| \ge 1$ to hold whenever z lies on J. For such z, therefore,

$$|f(z) - g(z)| = |-e^z| = e^{\operatorname{Re} z} \le e < |c| \le |c||z| = |g(z)|\,.$$

By Rouché's theorem f and g have an equal number of zeros in D_r. Because the only zero of g there is a simple zero at the origin, f has exactly one zero in D_r as well. Accordingly, $e^z = cz$ has precisely one solution in D_r. As this is true for every $r \geq 2$, this equation has a unique solution belonging to H. (N.B. We have not ruled out the possibility that $e^z = cz$ has additional solutions in the set $\mathbb{C} \sim H$. Such solutions do, in fact, exist.)

The establishment of Rouché's theorem has put us in a position to assemble a qualitatively accurate picture of the local mapping properties of analytic functions. Roughly speaking, what this picture reveals is that, in the vicinity of a point where an analytic function assumes the value w_0 with multiplicity m, its behavior very much mimics that of the function $f(z) = w_0 + z^m$ in the proximity of the origin. We give this statement a precise formulation under the title "Branched Covering Principle," but we warn the reader that this is by no means standard terminology in the literature.

Theorem 3.7. (Branched Covering Principle) *Suppose that a function f is analytic in an open set U, that z_0 is a point of U, and that f takes the value w_0 with multiplicty m at z_0. Let $r > 0$ be any number sufficiently small that the following conditions prevail: the closed disk $\overline{\Delta} = \overline{\Delta}(z_0, r)$ is contained in U, and the statements $f(z) \neq w_0$, $f'(z) \neq 0$ are true for every z in the set $\overline{\Delta} \sim \{z_0\}$. Define $s = s(r) > 0$ by $s = \min\{|f(z) - w_0| : z \in K(z_0, r)\}$. Then $G = \{z \in \Delta(z_0, r) : f(z) \in \Delta(w_0, s)\}$ is a domain. Moreover, for each point w of the punctured disk $\Delta^*(w_0, s)$ the set $E_w = \{z \in \Delta(z_0, r) : f(z) = w\}$ consists of exactly m points of G, at each of which f assumes the value w with multiplicity one.*

Proof. (Figure 9 illustrates the situation for $m = 3$.) The definition of f having a multiplicity at z_0 includes the assumption that f is not constant in

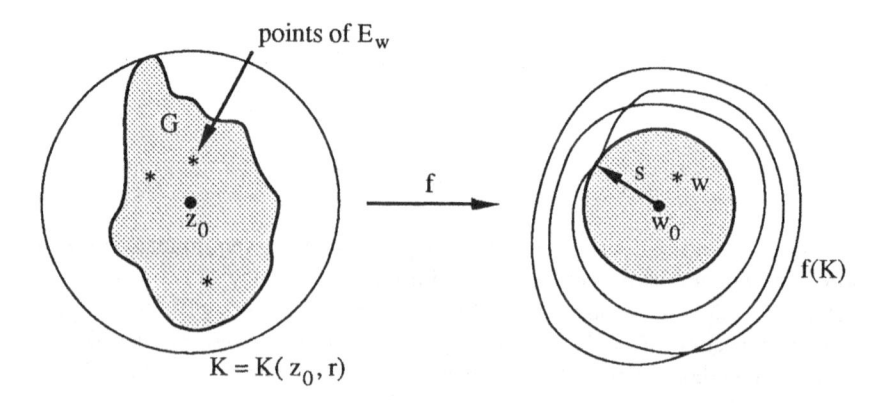

Figure 9.

3. The Residue Theorem and its Consequences

D, the component of U which contains z_0 — hence, that f' does not vanish identically in this domain. The discrete mapping theorem thus certifies that both $\{z \in D : f(z) = w_0\}$ and $\{z \in D : f'(z) = 0\}$ are discrete subsets of D. In particular, it is the case that for every suitably small $r > 0$ the disk $\overline{\Delta} = \overline{\Delta}(z_0, r)$ lies in D and the conditions $f(z) \neq w_0$ and $f'(z) \neq 0$ are satisfied at every point z of $\overline{\Delta} \sim \{z_0\}$. We focus on such an r. Since the function sending z to $|f(z) - w_0|$ is continuous and positive on the circle $K = K(z_0, r)$, it has a positive minimum value on K, a value we have labeled s. The continuity of f in $\Delta(z_0, r)$ implies that $G = \{z \in \Delta(z_0, r) : f(z) \in \Delta(w_0, s)\}$ describes an open set (Theorem II.2.5).

We now fix a point w belonging to the punctured disk $\Delta^*(w_0, s)$. We apply Rouché's theorem to the pair of functions $g(z) = f(z) - w_0$ and $h(z) = f(z) - w$ in the disk $\Delta = \Delta(z_0, r)$. Since

$$|g(z) - h(z)| = |f(z) - w_0 - (f(z) - w)| = |w - w_0| < s \leq |g(z)|$$

for every point z of $\partial \Delta = K$, we infer that, once multiplicity is accounted for, g and h have an equal number of zeros in Δ. By construction the function g has exactly m zeros there, all of them concentrated in a zero of order m at z_0. Consequently, h also has precisely m zeros in Δ — these necessarily located in $\Delta^*(z_0, r)$, for $w \neq w_0$. The fact that $h'(z) = f'(z) \neq 0$ for z in $\Delta^*(z_0, r)$ means that each of the m zeros of h in $\Delta(z_0, r)$ is a simple zero. In other words, $E_w = \{z \in \Delta(z_0, r) : f(z) = w\}$ contains exactly m points, at each of which f assumes the value w with multiplicity one. By the definition of G, E_w is a subset of G.

It remains only to demonstrate that the open set G is connected. For this, let V be any component of G. We show that z_0 must be a point of V. If this is true, then G has as its sole component the unique one containing z_0; i.e., G is connected. We shall assume, to the contrary, that z_0 does not belong to V, and then argue to a contradiction. Consider the function $k : \overline{V} \to \mathbb{C}$ given by $k(z) = [f(z) - w_0]/s$. This function is continuous, it is analytic in V, and, since z_0 is not a point of V and V lies in $\overline{\Delta}$, k is free of zeros in V. Owing to the definition of G, $|k(z)| < 1$ holds for all z in V. Thus, the continuity of k implies that $|k(z)| \leq 1$ throughout $\overline{V}$. It now follows from the fact that V is a component of G that $|k(z)| = 1$ whenever z is a point of ∂V. (N.B. If $|k(z)| < 1$ were to hold for some z on ∂V, then z would have to lie in Δ — z is certainly an element of $\overline{\Delta}$ and, in view of the definition of s, $|k(z)| \geq 1$ on the circle $K = K(z_0, r)$ — and by continuity there would be an open disk $\Delta(z, \rho)$ contained in Δ with the property that $|k(\zeta)| < 1$ for every ζ in $\Delta(z, \rho)$. This would stamp $\Delta(z, \rho)$ as a subset of G and so place this disk inside some component of G, a state of affairs incompatible with z being a boundary point of a component of G.) We conclude that the function $1/k$ is continuous on $\overline{V}$, analytic in V, and of modulus one everywhere on ∂V. The maximum principle — or, more precisely, Corollary V.3.11 — informs us that $1/|k(z)| \leq 1$ for every

z in V; i.e., $|k(z)| \geq 1$ in V. This obvious contradiction renders untenable the assumption that z_0 is not a member of V. Accordingly, G has but one component — hence, is a domain. ∎

The paradigm for Theorem 3.7 is found in the function $f(z) = z^m$ at $z_0 = 0$. In this example the conditions on r are met for every $r > 0$, we have $s = r^m$ and $G = \Delta(0, r)$, and for w in $\Delta^*(0, s)$ the points of E_w are just the mth-roots of w. In the general case described by the theorem the arrangement of points in the sets E_w need not be so regular, but f still maps the set $G \sim \{z_0\}$ in an m-to-one fashion onto the punctured disk $\Delta^*(w_0, s)$.

The restriction of the function f in Theorem 3.7 to the domain G belongs to a class of mappings that has great significance both for complex analysis and for topology, the class of "branched covering mappings." Although there are slight discrepancies in the way the term "branched covering" is used in the two subjects, complex analysts and topologists would concur here in describing the restriction of f to G as an *m-sheeted branched covering* of the disk $\Delta(w_0, s)$, *branched* (or *ramified*) in the case $m > 1$ over the point w_0. If $m > 1$, z_0 is the unique *branch point* of this particular covering, meaning the only point of G where the multiplicity of f exceeds one. The machinery of covering space theory, a part of topology, leads to an interesting refinement in the conclusion of the branched covering principle: *if $\Delta_0 = \Delta(0, \sqrt[m]{s})$, then it is possible to construct a univalent analytic function $\psi: \Delta_0 \to \mathbb{C}$ that satisfies $\psi(\Delta_0) = G, \psi(0) = z_0$, and $f[\psi(\zeta)] = w_0 + \zeta^m$ for every ζ in Δ_0* (Figure 10).

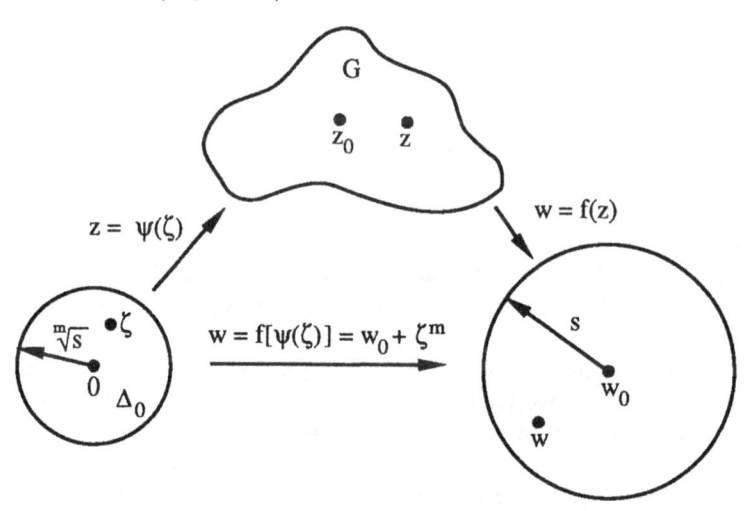

Figure 10.

It is not uncommon in mathematics to simplify a discussion by making an appropriate change of variables or, to put it differently, by choosing a coordinate system suited to that discussion. For instance, a correspondence $w = f(z) = f(x, y)$ might be more easily understood when viewed in polar coordinates, $w = f(r, \theta)$, following the change of variables $x = r\cos\theta$, $y = r\sin\theta$. The message of Figure 10 is that the mapping ψ can be used to install a new coordinate system in the domain G by assigning to a point z of G the coordinates of the point $\zeta = \xi + i\eta$ in Δ_0 for which $\psi(\zeta) = z$ and that, when expressed in terms of this transplanted coordinate system, the mapping $w = f(z)$ assumes the extremely simple form $w = f(\zeta) = w_0 + \zeta^m$. The mapping ψ is a particularly nice one — ψ is a "conformal mapping," the type of mapping that provides the subject matter for the next chapter — so nice, in fact, that almost every important qualitative geometric feature exhibited by f in G is inherited by the function $f \circ \psi$. Accepting this statement at face value and realizing that the geometry of $f \circ \psi$, which just maps ζ to $w_0 + \zeta^m$, is completely transparent, we recognize in the mapping ψ, whose existence is assured by covering space theory, the key to a total understanding of the local geometric structure of f near z_0.

A complex-valued function f whose domain-set includes the open set V in the complex plane is called an *open mapping of* V under the condition that $f(U)$ is an open set whenever U is an open subset of V. Not the least of the corollaries of the branched covering principle is:

Theorem 3.8. (Open Mapping Theorem) *If a function f is analytic and non-constant in a plane domain D, then f is an open mapping of D. In particular, $f(D)$ is a domain.*

Proof. Let U be an open set that is contained in D. We must verify that the set $f(U)$ is open. Given a point w_0 of $f(U)$ — say $w_0 = f(z_0)$, where z_0 belongs to U — we must produce an open disk $\Delta(w_0, s)$ that is a subset of $f(U)$. To this end, we simply choose $r > 0$ such that $\overline{\Delta} = \overline{\Delta}(z_0, r)$ lies in U and such that the requirements $f(z) \neq w_0$, $f'(z) \neq 0$ in the branched covering principle are fulfilled at every point z of $\overline{\Delta} \sim \{z_0\}$. The latter is possible because f is assumed to be analytic and non-constant in D. The aforementioned proposition certifies that every point of the disk $\Delta(w_0, s)$, where $s = \min\{|f(z) - w_0| : z \in K(z_0, r)\}$, is in the image of $\Delta(z_0, r)$ under f — hence, lies in $f(U)$. Thus, $f(U)$ is an open set. In particular, $f(D)$ is an open set. As this set is also connected (Theorem II.3.8), it is seen to be a domain. ∎

Another interesting observation based on the branched covering principle will be required for our discussion of conformal mappings in the succeeding chapter.

Theorem 3.9. *Let f be a function that is analytic in a domain D. If f is univalent in D, then $f'(z) \neq 0$ holds for every z in D.*

Proof. Suppose that $f'(z_0) = 0$ for some point z_0 of D. Being univalent in D, f is certainly non-constant there. Therefore, f assumes the value

$w_0 = f(z_0)$ with some multiplicity m at z_0. Because $f'(z_0) = 0$, $m \geq 2$. The branched covering principle shows that any point w sufficiently close to w_0, but different from it, has at least two preimages in D. This contradicts the univalence hypothesis. The conclusion: $f'(z) \neq 0$ must hold at each z in D. ∎

Theorem 3.9 should be contrasted with the situation in ordinary calculus where, for instance, the function $f(x) = x^3$ is a univalent differentiable function from $\mathbb{R}$ to itself, and yet $f'(0) = 0$. The converse of Theorem 3.9 is false. As an example, $f(z) = e^z$ is an entire function, $f'(z) = e^z$ is never zero, but f is far from univalent. On a local level, however, Theorem 3.9 does admit a converse.

Theorem 3.10. *Let f be a function that is analytic in a domain D. If z_0 is a point of D for which $f'(z_0) \neq 0$, then there is a subdomain G of D containing z_0 in which f is univalent.*

Proof. Since $f'(z_0) \neq 0$, f assumes the value $w_0 = f(z_0)$ with multiplicity $m = 1$ at z_0. Referring to the statement of Theorem 3.7 and taking $U = D$, we see that in the case $m = 1$ the domain G contains z_0 and is mapped by f in a univalent fashion onto the disk $\Delta(w_0, s)$. ∎

A final contribution to the circle of ideas surrounding the branched covering principle concerns inverses of analytic functions.

Theorem 3.11. (Inverse Function Theorem) *Suppose that D is a domain in the complex plane and that $f \colon D \to \mathbb{C}$ is a univalent analytic function. Then its inverse function $f^{-1} \colon f(D) \to D$ is also analytic.*

Proof. The open mapping theorem insures that $D' = f(D)$ is a domain. We first prove that f^{-1} is a continuous function. Let z_0 be a point of D', and let $w_0 = f^{-1}(z_0)$. Given $\epsilon > 0$ we shall produce a $\delta > 0$ for which $f^{-1}[\Delta(z_0, \delta)]$ is contained in $\Delta(w_0, \epsilon)$. Now $U = D \cap \Delta(w_0, \epsilon)$ is an open set in D, one that includes the point w_0, so by the open mapping theorem $U' = f(U)$ is an open subset of D' containing z_0. Plainly $f^{-1}(U') = U$. We select $\delta > 0$ so that $\Delta = \Delta(z_0, \delta)$ is contained in U'. Then $f^{-1}(\Delta)$ is a subset of U — hence, of $\Delta(w_0, \epsilon)$. This demonstrates that f^{-1} is continuous at each point of D'. It follows that $g = f^{-1}$ satisfies the hypotheses of Theorem 1.8 in the domain D'. We conclude on the basis of that result that f^{-1} is an analytic function. ∎

Rouché's theorem is pertinent to topics other than the local behavior of analytic functions. Evidence of this fact surfaces in the following convergence theorem of Adolf Hurwitz (1859-1919).

Theorem 3.12. (Hurwitz's Theorem) *Suppose that each function in a sequence $\langle f_n \rangle$ is analytic and zero-free in a domain D and that $f_n \to f$ normally in D. Then either f is free of zeros in D or it is identically zero there.*

Proof. The function f is analytic in D. Assume that $f(z_0) = 0$ for some point z_0 of D. We show that $f(z) = 0$ for every z in D. Suppose this not to be the case. The alternative is for z_0 to be an isolated zero of f. We can, accordingly, choose an $r > 0$ such that the closed disk $\overline{\Delta} = \overline{\Delta}(z_0, r)$ is contained in D and such that $f(z) \neq 0$ holds for every z on the circle $K = K(z_0, r)$. The quantity $|f(z)|$ attains a positive minimum value on K, call that value ϵ. Since $f_n \to f$ uniformly on K, there exists an index n with the property that $|f_n(z) - f(z)| < \epsilon \leq |f(z)|$ is true for every point z of K. Rouché's theorem dictates that f_n and f have the same number of zeros in $\Delta(z_0, r)$. Therefore, f_n has at least one zero there, contrary to hypothesis. The contradiction compels the conclusion that f does, indeed, vanish identically in D. ∎

The function $f_n(z) = z/n$ is analytic and zero-free in the domain $D = \mathbb{C} \sim \{0\}$. Quite clearly, $f_n \to 0$ normally in D. The second possibility allowed by the conclusion of Hurwitz's theorem is thus seen to be a real one. In the next chapter we shall have need of a noteworthy consequence of Hurwitz's theorem, a result with which we close this section.

Theorem 3.13. *Suppose that each function in a sequence $\langle f_n \rangle$ is analytic and univalent in a domain D and that $f_n \to f$ normally in D. Then either f is univalent in D or it is constant there.*

Proof. The function $f: D \to \mathbb{C}$ is analytic. Under the assumption that f is non-constant we verify that it is univalent in D. Fix z_0 in D. We show that $f(z) \neq f(z_0)$ when $z \neq z_0$. For this, consider the sequence of functions g_n defined in the domain $D_0 = D \sim \{z_0\}$ by $g_n(z) = f_n(z) - f_n(z_0)$. Because f_n is univalent in D, g_n is zero-free in D_0. Of course, $g_n \to g$ normally in D_0, where $g(z) = f(z) - f(z_0)$. Furthermore, g is not the zero function in D_0. (If it were, then f would be constant in D.) Hurwitz's theorem confirms that g is free of zeros in D_0; i.e., $f(z) \neq f(z_0)$ for $z \neq z_0$. As z_0 was an arbitrary point of D, the univalence of f is established. ∎

4 Function Theory on the Extended Plane

4.1 The Extended Complex Plane

Expressions like

$$\lim_{x \to x_0} f(x) = \infty \quad , \quad \lim_{x \to -\infty} g(x) = \ell \quad , \quad \lim_{x \to \infty} h(x) = -\infty$$

are commonplace in calculus. Such "limits involving infinity" provide us with valuable knowledge about the behavior of functions in situations that are not strictly covered by the primary definition of a limit, in which both the limit itself and the point at which it is taken are required to be real numbers. That definition can be modified, of course, so as to turn each of the above limit expressions into a meaningful and informative mathematical statement. A standard way to unify the treatments of ordinary limits and limits involving infinity is to work in the context of the "extended real line" obtained by attaching two "ideal boundary points," the points $-\infty$ and ∞, to the field $\mathbb{R}$ of real numbers. These added points are not to be thought of as "extra" real numbers. In fact, $-\infty < x < \infty$ holds by definition for all real numbers x. Moreover, the algebraic structures in $\mathbb{R}$ do not extend fully to $[-\infty, \infty]$: certain sums, products, and quotients containing the points at infinity can be consistently defined — e.g., $\infty + \infty = \infty$, $\infty \cdot (-\infty) = -\infty$, $1/\infty = 0$ — while others remain indeterminate — e.g., $\infty + (-\infty)$, $0 \cdot \infty$, $1/0$. Nevertheless, this expanded number system furnishes an excellent framework in which to discuss simultaneously all the variants of the limit concept suggested above.

An extended system similar to the one just described for the real numbers can be introduced for the complex numbers. There is, however, a crucial difference between the real and complex cases: the "extended complex plane" is formed by adjoining to $\mathbb{C}$ not two, but a single point at infinity. The disparity is tied up with the fact that $\mathbb{C}$ is an unorderable field, whereas $\mathbb{R}$ comes equipped with its standard ordering. (See Appendix A.) Thus, the *extended complex plane*, a set we symbolize by $\widehat{\mathbb{C}}$, consists of the complex plane $\mathbb{C}$ and an ideal point ∞ which, as part of its definition, is required not to be an element of $\mathbb{C}$. The plane $\mathbb{C}$ and the point ∞ fit together to form $\widehat{\mathbb{C}}$ in such a fashion that any sequence $\langle z_n \rangle$ of complex numbers without an accumulation point in $\mathbb{C}$ must tend to ∞ as a limit in $\widehat{\mathbb{C}}$. (This will be made precise when we discuss the topology of $\widehat{\mathbb{C}}$ several paragraphs hence.) The following algebraic rules are postulated for $\widehat{\mathbb{C}}$ to supplement those already in effect in $\mathbb{C}$:

(8.28) $\begin{cases} \infty \pm z = z \pm \infty = \infty & \text{for } z \text{ in } \mathbb{C}\ ; \\ \infty \cdot z = z \cdot \infty = \infty & \text{for } z \text{ in } \widehat{\mathbb{C}} \sim \{0\}\ ; \\ z/\infty = 0 & \text{for } z \text{ in } \mathbb{C}\ ; \\ z/0 = \infty & \text{for } z \text{ in } \widehat{\mathbb{C}} \sim \{0\}\ . \end{cases}$

Observe that none of the expressions $\infty + \infty$, $\infty - \infty$, ∞/∞, $0/0$, or $0 \cdot \infty$ is assigned meaning in $\widehat{\mathbb{C}}$.

4.2 The Extended Plane and Stereographic Projection

There is a traditional way of representing the extended complex plane $\widehat{\mathbb{C}}$ as a concrete geometric object. To describe it, we start with the sphere $S = \{(x_1, x_2, x_3) : x_1^2 + x_2^2 + x_3^2 = 1\}$ in three-dimensional euclidean space $\mathbb{R}^3$. By identifying the complex number $x_1 + ix_2$ with the point $(x_1, x_2, 0)$

4. Function Theory on the Extended Plane

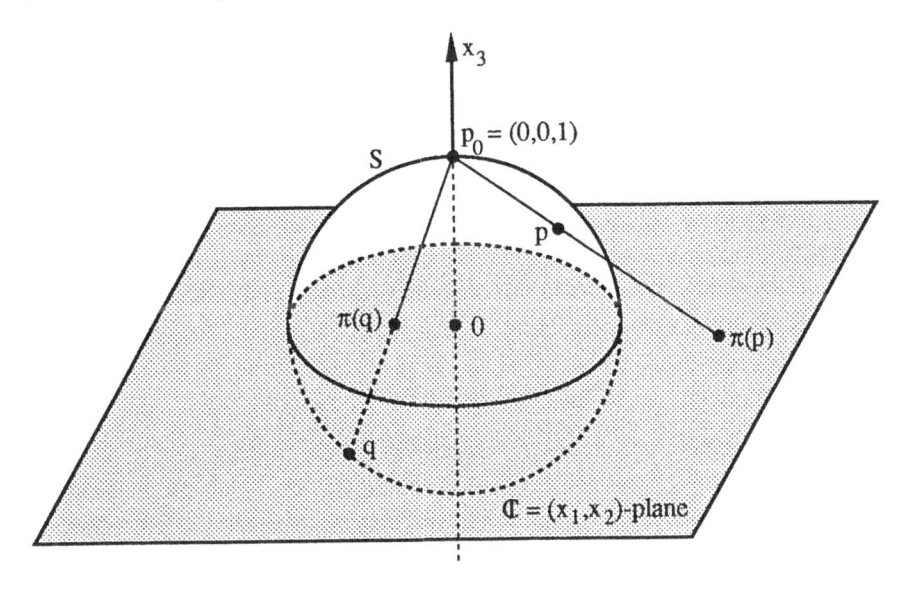

Figure 11.

we are free to think of $\mathbb{C}$ as sitting inside $\mathbb{R}^3$, masquerading as the (x_1, x_2)-plane. Having made this identification, we set $p_0 = (0, 0, 1)$ and define a function $\pi: S \sim \{p_0\} \to \mathbb{C}$ as follows: for p in S, $p \neq p_0$, $\pi(p)$ is the point of intersection of $\mathbb{C}$ with the ray in $\mathbb{R}^3$ that issues from p_0 and passes through p (Figure 11). The function π is called the *stereographic projection of $S \sim \{p_0\}$ onto* $\mathbb{C}$.

To determine an explicit formula for $\pi(p)$, notice that by definition $\pi(p) = p_0 + t(p - p_0)$, where t is the unique positive real number that forces the third coordinate of $p_0 + t(p - p_0)$ to vanish. Writing $p = (x_1, x_2, x_3)$, we obtain

$$p_0 + t(p - p_0) = (0, 0, 1) + t(x_1, x_2, x_3 - 1) = (tx_1, tx_2, 1 + t(x_3 - 1))$$

and see that $t = (1 - x_3)^{-1}$ is the number in question. Consequently, for $p = (x_1, x_2, x_3)$ in $S \sim \{p_0\}$ we have

$$(8.29) \quad \pi(p) = \left(\frac{x_1}{1 - x_3}, \frac{x_2}{1 - x_3}, 0 \right) = \frac{x_1}{1 - x_3} + i \left(\frac{x_2}{1 - x_3} \right).$$

Given a complex number $z = x + iy$, one checks easily that $z = \pi(p)$ for

$$p = \left(\frac{2x}{|z|^2 + 1}, \frac{2y}{|z|^2 + 1}, \frac{|z|^2 - 1}{|z|^2 + 1} \right)$$

and arrives at the fact that π has an inverse function $\pi^{-1}: \mathbb{C} \to S \sim \{p_0\}$, the function whose rule of correspondence is

$$(8.30) \quad \pi^{-1}(z) = \left(\frac{2x}{|z|^2 + 1}, \frac{2y}{|z|^2 + 1}, \frac{|z|^2 - 1}{|z|^2 + 1} \right).$$

The upshot of the preceding comments is that π establishes a one-to-one correspondence between $S \sim \{p_0\}$ and $\mathbb{C}$. Furthermore, formulas (8.29) and (8.30) make it evident that both π and π^{-1} are continuous functions. In the standard language of topology a continuous function with a continuous inverse is known as a *homeomorphism*. The domain-set and range of a homeomorphism are from a purely topological point of view indistinguishable sets. In our case, $S \sim \{p_0\}$ and $\mathbb{C}$ are thus seen via π to be *homeomorphic* (or *topologically equivalent*) sets. Remarking that $|\pi(p)| \to \infty$ as $p \to p_0$ and that $\pi^{-1}(z) \to p_0$ as $|z| \to \infty$, it seems only natural to extend π to a univalent function on S with range $\widehat{\mathbb{C}}$ by setting $\pi(p_0) = \infty$. Once this is done (and once the appropriate topological structure on $\widehat{\mathbb{C}}$ is in place), π becomes a continuous mapping from S to $\widehat{\mathbb{C}}$ that has a continuous inverse; i.e., π becomes a homeomorphism from S onto $\widehat{\mathbb{C}}$. Because of this beautiful correspondence between S and $\widehat{\mathbb{C}}$ under stereographic projection, it is topologically correct to visualize $\widehat{\mathbb{C}}$ as a sphere. Indeed, one of the common terms of reference to $\widehat{\mathbb{C}}$ describes it as the "Riemann sphere."

The stereographic projection mapping $\pi\colon S \to \widehat{\mathbb{C}}$ has many interesting geometric properties. At least a couple of these deserve a mention in passing. Thinking of p_0 as the north pole of S, we note that latitudinal circles on S project under π to circles in $\mathbb{C}$ centered at the origin, whereas longitudinal circles on S are mapped by π to lines in $\mathbb{C}$ passing through the origin — or, rather, to such lines with the point ∞ appended to them. In fact, it is a general phenomenon that a circle K situated on S will be transformed by π to a *circle in* $\widehat{\mathbb{C}}$, by which we mean either a true circle in $\mathbb{C}$ or a line in $\mathbb{C}$ with the point ∞ adjoined, and that circles in $\widehat{\mathbb{C}}$ are carried by π^{-1} to circles on S (Exercise 5.81). In addition, if two different circles on S intersect at a point p of $S \sim \{p_0\}$ at an angle θ — in order to be definite, use the non-obtuse angle of intersection — it can be demonstrated that the images of these circles under π intersect at the point $\pi(p)$ at the same angle θ. Consequently, we speak of π as a "conformal" (= angle preserving) mapping.

4.3 Functions in the Extended Setting

Just as we have declared that, when nothing is said to the contrary, the notation $f\colon A \to \mathbb{C}$ always indicates a function whose domain-set A lies in $\mathbb{C}$, we now establish the convention that, unless otherwise stated, $f\colon A \to \widehat{\mathbb{C}}$ will invariably signify a function having a subset A of $\widehat{\mathbb{C}}$ as its domain-set. The latter class of functions includes the former one, but it also embraces functions with ∞ located in their domain-sets or ranges or both. We have earlier agreed to take as the domain-set of a function f that is given by a formula, with no domain-set specified, the set of all points in $\mathbb{C}$ for which the formula expressing f has meaning. The analogous convention will ap-

4. Function Theory on the Extended Plane

ply in $\widehat{\mathbb{C}}$. Bearing in mind the extra algebraic rules we've adopted for $\widehat{\mathbb{C}}$, however, we remark that a formula which fails to make sense relative to the complex number system at a given point may well become meaningful at that point when viewed in the context of $\widehat{\mathbb{C}}$. Thus, a function presented by a formula typically has associated with it two distinct domain-sets, the first arising when the formula is interpreted as describing a complex-valued function of a complex variable, the second when ∞ is admitted as a candidate for the domain-set or range of the function in question. In order to distinguish between the two situations we shall refer to the domain-set in $\widehat{\mathbb{C}}$ of a formula-generated function as its *extended domain-set*. Take, for example, the function $f(z) = z^{-1}$. By our previous convention, its domain-set is $\mathbb{C} \sim \{0\}$. Its extended domain-set is $\widehat{\mathbb{C}}$, since according to (8.28) both $f(0) = 1/0 = \infty$ and $f(\infty) = 1/\infty = 0$ are well-defined in $\widehat{\mathbb{C}}$. In fact, $f(z) = z^{-1}$ defines a univalent mapping of $\widehat{\mathbb{C}}$ onto itself. The domain-set and the extended domain-set of the function $g(z) = e^z$, on the other hand, coincide (each is the plane $\mathbb{C}$) for e^∞ is an undefined expression in $\widehat{\mathbb{C}}$.

Let f be a rational function of z, say

(8.31) $$f(z) = \frac{a_0 + a_1 z + \cdots + a_n z^n}{b_0 + b_1 z + \cdots + b_m z^m}$$

with $a_n \neq 0$ and $b_m \neq 0$. We assume that f is in lowest terms, which demands that its numerator and denominator have no common polynomial factor of positive degree or, equivalently, that these functions have no common zero. (N.B. A rational function of z not in lowest terms can always be reduced to lowest terms and, in fact, there is an algorithm for carrying out the reduction that does not depend on finding the roots of its numerator and denominator.) As a formula defining a function in $\widehat{\mathbb{C}}$, (8.31) then has meaning for every z in $\mathbb{C}$. It assigns the value ∞ at each root of the denominator. Furthermore, by rewriting $f(z)$ for $z \neq 0$ in the form

$$f(z) = z^{n-m} \frac{\frac{a_0}{z^n} + \frac{a_1}{z^{n-1}} + \cdots + a_n}{\frac{b_0}{z^m} + \frac{b_1}{z^{m-1}} + \cdots + b_m},$$

we can also make sense of $f(\infty)$; namely, $f(\infty) = a_n/b_n$ if $n = m$, $f(\infty) = 0$ if $n < m$, and $f(\infty) = \infty$ if $n > m$. In other words, we can assert: *allowing for reduction to lowest terms, the extended domain-set of a rational function of z is $\widehat{\mathbb{C}}$*. As a special case, any polynomial function of z has extended domain-set $\widehat{\mathbb{C}}$ and, if it has positive degree, it takes the value ∞ at ∞.

Given functions $f: A \to \widehat{\mathbb{C}}$ and $g: B \to \widehat{\mathbb{C}}$ we can form the functions cf with a complex constant c, $f + g$, fg, and f/g, just as we did for complex-valued functions. In the extended case it is necessary to exercise a bit more caution with regard to the extended domain-sets of such algebraic combinations of f and g, for any one of these may be different from $A \cap B$.

The point is that in forming these functions we must avoid elements of $A \cap B$ at which indeterminacies such as $\infty + \infty$, $0 \cdot \infty$, etc., crop up. Assuming that $f(A)$ is contained in B, the composition $g \circ f$ is defined exactly as it was earlier. In particular, the notion of an inverse function carries over directly to the setting of $\widehat{\mathbb{C}}$.

4.4 Topology in the Extended Plane

Pains were taken in Chapter II to phrase each of the basic topological definitions in terms that would make the transition from the topology of $\mathbb{C}$ to that of $\widehat{\mathbb{C}}$ as smooth as possible. We now briefly retrace the main lines of development in Chapter II and indicate how to transfer the key ideas found there to the extended plane. Since most of the pertinent definitions were formulated using the open disks $\Delta(z,r)$ for z in $\mathbb{C}$ and $r > 0$, the first thing we must do is introduce the analogous "open disks centered at ∞." This presents no problem: for $r > 0$, we simply define

$$\Delta(\infty, r) = \{z : |z| > r^{-1}\} \cup \{\infty\} = \widehat{\mathbb{C}} \sim \overline{\Delta}(0, r^{-1}) .$$

The closed disk $\overline{\Delta}(\infty, r)$ and punctured disk $\Delta^*(\infty, r)$ are defined similarly.

As in the finite plane, a point z of $\widehat{\mathbb{C}}$ is called an *interior point* of a subset A of $\widehat{\mathbb{C}}$ if there exists an $r > 0$ such that the open disk $\Delta(z,r)$ is contained in A. A set U in $\widehat{\mathbb{C}}$ all of whose points are interior points of U is pronounced an *open subset* of $\widehat{\mathbb{C}}$. Thus, any open set in $\mathbb{C}$ is an open set in $\widehat{\mathbb{C}}$ as well. We declare a subset A of $\widehat{\mathbb{C}}$ to be *closed in* $\widehat{\mathbb{C}}$ provided $\widehat{\mathbb{C}} \sim A$ is an open set. A closed set A in $\mathbb{C}$ need not be closed in $\widehat{\mathbb{C}}$; in fact, a closed subset of $\mathbb{C}$ is closed in $\widehat{\mathbb{C}}$ if and only if the set in question is also bounded. Theorems II.1.1 and II.1.2 have obvious parallels in $\widehat{\mathbb{C}}$.

To say that a point z in $\widehat{\mathbb{C}}$ is a *boundary point* of a subset A of the extended plane means that for every $r > 0$ the disk $\Delta(z,r)$ has non-empty intersection with both A and $\widehat{\mathbb{C}} \sim A$. The *extended boundary* of A, a set we represent by $\widehat{\partial} A$, is made up of all such boundary points. If A is a set in $\mathbb{C}$, then $\widehat{\partial} A = \partial A$ when A is bounded and $\widehat{\partial} A = \partial A \cup \{\infty\}$ otherwise. The *extended closure* $\widehat{A}$ of a set A in $\widehat{\mathbb{C}}$ is defined by $\widehat{A} = A \cup \widehat{\partial} A$. Again, for a subset A of $\mathbb{C}$ we have $\widehat{A} = \overline{A}$ when A is bounded and $\widehat{A} = \overline{A} \cup \{\infty\}$ in the unbounded case. Both of the sets $\widehat{\partial} A$ and $\widehat{A}$ are closed in $\widehat{\mathbb{C}}$.

Let $\langle z_n \rangle$ be a sequence of $\widehat{\mathbb{C}}$. The statement that $\langle z_n \rangle$ is *convergent in* $\widehat{\mathbb{C}}$ with the point z_0 of $\widehat{\mathbb{C}}$ as its *limit* means this: corresponding to each $\epsilon > 0$ there exists an index $N = N(\epsilon)$ such that z_n belongs to $\Delta(z_0, \epsilon)$ for every $n \geq N$. Similarly, $\langle z_n \rangle$ has the point z_0 of $\widehat{\mathbb{C}}$ as an *accumulation point* if for every $\epsilon > 0$ the disk $\Delta(z_0, \epsilon)$ contains z_n for infinitely many values of n. For a sequence $\langle z_n \rangle$ in $\mathbb{C}$ we emphasize the distinction between "$\langle z_n \rangle$ is convergent" and "$\langle z_n \rangle$ is convergent in $\widehat{\mathbb{C}}$": the former requires that $\lim_{n \to \infty} z_n$ belong to $\mathbb{C}$, whereas the latter recognizes $\lim_{n \to \infty} z_n = \infty$ as

4. Function Theory on the Extended Plane

an acceptable alternative. Observe, too, that a sequence $\langle z_n \rangle$ in $\mathbb{C}$ which has no accumulation points in $\mathbb{C}$ necessarily satisfies $\lim_{n\to\infty} z_n = \infty$ in $\widehat{\mathbb{C}}$. The entire discussion of complex sequences in Chapter II demands little more than "fine tuning" when we transport it to $\widehat{\mathbb{C}}$. For instance, Theorem II.1.2, which describes the algebra of limits for convergent sequences, must undergo a slight adjustment in order to accomodate the rules of arithmetic peculiar to $\widehat{\mathbb{C}}$. One especially useful remark is that a sequence $\langle z_n \rangle$ in $\mathbb{C}$ has $\lim_{n\to\infty} z_n = \infty$ if and only if $\lim_{n\to\infty} 1/z_n = 0$.

A function $f: A \to \widehat{\mathbb{C}}$ is *continuous at a point* z_0 provided z_0 belongs to A and corresponding to each $\epsilon > 0$ there exists a $\delta = \delta(\epsilon, z_0) > 0$ with the property that

$$f[A \cap \Delta(z_0, \delta)] \subset \Delta(f(z_0), \epsilon) .$$

The only difference between this definition and the definition of continuity recorded in Chapter II is that we now admit $z_0 = \infty$ or $f(z_0) = \infty$ into the realm of possibilities. If f is continuous at every point of A, then f is proclaimed a *continuous function*. The rational functions of z, for example, constitute an important subclass of the continuous functions from $\widehat{\mathbb{C}}$ to $\widehat{\mathbb{C}}$. By simply writing $\widehat{\mathbb{C}}$ in place of $\mathbb{C}$ and substituting the words "extended complex plane" for "complex plane" we can transfer the bulk of the discussion of continuity from Chapter II to the present setting. A few points, these relating primarily to the algebraic conventions in $\widehat{\mathbb{C}}$ and their effects on the domain-sets of functions, demand minor cosmetic attention before they become valid in $\widehat{\mathbb{C}}$. This comment applies, for instance, to the counterpart of Theorem II.2.2 for the extended plane.

Modulo small technical details, again associated with the algebraic rules operative in $\widehat{\mathbb{C}}$, the treatment of limits of functions in the extended complex plane is also a retelling of the tale related in Chapter II concerning limits in the finite plane. Thus, if $f: A \to \widehat{\mathbb{C}}$ and if z_0 is a limit point of A in $\widehat{\mathbb{C}}$ (as in $\mathbb{C}$, this means that $A \cap \Delta^*(z_0, r) \neq \phi$ for every $r > 0$) we speak of f possessing a *limit in* $\widehat{\mathbb{C}}$ *at* z_0 if there exists a (necessarily unique) point w_0 of $\widehat{\mathbb{C}}$ with the following property: corresponding to each $\epsilon > 0$ there is a $\delta = \delta(\epsilon) > 0$ such that

$$f[A \cap \Delta^*(z_0, \delta)] \subset \Delta(w_0, \epsilon) .$$

We continue to refer to w_0 as the *limit of f at z_0* and to write $\lim_{z \to z_0} f(z) = w_0$ or $f(z) \to w_0$ as $z \to z_0$. In line with an earlier comment about sequences, we make a distinction, when dealing with a function $f: A \to \mathbb{C}$ and a finite limit point z_0 of A, between the statements "$\lim_{z \to z_0} f(z)$ exists" and "$\lim_{z \to z_0} f(z)$ exists in $\widehat{\mathbb{C}}$." In the first case, $\lim_{z \to z_0} f(z)$ must by definition be a complex number; in the second instance, $\lim_{z \to z_0} f(z) = \infty$

is an admissible possibility. Here are three simple examples:

$$\lim_{z \to 1} \frac{z^2 - 1}{(z-1)^2} = \lim_{z \to 1} \frac{z+1}{z-1} = \frac{2}{0} = \infty ;$$

$$\lim_{z \to \infty} (z^4 + z^2 + 1) = \lim_{z \to \infty} \left[z^4 \left(1 + \frac{1}{z^2} + \frac{1}{z^4} \right) \right]$$

$$= \lim_{z \to \infty} z^4 \lim_{z \to \infty} \left(1 + \frac{1}{z^2} + \frac{1}{z^4} \right) = \infty \cdot 1 = \infty ;$$

$$\lim_{z \to \infty} z \sin(z^{-1}) = \lim_{z \to 0} \frac{\sin z}{z} = 1 .$$

The first and second examples, which illustrate the algebraic rules governing limits $\widehat{\mathbb{C}}$, also exploit the continuity of rational functions of z; the third applies the useful fact that $\lim_{z \to \infty} f(z) = \lim_{z \to 0} f(z^{-1})$ when one of these limits is known to exist in $\widehat{\mathbb{C}}$.

The definitions of *connected set*, *domain*, and *compact set* carry over verbatim from $\mathbb{C}$ to $\widehat{\mathbb{C}}$. A non-empty open set U in $\widehat{\mathbb{C}}$ is again the disjoint union of domains, the *components* of U. An important remark is that any closed subset K of $\widehat{\mathbb{C}}$ — this includes $K = \widehat{\mathbb{C}}$ — is compact. Indeed, let $\langle z_n \rangle$ be a sequence in K. Then $\langle z_n \rangle$ has an accumulation point in $\widehat{\mathbb{C}}$, for either it has an accumulation point in $\mathbb{C}$ or $z_n \to \infty$. Being closed in $\widehat{\mathbb{C}}$, K contains all accumulation points of $\langle z_n \rangle$ in $\widehat{\mathbb{C}}$. The sequence $\langle z_n \rangle$ must, therefore, have an accumulation point in K. Theorems II.3.8 and II.4.6 have analogues in $\widehat{\mathbb{C}}$: if $f : A \to \widehat{\mathbb{C}}$ is a continuous function, if C is a connected subset of A, and if K is a compact subset of A, then $f(C)$ is a connected set and $f(K)$ is a compact set.

Finally, the notions of a *discrete subset of an open set*, a *discrete mapping*, and an *open mapping* are defined in the obvious ways that generalize the definitions given for the corresponding ideas in the finite plane situation.

4.5 Meromorphic Functions and the Extended Plane

Let U be an open set in $\widehat{\mathbb{C}}$. A complex-valued function f is called *meromorphic in U* subject to the condition that f have at each point z_0 of U no worse than a pole. The only difference between this and the previous definition offered for the concept is that the set U is now allowed to include the point ∞, in which event the new definition adds the requirement that f have an isolated singularity at ∞ in the sense of Section 2.6 and that this singularity be non-essential. For example: any rational function of z is meromorphic in $\widehat{\mathbb{C}}$; $f(z) = \tan z$ is meromorphic in $\mathbb{C}$, but not in $\widehat{\mathbb{C}}$, since it

4. Function Theory on the Extended Plane

does not have an isolated singularity at ∞; $g(z) = (z^2 + 1)^{-1} e^z$ is likewise meromorphic in the finite plane, but not the extended plane, for g has an essential singularity at ∞.

As soon as it is confirmed that a function f is meromorphic in an open subset U of $\widehat{\mathbb{C}}$, we shall enforce our announced policy of summarily ridding f of any removable singularities in U: when f has a removable singularity at a point z_0 of U — $z_0 = \infty$ is a possibility now — $f(z_0)$ automatically finds itself defined or redefined so as to insure that $f(z_0) = \lim_{z \to z_0} f(z)$. If we insist on working only with complex-valued functions, then no similar policy can be implemented for poles of f. If, however, we are willing to broaden our perspective and to admit functions taking values in $\widehat{\mathbb{C}}$, then something along these lines can be done. Namely, it follows from Theorem 2.4 that when z tends to a pole z_0 of f — $z_0 = \infty$ not excluded — $f(z) \to \infty$ with respect to the topology of $\widehat{\mathbb{C}}$. As a consequence, by defining $f(z_0) = \infty$ at each pole z_0 of f in the set U we extend the already modified function to a continuous function from U into $\widehat{\mathbb{C}}$. Motivated by these thoughts we issue a revamped "removal of singularities" policy: *whenever it is established that a function is meromorphic in an open subset U of $\widehat{\mathbb{C}}$, it is assumed that this function undergoes immediate modification to make it continuous in U as a function whose target-set is $\widehat{\mathbb{C}}$.*

We henceforth reserve the title *meromorphic function* for a function of the type $f: U \to \widehat{\mathbb{C}}$, where U is an open set in $\widehat{\mathbb{C}}$ and f is both continuous and meromorphic in U. Embedded in the foregoing discussion is the following characterization of a meromorphic function.

Theorem 4.1. *Let U be an open set in $\widehat{\mathbb{C}}$, and let $f: U \to \widehat{\mathbb{C}}$ be a continuous function. Then f is a meromorphic function if and only if the set $E = \{z \in U : z = \infty \text{ or } f(z) = \infty\}$ is a discrete subset of U and f is analytic in the open subset $U \sim E$ of $\mathbb{C}$.*

It is occasionally convenient to have a special designation for a function that is meromorphic in an open set U in $\widehat{\mathbb{C}}$ but has no poles in U. We refer to such a function as *holomorphic in U*. (If U lies in $\mathbb{C}$, then because of our removable singularity policy the phrase "holomorphic in U" is essentially synonymous with "analytic in U.") A *holomorphic function* is a meromorphic function without poles. There is one open set in which the holomorphic functions are especially easy to describe.

Theorem 4.2. *If $f: \widehat{\mathbb{C}} \to \widehat{\mathbb{C}}$ is a holomorphic function, then f is constant in $\widehat{\mathbb{C}}$.*

Proof. Since f is continuous in $\widehat{\mathbb{C}}$ and since $\widehat{\mathbb{C}}$ is a compact set, $f(\widehat{\mathbb{C}})$ is a compact set. Because f does not take the value ∞ anywhere, $f(\widehat{\mathbb{C}})$ is actually a compact set in $\mathbb{C}$ — hence, a bounded set. The restriction of f to $\mathbb{C}$ is, therefore, a bounded entire function. Liouville's theorem states that f is constant in $\mathbb{C}$. Owing to its continuity, f is then constant in $\widehat{\mathbb{C}}$. ∎

As was the case in the finite plane, when functions f and g are both meromorphic in an open subset U of the extended complex plane, so too are the functions f', $f + g$, fg, and, unless g is identically zero in some component of U, f/g. Along with Theorem 4.2 this observation comes into play in identifying all the functions that are meromorphic in $\widehat{\mathbb{C}}$.

Theorem 4.3. *If $f: \widehat{\mathbb{C}} \to \widehat{\mathbb{C}}$ is a meromorphic function, then f is a rational function of z.*

Proof. (We are already aware, of course, that rational functions of z are meromorphic.) By definition, $E = \{z \in \widehat{\mathbb{C}} : z = \infty \text{ or } f(z) = \infty\}$ is a discrete subset of $\widehat{\mathbb{C}}$. As $\widehat{\mathbb{C}}$ is compact, this can only happen if E is a finite set. Let $z_1, z_2, \ldots, z_p = \infty$ list the points of E, and let S_k denote the singular part of f at z_k. Then S_k is a rational function of z whose only singularities appear at z_k and ∞. For $k < p$ the singularity of S_k at ∞ is clearly removable: for such k, $S_k(z) \to 0$ as $z \to \infty$. The function $g = f - S_1 - S_2 \cdots - S_p$ is thus seen to be meromorphic in $\widehat{\mathbb{C}}$, where its only singularities are removable ones at $z_1, z_2, \ldots, z_p$; i.e., g is holomorphic in $\widehat{\mathbb{C}}$. Theorem 4.2 reveals that, once its singularities are removed, g is a constant function in $\widehat{\mathbb{C}}$ — say $g(z) = c$ for all z in $\widehat{\mathbb{C}}$, where c is a complex number. Consequently, $f = c + S_1 + S_2 + \cdots + S_p$ is seen to be a rational function of z. ∎

Many properties of analytic functions, especially those of a topological character, are enjoyed by meromorphic functions as well. The next theorem, which generalizes Theorem 1.5, affords a prime example of such a property.

Theorem 4.4. (Discrete Mapping Theorem) *If D is a domain in $\widehat{\mathbb{C}}$ and if $f: D \to \widehat{\mathbb{C}}$ is a non-constant meromorphic function, then f is a discrete mapping of D.*

Proof. Let w be a point of $\widehat{\mathbb{C}}$. We claim that $E = \{z \in D : f(z) = w\}$ is a discrete subset of D. If $w = \infty$, this follows from Theorem 4.1. We proceed under the assumption that w is a finite point. The set E does not have a limit point in the domain $G = \{z \in D : z \neq \infty, f(z) \neq \infty\}$, where f is analytic: if it did, then the discrete mapping theorem for analytic functions would imply that f is constant in G — hence, by continuity, constant in D, contrary to hypothesis. The only way left in which E could conceivably have a limit point in D would be for ∞ to belong to D and for E to have ∞ as a limit point. Assuming this to be so, we choose $r > 0$ with the property that the punctured disk $\Delta^* = \Delta^*(\infty, r)$ is contained in the domain G and select a sequence $\langle z_n \rangle$ of points in $E \cap \Delta^*$ converging to ∞. The continuity of f tells us that $f(\infty) = \lim_{n \to \infty} f(z_n) = w$, so f does not have a pole at ∞. It follows that f takes only finite values in $\Delta(\infty, r)$. The function $g: \Delta(0, r) \to \mathbb{C}$ given by $g(z) = f(z^{-1})$ is continuous, and due to the choice of r it is analytic in $\Delta^*(0, r)$. We infer using Riemann's extension theorem

4. Function Theory on the Extended Plane

that g is analytic in $\Delta(0,r)$. On the other hand, $g(\zeta_n) = w$ for $\zeta_n = z_n^{-1}$, and $\langle\zeta_n\rangle$ is a sequence in $\Delta^*(0,r)$ satisfying $\zeta_n \to 0$. In view of the discrete mapping property of non-constant analytic functions, it can only be the case that $g(z) = w$ for every z in $\Delta(0,r)$ or, equivalently, that $f(z) = w$ for every z in $\Delta(\infty, r)$. In particular, any point of $\Delta^*(\infty, r)$ is a limit point of the set E in G, something we have already ruled out. We conclude on the grounds of this contradiction that ∞ could not be a limit point of E in D. As a set in D without limit points in this domain, E is a discrete subset of D. ∎

Theorem 4.4 has the following corollary, which is a generalization of the principle of analytic continuation (Exercise 5.82).

Corollary 4.5. *If meromorphic functions f and g are both defined in a domain D in $\widehat{\mathbb{C}}$ and if $f(z) = g(z)$ for all z belonging to some subset A of D that has a limit point in D, then $f(z) = g(z)$ for every z in D.*

Theorem 4.4 also makes it easy to see that, under mild technical restrictions, the composition of meromorphic functions is again meromorphic.

Theorem 4.6. *Let $f: D \to \widehat{\mathbb{C}}$ be a non-constant meromorphic function, where D is a domain in $\widehat{\mathbb{C}}$, and let $g: U \to \widehat{\mathbb{C}}$ be a meromorphic function such that U contains $f(D)$. Then the function $g \circ f$ is meromorphic.*

Proof. Write $h = g \circ f$. Certainly h maps D continuously into $\widehat{\mathbb{C}}$. Given a point z_0 in D, we shall exhibit a punctured disk $\Delta^*(z_0, r)$ in which h is analytic. Because $\lim_{z \to z_0} h(z) = h(z_0)$ exists in $\widehat{\mathbb{C}}$, the function h could not have an essential singularity at z_0. It follows that h has no worse than a pole at z_0, an arbitrary point of D. By definition, this makes h a meromorphic function. Set $w_0 = f(z_0)$. Since the function g is meromorphic, we can select $s > 0$ so that g is analytic in $\Delta^*(w_0, s)$. Then, as f is meromorphic, we can pick $r > 0$ with the property that f is analytic in $\Delta^*(z_0, r)$. Furthermore, the fact that f is a continuous, discrete mapping of D into $\widehat{\mathbb{C}}$ means that r can be so chosen that f maps $\Delta(z_0, r)$ into $\Delta(w_0, s)$ and that $f(z) \neq w_0$ holds for every z in $\Delta^*(z_0, r)$. For such a choice of r the set $f[\Delta^*(z_0, r)]$ lies in $\Delta^*(w_0, s)$. The analyticity of the composition h in $\Delta^*(z_0, r)$ is then clear. ∎

If the function f in Theorem 4.6 were constant in D, then $g \circ f$ would also be constant there. This would trivially make $g \circ f$ meromorphic in D, save in one case: should the value of f be a pole of g, $g \circ f$ would be constantly infinite in D — and definitely not meromorphic.

Another significant feature that meromorphic functions inherit from analytic functions is the open mapping property.

Theorem 4.7. (Open Mapping Theorem) *If D is a domain in $\widehat{\mathbb{C}}$ and if $f: D \to \widehat{\mathbb{C}}$ is a non-constant meromorphic function, then f is an open mapping of D. In particular, $f(D)$ is a domain in $\widehat{\mathbb{C}}$.*

Proof. Let U be an open set in D. We must prove that $f(U)$ is an open set in $\widehat{\mathbb{C}}$: given z_0 in U, we must show that there is an $s > 0$ with the property that the disk $\Delta(w_0, s)$ is contained in $f(U)$, where $w_0 = f(z_0)$. For this we assume initially that $w_0 \neq \infty$. Choose $r > 0$ such that $\Delta = \Delta(z_0, r)$ is contained in U and such that f is analytic in the punctured disk $\Delta^*(z_0, r)$. If $z_0 \neq \infty$, then f is just a non-constant analytic function in Δ. In this event the open mapping theorem for analytic functions implies that $f(\Delta)$, a subset of $f(U)$, is an open set in $\mathbb{C}$ and insures the existence of the requisite s. If $z_0 = \infty$, then the function $g: \Delta(0, r) \to \mathbb{C}$ defined by $g(z) = f(z^{-1})$ is non-constant and analytic in $\Delta(0, r)$. Furthermore, $f(\Delta) = g[\Delta(0, r)]$. The open mapping theorem applied to g confirms that $f(\Delta)$ is open in this case, too, again implying that an s with the desired property exists. Finally, suppose that $w_0 = \infty$. Then $h = 1/f$ is a non-constant meromorphic function in D and $h(z_0) = 0$. Applying the previous considerations to h, we are guaranteed the existence of a radius s for which $\Delta(0, s)$ lies in $h(U)$. This, of course, places $\Delta(\infty, s)$ within $f(U)$. We are thus able to conclude that $f(U)$ is an open set. In particular, the set $f(D)$ is open. As the continuous image of a connected set, this set is also connected — i.e., $f(D)$ is a domain. ∎

With the aid of the open mapping theorem we can verify that the inverse of a univalent meromorphic function is meromorphic.

Theorem 4.8. (Inverse Function Theorem) *Suppose that D is a domain in $\widehat{\mathbb{C}}$ and that $f: D \to \widehat{\mathbb{C}}$ is a univalent meromorphic function. Then its inverse function $f^{-1}: f(D) \to D$ is also meromorphic.*

Proof. By Theorem 4.7, $D' = f(D)$ is a domain. The identical argument used in the proof of Theorem 3.11 to demonstrate the continuity of f^{-1} works to establish the continuity of the inverse in the present situation. The function f maps the domain $G = \{z \in D : z \neq \infty, f(z) \neq \infty\}$ in an analytic and univalent fashion to $G' = \{w \in D' : w \neq \infty, f^{-1}(w) \neq \infty\}$. Theorem 3.11 implies that f^{-1} is analytic in G'. Thus f^{-1} is analytic except for two eventual isolated singularities in D', the point ∞ if it belongs to D' and the point $f(\infty)$ if ∞ should lie in D. The continuity of f^{-1} rules out any chance that either of these potential singularities might be essential, so f^{-1} can have no worse than a pole at any point of D' — hence, is meromorphic in D'. ∎

A further quotable consequence of the open mapping theorem is a maximum principle for functions holomorphic in subdomains of $\widehat{\mathbb{C}}$.

Theorem 4.9. (Maximum Principle) *Let $f: D \to \mathbb{C}$ be a holomorphic function, where D is a domain in $\widehat{\mathbb{C}}$. Suppose that there exists a point z_0 of D with the property that $|f(z)| \leq |f(z_0)|$ for every z in D. Then f is constant in D.*

Proof. The point $f(z_0)$ is not an interior point of the set $f(D)$, a bounded

4. Function Theory on the Extended Plane

subset of $\mathbb{C}$: if $f(z_0)$ were an interior point of $f(D)$, there would obviously be elements w in $f(D)$ for which $|w| > |f(z_0)|$, in conflict with the hypothesis. Accordingly, $f(D)$ is not an open set. The open mapping theorem leaves only one alternative: f is constant in D. ∎

The branched covering principle can likewise be generalized to the extended plane setting. We state the theorem to round out the discussion in this section, but we leave its proof as an exercise (Exercise 5.83). Even the statement demands a few words of explanation, these concerning the interpretation of "multiplicity." Consider a meromorphic function f that is defined and non-constant in some disk $D = \Delta(z_0, r)$ in $\widehat{\mathbb{C}}$. Set $w_0 = f(z_0)$. Suppose first that $z_0 \neq \infty$. If $w_0 \neq \infty$, then we already know what it means to say that f assumes the value w_0 with multiplicity m at z_0, for f is then analytic and non-constant in a perhaps smaller open disk centered at z_0. Next, if $w_0 = \infty$, then f has a pole of some order m at z_0. In this case we say that f assumes the value ∞ with multiplicity m at z_0. Lastly, when $z_0 = \infty$ we say that f assumes the value w_0 with multiplcity m at z_0 if and only if the function g defined by $g(z) = f(z^{-1})$ assumes the value w_0 with multiplicity m at the origin. In terms of the Laurent expansion of f in a punctured disk $\Delta^*(\infty, r)$, the definition of f assuming the value w_0 with multiplicity m at ∞ translates to the following: when $w_0 \neq \infty$, f takes the value w_0 with multiplicity m at ∞ if and only if the Laurent expansion in question has the form

$$f(z) = w_0 + \frac{a_{-m}}{z^m} + \frac{a_{-m-1}}{z^{m+1}} + \cdots$$

with $a_{-m} \neq 0$; f takes the value ∞ with multiplicity m at ∞ if and only if this expansion looks like

$$f(z) = a_m z^m + \cdots + a_1 z + a_0 + \frac{a_{-1}}{z} + \frac{a_{-2}}{z^2} + \cdots$$

with $m \geq 1$ and $a_m \neq 0$.

Theorem 4.10. (Branched Covering Principle) *Suppose that $f: U \to \widehat{\mathbb{C}}$ is a meromorphic function, that z_0 is a point of U, and that f takes the value w_0 with multiplicity m at z_0. Let $r > 0$ be any number sufficiently small that the following conditions prevail: the closed disk $\overline{\Delta} = \overline{\Delta}(z_0, r)$ is contained in U, and the statements $f(z) \neq w_0$, $f(z) \neq \infty$, $f'(z) \neq 0$ are true for every z in the set $\overline{\Delta} \sim \{z_0\}$. Define $s = s(r) > 0$ to be the largest number for which $\Delta(w_0, s) \cap f(K) = \phi$, where K is the circle that bounds $\overline{\Delta}$. Then $G = \{z \in \Delta(z_0, r) : f(z) \in \Delta(w_0, s)\}$ is a domain. Moreover, for each point w of the punctured disk $\Delta^*(w_0, s)$ the set $E_w = \{z \in \Delta(z_0, r) : f(z) = w\}$ consists of exactly m points of G, at each of which f assumes the value w with multiplicity one.*

5 Exercises for Chapter VIII

5.1 Exercises for Section VIII.1

5.1. Determine the multiplicity with which f takes its value at z_0: (i) $f(z) = e^{z\cos z - z}$, $z_0 = 0$; (ii) $f(z) = z^{\text{Log } z}$, $z_0 = 1$; (iii) $f(z) = \text{Log}^2(\cos z)$, $z_0 = 2\pi$; (iv) $f(z) = \tan^2(1 + 2z^2 + z^4)$, $z_0 = i$; (v) $f(z) = (1 + z^2 - e^{z^2})^3$, $z_0 = 0$; (vi) $f(z) = z^2 - \text{Arctan}(z^2)$, $z_0 = 0$.

5.2. Let f be a function that is analytic and non-constant in a domain D, and let g be a function with those same properties in a domain D' that contains the range of f. Show that the multiplicity of the composition $g \circ f$ at any point z_0 of D is the product of the multiplicities of f at z_0 and g at $w_0 = f(z_0)$.

5.3. Evaluate the following limits: (i) $\lim_{z\to 0} z^{-2}(1 - \cos z)$; (ii) $\lim_{z\to 0} z^{-2}[1 - \cos(z^{3/2})]$; (iii) $\lim_{z\to 1}(z-1)^{-2}(z^{1/z} - z)$; (iv) $\lim_{z\to 2\pi}(1 - e^{iz})^{-2}\text{Log}(\cos z)$; (v) $\lim_{z\to 0}(1 + cz)^{1/z}$ for any complex number c; (vi) $\lim_{z\to 0}[(e^z - 1)^{-1} - z^{-1}]$.

5.4. Let E be a discrete subset of a plane domain D. Show that $D \sim E$ is a domain.

5.5. A function f is analytic in a domain D. It is known that for each point z of D there is a non-negative integer n_z with the property that $f^{(n_z)}(z) = 0$. Show that f must be a polynomial function of z in D. (*Hint.* Choose and fix a closed disk K in D. Feel free to use the information that K cannot be written as a union $K = \cup_{n=0}^{\infty} A_n$, where $A_0, A_1, A_2, \cdots$ are finite sets. Exercise III.6.16 is also pertinent.)

5.6. Let f be a function that is continuous on the closed disk $\overline{\Delta}(0, r)$ and analytic in $\Delta(0, r)$. Assuming that $f(z) = 0$ for every point z of the circle $K(0, r)$ having $\text{Im } z \geq 0$, prove that $f(z) = 0$ for every z in $\overline{\Delta}(0, r)$. (*Hint.* Consider $f(-z)$ along with $f(z)$.)

5.7. Let Q be a closed square centered at the origin, and let S be a side of Q (including the endpoints). If $f: Q^0 \to \mathbb{C}$ is a bounded analytic function with the property that $\lim_{z\to\zeta} f(z) = 0$ for every point ζ of S, then $f(z) = 0$ for every z in Q^0. Prove this.

5.8. If $f: \overline{\Delta}(0, 1) \to \mathbb{C}$ is a non-constant continuous function that is analytic in $\Delta(0, 1)$ and satisfies $|f(z)| = 1$ for every z on the circle $K(0, 1)$, demonstrate that f has the form

$$f(z) = c \prod_{k=1}^{r} \left(\frac{z - a_k}{1 - \overline{a}_k z}\right)^{m_k}$$

for z in $\overline{\Delta}(0, 1)$, where $a_1, a_2, \ldots, a_r$ are distinct points of the disk $\Delta(0, 1)$, $m_1, m_2, \ldots, m_r$ are positive integers, and c is a constant of unit modulus.

5. Exercises for Chapter VIII

(*Hint.* Begin by showing the f has at least one zero, but at most finitely many zeros, in $\Delta(0,1)$. Remember Exercise V.8.38 and also Exercise I.4.21.)

5.9. If a non-constant entire function f obeys the condition $|f(z)| = 1$ whenever $|z| = 1$, show that f must be of the type $f(z) = cz^m$, where m is a positive integer and c is a constant with $|c| = 1$. (*Hint.* Use Exercise 5.8 and the principle of analytic continuation.)

5.10. Let f be a non-constant entire function with the following property: there exist a pair of circles K and K' in the complex plane such that $f(K)$ is a subset of K'. Establish that f has the structure $f(z) = c(az+b)^m + d$, where m is a positive integer and where $a \neq 0, b, c \neq 0$, and d are constants.

5.11. Suppose that a function f is analytic in a domain D. Under the assumption that a branch g of the pth-root of f exists in D, prove that there are exactly p distinct branches of the pth-root of f in D, each having the form cg for some p^{th}-root of unity c. This result completes a cycle of ideas that started with Theorem III.4.2 and was continued in Exercise V.8.57.

5.12. Let $\sum_{n=0}^{\infty} a_n(z-z_0)^n$ be a Taylor series with a finite radius of convergence $\rho > 0$, and let f denote the function that is its sum in the disk $\Delta = \Delta(z_0, \rho)$. Show that there must be at least one point ζ of the circle $K(z_0, \rho)$ about which the following is true: for no $r > 0$ can f be extended to a function that is analytic in the set $\Delta \cup \Delta(\zeta, r)$. (A point ζ of this type is called a "singular point" for the given Taylor series. *Hint.* Assume that no such ζ exists and derive a contradiction.)

5.13. Assume that a function f is analytic in an open set U and that $f(z) = \sum_{n=0}^{\infty} a_n(z-z_0)^n$ is its Taylor series expansion about a point z_0 of U. Let Δ denote the disk of convergence of this series. Given that $U \cap \Delta$ is connected, show that one defines an analytic function $h: U \cup \Delta \to \mathbb{C}$ by setting $h(z) = f(z)$ for z in U and $h(z) = \sum_{n=0}^{\infty} a_n(z-z_0)^n$ for z in Δ. (Recall the comments following Theorem VII.3.4.)

5.14. If D is a bounded plane domain that contains the origin and if $f: D \to D$ is an analytic function that obeys the conditions $f(0) = 0$ and $f'(0) = 1$, then $f(z) = z$ for every z in D. Justify this statement. (*Hint.* Let $\Delta = \Delta(0, r)$ be contained in D. Show first that $f(z) = z$ for every z in Δ. If not, the Taylor series expansion of f about the origin would assume the form $f(z) = z + a_m z^m + O(z^{m+1})$ for some $m \geq 2$, where $a_m \neq 0$. Cauchy's estimate gives the bound $|a_m| \leq cr^{-m}$, with $c = \max\{|z|: z \in \overline{D}\}$. What would the Taylor series development in Δ of the k-fold composition $f_k = f \circ f \circ \cdots \circ f$ (k factors) look like? The correct answer to this question will produce a contradiction when $k \to \infty$.)

5.15. Let f be a branch of the inverse secant function in a domain D — i.e., $f: D \to \mathbb{C}$ is a continuous function and $\sec[f(z)] = z$ for every z in D. Show that f is an analytic function, that $f'(z) = \pm z^{-1}(z^2 - 1)^{-1/2}$

for every z in D (the sign need not remain constant in D) and that any branch g of the inverse secant in D has either the form $g = f + 2k\pi$ for some integer k or the form $g = -f + 2k\pi$ for some integer k.

5.16. Establish the following result of Guisseppe Vitali (1875-1932) concerning a family $\mathcal{F}$ of analytic functions which is normal in a domain D: if $\langle f_n \rangle$ is a sequence from $\mathcal{F}$ and if $\langle f_n(\zeta) \rangle$ converges for every point ζ belonging to a subset A of D that has a limit point in D, then $\langle f_n \rangle$ converges normally in D. (*Hint*. Recall Exercise VII.5.87.)

5.2 Exercises for Section VIII.2

5.17. Classify the singularity of f at z_0 and determine $\text{Res}(z_0, f)$: (i) $f(z) = \cot z, z_0 = k\pi$ for an integer k ; (ii) $f(z) = (z-1)^{-3} \cos(\pi z/2), z_0 = 1$; (iii) $f(z) = (z - \pi)^{-6} \sin^2 z, z_0 = \pi$; (iv) $f(z) = z^2 e^{-1/z^3}, z_0 = 0$; (v) $f(z) = (z+1)^4 \sin[\pi(z+1)^{-1}], z_0 = -1$; (vi) $f(z) = \text{Arctan}^{-2} z, z_0 = 0$; (vii) $f(z) = (z-1)^{-5} \text{Log}^2 z, z_0 = 1$; (viii) $f(z) = (z^2 + z) \cos(z^{-1}), z_0 = 0$.

5.18. Compute the singular part of f at z_0, and use it to obtain $\text{Res}(z_0, f)$: (i) $f(z) = z^2(z-1)^{-2}, z_0 = 1$; (ii) $f(z) = (z^2 + 1)^{-3}, z_0 = -i$; (iii) $f(z) = (1 - \cos z)^{-3} \sin(z^3), z_0 = 0$; (iv) $f(z) = z \text{Log}^{-3}(1 + z), z_0 = 0$; (v) $f(z) = (z^2 + 1)(e^{\pi z} + 1)^{-4}, z_0 = i$; (vi) $f(z) = [z^2 - \text{Arctan}(z^2)]^{-1}, z_0 = 0$; (vii) $f(z) = (z^z - 1)^{-3}, z_0 = 1$.

5.19. Suppose that a function f is analytic in the punctured plane $\mathbb{C} \sim \{0\}$, where it obeys the estimate $|f(z)| \leq c|z||\text{Log } z|$ for some constant $c > 0$. Demonstrate that $f(z) = 0$ for every z in $\mathbb{C} \sim \{0\}$. (*Hint*. Recall Exercise V.8.29.)

5.20. Let f have a pole of order m at a point z_0, and let g have a pole of order n at the same point. Show (i) that $f + g$ has either a removable singularity at z_0 or a pole of order not greater than $\max\{m, n\}$ at z_0; (ii) that fg has a pole of order $m + n$ at z_0; (iii) that g/f has a removable singularity at z_0 if $m \geq n$ and a pole of order $n - m$ at z_0 if $m < n$.

5.21. Given that a function f has a pole of order m at a point z_0, check that its derivative f' has a pole of order $m + 1$ at that point.

5.22. Let $f: \Delta(z_0, r) \to \mathbb{C}$ be a non-constant analytic function. If f takes the value w_0 with multiplicity m at z_0 and if g is a function with a pole of order n at w_0, verify that the composition $g \circ f$, which is well-defined in $\Delta^*(z_0, r)$ for sufficiently small r, has a pole of order mn at z_0.

5.23. Assuming that a function f is both analytic and even in a punctured disk $\Delta^* = \Delta^*(0, r)$, show that $\text{Res}(0, f) = 0$.

5.24. Let $r > 1$. If a function f is analytic in the disk $\Delta(0, r)$ except for a simple pole at the point $z_0 = 1$, verify that $\lim_{n\to\infty}(n!)^{-1}f^{(n)}(0) = -\text{Res}(1, f)$.

5.25. If a function f is analytic in $\Delta^*(z_0, r)$ with a simple pole at z_0 and if $I(s)$ is defined for $0 < s < r$ by $I(s) = \int_{\gamma_s} f(z)dz$, where $\gamma_s(t) = z_0 + se^{it}$ for $a \leq t \leq b$, then $I(s) \to i(b-a)\text{Res}(z_0, f)$ as $s \to 0$. Justify this statement.

5.26. Functions f and g are both analytic modulo isolated singularities in a domain D. Let E_f and E_g denote the singular sets of these functions in D, and let all the singularities in question be non-removable — i.e., assume that any removable singularities have been "removed." From the information that $f(z) = g(z)$ for all z belonging to a subset A of the domain $G = D \sim (E_f \cup E_g)$ and that A has a limit point in G, deduce that $E_f = E_g$ and that $f(z) = g(z)$ for every z in $D \sim E_f$.

5.27. If f and g are entire functions and if $|g(z)| \leq |f(z)|$ for every z in $\mathbb{C}$, demonstrate that $g(z) = cf(z)$ throughout the complex plane for some constant c.

5.28. Assuming that a function f is meromorphic in $\mathbb{C}$ and satisfies $|f(z)| = 1$ whenever $|z| = 1$, show that f admits a representation

$$f(z) = c \prod_{k=1}^{r} \left(\frac{z - a_k}{1 - \overline{a}_k z}\right)^{m_k} \prod_{\ell=1}^{s} \left(\frac{1 - \overline{b}_\ell z}{z - b_\ell}\right)^{n_\ell},$$

where c is a constant of modulus one, where $a_1, a_2, \ldots, a_r$ are the zeros of f in the disk $\Delta = \Delta(0,1)$ and $m_1, m_2, \ldots, m_r$ their respective orders, and where $b_1, b_2, \ldots, b_s$ are the poles of f in Δ and $n_1, n_2, \ldots, n_s$ their respective orders. (N.B. If f has no zeros in Δ, just leave out the first product; if f has no poles there, omit the the second product. *Hint.* Begin by showing that the above representation is valid in Δ. For this make use of Exercise V.8.38.)

5.29. If a function f is meromorphic in the whole complex plane and if there exist circles K and K' in the plane such that $f(K)$ is a subset of K', then f is necessarily a rational function of z. Support this claim.

5.30. Prove Theorem 2.5.

5.31. Suppose that a function f is analytic in some open set which contains the punctured closed disk $\overline{\Delta}(0,\delta) \sim \{0\}$ and that f has an isolated singularity at the origin. For $0 < p \leq 2$, set $I(p) = \lim_{\epsilon \to 0} \iint_{\epsilon \leq |z| \leq \delta} |f(z)|^p dx dy$. Certify that: (i) the singularity of f at 0 is removable if and only if $I(2) < \infty$; (ii) if the singularity is a pole of order m, then $I(p) < \infty$ for $0 < p < 2/m$ and $I(p) = \infty$ for $2/m \leq p \leq 2$; (iii) if $I(p) = \infty$ for every p in $(0, 2]$, then the singularity is essential. (*Hint for* (i). Show that $\iint_{\epsilon \leq |z| \leq \delta} |f(z)|^2 dx dy = 2\pi \sum_{n=-\infty}^{\infty} |a_n|^2 \int_{\epsilon}^{\delta} r^{2n+1} dr$, $f(z) = \sum_{n=-\infty}^{\infty} a_n z^n$

being the Laurent expansion of f in $\Delta^*(0,\delta)$.)

5.32. Let a function f be analytic and odd in $\Delta^* = \Delta^*(0,r)$ with an essential singularity at the origin. Assuming that Picard's theorem is true, check that either $f(\Delta^*) = \mathbb{C}$ or $f(\Delta^*) = \mathbb{C} \sim \{0\}$.

5.33. A function f is analytic in $\Delta^* = \Delta^*(0,r)$, has an essential singularity at the origin, and takes a real value at every point of $\Delta^* \cap \mathbb{R}$. Given that $f(\Delta^*)$ contains $\mathbb{R}$, prove by means of Picard's theorem that $f(\Delta^*) = \mathbb{C}$.

5.34. In each of the following cases the function f experiences an essential singularity at the origin. Granted the validity of Picard's theorem, decide whether $f[\Delta^*(0,1)] = \mathbb{C}$ or not: (i) $f(z) = (z^2 - z)e^{1/z}$; (ii) $f(z) = z\cos(z^{-1})$; (iii) $f(z) = e^{2/z} + 2e^{1/z} + 1$; (iv) $f(z) = z^{-1}\sin(z^{-1})$; (v) $f(z) = e^{1/z} - e^{-1/z}$; (vi) $f(z) = z^3 e^{1/z}\sin(z^{-1})$.

5.35. An analytic function $f\colon \Delta^*(0,1) \to \mathbb{C}$ obeys the conditions $f(z) \neq a$ and $f(z) \neq b$ for every z in $\Delta^*(0,1)$, where a and b are distinct complex numbers. Confirm that f can have no worse than a pole at the origin. (*Hint*. We may assume $a = 0$. (Why?) For $n = 1, 2, 3, \cdots$ define f_n in $\Delta^*(0,1)$ by $f_n(z) = f(n^{-1}z)$. Apply Theorem VII.4.6 to the sequence $\langle f_n \rangle$ in $D = \Delta^*(0,1)$. Remember that the terms of any sequence from $C(D)$ which converges normally in D are uniformly bounded on each compact set in D — in particular, are so on the circle $K(0, 1/2)$.)

5.36. Deduce Picard's theorem from Exercise 5.35.

5.37. If a function f has an isolated singularity at ∞, then $\text{Res}[\infty, f(z)] = -\text{Res}[0, z^{-2}f(z^{-1})]$. Corroborate this fact.

5.38. Assuming that a function f has no worse than a pole at ∞, show that $\text{Res}(\infty, f'/f) = \text{Res}(0, g'/g)$, where $g(z) = f(z^{-1})$.

5.39. Classify the singularity of f at ∞, and compute $\text{Res}(\infty, f)$: (i) $f(z) = z(1-z)^{-2}$; (ii) $f(z) = (1+z^2)(1-z^2)^{-1}$; (iii) $f(z) = z^3 \cos(z^{-1})$; (iv) $f(z) = z^{-3}\cos(\sqrt{z})$; (v) $f(z) = z^2(2z+1)(z^2+1)^{-1}$; (vi) $f(z) = \exp(z - z^{-1})$.

5.40. Let $f(z) = [1 + \cos(\sqrt{z})]e^z$ and let $\Delta^* = \Delta^*(\infty, r)$. Appealing to Picard's theorem if need be, determine $f(\Delta^*)$.

5.41. If f is an entire function with the property that $|f(z)| \to \infty$ as $|z| \to \infty$, verify that $f(\mathbb{C}) = \mathbb{C}$.

5.3 Exercises for Section VIII.3

5.42. Let γ be a Jordan contour in the complex plane, let D be the inside of the Jordan curve $|\gamma|$, and let f be a function that is analytic modulo isolated singularities in D and continuous in $\overline{D} \sim E$, where E is the singular set of f in D. Assuming Goursat's theorem (Theorem V.5.7), prove that

5. Exercises for Chapter VIII

$\int_\gamma f(z)dz = 2\pi i \sum_{z \in E} \text{Res}(z, f)$. (*Hint.* Show first that the number of non-removable singularities of f in D must be finite. Then adapt the proof given in the text for the residue theorem.)

5.43. A function f is analytic modulo isolated singularities in the complex plane and has an isolated singularity at ∞. Let E be the singular set of f in $\mathbb{C}$, let γ be a Jordan contour in $\mathbb{C} \sim E$, and let D^* denote the outside of the Jordan curve $|\gamma|$. Verify that

$$\int_\gamma f(z)\,dz = -2\pi i \left[\text{Res}(\infty, f) + \sum_{z \in E \cap D^*} \text{Res}(z, f) \right].$$

Use this information to deduce that $\text{Res}(\infty, f) + \sum_{z \in E} \text{Res}(z, f) = 0$ for any function f of the type described. (*Hint.* Choose $r > 0$ so that $\Delta(0, r)$ contains both E and $|\gamma|$. Consider the cycle $\sigma = (\gamma, -\beta)$, where $\beta(t) = re^{it}$ for $0 \le t \le 2\pi$.)

5.44. A function f is analytic modulo isolated singularities in a simply connected domain D, where its singular set is E. Show that f has a primitive in the domain $G = D \sim E$ if and only if $\text{Res}(z, f) = 0$ for all points z of E.

5.45. Evaluate the integrals: (i) $\int_{|z|=3} z^{-1}(z-1)^{-1}(z-2)^{-1}\,dz$;
(ii) $\int_{|z-e|=2}[(z-1)\,\text{Log}\,z]^{-1}\,dz$; (iii) $\int_{|z|=1}[(2z+1)^2\,\text{Arctan}(iz/2)]^{-1}\,dz$;
(iv) $\int_{\partial Q}\csc(z^3)\,dz$, where Q is the square with vertices $1+i, -1+i, -1-i$, and $1-i$; (v) $\int_\sigma z^{-2}\tan z\,dz$, where $\sigma = (2\gamma, 3\beta)$ is the cycle in which $\gamma(t) = 1 + 2e^{it}$ and $\beta(t) = 2 + e^{-it}$ for $0 \le t \le 2\pi$.

5.46. Evaluate $\int_\gamma z\csc^2 z\,dz$ with γ given by $\gamma = \gamma_1 + \gamma_2 + \gamma_3 + \gamma_4 + \gamma_5$, where $\gamma_1(t) = e^{it}$ for $0 \le t \le 3\pi$, $\gamma_2(t) = 2 + 3e^{it}$ for $\pi \le t \le 2\pi$, $\gamma_3(t) = 3 + 2e^{it}$ for $0 \le t \le 6\pi$, $\gamma_4(t) = -1 + 6e^{it}$ for $0 \le t \le \pi$, and $\gamma_5(t) = -3 + 4e^{it}$ for $\pi \le t \le 4\pi$.

5.47. Compute $\int_\gamma (z+\pi)(z-\pi)^{-1}(e^z+1)^{-1}\,dz$ with $\gamma = \beta + [-5\pi, 0]$, in which $\beta(t) = te^{it}$ for $0 \le t \le 5\pi$.

5.48. Evaluate: (i) $\int_{|z|=1} z^2 e^{i/z}\,dz$; (ii) $\int_{|z-1|=2} z^5 \sin(z^{-2})\,dz$;
(iii) $\int_{|z-i|=3/2}(z^2+1)^{-1}e^{\pi/z}\,dz$; (iv) $\int_\gamma z^{-1}\exp(z+z^{-1})\,dz$, where $\gamma(t) = \cos t + 2i\sin t$ for $0 \le t \le 4\pi$. (N.B. The answer in (iv) involves an infinite series.)

5.49. Certify that:

(i) $\int_0^{2\pi}(a + b\cos\theta)^{-1}\cos\theta\,d\theta = 2\pi b^{-1}[1 - a(a^2 - b^2)^{-1/2}]$ if $0 < b < a$;

(ii) $\int_0^{2\pi}(a + b\cos\theta)^{-2}\,d\theta = 2\pi a(a^2 - b^2)^{-3/2}$ if $0 < b < a$;

(iii) $\int_0^{2\pi}(ai + b\sin\theta)^{-1}\,d\theta = -2\pi i(a^2 + b^2)^{-1/2}$ if $a > 0$ and $b > 0$;

(iv) $\int_0^{2\pi}(a^2 + b^2\sin^2\theta)^{-1}\,d\theta = 2\pi a^{-1}(a^2 + b^2)^{-1/2}$ if $a > 0$ and $b > 0$;

(vi) $\int_0^{2\pi}(a^2\cos^2\theta+b^2\sin^2\theta)^{-2}\,d\theta = \pi(ab)^{-3}(a^2+b^2)$ if $a>0$ and $b>0$. (*Hint.* In each instance start with the case $b=1$.)

5.50. Evaluate: (i) $\int_{-\infty}^{\infty}(x^2+\pi^2)^{-2}\cos x\,dx$; (ii) $\int_0^{\infty}(x^2+9)^{-1}\sin^2(\pi x)\,dx$; (iii) $\int_{-\infty}^{\infty}(x^2+a^2)^{-1}(x^2+b^2)^{-1}\sin(\mu x)\sin(\nu x)\,dx$, $0<\mu\le\nu$, $0<a\le b$; (iv) $\int_0^{\infty}x^2(x^2+1)^{-2}\cos^2(\pi x)\,dx$; (v) $\int_{-\infty}^{\infty}(x^2+1)^{-1}(x^2+4)^{-2}\cos(\pi x)\,dx$; (vi) $\int_{-\infty}^{\infty}x(1+x^4)^{-1}\sin x\,dx$.

5.51. Let f be a function that is analytic modulo isolated singularities in the complex plane and has no singularities on the real axis. Assume that $|zf(z)|$ tends to a (finite) limit as $|z|\to\infty$. Demonstrate that for $c>0$

$$(8.32)\qquad \int_{-\infty}^{\infty}f(x)e^{icx}\,dx = 2\pi i\sum_{k=1}^{p}\mathrm{Res}[z_k,f(z)e^{icz}],$$

where $z_1,z_2,\ldots,z_p$ are the non-removable singularities of f in the half-plane $H=\{z:\mathrm{Im}\,z>0\}$. Part of the problem is to establish the convergence of the integral in (8.32). The stated assumptions imply the existence of positive constants r and M such that $|f(z)|\le M/|z|$ whenever $|z|\ge r$, but this inequality does not guarantee the integral's convergence. (*Hint.* All the non-removable singularities of f must lie in the disk $\Delta(0,r)$. Consider $\int_{\partial R}f(z)e^{icz}\,dz$, where R is the rectangle with vertices $a,b,b+i(b-a)$, and $a+i(b-a)$ for $a<-r$ and $b>r$.)

5.52. Let f be a proper rational function of z (i.e., the degree of the denominator in f exceeds that of the numerator) without any real poles. Certify that for $c>0$

$$\int_{-\infty}^{\infty}f(x)\cos(cx)\,dx = 2\pi i\sum_{k=1}^{p}\mathrm{Res}[z_k,f(z)e^{icz}]$$

if f is an even function, while

$$\int_{-\infty}^{\infty}f(x)\sin(cx)\,dx = 2\pi\sum_{k=1}^{p}\mathrm{Res}[z_k,f(z)e^{icz}]$$

if f is an odd function. Here $z_1,z_2,\ldots,z_p$ are the poles of f in the half-plane $H=\{z:\mathrm{Im}\,z>0\}$. (*Hint.* Apply Exercise 5.51.)

5.53. Find: (i) $\int_{-\infty}^{\infty}x(x^2+1)^{-1}\sin(\pi x)\,dx$; (ii) $\int_0^{\infty}x^3(x^2+\pi^2)^{-2}\sin x\,dx$; (iii) $\int_{-\infty}^{\infty}x^3(x^2+1)^{-1}(x^2+4)^{-1}\sin(2\pi x)\,dx$; (iv) $\int_0^{\infty}(x^2-i)^{-1}\cos x\,dx$.

5.54. By first integrating the function $f(z)=z^{-2}(1-e^{2iz})$ along the path γ that was introduced to work Example 3.9, calculate $\int_0^{\infty}t^{-2}\sin^2 t\,dt$. (*Hint.* Recall Exercise 5.25.)

5.55. Use the method of Example 3.10 to demonstrate that

$$\int_0^{\infty}t^{-\lambda}(t+b)^{-2}\,dt = \lambda\pi b^{-\lambda-1}\csc(\lambda\pi)$$

and
$$\int_0^\infty t^{-\lambda}(t+b)^{-1} \operatorname{Log} t \, dt = \pi b^{-\lambda} \csc(\lambda\pi)[\pi \cot(\lambda\pi) + \operatorname{Log} b]$$
when $0 < \lambda < 1$ and $b > 0$. (N.B. The first of these formulas can also be obtained by differentiating both sides of (8.21) with respect to the parameter b; differentiation of (8.21) with respect to λ will yield the second.)

5.56. Show that $\int_0^\infty (1+x^n)^{-1} \, dx = \pi n^{-1} \csc(\pi/n)$ for any integer $n \geq 2$. Conclude that $\int_{-\infty}^\infty (1+x^{2n})^{-1} \, dx = \pi n^{-1} \csc(\pi/2n)$ for every positive integer n. (*Hint*. Make a change of variable in Example 3.10 for some λ.)

5.57. Derive the formula $\int_0^\infty t^\lambda (t^2+b^2)^{-1} \, dt = (\pi/2) b^\lambda \sec(\lambda\pi/2)$, valid when $-1 < \lambda < 1$ and $b > 0$. (*Hint*. Let $0 < s < b$ and $b < r < \infty$. Integrate $f(z) = z^\lambda (z^2+b^2)^{-1}$ along $\gamma = [s,r] + \alpha + [-r,-s] - \beta$, where α and β are the paths defined on $[0,\pi]$ by $\alpha(t) = re^{it}$ and $\beta(t) = se^{it}$. Feel free to invoke Exercise 5.42.)

5.58. Verify that $\int_{-\infty}^\infty e^{\lambda t} \operatorname{sech} t \, dt = \pi \sec(\lambda\pi/2)$ for any complex number λ satisfying $-1 < \operatorname{Re} \lambda < 1$. (*Hint*. The estimate $|e^z + e^{-z}| \geq (1/2) e^{|x|}$ when $|x| \geq 1$ insures the convergence of the integral. Consider $\int_{\partial R} f(z) \, dz$, where $f(z) = e^{\lambda z} \operatorname{sech} z$ and R is the rectangle with vertices $-c, c, c + \pi i$ and $-c + \pi i$ for $c \geq 1$.)

5.59. Let p be a positive integer, and let $f(z) = z^{-2p}(e^z - 1)^{-1}$. If $Q = \{z : |x| \leq (2n+1)\pi, |y| \leq (2n+1)\pi\}$ for a positive integer n, show that
$$\int_{\partial Q} f(z) \, dz = 2\pi i \left(\frac{(-1)^{p-1} B_p}{(2p)!} + \frac{2(-1)^p}{(2\pi)^{2p}} \sum_{k=1}^n \frac{1}{k^{2p}} \right),$$
where B_p is the pth Bernoulli number (Exercise VII.5.62). Make use of this information to sum the series $\sum_{k=1}^\infty k^{-2p}$: $\sum_{k=1}^\infty k^{-2p} = 2^{-1}[(2p)!]^{-1}(2\pi)^{2p} B_p$. (*Hint*. For the last part notice that $|e^z - 1| \geq 1 - e^{-3\pi}$ for every point z of the set ∂Q.)

5.60. Prove the following generalized argument principle: if a function f is meromorphic in an open set U and if γ is a closed, piecewise smooth path in U that is homologous to zero in U and finds no zeros or poles of f on its trajectory, then
$$\frac{1}{2\pi i} \int_\gamma \frac{f'(z) \, dz}{f(z)} = \sum_{j=1}^r m_j n(\gamma, z_j) - \sum_{k=1}^s \ell_k n(\gamma, w_k),$$
where $z_1, z_2, \ldots, z_r$ lists all zeros of f for which $n(\gamma, z_j) \neq 0$, $m_1, m_2, \ldots, m_r$ being the respective orders of these zeros, while $w_1, w_2, \ldots, w_s$ are all the poles of f for which $n(\gamma, w_k) \neq 0$, $\ell_1, \ell_2, \ldots, \ell_s$ indicating their respective orders. Interpret an "empty" sum to mean zero.

5.61. If f, U, γ, and D meet all the conditions spelled out in Theorem 3.5 and if g is a function that is analytic in some open set which contains $\overline{D}$, where $z_1, z_2, \ldots, z_r$ lists all zeros of f for which $n(\gamma, z_j) \neq 0$, $m_1, m_2, \ldots, m_r$ being the respective orders of these zeros, while $w_1, w_2, \ldots, w_s$ are all the poles of f for which $n(\gamma, w_k) \neq 0$, $\ell_1, \ell_2, \ldots, \ell_s$ indicating their respective orders. Interpret an "empty" sum to mean zero.

5.61. If f, U, γ, and D meet all the conditions spelled out in Theorem 3.5 and if g is a function that is analytic in some open set which contains $\overline{D}$, verify that

$$\frac{1}{2\pi i} \int_\gamma \frac{g(z) f'(z)\, dz}{f(z)} = \sum_{j=1}^r m_j g(z_j) - \sum_{k=1}^s \ell_k g(w_k),$$

where $z_1, z_2, \ldots, z_r$ are the zeros of f in D and $m_1, m_2, \ldots, m_r$ their respective orders, and where $w_1, w_2, \ldots, w_s$ are the poles of f in D and $\ell_1, \ell_2, \ldots, \ell_s$ the orders of these poles. An empty sum is taken to mean zero.

5.62. Rouché's theorem for meromorphic functions reads: if D is the domain inside the trajectory of a Jordan contour in the complex plane, if f and g are functions that are meromorphic in some open set which contains $\overline{D}$, if both of these functions are free of poles on ∂D, and if the inequality $|f(z) - g(z)| < |f(z)| + |g(z)|$ holds at every point z of ∂D, then $Z_f - P_f = Z_g - P_g$. Here Z_f and Z_g are the numbers of zeros that f and g display in D, while P_f and P_g are the corresponding totals for poles, all zeros and poles being counted with consideration for multiplicity. Establish this result.

5.63. It follows from elementary considerations in calculus that the real solutions of the equation $\tan z = z$ can be listed as $z_0 = 0, \pm z_1, \pm z_2, \cdots$, where $k\pi < z_k < (2k+1)\pi/2$. Show that this equation has no other solutions in the complex plane. Then find the sum of the series $\sum_{k=1}^\infty z_k^{-2}$. (*Hint.* Let $Q = \{z : |x| < n\pi, |y| < n\pi\}$, where n is an arbitrary positive integer. For the first part compare zero-pole differences for the functions $f(z) = z - \tan z$ and $g(z) = z$ in the interior of Q. For the second part look at $\int_{\partial Q}[(z - \tan z)^{-1} - z^{-1}]dz$. Useful in both parts is the observation that $|\tan z| \leq 2$ on ∂Q. Verify this.)

5.64. Let c be a complex number with $|c| > e$. Confirm that for every positive integer n the equation $cz^n = e^z$ has n different solutions in the disk $\Delta(0, 1)$ and has no solutions apart from these in the half-plane $H = \{z : \operatorname{Re} z < 1\}$. (*Hint.* Recall Example 3.12.)

5.65. Let $f(z) = z^5 + 5z^3 - 1$. Verify that f has five simple zeros, three of these lying in the disk $\Delta(0, 1)$ and the remaining two located in the set $\Delta(0, 7/3) \sim \{z : |z| \leq 2, |\operatorname{Re} z| > 3/2\}$.

5.66. Let $f(z) = z^3 + 3a^2 z - 1$, where $a > 0$. Then f has a simple zero in the interval $(0, 1)$, but no additional real roots. (Why?) The two non-real roots of f are complex conjugates of one another. Let z_2 denote the non-real root of f whose imaginary part is positive. By combining Rouché's theorem with the Gauss-Lucas theorem (Exercise V.8.35) and using other elementary facts about polynomials, demonstrate that z_2 finds itself in the set $\{z: -(1/2) < x < 0, y > a(1-x), \max\{1, a\} < |z| < \sqrt{1 + 3a^2}\,\}$.

5.67. For $n = 1, 2, 3, \cdots$ set $f_n(z) = \sum_{k=0}^{n} (k!)^{-1} z^k$. If $r > 0$ is fixed, prove that there is an index $N = N(r)$ such that f_n is free of zeros in the disk $\Delta(0, r)$ once $n \geq N$.

5.68. Confirm that the roots of a polynomial depend continuously on its coefficients in the following sense: given $f(z) = a_0 + a_1 z + \cdots + a_n z^n$, a polynomial of degree $n \geq 1$ whose different roots are $z_1, z_2, \ldots, z_r$, and given $\epsilon > 0$, there exists a corresponding $\delta > 0$ such that any polynomial $g(z) = b_0 + b_1 z + \cdots + b_n z^n$ with coefficients which satisfy $|b_k - a_k| < \delta$ for $k = 0, 1, \ldots, n$ will have at least one root in each of the disks $\Delta(z_\ell, \epsilon)$ and all of its roots in the set $\cup_{\ell=1}^{r} \Delta(z_\ell, \epsilon)$.

5.69. Let $f(z) = a_0 + a_1 z + \cdots + a_N z^N$ be a polynomial of degree $N \geq 2$ with the property that $\sum_{n=2}^{N} n|a_n| \leq |a_1|$. Prove that f is univalent in the disk $\Delta = \Delta(0, 1)$. (*Hint.* Noting that $a_1 = 0$ is not compatible with these assumptions, fix z_0 in Δ and compare the number of zeros that the functions $g(z) = a_1(z - z_0)$ and $h(z) = f(z) - f(z_0)$ have in Δ.)

5.70. Suppose that a function f is analytic and non-constant in the disk $\Delta = \Delta(0, 1)$, where it has a Taylor series expansion $f(z) = \sum_{n=0}^{\infty} a_n z^n$ whose coefficients obey the condition $\sum_{n=2}^{\infty} n|a_n| \leq |a_1|$. Prove that f is univalent in Δ. (*Hint.* Use the preceding exercise.)

5.71. A fairly general version of Rouché's theorem states: if D is a bounded, simply connected domain in the complex plane, if f and g are functions that are continuous on $\overline{D}$ and analytic in D, and if the inequality $|f(z) - g(z)| < |f(z)| + |g(z)|$ is satisfied for every point z of ∂D, then f and g have the same number of zeros in D, always presuming that zeros are counted with due regard for multiplicity. Prove this under the added assumption that D can be represented in the fashion $D = \varphi(\Delta)$, where $\Delta = \Delta(0, 1)$ and $\varphi: \Delta \to \mathbb{C}$ is a univalent analytic function. Conclude that, in particular, this form of Rouché's theorem is true when D is a disk. (N.B. The "Riemann Mapping Theorem," to be established in the next chapter, insures that every domain D of the type under consideration here can be represented in the manner indicated. *Hint.* Use a uniform continuity argument to show that Theorem 3.6 is applicable to f and g in the domain $D_r = \varphi[\Delta(0, r)]$ for all r sufficiently close to 1.)

5.72. Assume that a function f is continuous on the closed disk $\overline{\Delta}(0, 1)$, is analytic in $\Delta(0, 1)$, and satisfies $|f(z)| \leq 1$ when $|z| = 1$. Show that f has at least one fixed point in $\overline{\Delta}(0, 1)$ and that, if f fixes no point of the circle $K(0, 1)$, it has exactly one fixed point in $\overline{\Delta}(0, 1)$.

5.73. Let $f: \Delta(z_0, r) \to \mathbb{C}$ be a non-constant analytic function. If a function

g has an essential singularity at the point $w_0 = f(z_0)$, demonstrate that the composition $g \circ f$, defined in $\Delta^*(z_0, r)$ for small r, exhibits an essential singularity at z_0.

5.74. Identify all the entire functions f with the property that $f \circ f = f$.

5.75. Let D and D' be bounded domains in the complex plane, and let $f: \overline{D} \to \mathbb{C}$ be a continuous function that is analytic in D. Under the assumptions that $f(D)$ is a subset of D' and that $f(\partial D)$ is a subset of $\partial D'$, verify that $f(D) = D'$, $f(\partial D) = \partial D'$, and $f(\overline{D}) = \overline{D}'$. (*Hint.* To start, assume that $G = f(D) \neq D'$, and argue to a contradiction. Recall Exercise II.5.25.)

5.76. Show that a univalent entire function f must be a sense-preserving similarity transformation — i.e., $f(z) = az + b$ for constants $a \neq 0$ and b. (*Hint.* Begin by ruling out the possibility that f has an essential singularity at ∞.)

5.77. Let f be an entire function with the property that $f(z)$ is real for every point z of the real axis, but for no other z. Prove that $f'(z) \neq 0$ holds for all real z. Infer from this that f is univalent on the real axis. Conclude with the aid of Picard's theorem — or otherwise — that $f(z) = az + b$ for some real constants $a \neq 0$ and b. (*Hint.* Assuming that $f'(z_0) = 0$ for some point z_0 of the real axis, put the branched covering principle to work and obtain a contradiction.)

5.78. Let $f: U \to \mathbb{C}$ be a univalent analytic function, and let γ be a Jordan contour in U with the property that the inside D of the Jordan curve $|\gamma|$ is contained in U. Demonstrate that for every point w of the domain $D' = f(D)$ the value of the inverse of f at w is given by the formula
$$f^{-1}(w) = \frac{1}{2\pi i} \int_\gamma \frac{z f'(z) \, dz}{f(z) - w}.$$

5.79. Suppose that $f: \Delta(z_0, r) \to \mathbb{C}$ is a univalent analytic function, that $w_0 = f(z_0)$, and that the disk $\Delta(w_0, s)$ is contained in the range of f. Verify that $|f'(z_0)| \geq s/r$ and that equality holds if and only if f takes the form $f(z) = w_0 + (cs/r)(z - z_0)$ in $\Delta(z_0, r)$, where c is a constant of unit modulus. (*Hint.* Consider the function $g: \Delta(0, 1) \to \mathbb{C}$ defined by $g(z) = r^{-1}[f^{-1}(w_0 + sz) - z_0]$.)

5.80. Let D be a plane domain, let $f: D \to \mathbb{C}$ be an analytic function, and let v be a continuous function that is subharmonic in some open set which contains the range of f. Demonstrate that the composition $w = v \circ f$ is subharmonic in D. (*Hint.* Assuming f to be non-constant — the assertion follows trivially otherwise — begin by showing that w is subharmonic in the domain $G = \{z \in D : f'(z) \neq 0\}$. Consult Exercises VI.4.36 and VI.4.39.)

5.4 Exercises for Section VIII.4

5.81. Confirm that under the stereographic projection $\pi: S \to \widehat{\mathbb{C}}$ circles on S are transformed to circles in $\widehat{\mathbb{C}}$ and that under π^{-1} circles in $\widehat{\mathbb{C}}$ go to circles on S. (*Hint.* A circle on S is the intersection of S with a plane in $\mathbb{R}^3$. Make use of Exercise I.4.19.)

5.82. Prove Corollary 4.5.

5.83. Prove Theorem 4.10.

5.84. Let $f: \mathbb{C} \to \widehat{\mathbb{C}}$ be a non-constant meromorphic function with the feature that $f(\lambda z) = f(z)$ for every z in $\mathbb{C}$, where λ is a constant. Prove that there must be a positive integer p for which $\lambda^p = 1$.

5.85. Let $f: \widehat{\mathbb{C}} \to \widehat{\mathbb{C}}$ be a non-constant rational function of z. Prove that $f(\widehat{\mathbb{C}}) = \widehat{\mathbb{C}}$. (*Hint.* It is enough to show that each such function must have a pole in $\widehat{\mathbb{C}}$. Why?)

5.86. Refine the preceding exercise as follows: if $f: \widehat{\mathbb{C}} \to \widehat{\mathbb{C}}$ is a non-constant rational function of z, then f assumes every value in $\widehat{\mathbb{C}}$ the same number of times, provided multiplicity is taken into account. To be more precise, if w is a point of $\widehat{\mathbb{C}}$ and if $z_1, z_2, \ldots, z_r$ are the different solutions in $\widehat{\mathbb{C}}$ of the equation $f(z) = w$, then the sum $m_1 + m_2 + \cdots + m_r$, where m_k is the multiplicity of f at z_k, is independent of w. (*Hint.* It suffices to check that, for arbitrary f of this type, these multiplicity sums are the same for $w = 0$ and $w = \infty$. Why?)

5.87. Let $f: \Delta(z_0, r) \to \widehat{\mathbb{C}}$ be a non-constant meromorphic function. If a function g has an isolated singularity at the point $w_0 = f(z_0)$, prove that the composition $g \circ f$, which makes sense in $\Delta^*(z_0, r)$ when r is sufficiently small, has the same kind of singularity at z_0 that g does at w_0. Of course, we are allowing $z_0 = \infty$ or $w_0 = \infty$ — or both — here.

5.88. A function f has an isolated singularity at a point z_0 of $\widehat{\mathbb{C}}$. Assuming that $\operatorname{Re} f$ is bounded above (or below) in some punctured disk $\Delta^*(z_0, r)$, show that the singularity of f at z_0 is removable. (*Hint.* Consider e^f.)

5.89. A function f is meromorphic in a punctured disk $\Delta^* = \Delta^*(z_0, r)$ in $\widehat{\mathbb{C}}$. Furthermore, it is known that the set of poles of f in Δ^* has z_0 as a limit point. Prove that the set $\widehat{\mathbb{C}} \sim f(\Delta^*)$ can have no interior points. (*Warning.* The function f does not have an isolated singularity at z_0, so the terms "removable singularity," "pole," and "essential singularity" are not applicable there.)

Chapter IX

Conformal Mapping

Introduction

Both from a geometric outlook and in terms of its plentiful applications one of the most appealing areas of complex analysis is the part devoted to the study of univalent analytic functions. Functions of this type are also known as "conformal mappings," for they are "angle-preserving" in a sense to be explained presently. The epoch-making event in the history of conformal mapping was the announcement by Riemann in his 1851 Göttingen dissertation of a remarkable fact: any simply connected domain that is properly contained in the complex plane can be transformed by a conformal mapping to any other domain of the same description. It is this stunning discovery of Riemann's that sits at the core of the present chapter. Before taking up the Riemann mapping theorem we spend a few words in an effort to make precise the notion of a conformal mapping, and we examine the relationship between conformality and analyticity. After this we discuss in some detail an important elementary class of conformal mappings, the Möbius transformations. We follow up our treatment of the mapping theorem itself with an introduction to the boundary behavior of conformal mappings. Finally, we determine the structure of a conformal mapping of a half-plane onto a general polygonal region.

Needless to say, there are many aspects of the theory of conformal mappings that we shall not even touch upon in this book. For instance, we restrict ourselves almost exclusively to the conformal mapping of simply connected domains, thereby avoiding the topological and analytical complications that arise in the multiply connected case. An excellent supplementary text on the subject matter of this chapter is the classic *Conformal Mapping* by Zeev Nehari (McGraw-Hill, New York, 1952). At a more advanced level a fine reference for the topic is *Univalent Functions* by Christian Pommerenke (Vandenhoeck-Ruprecht, Göttingen, 1975), as is Peter Duren's book with the same title (Springer-Verlag, New York, Berlin, 1983).

1 Conformal Mappings

1.1 Curvilinear Angles

Since a typical complex function is unlikely to map straight line segments to straight line segments — hence, to transform angles, understood in the strict sense, to other angles — in speaking of "angle-preserving transformations" we cannot take the expression literally without at the same time drastically limiting the body of admissible transformations. To arrive at an exceedingly rich class of angle-preserving mappings, however, we need only recast the concept in a curvilinear setting. The reader is no doubt aware that the standard way of measuring the angle at which two smooth curves intersect is simply to measure the angle (for definiteness, use the non-obtuse angle) formed by the tangent lines to the respective curves at their point of intersection. A suitably well-behaved function will transform one smooth curve to another, which suggests an informal working definition of the term "angle-preserving" that captures the essential spirit of the later technical definition: a complex-valued function of a complex variable is angle-preserving if it transforms each intersecting pair of smooth curves in its domain-set to a pair of smooth curves whose angle of intersection is the same as that of the original curves. In fact, we wish to be somewhat more careful than this, for we want to consider mappings that preserve angles not just in size, but also in orientation. As a prelude to that refinement, we establish some convenient notation and terminology.

Let z and w be non-zero complex numbers. We refer to the quantity $\theta(z, w) = \text{Arg}(w/z)$ as the *oriented angle from z to w*. Geometrically $\theta(z, w)$ is a measurement in $(-\pi, \pi]$ of the smaller of the two angles formed at the origin by the vectors representing z and w. It is positive if a vector that sweeps out this angle starting at "side" z and proceeding to "side" w moves in a counter-clockwise direction; if that motion is clockwise, $\theta(z, w)$ is negative. (See Figure 1. The preceding remarks do not apply in the situation where $\theta(z, w) = \pi$, for then the vectors in question form a straight line

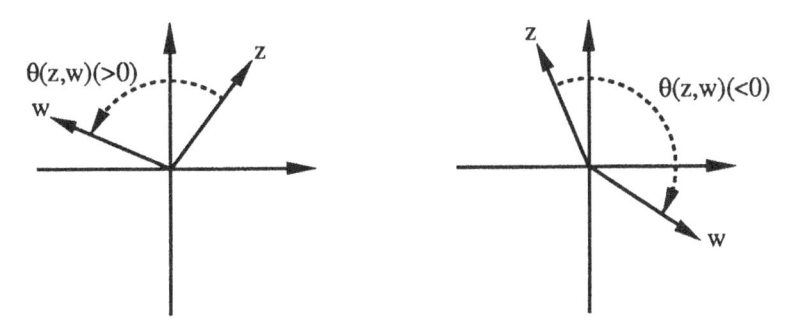

Figure 1.

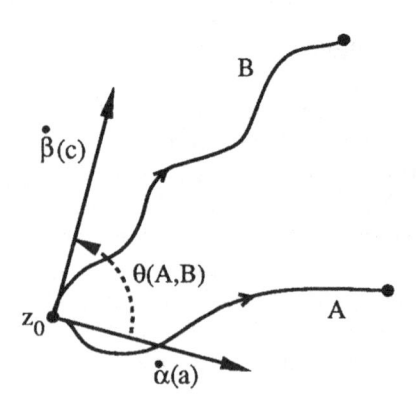

Figure 2.

segment passing through the origin, so no "smaller of the two angles" can be distinguished.) Observe that $\theta(w, z) = -\theta(z, w)$ and $\theta(\bar{z}, \bar{w}) = -\theta(z, w)$ except when $\theta(z, w) = \pi$, in which case both $\theta(w, z)$ and $\theta(\bar{z}, \bar{w})$ are π as well. Notice also that $\theta(cz, cw) = \theta(z, w)$ for any non-zero complex number c and that $\theta(rz, sw) = \theta(z, w)$ when r and s are positive real numbers.

As the name implies, the sides of a "curvilinear angle" are no longer required to be line segments, but instead are allowed to be more general smooth curves. To be precise, we introduce the expression *regular arc* to describe a set A that admits a representation of the sort $A = |\alpha|$, where $\alpha: [a, b] \to \mathbb{C}$ is a smooth, simple, non-closed path enjoying the added feature that $\dot{\alpha}(t) \neq 0$ for every t in $[a, b]$. Any such α is termed a *regular parametrization* of A. The points $\alpha(a)$ and $\alpha(b)$ are called the *endpoints* of A. Suppose that A and B are regular arcs which have one endpoint z_0 in common, but which are otherwise disjoint (Figure 2). In this event we say that A and B are the sides of a *curvilinear angle with vertex at* z_0, and we define $\theta(A, B)$, the *oriented angle from A to B*, by

(9.1) $$\theta(A, B) = \theta[\dot{\alpha}(a), \dot{\beta}(c)],$$

where $\alpha: [a, b] \to \mathbb{C}$ and $\beta: [c, d] \to \mathbb{C}$ are arbitrary regular parametrizations of A and B, respectively, having initial point z_0. (N.B. Definition (9.1) does not really depend on the regular parametrizations α and β chosen for A and B, since

$$\theta[\dot{\alpha}(a), \dot{\beta}(c)] = \theta\left[\frac{\dot{\alpha}(a)}{|\dot{\alpha}(a)|}, \frac{\dot{\beta}(c)}{|\dot{\beta}(c)|}\right]$$

and since the quantities $\dot{\alpha}(a)/|\dot{\alpha}(a)|$ and $\dot{\beta}(c)/|\dot{\beta}(c)|$ can be determined without reference to any parametrizations. For instance, it can be shown

1. Conformal Mappings 377

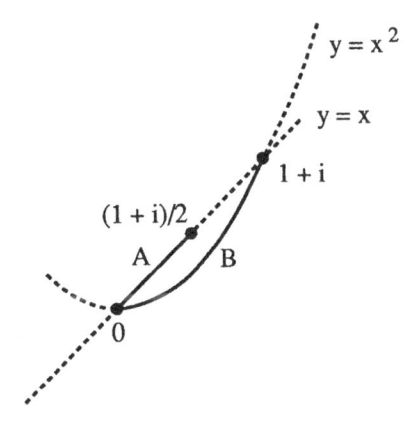

Figure 3.

that $\dot{\alpha}(a)/|\dot{\alpha}(a)|$ is just the limit of $(z - z_0)/|z - z_0|$ as z tends to z_0 along A. A similar statement holds for $\dot{\beta}(c)/|\dot{\beta}(c)|$.) As a simple illustration, consider the curvilinear angle pictured in Figure 3. Here the arc A lies on the line $y = x$, and B is a subset of the parabola $y = x^2$. We can use $\alpha(t) = t + it$ on the interval $[0, 1/2]$ and $\beta(t) = t + it^2$ on $[0, 1]$ as our regular parametrizations in computing

$$\theta(A, B) = \theta[\dot{\alpha}(0), \dot{\beta}(0)] = \theta(1+i, 1) = \text{Arg}[1/(1+i)]$$

$$= \text{Arg}[(1-i)/2] = -\pi/4 .$$

1.2 Diffeomorphisms

An object of considerable importance both for the immediate discussion and for subsequent developments in this chapter is the *Jacobian* J_f of a function $f = u + iv$. Assuming that U is an open set in the complex plane and that $f: U \to \mathbb{C}$ is a member of the class $C^1(U)$, J_f is the continuous real-valued function defined in U by

(9.2) $$J_f(z) = u_x(z)v_y(z) - u_y(z)v_x(z) ,$$

or, expressed in terms of z- and $\bar{z}$-derivatives,

(9.3) $$J_f(z) = |f_z(z)|^2 - |f_{\bar{z}}(z)|^2 .$$

In particular, if f happens to be differentiable at a point z_0 of U, then $f_z(z_0) = f'(z_0)$ and $f_{\bar{z}}(z_0) = 0$, with the result that

(9.4) $$J_f(z_0) = |f'(z_0)|^2 .$$

In saying what is required of a function for it to qualify as a conformal mapping we shall presuppose a certain amount of regularity on the part of the function in question; specifically, we shall insist that it be continuously differentiable in the real sense and have non-vanishing Jacobian. Thus, given a domain D in the complex plane, we are going to consider univalent functions $f: D \to \mathbb{C}$ that belong to the class $C^1(D)$ and obey the condition $J_f(z) \neq 0$ for every z in D. A mapping of this type is termed a *diffeomorphism* — or, more accurately, a C^1-*diffeomorphism* — *of D onto its range*. (N.B. While it is possible to introduce conformal mappings without imposing such strong preconditions on their smoothness, the resulting discussion becomes far too technical for a book of this level.) We remark that, if $f: D \to \mathbb{C}$ is a diffeomorphism, then either $J_f > 0$ everywhere in D or $J_f < 0$ throughout the domain. This follows from the fact that D is connected and $J_f: D \to \mathbb{R}$ is a continuous, zero-free function: the set $J_f(D)$ is a connected set of real numbers that does not contain zero — hence, $J_f(D)$ is either a subset of $(-\infty, 0)$ or a subset of $(0, \infty)$. When J_f is positive in D, the diffeomorphism f is called *orientation-preserving* (or *sense-preserving*); a diffeomorphism with negative Jacobian is said to be *orientation-reversing* (or *sense-reversing*). Notice that the conjugate $\bar{f}$ of a diffeomorphism $f: D \to \mathbb{C}$ is also a diffeomorphism, one for which

$$J_{\bar{f}} = -J_f .$$

Therefore $\bar{f}$ is orientation-reversing when f is orientation-preserving, and vice versa.

Suppose now that $f: D \to \mathbb{C}$ is a diffeomorphism and that A is a regular arc located in D. Then $f(A)$ is also a regular arc. To see this, choose a regular parametrization for A, say $\alpha(t) = x(t) + iy(t)$ on $[a, b]$. Since f is univalent and is a member of the class $C^1(D)$, $\beta(t) = f[\alpha(t)]$ defines a smooth, simple, non-closed path with $|\beta| = f(A)$. Could it be the case that $\dot{\beta}(t_0) = 0$ for some t_0 in $[a, b]$? Assuming that this were so, we could write $f = u + iv$ and employ the chain rule to get

$$0 = \dot{\beta}(t_0) = u_x(z_0)\dot{x}(t_0) + u_y(z_0)\dot{y}(t_0) + i\left[v_x(z_0)\dot{x}(t_0) + v_y(z_0)\dot{y}(t_0)\right] ,$$

where $z_0 = \alpha(t_0)$. It would follow that

(9.5) $\quad u_x(z_0)\dot{x}(t_0) + u_y(z_0)\dot{y}(t_0) = 0 , \quad v_x(z_0)\dot{x}(t_0) + v_y(z_0)\dot{y}(t_0) = 0 .$

The fact that $J_f(z_0) \neq 0$ would then permit us to solve (9.5) for $\dot{x}(t_0)$ and $\dot{y}(t_0)$. The result: $\dot{x}(t_0) = \dot{y}(t_0) = 0$. This would mean, however, that $\dot{\alpha}(t_0) = \dot{x}(t_0) + i\dot{y}(t_0) = 0$, contrary to the assumption that α is a regular parametrization of A. Accordingly, $\dot{\beta}(t) \neq 0$ must be true for every t in $[a, b]$, so β furnishes a regular parametrization of $f(A)$. A consequence of the preceding considerations is this: *a diffeomorphism $f: D \to \mathbb{C}$ transforms a curvilinear angle in D with sides A and B to a curvilinear angle in $D' = f(D)$ with sides $f(A)$ and $f(B)$* (Figure 4).

1. Conformal Mappings

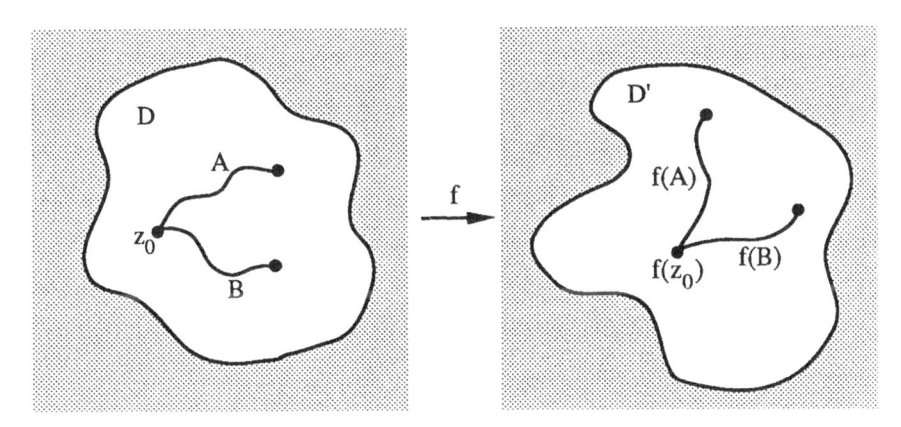

Figure 4.

1.3 Conformal Mappings

We are now ready to turn our earlier informal definition of an angle-preserving transformation into a proper one. As announced, we shall restrict attention to diffeomorphisms. A diffeomorphism $f: D \to \mathbb{C}$ is said to be *angle-preserving* (also called *isogonal*) *at a point z_0* of D under the condition that

$$|\theta(A, B)| = |\theta[f(A), f(B)]| \qquad (9.6)$$

whenever A and B are the sides of a curvilinear angle in D with vertex at z_0. It can be inferred from the soon-to-come Theorem 1.1 that condition (9.6) allows for two eventualities: if $J_f(z_0) > 0$, it reduces to

$$\theta(A, B) = \theta[f(A), f(B)] \qquad (9.7)$$

for all relevant A and B; if $J_f(z_0) < 0$, it becomes

$$\theta(A, B) = \theta[f(B), f(A)] . \qquad (9.8)$$

When (9.7) holds we say that f is *conformal at z_0*; in the alternative situation (9.8) we pronounce f *anti-conformal at z_0*. Thus, conformality demands that curvilinear angles be preserved not only in size, but also in sense. In the anti-conformal case, on the other hand, such angles are preserved in size, but their orientation gets reversed. It is not at all necessary, of course, for f to be angle-preserving at any point of D. Furthermore, since J_f does not undergo a change of sign in the domain D, it is not possible for f to be conformal at one point of D and anti-conformal at another. If the diffeomorphism f is conformal (anti-conformal) at every point of D, we call f a *conformal (anti-conformal) mapping of D*. For example, it is easily checked that a sense-preserving similarity transformation $f(z) = az + b$ —

remember that $a \neq 0$ here — is a conformal mapping of $\mathbb{C}$ onto itself. The canonical example of an anti-conformal mapping of the complex plane onto itself is found in the function $f(z) = \bar{z}$ (Exercise 6.5). A straightforward consequence of this last fact is that a diffeomorphism f is anti-conformal at a point z_0 if and only if its conjugate $\bar{f}$ is conformal there (Exercise 6.6).

The following theorem forges the link between the present discussion and the theory of analytic functions.

Theorem 1.1. *Let D be a domain in the complex plane, and let $f: D \to \mathbb{C}$ be a diffeomorphism. The following three statements concerning a point z_0 of D are equivalent: (i) f is differentiable at z_0; (ii) f is isogonal at z_0 and $J_f(z_0) > 0$; (iii) f is conformal at z_0.*

Proof. We first prove that (i) implies both (ii) and (iii). Assuming that f is differentiable at z_0, we infer from (9.4) that $J_f(z_0) = |f'(z_0)|^2 \geq 0$ and conclude, since $J_f(z_0) \neq 0$, that $J_f(z_0) > 0$ and $f'(z_0) \neq 0$. We establish the conformality of f at z_0, a property which encompasses isogonality at that point. Let A and B be the sides of a curvilinear angle in D with vertex at z_0. Choose regular parametrizations $\alpha: [a, b] \to \mathbb{C}$ for A and $\beta: [c, d] \to \mathbb{C}$ for B with $\alpha(a) = \beta(c) = z_0$. Then $\alpha_1(t) = f[\alpha(t)]$ and $\beta_1(t) = f[\beta(t)]$ provide regular parametrizations for $f(A)$ and $f(B)$, respectively. From the differentiability of f at z_0 it follows that

$$\dot{\alpha}_1(a) = f'[\alpha(a)]\dot{\alpha}(a) = f'(z_0)\dot{\alpha}(a)$$

and, similarly, $\dot{\beta}_1(c) = f'(z_0)\dot{\beta}(c)$. Because $f'(z_0) \neq 0$, we see that

$$\theta[f(A), f(B)] = \theta[\dot{\alpha}_1(a), \dot{\beta}_1(c)] = \theta[f'(z_0)\dot{\alpha}(a), f'(z_0)\dot{\beta}(c)]$$

$$= \theta[\dot{\alpha}(a), \dot{\beta}(c)] = \theta(A, B) ,$$

verifying the conformality of f at z_0.

We next derive (i) from (ii). Assume, therefore, that f is isogonal at z_0 and that $J_f(z_0) > 0$. As a diffeomorphism, the function f belongs to the class $C^1(D)$, which implies that it is differentiable in the real sense at each point of D (Theorem III.5.1). In particular, we are entitled to write

(9.9) $$f(z) = f(z_0) + c(z - z_0) + d(\bar{z} - \bar{z}_0) + E(z)$$

for z in D, where $c = f_z(z_0)$, $d = f_{\bar{z}}(z_0)$, and E is a function satisfying $|E(z)|/|z - z_0| \to 0$ as $z \to z_0$. The problem rests in showing that $d = 0$, a condition which, being a restatement of the Cauchy-Riemann relations at z_0, makes plain that f is differentiable at z_0 with $f'(z_0) = c$.

We fix $r > 0$ with the property that the closed disk $\bar{\Delta}(z_0, r)$ is contained in D. When $0 \leq \psi \leq \pi$, we let A_ψ denote the regular arc parametrized by $\alpha_\psi(t) = z_0 + te^{i\psi}$ for $0 \leq t \leq r$. If $0 < \psi \leq \pi$, then the arcs A_0 and A_ψ are the sides of a curvilinear angle — in fact, a true angle — in D with vertex

1. Conformal Mappings

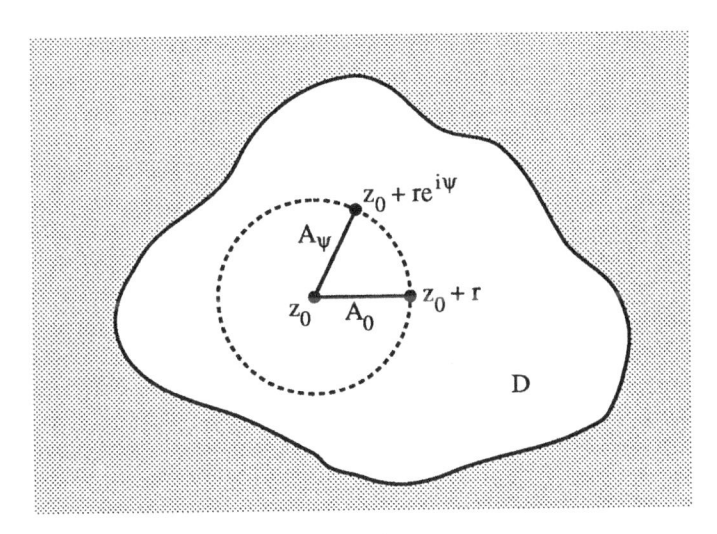

Figure 5.

at z_0 (Figure 5). Also, $\dot{\alpha}_\psi(0) = e^{i\psi}$, which implies that $\theta(A_0, A_\psi) = \psi$. The image arc $f(A_\psi)$ has the regular parametrization $\beta_\psi(t) = f[\alpha_\psi(t)] = f(z_0 + te^{i\psi})$ for $0 \leq t \leq r$. We use (9.9) to compute

$$\dot{\beta}_\psi(0) = \lim_{t \to 0+} \frac{f(z_0 + te^{i\psi}) - f(z_0)}{t}$$

$$= \lim_{t \to 0+} \left[ce^{i\psi} + de^{-i\psi} + \frac{e^{i\psi} E(z_0 + te^{i\psi})}{te^{i\psi}} \right] = ce^{i\psi} + de^{-i\psi}.$$

Note especially the implication that $ce^{i\psi} + de^{-i\psi} \neq 0$ for $0 \leq \psi \leq \pi$. We can thus safely write

$$\theta[f(A_0), f(A_\psi)] = \operatorname{Arg}\left(\frac{ce^{i\psi} + de^{-i\psi}}{c+d} \right)$$

when $0 < \psi \leq \pi$. Since f is isogonal at z_0, we draw the conclusion that

(9.10) $$\left| \operatorname{Arg}\left(\frac{ce^{i\psi} + de^{-i\psi}}{c+d} \right) \right| = \psi$$

for $0 \leq \psi \leq \pi$. Because $c\bar{d}e^{i\psi} + \bar{c}de^{-i\psi} = 2\operatorname{Re}(c\bar{d}e^{i\psi})$ is a real quantity and because — recall (9.3) — $|c|^2 - |d|^2 = J_f(z_0) > 0$, we learn that for such ψ

$$\operatorname{Im}\left(\frac{ce^{i\psi} + de^{-i\psi}}{c+d} \right) = \operatorname{Im}\left[\frac{(ce^{i\psi} + de^{-i\psi})(\bar{c} + \bar{d})}{|c+d|^2} \right]$$

$$= \frac{\operatorname{Im}\left(|c|^2 e^{i\psi} + c\bar{d}e^{i\psi} + \bar{c}de^{-i\psi} + |d|^2 e^{-i\psi} \right)}{|c+d|^2}$$

$$= \frac{\text{Im}(|c|^2 e^{i\psi} + |d|^2 e^{-i\psi})}{|c+d|^2} = \frac{(|c|^2 - |d|^2)\sin\psi}{|c+d|^2} \geq 0 \ .$$

The fact that the imaginary part of $(ce^{i\psi} + de^{-i\psi})/(c+d)$ is non-negative implies that the principal argument of this number lies in the interval $[0, \pi]$, so (9.10) simplifies to

$$(9.11) \qquad \text{Arg}\left(\frac{ce^{i\psi} + de^{-i\psi}}{c+d}\right) = \psi = \text{Arg}\, e^{i\psi} \ .$$

This means that, for every ψ in $[0, \pi]$, $(ce^{i\psi} + de^{-i\psi})/(c+d)$ is a positive real multiple of $e^{i\psi}$ — or, what amounts to the same thing, the quantity $(c + de^{-2i\psi})/(c+d)$ is real and positive. Should $d \neq 0$, however, the set of points $\{(c + de^{-2i\psi})/(c+d) \colon 0 \leq \psi \leq \pi\}$ would be a circle with center $c/(c+d)$ and radius $|d|/|c+d|$, a set definitely not contained in the positive real axis. The alternative: $d = 0$ and $f'(z_0)$ exists, as claimed.

The proof that (iii) implies (i) is basically the same as the one presented for the implication (ii)$\Rightarrow$(i), only simpler. Namely, conformality on the part of f at z_0 allows one to by-pass the computation leading from (9.10) to (9.11); (9.11) is a direct consequence of conformality at z_0. ∎

There is an analogue of Theorem 1.1 treating the anti-conformal case. We formulate it in an exercise (Exercise 6.8). The global version of Theorem 1.1 marks plane conformal mapping theory as a special topic in the theory of analytic functions.

Theorem 1.2. *Let D be a domain in the complex plane. A function $f \colon D \to \mathbb{C}$ is a conformal mapping of D if and only if f is a univalent analytic function.*

Proof. If f is a conformal mapping of D, then f is by definition a diffeomorphism, while Theorem 1.1 affirms that f is differentiable at each point of D; i.e., f is analytic in D. Conversely, if f is a univalent analytic function, then certainly f is a member of the class $C^1(D)$, and, in view of Theorem VIII.3.9, $J_f(z) = |f'(z)|^2 > 0$ throughout D. Consequently, f is seen to be a diffeomorphism. From Theorem 1.1 we learn that f is conformal at every point of D, making f a conformal mapping of D. ∎

It follows from Theorem 1.2 without further ado that the composition of conformal mappings is conformal. Coupled with Theorem VIII.3.11, Theorem 1.2 tells us that the inverse of a conformal mapping is likewise conformal.

Let D be a plane domain. Some authors pin the label "conformal mapping" on any analytic function $f \colon D \to \mathbb{C}$ that obeys the condition $f'(z) \neq 0$ for every z in D, the requirement of univalence being waived. (The exponential function $f(z) = e^z$ in $D = \mathbb{C}$ would be an example.) We prefer to characterize a mapping f of this type as *locally conformal* in D. Indeed,

1. Conformal Mappings 383

Theorem VIII.3.10 asserts that there exists corresponding to any point z_0 of D a subdomain G of D containing z_0 in which f is both analytic and univalent; i.e., to which the restriction of f is a conformal mapping according to our understanding of the term.

1.4 Some Standard Conformal Mappings

We compile here for handy reference a short list of examples of elementary conformal mappings.

EXAMPLE 1.1. The function $f(z) = (1-z)/(1+z)$ maps the disk $D = \Delta(0,1)$ conformally onto the half-plane $D' = \{w: \operatorname{Re} w > 0\}$ (Figure 6).

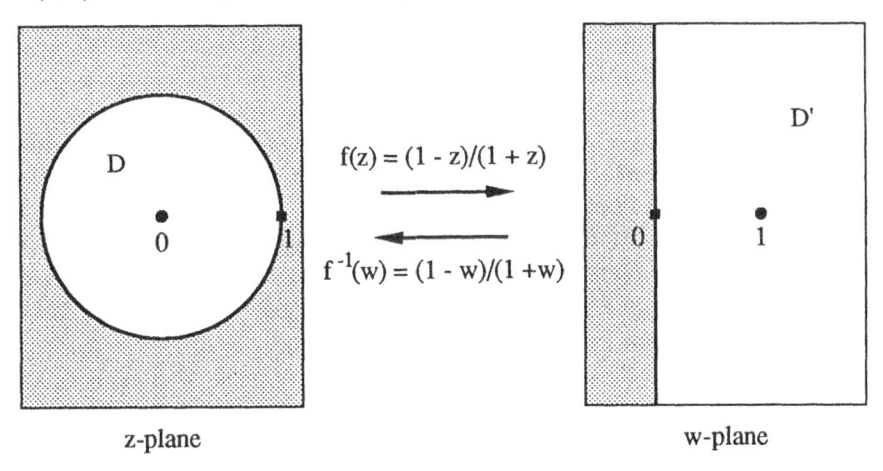

Figure 6.

The function f is analytic in D and maps D in a univalent fashion onto D'. (See Example I.3.4.) Theorem 1.2 certifies f as a conformal mapping of D onto D'. Note that $f^{-1} = f$ here, so f transforms D' conformally onto D as well.

EXAMPLE 1.2. Let $D = \mathbb{C} \sim (-\infty, 0]$ and let $0 < \lambda \leq 1$. The function $f: D \to \mathbb{C}$ defined by $f(z) = z^\lambda$ maps D conformally onto the angular region $D' = \{w: |\operatorname{Arg} w| < \lambda\pi\}$ (Figure 7).

For $0 < \lambda \leq 1$ and z in D we note that $|z^\lambda| = |z|^\lambda$ and $\operatorname{Arg}(z^\lambda) = \lambda \operatorname{Arg} z$. From these two facts it follows without difficulty that f is a univalent function with range D'. Since f is also plainly analytic, f is a conformal mapping of D onto D'. The inverse mapping $f^{-1}: D' \to D$ is given by $f^{-1}(w) = w^{1/\lambda}$. Observe that the restriction of f to $D \cap \Delta(0, r)$ gives a conformal mapping of this intersection onto $D' \cap \Delta(0, r^\lambda)$. (N.B. When $\lambda > 1$, $f(z) = z^\lambda$ is still analytic in D, but no longer univalent there (Ex-

384   IX. Conformal Mapping

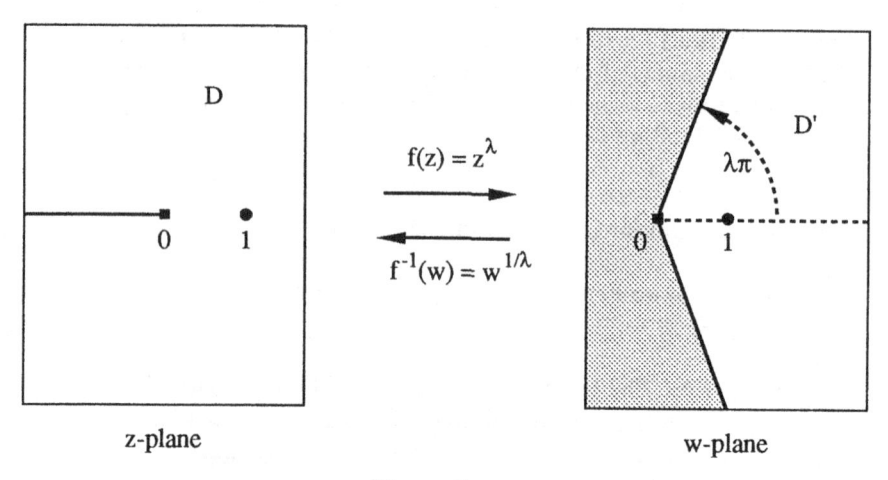

Figure 7.

ercise 6.10). Adhering to our definition of the concept, $f(z) = z^\lambda$ does not qualify as a conformal mapping of D for such λ. It does, however, meet the test to be a locally conformal mapping of D.)

EXAMPLE 1.3. If $-\infty \leq a < b \leq \infty$ and $-\pi \leq \alpha < \beta \leq \pi$, the exponential function $f(z) = e^z$ provides a conformal mapping of the domain $D = \{z : a < x < b, \alpha < y < \beta\}$ onto the region D' described by $D' = \{w : e^a < |w| < e^b,\ \alpha < \operatorname{Arg} w < \beta\}$ (Figure 8). The inverse $f^{-1} : D' \to D$ of this mapping is the restriction of the principal logarithm to D'; i.e., $f^{-1}(w) = \operatorname{Log} w$ for w in D'.

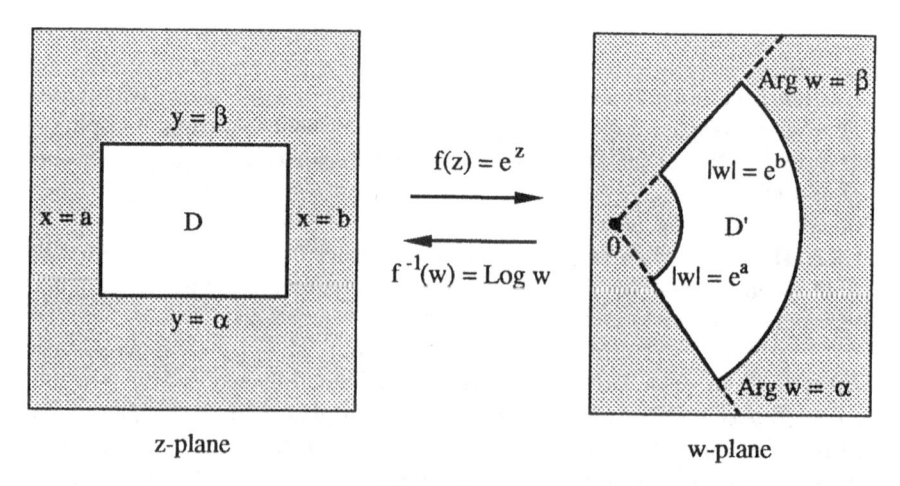

Figure 8.

1. Conformal Mappings

EXAMPLE 1.4. The function $f(z) = \sin z$ maps the infinite strip $D = \{z : |\operatorname{Re} z| < \pi/2\}$ conformally onto the domain $D' = \mathbb{C} \sim \{w : |\operatorname{Re} w| \geq 1$ and $\operatorname{Im} w = 0\}$. (See Figure 9. We refer the reader to Example III.3.7 for details.) Here $f^{-1} : D' \to D$ is the restriction to D' of the principal arcsine function.

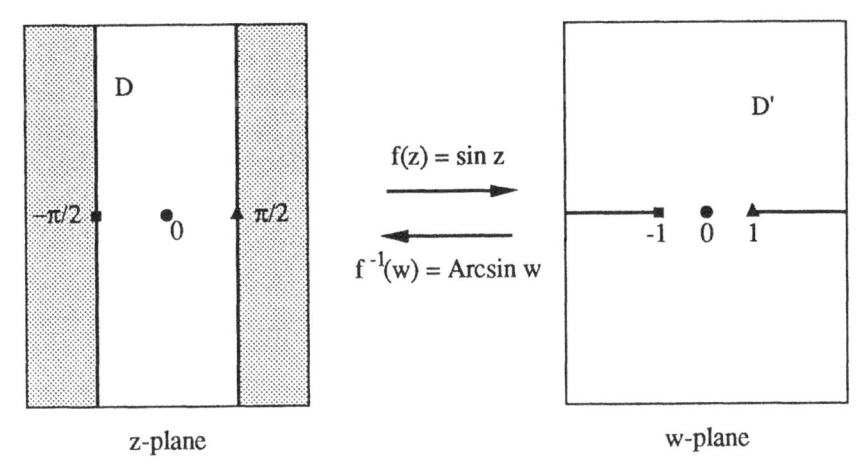

Figure 9.

EXAMPLE 1.5. The function $f(z) = \tan z$ is one that transforms the infinite strip $D = \{z : |\operatorname{Re} z| < \pi/2\}$ conformally to the domain $D' = \mathbb{C} \sim \{w : |\operatorname{Im} w| \geq 1$ and $\operatorname{Re} w = 0\}$. (See Figure 10. Recall, too, Example II.3.8.) The restriction of the principal arctangent function to D' furnishes the inverse mapping in this example.

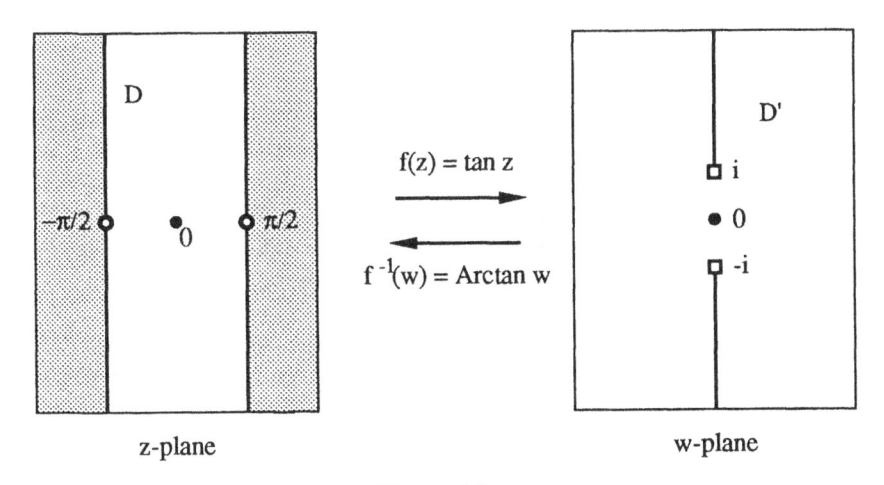

Figure 10.

Notice that $g(z) = i\sin z$ delivers a second function which maps the strip D conformally onto D', but maps it quite differently than f does. For example, f transforms the interval $(-\pi/2, \pi/2)$ to $(-\infty, \infty)$, whereas g maps $(-\pi/2, \pi/2)$ to the interval between $-i$ and i on the imaginary axis.

EXAMPLE 1.6. The restriction of the function $f(z) = (z + z^{-1})/2$ to the domain $D = \{z : |z| > 1\}$ gives a conformal mapping of D onto the domain $D' = \mathbb{C} \sim [-1, 1]$.

When $|z| = 1$, we observe that

$$f(z) = (z + z^{-1})/2 = (z + \bar{z})/2 = \operatorname{Re} z ,$$

so $f(z)$ is real and belongs to the interval $[-1,1] = \mathbb{C} \sim D'$. For w other than 1 or -1 the equation $(z + z^{-1})/2 = w$ has exactly two solutions, $z_1 = w + \sqrt{w^2 - 1}$ and $z_2 = w - \sqrt{w^2 - 1}$. Furthermore, if w is not a point of $[-1, 1]$, then by the preceding remark neither z_1 nor z_2 can have unit modulus. Since $|z_1||z_2| = |z_1 z_2| = 1$, one of these points must lie in D, the other in the disk $\Delta(0, 1)$. In fact, it can be verified that $|z_1| > 1$ for any w in D' such that $\operatorname{Re} w > 0$ or such that $\operatorname{Re} w = 0$ and $\operatorname{Im} w > 0$, whereas $|z_2| > 1$ for all other elements w of D'. (We remind the reader of Exercise I.4.38.) On the basis of the preceding information we can assert that the restriction of f to D is a univalent function whose range is D'. As f is analytic in D, this restriction is a conformal mapping of D onto D'. The inverse mapping is a bit awkward to write down here, its values being provided by z_1 at some points w of D' and z_2 at others, as just pointed out. (N.B. The same function f maps the punctured disk $\Delta^*(0, 1)$ conformally onto D'.)

The remaining two examples indicate how the foregoing elementary mappings can be used to construct more complicated conformal mappings.

EXAMPLE 1.7. Find a conformal mapping of the infinite strip $D = \{z : |\operatorname{Im} z| < \pi/2\}$ onto the disk $D' = \Delta(0, 1)$.

As suggested by Figure 11, we can arrive at such a mapping f by first transforming D with a conformal mapping f_1 to the half-space D_1, then choosing a conformal mapping f_2 of D_1 onto D', and taking $f = f_2 \circ f_1$. We actually carry this out by using $f_1(z) = e^z$ and $f_2(z) = (1-z)/(1+z)$, although other choices would certainly be possible. The composite function $f(z) = (1 - e^z)/(1 + e^z)$ is thus a conformal mapping of D onto D'.

EXAMPLE 1.8. Construct a conformal mapping of the infinite strip $D = \{z : 0 < \operatorname{Re} z < \pi\}$ onto the domain $D' = \{w : |\operatorname{Arg} w| < \pi/4\} \sim [1, \infty)$.

Again we build f in stages: $f = f_3 \circ f_2 \circ f_1$, where the component mappings meet the requirements implicit in Figure 12. Specifically, we take $f_1(z) = z/2$, $f_2(z) = \sin z$, and $f_3(z) = \sqrt{z}$. They compose to produce $f(z) = \sqrt{\sin(z/2)}$, a function that maps D conformally onto D'.

1. Conformal Mappings

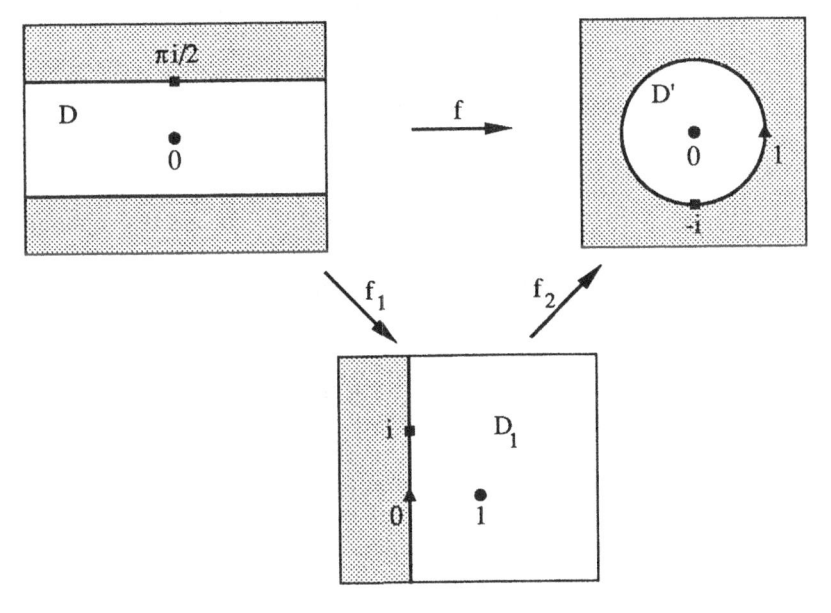

Figure 11.

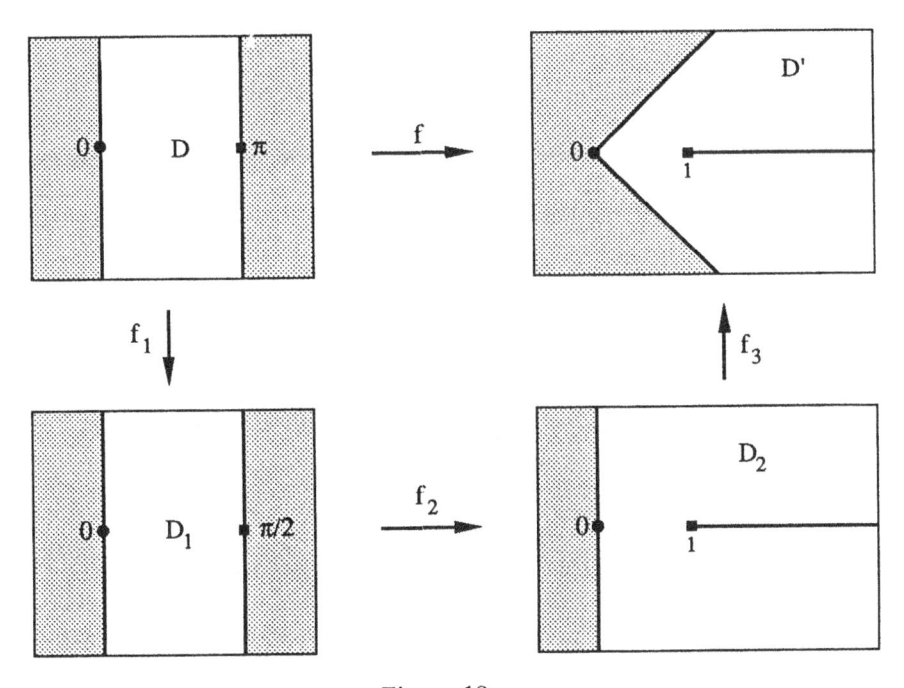

Figure 12.

1.5 Self-Mappings of the Plane and Unit Disk

Owing to Theorem 1.2 we can use the machinery of complex analysis to shed light on the structure of conformal mappings. The following two theorems illustrate this point. In the first we characterize the conformal mappings of the whole complex plane.

Theorem 1.3. *The conformal mappings $f: \mathbb{C} \to \mathbb{C}$ are the functions of the form $f(z) = az + b$, where a and b are complex numbers and $a \neq 0$. In particular, the range of any conformal mapping $f: \mathbb{C} \to \mathbb{C}$ is the entire complex plane.*

Proof. We already know that all functions of the stated description provide conformal mappings of the complex plane onto itself. We must show that there are no others. Theorem 1.2 tells us that a conformal mapping f of $\mathbb{C}$ is an entire function. As such, f admits a Taylor series expansion $f(z) = \sum_{n=0}^{\infty} a_n z^n$ valid throughout the complex plane. Since f is non-constant, $a_n \neq 0$ must be true for some $n > 0$, so the isolated singularity of f at ∞ is either an essential singularity or a pole. Could it be essential? Were this the case, the Casorati-Weierstrass theorem in concert with the open mapping theorem would imply that $f(U) \cap f(V) \neq \phi$, where $U = \Delta^*(\infty, 1)$ and $V = \Delta(0, 1)$. Indeed, the set $f(U)$ would be dense in $\mathbb{C}$, a condition that forces it to intersect the open set $f(V)$. Any point w in $f(U) \cap f(V)$ would clearly possess at least two preimages in $\mathbb{C}$, one in U and one in V. The resulting conflict with the univalence of f rules out an essential singularity at ∞. Therefore, f has a pole of some order m at ∞; i.e., the Taylor series expansion of f reduces to $f(z) = a_0 + a_1 z + \cdots + a_m z^m$, where $m \geq 1$ and where $a_m \neq 0$. If $m > 1$, then the fundamental theorem of algebra would imply that f', a polynomial of positive degree, would have a root. In light of Theorem VIII.3.9, the assumption that f is univalent would again be violated. As a consequence, $m = 1$ and f takes the form $f(z) = az + b$ with $a = a_1 \neq 0$ and $b = a_0$. ∎

The next result identifies the conformal self-mappings of the unit disk.

Theorem 1.4. *The functions that map the unit disk $D = \Delta(0, 1)$ conformally onto itself are the functions $f: D \to \mathbb{C}$ of the form*

(9.12) $$f(z) = e^{i\theta} \frac{z + c}{1 + \bar{c}z},$$

where θ is a real number and c is a complex number with $|c| < 1$.

Proof. We first check that each function of type (9.12) does represent a conformal self-mapping of D. Clearly $f = g \circ h$ in D, where $g(z) = e^{i\theta} z$ and $h(z) = (z + c)/(1 + \bar{c}z)$. The function g causes no problem: its effect is simply to rotate D about the origin. It plainly maps D conformally onto

1. Conformal Mappings

itself. The behavior of h is less transparent. Certainly the rational function h is analytic in D, for its denominator vanishes only at the point $-1/\bar{c}$, which lies outside D. A short computation — see Exercise I.4.21 — reveals that $|h(z)| < 1$ when $|z| < 1$. In other words, $h(D)$ is a subset of D. Next, we consider the function k defined by $k(z) = (z-c)/(1-\bar{c}z)$. As k has the same structure that h does, with $-c$ replacing c, it is also true that $k(D)$ is contained in D. Furthermore, easy calculations show that $k[h(z)] = z$ and $h[k(z)] = z$ for every z in D. (The function k is just the inverse of h.) This implies in the first place that h is univalent in D and secondly that

$$D = [h \circ k](D) = h[k(D)] \subset h(D) \subset D \,,$$

i.e., $h(D) = D$. Consequently, both g and h map D in a conformal manner onto itself. The composition $f = g \circ h$ does likewise.

It remains to be checked that an arbitrary conformal mapping f of D onto itself is of the kind described in (9.12). For this, we set $c = -f^{-1}(0)$ and look at the function $g : D \to D$ given by $g = f \circ k$, where as earlier $k : D \to D$ is the conformal mapping defined by $k(z) = (z-c)/(1-\bar{c}z)$. Then g, too, is a conformal self-mapping of D, one that has

$$g(0) = f[k(0)] = f(-c) = f[f^{-1}(0)] = 0 \,.$$

Schwarz's lemma (Theorem V.3.14) informs us that $|g'(0)| \leq 1$. The function g^{-1} is another conformal self-mapping of D that fixes the origin, so a second appeal to the Schwarz lemma delivers the inequality $1/|g'(0)| = |[g^{-1}]'(0)| \leq 1$. As a result, equality holds — $|g'(0)| = 1$. According to Schwarz's lemma this only happens when g has the structure $g(z) = e^{i\theta}z$ for some real number θ. Finally, recalling from the first part of this proof that the inverse of k is the function $h : D \to D$ with rule of correspondence $h(z) = (z+c)/(1+\bar{c}z)$, we conclude that $f = g \circ k^{-1} = g \circ h$ takes the form in (9.12). ∎

1.6 Conformal Mappings in the Extended Plane

For the sake of future discussions it will be convenient to have available the notion of a conformal mapping between domains in the extended complex plane. The most efficient way to introduce this concept (and the way best suited to our eventual needs) is to take a cue from Theorem 1.2 and to formulate the definition as follows: a function $f : D \to \widehat{\mathbb{C}}$, where D is a domain in $\widehat{\mathbb{C}}$, is a *conformal mapping of D* provided f is a univalent meromorphic function. (By Theorem 1.2 this reduces to the earlier definition of a conformal mapping if D and the range of f are subsets of $\mathbb{C}$. It follows from Theorem VIII.4.7, incidentally, that in the extended-plane setting the range of f is a domain in $\widehat{\mathbb{C}}$. The definition of a meromorphic function insures, of course, that f is continuous in the sense appropriate to mappings between sets in $\widehat{\mathbb{C}}$.) Since the requirement of univalence is imposed on f, there can be at most one point z_0 in D — there may be none — for which $f(z_0) = \infty$; i.e., f has at most one pole in D. If f does have a pole in D, it is necessarily a simple pole, for a meromorphic function is not univalent in the vicinity of a pole of order higher than one (Theorem VIII.4.10). The restriction of f to the finite domain $D_0 = \{z \in D : z \neq \infty, f(z) \neq \infty\}$ is both analytic and univalent — hence, is a conformal mapping of D_0 according to our original definition of the term. Theorem VIII.4.1 implies the following converse of this remark: if D is a domain in $\widehat{\mathbb{C}}$ and if $f: D \to \widehat{\mathbb{C}}$ is a univalent continuous function whose restriction to $D_0 = \{z \in D : z \neq \infty, f(z) \neq \infty\}$ is a conformal mapping of D_0, then f is a conformal mapping of D. As was the case in the finite plane, the composition of conformal mappings between domains in the extended plane is conformal, and the inverse of a conformal mapping is also conformal. By an *anti-conformal mapping* f of an extended-plane domain D we mean one whose conjugate $\bar{f}$ is a conformal mapping of D. Here we adopt the convention that $\overline{\infty} = \infty$.

A simple example of a conformal mapping $f: \widehat{\mathbb{C}} \to \widehat{\mathbb{C}}$ is the function $f(z) = z^{-1}$. (Remember: by our convention on extended domain-sets, $f(0) = \infty$ and $f(\infty) = 0$.) Being a rational function of z, f is meromorphic, and it is obviously univalent. This function also provides a useful conformal mapping of the disk $D = \Delta(0,1)$ onto its exterior $D' = \Delta(\infty, 1)$ and, since $f = f^{-1}$, of D' onto D.

The function $f(z) = z^{-1}$ is but one member of an extremely important class of conformal mappings, the class of *Möbius transformations*, named in honor of the geometer A.F. Möbius (1790-1868). These are the rational functions $f: \widehat{\mathbb{C}} \to \widehat{\mathbb{C}}$ that have the form

$$(9.13) \qquad f(z) = \frac{az+b}{cz+d},$$

where $a, b, c,$ and d are complex numbers with the property that $ad - bc \neq 0$. (Such functions are also known as *linear fractional transformations*.) As rational functions, Möbius transformations are meromorphic. The condition $ad - bc \neq 0$ makes certain that they are not constant functions in disguise. If $c = 0$, then $a \neq 0$ and (9.13) simplifies to an expression of the sort $f(z) = \alpha z + \beta$, with $\alpha = a/d \neq 0$; stated differently, when $c = 0$ the mapping f is a sense-preserving similarity transformation. In line with our extended domain-set convention we interpret (9.13) as saying that $f(\infty) = \infty$ in this case. If $c \neq 0$, then the Möbius transformation f has a pole at $-d/c$, so (9.13) is understood to include the assignment of value $f(-d/c) = \infty$. In this situation it is also implicit in (9.13) that $f(\infty) = \lim_{z \to \infty} f(z) = a/c$. To confirm that (9.13) does actually define a conformal self-mapping of $\widehat{\mathbb{C}}$ we still need to demonstrate that f is univalent. This is readily accomplished by writing down its inverse function:

$$f^{-1}(z) = \frac{dz-b}{-cz+a},$$

2. Möbius Transformations

as a straightforward calculation verifies (Exercises I.4.36 and I.4.48). Notice that f^{-1} is itself a Möbius transformation. We conclude, in particular, that $f(\widehat{\mathbb{C}}) = \widehat{\mathbb{C}}$. One checks without difficulty (Exercise I.4.35) that the composition of Möbius transformations is another function from that class.

Theorems 1.3 and 1.4 reveal that the conformal self-mappings of $\mathbb{C}$ and those of the disk $\Delta(0,1)$ are merely Möbius transformations of special types. The following proposition is a companion to those two results.

Theorem 1.5. *The conformal mappings* $f: \widehat{\mathbb{C}} \to \widehat{\mathbb{C}}$ *are the Möbius transformations. In particular, the range of any conformal mapping* $f: \widehat{\mathbb{C}} \to \widehat{\mathbb{C}}$ *is the entire extended plane.*

Proof. We have already observed that Möbius transformations map the extended plane conformally onto itself. Suppose that $f: \widehat{\mathbb{C}} \to \widehat{\mathbb{C}}$ is an arbitrary conformal mapping of $\widehat{\mathbb{C}}$. We define a function $h: \widehat{\mathbb{C}} \to \widehat{\mathbb{C}}$ as follows: if $f(\infty) = \infty$, then we set $h = f$; if $f(\infty) = w_0 \neq \infty$, then we take $h = g \circ f$, where $g(z) = 1/(z - w_0)$. Since g is a Möbius transformation mapping w_0 to ∞, h is in either case a conformal mapping of $\widehat{\mathbb{C}}$ that leaves the point ∞ fixed. The restriction of h to the finite plane $\mathbb{C}$ is then a conformal mapping of $\mathbb{C}$, which means in view of Theorem 1.3 that h has the form $h(z) = az + b$ with $a \neq 0$. We conclude that h is a Möbius transformation. Because either $f = h$ or $f = g^{-1} \circ h$, f is a Möbius transformation as well. ∎

2 Möbius Transformations

2.1 Elementary Möbius Transformations

In this section we take a closer look at the class of Möbius transformation, of which the previous section afforded us a brief glimpse. We shall discover that these mappings have quite a number of interesting and useful properties. To launch the discussion we observe that every Möbius transformation can be built up as the composition of extremely simple ones, mappings we refer to as *elementary Möbius transformations*. These fall into four categories, as follows: (i) a *translation* is a Möbius transformation of the form $f(z) = z + b$; (ii) a mapping of the type $f(z) = az$, where a is both real and positive, is called a *dilation* (or *homothety*) *with respect to the origin;* (iii) if $f(z) = az$ with $|a| = 1$, f is a *rotation about the origin;* (iv) the final elementary transformation is the *inversion* $f(z) = z^{-1}$. (N.B. The *identity transformation*, the Möbius transformation f defined by $f(z) = z$ for every z in $\widehat{\mathbb{C}}$, fits each of the descriptions (i), (ii), and (iii). To avoid such an overlap of categories some authors prefer to place the identity in a separate class of its own.) Suppose now that $f(z) = (az + b)/(cz + d)$ is an arbitrary Möbius transformation. If $c = 0$, then f is a similarity transformation, in which event f is the composition of elementary transformations of types (i), (ii), and (iii) (cf., Example I.3.1); if $c \neq 0$, then we can write

$$f(z) = \frac{a}{c} + \frac{bc - ad}{c^2} \frac{1}{z + (d/c)},$$

in which case it is again clear that f can be obtained as the composition of (five or fewer) mappings of elementary type. The value of the foregoing remarks is this: to establish that a property is shared by all Möbius transformations it suffices, first, to verify that every elementary transformation enjoys the property in question and, secondly, to check that the property is preserved under composition. Often the verification of a property for elementary transformations involves little or no work.

2.2 Möbius Transformations and Matrices

With a Möbius transformation $f(z) = (az + b)/(cz + d)$ we can associate the non-singular 2×2 complex matrix $A = \begin{bmatrix} a & b \\ c & d \end{bmatrix}$. Conversely, each matrix A of the stated character furnishes the coefficients of a Möbius transformation, one we symbolize by f_A. The correspondence $A \to f_A$ has the nice feature that $AB \to f_A \circ f_B$ and $A^{-1} \to (f_A)^{-1}$, where AB indicates ordinary matrix multiplication and A^{-1} signifies the inverse matrix to A (Exercise 6.24). Thanks to this observation, many computations involving Möbius transformations can be reduced to simple matrix calculations, at great savings of time and avoidance of ponderous functional notation. As an illustration of how matrices can expedite such calculations, let us compute $f = g \circ h \circ k$, where $g(z) = (z+1)/(z-1)$, $h(z) = z/(z+i)$, and $k(z) = (z - i)/z$. By the above comments we can express the composition as $f = f_A$ for

$$A = \begin{bmatrix} 1 & 1 \\ 1 & -1 \end{bmatrix} \begin{bmatrix} 1 & 0 \\ 1 & i \end{bmatrix} \begin{bmatrix} 1 & -i \\ 1 & 0 \end{bmatrix} = \begin{bmatrix} 2+i & -2i \\ -i & 0 \end{bmatrix},$$

which yields

$$f(z) = \frac{(2+i)z - 2i}{(-i)z} = \frac{(-1+2i)z + 2}{z}.$$

Though hardly overwhelming, the computation of f without resorting to matrices would be considerably less efficient.

The correspondence which sends a non-singular matrix A to the Möbius transformation f_A is definitely not one-to-one. Indeed, the fact that for $\lambda \neq 0$ the formulas

(9.14) $$f(z) = \frac{az + b}{cz + d}$$

and

2. Möbius Transformations

(9.15) $$f(z) = \frac{(\lambda a)z + \lambda b}{(\lambda c)z + \lambda d}$$

are alternate descriptions of one and the same function makes it clear that $f_A = f_{\lambda A}$ for every non-zero complex number λ. Presented with a Möbius transformation f in the form (9.14) we are free to choose λ for which $\lambda^2(ad - bc) = 1$ — namely, $\lambda = \pm(ad - bc)^{-1/2}$ — and to rewrite f as in (9.15). From this it becomes apparent that a Möbius transformation f always admits a representation of the type

(9.16) $$f(z) = \frac{az + b}{cz + d} \quad , \quad ad - bc = 1 \; .$$

When (9.16) holds we say that f is expressed in *normalized form*. The matrix $A = \begin{bmatrix} a & b \\ c & d \end{bmatrix}$, which then has determinant one, is called a *normalized matrix* corresponding to f. For instance, the functions $f(z) = 4z + 1$ and $g(z) = (z - 1)/(z + i)$ can be put into normalized form as follows:

$$f(z) = \frac{2z + (1/2)}{0z + (1/2)} \quad , \quad g(z) = \frac{\left(2^{-1/4}e^{-\pi i/8}\right)z - 2^{-1/4}e^{-\pi i/8}}{\left(2^{-1/4}e^{-\pi i/8}\right)z + 2^{-1/4}e^{-\pi i/8}i} \; .$$

Normalized matrices corresponding to the elementary transformations are:

(i) $\begin{bmatrix} 1 & b \\ 0 & 1 \end{bmatrix}$ for a translation $f(z) = z + b$;

(ii) $\begin{bmatrix} a^{1/2} & 0 \\ 0 & a^{-1/2} \end{bmatrix}$ for a transformation $f(z) = az$, which includes dilations ($a > 0$) and rotations ($|a| = 1$);

(iii) $\begin{bmatrix} 0 & i \\ i & 0 \end{bmatrix}$ for the inversion $f(z) = z^{-1}$.

Although at first glance it might seem that writing a Möbius transformation in normalized form only serves to complicate matters, for theoretical purposes it is really the opposite that is true: the use of normalized forms tends, in general, to simplify the treatment of these mappings. A perfect example of this is the discussion of fixed points found in Theorem 2.1. Even when restricted to matrices A with determinant one, the correspondence $A \rightarrow f_A$ fails to be one-to-one, for obviously $f_A = f_{-A}$. However, this is the only way in which it deviates from one-to-oneness: if $f_A = f_B$ and if $\det A = \det B = 1$, then either $A = B$ or $A = -B$ (Exercise 6.25).

2.3 Fixed Points

A giant step toward understanding the inner workings of a Möbius transformation f is taken with the realization that its set of fixed points, by which is meant the set of points z in $\widehat{\mathbb{C}}$ with the property that $f(z) = z$, has an extremely simple structure.

Theorem 2.1. *Let f be a Möbius transformation other than the identity. Then f has at most two fixed points. More precisely, if f is exhibited in normalized form as $f(z) = (az + b)/(cz + d)$, then f has exactly one fixed point when $a + d = \pm 2$ and exactly two fixed points otherwise.*

Proof. We may suppose that f is given to us in normalized form: $f(z) = (az + b)/(cz + d)$, with $ad - bc = 1$. Assume first that $c = 0$. Then $ad = 1$ and f is the similarity transformation $f(z) = \alpha z + \beta$, where $\alpha = a/d$ and $\beta = b/d$. Such a mapping always fixes ∞. In the event that $\alpha \neq 1$ it also fixes the point $-\beta/(\alpha - 1)$. Unless $\alpha = 1$ and $\beta = 0$ — this would make f the identity transformation — the equation $\alpha z + \beta = z$ has no other solutions, so f has no other fixed points. Notice that $\alpha = 1$ if and only if $a = d$. Since $(a + d)^2 = (a - d)^2 + 4ad = (a - d)^2 + 4$, $\alpha = 1$ holds if and only if $a + d = \pm 2$. In the case $c = 0$, therefore, f has exactly one fixed point when $a + d = \pm 2$ and exactly two such points otherwise.

We now consider the situation for non-zero c. In this case $f(-d/c) = \infty$ and $f(\infty) = a/c$, so neither $-d/c$ nor ∞ is fixed by f. Its only fixed points are the finite solutions z of the equation

$$\frac{az + b}{cz + d} = z ,$$

which, since we're concerned only with points where the denominator on the left is not zero, is equivalent to

$$cz^2 + (d - a)z - b = 0 .$$

Using the fact that $ad - bc = 1$, we find the roots of this quadratic equation to be

$$z = \frac{(a - d) \pm \sqrt{(a + d)^2 - 4}}{2c} .$$

Again when $c \neq 0$, the mapping f is seen to have a unique fixed point if $a + d = \pm 2$ and two fixed points if $a + d \neq \pm 2$. ∎

Möbius transformations f and g are said to be *conjugate* provided there exists a Möbius transformation h with the property that $g = h^{-1} \circ f \circ h$ or, equivalently, $h \circ g = f \circ h$. (N.B. The context will always make clear whether the term "conjugate" means the conjugate of a complex number, the conjugate of a harmonic function, or conjugate in the sense defined here for Möbius transformations.) Since $h[g(z)] = f[h(z)]$ is then true for every z in $\widehat{\mathbb{C}}$, it becomes clear that z is a fixed point of g precisely when $h(z)$ is fixed by f. In particular, conjugate Möbius transformations have the same number of fixed points. Many of the ideas and quantities naturally associated with Möbius transformations turn out to be invariant under

2. Möbius Transformations

conjugacy. Number of fixed points is the simplest example. We shall point out others as we proceed.

How much leeway exists for dictating the values of a Möbius transformation at specified points? The next two results provide the answer to this question.

Theorem 2.2. *Let (z_1, z_2, z_3) be an ordered triple of distinct points from $\widehat{\mathbb{C}}$. There exists a unique Möbius transformation f that has $f(z_1) = 1$, $f(z_2) = 0$, and $f(z_3) = \infty$.*

Proof. For the existence, assume initially that all three of the given points are finite. Then

$$(9.17) \qquad f(z) = \frac{(z_1 - z_3)(z - z_2)}{(z_1 - z_2)(z - z_3)}$$

is a Möbius transformation with the features sought. (We have not bothered to express f in the form $f(z) = (az + b)/(cz + d)$, but it is obvious that we could do so if pressed.) If one of the points z_1, z_2, or z_3 is ∞, we can obtain f by letting the appropriate term in (9.17) tend to ∞. To be explicit, this process yields

$$(9.18) \qquad \begin{cases} f(z) = \dfrac{z - z_2}{z - z_3} & \text{if } z_1 = \infty\,; \\[2mm] f(z) = \dfrac{z_1 - z_3}{z - z_3} & \text{if } z_2 = \infty\,; \\[2mm] f(z) = \dfrac{z - z_2}{z_1 - z_2} & \text{if } z_3 = \infty\,. \end{cases}$$

As for the uniqueness, suppose that both f and g are Möbius transformations which enjoy the stated property. Then $h = f^{-1} \circ g$ is a Möbius transformation that fixes z_1, z_2, and z_3. Theorem 2.1 implies that h is the identity transformation, which means that $g = (f^{-1})^{-1} = f$. This corroborates the uniqueness assertion. ∎

An immediate consequence of Theorem 2.2 is that, modulo the obvious restriction imposed by univalence, the values of a Möbius transformation can be prescribed arbitrarily at any three given points of $\widehat{\mathbb{C}}$.

Corollary 2.3. *Let (z_1, z_2, z_3) and (w_1, w_2, w_3) be ordered triples of distinct points from $\widehat{\mathbb{C}}$. There exists a unique Möbius transformation f that has $f(z_1) = w_1$, $f(z_2) = w_2$, and $f(z_3) = w_3$.*

Proof. We invoke Theorem 2.2 to choose Möbius transformations g and h for which $g(z_1) = 1$, $g(z_2) = 0$, and $g(z_3) = \infty$, while $h(w_1) = 1$, $h(w_2) = 0$, and $h(w_3) = \infty$. Then $f = h^{-1} \circ g$ is a Möbius transformation that maps z_1 to w_1, z_2 to w_2, and z_3 to w_3. Its uniqueness in this regard is confirmed by essentially the same argument used in the proof of Theorem 2.2 to demonstrate uniqueness there. ∎

2.4 Cross-ratios

In studying any class of mappings it is generally of value to identify quantities and structures that are preserved under application of those mappings. Later in this chapter we shall talk about a general "conformal invariant." For the time being we concentrate on three invariants — one numerical, two geometric — specific to the class of Möbius transformations. The first of these is the *cross-ratio* $[z_1, z_2, z_3, z_4]$ associated with an ordered quadruple (z_1, z_2, z_3, z_4) of distinct points from the extended complex plane. By definition, $[z_1, z_2, z_3, z_4]$ is the complex number $f(z_4)$, where f is the unique Möbius transformation that satisfies $f(z_1) = 1$, $f(z_2) = 0$, and $f(z_3) = \infty$. Referring to (9.17) and (9.18), we obtain the formula

$$(9.19) \qquad [z_1, z_2, z_3, z_4] = \frac{(z_1 - z_3)(z_2 - z_4)}{(z_1 - z_2)(z_3 - z_4)}$$

in case all four of these points are finite, whereas the situation in which one of the points is ∞ is covered by

$$(9.20) \qquad \begin{cases} [\infty, z_2, z_3, z_4] = \dfrac{z_2 - z_4}{z_3 - z_4} & , \quad [z_1, \infty, z_3, z_4] = \dfrac{z_1 - z_3}{z_4 - z_3}, \\[1em] [z_1, z_2, \infty, z_4] = \dfrac{z_2 - z_4}{z_2 - z_1} & , \quad [z_1, z_2, z_3, \infty] = \dfrac{z_1 - z_3}{z_1 - z_2}. \end{cases}$$

We stress that a cross-ratio is always a finite quantity different from 0 or 1. For example, we compute

$$[1, i, -1, -i] = \frac{2 \cdot (2i)}{(1-i)(-1+i)} = 2$$

and

$$[0, -1, i, \infty] = \frac{-i}{1} = -i \, .$$

The behavior of cross-ratios under Möbius transformations is the subject of the next theorem.

Theorem 2.4. *If f is a Möbius transformation, then $[z_1, z_2, z_3, z_4] = [f(z_1), f(z_2), f(z_3), f(z_4)]$ for every ordered quadruple (z_1, z_2, z_3, z_4) of distinct points from $\widehat{\mathbb{C}}$.*

Proof. Let g be the unique Möbius transformation that sends $f(z_1)$ to 1, $f(z_2)$ to 0, and $f(z_3)$ to ∞. Then $g \circ f$ transforms z_1 to 1, z_2 to 0, and z_3 to ∞. According to the definition of a cross-ratio

$$[z_1, z_2, z_3, z_4] = g \circ f(z_4) = g[f(z_4)] = [f(z_1), f(z_2), f(z_3), f(z_4)] \, ,$$

2. Möbius Transformations

as asserted. ∎

A slight refinement of Theorem 2.4 tells us exactly when two cross-ratios are equal.

Corollary 2.5. *Cross-ratios $[z_1, z_2, z_3, z_4]$ and $[w_1, w_2, w_3, w_4]$ are equal if and only if there exists a Möbius transformation f satisfying $f(z_j) = w_j$ for $j = 1, 2, 3, 4$.*

Proof. If f exists as described, then Theorem 2.4 certifies that the two cross-ratios agree. Conversely, assume that these cross-ratios are equal. By definition $[w_1, w_2, w_3, w_4] = g(w_4)$, where g is the unique Möbius transformation that has $g(w_1) = 1$, $g(w_2) = 0$, and $g(w_3) = \infty$. Now let f be the unique Möbius transformation for which $f(z_1) = w_1$, $f(z_2) = w_2$, and $f(z_3) = w_3$ (Corollary 2.3). On the basis of Theorem 2.4 we conclude that

$$g[f(z_4)] = [w_1, w_2, w_3, f(z_4)] = [z_1, z_2, z_3, z_4] = [w_1, w_2, w_3, w_4] = g(w_4).$$

Since g is univalent, this implies that $f(z_4) = w_4$, so $f(z_j) = w_j$ for $j = 1, 2, 3, 4$. ∎

Theorem 2.4 introduces a convenient way to represent the Möbius transformation f that maps three given points z_1, z_2, and z_3 to designated image points w_1, w_2, and w_3; namely, for $z \neq z_1, z_2, z_3$ the value $f(z)$ is completely determined by the relation

$$(9.21) \qquad [z_1, z_2, z_3, z] = [w_1, w_2, w_3, f(z)] \ .$$

(One implication of this formula is that Möbius transformations are the only self-mappings of $\widehat{\mathbb{C}}$ which preserve cross-ratios.) For many purposes it is far preferable to work with (9.21) than with some cumbersome formula for f. Even if a formula for f is the objective, (9.21) often provides the most efficient route to it, as the following example illustrates.

EXAMPLE 2.1. Determine the Möbius transformation f that maps i to ∞, 0 to 1, and ∞ to $-i$.

For $z \neq i, 0, \infty$ we appeal to (9.21) and write

$$[i, 0, \infty, z] = [\infty, 1, -i, f(z)] \ ,$$

which reduces with the help of (9.20) to

$$-iz = \frac{1 - f(z)}{-i - f(z)} \ .$$

Solving for $f(z)$ we arrive at the formula

$$f(z) = \frac{z + 1}{iz + 1} \ .$$

2.5 Circles in the Extended Plane

From a purely geometric standpoint the most important class of objects preserved by Möbius transformations is the class of circles in the extended complex plane. We remind the reader that the expression "circle in $\widehat{\mathbb{C}}$" refers to a subset K of $\widehat{\mathbb{C}}$ that is either an honest-to-god circle in the complex plane or a "circle through ∞," meaning a set of the type $K = L \cup \{\infty\}$, where L is a straight line in $\mathbb{C}$. Thus, through any triad of points in $\widehat{\mathbb{C}}$ there passes one and only one circle in $\widehat{\mathbb{C}}$. It is easily checked that the locus of points z in $\mathbb{C}$ which satisfy an equation

(9.22) $A|z|^2 + Bz + \overline{B}\overline{z} + C = 0 \qquad (A \text{ and } C \text{ real}, |B|^2 - AC > 0)$

is either a circle ($A \neq 0$) or a line ($A = 0$). Conversely, every circle or line in the complex plane can be described by an equation of this type. (Recall Exercise I.4.19.) We can therefore regard (9.22) as the general equation of a circle in $\widehat{\mathbb{C}}$, provided we make the convention that ∞ gets included in the locus of points satisfying (9.22) if and only if $A = 0$. After these preparatory remarks, we are set to prove:

Theorem 2.6. *Möbius transformations map circles in $\widehat{\mathbb{C}}$ to circles in $\widehat{\mathbb{C}}$.*

Proof. Since an arbitrary composition of mappings with the property of transforming circles in $\widehat{\mathbb{C}}$ to circles in $\widehat{\mathbb{C}}$ will again exhibit this property, it is enough to check that the image of K, an arbitrary circle in $\widehat{\mathbb{C}}$, under every elementary transformation is a circle in $\widehat{\mathbb{C}}$. If f is a translation or dilation or rotation, then it is perfectly obvious that $f(K)$ is another circle in $\widehat{\mathbb{C}}$. It is only for the inversion $f(z) = z^{-1}$ that the situation remains unclear and needs to be discussed. Choose an equation $A|z|^2 + Bz + \overline{B}\overline{z} + C = 0$ for K. Let complex numbers z and w be related by $w = z^{-1}$. (In particular, $z \neq 0$ and $w \neq 0$.) Then z lies on K if and only if — substitute w^{-1} for z in the equation of K — w obeys the condition

$$\frac{A}{|w|^2} + \frac{B}{w} + \frac{\overline{B}}{\overline{w}} + C = 0$$

or, equivalently, w is on the locus determined by

$$C|w|^2 + \overline{B}w + B\overline{w} + A = 0 \, .$$

The last line is the equation of another circle in $\widehat{\mathbb{C}}$, call it $\widetilde{K}$. The point $z = 0$ belongs to K precisely when $C = 0$, which occurs by the convention adopted above exactly when $w = \infty = 1/0$ belongs to $\widetilde{K}$. Similarly, $z = \infty$ is on K if and only if $w = 0 = 1/\infty$ is on $\widetilde{K}$. In summary, a point z of $\widehat{\mathbb{C}}$ is a point of K if and only if $w = z^{-1} = f(z)$ is a point of $\widetilde{K}$; i.e., $f(K) = \widetilde{K}$, a circle in $\widehat{\mathbb{C}}$. ∎

2. Möbius Transformations

EXAMPLE 2.2. Identify the images of the circles $K = \mathbb{R} \cup \{\infty\}$ and $K' = K(0,1)$ under the Möbius transformation $f(z) = (z+i)/(z+1)$.

In view of Theorem 2.6 the set $f(K)$ is the circle in $\widehat{\mathbb{C}}$ that passes through the points $f(1) = (1+i)/2$, $f(0) = i$, and $f(-1) = \infty$ (Figure 13). Accordingly, $f(K)$ can be described by the (real) equation $y = -x+1$. The image of K' is likewise a circle in $\widehat{\mathbb{C}}$. It passes through $f(1) = (1+i)/2$ and $f(-1) = \infty$. By conformality $f(K')$ intersects $f(K)$ orthogonally at $(1+i)/2$, for K' is perpendicular to K at the point 1. This is enough information to identify $f(K')$ as the circle in $\widehat{\mathbb{C}}$ with equation $y = x$.

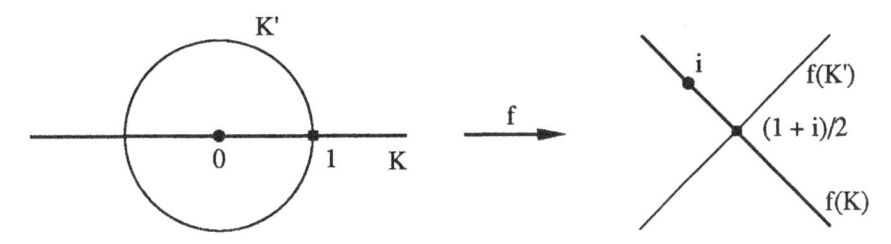

Figure 13.

2.6 Reflection and Symmetry

A third noteworthy invariant for Möbius transformations is the property of symmetry with respect to a circle in $\widehat{\mathbb{C}}$. In order to explain this notion we associate with each circle K in $\widehat{\mathbb{C}}$ a mapping $\rho_K : \widehat{\mathbb{C}} \to \widehat{\mathbb{C}}$ called the *reflection in K*. This we do as follows. Suppose first that K is the "true" circle in $\mathbb{C}$ with center z_0 and radius r. In this case we define ρ_K by

$$(9.23) \qquad \rho_K(z) = z_0 + \frac{r^2}{\overline{z} - \overline{z_0}} = \frac{z_0 \overline{z} + r^2 - |z_0|^2}{\overline{z} - \overline{z_0}}$$

for $z \neq z_0, \infty$, while $\rho_K(z_0) = \infty$ and $\rho_K(\infty) = z_0$. Put in geometric terms, ρ_K transforms a point z different from z_0 and ∞ to the unique point z^* that lies on the ray through z issuing from z_0 and that obeys the condition $|z - z_0||z^* - z_0| = r^2$ (Exercise 6.35). Figure 14 suggests how one can geometrically construct $\rho_K(z)$ from z when K is a genuine circle. Next, if $K = L \cup \{\infty\}$ for a line L with equation $Bz + \overline{B}\overline{z} + C = 0$ (here C is real and $B \neq 0$) then ρ_K is given by

$$(9.24) \qquad \rho_K(z) = (-\overline{B}/B)\overline{z} - (C/B)$$

for $z \neq \infty$ and $\rho_K(\infty) = \infty$. In this situation ρ_K acts on the plane in the way one ordinarily thinks of a reflection acting: it sends a point $z(\neq \infty)$ to its mirror image with respect to K; i.e., to the point z^* such that K

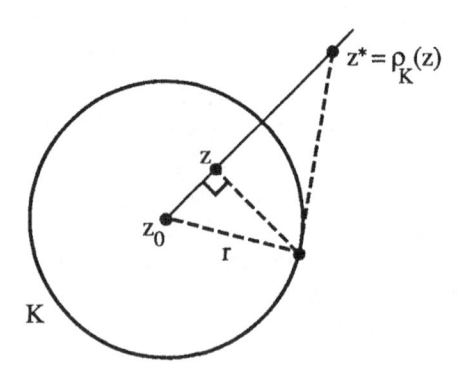

Figure 14.

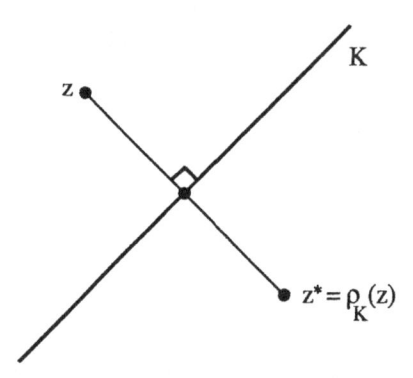

Figure 15.

is the perpendicular bisector of the line segment with endpoints z and z^*. (See Figure 15. See, too, Exercise 6.35.) In both (9.23) and (9.24) we observe that ρ_K fixes every point of K and that $\rho_K \circ \rho_K$ is the identity transformation of $\widehat{\mathbb{C}}$, from which it follows that $\rho_K^{-1} = \rho_K$.

It is evident from (9.23) and (9.24) that every reflection ρ_K belongs to the class of functions with the general structure

(9.25) $$f(z) = \frac{a\bar{z} + b}{c\bar{z} + d},$$

where $a, b, c,$ and d are complex numbers for which $ad - bc \neq 0$. Stated differently, f is a function of the form $f = g \circ \rho$, where g is a Möbius transformation and ρ is the reflection in the real axis, $\rho(z) = \bar{z}$. (N.B. Many books retain the name Möbius transformation for a function of this kind, but in keeping with other usage in this book we prefer to describe such a function as an *anti-Möbius transformation*. It should perhaps be emphasized that not every anti-Möbius transformation is a reflection. See Exercise 6.37.) Since ρ is the prototypical anti-conformal self-mapping of $\widehat{\mathbb{C}}$

2. Möbius Transformations

and since Möbius transformations are the conformal mappings of $\widehat{\mathbb{C}}$, it follows that anti-Möbius transformations map $\widehat{\mathbb{C}}$ anti-conformally onto itself. Indeed, it follows from Theorem 1.5 and the definition of anti-conformality in $\widehat{\mathbb{C}}$ that an arbitrary anti-conformal mapping of $\widehat{\mathbb{C}}$ onto itself has the structure (9.25). As might be expected, functions f that fit the description in (9.25) have many properties in common with Möbius transformations. For instance, if K is a circle in $\widehat{\mathbb{C}}$, then $f(K)$ is again a circle in $\widehat{\mathbb{C}}$. It is readily verified that by composing a pair of anti-Möbius transformations one produces a Möbius transformation. More generally, any composition of Möbius transformations and anti-Möbius transformations that involves an even number of factors of the latter sort will be a Möbius transformation.

Suppose now that K is a circle in $\widehat{\mathbb{C}}$. Points z and z^* of $\widehat{\mathbb{C}} \sim K$ are said to be *symmetric with respect to K* if $z^* = \rho_K(z)$ — hence, also, $z = \rho_K(z^*)$. The next theorem establishes the symmetry principle for Möbius transformations that was hinted at earlier.

Theorem 2.7. *If f is a Möbius transformation and if K is a circle in $\widehat{\mathbb{C}}$, then*

$$(9.26) \qquad f \circ \rho_K = \rho_{f(K)} \circ f \; .$$

In particular, if z and z^* are points that are symmetric with respect to K, then $f(z)$ and $f(z^*)$ are symmetric with respect to $f(K)$.

Proof. Statement (9.26) is equivalent to the assertion that the function $g = f^{-1} \circ \rho_{f(K)} \circ f \circ \rho_K$ is the identity transformation of $\widehat{\mathbb{C}}$. (Don't forget: $\rho_{f(K)}^{-1} = \rho_{f(K)}$.) As the composition of Möbius transformations and an even number of reflections, g is a Möbius transformation. Furthermore, it is evident that g fixes every point of K. Theorem 2.1 implies that g must fix every point of $\widehat{\mathbb{C}}$; i.e., g is the identity transformation, so (9.26) holds.

If z and z^* are symmetric with respect to K, then by (9.26) the points $w = f(z)$ and $w^* = f(z^*)$ satisfy

$$\rho_{f(K)}(w) = \rho_{f(K)}[f(z)] = f[\rho_K(z)] = f(z^*) = w^* \; ,$$

revealing that w and w^* are symmetric with respect to $f(K)$. ∎

The proof given for Theorem 2.7 would have worked just as well had f been a mapping of type (9.25) instead of a Möbius transformation. The conclusions of the theorem thus remain valid when f is an anti-Möbius transformation. Here is a simple application of the theorem.

EXAMPLE 2.3. Let K be the circle in $\widehat{\mathbb{C}}$ that passes through the three points z_1, z_2, and z_3. Show that the reflection ρ_K satisfies

$$(9.27) \qquad \overline{[z_1, z_2, z_3, z]} = [z_1, z_2, z_3, \rho_K(z)]$$

for $z \neq z_1, z_2, z_3$. (Compare this with (9.21).)

Let f be the Möbius transformation that maps z_1 to 1, z_2 to 0, and z_3 to ∞. Then $f(K) = \mathbb{R} \cup \{\infty\}$. Moreover, for z other than z_1, z_2, or z_3 we infer from the very definition of a cross-ratio that $[z_1, z_2, z_3, z] = f(z)$ and $[z_1, z_2, z_3, \rho_K(z)] = f[\rho_K(z)]$. If z lies on K, then $\rho_K(z) = z$ and (9.27) reduces to the statement $\overline{f(z)} = f(z)$, which is true because $f(z)$ is real in this case. If z lies off K, then by Theorem 2.7 the points $f(z)$ and $f[\rho_K(z)]$ must be symmetric with respect to $\mathbb{R} \cup \{\infty\}$, which is another way to say that $\overline{f(z)} = f[\rho_K(z)]$. Again, relation (9.27) is seen to be in force.

2.7 Classification of Möbius Transformations

Theorem 2.1 is the starting point for a standard classification of the non-identity Möbius transformations into four types. To a large extent (but not entirely) the classification parallels the breakdown of the elementary transformations into their categories.

A Möbius transformation with exactly one fixed point is known as a *parabolic transformation*. Let f be such a mapping, and let z_0 be its fixed point. If $z_0 = \infty$ the proof of Theorem 2.1 reveals that f must be a translation:

$$(9.28) \qquad f(z) = z + b ,$$

where $b = f(0) \neq 0$. Next, under the assumption that $z_0 \neq \infty$, we conjugate f by the transformation $h(z) = z_0 + z^{-1} = (z_0 z + 1)/z$ — h is chosen because it is the simplest Möbius transformation sending ∞ to z_0 — to form $g = h^{-1} \circ f \circ h$. Since f fixes only $z_0 = h(\infty)$, g has ∞ for its sole fixed point, so it is also parabolic. By our previous comments g has the form $g(z) = z + b$ for some non-zero complex number b. (In fact, b can be computed directly from f —

$$b = \lim_{z \to z_0} \left[\frac{1}{f(z) - z_0} - \frac{1}{z - z_0} \right]$$

— as (9.29) makes plain.) Observing that $h^{-1}(z) = (z - z_0)^{-1}$ and exploiting the fact that $h^{-1}[f(z)] = g[h^{-1}(z)]$ for every z in $\widehat{\mathbb{C}}$, we find that

$$(9.29) \qquad \frac{1}{f(z) - z_0} = b + \frac{1}{z - z_0} .$$

Equation (9.29) is quickly solved for $f(z)$ to yield

$$(9.30) \qquad f(z) = z_0 + \frac{z - z_0}{1 + b(z - z_0)}$$

as the general form of a parabolic transformation that fixes the finite point z_0. (A reminder: it is required that $b \neq 0$ here.) We remark that a mapping f of type (9.30) takes ∞ to $z_0 + b^{-1}$ and $z_0 - b^{-1}$ to ∞, which means that

2. Möbius Transformations

z_0, $f(\infty)$, and $f^{-1}(\infty)$ are collinear points in the finite plane. Let it be further noted that a Möbius transformation f is parabolic if and only if it is conjugate to a non-trivial translation g (in fact, one can take $g(z) = z+1$).

From (9.28) and (9.30) it is possible to read off immediately the *forward iterates* $f_2 = f \circ f$, $f_3 = f \circ f \circ f, \cdots$ and *backward iterates* $f_{-1} = f^{-1}$, $f_{-2} = f^{-1} \circ f^{-1}, f_{-3} = f^{-1} \circ f^{-1} \circ f^{-1}, \cdots$ of f: for any integer n it is true that

$$f_n(z) = z + nb \tag{9.31}$$

if $z_0 = \infty$, while

$$f_n(z) = z_0 + \frac{z - z_0}{1 + nb(z - z_0)} \tag{9.32}$$

if $z_0 \neq \infty$. (Implicit here is that $f_1 = f$. Also, f_0 is an alias for the identity transformation.) This assertion is essentially obvious in the former case and can be verified in the latter one by repeated application of (9.29). For instance, we have

$$\frac{1}{f_2(z) - z_0} = \frac{1}{f[f(z)] - z_0} = b + \frac{1}{f(z) - z_0} = 2b + \frac{1}{z - z_0},$$

and solving for $f_2(z)$ produces (9.32) with $n = 2$. Similarly, (9.29) gives

$$-b + \frac{1}{z - z_0} = -b + \frac{1}{f[f^{-1}(z)] - z_0} = \frac{1}{f^{-1}(z) - z_0} = \frac{1}{f_{-1}(z) - z_0},$$

which leads to the confirmation of (9.32) for $n = -1$. It should be pointed out that, except for $n = 0$, all the iterates f_n of f are likewise parabolic with the same fixed point z_0 and that, as $|n| \to \infty$, $f_n(z) \to z_0$ for every z in $\widehat{\mathbb{C}}$.

We now turn our attention to a Möbius transformation f that has exactly two fixed points. We label these points z_1 and z_2, establishing the notational convention for the present discussion and for similar situations later on that $z_1 \neq \infty$. If $z_2 = \infty$, then the proof of Theorem 2.1 shows that f has the appearance $f(z) = \kappa z + \beta$, in which $\kappa \neq 0, 1$. Since $f(z_1) = z_1$, it follows that $\beta = z_1 - \kappa z_1$. This allows us to rewrite f in the manner $f(z) = z_1 + \kappa(z - z_1)$. In particular, the relationship between z, $f(z)$, and z_1 is expressed by

$$f(z) - z_1 = \kappa(z - z_1) \tag{9.33}$$

for some complex number κ different from 0 and 1. If $z_2 \neq \infty$, we select a Möbius transformation h that sends 0 to z_1 and ∞ to z_2 — $h(z) = (z_2 z - z_1)/(z - 1)$ is our choice — and build the conjugate transformation $g = h^{-1} \circ f \circ h$. Then g has 0 and ∞ as its only fixed points, so $g(z) = \kappa z$ for some constant κ other than 0 or 1. Because $h^{-1}(z) = (z - z_1)/(z - z_2)$

and $h^{-1}[f(z)] = g[h^{-1}(z)]$ for every z in $\widehat{\mathbb{C}}$, we learn that f satisfies

(9.34) $$\frac{f(z) - z_1}{f(z) - z_2} = \kappa \left(\frac{z - z_1}{z - z_2} \right).$$

The constant κ appearing in (9.33) and (9.34) is known as a *multiplier* for the transformation f; (9.33) and (9.34) represent f in so-called *multiplier-fixed point format*. In the case of (9.34) such a representation is not uniquely determined, for it clearly depends on which fixed point we elect to designate as z_1. Reversing the labels on the fixed points, however, has only the effect of replacing the multiplier κ in (9.34) by κ^{-1}. (It follows that in all cases the quantity $\kappa + \kappa^{-1}$ is independent of the specific designations of z_1 and z_2.) Notice that we can always use the formula $\kappa = f'(z_1)$ to determine a multiplier for f. This is clear from (9.33) when $z_2 = \infty$ and follows from (9.34) via the calculation

$$\kappa = \lim_{z \to z_1} \left[\frac{f(z) - z_1}{z - z_1} \frac{z - z_2}{f(z) - z_2} \right]$$

$$= f'(z_1) \frac{z_1 - z_2}{f(z_1) - z_2} = f'(z_1) \frac{z_1 - z_2}{z_1 - z_2} = f'(z_1)$$

when $z_2 \neq \infty$. In the latter instance a similar computation would produce an alternate description of the κ in (9.34) — namely, $\kappa = 1/f'(z_2)$.

It has already been remarked that in the case $z_2 = \infty$ the transformation f takes the form

(9.35) $$f(z) = z_1 + \kappa(z - z_1).$$

Under the assumption that both z_1 and z_2 are finite we can solve (9.34) for $f(z)$ and thereby extract from it the expression

(9.36) $$f(z) = \frac{(z_1 - \kappa z_2)z + (\kappa - 1)z_1 z_2}{(1 - \kappa)z + \kappa z_1 - z_2}$$

for a general Möbius transformation that leaves fixed the two finite points z_1 and z_2. In practice, one should stress, it is frequently more convenient to work with (9.34) than with (9.36). It will prove useful for later considerations to register the fact that the function in (9.36) transforms ∞ to $(z_1 - \kappa z_2)/(1 - \kappa)$ and $(z_2 - \kappa z_1)/(1 - \kappa)$ to ∞.

There is no difficulty in passing from (9.33) and (9.34) to the corresponding representations for the iterate f_n of f, just as we did in the case of a parabolic transformation: for any integer n,

(9.37) $$f_n(z) - z_1 = \kappa^n(z - z_1)$$

if $z_2 = \infty$, while

2. Möbius Transformations

(9.38) $$\frac{f_n(z) - z_1}{f_n(z) - z_2} = \kappa^n \left(\frac{z - z_1}{z - z_2} \right)$$

if $z_2 \neq \infty$. For example, when $n = 2$ the verification of (9.38) reads

$$\frac{f_2(z) - z_1}{f_2(z) - z_2} = \frac{f[f(z)] - z_1}{f[f(z)] - z_2} = \kappa \left(\frac{f(z) - z_1}{f(z) - z_2} \right) = \kappa^2 \left(\frac{z - z_1}{z - z_2} \right).$$

From (9.37) and (9.38) it is a short step to a closed form expression for f_n:

(9.39) $$f_n(z) = \begin{cases} z_1 + \kappa^n(z - z_1) & \text{if } z_2 = \infty\,; \\ \dfrac{(z_1 - \kappa^n z_2)z + (\kappa^n - 1)z_1 z_2}{(1 - \kappa^n)z + \kappa^n z_1 - z_2} & \text{if } z_2 \neq \infty\,. \end{cases}$$

Several interesting phenomena are discernible in (9.39). First of all, in the situation where $|\kappa| \neq 1$ one of the fixed points of f has the property that $f_n(z)$ tends to this fixed point as $n \to \infty$ whenever z is an element of $\widehat{\mathbb{C}}$ different from the other fixed point of f. The fixed point of f that enjoys this property is named its *attracting fixed point*; the remaining fixed point is its *repelling fixed point*. Recalling that it is $\kappa = f'(z_1)$ which appears in (9.39), we can elicit from (9.39) the information that the fixed point z_1 is attracting when $|f'(z_1)| < 1$ and repelling when $|f'(z_1)| > 1$. A fixed point is plainly attracting for f if and only if it is repelling for f^{-1}. In the case of a Möbius transformation f for which $|\kappa| = 1$ no such distinction between its fixed points can be drawn. What does present itself as a possibility when $|\kappa| = 1$, however, is that $\kappa^n = 1$ for some integer n, in which event the iterate f_n of f reduces to the identity transformation. The smallest positive integer n with this property is then called *the order of f*. The multipliers κ for which this behavior is exhibited are those of the form $\kappa = e^{2\pi i \theta}$ with θ a rational number.

The classification of a Möbius transformation with exactly two fixed points hinges on the character of the multiplier κ in (9.33) or (9.34). (Recall: $\kappa \neq 1$.) If κ is real and positive, then f is termed a *hyperbolic transformation*; when $|\kappa| = 1$, we say that f is an *elliptic transformation*; in all remaining cases f is pronounced a *loxodromic transformation*. It is not difficult to verify — this was partially done in the derivation of (9.34) — that f is hyperbolic if and only if it is conjugate to a non-trivial dilation with respect to the origin, and elliptic if and only if it is conjugate to a non-trivial rotation about the origin (Exercise 6.47). A loxodromic transformation, on the other hand, can never be conjugated to one of the elementary transformations. When a Möbius transformation is presented in normalized form, its type can be detected instantly.

Theorem 2.8. *Let $f(z) = (az + b)/(cz + d)$ be a Möbius transformation that is not the identity and is displayed in normalized form. Then f is* *parabolic if and only if $a + d = \pm 2$; f is hyperbolic if and only if $a + d$ is real with $|a+d| > 2$; f is elliptic if and only if $a+d$ is real with $|a+d| < 2$; f is loxodromic if and only if $a + d$ is non-real.*

Proof. It is an immediate consequence of Theorem 2.1 that f is a parabolic transformation precisely in the circumstance that $a + d = \pm 2$. It remains to treat the case when f has two fixed points z_1 and z_2, where we continue to observe the convention that $z_1 \neq \infty$. Let κ be a multiplier for f. We claim first that

$$\kappa + \kappa^{-1} = (a + d)^2 - 2 . \tag{9.40}$$

If $z_2 = \infty$, then $c = 0$, $ad = 1$, and $\kappa = f'(z_1) = a/d$. Therefore,

$$\kappa + \kappa^{-1} = \frac{a}{d} + \frac{d}{a} = \frac{a^2 + 2ad + d^2 - 2ad}{ad} = (a+d)^2 - 2 . \tag{9.41}$$

If $z_2 \neq \infty$, then $c \neq 0$ and we may assume that $\kappa = f'(z_1) = 1/f'(z_2)$. In this case $\kappa + \kappa^{-1} = f'(z_1) + f'(z_2)$. Using the fact that $ad - bc = 1$, one checks that

$$f'(z) = \frac{1}{(cz+d)^2} .$$

Moreover, as the proof of Theorem 2.1 indicates, we can identify the fixed points of f as

$$z_1 = \frac{(a-d) + \sqrt{(a+d)^2 - 4}}{2c} , \quad z_2 = \frac{(a-d) - \sqrt{(a+d)^2 - 4}}{2c} .$$

It follows that

$$\kappa + \kappa^{-1} = f'(z_1) + f'(z_2) = \frac{1}{(cz_1 + d)^2} + \frac{1}{(cz_2 + d)^2}$$

$$= \frac{4}{[(a+d) + \sqrt{(a+d)^2 - 4}]^2} + \frac{4}{[(a+d) - \sqrt{(a+d)^2 - 4}]^2}$$

$$= \frac{4[(a+d) - \sqrt{(a+d)^2 - 4}]^2 + 4[(a+d) + \sqrt{(a+d)^2 - 4}]^2}{\{[(a+d) + \sqrt{(a+d)^2 - 4}][(a+d) - \sqrt{(a+d)^2 - 4}]\}^2}$$

$$= \frac{16(a+d)^2 - 32}{16} = (a+d)^2 - 2 ,$$

which combined with (9.41) completes the verification of (9.40).

We proceed to write κ in polar form: $\kappa = re^{i\theta}$, with $\theta = \operatorname{Arg} \kappa$. For κ to be both real and positive, it is required that $\theta = 0$. In this case $r = 1$ is ruled out, as $\kappa \neq 1$. Therefore, a hyperbolic transformation f has

$$(a+d)^2 = 2 + \kappa + \kappa^{-1} = 2 + r + r^{-1} ;$$

2. Möbius Transformations

i.e., $(a+d)^2$ is real and positive. Furthermore, $(a+d)^2 > 4$, because when r varies over $(0, \infty)$ the quantity $r + r^{-1}$ attains its minimum value of 2 only for $r = 1$. Thus, for f to be hyperbolic it is necessary that $a + d$ be real and satisfy $|a+d| > 2$. Next, suppose that f is elliptic. Then $|\kappa| = 1$. Since $\kappa \neq 1$, this only happens when $r = 1$ and $\theta \neq 0$. The result is that

$$(a+d)^2 = 2 + \kappa + \kappa^{-1} = 2 + e^{i\theta} + e^{-i\theta} = 2 + 2\cos\theta < 4 \, ,$$

which implies that in the elliptic case $a + d$ is real and has $|a+d| < 2$. Looking at the remaining cases — $\theta \neq 0$ and $r \neq 1$ — we see that the quantity $\kappa + \kappa^{-1}$ is not real when $0 < |\theta| < \pi$ and belongs to the interval $(-\infty, -2)$ when $\theta = \pi$. Under these conditions $(a+d)^2$ is found to be a member of the set $\mathbb{C} \sim [0, \infty)$. As a consequence, $a + d$ must be non-real for loxodromic f. Since our list of categories of non-identity Möbius transformations is exhaustive and these categories are mutually exclusive, and since the same is true of the list of possibilities for $a+d$ in the statement of Theorem 2.8, the proof is complete. ∎

It is an easy consequence of Theorem 2.8 that conjugate Möbius transformations always fall into the same class (Exercise 6.48). We apply Theorem 2.8 and the discussion preceding it to a pair of examples.

EXAMPLE 2.4. Classify and then describe the important properties of the transformation $f(z) = (4z+6)/(2z+4)$.

We begin by rewriting f in normalized form: $f(z) = (2z+3)/(z+2)$. Here $a + d = 4$, so on the basis of Theorem 2.8 we can say that f is hyperbolic. Its fixed points are $z_1 = \sqrt{3}$ and $z_2 = -\sqrt{3}$. To find a multiplier for f we compute

$$\kappa = f'(z_1) = \frac{1}{(z_1+2)^2} = \frac{1}{7+4\sqrt{3}} \, .$$

Since $|f'(z_1)| < 1$, z_1 is the attracting fixed point of f and z_2 is its repelling fixed point.

EXAMPLE 2.5. Classify the transformation $f(z) = [(1+i)z - i]/z$, and indicate any special features that it might possess.

In normalized form this mapping becomes

$$f(z) = \frac{\sqrt{2}\,z - e^{\pi i/4}}{e^{-\pi i/4}\,z + 0} \, .$$

Thus $a + d = \sqrt{2}$, which stamps f as elliptic. It has fixed points $z_1 = 1$ and $z_2 = i$. For a multiplier we can use $\kappa = f'(1) = i$. Noting that $\kappa^4 = 1$, whereas $\kappa^n \neq 1$ for $1 \leq n \leq 3$, we infer that f is a transformation of order four.

2.8 Invariant Circles

Save for those loxodromic transformations whose multipliers are not real, every non-identity Möbius transformation f has a characteristic family of *invariant circles*. This name applies to any circle K in $\widehat{\mathbb{C}}$ having the property that $f(K) = K$. With the help of its invariant circles it is possible to draw a very instructive picture detailing how the transformation acts on $\widehat{\mathbb{C}}$. We shall carry this out for f, in turn, of parabolic, hyperbolic, and elliptic type. In each instance we deal first with an elementary transformation, where the geometry is quite obvious. The treatment of the general transformation in each category is reduced to the elementary case by conjugation, using the principle that K is an invariant circle for $h^{-1} \circ f \circ h$ if and only if $h(K)$ is an invariant circle for f (Exercise 6.52). Invariant circles, however, paint only half the picture. To complete it we track the image under f and its sundry (forward and backward) iterates f_n of a circle $\widetilde{K}_0$ selected from the family dual to the invariant circles of f, the latter being the family of circles $\widetilde{K}$ in $\widehat{\mathbb{C}}$ that are perpendicular to every f-invariant circle. We repeat for emphasis that a loxodromic transformation with a multiplier that is not a negative real number does not have any invariant circles (Exercise 6.56).

Parabolic Case. Suppose that f is a parabolic transformation with fixed point z_0. If $z_0 = \infty$, then f is an elementary transformation — a translation, to be exact — say $f(z) = z + b$ with $b \neq 0$. It is evident that $K_0 = L_0 \cup \{\infty\}$, where L_0 is the line through 0 and $b = f(0)$, is an invariant circle for f, as is $K = L \cup \{\infty\}$ for any line L parallel to L_0. These are readily seen to be the only circles in $\widehat{\mathbb{C}}$ invariant under f. If $\widetilde{L}_0$ denotes the line perpendicular to L_0 at the origin, then the images of $\widetilde{K}_0 = \widetilde{L}_0 \cup \{\infty\}$ under f and its iterates partition the plane into parallel strips. The geometry of this mapping is captured by Figure 16, where the arrows indicate the direction in which points move along invariant circles under the application of f. One message which this diagram (and later diagrams akin to it) seeks to convey is that f maps a shaded region onto an adjacent unshaded region, and vice versa.

If z_0 is a finite point, we consider $g = h^{-1} \circ f \circ h$, where h is a Möbius transformation taking ∞ to z_0. As g is parabolic and fixes ∞, we know the nature of its invariant circles. In particular, if we choose one invariant circle K_0^* for g, then we can characterize all remaining g-invariant circles as the circles in $\widehat{\mathbb{C}}$ that intersect K_0^* only at ∞, the fixed point of g. It follows from Exercise 6.52 that, once we pin down a single invariant circle K_0 for f, we shall have determined all invariant circles for this transformation — namely, K_0 and every circle K in $\widehat{\mathbb{C}}$ that intersects K_0 only at z_0. Recalling (9.30) we write,

$$f(z) = z_0 + \frac{z - z_0}{1 + b(z - z_0)},$$

2. Möbius Transformations 409

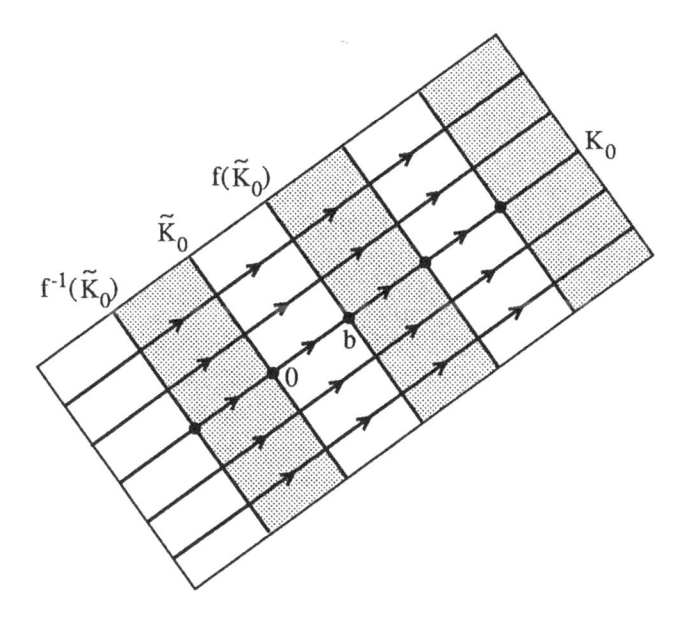

Figure 16.

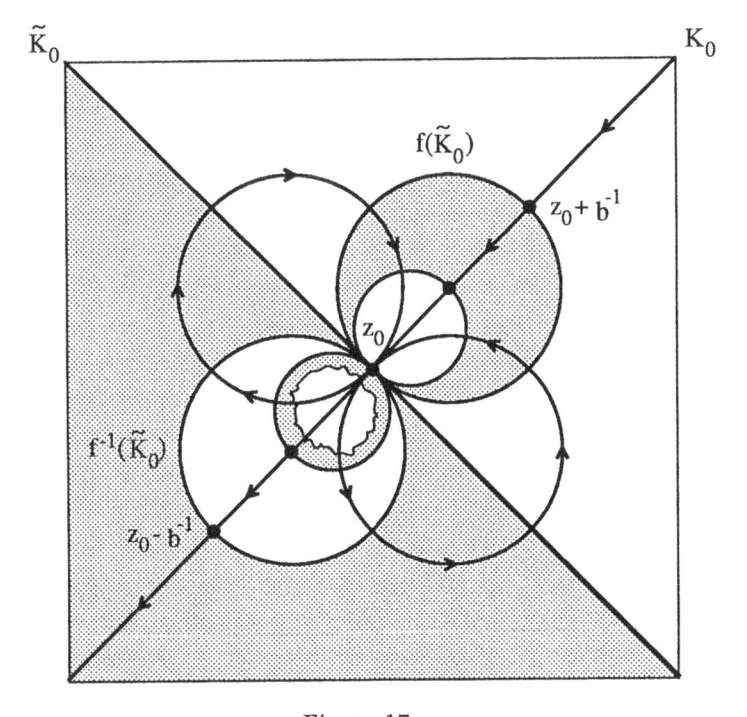

Figure 17.

where $b \neq 0$ is given by

$$b = \lim_{z \to z_0} \left[\frac{1}{f(z) - z_0} - \frac{1}{z - z_0} \right].$$

As observed at the time, the points z_0, $f(\infty) = z_0 + b^{-1}$, and $f^{-1}(\infty) = z_0 - b^{-1}$ lie on a line L_0 in the finite plane, so $K_0 = L_0 \cup \{\infty\}$ must be an invariant circle for f. In light of the foregoing remarks its other invariant circles are just the authentic circles in $\mathbb{C}$ that are tangent to L_0 at z_0. We direct the reader to Figure 17, the analogue of Figure 16 for a parabolic transformation whose fixed point z_0 is finite. The direction of the arrows is dictated by the fact that f transforms $z_0 + b^{-1}$ to $z_0 + (2b)^{-1}$, a point situated between z_0 and $z_0 + b^{-1}$. Taking $\widetilde{K}_0 = \widetilde{L}_0 \cup \{\infty\}$, where $\widetilde{L}_0$ is the line orthogonal to L_0 at z_0, we observe that here the images of $\widetilde{K}_0$ under f and its iterates partition the set $\mathbb{C} \sim \{z_0\}$ into (with two unbounded exceptions) crescent-shaped pieces.

Hyperbolic Case. Let f be a hyperbolic transformation with fixed points z_1 and z_2. In the situation where $z_1 = 0$ and $z_2 = \infty$ the mapping is a simple dilation, $f(z) = \kappa z$ with κ positive and different from 1. The circles invariant under f are easily described: they are the circles in $\widehat{\mathbb{C}}$ that pass through both fixed points of f; i.e., the circles $K = L \cup \{\infty\}$, in which L is a line through the origin. A convenient member of the perpendicular family is the circle $\widetilde{K}_0 = K(0,1)$. Under iteration of f, $\widetilde{K}_0$ and its images subdivide the punctured plane $\mathbb{C} \sim \{0\}$ into concentric annular regions. (See Figure 18. There we have assumed that $\kappa < 1$, which makes the origin the attracting fixed point of f.)

In the general hyperbolic case, we conclude by means of the conjugation principle that the family of invariant circles of f still consists of all the circles in $\widehat{\mathbb{C}}$ passing through both z_1 and z_2. (Here we would choose the conjugating Möbius transformation h so as to send 0 to z_1 and ∞ to z_2.) Figure 19 illustrates the basic geometry of f when both z_1 and z_2 are finite and z_1 is the attracting fixed point. In this situation $K_0 = L_0 \cup \{\infty\}$, with L_0 the line through z_1 and z_2, is one of the invariant circles for f. As a circle from the perpendicular family we can take $\widetilde{K}_0 = \widetilde{L}_0 \cup \{\infty\}$, $\widetilde{L}_0$ being the line perpendicular to L_0 at $(z_1 + z_2)/2$. Incidentally, notice that the fixed points z_1 and z_2 of a hyperbolic transformation f are always symmetric with respect to any circle in $\widehat{\mathbb{C}}$ that is perpendicular to every f-invariant circle. This is clear when $z_1 = 0$ and $z_2 = \infty$; through conjugation it then follows from conformality and the symmetry principle (Theorem 2.7) for an arbitrary hyperbolic transformation.

Elliptic Case. Consider, finally, an elliptic transformation f with fixed points z_1 and z_2. In case $z_1 = 0$ and $z_2 = \infty$ we can write $f(z) = e^{i\theta}z$, where $\theta \neq 0$ and $-\pi < \theta \leq \pi$. One obvious class of circles preserved by such a rotation consists of the true circles centered at the origin or, described in terms of Möbius invariant concepts, the circles in $\widehat{\mathbb{C}}$ with respect to which the fixed points of f are symmetric. (Unless $\theta = \pi$, f has no other invariant

2. Möbius Transformations

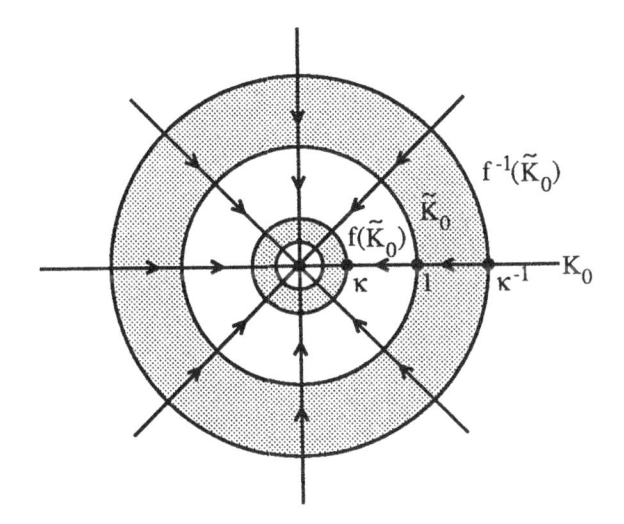

Figure 18.

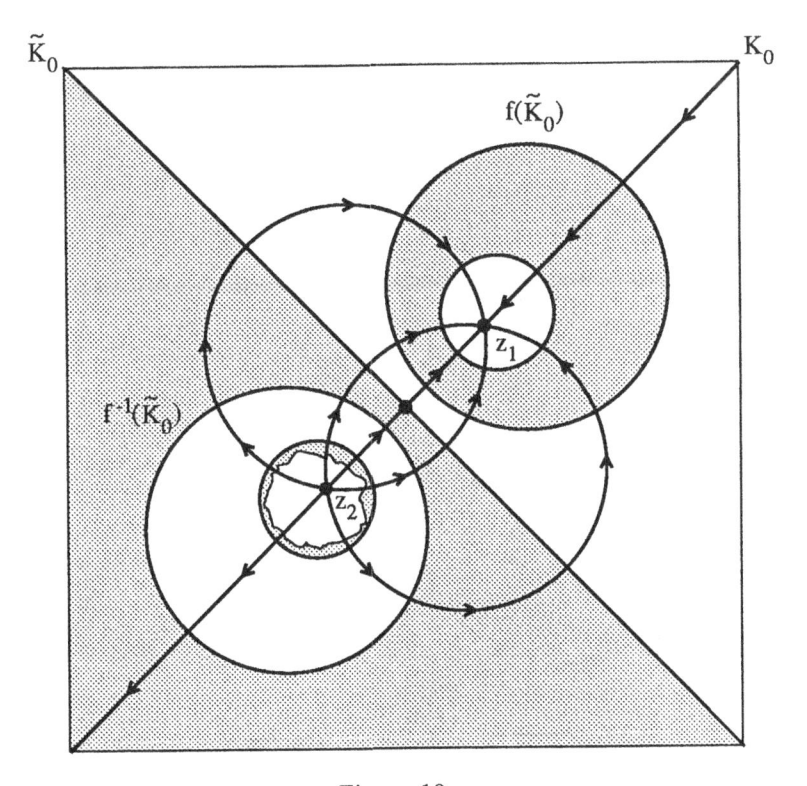

Figure 19.

412   IX. Conformal Mapping

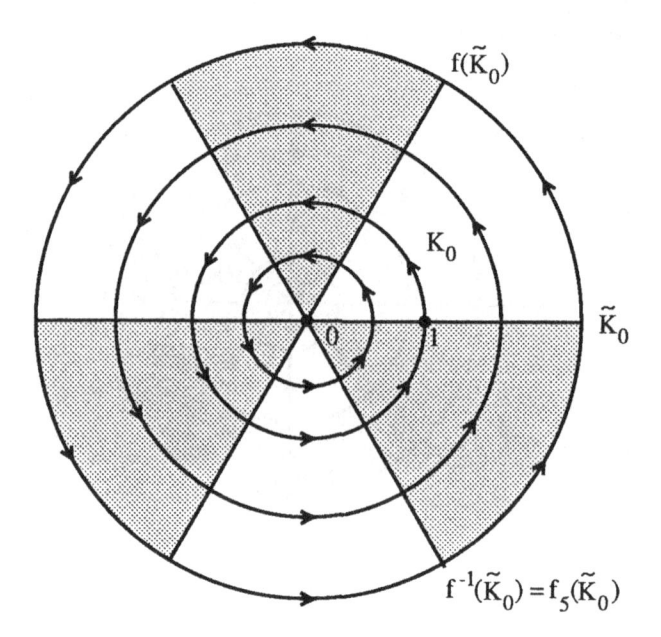

Figure 20.

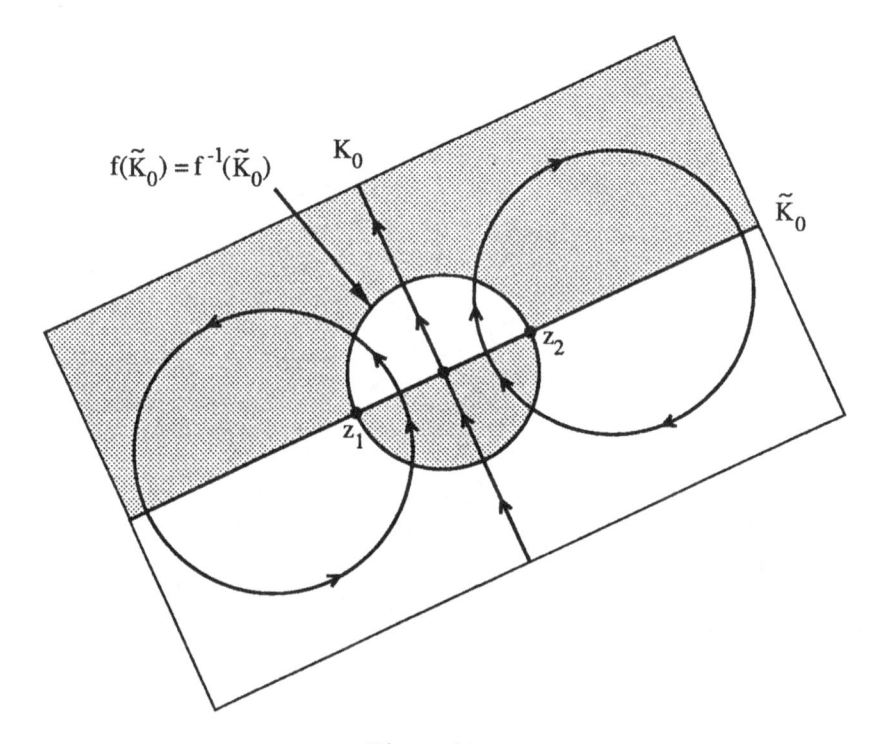

Figure 21.

2. Möbius Transformations

circles. When $\theta = \pi$, which means that $f(z) = -z$, there is an additional family of f-invariant circles: in this special case any circle in $\widehat{\mathbb{C}}$ that passes through both 0 and ∞ is also invariant under f.) If $K = \widetilde{L} \cup \{\infty\}$ is a circle in $\widehat{\mathbb{C}}$ that passes through both of these fixed points, then $f(\widetilde{K})$ is a second circle of the same kind and the oriented angle from $\widetilde{K}$ to $f(\widetilde{K})$ at the origin is θ. This statement applies, in particular, to $\widetilde{K}_0 = \mathbb{R} \cup \{\infty\}$. The images of $\widetilde{K}_0$ under f and its iterates do not typically partition the plane in any reasonable way; indeed, they normally seem to "fill up" the plane in a very chaotic fashion. An exception occurs when $\theta = 2\pi q$ for a rational number q; i.e., when f has finite order. In this case the images of $\widetilde{K}_0$ subdivide the plane into finitely many sectors, as pictured in Figure 20 for $\theta = \pi/3$.

Via conjugation we see that the general elliptic transformation f leaves invariant exactly those circles in $\widehat{\mathbb{C}}$ with respect to which its fixed points z_1 and z_2 are symmetric, supplemented when the multiplier of f is -1 by the circles in $\widehat{\mathbb{C}}$ that go through both z_1 and z_2. It maps an arbitrary circle passing through both fixed points to a new circle through those points. At z_1 — remember that by our notational convention $z_1 \neq \infty$ — this new circle intersects the original one at an angle $\theta = \text{Arg}\, f'(z_1)$. Figure 21 depicts the situation for two finite fixed points when $\theta = \pi/2$. Of course, if both fixed points are finite, then the perpendicular bisector L_0 of the segment between z_1 and z_2 is part of an invariant circle for f, that being $K_0 = L_0 \cup \{\infty\}$. A natural choice of circle $\widetilde{K}_0$ from the orthogonal family is then found in $\widetilde{K}_0 = \widetilde{L}_0 \cup \{\infty\}$, where $\widetilde{L}_0$ is the line through z_1 and z_2.

By way of transition from the topic of Möbius transformations back to the subject of general conformal mappings, we close this section with two examples that demonstrate how Möbius transformations and the concepts associated with them enhance our ability to construct conformal mappings.

EXAMPLE 2.6. Determine a conformal mapping of the open half-disk $D = \{z : |z| < 1, \text{Im}\, z > 0\}$ onto the full disk $D' = \Delta(0,1)$ (Figure 22).

We start by subjecting D to a Möbius transformation f_1 that sends 1 to 0 and -1 to ∞. (Reason: f_1 will transform D to a domain bounded by a pair of rays emanating from the origin. Such a domain is not hard to map to D'.) We choose $f_1(z) = (1-z)/(1+z)$, a mapping we have worked with before when we used it to map D' onto the half-plane $D'' = \{z : \text{Re}\, z > 0\}$. Now $f_1(0) = 1$, which implies that f_1 transforms $\mathbb{R} \cup \{\infty\}$ to the circle in $\widehat{\mathbb{C}}$ through 0, 1, and ∞ — hence, to itself. It maps the interval $(-1,1)$ to the interval $(0,\infty)$. The symmetry property of Möbius transformations implies that the domains $f_1(D)$ and $f_1[\Delta(0,1) \sim \overline{D}]$ are symmetric with respect to the real axis. It follows that $f_1(D)$ is one of the two quarter planes into which $(0,\infty)$ separates D''. Lastly, since the curvilinear angle whose sides are the interval $A = [0,1]$ and the circular arc $B = \{z : |z| = 1, \text{Im}\, z \geq 0\}$ has $\theta(A, B) = -\pi/2$, the conformality of f_1 tells us that $\theta[f_1(A), f_1(B)] = -\pi/2$. This added information is enough to nail

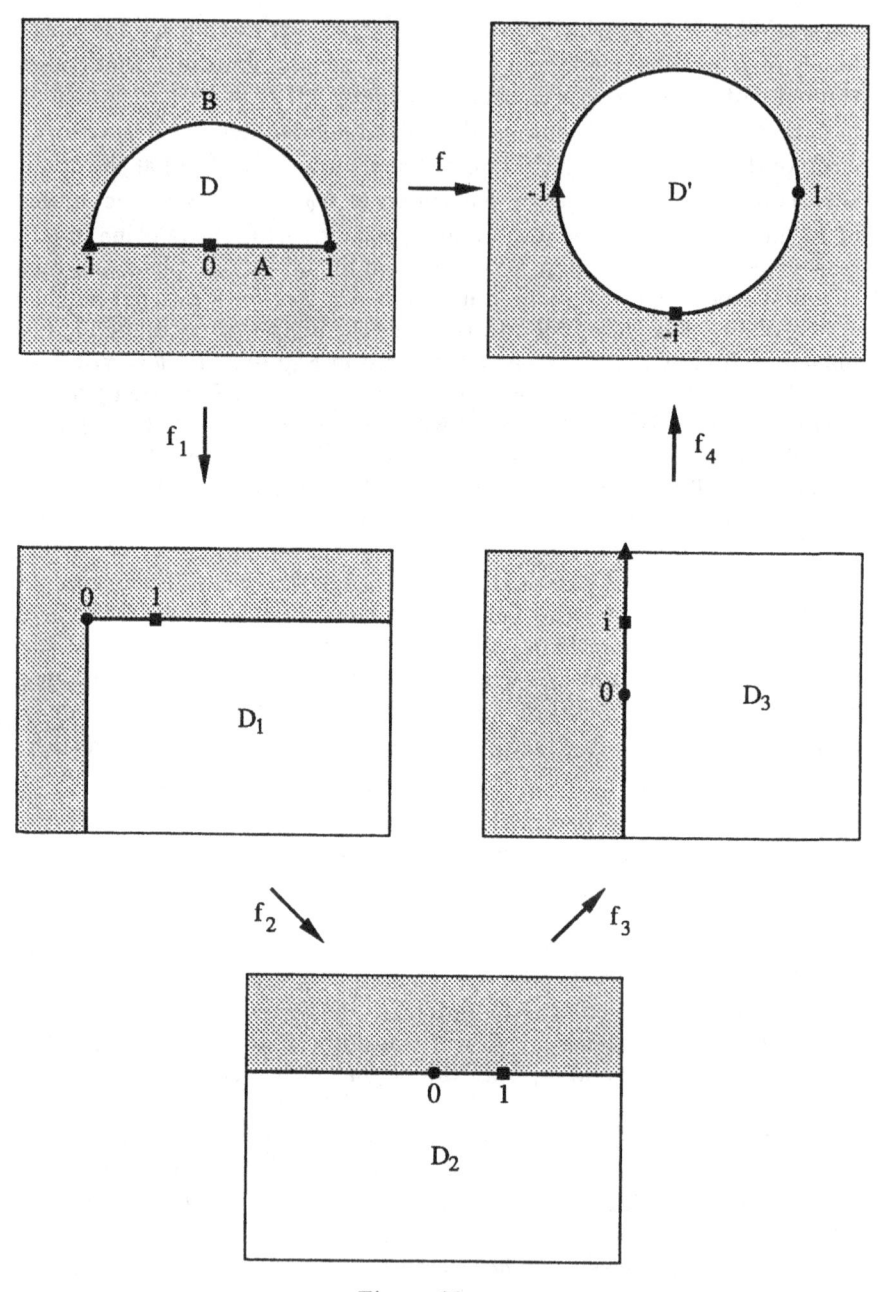

Figure 22.

2. Möbius Transformations

down the image of D under f_1: $D_1 = f_1(D) = \{z : \operatorname{Re} z > 0, \operatorname{Im} z < 0\}$. We proceed with the construction of a mapping f in the manner proposed by Figure 22. We take $f = f_4 \circ f_3 \circ f_2 \circ f_1$ with f_1 as above, $f_2(z) = z^2$, $f_3(z) = iz$, and $f_4 = f_1$. Evaluation of this composition at a point z of D yields the formula

$$f(z) = \frac{(1+z)^2 - i(1-z)^2}{(1+z)^2 + i(1-z)^2}$$

for a conformal mapping of D onto D'.

EXAMPLE 2.7. Find a conformal mapping of the region D pictured in Figure 23 onto the strip $D' = \{z : |\operatorname{Im} z| < \pi/2\}$.

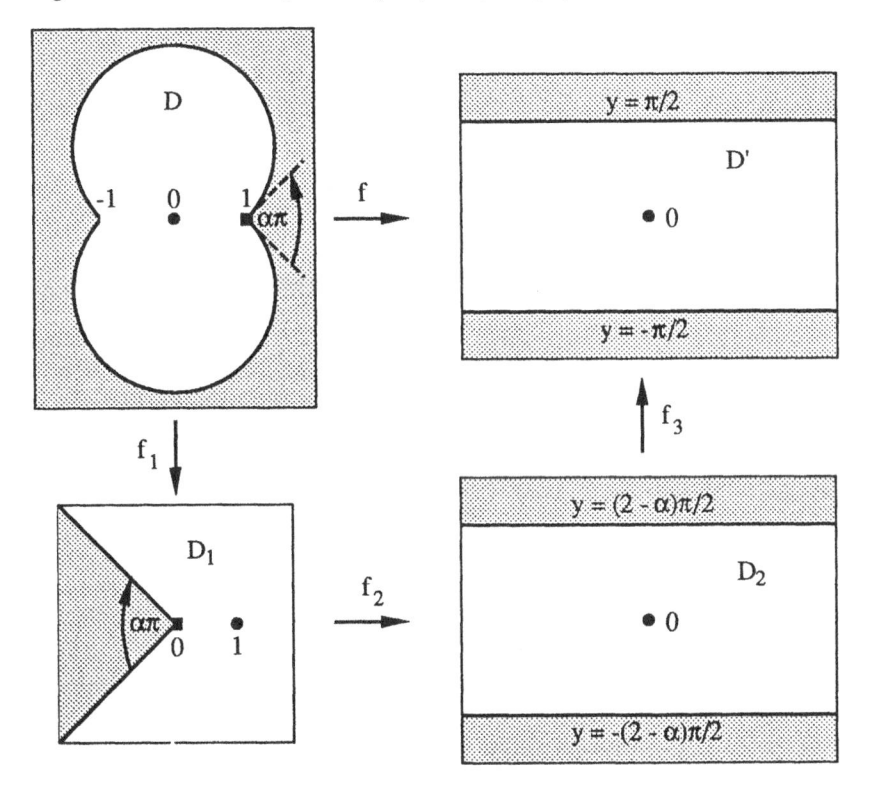

Figure 23.

Figure 23 is intended to indicate that the domain D is the union of two disks of the same radius whose bounding circles intersect in an angle $\alpha\pi, 0 < \alpha < 1$, at the points 1 and -1. In particular, D is symmetric with respect to the real axis. As in the previous example, the first step here is to "straighten out" the boundary of D. Once again this is accomplished by means of the mapping $f_1(z) = (1-z)/(1+z)$. The domain $D_1 = f_1(D)$ is symmetric with respect to the real axis, for D itself has this property and the Möbius transformation f_1 leaves the set $\mathbb{R} \cup \{\infty\}$ invariant. Also, D_1 is bounded by two rays issuing from the origin. Because the curvilinear angle exterior to D formed at the point 1 by the two circular arcs that compose ∂D has magnitude $\alpha\pi$, the same must be true of the angle exterior to D_1 formed by the aforementioned rays at the origin. Since f_1 maps the interval $(-1,1)$ to $(0,\infty)$, the latter interval lies in D_1. The domain D_1 is now completely determined: $D_1 = \{z : |\operatorname{Arg} z| < (2-\alpha)\pi/2\}$. A mapping f of D onto D' is obtained in the form $f = f_3 \circ f_2 \circ f_1$ suggested by Figure 23. Here f_1 is as announced, $f_2(z) = \operatorname{Log} z$, and $f_3(z) = z/(2-\alpha)$. Therefore,

$$f(z) = \frac{1}{2-\alpha} \operatorname{Log}\left(\frac{1-z}{1+z}\right)$$

is a mapping that does the required job.

3 Riemann's Mapping Theorem

3.1 Preparations

It is a noteworthy item in the lore of complex function theory that the proof which Riemann offered for his mapping theorem was flawed, resting as it did on the faulty hypothesis that a certain minimization problem in the calculus of variations is guaranteed to have a solution. In all fairness to Riemann, few of his contemporaries would have batted an eye at this assumption: the existence of solutions to such problems was taken largely for granted by mathematicians of the day. Not until Weierstrass appeared on the scene and spearheaded a campaign to enforce stricter standards of rigor in analysis did existence questions of this type begin to attract closer attention. Furthermore, the basic features of Riemann's approach were eventually salvaged by David Hilbert (1862-1943) and Richard Courant (1888-1972), albeit some fifty years after Riemann's original work. In the interim other proofs for the mapping theorem were devised. The argument we present is based on Montel's normal family theorem and an idea of Paul Koebe (1882-1945), to whom this particular proof is usually credited. For convenience we include the more technical aspects of the proof in three preparatory lemmas. The first of these informs us that simple connectivity, as we have defined the term, is preserved under conformal mapping.

Lemma 3.1. *Let $f: D \to \mathbb{C}$ be a conformal mapping, where D is a simply connected domain in the complex plane. Then $D' = f(D)$ is also a simply connected domain.*

Proof. If $D' = \mathbb{C}$, the conclusion is obvious. We assume, therefore, that $D' \neq \mathbb{C}$. Since f is a non-constant analytic function, D' is certainly a

3. Riemann's Mapping Theorem

domain. We must prove that $n(\beta, w) = 0$ whenever w is a point of $\mathbb{C} \sim D'$ and β is a closed, piecewise smooth path in D'. Fix w and β as described, say $\beta\colon [a, b] \to \mathbb{C}$. Define $\gamma\colon [a, b] \to \mathbb{C}$ by $\gamma = f^{-1} \circ \beta$, so that $\beta(t) = f[\gamma(t)]$. Then γ is a closed, piecewise smooth path in D. As D is simply connected, γ is homologous to zero in this domain. Because w does not belong to D', the function $f'/(f - w)$ is analytic in D. Cauchy's theorem thus leads to

$$0 = \int_\gamma \frac{f'(z)\,dz}{f(z) - w} = \int_a^b \frac{f'[\gamma(t)]\dot\gamma(t)\,dt}{f[\gamma(t)] - w} = \int_a^b \frac{\dot\beta(t)\,dt}{\beta(t) - w}$$

$$= \int_\beta \frac{d\zeta}{\zeta - w} = 2\pi i\, n(\beta, w)\,,$$

which confirms that $n(\beta, w) = 0$. ∎

Our second preliminary lemma states that a simply connected subdomain of $\mathbb{C}$ which is not the whole plane can always be transformed conformally to a domain that is contained in the unit disk.

Lemma 3.2. *Let D be a simply connected domain in the complex plane, $D \neq \mathbb{C}$, and let z_0 be a point of D. There is a conformal mapping $f\colon D \to \mathbb{C}$ with the following properties: (i) the domain $f(D)$ is contained in $\Delta = \Delta(0, 1)$; (ii) $f(z_0) = 0$ and $f'(z_0) > 0$.*

Proof. We shall exhibit a mapping f with the stated properties as the composition $f = f_5 \circ f_4 \circ f_3 \circ f_2 \circ f_1$ of five relatively simple conformal mappings, which we now proceed to describe. To start, we choose a point b of $\mathbb{C} \sim D$ and define $f_1\colon D \to \mathbb{C}$ by $f_1(z) = z - b$. Since D is a proper subdomain of $\mathbb{C}$, the selection of b is possible. The mapping f_1 merely translates D to a simply connected domain $D_1 = f_1(D)$ that does not contain the origin. For the mapping f_2 we take any branch of $\log z$ in D_1. That such a branch f_2 exists is guaranteed by Theorem V.6.2. Moreover, we know that f_2 is a univalent analytic function. Fix a point w_0 in the domain $D_2 = f_2(D_1)$, together with a radius $r > 0$ for which the closed disk $\overline{\Delta}(w_0, r)$ lies in D_2. Setting $\widetilde{w}_0 = w_0 + 2\pi i$, we observe that $\overline{\Delta}(\widetilde{w}_0, r)$ and D_2 are necessarily disjoint. Indeed, should there be a point $\widetilde{w}$ in $\overline{\Delta}(\widetilde{w}_0, r) \cap D_2$, then on the one hand $\widetilde{w}$ would be of the form $\widetilde{w} = f_2(\widetilde{z})$ for some $\widetilde{z}$ in D_1, while on the other we could represent $\widetilde{w}$ as $\widetilde{w} = w + 2\pi i$ for some w in $\overline{\Delta}(w_0, r)$. Of course, it would also be true that $w = f_2(z)$ with z a point of D_1, implying that

$$\widetilde{z} = e^{f_2(\widetilde{z})} = e^{\widetilde{w}} = e^{w + 2\pi i} = e^w = e^{f_2(z)} = z\,.$$

It would follow that $w = f_2(z) = f_2(\widetilde{z}) = \widetilde{w} = w + 2\pi i$, an obvious contradiction. We conclude that $\overline{\Delta}(\widetilde{w}_0, r) \cap D_2 = \phi$, as claimed. Accordingly, $|z - \widetilde{w}_0| > r$ holds for every z in D_2. This fact implies that the image $D_3 = f_3(D_2)$ of D_2 under the Möbius transformation $f_3(z) = r/(z - \widetilde{w}_0)$ is a subset of Δ. The function $f_3 \circ f_2 \circ f_1$ thus furnishes a conformal mapping of D whose range is contained in Δ.

Next, set $c = f_3 \circ f_2 \circ f_1(z_0)$. According to Theorem 1.4 the function $f_4: \Delta \to \Delta$ given by $f(z) = (z-c)/(1-\bar{c}z)$ maps Δ conformally onto itself. It carries the point c to the origin. The composition $f_4 \circ f_3 \circ f_1 \circ f_1$ then maps D conformally inside Δ and transforms z_0 to the origin.

Finally, $d = (f_4 \circ f_3 \circ f_2 \circ f_1)'(z_0) \neq 0$ (Theorem VIII.3.9). Write $u = e^{-i \operatorname{Arg} d}$ and define f_5 in Δ by $f_5(z) = uz$. Then $f = f_5 \circ f_4 \circ f_3 \circ f_2 \circ f_1$ provides a conformal mapping of D onto a subdomain of Δ. It has $f(z_0) = 0$ and $f'(z_0) = f_5'(0)(f_4 \circ f_3 \circ f_2 \circ f_1)'(z_0) = ud = |d| > 0$. In short, f is a conformal mapping of D that enjoys properties (i) and (ii). ∎

The final lemma in the present sequence articulates the idea of Koebe that was alluded to earlier.

Lemma 3.3. *Suppose that D is a simply connected domain in the complex plane, $D \neq \mathbb{C}$, that z_0 is a point of D, and that $f: D \to \mathbb{C}$ is a conformal mapping which exhibits properties (i) and (ii) in Lemma 3.2. Assume that $f(D) \neq \Delta$. Then there exists a conformal mapping $g: D \to \mathbb{C}$ that is also endowed with these two properties, but has $g'(z_0) > f'(z_0)$.*

Proof. Write $D_0 = f(D)$. We again obtain g in the form of a composition $g = g_3 \circ g_2 \circ g_1 \circ f$, where the mappings g_1, g_2, and g_3 are constructed as follows. First, select a point b in $\Delta \sim D_0$. Since $0 = f(z_0)$ is an element of D_0, $b \neq 0$. Theorem 1.4 tells us that the Möbius transformation defined by $g_1(z) = (z - b)/(1 - \bar{b}z)$ maps Δ conformally onto itself — hence, maps the simply connected subdomain D_0 of Δ onto another such domain $D_1 = g_1(D_0)$. The domain D_1 does not contain the origin $(= g_1(b))$, but it does contain the point $-b = g_1(0)$. (See Figure 24.) Direct calculation shows that $g_1'(0) = 1 - |b|^2$.

Next, Theorem V.6.2 certifies the existence of a branch of $\log z$ in D_1. We choose one and call it L. More to the point, there is associated with L a branch g_2 of the square root function in D_1 — namely, the one given by $g_2(z) = \exp[L(z)/2]$. Then $|g_2(z)| = \sqrt{|z|} < 1$ for z in D_1, and g_2 is univalent in D_1: if $g_2(z) = g_2(\tilde{z})$, then $z = [g_2(z)]^2 = [g_2(\tilde{z})]^2 = \tilde{z}$. In other words, g_2 is a conformal mapping of D_1 onto some other simply connected domain $D_2 = g_2(D_1)$ lying in Δ (Figure 24). We write $c = g_2(-b)$, a point of D_2, and observe that $g_2'(-b) = 1/[2g_2(-b)] = 1/(2c)$.

Lastly, take g_3 to be the conformal self-mapping of Δ defined by $g_3(z) = u(z-c)/(1-\bar{c}z)$, with $u = \exp[i \operatorname{Arg} c]$. The domain $D_3 = g_3(D_2)$ is contained in Δ, the origin $(= g_3(c))$ belongs to D_3, and, as a simple computation reveals, $g_3'(c) = u/(1 - |c|^2)$.

The composite mapping $g = g_3 \circ g_2 \circ g_1 \circ f$ transforms D conformally to D_3, sends z_0 to the origin, and has

$$g'(z_0) = g_3'(c) g_2'(-b) g_1'(0) f'(z_0) = \left(\frac{u}{1 - |c|^2}\right)\left(\frac{1}{2c}\right)(1 - |b|^2) f'(z_0)$$

3. Riemann's Mapping Theorem

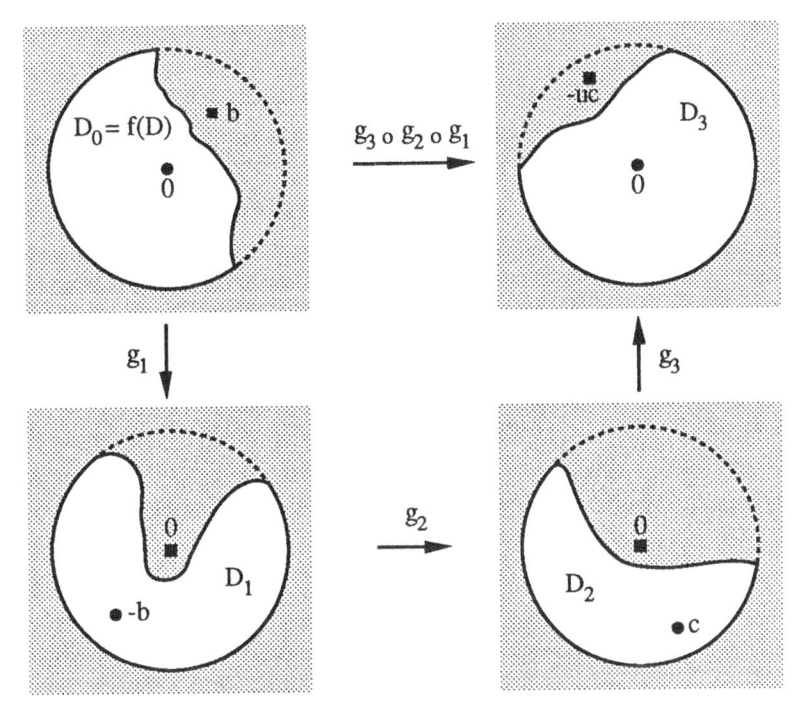

Figure 24.

$$= \left(\frac{1+|c|^2}{2|c|}\right) f'(z_0) > f'(z_0) \, ,$$

since $u/c = 1/|c|$, $|c|^2 = |g_2(-b)|^2 = |b|$, and $1 + |c|^2 > 2|c|$. ∎

3.2 The Mapping Theorem

We now move to the statement and proof of Riemann's mapping theorem. We preface the discussion with several remarks. First, we present the result in a normalized situation, requiring that the disk $\Delta = \Delta(0,1)$ be the image domain of the conformal mapping whose existence we establish. Nowadays this is quite standard, but it is not the way Riemann formulated his theorem. Secondly, just as Riemann sought to do, we arrive at a mapping by solving an extremal problem. The beauty of the Koebe-Montel approach, however, is that by avoiding the calculus of variations it skirts pitfalls like the one which was Riemann's misfortune. Thirdly, the conditions used in Theorem 3.4 to provide for the uniqueness of the mapping differ from those found in Riemann's own version of the theorem. All things considered, Theorem 4.11, which deals with domains bounded by Jordan curves, is more faithful to the original statement of the mapping theorem than are the results in the present section.

Theorem 3.4. (Riemann's Mapping Theorem) *Suppose that D is a simply connected domain in the complex plane, $D \neq \mathbb{C}$, and that z_0 is a point of D. There exists a unique conformal mapping f of D onto the disk $\Delta = \Delta(0,1)$ satisfying the conditions $f(z_0) = 0$ and $f'(z_0) > 0$.*

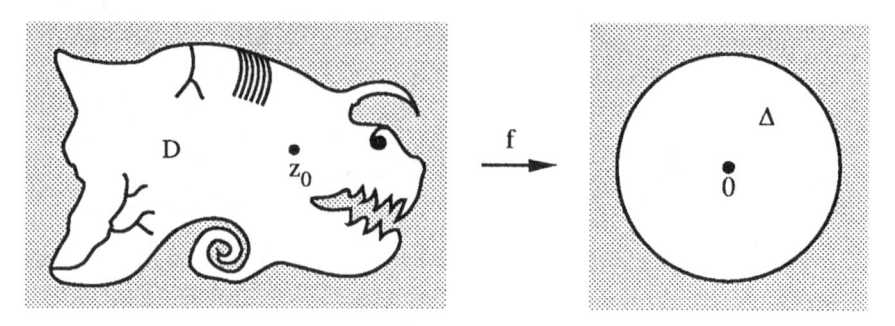

Figure 25.

Proof. Let $\mathcal{F}$ designate the family of all functions $f: D \to \mathbb{C}$ that meet the following specifications: f is a conformal mapping of D onto a subdomain of Δ, $f(z_0) = 0$, and $f'(z_0) > 0$. Lemma 3.3 insures that the family $\mathcal{F}$ is not empty. Suppose $r > 0$ has the property that $\Delta(z_0, r)$ lies in D. Cauchy's estimate (Theorem V.3.6) shows that $f'(z_0) = |f'(z_0)| \leq r^{-1}$ holds for every member f of $\mathcal{F}$. We infer that $\{f'(z_0): f \in \mathcal{F}\}$ is a bounded set of positive real numbers. As such, this set has a least upper bound, which we denote by ℓ. For each positive integer n we can select a function f_n in $\mathcal{F}$ for which $\ell - n^{-1} \leq f_n'(z_0) \leq \ell$. Because the family $\mathcal{F}$ is manifestly locally bounded in D, Montel's theorem empowers us to extract from the sequence $\langle f_n \rangle$ a subsequence $\langle f_{n_k} \rangle$ that converges normally in D to a function f. This limit function is analytic in D. Also, $f(z_0) = \lim_{k \to \infty} f_{n_k}(z_0) = 0$ and $f'(z_0) = \lim_{k \to \infty} f_{n_k}'(z_0) = \ell > 0$. In particular, f is non-constant in D. Corollary VIII.3.13 asserts that f is univalent in this domain. Obviously $f(D)$ lies in $\overline{\Delta}$. By the open mapping theorem $f(D)$ is an open set and so must actually be a subset of Δ. The function f, therefore, is a member of the family $\mathcal{F}$. If it could be demonstrated that $f(D) = \Delta$, the existence portion of the proof would be complete. Were it true, however, that $f(D) \neq \Delta$, then Lemma 3.3 would enable us to produce a member of $\mathcal{F}$ whose derivative at z_0 exceeds $f'(z_0) = \ell$. Given the definition of ℓ, this is not possible. As a result, f must map D conformally onto Δ.

To address the question of uniqueness, assume that g is a second conformal mapping of D onto Δ which has $g(z_0) = 0$ and $g'(z_0) > 0$. Consider the function $\varphi: \Delta \to \Delta$, $\varphi = g \circ f^{-1}$. This function provides a conformal self-mapping of Δ that fixes the origin and satisfies $\varphi'(0) = g'(z_0)/f'(z_0) > 0$. Theorem 1.4 implies that the only conformal mappings φ of Δ onto itself

3. Riemann's Mapping Theorem

with $\varphi(0) = 0$ are the rotations about the origin. The only such rotation with $\varphi'(0) > 0$ is the trivial rotation given by $\varphi(z) = z$. We conclude that $g(z) = \varphi[f(z)] = f(z)$ holds for every z in D; i.e., the alledged second mapping with the stated properties turns out to be nothing of the sort. ∎

Theorem 3.4 makes it possible to transform any simply connected proper subdomain D of $\mathbb{C}$ conformally to any other such domain. The mapping is uniquely determined once $f(z_0)$ and $\mathrm{Arg}[f'(z_0)]$ are specified for some point z_0 of D.

Theorem 3.5. *Let D and D' be simply connected domains in $\mathbb{C}$, neither the whole complex plane. Corresponding to given z_0 in D, z_0' in D', and θ_0 in $(-\pi, \pi]$ there exists a unique conformal mapping f of D onto D' that obeys the conditions $f(z_0) = z_0'$ and $\mathrm{Arg}[f'(z_0)] = \theta_0$.*

Proof. Let g be the conformal mapping of D onto $\Delta = \Delta(0,1)$ satisfying $g(z_0) = 0$ and $g'(z_0) > 0$, and let h be the corresponding mapping for the pair D' and z_0'. Then the function $f: D \to D'$ defined by $f(z) = h^{-1}[e^{i\theta_0} g(z)]$ gives a conformal mapping of D onto D' for which $f(z_0) = z_0'$ and $f'(z_0) = e^{i\theta_0} g'(z_0)/h'(z_0')$. Since $g'(z_0) > 0$ and $h'(z_0') > 0$, $\mathrm{Arg}[f'(z_0)] = \theta_0$. To establish the uniqueness of f, consider an arbitrary conformal mapping f_0 of D onto D' with the prescribed features. Then $\varphi = g \circ f^{-1} \circ f_0 \circ g^{-1}$ is a conformal self-mapping of Δ which satisfies $\varphi(0) = 0$ and

$$\varphi'(0) = g'(0) \cdot \frac{1}{f'(z_0)} \cdot f_0'(z_0) \cdot \frac{1}{g'(0)} = \frac{f_0'(z_0)}{f'(z_0)} = \frac{|f_0'(z_0)|}{|f'(z_0)|} > 0 \, ,$$

since $\mathrm{Arg}[f_0'(z_0)] = \mathrm{Arg}[f'(z_0)]$. As in the proof of Theorem 3.4 we conclude that $\varphi(z) = z$ for all z in Δ, from which it follows easily that $f_0(z) = f(z)$ for every z in D. ∎

Theorems 3.4 and 3.5 fall under the heading of pure existence theorems, for they do not tell us how to construct in any explicit way the mappings whose existence they grant. In Section 5 we shall look at the problem of mapping a half-plane conformally to a polygon, a situation in which this state of affairs can be remedied to some extent: there one can represent the mapping by an elementary integral formula. The catch is that the formula contains a number of parameters which are not easily computed unless the target polygon displays a good deal of symmetry. Thus, even mappings of half-planes to polygons remain somewhat elusive creatures.

With the aid of the Riemann mapping theorem we can at last lay to rest any lingering confusion about what it means for a plane domain to be simply connected. We now show that the standard topological definition of the concept and the definition we have adopted are equivalent. We also take the occasion to introduce yet another characterization of such domains, one that involves neither homotopy nor winding numbers.

Theorem 3.6. *The following statements about a domain D in $\mathbb{C}$ are equivalent:*

(i) D is simply connected;

(ii) every closed path in D is contractible in D;

(iii) the complement of D in the extended plane $\widehat{\mathbb{C}}$ is connected.

Proof of (ii)$\Rightarrow$(i). This implication is a restatement of Theorem V.7.4.

Proof of (i)$\Rightarrow$(ii). Assume that D is simply connected, and let $\gamma\colon [a,b] \to \mathbb{C}$ be a closed path in D. If $D = \mathbb{C}$, then $H(t,s) = (1-s)\gamma(t)$ gives a free homotopy in D between γ and the constant path β defined on $[a,b]$ by $\beta(t) = 0$. If $D \neq \mathbb{C}$, then we can choose a conformal mapping f of D onto $\Delta = \Delta(0,1)$. In this instance $H(t,s) = f^{-1}\{(1-s)f[\gamma(t)]\}$ supplies a free homotopy in D from γ to the constant path β given by $\beta(t) = f^{-1}(0)$ for $a \leq t \leq b$. In either case γ is seen to be contractible in D.

Proof of (iii)$\Rightarrow$(i). We assume $D \neq \mathbb{C}$, for only in this case is the implication unclear. Let γ be an arbitrary closed and piecewise smooth path in D. Under the assumption that $\widehat{\mathbb{C}} \sim D$ is connected we are asked to prove that γ is homologous to zero in D; i.e., $n(\gamma, z) = 0$ for every z in $\mathbb{C} \sim D$. Let D_0 denote the component of $\widehat{\mathbb{C}} \sim |\gamma|$ that contains the point ∞. Since $|\gamma|$ lies in D, $\widehat{\mathbb{C}} \sim D$ is contained in $\widehat{\mathbb{C}} \sim |\gamma|$. Furthermore, because $\widehat{\mathbb{C}} \sim D$ is assumed to be a connected set, it must actually be contained in some component of $\widehat{\mathbb{C}} \sim |\gamma|$. As ∞ is one point that D_0 and $\widehat{\mathbb{C}} \sim D$ have in common, that unnamed component can be none other than D_0. In particular, $\mathbb{C} \sim D$ is a subset of $D_0 \sim \{\infty\}$, which is just the unbounded component of $\mathbb{C} \sim |\gamma|$. By Lemma V.2.1(ii), $n(\gamma, z) = 0$ for every z in $\mathbb{C} \sim D$.

Proof of (i)$\Rightarrow$(iii). We assume that D is simply connected and prove that $\widehat{\mathbb{C}} \sim D$ is connected. If $D = \mathbb{C}$, then $\widehat{\mathbb{C}} \sim D = \{\infty\}$ is certainly a connected set. We proceed under the assumption that $D \neq \mathbb{C}$, which allows us to choose a conformal mapping f of $\Delta = \Delta(0,1)$ onto D. For $0 < r < 1$ let D_r be the image of the disk $\Delta(0,r)$ under f. We show initially that the set $\widehat{\mathbb{C}} \sim D_r$ is connected. This conclusion is essentially trivial if one is prepared to invoke the Jordan curve theorem, but a direct proof is also possible. In the interest of keeping the present discussion self-contained — but at the risk of making it seem slightly long-winded — we present the latter argument.

Fix r in $(0,1)$ and write $K = f[\overline{\Delta}(0,r)]$. The set K is a compact subset of D. The first step in demonstrating that $\widehat{\mathbb{C}} \sim D_r$ is connected is to show that the open set $\mathbb{C} \sim K$ is connected. Setting this as our immediate objective, we remark that $f[\Delta \sim \overline{\Delta}(0,r)]$ is a connected subset of $\mathbb{C} \sim K$ and thus lies in some component — label it G — of $\mathbb{C} \sim K$. We claim that, in fact, $\mathbb{C} \sim K = G$. If so, $\mathbb{C} \sim K$ is definitely connected. Let w_0 be a point of $\mathbb{C} \sim K$. The compactness of K and the continuity of the function $d(w) = |w - w_0|$ insure the existence of a point w_0' of K with the property that $|w_0' - w_0| = \min\{|w - w_0| : w \in K\}$. Being a point of K, w_0' belongs to D. Let Δ_0 be an open disk centered at w_0' and contained in D. Then $\Delta_0 \sim K$ is contained in $f[\Delta \sim \overline{\Delta}(0, r)]$ and, for this reason, is contained in G. Consider $S = L \sim \{w_0'\}$, where L is the line segment with endpoints w_0 and w_0'. The set S is connected, it lies in $\mathbb{C} \sim K$, and it intersects Δ_0. It thus meets $\Delta_0 \sim K$ — hence, meets G. On the other hand, S is contained in some component of $\mathbb{C} \sim K$, a component that we can now assert to be G. In particular, the point w_0, an arbitrary point in $\mathbb{C} \sim K$, belongs to G; i.e., $\mathbb{C} \sim K = G$. It is then elementary to check that $\partial G = \partial D_r$ and to conclude that $\widehat{\mathbb{C}} \sim D_r = (\widehat{\mathbb{C}} \sim \overline{D}_r) \cup \partial D_r = G \cup \partial G \cup \{\infty\} = \overline{G} \cup \{\infty\} = \widehat{G}$, the closure of G in $\widehat{\mathbb{C}}$. Since the closure in $\widehat{\mathbb{C}}$ of a connected subset of $\widehat{\mathbb{C}}$ is again connected, $\widehat{\mathbb{C}} \sim D_r$ is connected.

To complete the proof of (i)⇒(iii), set $D_n = D_{r_n}$, where $r_n = 1 - 2^{-n}$ for $n = 1, 2, 3, \cdots$. Then $D_1 \subset D_2 \subset D_3 \subset \cdots$ and $D = \bigcup_{n=1}^{\infty} D_n$. As a result, $F = \widehat{\mathbb{C}} \sim D = \bigcap_{n=1}^{\infty} F_n$, in which $F_n = \widehat{\mathbb{C}} \sim D_n$. Each F_n is a compact, connected set in $\widehat{\mathbb{C}}$, and $F_1 \supset F_2 \supset F_3 \supset \cdots$. Suppose that F were disconnected. Then there would exist disjoint open sets U and V in $\widehat{\mathbb{C}}$ with $U \cap F \neq \phi$, $V \cap F \neq \phi$, and F contained in $U \cup V$. The sequence $F_1 \sim (U \cup V), F_2 \sim (U \cup V), F_3 \sim (U \cup V), \cdots$ would be a non-increasing sequence of compact sets in $\widehat{\mathbb{C}}$ with empty intersection. In view of Cantor's theorem (Theorem II.4.5) — or, to be more precise, its analogue in $\widehat{\mathbb{C}}$ — $F_n \sim (U \cup V) = \phi$ would have to be true for some n. For such an n we would find F_n contained in the union of two disjoint open sets, both of which intersect F_n. At this point we would run into a contradiction, for we already know that F_n is connected. The contradiction forces us to conclude that $F = \widehat{\mathbb{C}} \sim D$ is, after all, a connected set. ∎

We have confined our discussion in this section to the conformal mapping of simply connected domains. The existence of conformal mappings between domains of more complicated topological structure is a trickier business and will not be considered in this book.

4 The Carathéodory-Osgood Theorem

4.1 Topological Preliminaries

Suppose that D and D' are simply connected plane domains, neither of which is the entire complex plane. We are now in possession of the knowledge that conformal mappings f from D onto D' exist. Moreover, the function theory we have so far developed secures for us a reasonably good hold on the local geometric and analytical properties of such a mapping near any point of D. Left wide open to speculation is its behavior at the boundary of D: What can be said about $f(\zeta)$ as ζ approaches a point z of ∂D? Framed in such generality this question has no simple answer. Indeed, there are many facets of the problem that are not yet fully understood and are the subjects of continuing research. There is, however, one special set of circumstances in which, from a strictly topological viewpoint, the matter can be completely settled, thanks to a beautiful theorem discovered independently by Constantin Carathéodory (1873-1950) and William F. Osgood (1884-1943): if each of the domains D and D' is the inside of a plane Jordan curve, then f extends in a unique way to a continuous and univalent function $\tilde{f}$ that maps $\overline{D}$ onto $\overline{D}'$. It turns out that under these conditions $\tilde{f}^{-1}: \overline{D}' \to \overline{D}$ will automatically be continuous as well, making $\tilde{f}$ a homeomorphism of $\overline{D}$ onto $\overline{D}'$. (Recall: a function $h: A \to B$ between sets A and B in the complex plane — or, more generally, in the extended complex plane — is called a *homeomorphism of A onto B* if h is univalent, if its range is B, and if both h and h^{-1} are continuous.) This section is devoted to a proof, modulo one technical detail, of the Carathéodory-Osgood theorem and to a discussion of several of the theorem's many consequences. The proof that we give relies on estimates for an important conformal invariant, of interest in its own right, which is introduced in Subsection 4.4.

As we did in the case of the Riemann mapping theorem, we shall "normalize" our discussion of the extension problem, this time by considering initially only conformal mappings $f: \Delta \to \mathbb{C}$, where $\Delta = \Delta(0,1)$. Since we want to allow for the possibility of $f(\Delta)$ being unbounded, we shall work in the extended plane $\widehat{\mathbb{C}}$ when attempting to extend f. (We remind the reader that the notations $\widehat{A}$ and $\widehat{\partial} A$ indicate the closure in $\widehat{\mathbb{C}}$ and boundary in $\widehat{\mathbb{C}}$, respectively, of a subset A of the extended plane. In particular, for a domain D in the finite plane, $\widehat{D} = \overline{D}$ and $\widehat{\partial} D = \partial D$ if D is bounded, whereas $\widehat{D} = \overline{D} \cup \{\infty\}$ and $\widehat{\partial} D = \partial D \cup \{\infty\}$ in the unbounded case.) As a beginning step we summarize in the form of a lemma some essential topological background information.

Lemma 4.1. *Let $\Delta = \Delta(0,1)$, let $f: \Delta \to \mathbb{C}$ be a continuous function, and let $D = f(\Delta)$. Assume that $\lim_{\zeta \to z} f(\zeta)$ exists in $\widehat{\mathbb{C}}$ for every point z of $\partial \Delta$. Then the function $\tilde{f}: \overline{\Delta} \to \widehat{\mathbb{C}}$ defined by*

$$\tilde{f}(z) = \begin{cases} f(z) & \text{if } z \in \Delta, \\ \lim_{\zeta \to z} f(\zeta) & \text{if } z \in \partial \Delta, \end{cases}$$

is the unique extension of f to a continuous function with domain-set $\overline{\Delta}$. Furthermore, $\tilde{f}(\overline{\Delta}) = \widehat{D}$. If, in addition, $\tilde{f}$ is univalent, then $\tilde{f}$ is a homeomorphism of $\overline{\Delta}$ onto $\widehat{D}$.

Proof. The relation $\tilde{f}(z) = \lim_{\zeta \to z} f(\zeta)$ is actually true for every point z

4. The Carathéodory-Osgood Theorem

of $\overline{\Delta}$, by definition if z lies on $\partial \Delta$ and by the assumed continuity of f if z belongs to Δ. What needs to be checked first is that $\widetilde{f}$, which is obviously continuous at all points of Δ, is continuous at every point of $\partial \Delta$. Fix such a point, say z_0. We consider an arbitrary sequence $\langle z_n \rangle$ in $\overline{\Delta}$ with the property that $z_n \to z_0$ and verify that $\widetilde{f}(z_n) \to \widetilde{f}(z_0)$. Since $\widetilde{f}(z_n) = \lim_{\zeta \to z_n} f(\zeta)$, we are at liberty to choose a point ζ_n in Δ satisfying $|\zeta_n - z_n| < n^{-1}$ for which it is also true that $|f(\zeta_n) - \widetilde{f}(z_n)| < n^{-1}$ if $\widetilde{f}(z_n) \neq \infty$ and $|f(\zeta_n)| > n$ if $\widetilde{f}(z_n) = \infty$. Then

$$|\zeta_n - z_0| \leq |\zeta_n - z_n| + |z_n - z_0| \leq \frac{1}{n} + |z_n - z_0| \to 0$$

as $n \to \infty$; i.e., $\zeta_n \to z_0$. Since $\langle \zeta_n \rangle$ is a sequence in Δ, it follows from the definition of $\widetilde{f}(z_0)$ that $f(\zeta_n) \to \widetilde{f}(z_0)$. If $\widetilde{f}(z_0) \neq \infty$, it must be so that $|f(\zeta_n)| < n$ — hence, that $\widetilde{f}(z_n) \neq \infty$ — once n is suitably large. For large n, therefore, we have

$$|\widetilde{f}(z_n) - \widetilde{f}(z_0)| \leq |\widetilde{f}(z_n) - f(\zeta_n)| + |f(\zeta_n) - \widetilde{f}(z_0)| \leq \frac{1}{n} + |f(\zeta_n) - \widetilde{f}(z_0)| \to 0,$$

showing that $\widetilde{f}(z_n) \to \widetilde{f}(z_0)$ in the case $\widetilde{f}(z_0) \neq \infty$. If $\widetilde{f}(z_0) = \infty$, then by construction $|f(\zeta_n)| \to \infty$. Either $\widetilde{f}(z_n) = \infty$ or

$$|\widetilde{f}(z_n)| \geq |f(\zeta_n)| - \frac{1}{n},$$

which facts make it evident that $\widetilde{f}(z_n) \to \infty = \widetilde{f}(z_0)$ in this case, too. Having checked the continuity of $\widetilde{f}$ at all points where it could conceivably be in doubt, we can assert that $\widetilde{f}$ is a continuous mapping of $\overline{\Delta}$ into $\widehat{\mathbb{C}}$. The very definition of continuity makes it clear that $\widetilde{f}$ is the only possible continuous function from $\overline{\Delta}$ into $\widehat{\mathbb{C}}$ which agrees with f in Δ.

We next prove that $\widetilde{f}(\overline{\Delta}) = \widehat{D}$. It follows almost immediately from the definition of $\widetilde{f}$ that $\widetilde{f}(\overline{\Delta})$ is a subset of $\widehat{D}$. As to the opposite containment, let w_0 belong to the set $\widehat{D}$. We must produce a point z_0 of $\overline{\Delta}$ for which $\widetilde{f}(z_0) = w_0$. To this end, we pick a sequence $\langle w_n \rangle$ in D such that $w_n \to w_0$. Because $D = f(\Delta)$, we are then entitled to choose a point z_n in Δ with $w_n = f(z_n) = \widetilde{f}(z_n)$. Because the set $\overline{\Delta}$ is compact, we can extract from the sequence $\langle z_n \rangle$ a subsequence $\langle z_{n_k} \rangle$ with the property that $z_{n_k} \to z_0$, some point of $\overline{\Delta}$. Finally, the continuity of $\widetilde{f}$ at z_0 gives

$$w_0 = \lim_{k \to \infty} w_{n_k} = \lim_{k \to \infty} \widetilde{f}(z_{n_k}) = \widetilde{f}(z_0),$$

as desired. Thus, $\widetilde{f}(\overline{\Delta}) = \widehat{D}$.

Assuming that $\widetilde{f}$ is univalent, we can go further and state that its inverse $\widetilde{f}^{-1}: \widehat{D} \to \overline{\Delta}$ is another continuous function. If $\widetilde{f}^{-1}$ fails to be continuous at some point w_0 of $\widehat{D}$, then there is a sequence $\langle w_n \rangle$ in $\widehat{D}$ such that $w_n \to w_0$, whereas $z_n = \widetilde{f}^{-1}(w_n) \not\to z_0 = \widetilde{f}^{-1}(w_0)$. The last fact and the compactness of $\overline{\Delta}$ imply that $\langle z_n \rangle$ has some subsequence $\langle z_{n_k} \rangle$ with the property that $z_{n_k} \to z_0'$, where z_0' is a point of $\overline{\Delta}$ different from z_0. On the other hand, the continuity of $\widetilde{f}$ tells us that

$$\widetilde{f}(z_0') = \lim_{k \to \infty} \widetilde{f}(z_{n_k}) = \lim_{k \to \infty} w_{n_k} = w_0 = \widetilde{f}(z_0),$$

contradicting the univalence of $\widetilde{f}$. Accordingly, if the extension $\widetilde{f}$ is univalent, then it is a homeomorphism of $\overline{\Delta}$ onto $\overline{D}$. ∎

4.2 Double Integrals

In anticipation of the approaching material dealing with the conformal modulus of a path family, we review some basic facts about integrals of the type $\iint_S \rho(z)\,dx\,dy$, in which S is a subset of $\mathbb{C}$ and ρ is a real-valued function that is continuous on S. In multi-variable calculus courses it is typically shown that such an integral is meaningful when S has the form $S = \overline{D}$, where D is a bounded plane domain with the property that ∂D can be expressed as the union of a finite number of regular arcs. For want of a better term we call a set S of this kind a "standard integration region." For instance, a closed rectangle $S = \{z : a \leq x \leq b, c \leq y \leq d\}$ is a standard integration region and, in this instance, the double integral can be computed by means of iterated integrations:

(9.42)
$$\iint_S \rho(z)\,dx\,dy = \int_a^b \left\{ \int_c^d \rho(x+iy)\,dy \right\} dx$$
$$= \int_c^d \left\{ \int_a^b \rho(x+iy)\,dx \right\} dy .$$

A second example is a set of the form $S = \{re^{i\theta} : a \leq r \leq b, \alpha \leq \theta \leq \beta\}$ with $0 \leq a < b < \infty$ and $0 < \beta - \alpha \leq 2\pi$. Here it is often convenient to make a switch to polar coordinates for the evaluation:

(9.43)
$$\iint_S \rho(z)\,dx\,dy = \int_a^b \left\{ \int_\alpha^\beta \rho(re^{i\theta})\,d\theta \right\} r\,dr$$
$$= \int_\alpha^\beta \left\{ \int_a^b \rho(re^{i\theta})\,r\,dr \right\} d\theta .$$

This shift to polar coordinates is a special case of a more general "change of variables" formula: if $f : D \to \mathbb{C}$ is a diffeomorphism, if S is a standard integration region that is contained in D, and if $\rho : f(S) \to \mathbb{R}$ is a continuous

4. The Carathéodory-Osgood Theorem

function, then

$$(9.44) \qquad \iint_{f(S)} \rho(w)\,du\,dv = \iint_S \rho[f(z)]|J_f(z)|\,dx\,dy\ .$$

We notice especially that

$$(9.45) \qquad \iint_{f(S)} \rho(w)\,du\,dv = \iint_S \rho[f(z)]|f'(z)|^2\,dx\,dy$$

when f is a conformal mapping. Implicit in (9.44) and (9.45) is the fact that $f(S)$ is itself a standard integration region. To avoid confusion, we have indicated the variables of integration in $f(S)$ by u and v. Thus one is to think of making the change of variable $w = f(z)$ in the integral $\iint_{f(S)} \rho(w)\,du\,dv$, thereby transforming it into $\iint_S \rho[f(z)]|J_f(z)|\,dx\,dy$.

We shall also need to deal with "improper" double integrals of the sort $\iint_G \rho(z)\,dx\,dy$, where G is a "general integration region" in the complex plane — we use this expression to describe a set G that can be written in the form $G = \bigcup_{n=1}^{\infty} S_n$, in which $S_1 \subset S_2 \subset S_3 \subset \cdots$ is a sequence of standard integration regions — and where $\rho: G \to [0, \infty)$ is a continuous function. (A plane domain G is an example of a general integration region. Of course, any standard integration region also qualifies as a general integration region.) The definition of the integral in this situation reads

$$(9.46) \qquad \iint_G \rho(z)\,dx\,dy = \sup\left\{ \iint_S \rho(z)\,dx\,dy\colon S \text{ stand. integ. reg.}, S \subset G \right\}.$$

We must allow here for the possibility that $\iint_G \rho(z)\,dx\,dy = \infty$, a state of affairs which arises when the set of numbers on the right-hand side of (9.46) is unbounded. The change of variables formula carries over to this setting and permits one to conclude, for instance, that

$$(9.47) \qquad \iint_{f(G)} \rho(w)\,du\,dv = \iint_G \rho[f(z)]|f'(z)|^2\,dx\,dy$$

in the event $f: D \to \mathbb{C}$ is a conformal mapping of a domain D which contains G and $\rho: f(G) \to [0, \infty)$ is a continuous function.

4.3 Conformal Modulus

Let G be a general integration region. A continuous function $\rho: G \to [0, \infty)$ will be called a *density* in G. We can regard such a function as a vehicle for setting up in G a new system for measuring lengths and areas, as follows: the "ρ-length" $\ell_\rho(\gamma)$ of piecewise smooth path γ in G is given by

$$\ell_\rho(\gamma) = \int_\gamma \rho(z)|dz| ,$$

while the "ρ-area" $A_\rho(G')$ of any general integration region G' that is a subset of G is defined by

$$A_\rho(G') = \iint_{G'} [\rho(z)]^2 \, dx dy .$$

(N.B. When $\rho(z) = 1$ throughout G these reduce to the ordinary length $\ell(\gamma)$ of γ and area $A(G')$ of G'.) One is reminded of a situation in physics where G might serve as a model for a thin plate composed of some non-homogeneous substance and ρ would specify the "linear mass density" (= mass per unit length) at points of the plate. In this analogy, $\ell_\rho(\gamma)$ would represent the mass of a "wire" formed from the trajectory of γ and $A_\rho(G')$ would give the mass of the "subplate" G' of G.

We are going to consider configurations of sets E, F, and G, where E and F are non-empty, disjoint subsets of an integration region G (Figure 26). The notation $[E, F: G]$ is used to symbolize a configuration of this

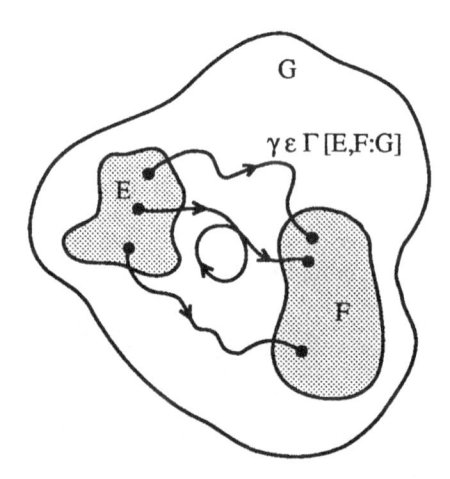

Figure 26.

kind. With each such configuration we shall associate an extended real number $M[E, F: G]$, the *conformal modulus* of $[E, F: G]$, in such a way that this quantity remains invariant under conformal mapping. Our method for doing this was introduced by Lars Ahlfors and Arne Beurling (1905-1986) in their landmark paper "Conformal invariants and function-theoretic null

4. The Carathéodory-Osgood Theorem

sets" in *Acta Mathematica*, Vol. 83, 1950. The invariant they defined there has subsequently developed into one of the major theoretical tools at the disposal of researchers in complex analysis. Let $\Gamma[E, F:G]$ designate the family of all piecewise smooth paths in G having initial point in E and terminal point in F. To define $M[E, F:G]$ we look at densities ρ in G with the extra feature that $\ell_\rho(\gamma) \geq 1$ for every path γ of $\Gamma[E, F:G]$. Such a function ρ is termed an *admissible density* for the configuration $[E, F:G]$. The collection of all densities in G that are admissible for $[E, F:G]$ is denoted by $\text{Adm}[E, F:G]$. (*Warning*: there are situations in which no admissible densities exist; i.e., for which $\text{Adm}[E, F:G] = \phi$.) We define

$$(9.48) \qquad M[E, F:G] = \inf\{A_\rho(G) : \rho \in \text{Adm}[E, F:G]\}$$

if $\text{Adm}[E, F:G] \neq \phi$, the infimum being taken in the extended real interval $[0, \infty]$, and set $M[E, F:G] = \infty$ otherwise. In plain words, $M[E, F:G]$ is the smallest area a density ρ could assign to G, given that ρ is constrained to assign a length no less than one to every path in the family $\Gamma[E, F:G]$. (N.B. In the paper cited Ahlfors and Beurling found it more convenient to work with $\lambda[E, F:G] = 1/M[E, F:G]$, a quantity they dubbed the *extremal length* of $\Gamma[E, F:G]$, than with $M[E, F:G]$. The number $\lambda[E, F:G]$ is also known as the *extremal distance between E and F in G*.) Since $-\gamma$ belongs to $\Gamma[F, E:G]$ if and only if γ belongs to $\Gamma[E, F:G]$ and since $\int_{-\gamma} \rho(z)\,|dz| = \int_\gamma \rho(z)\,|dz|$, it is clear that $M[E, F:G] = M[F, E:G]$.

It is not easy to gather from this admittedly abstruse definition what information the number $M[E, F:G]$ might encode, to say nothing of the relevance of that information for the theory of analytic functions. Nor shall we make any serious effort to explain this definition by delving into the ideas that underlie and motivate it. Instead, we hope that by seeing the concept in action in the proof of the Carathéodory-Osgood theorem the reader will gain an appreciation for the power and utility of this invariant, even though many aspects of it remain more than a little hazy. Perhaps the following example, which illustrates some common techniques for computing and estimating conformal moduli, may serve to shed a bit of light on the definition itself.

EXAMPLE 4.1. For $a > 0$ and $b > 0$ let G be the closed rectangle with vertices $0, a, a + ib$, and ib (Figure 27). If $E = \{x : 0 \leq x \leq a\}$ and $F = \{x + ib : 0 \leq x \leq a\}$, show that $M[E, F:G] = a/b$.

We first demonstrate that $A_\rho(G) \geq a/b$ for every density ρ in $\text{Adm}[E, F:G]$. It will follow immediately that $M[E, F:G] \geq a/b$. Fix such a ρ. For each x satisfying $0 \leq x \leq a$ the path $\gamma_x : [0, b] \to \mathbb{C}$ defined by $\gamma_x(y) = x + iy$ is certainly a member of $\Gamma[E, F:G]$. By the definition of an admissible density, it must be the case that

$$1 \leq \int_{\gamma_x} \rho(z)|dz| = \int_0^b \rho(x + iy)\,dy \ .$$

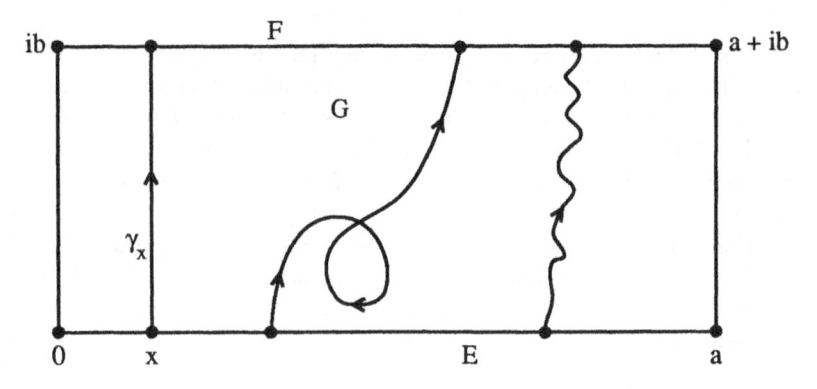

Figure 27.

Consequently,
$$1 \le \left[\int_0^b \rho(x+iy)\,dy\right]^2 \le \left\{\int_0^b [\rho(x+iy)]^2\,dy\right\}\left\{\int_0^b dy\right\}$$
$$= b\int_0^b [\rho(x+iy)]^2\,dy\ .$$

Here we have appealed to a classical inequality for Riemann integrals, the Cauchy-Schwarz inequality:

(9.49) $$\left[\int_c^d f(y)g(y)\,dy\right]^2 \le \left\{\int_c^d [f(y)]^2\,dy\right\}\left\{\int_c^d [g(y)]^2\,dy\right\}\ ,$$

if f and g are real-valued functions that are Riemann integrable over $[c,d]$. In our case $f(y) = \rho(x+iy)$ and $g(y) = 1$ on $[0,b]$. We infer that the inequality
$$\frac{1}{b} \le \int_0^b [\rho(x+iy)]^2\,dy$$
is valid for each x in $[0,a]$. As a result of (9.42),
$$\frac{a}{b} = \int_0^a \frac{dx}{b} \le \int_0^a \left\{\int_0^b [\rho(x+iy)]^2\,dy\right\}dx$$
$$= \iint_G [\rho(z)]^2\,dx\,dy = A_\rho(G)\ .$$

Therefore, the lower bound $A_\rho(G) \ge a/b$ is valid for every admissible density ρ — hence, $M[E,F:G] \ge a/b$.

4. The Carathéodory-Osgood Theorem

To confirm that equality actually holds we consider a particular density $\rho_0: G \to [0, \infty)$, the one given by $\rho_0(z) = b^{-1}$. Since $\ell(\gamma) \geq b$ is without a doubt true of any path γ in $\Gamma[E, F: G]$, we see that

$$\ell_{\rho_0}(\gamma) = \int_\gamma \rho_0(z)\,|dz| = \int_\gamma \frac{|dz|}{b} = \frac{\ell(\gamma)}{b} \geq 1$$

for each such γ, which places ρ_0 in $\mathrm{Adm}[E, F: G]$. This means that

$$M[E, F: G] \leq A_{\rho_0}(G) = \iint_G [\rho_0(z)]^2\,dx\,dy = \iint_G \frac{dx\,dy}{b^2} = \frac{ab}{b^2} = \frac{a}{b}\,.$$

Coupled with the inequality $M[E, F: G] \geq a/b$, this gives $M[E, F: G] = a/b$.

The property that ultimately turns $M[E, F: G]$ into a useful commodity is its invariance under conformal mapping. The precise formulation of that invariance is the purpose of the next theorem.

Theorem 4.2. *Suppose that D is a domain in the complex plane and that $f: D \to \mathbb{C}$ is a conformal mapping. Then*

(9.50) $$M[E, F: G] = M[f(E), f(F): f(G)]$$

for any configuration $[E, F: G]$ with G contained in D.

Proof. Write $M = M[E, F: G]$ and $M' = M[f(E), f(F): f(G)]$. It is enough to verify that $M \leq M'$, for the reverse inequality can be deduced from this one simply by applying it to f^{-1}. We may further assume that $M' < \infty$, since the inequality $M \leq M'$ holds trivially otherwise. Let $\tilde{\rho}$ be an admissible density for the configuration $[f(E), f(F): f(G)]$. We define a density ρ in G by $\rho(z) = \tilde{\rho}[f(z)]|f'(z)|$. The claim is that ρ constitutes an admissible density for $[E, F: G]$. To see this, let $\gamma: [a, b] \to \mathbb{C}$ be a path in $\Gamma[E, F: G]$. Then $\beta = f \circ \gamma$ is a path belonging to the family $\Gamma[f(E), f(F): f(G)]$. Since $\tilde{\rho}$ is a member of $\mathrm{Adm}[f(E), f(F): f(G)]$, we obtain

$$1 \leq \int_\beta \tilde{\rho}(z)\,|dz| = \int_a^b \tilde{\rho}[\beta(t)]|\dot{\beta}(t)|\,dt = \int_a^b \tilde{\rho}\{f[\gamma(t)]\}|f'[\gamma(t)]||\dot{\gamma}(t)|\,dt$$

$$= \int_a^b \rho[\gamma(t)]|\dot{\gamma}(t)|\,dt = \int_\gamma \rho(z)\,|dz|\,,$$

which demonstrates that ρ belongs to $\mathrm{Adm}[E, F: G]$. Accordingly, M does not exceed $A_\rho(G)$. The change of variable formula (9.47) for double integrals then leads to

$$M \leq A_\rho(G) = \iint_G [\rho(z)]^2\,dx\,dy = \iint_G \{\tilde{\rho}[f(z)]\}^2|f'(z)|^2\,dx\,dy$$

$$= \iint_{f(G)} [\tilde{\rho}(w)]^2 \, du dv = A_{\tilde{\rho}}[f(G)] .$$

Because $\tilde{\rho}$ was an arbitrary admissible density for $[f(E), f(F): f(G)]$, we are able to conclude that $M \leq M'$, as desired. ∎

In every application that we intend to make of Theorem 4.2 in this book we shall have $G = D$, but the reader must not get the impression that the utility of the theorem is limited to that case.

In general it is quite difficult to determine the quantity $M[E, F: G]$ exactly. One must ordinarily make do with upper or lower bounds for the modulus. Fortunately, such bounds are not exceedingly hard to come by and are all that is needed for many applications of the invariant. This point is dramatized by the proof we give of the Carathéodory-Osgood theorem, the crux of which resides in the modulus estimates afforded by the following sequence of lemmas. The first of these describes a situation in which we are assured that a certain modulus cannot be unduly small.

Lemma 4.3. *Let $\Delta = \Delta(0,1)$, let $E = \overline{\Delta}(0,1/2)$, and let $z_1 = e^{i\theta_1}$ and $z_2 = e^{i\theta_2}$, where $0 \leq \theta_1 < \theta_2 < 2\pi$. Then*

(9.51) $$M[E, F: \Delta] \geq \min\{\theta_2 - \theta_1, 2\pi - (\theta_2 - \theta_1)\}$$

for every connected set F in $\Delta \sim E$ with the property that both z_1 and z_2 belong to $\overline{F}$ (Figure 28).

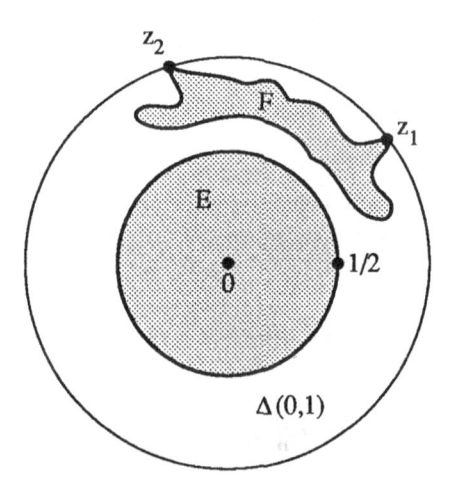

Figure 28.

Proof. By the action of performing a preliminary rotation and using the conformal invariance of the modulus, we reduce the proof to consideration of the case $z_1 = 1$ and $z_2 = e^{i\theta_0}$, where $0 < \theta_0 = \theta_2 - \theta_1 < 2\pi$. Fix a

4. The Carathéodory-Osgood Theorem

connected set F in $\Delta \sim E$ that has both 1 and $e^{i\theta_0}$ as points of its closure. For $0 \leq \theta < 2\pi$, let $S_\theta = \{re^{i\theta} : 1/2 \leq r < 1\}$. We start the argument with the observation that either F intersects S_θ for every θ in $(0, \theta_0)$ or it intersects S_θ for every θ in $(\theta_0, 2\pi)$. If not, there would exist angles ψ in $(0, \theta_0)$ and φ in $(\theta_0, 2\pi)$ such that both S_ψ and S_φ are disjoint from F. The open set $\Delta \sim (E \cup S_\psi \cup S_\varphi)$ would then consist of two components U and V, one of which has 1 as a boundary point and the other of which has $e^{i\theta_0}$ on its boundary. Due to the location of ψ and φ, U and V would be disjoint open sets which both meet F (remember that 1 and $e^{i\theta_0}$ belong to $\overline{F}$) and which include F in their union, a situation at odds with the hypothesis that F is a connected set. We shall proceed under the assumption that S_θ intersects F for every θ in $(0, \theta_0)$ and derive for $M[E, F : \Delta]$ the estimate

$$(9.52) \qquad M[E, F : \Delta] \geq \theta_0 .$$

In the other case, similar reasoning would yield

$$(9.53) \qquad M[E, F : \Delta] \geq 2\pi - \theta_0 .$$

Together (9.52) and (9.53) imply that

$$M[E, F : \Delta] \geq \min\{\theta_0, 2\pi - \theta_0\} ,$$

as announced in (9.51).

In verifying (9.52) we may assume that $M[E, F : \Delta] < \infty$, there being nothing to prove should this modulus be infinite. Let ρ be an arbitrary admissible density for $[E, F : \Delta]$. It is our task to demonstrate that $A_\rho(\Delta) \geq \theta_0$. To arrive at this lower bound for $A_\rho(\Delta)$ we first choose for each θ in $(0, \theta_0)$ a number b_θ in $(1/2, 1)$ with the feature that $b_\theta e^{i\theta}$ is a point of F. Such a choice is possible because by assumption S_θ does intersect F. The path $\gamma_\theta : [1/2, b_\theta] \to \mathbb{C}$ defined by $\gamma_\theta(r) = re^{i\theta}$ qualifies as a member of $\Gamma[E, F : \Delta]$, with the result that

$$(9.54) \qquad 1 \leq \int_{\gamma_\theta} \rho(z) |dz| = \int_{1/2}^{b_\theta} \rho(re^{i\theta}) \, dr$$

by the definition of an admissible density. To certify that $A_\rho(\Delta) \geq \theta_0$ it is enough to check that

$$(9.55) \qquad A_\rho(\Delta) \geq (1 - \epsilon)^2 (\beta - \alpha)$$

whenever $0 < \alpha < \beta < \theta_0$ and $0 < \epsilon < 1$, for we then get $A_\rho(\Delta) \geq \theta_0$ by letting $\beta \to \theta_0$, $\alpha \to 0$, and $\epsilon \to 0$ on the right-hand side of (9.55). The argument that proves (9.55) is quite technical. We apologize for the technicality, but we find it necessary in order to avoid complications associated with iterated improper Riemann integrals. (A much shorter derivation of (9.52) — indeed, a much cleaner treatment of conformal moduli overall — is available in the context of Lebesgue integration.)

Fix α, β, and ϵ as indicated. We use (9.54) to establish the existence of a number b in $(1/2, 1)$ with the property that the inequality

$$\text{(9.56)} \qquad \int_{1/2}^{b} \rho(re^{i\theta})\,dr \geq 1 - \epsilon$$

is true for every θ in the interval $[\alpha, \beta]$. Suppose that no such b existed. In particular, taking $b = b_n = 1 - 2^{-n-1}$ for a positive integer n would not meet the requirement, so we could find an angle θ_n in $[\alpha, \beta]$ for which

$$\int_{1/2}^{b_n} \rho(re^{i\theta_n})\,dr < 1 - \epsilon \, .$$

The sequence $\langle \theta_n \rangle$ in $[\alpha, \beta]$ thus generated would have an accumulation point in that interval. Pick one and call it θ. By passing to subsequences and relabeling, if necessary, we could assume that $\theta_n \to \theta$. Now $b_n \to 1$, implying that for large n we would have $b_n \geq b_\theta$, the number appearing in (9.54) for our special θ. Since ρ is non-negative, we see that

$$\int_{1/2}^{b_\theta} \rho(re^{i\theta_n})\,dr \leq \int_{1/2}^{b_n} \rho(re^{i\theta_n})\,dr < 1 - \epsilon$$

would hold for any such n. The continuity of ρ would allow us to infer that $\rho(re^{i\theta_n}) \to \rho(re^{i\theta})$ uniformly in r on the interval $[1/2, b_\theta]$, which would have the consequence that

$$\int_{1/2}^{b_\theta} \rho(re^{i\theta})\,dr = \lim_{n \to \infty} \int_{1/2}^{b_\theta} \rho(re^{i\theta_n})\,dr \leq 1 - \epsilon \, .$$

This is clearly in conflict with (9.54). The only way to avert this contradiction is to accept the existence of b with property (9.56).

We now select and fix b in $(1/2, 1)$ such that (9.56) is valid for every θ in $[\alpha, \beta]$. For each such θ we conclude with the help of the Cauchy-Schwarz inequality (9.49) that

$$(1 - \epsilon)^2 \leq \left[\int_{1/2}^{b} \rho(re^{i\theta})\,dr \right]^2 = \left[\int_{1/2}^{b} \rho(re^{i\theta})\, r^{1/2} r^{-1/2}\,dr \right]^2$$

$$\leq \left\{ \int_{1/2}^{b} [\rho(re^{i\theta})]^2\, r\,dr \right\} \left\{ \int_{1/2}^{b} \frac{dr}{r} \right\} = \mathrm{Log}(2b) \int_{1/2}^{1} [\rho(re^{i\theta})]^2\, r\,dr \, .$$

Since $\mathrm{Log}(2b) < \mathrm{Log}\,2 < 1$, we can thus be sure that

$$\text{(9.57)} \qquad \int_{1/2}^{b} [\rho(re^{i\theta})]^2\, r\,dr \geq (1 - \epsilon)^2$$

4. The Carathéodory-Osgood Theorem

when $\alpha \leq \theta \leq \beta$. Finally, let $S = \{re^{i\theta} : 1/2 \leq r \leq b, \alpha \leq \theta \leq \beta\}$, a standard integration region in Δ. After a shift to polar coordinates, we deduce from (9.43) and (9.57) that

$$A_\rho(\Delta) \geq A_\rho(S) = \iint_S [\rho(z)]^2 \, dxdy = \int_\alpha^\beta \left\{ \int_{1/2}^b [\rho(re^{i\theta})]^2 \, rdr \right\} d\theta$$

$$\geq \int_\alpha^\beta (1-\epsilon)^2 \, d\theta = (1-\epsilon)^2(\beta - \alpha) \, .$$

Inequality (9.55) is thereby confirmed, and with that the proof of the lemma is complete. ∎

The key feature of the lower bound in (9.51) is that it depends only on the points z_1 and z_2 (or, to be more precise, on $|z_1 - z_2|$), not on the connected set F. As stated earlier, there are circumstances under which a configuration $[E, F: G]$ can have $M[E, F: G] = \infty$. The next lemma presents one set of circumstances where this happens.

Lemma 4.4. *Let $H = \{z : \operatorname{Im} z > 0\}$. Then*

(9.58) $$M[E, F: H] = \infty$$

for any disjoint pair of connected sets E and F in H with the property that the origin belongs to both $\overline{E}$ and $\overline{F}$.

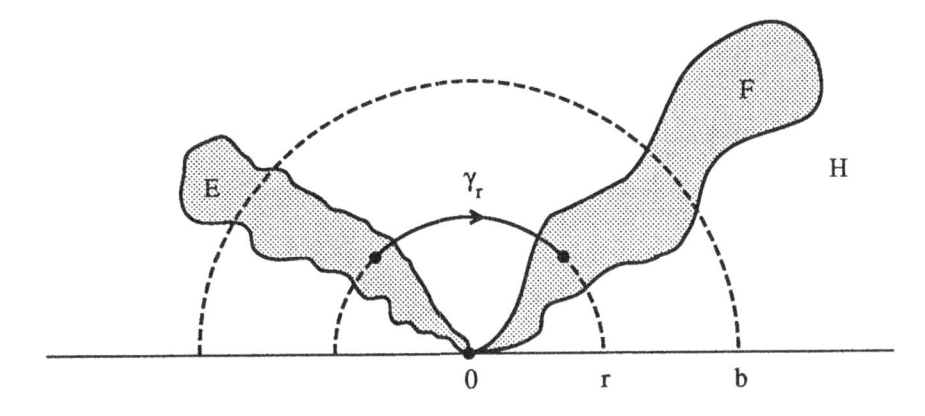

Figure 29.

Proof. Let E and F be sets as indicated (Figure 29), and let ρ be an arbitrary admissible density for the configuration $[E, F: H]$. (N.B. If no such densities ρ exist, (9.58) is true by definition.) We must show that $A_\rho(H) = \infty$. We fix a radius $b > 0$ so that both E and F contain points outside the disk $\overline{\Delta}(0, b)$ and prove that the inequality

(9.59) $$A_\rho(H) \geq \frac{1}{\pi} \log \frac{b}{a}$$

holds whenever $0 < a < b$. This is enough to finish the proof of the lemma, for by letting $a \to 0$ in (9.59) we deduce that $A_\rho(H) = \infty$. The initial step in the verification of (9.59) is the remark that for each r in the interval $(0, b]$ the semi-circle $S_r = \{re^{i\theta} : 0 < \theta < \pi\}$ has to intersect both E and F. Suppose, for instance, that S_r did not meet E. Then $U = \Delta(0, r)$ and $V = H \sim \overline{\Delta}(0, r)$ would be disjoint open sets satisfying $E \cap U \neq \phi$ (by reason of the fact that 0 belongs to $\overline{E}$), $E \cap V \neq \phi$ (by the choice of b), and also $E \subset U \cup V$. This would violate the connectedness of E. Similarly, S_r must intersect F.

Given r in $(0, b]$ we are now free to pick angles φ_r and ψ_r from $(0, \pi)$ in such a way that $re^{i\varphi_r}$ lies in E and $re^{i\psi_r}$ in F. Define a path γ_r as follows: if $\varphi_r < \psi_r$ then $\gamma_r(\theta) = re^{i\theta}$ for $\varphi_r \leq \theta \leq \psi_r$; if $\psi_r < \varphi_r$, then $\gamma_r(\theta) = re^{i(\varphi_r + \psi_r - \theta)}$ for $\psi_r \leq \theta \leq \varphi_r$. (See Figure 29.) The path γ_r belongs to $\Gamma[E, F : H]$. Owing to the admissibility of ρ for $[E, F : H]$, we know that

$$1 \leq \int_{\gamma_r} \rho(z) |dz| = \left| \int_{\varphi_r}^{\psi_r} \rho(re^{i\theta}) r d\theta \right| .$$

In summary, for each r in $(0, b]$ we can fix a subinterval $[\alpha_r, \beta_r]$ of $(0, \pi)$ — take $[\varphi_r, \psi_r]$ if $\varphi_r < \psi_r$ and $[\psi_r, \varphi_r]$ if $\psi_r < \varphi_r$ — with the property that

$$(9.60) \qquad \int_{\alpha_r}^{\beta_r} \rho(re^{i\theta}) r d\theta \geq 1 .$$

We shall derive (9.59) from (9.60) in much the same way we obtained (9.55) from (9.54) in the proof of Lemma 4.3. Once again a rather technical step is necessary if we are to argue rigorously without at the same time being forced out of the confines of normal Riemann integration.

We fix a in $(0, b)$ and ϵ in $(0, 1)$. We contend that there exists an angle δ in $(0, \pi/2)$ for which

$$(9.61) \qquad \int_{\delta}^{\pi - \delta} \rho(re^{i\theta}) r d\theta \geq 1 - \epsilon$$

whenever $a \leq r \leq b$. If not, there would exist sequences $\langle \delta_n \rangle$ in $(0, \pi/2)$ and $\langle r_n \rangle$ in $[a, b]$ such that $\delta_n \to 0$, but such that

$$\int_{\delta_n}^{\pi - \delta_n} \rho(r_n e^{i\theta}) r_n d\theta < 1 - \epsilon .$$

Replacing the original sequences by subsequences if necessary, we could assume that $r_n \to r$, a number in $[a, b]$. For large n the interval $[\delta_n, \pi - \delta_n]$ would contain the interval $[\alpha_r, \beta_r]$ that was chosen for this particular r in order to achieve (9.60). For large n, therefore, we would have

$$\int_{\alpha_r}^{\beta_r} \rho(r_n e^{i\theta}) r_n d\theta \leq \int_{\delta_n}^{\pi - \delta_n} \rho(r_n e^{i\theta}) r_n d\theta < 1 - \epsilon .$$

4. The Carathéodory-Osgood Theorem

Because the convergence of $\rho(r_n e^{i\theta})r_n$ to $\rho(re^{i\theta})r$ would be uniform on $[\alpha_r, \beta_r]$, we could infer that

$$\int_{\alpha_r}^{\beta_r} \rho(re^{i\theta})\, r d\theta = \lim_{n\to\infty} \int_{\alpha_r}^{\beta_r} \rho(r_n e^{i\theta})\, r_n d\theta \leq 1 - \epsilon,$$

which would contradict (9.60). The only option left open is for $(0, \pi/2)$ to contain a δ with property (9.61). We choose such a δ and fix it for the rest of the proof.

Applying the Cauchy-Schwarz inequality to (9.61) we discover that

$$(1-\epsilon)^2 \leq r^2 \left[\int_\delta^{\pi-\delta} \rho(re^{i\theta})\, d\theta \right]^2 \leq r^2 \left\{ \int_\delta^{\pi-\delta} [\rho(re^{i\theta})]^2\, d\theta \right\} \left\{ \int_\delta^{\pi-\delta} 1\, d\theta \right\}$$

$$= (\pi - 2\delta) r^2 \int_\delta^{\pi-\delta} [\rho(re^{i\theta})]^2\, d\theta \leq \pi r^2 \int_\delta^{\pi-\delta} [\rho(re^{i\theta})]^2\, d\theta$$

for every r in $[a, b]$, so that

(9.62) $$\frac{(1-\epsilon)^2}{\pi r} \leq \int_\delta^{\pi-\delta} [\rho(re^{i\theta})]^2\, r d\theta$$

for such r. The set $S = \{re^{i\theta} : a \leq r \leq b, \delta \leq \theta \leq \pi - \delta\}$ is a standard integration region in H. By changing to polar coordinates and appealing to (9.43) and (9.62) we are put in a position to conclude that

$$A_\rho(H) \geq A_\rho(S) = \iint_S [\rho(z)]^2\, dxdy = \int_a^b \left\{ \int_\delta^{\pi-\delta} [\rho(re^{i\theta})]^2\, r d\theta \right\} dr$$

$$\geq \int_a^b \frac{(1-\epsilon)^2\, dr}{\pi r} = \frac{(1-\epsilon)^2}{\pi} \operatorname{Log} \frac{b}{a}.$$

Because the estimate

$$A_\rho(H) \geq \frac{(1-\epsilon)^2}{\pi} \operatorname{Log} \frac{b}{a}$$

holds for every ϵ in $(0, 1)$, we can let $\epsilon \to 0$ to obtain (9.59). As observed, this finishes the proof. ∎

Our third lemma introduces one of the rare configurations whose conformal modulus can be calculated exactly and expressed in elementary terms.

Lemma 4.5. *Suppose that z_0 is a point in the complex plane and that $0 < r_0 < r_1 < \infty$. If $E = \overline{\Delta}(z_0, r_0)$ and $F = \mathbb{C} \sim \Delta(z_0, r_1)$, then*

(9.63) $$M[E, F : \mathbb{C}] = 2\pi \left(\operatorname{Log} \frac{r_1}{r_0} \right)^{-1}.$$

Proof. We treat the case $z_0 = 0$. The general case can be reduced to this special one by making a translation and invoking the conformal invariance of the modulus. Let ρ be an arbitrary admissible density for the configuration $[E, F: \mathbb{C}]$. We begin the proof by showing that

$$(9.64) \qquad A_\rho(\mathbb{C}) \geq 2\pi \left(\operatorname{Log} \frac{r_1}{r_0} \right)^{-1}.$$

For each θ in $[0, 2\pi]$ the path γ_θ defined on $[r_0, r_1]$ by $\gamma_\theta(r) = re^{i\theta}$ is a member of $\Gamma[E, F: \mathbb{C}]$, from which it follows that

$$1 \leq \int_{\gamma_\theta} \rho(z)\, |dz| = \int_{r_0}^{r_1} \rho(re^{i\theta})\, dr.$$

An application of (9.49) produces the inequality

$$1 \leq \left[\int_{r_0}^{r_1} \rho(re^{i\theta})\, dr \right]^2 \leq \left\{ \int_{r_0}^{r_1} [\rho(re^{i\theta})]^2 \, r\, dr \right\} \left\{ \int_{r_0}^{r_1} \frac{dr}{r} \right\}$$

$$= \left(\operatorname{Log} \frac{r_1}{r_0} \right) \int_{r_0}^{r_1} [\rho(re^{i\theta})]^2 \, r\, dr$$

for every θ in $[0, 2\pi]$. If $S = \{z : r_0 \leq |z| \leq r_1\}$, then

$$A_\rho(\mathbb{C}) \geq A_\rho(S) = \iint_S [\rho(z)]^2 \, dx\, dy = \int_0^{2\pi} \left\{ \int_{r_0}^{r_1} [\rho(re^{i\theta})]^2 \, r\, dr \right\} d\theta$$

$$\geq \int_0^{2\pi} \left(\operatorname{Log} \frac{r_1}{r_0} \right)^{-1} d\theta = 2\pi \left(\operatorname{Log} \frac{r_1}{r_0} \right)^{-1},$$

which establishes (9.64). Because the density ρ in $\operatorname{Adm}[E, F: \mathbb{C}]$ was arbitrary, (9.64) implies that

$$(9.65) \qquad M[E, F: \mathbb{C}] \geq 2\pi \left(\operatorname{Log} \frac{r_1}{r_0} \right)^{-1}.$$

It remains to prove that the inequality sign in (9.65) can also be reversed. We accomplish this by first showing that

$$(9.66) \qquad M[E, F: \mathbb{C}] \leq 2\pi \left(\operatorname{Log} \frac{s_1}{s_0} \right) \left(\operatorname{Log} \frac{r_1}{r_0} \right)^{-2}$$

whenever $0 < s_0 < r_0$ and $r_1 < s_1 < \infty$. Fix s_0 and s_1 obeying these conditions. Let $h: [0, \infty) \to [0, \infty)$ be the continuous function defined as follows: $h(r) = 0$ if $r < s_0$ or $r > s_1$; $h(r) = [\operatorname{Log}(r_1/r_0)]^{-1}$ if $r_0 \leq r \leq r_1$; h is linear on each of the intervals $[s_0, r_0]$ and $[r_1, s_1]$. We can then define

4. The Carathéodory-Osgood Theorem

a continuous function $\rho: \mathbb{C} \to [0, \infty)$ by $\rho(z) = |z|^{-1} h(|z|)$ if $z \neq 0$ and $\rho(0) = 0$. We maintain that ρ is an admissible density for $[E, F: \mathbb{C}]$; i.e., we claim that $\int_\gamma \rho(z) |dz| \geq 1$ for every path γ belonging to $\Gamma[E, F: \mathbb{C}]$. Let $\gamma: [a, b] \to \mathbb{C}$ be such a path. Determine numbers c and d by the following prescription:

$$c = \max\{t : t \in [a, b], |\gamma(t)| \leq r_0\} \quad , \quad d = \min\{t : t \in [c, b], |\gamma(t)| \geq r_1\} .$$

The fact that γ is a continuous function on $[a, b]$ having $|\gamma(a)| \leq r_0$ and $|\gamma(b)| \geq r_1$ makes certain that c and d are well-defined, that $|\gamma(c)| = r_0$ and $|\gamma(d)| = r_1$, and that $r_0 \leq |\gamma(t)| \leq r_1$ when $c \leq t \leq d$. In particular, $\rho[\gamma(t)] = [|\gamma(t)| \operatorname{Log}(r_1/r_0)]^{-1}$ for t in $[c, d]$. Therefore,

$$\int_\gamma \rho(z) |dz| = \int_a^b \rho[\gamma(t)] |\dot\gamma(t)| \, dt \geq \int_c^d \rho[\gamma(t)] |\dot\gamma(t)| \, dt$$

$$= \left(\operatorname{Log} \frac{r_1}{r_0}\right)^{-1} \int_c^d \frac{|\dot\gamma(t)| \, dt}{|\gamma(t)|} = \left(\operatorname{Log} \frac{r_1}{r_0}\right)^{-1} \int_c^d \frac{|\overline{\gamma(t)}\dot\gamma(t)| \, dt}{|\gamma(t)|^2}$$

$$\geq \left(\operatorname{Log} \frac{r_1}{r_0}\right)^{-1} \int_c^d \frac{\operatorname{Re}[\overline{\gamma(t)}\dot\gamma(t)] \, dt}{|\gamma(t)|^2} = \left(\operatorname{Log} \frac{r_1}{r_0}\right)^{-1} \int_c^d \frac{d}{dt} \{\operatorname{Log} |\gamma(t)|\} dt$$

$$= \left(\operatorname{Log} \frac{r_1}{r_0}\right)^{-1} \left[\operatorname{Log} |\gamma(t)|\right]_c^d = \left(\operatorname{Log} \frac{r_1}{r_0}\right)^{-1} \left(\operatorname{Log} \frac{r_1}{r_0}\right) = 1,$$

putting ρ in $\operatorname{Adm}[E, F: \mathbb{C}]$. Remembering that $\rho(z) = 0$ when $|z| \leq s_0$ or $|z| \geq s_1$ and noticing that $\rho(z) \leq [|z| \operatorname{Log}(r_1/r_0)]^{-1}$ for all z, we deduce:

$$M[E, F: \mathbb{C}] \leq A_\rho(\mathbb{C}) = \iint_\mathbb{C} [\rho(z)]^2 \, dx dy$$

$$= \iint_{s_0 \leq |z| \leq s_1} [\rho(z)]^2 \, dx dy \leq \left(\operatorname{Log} \frac{r_1}{r_0}\right)^{-2} \iint_{s_0 \leq |z| \leq s_1} \frac{dx dy}{|z|^2}$$

$$= \left(\operatorname{Log} \frac{r_1}{r_0}\right)^{-2} \int_0^{2\pi} \left\{\int_{s_0}^{s_1} \frac{r dr}{r^2}\right\} d\theta = 2\pi \left(\operatorname{Log} \frac{s_1}{s_0}\right) \left(\operatorname{Log} \frac{r_1}{r_0}\right)^{-2} .$$

This gives (9.66). Letting $s_0 \to r_0$ and $s_1 \to r_1$ in (9.66) leads to

$$M[E, F: \mathbb{C}] \leq 2\pi \left(\operatorname{Log} \frac{r_1}{r_0}\right)^{-1} .$$

In tandem with (9.65) the last inequality yields (9.63). ∎

The final lemma in the present series is a monotonicity principle for conformal moduli.

Lemma 4.6. *Suppose that configurations* $[E_1, F_1: G_1]$ *and* $[E_2, F_2: G_2]$ *satisfy* $E_1 \subset E_2, F_1 \subset F_2,$ *and* $G_1 \subset G_2$. *Then*

(9.67) $$M[E_1, F_1: G_1] \leq M[E_2, F_2: G_2] .$$

Proof. We may assume that $M[E_2, F_2: G_2] < \infty$, for (9.67) is a trivial statement otherwise. If ρ is an admissible density for $[E_2, F_2: G_2]$, then its restriction to G_1 is obviously an admissible density for $[E_1, F_1: G_1]$. Consequently,
$$M[E_1, F_1: G_1] \leq A_\rho(G_1) \leq A_\rho(G_2) .$$
Since ρ in $\mathrm{Adm}[E_2, F_2: G_2]$ was arbitrary, (9.67) follows. ∎

4.4 Extending Conformal Mappings of the Unit Disk

It is definitely not the case that an arbitrary conformal mapping f of the unit disk $\Delta = \Delta(0,1)$ into $\mathbb{C}$ will extend to a continuous mapping of $\overline{\Delta}$ into $\widehat{\mathbb{C}}$. Such an extension will exist only when the boundary of the domain $D = f(\Delta)$ is suitably regular. In overly simplified terms the existence of an extension is predicated on a negative answer to the question: Does the boundary of D chop any small open disks centered at boundary points of D into infinitely many pieces? We can make this vague statement precise by introducing the following notion: a plane domain D is *finitely connected along its boundary* if corresponding to each point z of $\widehat{\partial}D$ and each $r > 0$ there exists an s in the interval $(0, r)$ such that $D \cap \Delta(z, s)$ intersects at most finitely many components of the open set $D \cap \Delta(z, r)$. As a rule, both the size of s and the number of components of $D \cap \Delta(z, r)$ that $D \cap \Delta(z, s)$ meets may vary wildly with z and r. In the nicest of situations it will be true that for every point z on $\widehat{\partial}D$ and every $r > 0$ a number s in $(0, r)$ corresponding to z and r can be found for which $D \cap \Delta(z, s)$ intersects — hence, is contained in — exactly one component of $D \cap \Delta(z, r)$. When the latter happens we say that D is *locally connected along its boundary*. In Figure 30, D_1 is locally connected along its boundary, D_2 and D_3 are finitely connected along their respective boundaries, but D_4 is not finitely connected along its boundary. For instance, the defining condition is violated by D_4 for small r at the point z_0 indicated. We emphasize that for an unbounded domain D to be finitely (or locally) connected along its boundary the pertinent condition must be satisfied at $z = \infty$, as well as at all points of ∂D. Just to illustrate this, in Figure 30 the quarter plane $D_5 = \{z: x > 0, y > 0\}$ is locally connected along its boundary, and the strip $D_6 = \{z: |y| < 1\}$ is finitely connected along its boundary. The fact that $D_6 \cap \Delta(\infty, r) = \{z \in D_6 : |z| > r^{-1}\}$ has two components whenever $0 < r < 1$ prevents D_6 from being locally connected along its boundary, the required behavior breaking down at ∞.

4. The Carathéodory-Osgood Theorem

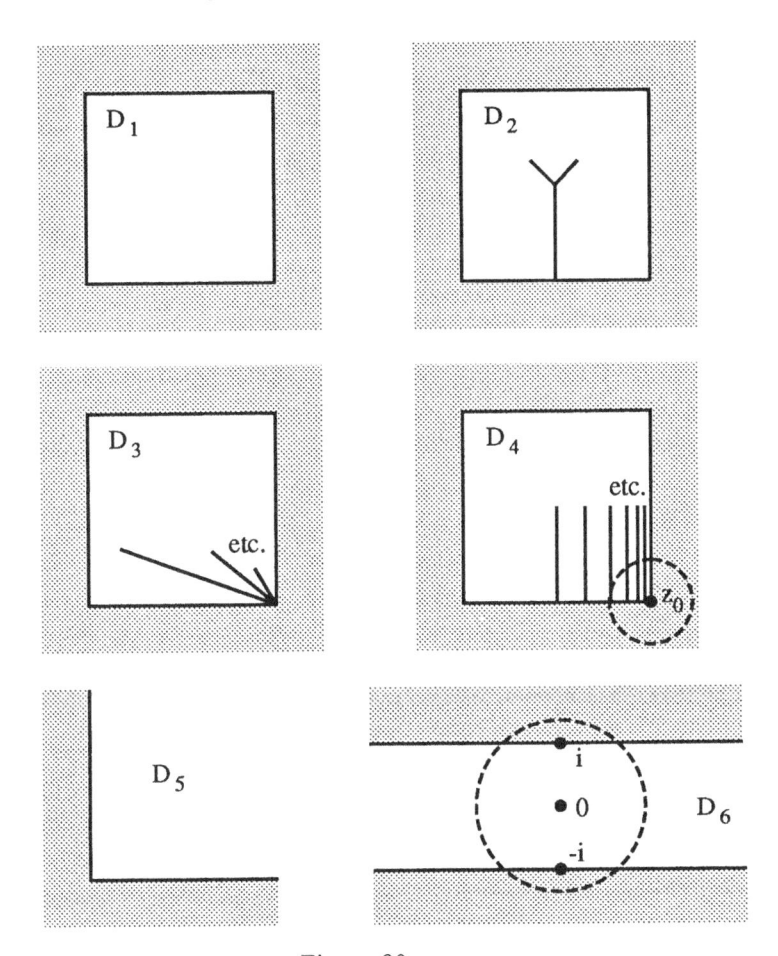

Figure 30.

The preceding definitions bring us to the point where we can establish our first extension result. To the best of the author's knowledge the theorem, as stated here, is due to Jussi Väisälä and Raimo Näkki.

Theorem 4.7. *Let f be a conformal mapping of the disk $\Delta = \Delta(0,1)$ onto a domain D in $\mathbb{C}$. Then f can be extended to a continuous mapping $\tilde{f}$ of $\overline{\Delta}$ onto $\widehat{D}$ if and only if D is finitely connected along its boundary.*

Proof. Assume first that D is finitely connected along its boundary. Owing to Lemma 4.1 the existence of $\tilde{f}$ will be established if it can be shown that $\lim_{\zeta \to z} f(\zeta)$ exists in $\widehat{\mathbb{C}}$ for each point z of $\partial \Delta$. We argue to this conclusion indirectly; i.e., we suppose that $\lim_{\zeta \to z_0} f(\zeta)$ fails to exist in $\widehat{\mathbb{C}}$ for some point z_0 of $\partial \Delta$ and derive a contradiction. Choose a sequence $\langle z_n \rangle$ in Δ such that $z_n \to z_0$. If $w_n = f(z_n)$, then due to the compactness of $\widehat{D}$ we may assume, possibly after passing to a subsequence and doing some relabeling, that $w_n \to w_0$, a point of $\widehat{D}$. In fact, w_0 must belong to $\widehat{\partial}D$. (Otherwise w_0 would be a point of D, $f^{-1}(w_0)$ would lie in Δ, and the continuity of f^{-1} at w_0 would lead to an immediate contradiction — namely, that

$$z_0 = \lim_{n\to\infty} z_n = \lim_{n\to\infty} f^{-1}(w_n) = f^{-1}(w_0) \in \Delta \;.)$$

The assumption that $\lim_{\zeta \to z_0} f(\zeta)$ fails to exist means there has to be a second sequence $\langle z'_n \rangle$ in D such that $z'_n \to z_0$, yet $w'_n = f(z'_n) \not\to w_0$. Once again passage to an appropriate subsequence allows us to assume that $w'_n \to w'_0$, where w'_0 lies on $\widehat{\partial}D$ and $w'_0 \neq w_0$. We select and fix $r > 0$ for which the disks $\overline{\Delta}(w_0, 2r)$ and $\overline{\Delta}(w'_0, 2r)$ are disjoint.

Because D is finitely connected along its boundary, it is possible to pick an s in $(0, r)$ with the property that $D \cap \Delta(w_0, s)$ meets only a finite number of components of $D \cap \Delta(w_0, r)$. Since w_n lies in $D \cap \Delta(w_0, s)$ for all large n, we infer that at least one of those finitely many components must contain w_n for infinitely many values of n. Stated differently, it is possible to extract a subsequence $\langle w_{n_k} \rangle$ from $\langle w_n \rangle$ in such a way that w_{n_k} belongs to a fixed component E' of $D \cap \Delta(w_0, r)$ for every k. We can likewise single out a component F' of $D \cap \Delta(w'_0, r)$ and a subsequence $\langle w'_{m_k} \rangle$ of $\langle w'_n \rangle$ such that w'_{m_k} is a point of F' for every k. Note that $E = f^{-1}(E')$ and $F = f^{-1}(F')$ are then disjoint connected sets in Δ.

We now examine the configurations $[E', F': D]$ and $[E, F: \Delta]$. At least one of the points w_0 or w'_0 must be a finite point — say $w_0 \neq \infty$. By construction E' is contained in $\widetilde{E} = \overline{\Delta}(w_0, r)$ and F' in $\widetilde{F} = \mathbb{C} \sim \Delta(w_0, 2r)$. According to Lemmas 4.6 and 4.5

$$M[E', F': D] \leq M[\widetilde{E}, \widetilde{F}: \mathbb{C}] = \frac{2\pi}{\mathrm{Log}\, 2} < \infty \;.$$

Next, $z_{n_k} = f^{-1}(w_{n_k})$ is an element of E for all k. Since $z_{n_k} \to z_0$ it follows that z_0 is a point of $\overline{E}$. For similar reasons z_0 lies in $\overline{F}$. Let h be a Möbius transformation that maps Δ to $H = \{z : \mathrm{Im}\, z > 0\}$ and sends z_0 to 0. The sets $h(E)$ and $h(F)$ are disjoint connected subsets of H having the origin as a common point of their closures. Theorem 4.2 and Lemma 4.4 tell us that

$$M[E, F: \Delta] = M[h(E), h(F): H] = \infty \;.$$

But then $M[E, F: \Delta] \neq M[E', F': D]$, which contravenes the conformal invariance of the modulus. This contradiction arose from the assumption that $\lim_{\zeta \to z_0} f(\zeta)$ failed to exist in $\widehat{\mathbb{C}}$. Thus, when all is said and done, the limit in question must exist. The existence of the extension $\widetilde{f}$ is now promised by Lemma 4.1.

The proof of the converse is also by contradiction. We suppose that the extension $\widetilde{f}$ exists, but that D is not finitely connected along its boundary. There must then exist a point w_0 of $\widehat{\partial}D$ and a number $r > 0$ such that for

4. The Carathéodory-Osgood Theorem

each s in $(0, r)$ the set $D \cap \Delta(w_0, s)$ intersects infinitely many components of $D \cap \Delta(w_0, r)$. It follows easily that we can manufacture a sequence $\langle w_n \rangle$ in D with the properties that $w_n \to w_0$ and that for $n \neq m$ the points w_n and w_m lie in different components of $D \cap \Delta(w_0, r)$. Let $z_n = f^{-1}(w_n)$. Passing to a subsequence, if need be, we may assume that $z_n \to z_0$, necessarily a point of $\partial \Delta$. Since $\tilde{f}$ is continuous at z_0 and has

$$\tilde{f}(z_0) = \lim_{n \to \infty} f(z_n) = \lim_{n \to \infty} w_n = w_0 ,$$

there exists a $\delta > 0$ for which it is true that $\tilde{f}[\overline{\Delta} \cap \Delta(z_0, \delta)]$ is contained in $\Delta(w_0, r)$. The set $C = \Delta \cap \Delta(z_0, \delta)$ is connected, and $\tilde{f}(C) = f(C)$ is contained in $D \cap \Delta(w_0, r)$. Itself a connected set, $f(C)$ must therefore be a subset of some component of $D \cap \Delta(w_0, r)$. On the other hand, z_n is a point of C for all large n, which implies that $f(C)$ contains w_n for all large n, in disagreement with the fact that the points w_n all lie in different components of $D \cap \Delta(w_0, r)$. This contradiction shows that D has to be finitely connected along its boundary. ∎

Theorem 4.7 informs us, for example, that a conformal mapping of $\Delta = \Delta(0, 1)$ onto any one of the domains D_1, D_2, D_3, D_5, or D_6 in Figure 30 extends continuously to $\overline{\Delta}$, whereas a conformal mapping of Δ onto D_4 definitely does not admit a continuous extension to $\overline{\Delta}$.

Relying on slightly more sophisticated topological information, Carathéodory was able to prove the following variant of Theorem 4.7: *a conformal mapping f of Δ onto D can be extended to a continuous mapping $\tilde{f}$ of $\overline{\Delta}$ onto $\widehat{D}$ if and only if ∂D is a locally connected subset of $\mathbb{C}$.* (To say that a set A in $\mathbb{C}$ is *locally connected* means this: corresponding to any point z of A and any $r > 0$ there exist an s in $(0, r)$ and a connected set C satisfying $A \cap \Delta(z, s) \subset C \subset A \cap \Delta(z, r)$.) A proof of this result of Carathéodory's can be found in the book of Pommerenke cited in the introduction to this chapter.

The analogue of Theorem 4.7 in which we insist that the extension $\tilde{f}$ be a homeomorphism merely replaces the requirement that D be finitely connected along its boundary with the more stringent demand that D be locally connected along its boundary. (*Warning*: Do not confuse "D is locally connected along its boundary" with "∂D is locally connected.")

Theorem 4.8. *Let f be a conformal mapping of the disk $\Delta = \Delta(0, 1)$ onto a domain D in $\mathbb{C}$. Then f can be extended to a homeomorphism $\tilde{f}$ of $\overline{\Delta}$ onto $\widehat{D}$ if and only if D is locally connected along its boundary.*

Proof. Attacking the sufficiency first, we assume that D is locally connected along its boundary. Theorem 4.7 certifies that f can be extended to a continuous function $\tilde{f}$ mapping $\overline{\Delta}$ onto $\widehat{D}$. If we are able to demonstrate that $\tilde{f}$ is univalent, Lemma 4.1 will insure that $\tilde{f}$ is a homeomorphism. We show, in fact, that a breakdown in univalence on the part of $\tilde{f}$ leads to a contradiction. Since f is univalent and since $\tilde{f}(\partial\Delta)$ is contained in $\hat{\partial}D$, the only way that $\tilde{f}$ can fail to be univalent is for $\tilde{f}(z_0)$ to coincide with $\tilde{f}(z_0')$ for a pair of distinct points z_0 and z_0' of $\partial\Delta$. Suppose this occurs for $z_0 = e^{i\theta_1}$ and $z_0' = e^{i\theta_2}$, where $0 \leq \theta_1 < \theta_2 < 2\pi$. Write $w_0 = \tilde{f}(z_0) = \tilde{f}(z_0')$, $E = \overline{\Delta}(0, 1/2)$, $E' = f(E)$, and $m = \min\{\theta_2 - \theta_1, 2\pi - (\theta_2 - \theta_1)\}$. The set E' is a compact set in D and w_0, a point of $\hat{\partial}D$, does not belong to E'. We can therefore choose $r_1 > 0$ with the property that E' and $\overline{\Delta}(w_0, r_1)$ are disjoint. Next, we pick r_0 in $(0, r_1)$ for which $2\pi[\mathrm{Log}(r_1/r_0)]^{-1} < m$. Finally, exploiting the fact that D is locally connected along its boundary, we select s in $(0, r_0)$ such that $D \cap \Delta(w_0, s)$ lies in a component F' of $D \cap \Delta(w_0, r_0)$. If $w_0 \neq \infty$, the set E' is contained in $\widetilde{E} = \mathbb{C} \sim \Delta(w_0, r_1)$ and F' lies in $\widetilde{F} = \overline{\Delta}(w_0, r_0)$; if $w_0 = \infty$, on the other hand, E' and F' are subsets of $\widetilde{E} = \overline{\Delta}(0, r_1^{-1})$ and $\widetilde{F} = \mathbb{C} \sim \Delta(0, r_0^{-1})$, respectively. In either case, Lemmas 4.6 and 4.5 imply that

$$(9.68) \qquad M[E', F': D] \leq M[\widetilde{E}, \widetilde{F}: \mathbb{C}] = 2\pi \left(\mathrm{Log}\frac{r_1}{r_0}\right)^{-1} < m \ .$$

The set $F = f^{-1}(F')$ is a connected set in $\Delta \sim E$. If $\langle z_n \rangle$ is a sequence in Δ such that $z_n \to z_0$, then from the continuity of $\tilde{f}$ we infer that $w_n = f(z_n) \to \tilde{f}(z_0) = w_0$. This dictates that w_n belong to $D \cap \Delta(w_0, s)$ — and so to F' — once n is suitably large. Accordingly, as soon as n gets sufficiently large, z_n is a point of F. Since $z_n \to z_0$, z_0 has to be a point of $\overline{F}$. A similar argument reveals that z_0' is also a point of $\overline{F}$. Lemma 4.3 and Theorem 4.2 then combine to give

$$m \leq M[E, F: \Delta] = M[E', F': D] \ ,$$

contradicting (9.68). The upshot of the contradiction: $\tilde{f}$ is univalent — hence, in view of Lemma 4.1, is a homeomorphism of $\overline{\Delta}$ onto $\hat{D}$.

As for the converse, assume that f admits a homeomorphic extension $\tilde{f}$ to $\overline{\Delta}$. Let w_0 be a point of $\hat{\partial}D$ and let $r > 0$. Setting $z_0 = \tilde{f}^{-1}(w_0)$, we use the continuity of $\tilde{f}$ to choose $\delta > 0$ so that $\tilde{f}[\overline{\Delta} \cap \Delta(z_0, \delta)]$ is contained in $\Delta(w_0, r)$. The set $f[\Delta \cap \Delta(z_0, \delta)]$ is a connected set in $D \cap \Delta(w_0, r)$ and, for this reason, lies inside some component C of $D \cap \Delta(w_0, r)$. We now claim that, for sufficiently small s in $(0, r)$, $D \cap \Delta(w_0, s)$ is a subset of C. The alternative to this is the existence of a sequence $\langle w_n \rangle$ in $D \sim C$ such that $w_n \to w_0$. Write $z_n = f^{-1}(w_n)$. We can extract from $\langle z_n \rangle$ a convergent subsequence $\langle z_{n_k} \rangle$ whose limit z_0' is a point of $\overline{\Delta}$. Again by continuity

$$\tilde{f}(z_0') = \lim_{k \to \infty} f(z_{n_k}) = \lim_{k \to \infty} w_{n_k} = w_0 \ .$$

From the univalence of $\tilde{f}$ we deduce that z_0' and z_0 are one and the same point. This implies, however, that z_{n_k} belongs to $\Delta \cap \Delta(z_0, \delta)$ once k is

4. The Carathéodory-Osgood Theorem 445

suitably large. For such k the point $w_{n_k} = f(z_{n_k})$ belongs to C, contrary to the selection of $\langle w_n \rangle$ as a sequence in $D \sim C$. In other words, given w_0 in $\widehat{\partial} D$ and $r > 0$, we have demonstrated that $D \cap \Delta(w_0, s)$ meets only one component of $D \cap \Delta(w_0, r)$, provided s in $(0, r)$ is taken appropriately small. Thus, D is locally connected along its boundary. ∎

4.5 Jordan Domains

Let D be a simply connected domain in $\mathbb{C}$, not the entire plane. Assume that D is locally connected along its boundary. Define a function $\gamma: [0, 2\pi] \to \widehat{\mathbb{C}}$ by $\gamma(t) = \widetilde{f}(e^{it})$, where $\widetilde{f}$ is the homeomorphic extension to $\overline{\Delta}$ of any conformal mapping f that transforms $\Delta = \Delta(0, 1)$ to D. Then γ is a simple and closed path in $\widehat{\mathbb{C}}$, one whose trajectory is $\widehat{\partial} D$, so $\widehat{\partial} D$ is a Jordan curve in $\widehat{\mathbb{C}}$. A domain D in $\widehat{\mathbb{C}}$ with the property that $\widehat{\partial} D$ is a Jordan curve in $\widehat{\mathbb{C}}$ is called a *Jordan domain*. What we have just observed is this: if D is a simply connected domain in $\mathbb{C}$, $D \neq \mathbb{C}$, and if D is locally connected along its boundary, then D is a Jordan domain. The final link needed to complete the chain of proof for the Carathéodory-Osgood extension theorem is the converse of the preceding statement: if D is a Jordan domain in $\mathbb{C}$, then D is simply connected and is locally connected along its boundary. This converse is known to be true. Its standard proof is a by-product of the considerations that enter into the proof of the Jordan curve theorem. We shall not attempt to reproduce the proof, for to do so would divert us from our real mission in this chapter, which is to study conformal mappings. We shall merely accept the result as an established fact, just as we have done all along with the Jordan curve theorem itself, and go on. For the record we should point out that the Jordan curve theorem remains valid in $\widehat{\mathbb{C}}$: if J is a Jordan curve in $\widehat{\mathbb{C}}$, then $\widehat{\mathbb{C}} \sim J = D \cup D^*$, where D and D^* are disjoint domains in $\widehat{\mathbb{C}}$ having $\widehat{\partial} D = \widehat{\partial} D^* = J$. This implies, for instance, that a Jordan domain D in $\mathbb{C}$ is simply connected, because the set $\widehat{\mathbb{C}} \sim D$ is connected (Theorem 3.6). Indeed, if we apply the Jordan curve theorem to $J = \widehat{\partial} D$, we find that $\widehat{\mathbb{C}} \sim D = \widehat{D}^*$, a connected set. Making allowances for the preceding topological details, we have assembled all the components in the featured result of this section.

Theorem 4.9. (Carathéodory-Osgood Theorem) *A conformal mapping f of $\Delta = \Delta(0, 1)$ onto a domain D in $\mathbb{C}$ can be extended to a homeomorphism of $\overline{\Delta}$ onto $\widehat{D}$ if and only if D is a Jordan domain.*

Having formally stated the Carathéodory-Osgood theorem, we wish to stress that in concrete situations Theorem 4.8, which amounts to the Carathéodory-Osgood theorem stripped of a few purely topological trappings, provides a perfectly worthy substitute for the higher powered result. In particular, it is not difficult to show that Theorem 4.8 applies to any bounded domain D in the complex plane whose boundary ∂D is the trajectory of a simple, closed, piecewise smooth path $\gamma:[a,b] \to \mathbb{C}$ with the property that $\dot\gamma(t) \neq 0$ for every t in $[a,b]$. (At possible points of non-differentiability this condition is interpreted to mean that both one-sided derivatives of γ are non-zero.) The domains falling under this description include, for instance, all bounded domains D for which ∂D is the trajectory of a simple and closed polygonal path.

For reference purposes we record the following useful corollary of Theorem 4.9.

Corollary 4.10. *Let f be a conformal mapping of a plane Jordan domain D onto another such domain D'. Then f extends to a homeomorphism $\widetilde{f}$ of $\widehat{D}$ onto $\widehat{D}'$.*

Proof. We use the Riemann mapping theorem to choose a conformal mapping g of $\Delta = \Delta(0,1)$ onto D. Then $h = f \circ g$ maps Δ conformally to D'. By Theorem 4.9 both g and h have homeomorphic extensions to $\overline{\Delta}$, call them $\widetilde{g}$ and $\widetilde{h}$. We see that $\widetilde{h} \circ \widetilde{g}^{-1}$ is a homeomorphism of $\widehat{D}$ onto $\widehat{D}'$ that extends f. ∎

As we saw in Theorem 3.2, a conformal mapping f of one simply connected plane domain $D(\neq \mathbb{C})$ onto another such domain D' is uniquely determined once $f(z_0)$ in D' and $\text{Arg}[f'(z_0)]$ are prescribed at some point z_0 of D. When D and D' are both Jordan domains, the Carathéodory-Osgood theorem opens the door to other means for uniquely specifying a conformal mapping between D and D'. The first of these occurs in the situation suggested by Figure 31. As a matter of fact, the associated theorem reproduces almost verbatim — concessions are made to present-day terminology — Riemann's own statement of his mapping theorem.

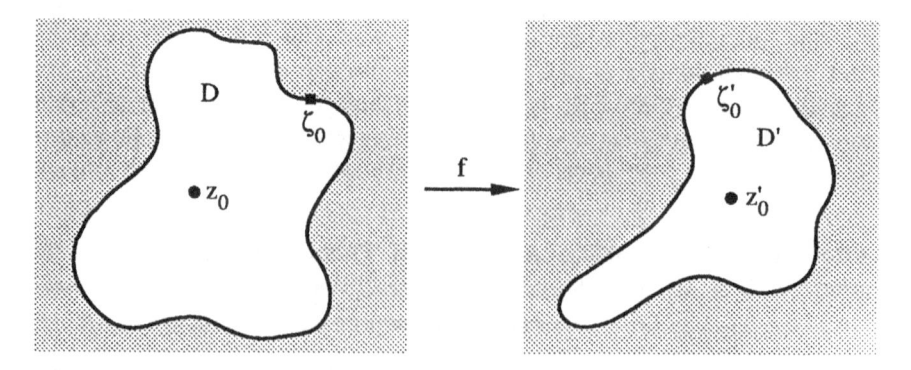

Figure 31.

Theorem 4.11. *Let D and D' be Jordan domains in the complex plane. Corresponding to any given points z_0 in D, ζ_0 on $\widehat{\partial}D$, z_0' in D', and ζ_0'*

4. The Carathéodory-Osgood Theorem

on $\widehat{\partial D'}$ there is a unique homeomorphism f of $\widehat{D}$ onto $\widehat{D}'$ that maps D conformally onto D' and meets the conditions $f(z_0) = z'_0$ and $f(\zeta_0) = \zeta'_0$.

Proof. Let φ be any conformal mapping of $\Delta = \Delta(0,1)$ onto D taking 0 to z_0, and let $\widetilde{\varphi}$ be its homeomorphic extension to $\overline{\Delta}$. If $\widetilde{\varphi}^{-1}(\zeta_0) = e^{i\theta_0}$, then $g(z) = \widetilde{\varphi}(e^{i\theta_0}z)$ defines a homeomorphism g of $\overline{\Delta}$ onto $\widehat{D}$ that transforms Δ conformally to D with $g(0) = z_0$ and $g(1) = \zeta_0$. Similarly, one constructs a homeomorphism h of $\overline{\Delta}$ onto $\widehat{D}'$ that maps Δ to D' in a conformal fashion and satisfies $h(0) = z'_0$, $h(1) = \zeta'_0$. The function $f = h \circ g^{-1}$ furnishes a homeomorphism of $\widehat{D}$ onto $\widehat{D}'$ that has all the desired features. Furthermore, f is the unique such mapping. If, namely, f_0 is any homeomorphism of $\widehat{D}$ onto $\widehat{D}'$ that maps D conformally onto D' and that sends z_0 to z'_0 and ζ_0 to ζ'_0, then $h^{-1} \circ f_0 \circ g$ is a homeomorphism of $\overline{\Delta}$ onto itself that is conformal in Δ and fixes both 0 and 1. The only conformal self-mappings of Δ fixing the origin are rotations about the origin; the only such rotation also fixing 1 is the identity mapping. Thus $h^{-1} \circ f_0 \circ g(z) = z$ for every z in $\overline{\Delta}$. This implies that $f_0(z) = h \circ g^{-1}(z) = f(z)$ for all points z of $\widehat{D}$ — and so confirms the uniqueness of f. ∎

4.6 Oriented Boundaries

A second procedure for specifying in a unique way a conformal mapping of one plane Jordan domain D onto another is to dictate the values that the mapping should take — or, more accurately, the values that its homeomorphic extension to $\widehat{D}$ should take — at three given boundary points of D. Before this pronouncement can be incorporated into a theorem, a few words must be said concerning the notion of "orientation." Let D be a Jordan domain in $\mathbb{C}$, and let $\gamma:[a,b] \to \widehat{\mathbb{C}}$ be a simple, closed path that parametrizes $\widehat{\partial}D$. The intuitive meaning of the statement "γ is positively oriented with respect to D" is that, as t increases from a to b and $\gamma(t)$ traverses $\widehat{\partial}D$, D "stays to the left" of $\gamma(t)$. In the case where the domain D is bounded (we have heretofore referred to D as the "inside" of the Jordan curve $J = \partial D$ in this situation) and γ is piecewise smooth, we were able in Chapter V to turn this intuitive idea into a precise mathematical definition, always taking for granted the conclusions of the Jordan curve theorem: γ is positively oriented with respect to D — at the time we simply said that γ was positively oriented — if and only if $n(\gamma, z) = 1$ for every z in D. By revamping our treatment of winding numbers, a process that would entail some non-elementary topology, it would be possible to extend this definition to cover general D and γ. We shall again resist the temptation to go into detail on this topic, preferring to keep our attention centered on complex analysis rather than letting the focus shift to topology. The key fact we shall need — and shall use without proof — is this: *if f maps* the disk $\Delta = \Delta(0,1)$ *conformally onto a Jordan domain D in* $\mathbb{C}$, *if* $\tilde{f}$ *is the homeomorphic extension of f to* $\overline{\Delta}$, *and if* γ *is the parametrization of* $\hat{\partial}D$ *defined on* $[0, 2\pi]$ *by* $\gamma(t) = \tilde{f}(e^{it})$, *then* γ *is positively oriented with respect to D.* This statement is rendered at least plausible by the following argument. For $0 < r < 1$ we define a path γ_r on $[0, 2\pi]$ by $\gamma_r(t) = f(re^{it})$. If z is a point of D and if $|f^{-1}(z)| < r < 1$, then the argument principle lets us know that

$$n(\gamma_r, z) = \frac{1}{2\pi i} \int_0^{2\pi} \frac{f'(re^{it}) i r e^{it}\, dt}{f(re^{it}) - z} = \frac{1}{2\pi i} \int_{|\zeta|=r} \frac{f'(\zeta)\, d\zeta}{f(\zeta) - z} = 1\ .$$

Whatever the exact definition of $n(\gamma, z)$ might be, it is not unreasonable to expect that $n(\gamma, z) = \lim_{r\to 1} n(\gamma_r, z)$. Thus, $n(\gamma, z) = 1$ ought to hold for every z in D.

An ordered triple $(\zeta_0, \zeta_1, \zeta_2)$ of distinct boundary points of a Jordan domain D in $\mathbb{C}$ is called *positively oriented relative to D* if, when starting at ζ_0 and traversing $\hat{\partial}D$ according to any simple parametrization that is positively oriented with respect to D, one passes through the given points in the order $\zeta_0, \zeta_1, \zeta_2, \zeta_0$ (i.e., if a simple parametrization $\gamma: [a, b] \to \hat{\partial}D$ is positively oriented with respect to D and has $\gamma(a) = \gamma(b) = \zeta_0$, it is required that the points t_1 and t_2 for which $\gamma(t_1) = \zeta_1$ and $\gamma(t_2) = \zeta_2$ satisfy $t_1 < t_2$.) If the points are encountered in the order $\zeta_0, \zeta_2, \zeta_1, \zeta_0$ for such a parametrization, the triple $(\zeta_0, \zeta_1, \zeta_2)$ is declared to be *negatively oriented relative to D*. (See Figure 32. The domain D_1 there is intended to be an unbounded Jordan domain.) We are now prepared to state:

Theorem 4.12. *Let D and D' be Jordan domains in the complex plane. Corresponding to any ordered triples of distinct points $(\zeta_0, \zeta_1, \zeta_2)$ from* $\hat{\partial}D$

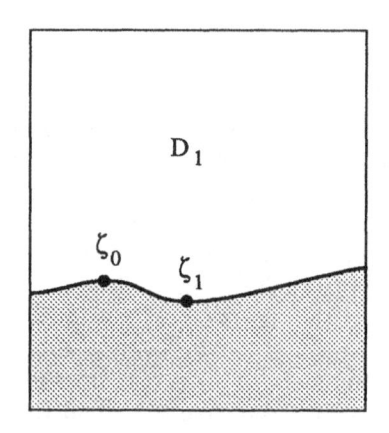

$(\zeta_0, \zeta_1, \infty)$ positively oriented relative to D_1

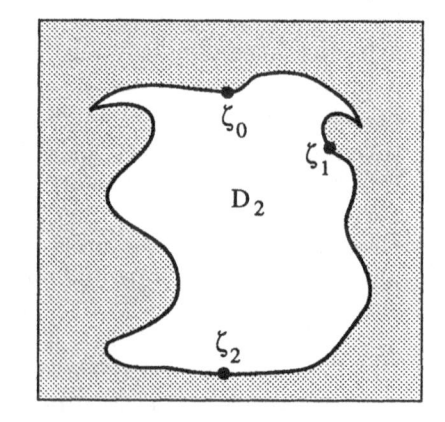

$(\zeta_0, \zeta_1, \zeta_2)$ negatively oriented relative to D_2

Figure 32.

4. The Carathéodory-Osgood Theorem

and $(\zeta_0', \zeta_1', \zeta_2')$ from $\widehat{\partial} D'$ that exhibit like orientations relative to D and D', respectively, there exists a unique homeomorphism f of $\widehat{D}$ onto $\widehat{D}'$ that maps D conformally onto D' and has $f(\zeta_j) = \zeta_j'$ for $j = 0, 1, 2$.

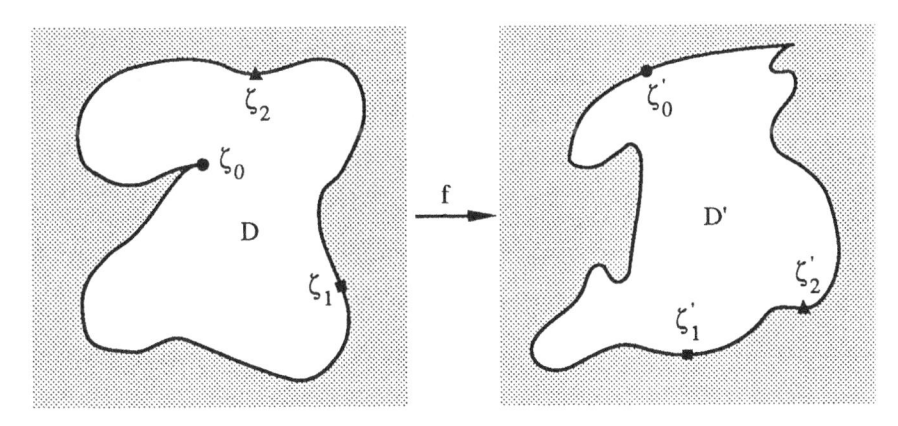

Figure 33.

Proof. We may assume that both triples are positively oriented. (If not, consider $(\zeta_0, \zeta_2, \zeta_1)$ and $(\zeta_0', \zeta_2', \zeta_1')$ instead of the original triples.) Let $\Delta = \Delta(0,1)$, let g be a homeomorphism of $\overline{\Delta}$ onto $\widehat{D}'$ that maps Δ conformally onto D and takes 1 to ζ_0, and let h be a homeomorphism of $\overline{\Delta}$ onto $\widehat{D}'$ that transforms Δ conformally to D' and has $h(1) = \zeta_0'$. Because the parametrization γ of $\widehat{\partial} D$ given by $\gamma(t) = g(e^{it})$ for $0 \le t \le 2\pi$ is positively oriented with respect to D and has initial point $\gamma(0) = g(1) = \zeta_0$ and because $(\zeta_0, \zeta_1, \zeta_2)$ is positively oriented relative to D, we can write $\zeta_1 = g(e^{i\theta_1})$ and $\zeta_2 = g(e^{i\theta_2})$, where $0 < \theta_1 < \theta_2 < 2\pi$. For similar reasons we can express $\zeta_1' = h(e^{i\psi_1})$ and $\zeta_2' = h(e^{i\psi_2})$, with $0 < \psi_1 < \psi_2 < 2\pi$. Next, let φ be the unique Möbius transformation that sends 1 to 1, $e^{i\theta_1}$ to $e^{i\psi_1}$, and $e^{i\theta_2}$ to $e^{i\psi_2}$. Then φ obviously maps the circle $K = K(0,1)$ to itself. It follows that either $\varphi(\Delta) = \Delta$ or $\varphi(\Delta) = \widehat{\mathbb{C}} \sim \overline{\Delta}$. If the latter were true, φ would have a single simple pole in Δ and be free of zeros there. The argument principle tells us that the parametrization of K given by $\beta(t) = \varphi(e^{it})$ for $0 \le t \le 2\pi$ would then have

$$n(\beta, 0) = \frac{1}{2\pi i} \int_0^{2\pi} \frac{\varphi'(e^{it}) i e^{it} dt}{\varphi(e^{it})} = \frac{1}{2\pi i} \int_{|z|=1} \frac{\varphi'(z)\, dz}{\varphi(z)} = -1,$$

implying that β would be negatively oriented with respect to Δ. This would force the triple $(\beta(0), \beta(\theta_1), \beta(\theta_2)) = (1, e^{i\psi_1}, e^{i\psi_2})$ to be negatively oriented relative to Δ, which it plainly isn't. We infer from these remarks that $\varphi(\overline{\Delta}) = \overline{\Delta}$ must be true and conclude that $f = h \circ \varphi \circ g^{-1}$ is a homeomorphism blessed with the requisite properties.

To establish the uniqueness of f, suppose that f_0 is an arbitrary homeomorphism of $\widehat{D}$ onto $\widehat{D}'$ meeting all the stated requirements. Then $\varphi_0 = g^{-1} \circ f_0^{-1} \circ f \circ g$ is a homeomorphism of $\overline{\Delta}$ onto itself. It maps Δ conformally to itself and fixes the points 1, $e^{i\theta_1}$, and $e^{i\theta_2}$. Theorem 1.4 implies that φ_0 is the restriction to $\overline{\Delta}$ of a Möbius transformation. Since φ_0 has three fixed points, that Möbius transformation is necessarily the identity transformation. In particular, $\varphi_0(z) = z$ for every z in $\overline{\Delta}$. From this it can be deduced that $f_0 = f$, which is the uniqueness assertion desired. ∎

The results in this section have many applications. We conclude the section with one sample — to wit, we use the Carathéodory-Osgood theorem to prove that any bounded Jordan domain is regular for the Dirichlet problem. This allows us to make good on a promise issued in Chapter VI.

Theorem 4.13. *Any bounded Jordan domain D in the complex plane is regular for the Dirichlet problem.*

Proof. Consider a continuous function $h: \partial D \to \mathbb{R}$. It is our job to produce a continuous function $u: \overline{D} \to \mathbb{R}$ that is harmonic in D and satisfies $u = h$ on ∂D. Let $\Delta = \Delta(0, 1)$, and let f be a homeomorphism of $\overline{D}$ onto $\overline{\Delta}$ that transforms D conformally to Δ. The function $h_0 = h \circ f^{-1}$ is continuous on $\partial \Delta$. By Theorem VI.3.1 there exists a continuous function $u_0: \overline{\Delta} \to \mathbb{R}$ that is harmonic in Δ and agrees with h_0 on $\partial \Delta$. Finally, $u = u_0 \circ f$ is continuous on $\overline{D}$, $u = h$ on ∂D, and a direct calculation — see Exercise VI.4.5 — demonstrates that $u_{xx} + u_{yy} = 0$ in D; i.e., u is harmonic in D. ∎

The argument just presented also works for an unbounded Jordan domain D, provided the boundary data h is continuous not just on ∂D, but also at ∞. Since the Dirichlet problem in $\Delta(0, 1)$ has a concrete solution given by the Poisson integral formula, the proof of Theorem 4.13 gives rise to an integral formula representing the solution of the Dirichlet problem in any Jordan domain for which the mapping f can be explicitly determined.

5 Conformal Mappings onto Polygons

5.1 Polygons

Let $\gamma = [z_1, z_2] + [z_2, z_3] + \cdots + [z_{n-1}, z_n] + [z_n, z_1]$ be a simple and closed polygonal path in the complex plane endowed with the extra feature that for $j = 1, 2, \ldots, n - 1$ the points z_j, z_{j+1}, and z_{j+2} are not collinear. (To make this statement meaningful for $j = n - 1$, set $z_{n+1} = z_1$.) Under these conditions we describe $P = \overline{D}$, where D is the inside of the Jordan curve $J = |\gamma|$, as the *closed polygon with vertices* $z_1, z_2, \ldots, z_n$. (See Figure 34.) We shall always tacitly assume that the vertices of P are listed in an order which causes the path γ to be positively oriented with respect to D. If λ_j is

5. Conformal Mappings onto Polygons

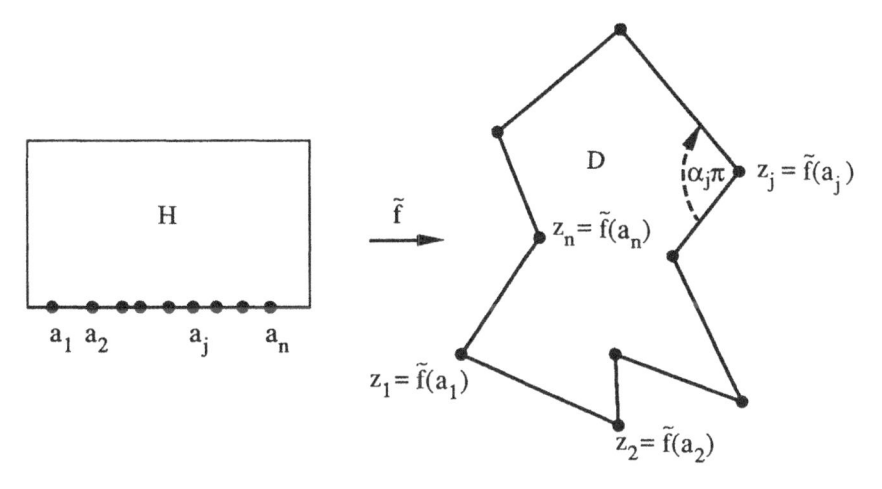

Figure 34.

the line segment joining z_j and z_{j+1}, then $\lambda_1, \lambda_2, \ldots, \lambda_n$ are the *sides* of P. The (unoriented) interior angle of P at the vertex z_j will be written $\alpha_j \pi$, with $0 < \alpha_j < 2$. (N.B. The requirements we have imposed on γ insure that z_j is an "authentic" vertex of P, meaning that $\alpha_j \neq 1$.) It is a fact of elementary geometry that

(9.69) $$\sum_{j=1}^{n} \alpha_j = n - 2 .$$

In this section we study conformal mappings of the half-plane $H = \{z : \operatorname{Im} z > 0\}$ onto the interior D of such a closed polygon P. It is our goal to obtain a concrete representation for such a mapping, a representation in the shape of a famous integral formula that bears the names of H.A. Schwarz and E.B. Christoffel (1829-1900). (As it is an easy matter to pass conformally from the disk $\Delta = \Delta(0, 1)$ to H by means of a Möbius transformation, we shall also determine the structure of a conformal mapping from Δ onto D.) If f maps H conformally onto D, then f can be extended to a homeomorphism $\tilde{f}$ of $\hat{H}$ onto P. We denote by $a_1, a_2, \ldots, a_n$ the preimages of the vertices $z_1, z_2, \ldots, z_n$ of P under $\tilde{f}$. We shall insist that the labeling of vertices be done in such a way that $-\infty < a_1 < a_2 < \cdots < a_n \leq \infty$, as in Figure 34.

5.2 The Reflection Principle

Underlying our derivation of the Schwarz-Christoffel formula is a method, due to Schwarz, of extending certain analytic functions from their given domain-sets to larger ones through a process of reflection. As a point of historical interest, we might add that Schwarz originally developed his "reflection principle" with the study of conformal mappings of H to polygons as the motivation. Our treatment of this result begins with the following refinement of Lemma V.1.1.

Lemma 5.1. *Let R be a closed rectangle in the complex plane, and let $f: R \to \mathbb{C}$ be a continuous function that is analytic in the interior of R. Then $\int_{\partial R} f(z)\, dz = 0$.*

Proof. The argument we make is a variation on the one used to prove Lemma V.1.2. Given $\epsilon > 0$, we verify that

$$\left| \int_{\partial R} f(z)\, dz \right| \leq 4\epsilon L, \tag{9.70}$$

in which L is the perimeter of R. Letting $\epsilon \to 0$ in (9.70) we arrive at the stated conclusion. In view of the uniform continuity of f on R, we can choose an integer $n > 2$ such that $|f(z_1) - f(z_2)| < \epsilon$ for all points z_1 and z_2 of R that satisfy $|z_1 - z_2| < d/n$. Here d is the length of the diagonal of R. As in the proof of Lemma V.1.2, we subdivide R into n^2 congruent rectangles $R_{k\ell}$ and observe that

$$\left| \int_{\partial R} f(z)\, dz \right| = \left| \sum_{k,\ell} \int_{\partial R_{k\ell}} f(z)\, dz \right| \leq \sum_{k,\ell} \left| \int_{\partial R_{k\ell}} f(z)\, dz \right|. \tag{9.71}$$

The rectangles $R_{k\ell}$ fall into two classes: $(n-2)^2$ of them lie in the interior of R; each of the remaining $4(n-1)$ rectangles has at least one side contained in ∂R. For any of the "interior" rectangles $R_{k\ell}$ we have $\int_{\partial R_{k\ell}} f(z)\, dz = 0$ by Lemma V.1.1. For a "border" rectangle $R_{k\ell}$ with center $z_{k\ell}$ we get

$$\left| \int_{\partial R_{k\ell}} f(z)\, dz \right| = \left| \int_{\partial R_{k\ell}} f(z) - \int_{\partial R_{k\ell}} f(z_{k\ell})\, dz \right|$$

$$\leq \int_{\partial R_{k\ell}} |f(z) - f(z_{k\ell})|\, |dz| < \frac{\epsilon L}{n}$$

by reason of the fact that $|z - z_{k\ell}| < d/n$ is true for every z in $\partial R_{k\ell}$ — this forces $|f(z) - f(z_{k\ell})| < \epsilon$ to hold for such z — and that $R_{k\ell}$ has perimeter L/n. Because the "border" rectangles are $4(n-1)$ in number, (9.71) leads to

$$\left| \int_{\partial R} f(z)\, dz \right| \leq \frac{4(n-1)\epsilon L}{n} \leq 4\epsilon L,$$

confirming (9.70). ∎

The next lemma presents the reflection principle of Schwarz in its simplest setting.

5. Conformal Mappings onto Polygons

Lemma 5.2. *Let D be a plane domain that is symmetric about the real axis, let G and G^* be the components of $D \sim \mathbb{R}$, and let $I = D \cap \mathbb{R}$. If $f: G \cup I \to \mathbb{C}$ is a continuous function that is analytic in G and real-valued on I, then the function $F: D \to \mathbb{C}$ defined by*

$$F(z) = \begin{cases} f(z) & \text{if } z \in G \cup I, \\ \overline{f(\bar{z})} & \text{if } z \in G^*, \end{cases}$$

is an analytic function.

Proof. (Recall Figure VI.2.) The function F is analytic in G by hypothesis. It is also analytic in G^*, for when z_0 lies in G^* we have

$$\lim_{z \to z_0} \frac{F(z) - F(z_0)}{z - z_0} = \lim_{z \to z_0} \frac{\overline{f(\bar{z})} - \overline{f(\bar{z_0})}}{z - z_0} = \lim_{z \to z_0} \overline{\left(\frac{f(\bar{z}) - f(\bar{z_0})}{\bar{z} - \bar{z_0}}\right)} = \overline{f'(\bar{z_0})} ;$$

i.e., F is differentiable at z_0 with $F'(z_0) = \overline{f'(\bar{z_0})}$. The real-valuedness and continuity of f at the points of I insure that F, too, is continuous at these points. Therefore, F is at least a continuous function. Morera's theorem will testify to the analyticity of F if only it can be demonstrated that $\int_{\partial R} F(z)\,dz = 0$ for each closed rectangle R in D whose sides are parallel to the coordinate axes. Fix such a rectangle R. If R is contained in either $G \cup I$ or $G^* \cup I$, then the vanishing of the integral in question is a direct consequence of Lemma 5.1. In all other cases R is partitioned by I into two smaller rectangles R' and R'', where R' is located in $G \cup I$ and R'' in $G^* \cup I$. Thus, by the preceding remark,

$$\int_{\partial R} F(z)\,dz = \int_{\partial R'} F(z)\,dz + \int_{\partial R''} F(z)\,dz = 0 + 0 = 0 \,.$$

The invocation of Morera's theorem completes the proof. ∎

We generalize Lemma 5.2 as follows.

Theorem 5.3. (Schwarz Reflection Principle) *Let K and $\widetilde{K}$ be circles in $\widehat{\mathbb{C}}$ with associated reflections ρ and $\widetilde{\rho}$, let D be a domain in $\mathbb{C}$ that is symmetric about K (i.e., $\rho(D) = D$), let G and G^* be the components of $D \sim K$, and let $I = D \cap K$. If $f: G \cup I \to \mathbb{C}$ is a continuous function that is analytic in G and maps I to a subset of $\widetilde{K}$, then the function $F: D \to \widehat{\mathbb{C}}$ defined by*

$$F(z) = \begin{cases} f(z) & \text{if } z \in G \cup I, \\ \widetilde{\rho} \circ f \circ \rho(z) & \text{if } z \in G^*, \end{cases}$$

is a meromorphic function. In fact, F is an analytic function except when $\widetilde{K}$ is a true circle whose center belongs to the range of f, in which case the only poles of F are found at the points z of G^* with the property that $f[\rho(z)]$ is the center of $\widetilde{K}$.

Proof. Since the function $\tilde{\rho} \circ f \circ \rho$ is continuous in $G^* \cup I$ and coincides with f on I, we see easily that the function F is continuous. It only assumes the value ∞ when $\tilde{K}$ is a true circle with its center in the range of f. In this event $F(z) = \infty$ at precisely those points z of G^* for which $f[\rho(z)]$ is the center of $\tilde{K}$. These are the only places where F could possibly have poles. We now fix an arbitrary open subarc I' of I such that $I' \neq I$ and $f(I') \neq \tilde{K}$. We confirm that F is meromorphic in the domain $D' = G \cup I' \cup G^*$. The establishment of this fact for every such set I' clearly suffices to complete the proof of the theorem. Write $K_0 = \mathbb{R} \cup \{\infty\}$. We select Möbius transformations g and h with the following properties: $g(K) = K_0$; $h(K_0) = \tilde{K}$; ∞ does not lie in $I_0 = g(I')$; $h(\infty)$ does not lie in $f(I')$. By Theorem 2.7 the plane domain $D_0 = g(D')$ is symmetric about the real axis. The components of $D_0 \sim \mathbb{R}$ are $G_0 = g(G)$ and $G_0^* = g(G^*)$, while $D_0 \cap \mathbb{R} = I_0$. Furthermore, the function $f_0: G_0 \cup I_0 \to \mathbb{C}$ given by $f_0(z) = h^{-1} \circ f \circ g^{-1}(z)$ is continuous, it is analytic in G_0, and it is real-valued on I_0. We can appeal to Lemma 5.2 and assert that the function $F_0: D_0 \to \mathbb{C}$ defined by

$$F_0(z) = \begin{cases} f_0(z) & \text{if } z \in G_0 \cup I_0 \,, \\ \overline{f_0(\bar{z})} & \text{if } z \in G_0^* \,, \end{cases}$$

is analytic. It follows that $h \circ F_0 \circ g$ is a meromorphic function with domain-set D'. We claim that in D' this function and F are one and the same. It is immediate from our definition of f_0 that $h \circ F_0 \circ g = f = F$ in the set $G \cup I'$. What happens in G^*? To answer this question we once more utilize information given to us by Theorem 2.7: $g[\rho(z)] = \overline{g(z)}$ and $\tilde{\rho}[h(z)] = h(\bar{z})$ for every z in $\widehat{\mathbb{C}}$. We learn from it that

$$h \circ F_0 \circ g(z) = h\overline{\{f_0[\overline{g(z)}]\}} = [(\tilde{\rho} \circ h) \circ f_0 \circ (g \circ \rho)](z)$$

$$= [\tilde{\rho} \circ (h \circ f_0 \circ g) \circ \rho](z) = \tilde{\rho} \circ f \circ \rho(z) = F(z)$$

for every z in G^*. Agreeing with $h \circ F_0 \circ g$ in D', the function F is meromorphic in this set, as maintained. ∎

The function F constructed under the conditions detailed in Theorem 5.3 is called the *continuation of f across I by reflection*. The basic reflection principle that we have elected to state here admits further generalizations. A couple of these are indicated in the exercises.

5.3 The Schwarz-Christoffel Formula

Theorem 5.3 equips us with just the tool we need to expose the structure of a conformal mapping from the upper half-plane H to a polygonal region.

5. Conformal Mappings onto Polygons

In an effort to streamline our presentation of the theorem that does this we collect some technical details in two preliminary lemmas.

Lemma 5.4. *Suppose that f is a conformal mapping of the half-plane $H = \{z : \operatorname{Im} z > 0\}$ onto the interior D of a closed polygon P in the complex plane and that $-\infty < a_1 < a_2 < \cdots < a_n \leq \infty$ is the list of points transformed to the vertices of P by $\tilde{f}$, the homeomorphic extension of f to $\widehat{H}$. Then the function f''/f' can be extended to a function g that is analytic in the domain $G = \mathbb{C} \sim \{a_1, a_2, \ldots, a_n\}$.*

Proof. Let $I_1 = (-\infty, a_1), \ldots, I_n = (a_{n-1}, a_n)$, and $I_{n+1} = (a_n, \infty)$ be the open intervals into which the real axis splits upon removal of the points $a_1, a_2, \ldots, a_n$. (In the event that $a_n = \infty$, disregard I_{n+1}.) The image of I_j under $\tilde{f}$ is contained in a side of P — hence, in a line L_j. Let ρ_j denote reflection in L_j, and let ρ be reflection in the real axis. Using the Schwarz reflection principle we can continue $\tilde{f}$ across I_j to define an analytic function F_j in the domain $D_j = H \cup I_j \cup H^*$. Here $H^* = \{z : \operatorname{Im} z < 0\}$, a set in which F_j is given by

$$(9.72) \qquad F_j = \rho_j \circ f \circ \rho.$$

The function F_j is clearly univalent in both H and H^*. (It might fail to be univalent in D_j, however, for nothing in our hypotheses rules out the eventuality that the two sets $F_j(H)$ and $F_j(H^*)$ overlap.) In particular, $F_j'(z) \neq 0$ holds for every z in $H \cup H^*$ (Theorem VIII.3.9). It is also true that $F_j'(z) \neq 0$ if z belongs to I_j. To see this, fix a point z_0 of I_j. Then $w_0 = F_j(z_0) = \tilde{f}(z_0)$ is a point of $\partial D \cap L_j$, but w_0 is not a vertex of P. We can therefore choose $s > 0$ so that the set $D \cap \Delta(w_0, s)$ is a half-disk, one of the two components of $\Delta(w_0, s) \sim L_j$. Next, pick $r > 0$ small enough that $\mathbb{R} \cap \Delta(z_0, r)$ is a subset of I_j and that $\tilde{f}[H \cap \Delta(z_0, r)]$ is contained in $D \cap \Delta(w_0, s)$. The set $F_j[H \cap \Delta(z_0, r)]$ thus lies on one side of the line L_j, and its reflection in L_j — i.e., the set $F_j[H^* \cap \Delta(z_0, r)]$ — lies on the opposite side of L_j (Figure 35). This insures that F_j is univalent in $\Delta(z_0, r)$ and, as a result, that $F_j'(z_0) \neq 0$. An implication of the above considerations is that the function $g_j = F_j''/F_j'$ is analytic in D_j.

How are F_j and F_k related in H^* when $j \neq k$? The function F_j maps H^* conformally onto $\rho_j(D)$. We use (9.72) to compute $F_k \circ F_j^{-1}$ in $\rho_j(D)$:

$$F_k \circ F_j^{-1} = (\rho_k \circ f \circ \rho) \circ (\rho_j \circ f \circ \rho)^{-1} = \rho_k \circ f \circ \rho \circ \rho^{-1} \circ f^{-1} \circ \rho_j^{-1} = \rho_k \circ \rho_j.$$

(Remember: $\rho_j^{-1} = \rho_j$.) As the composition of two reflections in lines, $\rho_k \circ \rho_j$ has the form $\rho_k \circ \rho_j(z) = az + b$, where $a \neq 0$ and b are constants (they do depend on j and k, of course). Accordingly, we can write

$$F_k \circ F_j^{-1}(z) = az + b$$

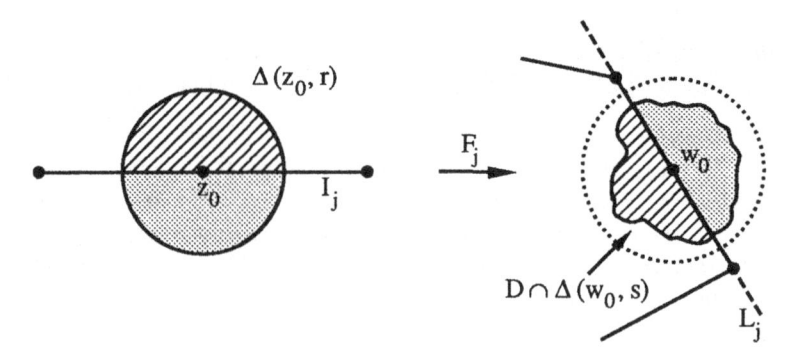

Figure 35.

for all z in $\rho_j(D)$ or, equivalently,

$$F_k(z) = aF_j(z) + b$$

for every z in H^*. The last relation leads to the conclusion that $g_k = F_k''/F_k' = F_j''/F_j' = g_j$ in H^* when $j \neq k$. If, as a consequence, we define a function g in the domain $G = \mathbb{C} \sim \{a_1, a_2, \ldots, a_n\} = \bigcup_j D_j$ by setting $g(z) = g_j(z)$ for z in D_j, then we succeed in creating an analytic function that coincides with f''/f' in H. ∎

The next lemma clarifies the nature of the the singularities of the function g just constructed.

Lemma 5.5. *Under the conditions of Lemma 5.4, let the interior angle of P at the vertex $\tilde{f}(a_j)$ be $\alpha_j \pi$. If $a_j \neq \infty$, the singularity of the function g at a_j is a simple pole with residue $\alpha_j - 1$. Furthermore, $g(z) \to 0$ as $z \to \infty$.*

Proof. To conserve on notation we write $a = a_j$ and $\alpha = \alpha_j$. In the first part of the proof we assume that $a \neq \infty$. Let $h(z) = g(z) - (\alpha - 1)(z - a)^{-1}$ for z in the domain G. We shall show that in some punctured disk $\Delta^*(a, \delta)$ this function h can be expressed in the form

(9.73) $$h(z) = \frac{(\alpha + 1)f_0'(z) + (z - a)f_0''(z)}{\alpha f_0(z) + (z - a)f_0'(z)},$$

where f_0 is a function that is both analytic and zero-free in the full disk $\Delta(a, \delta)$. Since the right-hand side of (9.73) plainly approaches the finite limit $(\alpha + 1)f_0'(a)/\alpha f_0(a)$ as $z \to a$, it becomes evident that h has a removable singularity at a. The singularity of g at a is then immediately recognized to be a simple pole at which g has residue $\alpha - 1$.

If $c \neq 0$ and d are complex constants, then $\varphi = cf + d$ also maps H conformally to a polygonal region, one similar to D. Its homeomorphic extension to $\bar{H}$ is $\tilde{\varphi} = c\tilde{f} + d$, so $\tilde{\varphi}$ also takes the points $a_1, a_2, \ldots, a_n$ to the vertices of its image polygon. It follows that Lemma 5.4 can be applied to

5. Conformal Mappings onto Polygons

φ. The result is a function that is analytic in D and coincides with $\varphi''/\varphi' = f''/f'$ in H. By Corollary VIII.1.6 the only analytic extension of f''/f' to G is g, the function constructed in the proof of Lemma 5.4. In other words, it makes no difference whether we apply Lemma 5.4 to f or to φ — we obtain the same function g in either case. Consequently, in verifying (9.73) we are allowed to make the simplifying assumptions that the extension $\tilde{f}$ of f to $\widehat{H}$ satisfies $\tilde{f}(a) = 0$ and that the interior angle of P at the origin is bisected by the positive real axis, as in Figure 36. (If not, replace f by $cf + d$ for suitable c and d.) Under these assumptions we can choose and fix s for which $D \cap \Delta(0,s) = \{w \in \Delta(0,s) : |\operatorname{Arg} w| < \alpha\pi/2\}$. We can then select r with the property that $\tilde{f}[\widehat{H} \cap \Delta(a,r)]$ is a subset of $\Delta(0,s)$. Finally, we can define a function $\psi : \widehat{H} \cap \Delta(a,r) \to \mathbb{C}$ as follows: $t = \psi(z) = [\tilde{f}(z)]^{1/\alpha}$. This function is a homeomorphism that maps $H \cap \Delta(a,r)$ conformally onto a domain contained in the half-plane $\{t : \operatorname{Re} t > 0\}$. It transforms the interval $I = (a - r, a + r)$ to an open interval on the imaginary axis in the t-plane (Figure 36).

We now continue the function ψ across I by reflection. Through this process we produce a conformal mapping $\Psi : \Delta(a,r) \to \mathbb{C}$ with the property that $f(z) = [\Psi(z)]^\alpha$ for every z in $H \cap \Delta(a,r)$. As a function analytic in $\Delta(a,r)$ whose only zero there is a simple one at a, Ψ can be rewritten in the

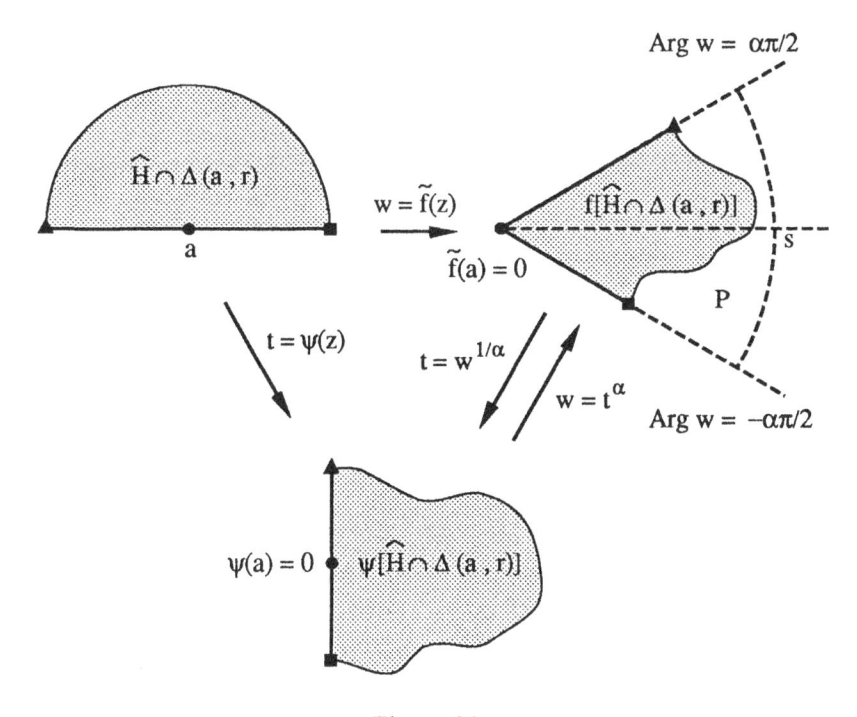

Figure 36.

form $\Psi(z) = (z-a)\Psi_0(z)$, where Ψ_0 is a function that is both analytic and free of zeros in $\Delta(a,r)$. Theorem V.6.2 insures the existence of branches of $\log \Psi_0(z)$ in $\Delta(a,r)$. We pick one and label it L. In fact, after possibly adjusting our original choice for L by adding to it an integral multiple of $2\pi i$, we may assume that $L(z) = \text{Log } \Psi(z) - \text{Log}(z-a)$ whenever z belongs to $H \cap \Delta(a,r)$. (Reason: the restriction of L to $H \cap \Delta(a,r)$ is a branch of $\log \Psi_0(z)$ in $H \cap \Delta(a,r)$, as is the function $\widetilde{L}$ given in that half-disk by $\widetilde{L}(z) = \text{Log } \Psi(z) - \text{Log}(z-a)$, conditions which imply that L and $\widetilde{L}$ differ in $H \cap \Delta(a,r)$ by an integral multiple of $2\pi i$.) For z in $H \cap \Delta(a,r)$ we can thus assert that

$$f(z) = [\Psi(z)]^\alpha = e^{\alpha \text{ Log } \Psi(z)} = e^{\alpha[\text{Log } \Psi(z) - \text{Log}(z-a)] + \alpha \text{ Log}(z-a)}$$

$$= (z-a)^\alpha e^{\alpha L(z)}.$$

Note that the function f_0 defined by $f_0(z) = e^{\alpha L(z)}$ is actually analytic in the whole disk $\Delta(a,r)$, where it has no zeros. The purpose of these rather drawn out deliberations was to arrive at the representation $f(z) = (z-a)^\alpha f_0(z)$ for the function f in $H \cap \Delta(a,r)$. It leads by direct calculation to a representation for the function h in $H \cap \Delta(a,r)$ — namely,

$$h(z) = \frac{f''(z)}{f'(z)} - \frac{\alpha-1}{z-a} = \frac{(\alpha+1)f_0'(z) + (z-a)f_0''(z)}{\alpha f_0(z) + (z-a)f_0'(z)}.$$

Since the denominator in the last expression does not vanish at the point a, we can choose δ in $(0,r)$ so that the function defined by this formula is analytic in the disk $\Delta(a,\delta)$. Because that function and h coincide in $H \cap \Delta(a,r)$ and because both are analytic in $\Delta^*(a,\delta)$, they must, according to the principle of analytic continuation, agree everywhere in $\Delta^*(a,\delta)$. This completes the verification of (9.73) and, with it, the first half of the proof.

It remains to pin down the behavior of $g(z)$ as $z \to \infty$. For this, we consider the function $f_1: H \to \mathbb{C}$ given by $f_1(z) = f(-z^{-1})$. (N.B. The relation $f(z) = f_1(-z^{-1})$ is equally valid.) Then f_1 is a second conformal mapping of H onto D, one satisfying $\widetilde{f}_1(0) = \widetilde{f}(\infty)$. Let g_1 be constructed from f_1 in the same way g was from f. Straightforward computation yields

$$g(z) = \frac{f''(z)}{f'(z)} = -\frac{2}{z} + \frac{f_1''(-z^{-1})}{z^2 f_1'(-z^{-1})}$$

for z in H; i.e., we conclude that in H

(9.74) $$g(z) = -\frac{2}{z} + \frac{1}{z^2} g_1\left(-\frac{1}{z}\right).$$

Each side of (9.74) describes a function that is analytic in the domain $U = \mathbb{C} \sim \{0, a_1, \ldots, a_n\}$. As these functions are the same in H, Corollary VIII.1.6 confirms that (9.74) remains valid throughout U. Two cases must

5. Conformal Mappings onto Polygons

now be distinguished. First, if $\widetilde{f}_1(0)$ is not a vertex of P, then g_1 is analytic in the vicinity of the origin, so $g_1(-z^{-1}) \to g_1(0) \neq \infty$ as $z \to \infty$. Coupled with (9.74) this implies that $g(z) \to 0$ as $z \to \infty$. If, on the other hand, $\widetilde{f}_1(0)$ happens to be a vertex of P, then by the first part of the lemma g_1 has a simple pole at the origin. As a result, $zg_1(z) \to \ell$, a finite limit, as $z \to 0$. Thus $z^{-1}g_1(-z^{-1}) \to -\ell$ as $z \to \infty$. In this instance, too, it follows from (9.74) that $g(z) \to 0$ as $z \to \infty$. ∎

We now come to the main result of this section.

Theorem 5.6. (Schwarz-Christoffel Formula) *Let f be a conformal mapping of the half-plane $H = \{z : \operatorname{Im} z > 0\}$ onto the interior D of a closed polygon P in the complex plane, let $-\infty < a_1 < a_2 < \cdots < a_n \leq \infty$ be the list of points transformed to the vertices of P by the homeomorphic extension $\widehat{f}$ of f to $\widehat{H}$, and let $\alpha_j \pi$ be the interior angle of P at the vertex $\widehat{f}(a_j)$. There exist constants A and B such that for every z in H*

$$(9.75) \quad f(z) = A \int_i^z (\zeta - a_1)^{\alpha_1 - 1}(\zeta - a_2)^{\alpha_2 - 1} \cdots (\zeta - a_n)^{\alpha_n - 1}\, d\zeta + B$$

if $a_n \neq \infty$, while

$$(9.76) \quad f(z) = A \int_i^z (\zeta - a_1)^{\alpha_1 - 1}(\zeta - a_2)^{\alpha_2 - 1} \cdots (\zeta - a_{n-1})^{\alpha_{n-1} - 1}\, d\zeta + B$$

if $a_n = \infty$.

Proof. Let g be the function constructed from f as in Lemma 5.4. We consider the function h defined in the domain-set G of g by

$$h(z) = g(z) - \sum_{j=1}^{n} \frac{\alpha_j - 1}{z - a_j}$$

if $a_n \neq \infty$, and

$$h(z) = g(z) - \sum_{j=1}^{n-1} \frac{\alpha_j - 1}{z - a_j}$$

if $a_n = \infty$. Then h is analytic in G. On the basis of Lemma 5.5 we can assert that the singularities of h at the points $a_1, a_2, \ldots, a_n$, and ∞ are all removable. We remove them and thereby turn h into a holomorphic function on $\widehat{\mathbb{C}}$. By Theorem VIII.4.2, h is constant on $\widehat{\mathbb{C}}$. Since $h(\infty) = 0$, we conclude that $h(z) = 0$ for every z in $\mathbb{C}$. We arrive in this way at an explicit formula for g.

In the event that $a_n \neq \infty$, we choose a branch ℓ of $\log f'(z)$ in H and observe that

$$\ell'(z) = \frac{f''(z)}{f'(z)} = g(z) = \sum_{j=1}^{n} \frac{\alpha_j - 1}{z - a_j} = \frac{d}{dz}\sum_{j=1}^{n} (\alpha_j - 1) \operatorname{Log}(z - a_j)$$

in this half-plane. (Such a function ℓ exists because f' is analytic and zero-free in H.) In H, therefore,

$$\ell(z) = \sum_{j=1}^{n}(\alpha_j - 1)\operatorname{Log}(z - a_j) + C$$

for some constant C. Exponentiation results in

$$f'(z) = e^{\ell(z)} = A(z-a_1)^{\alpha_1-1}(z-a_2)^{\alpha_2-1}\cdots(z-a_n)^{\alpha_n-1},$$

where $A = e^C$. Integration then produces (9.75) with $B = f(i)$. The argument in the case $a_n = \infty$ involves only minute changes and leads to (9.76) in place of (9.75). ∎

We accent the fact that all powers occurring in (9.75) and (9.76) are principal powers. Since the integrands in these formulas are continuous functions in $\overline{H} \sim \{a_1, a_2, \ldots, a_n\}$, it follows without any fuss that the integrals which appear there evaluate to $\widetilde{f}(z)$ for every point z of the set $\mathbb{R} \sim \{a_1, a_2, \ldots, a_n\}$. As a matter of fact, even more is true: if $z = a_j$ for some j or if $z = \infty$, then the corresponding integral on the right-hand side of (9.75) or (9.76) still has the value $\widetilde{f}(z)$, although the integral in question becomes improper when $z = a_j$ for a value of j such that the corresponding α_j is less than 1 or when $z = \infty$. (In the latter case use $\gamma(t) = ti$, $1 \leq t \leq \infty$, for the path of integration.) The choice of i as the initial point of integration in (9.75) and (9.76) was quite arbitrary. Any other point of $\overline{H}$ would serve just as well, the only thing affected by a change being the constant B. For instance, one frequently encounters these formulas rewritten as

(9.77) $\quad f(z) = A\int_0^z (\zeta - a_1)^{\alpha_1-1}(\zeta - a_2)^{\alpha_2-1}\cdots(\zeta - a_n)^{\alpha_n-1}\,d\zeta + B$

when $a_n \neq \infty$, and

(9.78) $\quad f(z) = A\int_0^z (\zeta - a_1)^{\alpha_1-1}(\zeta - a_2)^{\alpha_2-1}\cdots(\zeta - a_{n-1})^{\alpha_{n-1}-1}\,d\zeta + B$

when $a_n = \infty$. It should be pointed out that the integrals which turn up in conjunction with the Schwarz-Christoffel formula can seldom be evaluated explicitly in terms of elementary functions. They are, however, amenable to numerical techniques of integration, a fact that frees the way to numerical approximation of the conformal mapping f.

Theorem 5.6 actually covers a situation somewhat more general than we claimed. Apart from a bit of tinkering with details, the proof that we've given can be used to establish the validity of the Schwarz-Christoffel formula for a conformal mapping f of H onto any bounded domain D whose boundary ∂D is the trajectory of a closed — but not necessarily simple — polygonal path $\gamma = [z_1, z_2] + [z_2, z_3] + \cdots + [z_{n-1}, z_n] + [z_n, z_1]$ with the

5. Conformal Mappings onto Polygons

property that for $j = 1, 2, \ldots, n-1$ the points z_j, z_{j+1}, and z_{j+2} are not collinear. (As earlier, put $z_{n+1} = z_1$.) This description permits points of ∂D to be "repeated vertices." The domain pictured in Figure 37 offers an example. In this more general setting f still admits a continuous extension

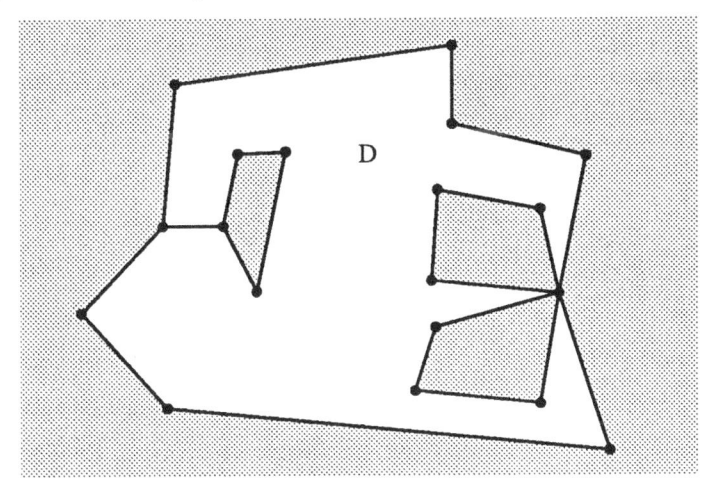

Figure 37.

$\widetilde{f}$ to $\widehat{H}$, though $\widetilde{f}$ will not be a homeomorphism in the presence of repeated vertices, and we can once more list the points mapped by $\widetilde{f}$ to the vertices in the manner $-\infty < a_1 < a_2 < \cdots < a_n \leq \infty$. We continue to express the angle at $\widetilde{f}(a_j)$ interior to D as $\alpha_j \pi$. Should $\widetilde{f}(a_j)$ happen to be a repeated vertex, we must take care that $\alpha_j \pi$ measures the particular interior angle corresponding to a_j, meaning the one whose sides are the images under $\widetilde{f}$ of $[a_{j-1}, a_j]$ and $[a_j, a_{j+1}]$. (So that this will make sense for $j = 1$ and $j = n$, we set $[a_0, a_1] = [-\infty, a_1]$ and agree that $[a_n, a_{n+1}] = [a_n, \infty]$ if $a_n \neq \infty$, while $[a_n, a_{n+1}] = [-\infty, a_1]$ if $a_n = \infty$.) Subject to these provisions Theorem 5.6 remains true as it stands for mappings to such generalized polygons.

While Theorem 5.6 does accurately convey the structure of a conformal mapping f of H onto the interior of a given closed polygon P, it by no means hands one a foolproof method for describing f explicitly, even in integral form. A snag often develops in trying to get a hold on the points $a_1, a_2, \ldots, a_n$ that are mapped by $\widetilde{f}$ to the vertices of P, something we must certainly do to make the Schwarz-Christoffel formula truly explicit. (After these points are identified, the computation of the constants A and B is usually quite manageable.) In light of Theorem 4.12 we are at liberty in constructing such a mapping f to specify the points that $\widetilde{f}$ transforms to any three selected vertices of P, modulo constraints imposed by orientation. As an example of this kind of normalization, a triple $(\zeta_0, \zeta_1, \zeta_\infty)$ of consecutive vertices of P can be chosen, one that displays the positive orientation relative to the interior of P, and it can then be required that

$\widetilde{f}$ take 0 to ζ_0, 1 to ζ_1, and ∞ to ζ_∞. For this mapping f we thus have $a_n = \infty$, $a_{n-1} = 1$, and $a_{n-2} = 0$. With that all flexibility in the matter comes to an end, for the values of $a_1, a_2, \ldots, a_{n-3}$ are dictated by the choices of a_{n-2}, a_{n-1}, and a_n. It can be a formidable problem to ascertain the values of these remaining a_j. Indeed, unless P is blessed with a considerable amount of symmetry, their exact determination is rarely possible. The role of symmetry is exemplified by the construction of a conformal mapping of H to a square in Example 5.2. Before this, we consider a more straightforward application of the Schwarz-Christoffel formula.

EXAMPLE 5.1. Find a conformal mapping of $H = \{z : \operatorname{Im} z > 0\}$ onto the interior D of the closed triangle T with vertices 0, 1, and ic, where $\operatorname{Im} c > 0$ (Figure 38).

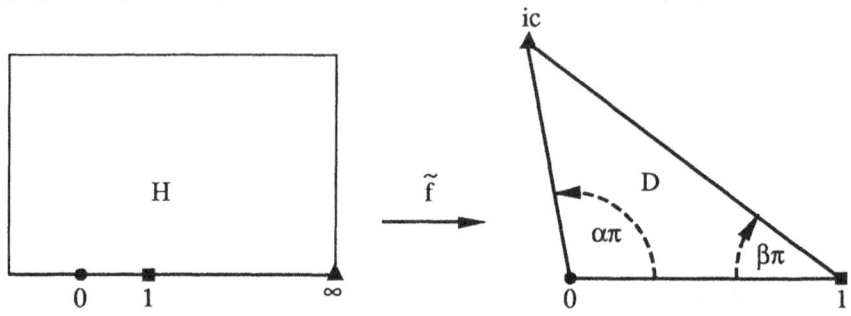

Figure 38.

Let $\alpha\pi$ and $\beta\pi$ be the interior angles of T at the vertices 0 and 1, respectively. We shall obtain the conformal mapping f of H onto D whose homeomorphic extension $\widetilde{f}$ to $\widehat{H}$ sends 0 to 0, 1 to 1, and ∞ to ic. In view of (9.78) and other comments made in the aftermath of Theorem 5.6, we can express $\widetilde{f}(z)$ for z belonging to $\widehat{H}$ in integral form,

$$\widetilde{f}(z) = A \int_0^z \zeta^{\alpha-1}(\zeta-1)^{\beta-1} \, d\zeta + B ,$$

for certain constants A and B. Since $\widetilde{f}(0) = 0$ and since the integral here tends to zero when z does, we must have $B = 0$; since $\widetilde{f}(1) = 1$, we then see that

$$A = \frac{1}{\int_0^1 \zeta^{\alpha-1}(\zeta-1)^{\beta-1} \, d\zeta} .$$

Therefore, one conformal mapping of H onto D is given by

$$f(z) = \frac{\int_0^z \zeta^{\alpha-1}(\zeta-1)^{\beta-1} \, d\zeta}{\int_0^1 \zeta^{\alpha-1}(\zeta-1)^{\beta-1} \, d\zeta} .$$

Mapping H conformally to a square offers more of a challenge.

5. Conformal Mappings onto Polygons

EXAMPLE 5.2. Determine a conformal mapping of $H = \{z: \operatorname{Im} z > 0\}$ onto the interior D of the closed square Q with vertices $0, 1, 1+i$, and i.

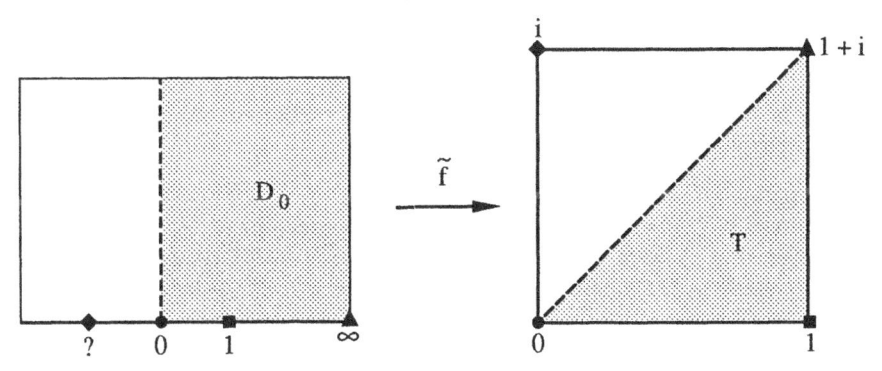

Figure 39.

We shall use Theorem 5.6 to identify the unique conformal mapping f of H onto D whose homeomorphic extension $\widetilde{f}$ to $\widehat{H}$ satisfies $\widetilde{f}(0) = 0$, $\widetilde{f}(1) = 1$, and $\widetilde{f}(\infty) = 1 + i$. Before we can apply the theorem, however, we must first settle the question: Which point of the interval $(-\infty, 0)$ gets mapped by $\widetilde{f}$ to i? We profit by the symmetry of Q to arrive at the answer. Consider the quarter plane $D_0 = \{z: \operatorname{Re} z > 0, \operatorname{Im} z > 0\}$ and the triangle T with vertices $0, 1$, and $1 + i$ (Figure 39). Let g be the unique conformal mapping of D_0 onto the interior of T whose homeomorphic extension $\widetilde{g}$ to $\widehat{D}_0$ has $\widetilde{g}(0) = 0$, $\widetilde{g}(1) = 1$, and $\widetilde{g}(\infty) = 1+i$. The mapping $\widetilde{g}$ transforms the positive imaginary axis I to the diagonal of Q with endpoints 0 and $1+i$. We can continue $\widetilde{g}$ across I by reflection. Because a square is symmetric about its diagonals, the reflection process generates a homeomorphism G of $\widehat{H}$ onto Q that carries H conformally onto D and satisfies $G(0) = 0$, $G(1) = 1$, $G(\infty) = 1 + i$, and $G(-1) = i$. The uniqueness statement in Theorem 4.12 makes it clear that $G = \widetilde{f}$. In particular, we learn that $\widetilde{f}(-1) = i$. We are now able to invoke (9.78) and obtain an integral representation for $\widetilde{f}$:

$$\widetilde{f}(z) = A \int_0^z \zeta^{-1/2}(\zeta - 1)^{-1/2}(\zeta + 1)^{-1/2}\, d\zeta + B$$

for z in $\widehat{H}$. The conditions $\widetilde{f}(0) = 0$ and $\widetilde{f}(1) = 1$ allow us to evaluate the constants A and B. This results in the identification of

$$f(z) = \frac{\int_0^z \zeta^{-1/2}(\zeta - 1)^{-1/2}(\zeta + 1)^{-1/2}\, d\zeta}{\int_0^1 \zeta^{-1/2}(\zeta - 1)^{-1/2}(\zeta + 1)^{-1/2}\, d\zeta}$$

as a conformal mapping mapping of H onto D.

Theorem 5.6 admits a multitude of extensions and refinements. As most of these require for their development ideas that go well beyond the elementary arguments which led to Theorem 5.6, we say nothing about them here. (Some can be found in the book of Nehari mentioned at the opening of this chapter.) The one exception we make in this regard is the derivation of an integral representation for a conformal mapping of the disk $\Delta(0,1)$ onto the interior of a polygon — not barring one with repeated vertices — since this representation can be extracted painlessly enough from Theorem 5.6.

Theorem 5.7. *Let f be a conformal mapping of the disk $\Delta = \Delta(0,1)$ onto the interior D of a closed polygon P in the complex plane, let $b_1, b_2, \ldots, b_n$ be a list of the points of $\partial \Delta$ that are transformed to the vertices of P by $\widetilde{f}$, the continuous extension of f to $\overline{\Delta}$, and let $\alpha_j \pi$ be the interior angle of P at the vertex $\widetilde{f}(b_j)$. There exist constants A and B such that*

$$f(z) = A \int_0^z \left(\frac{b_1 - \zeta}{1+\zeta}\right)^{\alpha_1 - 1} \left(\frac{b_2 - \zeta}{1+\zeta}\right)^{\alpha_2 - 1} \cdots \left(\frac{b_n - \zeta}{1+\zeta}\right)^{\alpha_n - 1} \frac{d\zeta}{(1+\zeta)^2} + B$$

for every z in Δ.

Proof. For convenience in the proof we assume the labeling of vertices has been arranged so that $-\pi < \operatorname{Arg} b_1 < \operatorname{Arg} b_2 < \cdots < \operatorname{Arg} b_n \leq \pi$. Noting that the Möbius transformation $\varphi(z) = (1+iz)/(1-iz)$ maps the upper half-plane $H = \{z : \operatorname{Im} z > 0\}$ onto Δ, we see that $g = f \circ \varphi$ delivers a conformal mapping of H onto D. Since $\varphi^{-1}(z) = i[(1-z)/(1+z)]$, the points mapped to the vertices of P under the extension $\widetilde{g} = \widetilde{f} \circ \varphi$ of g to $\widehat{H}$ are the points $-\infty < a_1 < a_2 < \cdots < a_n \leq \infty$ given by

$$a_j = \varphi^{-1}(b_j) = i\left(\frac{1-b_j}{1+b_j}\right).$$

Also, from the relation $f = g \circ \varphi^{-1}$ we infer that

(9.79) $$f'(z) = g'[\varphi^{-1}(z)][\varphi^{-1}]'(z) = -\frac{2ig'[\varphi^{-1}(z)]}{(1+z)^2}.$$

Suppose initially that $\operatorname{Arg} b_n < \pi$; i.e., $b_n \neq -1$. Then $a_n \neq \varphi^{-1}(-1) = \infty$. On the strength of Theorem 5.6 we can assert that g' takes the form

(9.80) $$g'(z) = A_0(z-a_1)^{\alpha_1 - 1}(z-a_2)^{\alpha_2 - 1} \cdots (z-a_n)^{\alpha_n - 1}$$

for some constant A_0.

We now fix j. A simple calculation reveals that

$$\varphi^{-1}(z) - a_j = \frac{2i(b_j - z)}{(1+b_j)(1+z)}.$$

The Möbius transformation $\psi(z) = (b_j - z)/(1+z)$ maps the circle $K = K(0,1)$ to $\widetilde{K} = L \cup \{\infty\}$, where L is a line through the origin. Since 0 and

5. Conformal Mappings onto Polygons

∞ are symmetric with respect to K, $b_j = \psi(0)$ and $-1 = \psi(\infty)$ must be symmetric with respect to $\tilde{K}$. This remark enables us to identify L as the perpendicular bisector of the segment joining b_j and -1. It follows easily that $\psi(\Delta)$ is one of the two open half-planes bounded by L — namely, the one that contains b_j. Since the complementary half-plane contains the real interval $(-\infty, 0]$, $\psi(\Delta)$ is disjoint from $(-\infty, 0]$. The consequence of this observation is that $\ell(z) = \text{Log}[(b_j - z)/(1 + z)]$ defines a function that is analytic in Δ. The function ℓ_0 given by $\ell_0(z) = \text{Log}[\varphi^{-1}(z) - a_j]$ is likewise analytic in Δ. Furthermore, in this disk

$$e^{\ell_0(z) - \ell(z)} = \frac{[\varphi^{-1}(z) - a_j](1 + z)}{(b_j - z)} = \frac{2i}{1 + b_j},$$

a constant, from which one concludes without difficulty that ℓ_0 and ℓ also differ by a constant in Δ, say $\ell_0(z) = \ell(z) + \gamma_j$. As a result,

$$[\varphi^{-1}(z) - a_j]^{\alpha_j - 1} = e^{(\alpha_j - 1)\ell_0(z)} = e^{(\alpha_j - 1)\ell(z) + (\alpha_j - 1)\gamma_j}$$

$$= c_j \left(\frac{b_j - z}{1 + z}\right)^{\alpha_j - 1}$$

for every z in Δ, where $c_j = e^{(\alpha_j - 1)\gamma_j}$. This is true for $j = 1, 2, \ldots, n$.

Referring to (9.79) and (9.80) we come to the conclusion that, at least when $b_n \neq -1$,

$$(9.81) \quad f'(z) = \frac{A}{(1 + z)^2} \left(\frac{b_1 - z}{1 + z}\right)^{\alpha_1 - 1} \left(\frac{b_2 - z}{1 + z}\right)^{\alpha_2 - 1} \cdots \left(\frac{b_n - z}{1 + z}\right)^{\alpha_n - 1}$$

for every z in Δ, where A is some constant. Should $b_n = -1$, Theorem 5.6 would give rise to an expression for $f'(z)$ similar to (9.81), but with the last factor in the product deleted. As a matter of fact, however, (9.81) still applies when $b_n = -1$, for in that event the last factor reduces to $(-1)^{\alpha_n - 1}$, a constant that can be absorbed into A. Integration of (9.81) results in the stated formula for f with $B = f(0)$. ∎

We again stress that the powers appearing in Theorem 5.7 are principal powers. In many books the conclusion of this result is reported as

$$(9.82) \quad f(z) = A \int_0^z (b_1 - \zeta)^{\alpha_1 - 1}(b_2 - \zeta)^{\alpha_2 - 1} \cdots (b_n - \zeta)^{\alpha_n - 1} d\zeta + B,$$

a formula obtained when, with the aid of (9.69), one formally simplifies the integrand found in Theorem 5.7. Unfortunately, if the powers present in (9.82) are interpreted as principal powers, consistent with the notational conventions in force in this book, the integrand in this formula may have discontinuities in Δ. Formula (9.82) is valid, of course, but only with the understanding that the notation $(b_j - z)^{\alpha_j - 1}$ is merely being employed as shorthand to indicate some branch of the $(\alpha_j - 1)$-power of the analytic function $p_j(z) = b_j - z$ in Δ.

6 Exercises for Chapter IX

6.1 Exercises for Section IX.1

6.1. Assuming that $f:U \to V$ belongs to the class $C^1(U)$ and $g:V \to \mathbb{C}$ is a member of $C^1(V)$, check that $J_{g \circ f}(z) = J_g[f(z)]J_f(z)$ holds for every point z of U.

6.2. Let $f:\mathbb{C} \to \mathbb{C}$ be defined by $f(z) = x^3 + iy^3$. Is f a univalent C^1-function? Is f a diffeomorphism?

6.3. Let c and d be complex numbers such that $|c| \neq |d|$. Show that $f(z) = cz + d\bar{z}$ defines a diffeomorphism of $\mathbb{C}$ onto itself, and determine f^{-1}. Under what conditions on c and d is f orientation-preserving?

6.4. If $D = \mathbb{C} \sim \{0\}$, show that for any non-zero real number λ the function $f(z) = |z|^{\lambda-1}z$ furnishes a diffeomorphism of D onto itself. For which λ is f orientation-preserving? What is f^{-1}? (*Hint.* Use (9.3) in computing J_f.)

6.5. Working from the definitions of conformality and anti-conformality, verify that a sense-preserving similarity transformation $f(z) = az + b$ gives a conformal mapping of the complex plane onto itself, but that a sense-reversing similarity $g(z) = a\bar{z} + b$ maps the plane anti-conformally onto itself.

6.6. Let $f:D \to \mathbb{C}$ be a diffeomorphism. Confirm that f is anti-conformal at a point z_0 of D if and only if its conjugate $\bar{f}$ is conformal at that point. Conclude that f is an anti-conformal mapping of D if and only if $\bar{f}$ is a conformal mapping of D.

6.7. For $j = 1, 2, \ldots, n$ let $f_j:D_j \to \mathbb{C}$ be either a conformal or an anti-conformal mapping of a domain D_j. Assuming that $f_j(D_j)$ is contained in D_{j+1} for $j = 1, 2, \ldots, n-1$, check that the composition $f = f_n \circ f_{n-1} \circ \cdots \circ f_1$ is likewise either conformal or anti-conformal. Show that f is conformal if and only if the number of anti-conformal factors in its make-up is even.

6.8. Let D be a domain in the complex plane, and let $f:D \to \mathbb{C}$ be a diffeomorphism. Prove that the following statements concerning a point z_0 of D are equivalent: (i) $\bar{f}$ is differentiable at z_0; (ii) f is isogonal at z_0 and $J_f(z_0) < 0$; (iii) f is anti-conformal at z_0.

6.9. Let $f:D \to \mathbb{C}$ be a diffeomorphism, and let z_0 be a point of D. Given the knowledge that $\lim_{z \to z_0} |f(z) - f(z_0)|/|z - z_0|$ exists, deduce that f is either conformal or anti-conformal at z_0. (*Hint.* Exercise III.6.65 is pertinent to this exercise.)

6.10. Let $D = \mathbb{C} \sim (-\infty, 0]$ and let $1 < \lambda < \infty$. Show that the function $f:D \to \mathbb{C}$ defined by $f(z) = z^\lambda$ is not a univalent function, but is a locally conformal mapping of D. What is its range?

6.11. Find a conformal mapping of the half-plane $D = \{z : x > 0\}$ onto the domain $D' = \{w : |\operatorname{Arg} w| < \lambda\pi\}$, where $0 < \lambda \leq 1$.

6.12. Construct a conformal mapping of the domain $D = \{z : x > |y|\}$ onto the disk $D' = \Delta(0,1)$.

6.13. Determine a conformal mapping of the semi-infinite strip $D = \{z : x < 0, 0 < y < 1\}$ onto the quarter-disk $D' = \{w : |w| < 1, v > |u|\}$.

6.14. Construct a conformal mapping of $D = \{z : |x| + |y| < \sqrt{2}/2\}$ onto the domain $D' = \{w : 1 < |w| < e^\pi, v > 0\}$.

6.15. Find a conformal mapping of the strip $D = \{z : |\operatorname{Re} z| < \pi/2\}$ onto itself that transforms the real interval $(-\pi/2, \pi/2)$ to the full imaginary axis.

6.16. Build a conformal mapping of the disk $D = \Delta(0,1)$ onto the domain $D' = \{w : u^2 - v^2 < 1\}$. (*Hint.* Remember Example III.3.7.)

6.17. Exhibit a conformal mapping of the disk $D = \Delta(0,1)$ onto the domain D' inside the cardioid whose polar equation is $r = 1 + \cos\theta$. (*Hint.* Recall Exercises I.4.55 and VIII.5.75.)

6.18. Find a conformal mapping of the half-plane $D = \{z : y > 0\}$ onto the domain $D' = \{w : u < v^2\}$. (*Hint.* Recall Exercise I.4.53.)

6.19. Display a conformal mapping of the strip $D = \{z : |y| < \pi/2\}$ onto the domain $D' = \{w : v < u^2\}$.

6.20. Construct a conformal mapping f of the disk $D = \Delta(0,1)$ onto the domain $D' = \mathbb{C} \sim (-\infty, -1/4]$ with $f(0) = 0$ and $f'(0) = 1$.

6.21. Let $D = \Delta(0,1)$, and let D' be the component of the open set $\{w : |w^2 - 1| < 1\}$ that contains the point -1. Find a conformal mapping of D onto D'. (*Hint.* Recall Exercise V.8.40.)

6.22. Let $a > 0$ and $b > 0$. Determine a conformal mapping of the domain $D = \Delta(\infty, 1)$ onto $D' = \widehat{\mathbb{C}} \sim \{w : (u/a)^2 + (v/b)^2 \leq 1\}$. (*Hint.* The function $f(z) = cz + dz^{-1}$ for properly chosen c and d will do the job.)

6.2 Exercises for Section IX.2

6.23. If $f(z) = z/(z+1)$, calculate $g = f \circ f \circ \cdots \circ f$ (n terms), the nth iterate of f. (*Hint:* $f = f_A$ for $A = ?$)

6.24. Confirm that under the correspondence $A \to f_A$ of 2×2 non-singular matrices to Möbius transformations it is true that $AB \to f_A \circ f_B$ and $A^{-1} \to (f_A)^{-1}$.

6.25. Let A and B be 2×2 matrices such that $\det A = \det B = 1$ and $f_A = f_B$. Show that either $A = B$ or $A = -B$.

6.26. Express f in normalized form, and identify its fixed points: (i) $f(z) =$

$(z-3i)/(iz-1)$; (ii) $f(z) = (4z+i)/(4iz)$; (iii) $f(z) = (3z-4i)/(5iz+7)$.

6.27. Check that, when written in normalized form, a Möbius transformation f that maps the disk $\Delta(0,1)$ onto itself has the appearance $f(z) = (az+b)/(\bar{b}z+\bar{a})$, where $|a|^2 - |b|^2 = 1$.

6.28. Calculate the cross-ratios: (i) $[1, i, -1, -i]$; (ii) $[1, -1, i, -i]$; (iii) $[0, \infty, 1+i, 1-i]$; (iv) $[0, a, b, \infty]$ for finite non-zero a and b, $a \neq b$; (v) $[c, c^{-1}, 1, 0]$ for finite $c \neq 0, \pm 1$; (vi) $[z, z^{-1}, \bar{z}, \infty]$ for finite, non-real z, $|z| \neq 1$; (vii) $[-1, 1, w, w^{-1}]$ for $w \neq \pm 1$; (viii) $[e^{i\theta}, -e^{-i\theta}, -e^{i\theta}, e^{-i\theta}]$ for $\theta \in (0, \pi/2)$.

6.29. Let $\lambda = [z_1, z_2, z_3, z_4]$. Express in terms of λ the various cross-ratios that arise when the points z_1, z_2, z_3, and z_4 are permuted. (N.B. There are twenty-four possible permutations to deal with, but only six different values turn up among the associated cross-ratios. *Hint.* It suffices to consider $z_1 = 1, z_2 = 0$, and $z_3 = \infty$. Why?)

6.30. Substantiate the following claim: four distinct points z_1, z_2, z_3, and z_4 in $\widehat{\mathbb{C}}$ lie on a circle in $\widehat{\mathbb{C}}$ if and only if the cross-ratio $[z_1, z_2, z_3, z_4]$ is real. (*Hint.* Reduce the problem to the case $z_1 = 1, z_2 = 0$, and $z_3 = \infty$.)

6.31. Refine the preceding exercise as follows: four distinct points z_1, z_2, z_3, and z_4 in $\widehat{\mathbb{C}}$ lie on a circle in $\widehat{\mathbb{C}}$ and are so arranged on that circle that z_1 and z_3 separate z_2 and z_4 if and only if the condition

$$|[z_1, z_3, z_2, z_4]| + |[z_3, z_1, z_2, z_4]| = 1$$

is fulfilled. (*Hint.* What does this condition boil down to when $z_1 = 0, z_3 = 1$, and $z_4 = \infty$?)

6.32. Let z_1, z_2, z_3, and z_4 be distinct points of $\widehat{\mathbb{C}}$, finite except possibly for z_4, let K be the circle in $\widehat{\mathbb{C}}$ on which z_1, z_2, and z_4 lie, and let $\widetilde{K}$ be the circle in $\widehat{\mathbb{C}}$ that passes through z_1, z_3, and z_4. Confirm that $\text{Arg}([z_1, z_2, z_3, z_4]) = \theta(A, \widetilde{A})$, where A is the arc of K from z_1 to z_2 that does not contain z_4 and $\widetilde{A}$ is the arc of $\widetilde{K}$ from z_1 to z_3 that misses z_4.

6.33. Construct a Möbius transformation f with the specified effect: (i) f sends i to 1, 1 to ∞, and ∞ to i ; (ii) f transforms $K = K(i,1)$ to $\widetilde{K} = \{w : (1+i)w + (1-i)\bar{w} = 0\}$; (iii) f takes $K = \mathbb{R} \cup \{\infty\}$ to $\widetilde{K} = K(0,1)$, leaving the point -1 fixed; (iv) f maps $D = \Delta(0,1)$ to $D' = \{w : \text{Im}\, w > 0\}$, with $f(0) = 1 + 2i$; (v) f maps $K(0,1)$ to itself and $K(1/4, 1/4)$ to $K(0, r)$ for some $r < 1$.

6.34. Let $f(z) = (az+b)/(cz+d)$ be a normalized Möbius transformation for which $c \neq 0$. The circle $K(f) = \{z : |cz+d| = 1\}$ is called the *isometric circle* of f for the reason that f preserves the distance between points of $K(f)$ — i.e., $|f(z_2) - f(z_1)| = |z_2 - z_1|$ whenever z_1 and z_2 lie on $K(f)$. Certify that this is so. Show also that f transforms $K(f)$ to $K(f^{-1})$. To what set does f map the disk inside of $K(f)$?

6.35. Let K be a circle in $\widehat{\mathbb{C}}$. Certify that the reflection ρ_K defined by (9.23) (respectively, (9.24)) has the geometric effect described in the text.

6.36. How do cross-ratios behave under anti-Möbius transformations?

6.37. Show that the fixed point set of an anti-Möbius transformation f either contains at most two points or is a circle K in $\widehat{\mathbb{C}}$. Furthermore, prove that the latter occurs if and only if $f = \rho_K$, the reflection in K.

6.38. Let K be a circle in $\widehat{\mathbb{C}}$, and let z and w be different points of $\widehat{\mathbb{C}} \sim K$. Prove that z and w are symmetric with respect to K if and only if every circle in $\widehat{\mathbb{C}}$ that passes through both z and w is perpendicular to K. (*Hint.* What is the situation when $K = \mathbb{R} \cup \{\infty\}$?)

6.39. If z_1, z_2, and z_3 are distinct points of $\widehat{\mathbb{C}}$, verify that through z_1 passes one and only one circle in $\widehat{\mathbb{C}}$ with respect to which z_2 and z_3 are symmetric.

6.40. Let K and K' be circles in $\widehat{\mathbb{C}}$. If z_0 belongs to K, z_0' to K', w_0 to $\widehat{\mathbb{C}} \sim K$, and w_0' to $\widehat{\mathbb{C}} \sim K'$, then there is a unique Möbius transformation that maps K to K', z_0 to z_0', and w_0 to w_0'. Verify this.

6.41. Let $f(z) = (az+b)/(cz+d)$ be a normalized Möbius transformation, and let $H = \{z : \operatorname{Im} z > 0\}$. Demonstrate that $f(H) = H$ if and only if a, b, c, and d are real numbers. (*Hint.* For the sufficiency, look at $\operatorname{Im} f(z)$. For the necessity, consider $g(z) = \overline{f(\bar{z})}$.)

6.42. Show that the conformal self-mappings f of the half-plane $H = \{z : \operatorname{Im} z > 0\}$ are precisely the restrictions to H of the Möbius transformations that, when written in normalized form, have real coefficients. (*Hint.* Let $f : H \to H$ be a conformal mapping of H onto itself. Show first that f is the restriction to H of a Möbius transformation by considering $g = h^{-1} \circ f \circ h$ in $\Delta(0,1)$, where $h(z) = (z+i)/(iz+1)$.)

6.43. Demonstrate that any univalent function $f : \widehat{\mathbb{C}} \to \widehat{\mathbb{C}}$ that preserves the class of circles in $\widehat{\mathbb{C}}$ — i.e., if K is a circle in $\widehat{\mathbb{C}}$, then so is $f(K)$ — is either a Möbius transformation or an anti-Möbius transformation. (*Hint.* Start by looking at a function f of the above kind that also fixes $0, 1$, and ∞. Verify that f is either the identity mapping of $\widehat{\mathbb{C}}$ or the reflection in the real axis. Try first to pin down the images under f of the real and imaginary axes, the line with equation $x = 1$, and the circle $K(1/2, 1/2)$. Assume, if need be, that f is continuous.)

6.44. Let S be the standard unit sphere in $\mathbb{R}^3$, and let $\pi : S \to \widehat{\mathbb{C}}$ be stereographic projection. If $F : S \to S$ is the mapping created by reflecting S in a plane P that passes through its center, then the mapping $f = \pi \circ F \circ \pi^{-1}$ of $\widehat{\mathbb{C}}$ is just the reflection ρ_K, where $K = \pi(S \cap P)$. Prove this. (*Hint.* Use Exercises 6.41 and 6.37 along with Exercise VIII.5.81. Notice that f could not be a Möbius transformation. Why not?)

6.45. As in the preceding exercise, let $\pi\colon S \to \widehat{\mathbb{C}}$ be the stereographic projection, but now let $F\colon S \to S$ be the antipodal map; i.e., $F(p) = -p$. Check that $f = \pi \circ F \circ \pi^{-1}$ is another anti-Möbius transformation — namely, $f(z) = -1/\bar{z}$.

6.46. Express f in normalized form and classify it; if f is non-parabolic, represent it in multiplier-fixed point format: (i) $f(z) = (z+i)/(iz+1)$; (ii) $f(z) = (2z+1)/(2z+3)$; (iii) $f(z) = (2z-1)/(2z-3)$; (iv) $f(z) = z/(iz+1)$.

6.47. Confirm that a Möbius transformation f is parabolic if and only if it is conjugate to the translation $g(z) = z + 1$, that f is hyperbolic if and only if it is conjugate to a non-trivial dilation with respect to the origin, and that f is elliptic if and only if it is conjugate to a non-trivial rotation about the origin.

6.48. Verify that conjugate Möbius transformations f and g always fall into the same category in the standard classification.

6.49. Assume that f is a Möbius transformation with exactly two fixed points z_1 and z_2. Let g be a Möbius transformation with the property that $g \circ f = f \circ g$. Show that either g fixes z_1 and z_2 or it interchanges these points, the latter case only being possible if g has order two.

6.50. Let f and g be Möbius transformations different from the identity. Assuming that f and g share the same set of fixed points, verify that $g \circ f = f \circ g$. Conversely, prove that the condition $g \circ f = f \circ g$, supplemented by the information that neither f nor g has order two, implies that f and g fix exactly the same points. Show by way of example that $g \circ f = f \circ g$ need not force a coincidence of fixed points when at least one of the transformations has order two. (*Hint.* First treat the case where f is parabolic. In the case of non-parabolic f Exercise 6.47 is relevant.)

6.51. Let S be the standard unit sphere in $\mathbb{R}^3$, let $F\colon S \to S$ be a rotation of S about an axis through the origin, and let $f = \pi \circ F \circ \pi^{-1}$, where $\pi\colon S \to \widehat{\mathbb{C}}$ is stereographic projection. Show that f is a Möbius transformation, one which appears in normalized form as $f(z) = (az - \bar{b})/(bz + \bar{a})$ with $|a|^2 + |b|^2 = 1$. Conversely, verify that any Möbius transformation which owns such a normalized form is generated in the above way by a rotation of S. (*Hint.* For topological reasons the mapping f must be sense-preserving. Use this information, without proof if necessary, in showing that f is a Möbius transformation. Then check that, unless f reduces to the identity transformation of $\widehat{\mathbb{C}}$, f is necessarily an elliptic transformation whose fixed points z_1 and z_2 are related by $z_2 = (\bar{z}_1)^{-1}$ and whose multiplier can be taken to be $\kappa = e^{i\theta}$, where θ is the rotation angle of the mapping F.)

6.52. Let f and g be conjugate Möbius transformations, say $g = h^{-1} \circ f \circ h$ for a Möbius transformation h. Verify that K is a g-invariant circle in $\widehat{\mathbb{C}}$ if and only if $h(K)$ is invariant under f.

6.53. Let $f(z) = z + b$, where $b \neq 0$, and let K be an f-invariant circle in $\widehat{\mathbb{C}}$. Give a careful proof that $K = L \cup \{\infty\}$, in which L is a line parallel to the one through 0 and b.

6.54. Given that $f(z) = \kappa z$, where $\kappa > 0$ and $\kappa \neq 1$, confirm that the invariant circles for f are the circles in $\widehat{\mathbb{C}}$ that pass through both 0 and ∞.

6.55. Let $f(z) = e^{i\theta}z$, where $\theta \neq 0$ and $-\pi < \theta \leq \pi$, and let K be an invariant circle for f. Assuming that $\theta \neq \pi$, prove that K must be a genuine circle centered at the origin. In case $\theta = \pi$, show that K is either a circle centered at the origin or a circle in $\widehat{\mathbb{C}}$ of the kind $L \cup \{\infty\}$, where L is a line through the origin.

6.56. It was claimed in the text that a loxodromic Möbius transformation f has no invariant circles, except when it has a negative real multiplier κ. Substantiate this claim. In addition, show that when $\kappa < 0$ the f-invariant circles are the circles in $\widehat{\mathbb{C}}$ that pass through both the fixed points of f.

6.57. Sketch a representative set of invariant circles for f: (i) $f(z) = (2z - 1)/z$; (ii) $f(z) = (2z - 1)/(-z + 1)$; (iii) $f(z) = (2z - 3)/(z - 1)$.

6.58. Let $\widetilde{K}_0$ be a circle in $\widehat{\mathbb{C}}$, and let f be the reflection in $\widetilde{K}_0$. Prove that a circle K in $\widehat{\mathbb{C}}$ is an invariant circle for f if and only if K belongs to the family of circles in $\widehat{\mathbb{C}}$ that are orthogonal to $\widetilde{K}_0$. (*Hint.* Reduce the problem to the case $\widetilde{K}_0 = \mathbb{R} \cup \{\infty\}$.)

6.59. Construct a conformal mapping of the open quarter-disk $D = \{z : |z| < 1, x > 0, y > 0\}$ onto the open quadrant $D' = \{w : u > 0, v > 0\}$.

6.60. Let $D = \Delta_1 \cap \Delta_2$, where Δ_1 and Δ_2 are the open disks of equal radius whose bounding circles meet an angle (exterior to D) $\alpha\pi, 0 < \alpha < 1$, at the points i and $-i$. Construct a conformal mapping of D onto $D' = \Delta(0, 1)$.

6.61. Let $D = \Delta(0, 1) \sim [\overline{\Delta}(-1+i, 1) \cup \overline{\Delta}(-1-i, 1)]$. Map D conformally to the half-plane $D' = \{w : \operatorname{Im} w > 0\}$. (*Hint.* Start a construction by sending 1 to 0, 0 to 1, and -1 to ∞ with a Möbius transformation.)

6.62. Let $D = \Delta_1 \cap \Delta_2 \cap \Delta_3$, where Δ_1, Δ_2, and Δ_3 are open disks of the same radius whose boundary circles K_1, K_2, and K_3 obey the following conditions: K_1 intersects K_2 orthogonally at $i\sqrt{3}$, K_1 intersects K_3 orthogonally at -1, and K_2 intersects K_3 orthogonally at 1. Determine a conformal mapping of D onto the half-plane $D' = \{w : \operatorname{Im} w > 0\}$.

6.3 Exercises for Section IX.3

6.63. Suppose that f is a conformal mapping of a plane domain D onto $\Delta(0, 1)$; assume it to be known that $f(z_0) = 0$ and $f'(z_0) = i$ at some specified point z_0 of D. Construct from f the conformal mapping g of D onto $\Delta(0,1)$ that obeys the conditions $g(z_0) = 1/2$ and $g'(z_0) > 0$. Find $g'(z_0)$.

6.64. If D is a simply connected plane domain, $D \neq \mathbb{C}$, and if z_0 is a point of D, then there is a unique radius $r > 0$ such that a conformal mapping f of D onto $\Delta(0,r)$ exists which satisfies $f(z_0) = 0$ and $f'(z_0) = 1$. Establish this fact.

6.65. Let $D(\neq \mathbb{C})$ be a simply connected plane domain that contains the origin and is symmetric with respect to it — the latter condition requires that $-z$ belong to D whenever z does — and let f be a conformal mapping of D onto $\Delta(0,1)$ with $f(0) = 0$. Demonstrate that f is necessarily an odd function. (*Hint.* Exploit the uniqueness assertion in the Riemann mapping theorem.)

6.66. Let $D(\neq \mathbb{C})$ be a simply connected plane domain that is symmetric with respect to a line L through the origin (i.e., $\rho(D) = D$, where ρ is the reflection in $K = L \cup \{\infty\}$), let z_0 be a point of $L \cap D$, and let f be the conformal mapping of D onto $\Delta(0,1)$ that has $f(z_0) = 0$ and $f'(z_0) > 0$. Show that f is symmetric about L, in the sense that $f[\rho(z)] = \rho[f(z)]$ for every point z of D. Using this information, conclude that $L \cap D$ is an open interval whose image under f is the set $L \cap \Delta(0,1)$. (*Hint.* Look at $g = \rho \circ f \circ \rho$ in D.)

6.67. The domain D inside the ellipse with equation $(x/a)^2 + (y/b)^2 = 1$ is mapped conformally onto $\Delta(0,1)$ by a function f. Given the information that $f(0) = 0$ and $f'(0) > 0$, ascertain the images under f of the (open) axes of the ellipse.

6.68. Let D be the interior of the square with vertices $1, i, -1$, and $-i$. Given that f maps D conformally onto $\Delta(0,1)$ with $f(0) = 0$ and $f'(0) < 0$, identify the images under f of the (open) diagonals of D and of the two (open) line segments that join the midpoints of its opposite sides.

6.69. If D is the interior of the triangle whose vertices are $1, e^{2\pi i/3}$, and $e^{4\pi i/3}$, determine the images of the (open) altitudes of D under the function f that maps D conformally onto $\Delta(0,1)$, fixes 0, and has $\operatorname{Arg} f'(0) = \pi/2$.

6.70. A domain D in $\widehat{\mathbb{C}}$ is called simply connected if either of the statements (ii) or (iii) in Theorem 3.6 is true of D. (The interpretation of "homologous in D" would require some serious reworking before our originally adopted definition of simple connectivity would make sense for domains that contain the point ∞.) Let D and D' be simply connected domains in $\widehat{\mathbb{C}}$ such that each of the sets $\widehat{\mathbb{C}} \sim D$ and $\widehat{\mathbb{C}} \sim D'$ contains at least two points, let z_0 be a point of D, and let z'_0 belong to D'. Establish the existence of a conformal mapping of D onto D' with $f(z_0) = z'_0$. Is the same true if both $\widehat{\mathbb{C}} \sim D$ and $\widehat{\mathbb{C}} \sim D'$ are one-point sets? What if $\widehat{\mathbb{C}} \sim D$ contains at least two points but $\widehat{\mathbb{C}} \sim D'$ has fewer than two elements?

6.4 Exercises for Section IX.4

6.71. Under the assumption that $\ell(\gamma) \geq c > 0$ for every path γ in $\Gamma[E, F:G]$, obtain the estimate $M[E, F:G] \leq c^{-2}A(G)$.

6.72. Let $G = \{z : r_0 \leq |z - z_0| \leq r_1\}$ with $0 < r_0 < r_1 < \infty$, let $E = K(z_0, r_0)$, and let $F = K(z_0, r_1)$. Verify that $M[E, F:G] = 2\pi[\text{Log}(r_1/r_0)]^{-1}$.

6.73. If $G = \overline{\Delta}(z_0, r)$, $E = \{z_0\}$, and $F = K(z_0, r)$, then $M[E, F:G] = 0$. Prove this.

6.74. Suppose that f is the reflection in the real axis. Demonstrate that $M[E, F:G] = M[f(E), f(F): f(G)]$ for any configuration $[E, F:G]$. Employ this information to check that the modulus $M[E, F:G]$ is also invariant under anti-conformal mappings.

6.75. Let L be a line in the complex plane, and let G and G^* be the complementary closed half-spaces determined by L. Show that

$$M[E_1, F:G] + M[E_2, F:G^*] \leq M[E_1 \cup E_2, F:\mathbb{C}]$$

for any configurations $[E_1, F:G]$ and $[E_2, F:G^*]$ with F contained in L.

6.76. Suppose that E lies in the interior of $G = \{z : \text{Im } z \geq 0\}$ and that E^* is the reflection of E in the real axis. Certify that $M[E \cup E^*, \mathbb{R}:\mathbb{C}] = 2M[E, \mathbb{R}:G]$. (*Hint*. Exercises 6.72 and 6.73 lead to an inequality in one direction. There is an easy way to construct an admissible density $\widetilde{\rho}$ for $[E \cup E^*, \mathbb{R}:\mathbb{C}]$ from an admissible density ρ for $[E, \mathbb{R}:G]$. This gives an equality in the other direction.)

6.77. Suppose that E lies in the interior of $G = \{z : \text{Im } z \geq 0\}$ and that E^* is the reflection of E in the real axis. Show that $M[E, \mathbb{R}:G] = 2M[E, E^*:\mathbb{C}]$. (*Hint*. To get an inequality in one direction consider $\rho \in \text{Adm}[E, \mathbb{R}:G]$, define $\widetilde{\rho}$ in $\mathbb{C}$ by $\widetilde{\rho}(z) = \rho(z)$ if $z \in G$ and $\widetilde{\rho}(z) = \rho(\overline{z})$ if $z \in G^*$, and show that $\widetilde{\rho}/2 \in \text{Adm}[E, E^*:\mathbb{C}]$. Conversely, given $\rho \in \text{Adm}[E, E^*:\mathbb{C}]$, define $\widetilde{\rho}: G \to [0, \infty)$ by $\widetilde{\rho}(z) = \rho(z) + \rho(\overline{z})$ and confirm that $\widetilde{\rho} \in \text{Adm}[E, \mathbb{R}:G]$. This will produce an inequality in the opposite direction.)

6.78. Assume that D is a domain in the complex plane and that Γ is an arbitrary family of piecewise smooth paths in D. Define an extended real number $M[\Gamma; D]$ as follows: $M[\Gamma:D] = \inf\{A_\rho(D) : \rho \in \text{Adm}[\Gamma:D]\}$ if $\text{Adm}[\Gamma:D] \neq \phi$ and $M[\Gamma:D] = \infty$ otherwise. Here, as one might anticipate, $\text{Adm}[\Gamma:D]$ indicates the class of densities ρ in D such that $\int_\gamma \rho(z)|dz| \geq 1$ for every path γ in Γ. (This is obviously a generalization of the conformal modulus of a configuration $[E, F:D]$, one in which the paths involved are no longer constrained to begin and end in preordained sets.) Show that $M[\Gamma\colon D]$ is conformally invariant: if $f\colon D \to \mathbb{C}$ is a conformal mapping, then $M[\Gamma\colon D] = M[f(\Gamma)\colon f(D)]$, where $f(\Gamma) = \{f \circ \gamma : \gamma \in \Gamma\}$.

6.79. A plane domain D is mapped conformally by f onto $\Delta(0,1)$. One is told that $|f'(z)| \geq c$ throughout D, where $c > 0$ is a constant. Derive the bound $M[\Gamma\colon D] \leq \pi(cL)^{-2}$ for any family Γ of piecewise smooth paths in D, in which $L = \inf\{\ell(\gamma) : \gamma \in \Gamma\}$.

6.80. Let $D = \{z : r_0 < |z| < r_1\}$, where $0 \leq r_0 < r_1 \leq \infty$, and let Γ be the collection of all closed, piecewise smooth paths γ in D with the property that $n(\gamma, 0) \neq 0$. Demonstrate that $M[\Gamma\colon D] = (2\pi)^{-1}\operatorname{Log}(r_1/r_0)$ if $0 < r_0 < r_1 < \infty$, while $M[\Gamma\colon D] = \infty$ if $r_0 = 0$ or $r_1 = \infty$.

6.81. If f is a conformal mapping of an annulus $D = \{z : r < |z| < 1\}$ onto an annulus $D' = \{w : s < |w| < 1\}$, where $0 \leq r < 1$ and also $0 \leq s < 1$, then necessarily $r = s$. Prove this. (*Hint*. Let Γ be the family of closed, piecewise smooth paths γ in D with the property that $n(\gamma, 0) \neq 0$. To start, show that any path $\beta = f \circ \gamma$, with γ a member of Γ, must have $n(\beta, 0) \neq 0$. If not, β would be homologous to zero in D'. Why? And why is this unacceptable?)

6.82. Show that any domain D in $\widehat{\mathbb{C}}$ such that ∂D consists of two disjoint circles in $\widehat{\mathbb{C}}$ can be mapped by a Möbius transformation to $D' = \{z : r < |z| < 1\}$ for a unique r in the interval $(0,1)$.

6.83. Given that D is a Jordan domain in the complex plane and that E and F are disjoint connected subsets of D with the feature that $\widehat{E}$ and $\widehat{F}$ have a point of ∂D in common, prove that $M[E, F\colon D] = \infty$. Show that, if $D = \{z : 0 < y < 1\}$, $E = \{z \in D : x \geq 1\}$, and $F = \{z \in D : x \leq -1\}$, then $M[E, F\colon D] < \infty$, even though $\infty \in \widehat{E} \cap \widehat{F}$. Are these two results in conflict?

6.84. Modify the proof of Lemma 4.3 to upgrade that result as follows: if $\Delta = \Delta(0,1)$ and $E = \overline{\Delta}(0, 1/2)$, if z_1 and z_2 are distinct points of $\partial\Delta$, and if $0 < r < |z_1 - z_2|/2$, then there is a constant $c > 0$ — it will depend on $|z_1 - z_2|$ and r — with the property that $M[E, F\colon \Delta] \geq c$ for every connected subset F of $\Delta \sim E$ which intersects both $\Delta(z_1, r)$ and $\Delta(z_2, r)$.

6.85. Let f be a conformal mapping of a bounded domain D onto $\Delta = \Delta(0, 1)$. Establish the following extension result of Raimo Näkki: the function f extends to a continuous mapping $\widetilde{f}$ of $\overline{D}$ onto $\overline{\Delta}$ if and only if D has the property (∗) $\inf\{\ell(\gamma) : \gamma \in \Gamma[E, F\colon D]\} = 0$ whenever E and F are disjoint connected sets in D for which $\overline{E} \cap \overline{F} \cap \partial D \neq 0$. (N.B. The domain $D = \Delta \sim [0, 1)$, for example, does not have the aforementioned property. *Hint*. For the necessity use Exercise 6.69 and Lemma 4.4. As to the sufficiency, it is enough to check that $\lim_{\zeta \to z} f(\zeta)$ exists for every point z of ∂D, for an analogue of Lemma 4.1 then applies. Suppose this not to be so at some z and argue to a contradiction. Use Exercise 6.81 and the following consequence of (∗): if $E_0 = f^{-1}[\overline{\Delta}(0,1/2)]$, if E and F are disjoint connected sets in $D \sim f^{-1}[\overline{\Delta}(0,3/4)]$ with $\overline{E} \cap \overline{F} \cap \partial D \neq \phi$, and if $c > 0$, then there must be a path γ in $\Gamma[E, F: D]$ for which $M[E_0, F_0: D] < c$, where $F_0 = |\gamma|$.)

6.86. Let D be the domain inside the cardioid whose polar equation is $r = 2 + 2\cos\theta$. Find an integral representation for the solution of the Dirichlet problem in D with boundary data h.

6.5 Exercises for Section IX.5

6.87. A function f is analytic in a Jordan domain D in the complex plane. It is known that $\lim_{\zeta \to z} f(\zeta) = 0$ for all points z belonging to C, a connected subset of $\widehat{\partial D}$ which contains more than one point. Deduce that $f(z) = 0$ for every z in D. (*Hint.* First do the case where D is a half-plane.)

6.88. Let $D = \{z: 0 < x < 1, y > 0\}$, and let $f: \overline{D} \to \mathbb{C}$ be a bounded continuous function that is analytic in D and real-valued on the interval $(0, 1)$. Assume that $|f(iy)| \leq m_0$ and $|f(1 + iy)| \leq m_1$ for all $y \geq 0$, where m_0 and m_1 are constants. Prove that $|f(x+iy)| \leq m_0^{1-x} m_1^x$ holds whenever $0 < x < 1$ and $y \geq 0$.

6.89. A plane domain D is symmetric about the real axis, and G is one of the components of $D \sim \mathbb{R}$. Given that a function $f = u + iv$ is analytic in G and obeys the condition $\lim_{\zeta \to z} u(\zeta) = 0$ for every point z of $I = D \cap \mathbb{R}$, establish the existence of an analytic function $F: D \to \mathbb{R}$ that coincides with f in G. (*Hint.* First prove that $\lim_{\zeta \to z} f(\zeta)$ exists for every z in I. For this use Theorem VI.3.4 and the fact that harmonic functions always have harmonic conjugates in disks.)

6.90. Make the following changes in the hypotheses of Theorem 5.3: assume that D is now a domain in $\widehat{\mathbb{C}}$ and that $f: G \cup I \to \widehat{\mathbb{C}}$ is a continuous function which is meromorphic in G. Keeping the remaining hypotheses as they stand, prove that the function F defined there is still meromorphic.

6.91. Suppose that $D = \{z: 0 < x < a, 0 < y < b\}$ and that $D' = \{w: 0 < u < c, 0 < v < d\}$. Establish the fact that there is a conformal mapping f of D onto D' whose homeomorphic extension $\widetilde{f}$ to $\overline{D}$ satisfies $\widetilde{f}(0) = 0$, $\widetilde{f}(a) = c$, $\widetilde{f}(ib) = id$, and $\widetilde{f}(a+ib) = c+id$ if and only if $a/b = c/d$. (*Hint.* Assuming that a mapping f of the given description exists, prove that it can be extended to a conformal self-mapping of the complex plane.)

6.92. Let $D = \{z: |z| > 1, |x| < 1, y > 0\}$, let $H = \{w: v > 0\}$, and let f be the homeomorphism of $\widehat{D}$ onto $\widehat{H}$ that maps D conformally onto H and satisfies $f(-1) = -1$, $f(1) = 1$, and $f(\infty) = \infty$. Show that f can be extended to an analytic function $F: H \to \mathbb{C}$ whose range is $\mathbb{C} \sim \{1, -1\}$.

6.93. Let f be a conformal mapping of $H = \{z: \text{Im } z > 0\}$ onto the interior of a closed polygon P, and let z_0 be a point of ∂H. We can see from the proof of Lemma 5.4 that, if $\widetilde{f}(z_0)$ is not a vertex of P, then $\lim_{z \to z_0} |f'(z)|$ exists and is not zero. Show that $\lim_{z \to z_0} |f'(z)| = \infty$ in case $\widetilde{f}(z_0)$ is a vertex of P at which the interior angle is less than π, whereas $\lim_{z \to z_0} |f'(z)| = 0$ if $\widetilde{f}(z_0)$ is a vertex of P at which the interior angle is more than π. Verify, also, that $|f'(z)| \to 0$ as $z \to \infty$ through H. (*Hint.* For the last part remember relation (9.69).)

6.94. Referring to Example 5.2, show that $\widetilde{f}$ maps the arc $J = \widehat{H} \cap K(0,1)$ to the diagonal of Q which joins 1 and i. From this determine the point of H that gets mapped by f to the center of Q. (*Hint.* Consider $g = \rho_1 \circ f \circ \rho$ for suitable reflections ρ and ρ_1.)

6.95. Construct a conformal mapping of $H = \{z: \text{Im } z > 0\}$ onto the interior of the closed rhombus P that has 1 and -1 as opposite vertices, the interior angle at each being $\alpha\pi$ for $0 < \alpha < 1$.

6.96. Let P be the regular polygon whose vertices are the nth-roots of unity, where $n \geq 3$. Find an integral representation for the conformal mapping f of $\Delta(0,1)$ onto the interior of P such that $f(0) = 0$ and $\widetilde{f}(1) = 1$. What are the points of $K(0,1)$ that get mapped by $\widetilde{f}$ to the midpoints of the sides of P?

6.97. Let $f(z) = \int_0^z (\zeta - a)^{-1/2}(\zeta - b)^{-1/2}(\zeta - c)^{-1/2} d\zeta$, where $a < b < c$ are real numbers. Show directly that f maps $H = \{z: \text{Im } z > 0\}$ conformally onto a rectangle with vertices $f(a), f(b), f(c)$, and $f(\infty)$. (*Hint.* Start by examining $\text{Arg } \dot{\gamma}(t)$, where $\gamma(t) = f(t)$ for real t.)

6.98. Let $D = \{z: |x| < 1, |y| < c\}$, where $c > 0$. Demonstrate the existence and uniqueness of a number k in the interval $(0,1)$ — you will not be able to identify k unless $c = 1$ — with the property that there is a conformal mapping f of $H = \{z: \text{Im } z > 0\}$ onto D for which $\widetilde{f}(-1) = 1+ic$, $\widetilde{f}(-k) = -1+ic$, $\widetilde{f}(k) = -1-ic$, and $\widetilde{f}(1) = 1-ic$. Thus f has the structure $f(z) = A \int_0^z (\zeta - 1)^{-1/2}(\zeta + 1)^{-1/2}(\zeta - k)^{-1/2}(\zeta + k)^{-1/2} d\zeta + B$, where A and B are constants. What point does f map to the center of D? What is k when $c = 1$? (*Hint.* Seek f in the form $f = g^{-1} \circ h$, where h is a Möbius transformation that maps H onto $\Delta(0,1)$ and g is the conformal mapping of D onto $\Delta(0,1)$ with $g(0) = 0$ and $g'(0) > 0$. Recall Exercises 6.66 and 6.28(viii).)

6.99. Let D be a Jordan domain in $\mathbb{C}$ and let (z_1, z_2, z_3, z_4) be a quadruple of distinct points from $\widehat{\partial} D$ that is positively oriented relative to D. Demonstrate that there is a unique number $M > 0$ with the following property: there exists a homeomorphism f of $\widehat{D}$ onto $\{z: 0 \leq x \leq M, 0 \leq y \leq 1\}$ that is conformal in D and maps z_1 to 0, z_2 to M, z_3 to $M+i$, and z_4 to i.

Chapter X

Constructing Analytic Functions

Introduction

It is our aim with the present chapter to impress upon the reader just how great the wealth of functions is to which the ideas related in this book apply. The theorem of Mittag-Leffler, Weierstrass's theorem, and the method of analytic continuation generate a supply of analytic and meromorphic functions rich beyond one's wildest expectations. At the same time the chapter is intended to serve as a gateway to more sophisticated topics in complex analysis. The subjects introduced here come to full fruition in such diverse areas as the theory of entire functions, the theory of value distribution for meromorphic functions, analytic number theory, and the theory of Riemann surfaces, to name but a few. Hopefully the reader will find in the contents of this chapter a stimulus for continuing the study of complex function theory at more advanced levels.

1 The Theorem of Mittag-Leffler

1.1 Series of Meromorphic Functions

The material in this section demands that we deal with function series $\sum_{n=1}^{\infty} f_n$ (and $\sum_{n=-\infty}^{\infty} f_n$) whose terms are meromorphic in a plane open set U. Since the value ∞ is allowed to appear in the range of any or all f_n, the notions of convergence we have heretofore considered do not strictly apply to such a series. The concept of normal convergence, in particular, requires some adjustment. That adjustment is conventionally made as follows: a series of functions $\sum_{n=1}^{\infty} f_n$ whose terms are meromorphic in an open set U is said to *converge normally in U* if corresponding to each compact subset K of U there exists an index N such that no term f_n with

$n > N$ has a pole in K and such that the truncated series $\sum_{n=N+1}^{\infty} f_n$ converges uniformly on K. (N.B. As was the case for the earlier definition of normal convergence, to insure this behavior it is sufficient that the stated condition be satisfied for every closed disk K in U.) A doubly infinite series $\sum_{n=-\infty}^{\infty} f_n$ of functions meromorphic in U converges normally in U provided both of the series $\sum_{n=0}^{\infty} f_n$ and $\sum_{n=1}^{\infty} f_{-n}$ are normally convergent there. It goes without saying that this generalized understanding of normal convergence reduces to the original concept for series whose terms are actually analytic in U. To illustrate the new interpretation of normal convergence, recall Example VII.3.2. The argument presented there shows that $\sum_{n=-\infty}^{\infty} (z-n)^{-2}$, when viewed as a series of meromorphic functions, is normally convergent in the whole complex plane. By the old definition it was only regarded as normally convergent in $\mathbb{C} \sim \{0, \pm 1, \pm 2, \cdots\}$.

If a series $\sum_{n=1}^{\infty} f_n$ of functions that are meromorphic in an open set U converges normally there, then the set $E = \bigcup_{n=1}^{\infty} E_n$, where E_n is the set of poles of f_n in U, is a discrete subset of U. Indeed, the definition of normal convergence and the fact that each individual set E_n is a discrete subset of U imply that any closed disk $\overline{\Delta}(z_0, r)$ in U intersects E in at most finitely many points, eliminating z_0 as a potential limit point of E. All of the terms f_n are analytic in the open set $V = U \sim E$, and the series $\sum_{n=1}^{\infty} f_n$ converges normally in V. As a result, the function $f: V \to \mathbb{C}$ defined by $f(z) = \sum_{n=1}^{\infty} f_n(z)$ is analytic. It has an isolated singularity at every point of E. What is the nature of these singularities? Given z_0 in E, we can choose a closed disk $K = \overline{\Delta}(z_0, r)$ in U with the property that $K \cap E = \{z_0\}$. According to the definition of normal convergence we can then fix an index N such that f_n is free of poles in K — hence, is analytic in the open disk $\Delta = \Delta(z_0, r)$ — as soon as $n > N$ and such that the series $\sum_{n=N+1}^{\infty} f_n$ converges uniformly on K. It follows that $g(z) = \sum_{n=N+1}^{\infty} f_n(z)$ defines an analytic function in Δ. In the punctured disk $\Delta^* = \Delta^*(z_0, r)$ the function f admits the representation $f = f_1 + f_2 + \cdots + f_N + g$. Because each of the finitely many functions $f_1, f_2, \ldots, f_N$, and g has no worse than a pole at z_0, the same is true of their sum f. The conclusion: at each point of U the function f has at worst a pole, so f is meromorphic in U. By convention, f is automatically assigned values at the points of E so as to extend it to a continuous function from U into $\widehat{\mathbb{C}}$. We continue to speak of f as the *sum* of the series $\sum_{n=1}^{\infty} f_n$ in U and to write $f(z) = \sum_{n=1}^{\infty} f_n(z)$ for all z in U, despite the fact that $\sum_{n=1}^{\infty} f_n(z_0)$ is technically undefined when z_0 belongs to E: for such z_0 the value $f(z_0)$ is officially given by $f(z_0) = \lim_{z \to z_0} \sum_{n=1}^{\infty} f_n(z)$, the limit being taken in the extended plane $\widehat{\mathbb{C}}$. We summarize the preceding remarks in a theorem.

Theorem 1.1. *Suppose that each term in a function series $\sum_{n=1}^{\infty} f_n$ is meromorphic in an open set U and that the series converges normally in U, with the function f as its sum. Then f is meromorphic in U. Any pole of f is a pole of f_n for at least one value, but at most finitely many values, of*

1. The Theorem of Mittag-Leffler

n. Furthermore, $f^{(k)} = \sum_{n=1}^{\infty} f_n^{(k)}$ in U for every positive integer k. Each of these derived series converges normally in U.

Proof. Only the statements concerning the series of derivatives are still in need of justification. Fix a positive integer k. We first establish that the series of meromorphic functions $\sum_{n=1}^{\infty} f_n^{(k)}$ converges normally in U. For this, let $K = \overline{\Delta}(z_0, r)$ be an arbitrary closed disk in U. We must prove that there is an index N for which $f_n^{(k)}$ is without poles in K whenever $n > N$ and for which the series $\sum_{n=N+1}^{\infty} f_n^{(k)}$ is uniformly convergent on K. Choose s satisfying $s > r$ such that $K_1 = \overline{\Delta}(z_0, s)$ is still contained in U. Since $\sum_{n=1}^{\infty} f_n$ is given to be normally convergent in U, we are guaranteed the existence of an index N such that no f_n with $n > N$ exhibits a pole in K_1 and such that the series $\sum_{n=N+1}^{\infty} f_n$ converges uniformly on K_1. It follows that $\sum_{n=N+1}^{\infty} f_n$ is a normally convergent series of analytic functions in the disk $\Delta = \Delta(z_0, s)$. Denote its sum there by g. When $n > N$ the function $f_n^{(k)}$, being itself analytic in Δ, has no poles in K. Also, in view of Theorem VII.3.2 the series $\sum_{n=N+1}^{\infty} f_n^{(k)}$ converges normally in Δ — hence, converges uniformly on K — where its sum is $g^{(k)}$. Notice that in Δ we have $f = f_1 + f_2 + \cdots + f_N + g$, from which we infer that in this disk

$$f^{(k)} = f_1^{(k)} + f_2^{(k)} + \cdots + f_N^{(k)} + g^{(k)}$$

$$= f_1^{(k)} + f_2^{(k)} + \cdots + f_N^{(k)} + \sum_{n=N+1}^{\infty} f_n^{(k)} = \sum_{n=1}^{\infty} f_n^{(k)}.$$

Therefore, we have shown that $\sum_{n=1}^{\infty} f_n^{(k)}$ is normally convergent in U and in the process demonstrated that each point of U is the center of an open disk in which $f^{(k)} = \sum_{n=1}^{\infty} f_n^{(k)}$, a fact which implies that $f^{(k)} = \sum_{n=1}^{\infty} f_n^{(k)}$ throughout U. ∎

An obvious analogue of Theorem 1.1 is valid for doubly infinite series.

1.2 Constructing Meromorphic Functions

The basic global existence theorem for meromorphic functions was discovered by the dynamic Swedish mathematician and entrepreneur Gösta Mittag-Leffler (1846-1927), whose mathematical legacy also includes the journal *Acta Mathematica*, which he founded and which remains to this day one of the world's most prestigious publications devoted to mathematical research. In stating Mittag-Leffler's result we use the terminology *rational singular part at z_0* to describe a rational function S of the type

$$S(z) = \frac{a_{-m}}{(z-z_0)^m} + \frac{a_{-m+1}}{(z-z_0)^{m-1}} + \cdots + \frac{a_{-1}}{z-z_0},$$

in which $a_{-m} \neq 0$; i.e., S has exactly one pole in $\widehat{\mathbb{C}}$, it located at the point z_0, and $S(\infty) = 0$.

Let $E = \{z_1, z_2, z_3, \cdots\}$ be a discrete subset of an open set U in $\mathbb{C}$. (N.B. Implicit here is the fact that the elements of a discrete subset of U can be listed in a finite or infinite sequence, which for present purposes we always assume to be univalent — $z_n \neq z_m$ when $n \neq m$.) Assume that at each point z_n of E we are presented with a rational singular part S_n. Does there exist a meromorphic function in U whose set of poles in this set coincides with E and whose singular part at z_n is S_n for $n = 1, 2, \cdots$? Mittag-Leffler's theorem provides an affirmative answer to this question. Of course, when $E = \{z_1, z_2, z_3, \ldots, z_p\}$ is just a finite subset of U, it is no great feat to manufacture such a function: $f = S_1 + S_2 + \cdots + S_p$ does the job. In the more interesting case where E is infinite the natural temptation is to hope that $f = \sum_{n=1}^{\infty} S_n$ will meet the requirements, as indeed it will when this series happens to converge normally in U. Unfortunately, one cannot in general count on that convergence. Mittag-Leffler's insight was to realize that, upon subtracting from S_n an appropriate rational function R_n whose poles are outside U, one can produce a series $\sum_{n=1}^{\infty} (S_n - R_n)$ that does converge normally in U. The function $f = \sum_{n=1}^{\infty} (S_n - R_n)$ is then meromorphic in U, it is analytic in $U \sim E$, and it has a pole with the prescribed singular part at each point of E. (The last statement follows from the discussion preceding Theorem 1.1, which confirms that $f - S_n = -R_n + \sum_{k \neq n} (S_k - R_k)$ is analytic in a suitably small punctured disk centered at z_n. Remember: for $k \neq n$ the function $S_k - R_k$ is analytic in the neighborhood of z_n.) A collection of functions R_n for which Mittag-Leffler's procedure is successful is called a set of "convergence inducing summands" for the problem.

Theorem 1.2. (Mittag-Leffler's Theorem) *Let $E = \{z_1, z_2, z_3, \cdots\}$ be a discrete subset of an open set U in the complex plane, and for $n = 1, 2, 3, \cdots$ let S_n be a rational singular part at z_n. There exists a meromorphic function $f: U \to \widehat{\mathbb{C}}$ that has E as its set of poles and that for $n = 1, 2, 3, \cdots$ has singular part S_n at z_n. Any two functions fitting this description differ by a function that is analytic in U.*

Proof. Only when E is an infinite set does the existence of such a function remain in doubt, so it is an infinite set E that we consider in the proof. For technical reasons we shall assume that E does not include the origin. (This entails no loss of generality: if the origin does belong to E — say $z_1 = 0$ — then the function $f = S_1 + \widetilde{f}$, where $\widetilde{f}$ is a solution of the same problem for the set $\widetilde{E} = \{z_2, z_3, z_4, \cdots\}$ with the singular part S_n still specified at z_n for $n = 2, 3, 4, \cdots$, has all the features demanded.) Let δ_n designate the radius of the largest open disk centered at z_n that is contained in U. We distinguish two preliminary cases.

Case 1. $|z_n|\delta_n \geq 1$ for every n. (N.B. If $U = \mathbb{C}$, then $\delta_n = \infty$ for

1. The Theorem of Mittag-Leffler

every n. By assumption $z_n \neq 0$, so $|z_n|\delta_n = \infty$ for every n. Hence, Case 1 is actually the general case when U is the whole complex plane.) We claim that in this situation $|z_n| \to \infty$ as $n \to \infty$. If not, we would be able to extract from the sequence $\langle z_n \rangle$ a subsequence $\langle z_{n_k} \rangle$ that converges to some finite accumulation point z_0 of $\langle z_n \rangle$. Should $U = \mathbb{C}$, this would instantly contradict the fact that E is a discrete subset of U. Assuming that $U \neq \mathbb{C}$, we know from the definition of δ_n and the assumption in Case 1 that

$$|z_n - z| \geq \delta_n \geq \frac{1}{|z_n|}$$

for every $n \geq 1$ and every z in $\mathbb{C} \sim U$. Accordingly, we would obtain

$$|z_0 - z| = \lim_{k \to \infty} |z_{n_k} - z| \geq \lim_{k \to \infty} \frac{1}{|z_{n_k}|} = \frac{1}{|z_0|}$$

for every z in $\mathbb{C} \sim U$, which both rule out $z_0 = 0$ and place z_0 in the set U. This would again mark z_0 as a limit point of E in U, another violation of discreteness. Therefore, $|z_n| \to \infty$ as declared. We shall see that the condition $|z_n| \to \infty$ leads to an even stronger conclusion than the one promised by the theorem.

The rational function S_n, whose unique pole is at z_n, is analytic in the disk $\Delta_n = \Delta(0, |z_n|)$. It follows that S_n can be represented in Δ_n as the sum of a normally convergent Taylor series centered at the origin, $S_n(z) = \sum_{k=0}^{\infty} a_k^{(n)} z^k$. As $d \to \infty$, $\sum_{k=0}^{d} a_k^{(n)} z^k \to S_n(z)$ uniformly on the closed disk $K_n = \overline{\Delta}(0, |z_n|/2)$. For this reason we are at liberty to select and fix an index $d(n)$ such that the polynomial $R_n(z) = \sum_{k=0}^{d(n)} a_k^{(n)} z^k$ satisfies $|S_n(z) - R_n(z)| < 2^{-n}$ for every z in K_n. We do this for $n = 1, 2, 3, \cdots$.

We now assert that the series $\sum_{n=1}^{\infty} (S_n - R_n)$ is normally convergent in $\mathbb{C}$. If so, $f = \sum_{n=1}^{\infty} (S_n - R_n)$ will define a function that is meromorphic in $\mathbb{C}$ — not just meromorphic in U, which is all we originally asked for — having the prescribed poles and singular parts. Let K be an arbitrary compact set in $\mathbb{C}$. Since $|z_n| \to \infty$, we can fix an index N with the property that K is contained in the disk K_n whenever $n > N$. If $n > N$, then the only pole in $\mathbb{C}$ of the function $S_n - R_n$, the one at z_n, is not a point of K. Also, for such n the estimate $|S_n(z) - R_n(z)| < 2^{-n}$ holds throughout K. On the basis of the Weierstrass M-test we can pronounce the series $\sum_{n=N+1}^{\infty}(S_n - R_n)$ uniformly convergent on K. We have thus established the normal convergence of $\sum_{n=1}^{\infty}(S_n - R_n)$ in $\mathbb{C}$. With this we have completed the proof of the theorem in Case 1, obtaining a slightly better result than the theorem stated.

Case 2: $|z_n|\delta_n < 1$ for every n. To handle Case 2 we first observe that its defining condition implies that $\delta_n \to 0$ as $n \to \infty$. Suppose this not to be so. Then there exists an $\epsilon > 0$ such that $\delta_n \geq \epsilon$ is true for infinitely many n. In other words, the sequence $\langle \delta_n \rangle$ has a subsequence $\langle \delta_{n_k} \rangle$ such that $\delta_{n_k} \geq \epsilon$ is satisfied for every k. Thus $|z_{n_k}| < \delta_{n_k}^{-1} < \epsilon^{-1}$ holds for all

k, making $\langle z_{n_k} \rangle$ a bounded sequence. By passing to a further subsequence and relabeling, if need be, we may assume that $\langle z_{n_k} \rangle$ is convergent. Let z_0 be its limit. For every $k \geq 1$ and every z in $\mathbb{C} \sim U$ it is then the case that

$$|z_{n_k} - z| \geq \delta_{n_k} \geq \epsilon .$$

For all such z, therefore,

$$|z_0 - z| = \lim_{k \to \infty} |z_{n_k} - z| \geq \epsilon > 0 ,$$

which means that z_0 is a point of U. As z_0 is also a limit point of E, a discrete subset of U, we have reached a contradiction. We are forced to conclude that $\delta_n \to 0$ in Case 2.

Since $\delta_n < \infty$ in Case 2, we can choose and fix for each n a point ζ_n of ∂U with the property that $|\zeta_n - z_n| = \delta_n$. (If no such ζ_n existed, $\Delta(z_n, \delta_n)$ would not be the largest open disk in U centered at z_n.) Because the function S_n is analytic in the annulus $G_n = \{z : \delta_n < |z - \zeta_n| < \infty\}$, it can be expanded there in a Laurent series centered at ζ_n. The structure of S_n permits us to extract this expansion from the identity

$$\frac{1}{z - z_n} = \sum_{k=1}^{\infty} \frac{(z_n - \zeta_n)^{k-1}}{(z - \zeta_n)^k} ,$$

valid for all z in G_n, by repeatedly differentiating it and by taking the appropriate linear combination of the resulting derived series, a process which reveals that the Laurent expansion of S_n in G_n has the form $S_n(z) = \sum_{k=1}^{\infty} a_k^{(n)} (z - \zeta_n)^{-k}$. Furthermore, this series converges uniformly on the set $A_n = \{z : 2\delta_n \leq |z - \zeta_n| < \infty\}$. This entitles us to select and fix an index $d(n)$ for which the rational function $R_n(z) = \sum_{k=1}^{d(n)} a_k^{(n)} (z - \zeta_n)^{-k}$ meets the following specification: $|S_n(z) - R_n(z)| < 2^{-n}$ is true for every z in A_n. Again, this is to be done for $n = 1, 2, 3, \cdots$. Notice that R_n is analytic in U, for its only pole ζ_n lies on ∂U.

To cap off the proof of the theorem in Case 2 we need only verify that the series $\sum_{n=1}^{\infty}(S_n - R_n)$ converges normally in U, for then $f = \sum_{n=1}^{\infty}(S_n - R_n)$ is a function blessed with all the properties insisted upon in the theorem. The poles of $S_n - R_n$ — namely, z_n and ζ_n — are not elements of A_n. If K is any compact subset of U, then there is a number $\delta > 0$ such that $|z - \zeta| \geq \delta$ holds for every z in K and every ζ in $\mathbb{C} \sim U$ (Lemma II.4.3). Since $2\delta_n \to 0$ as $n \to \infty$, it follows easily that there is an index N such that K is a subset of A_n once $n > N$. For such n the function $S_n - R_n$ is free of poles in K and obeys the estimate $|S_n(z) - R_n(z)| < 2^{-n}$ everywhere in K. By the Weierstrass M-test the series $\sum_{n=N+1}^{\infty}(S_n - R_n)$ is uniformly convergent on K. This certifies $\sum_{n=1}^{\infty}(S_n - R_n)$ as normally convergent in U and so finishes the proof of the theorem in Case 2.

Finally, we come to an infinite discrete subset $E = \{z_1, z_2, z_3, \cdots\}$ of U not covered by either Case 1 or Case 2. Set $J = \{n : |z_n|\delta_n \geq 1\}$,

1. The Theorem of Mittag-Leffler

$E_1 = \{z_n : n \in J\}$, and $E_2 = E \sim E_1$. Then E_1 and E_2 are non-empty, disjoint sets whose union is E. Moreover, E_1 is either a finite set or a set to which Case 1 applies, whereas E_2 either is finite or falls within the scope of Case 2. As a consequence, there exists for $j = 1, 2$ a meromorphic function $f_j : U \to \widehat{\mathbb{C}}$ that has pole-set E_j, at each point of which f_j has the singular part specified for that point in the data for E. The function $f = f_1 + f_2$ is meromorphic in U, its set of poles there is $E_1 \cup E_2 = E$, and it has at each point of E the prescribed singular part. If g is any other function with the same properties, then $f - g$ is analytic in $U \sim E$ and has a removable singularity at every point of E. Upon removal of these singularities, $f - g$ becomes analytic in U. The final assertion of the theorem is thus clear. ∎

Mittag-Leffler's theorem remains true essentially as stated — in the last sentence the work "analytic" must be replaced by "holomorphic" — for an open set U in the extended complex plane $\widehat{\mathbb{C}}$. By a *rational singular part at* ∞, in case ∞ is a point of E, is meant a polynomial function of z having positive degree and having a zero at the origin. The proof when U is a set containing ∞ reduces more or less to the proof of Case 2 in the argument presented. We leave the details as an exercise (Exercise 4.1).

There are many situations in which sets of convergence inducing summands for Mittag-Leffler's procedure can be written down explicitly. We indicate one instance of this. Suppose, namely, that $E = \{z_1, z_2, z_3, \cdots\}$ is an infinite discrete subset of $\mathbb{C}$, which for convenience is assumed not to have the origin as an element. A meromorphic function $f : \mathbb{C} \to \widehat{\mathbb{C}}$ is desired that has a simple pole with residue one at each of the points z_n and no other poles. If the series $\sum_{n=1}^{\infty} |z_n|^{-1}$ converges, then

(10.1) $$f(z) = \sum_{n=1}^{\infty} \frac{1}{z - z_n}$$

already fills the bill; i.e., no convergence inducing modifications are needed. The reason: since $|z_n| \to \infty$ as $n \to \infty$ and since for any given compact subset K of $\mathbb{C}$ the inequality

$$\frac{1}{|z - z_n|} \leq \frac{2}{|z_n|}$$

holds for every z in K as soon as n is large enough that $|z_n|$ exceeds the number $2 \max\{|z| : z \in K\}$, it is evident that the series of meromorphic function in (10.1) is normally convergent in $\mathbb{C}$. If $\sum_{n=1}^{\infty} |z_n|^{-p}$ is convergent for an integer $p > 1$, then the function

(10.2) $$f(z) = \sum_{n=1}^{\infty} \left(\frac{1}{z - z_n} + \frac{1}{z_n} + \frac{z}{z_n^2} + \cdots + \frac{z^{p-2}}{z_n^{p-1}} \right)$$

meets the demand. Indeed, for any compact set K in $\mathbb{C}$ and any n sufficiently large that $c = \max\{|z| : z \in K\} \le |z_n|/2$ we obtain the estimate

$$\text{(10.3)} \qquad \left| \frac{1}{z-z_n} + \frac{1}{z_n} + \frac{z}{z_n^2} + \cdots + \frac{z^{p-2}}{z_n^{p-1}} \right| \le \frac{2c^{p-1}}{|z_n|^p}$$

for every z in K, as follows easily from the observation that

$$\frac{1}{z-z_n} + \frac{1}{z_n} + \frac{z}{z_n^2} + \cdots + \frac{z^{p-2}}{z_n^{p-1}} = \frac{1}{z-z_n} + \frac{1}{z_n} \frac{1-(z/z_n)^{p-1}}{1-(z/z_n)} = \frac{z^{p-1}}{z_n^{p-1}(z-z_n)}.$$

Coupled with the convergence of $\sum_{n=1}^{\infty} |z_n|^{-p}$, (10.3) implies that the series in (10.2) converges normally in the whole plane. It is worth noting that the functions in (10.1) and (10.2) do not depend on the particular way we elect to list the elements of E: since the defining series are absolutely convergent in $\mathbb{C} \sim E$, their terms can be rearranged at will (rearrangement is tantamount to relabeling the points of E) without affecting the sum.

A classic example of (10.2) is the function

$$f(z) = \frac{1}{z} + \sum_{|n|\ge 1} \left(\frac{1}{z-n} + \frac{1}{n} \right),$$

whose singularities in $\mathbb{C}$ are simple poles with residue one at all integers n. (Here we apply (10.2) to $E = \{\pm 1, \pm 2, \cdots\}$ with $p = 2$. The use of (10.2) is legitimate, because $\sum_{n=1}^{\infty} |z_n|^{-2} = 2\sum_{n=1}^{\infty} n^{-2} < \infty$. We have also thrown in a pole at the origin for good measure.) As luck would have it, we already know a function with these qualifications, the function $g(z) = \pi \cot(\pi z)$. What is the relationship between f and g? The next example furnishes the answer to this question.

EXAMPLE 1.1. Show that

$$\text{(10.4)} \qquad \pi \cot(\pi z) = \frac{1}{z} + \sum_{|n|\ge 1} \left(\frac{1}{z-n} + \frac{1}{n} \right)$$

for every complex number z.

Since both sides of (10.4) describe meromorphic functions in $\mathbb{C}$ having poles — hence, taking the value ∞ — at all integers, we need only concern ourselves with verifying (10.4) for non-integral complex numbers z. Fix such a point z and consider the function h defined by

$$h(\zeta) = \frac{\pi \cot(\pi \zeta)}{\zeta(\zeta - z)} - \frac{1}{\zeta^2(\zeta - z)}.$$

This function is meromorphic in $\mathbb{C}$. It has a simple pole at the point z, with

$$\text{Res}(z, h) = \frac{\pi \cot(\pi z)}{z} - \frac{1}{z^2},$$

1. The Theorem of Mittag-Leffler

and a simple pole at each non-zero integer n, for which

$$\text{Res}(n, h) = \frac{1}{n(n-z)} = -\frac{1}{z}\left(\frac{1}{z-n} + \frac{1}{n}\right).$$

The singularity of h at the origin is removable. Let N be any integer satisfying $N > |z|$, and let $Q_N = \{z : |x| \leq N + (1/2), |y| \leq N + (1/2)\}$. An application of the residue theorem yields

(10.5) $\quad \dfrac{1}{2\pi i} \displaystyle\int_{\partial Q_N} h(\zeta)\, d\zeta = \dfrac{\pi \cot(\pi z)}{z} - \dfrac{1}{z^2} - \dfrac{1}{z} \displaystyle\sum_{1 \leq |n| \leq N} \left(\dfrac{1}{z-n} + \dfrac{1}{n}\right).$

We claim that $\int_{\partial Q_N} h(\zeta)\, d\zeta \to 0$ as $N \to \infty$. Once this is demonstrated, we shall be in a position to let $N \to \infty$ in (10.5) and conclude that

$$\frac{\pi \cot(\pi z)}{z} - \frac{1}{z^2} - \frac{1}{z}\sum_{|n|\geq 1}\left(\frac{1}{z-n} + \frac{1}{n}\right) = 0,$$

from which (10.4) follows immediately.

We observe that

(10.6) $\quad |\cot(\pi \zeta)| \leq m = \dfrac{1 + e^{-3\pi}}{1 - e^{-3\pi}}$

for every ζ in ∂Q_N. To see this, observe first that for $\zeta = N + (1/2) + it$, where t is an arbitrary real number,

$$|\cot(\pi\zeta)| = \left|\cot(\pi N + 2^{-1}\pi + i\pi t)\right| = |\tan(i\pi t)|$$

$$= \left|\frac{e^{-\pi t} - e^{\pi t}}{e^{-\pi t} + e^{\pi t}}\right| \leq \frac{e^{-\pi t} + e^{\pi t}}{e^{-\pi t} + e^{\pi t}} = 1 < m.$$

Similarly, if $\zeta = t + Ni + (i/2)$ for any real t, then

$$|\cot(\pi\zeta)| = \left|\frac{e^{i\pi\zeta} + e^{-i\pi\zeta}}{e^{i\pi\zeta} - e^{-i\pi\zeta}}\right| = \left|\frac{e^{i2\pi\zeta} + 1}{e^{i2\pi\zeta} - 1}\right| \leq \frac{1 + |e^{i2\pi\zeta}|}{1 - |e^{i2\pi\zeta}|}$$

$$= \frac{1 + e^{-(2N+1)\pi}}{1 - e^{-(2N+1)\pi}} \leq \frac{1 + e^{-3\pi}}{1 - e^{-3\pi}} = m.$$

Since $|\cot(\pi\zeta)| = |\cot(-\pi\zeta)|$, the same bound is valid on the other two sides of Q_N, which confirms (10.6). As $|\zeta| \geq N$ for ζ on ∂Q_N, we retrieve from (10.6) the estimate

$$|h(\zeta)| \leq \frac{m\pi}{N(N-|z|)} + \frac{1}{N^2(N-|z|)} = \frac{m\pi N + 1}{N^2(N-|z|)}$$

for all such ζ. Because Q_N has perimeter $8N + 4$,

$$\left| \int_{\partial Q_N} h(\zeta)\, d\zeta \right| \leq \int_{\partial Q_N} |h(\zeta)||d\zeta| \leq \frac{(8N+4)(m\pi N + 1)}{N^2(N - |z|)} \to 0$$

as $N \to \infty$, just what we needed to know.

The beautiful identity (10.4) spawns a number of identities of the same general type. Here is one sample. Others can be found in the exercises.

EXAMPLE 1.2. Verify that

(10.7) $$\frac{\pi^2}{\sin^2(\pi z)} = \sum_{n=-\infty}^{\infty} \frac{1}{(z-n)^2}$$

for every complex number z.

Owing to Theorem 1.1 a series representation for the derivative of $f(z) = \pi \cot(\pi z)$ can be arrived at by differentiating the right-hand side of (10.4) term by term, a step that results in

$$-\pi^2 \csc^2(\pi z) = -\frac{1}{z^2} - \sum_{|n|\geq 1} \frac{1}{(z-n)^2} = -\sum_{n=-\infty}^{\infty} \frac{1}{(z-n)^2}$$

for every complex number z. This is plainly equivalent to (10.7).

1.3 The Weierstrass $\wp$-function

We could not abandon the present topic without saying at least a few words about one of the most significant functions to issue directly from the circle of ideas surrounding Mittag-Leffler's theorem, the so-called "$\wp$-function" of Weierstrass, which turns up in a host of advanced mathematical discussions. Its construction begins with a pair of non-zero complex numbers ω_1 and ω_2 whose ratio $\tau = \omega_2/\omega_1$ is not real. The set $\Omega = \Omega(\omega_1, \omega_2)$ consisting of all complex numbers ω of the form $\omega = n\omega_1 + m\omega_2$, where n and m are integers, is called the *lattice generated by ω_1 and ω_2* (Figure 1). It is not difficult to see that Ω is a discrete subset of the complex plane. We use Ω^* to signify the set $\Omega \sim \{0\}$. The *Weierstrass function corresponding to the lattice Ω* is the function $\wp$ defined on $\mathbb{C}$ by

(10.8) $$\wp(z) = \frac{1}{z^2} + \sum_{\omega \in \Omega^*} \left[\frac{1}{(z-\omega)^2} - \frac{1}{\omega^2} \right].$$

The expanded notation $\wp_\Omega$ is sometimes used for this function, especially in situations where two or more different lattices are under consideration

1. The Theorem of Mittag-Leffler

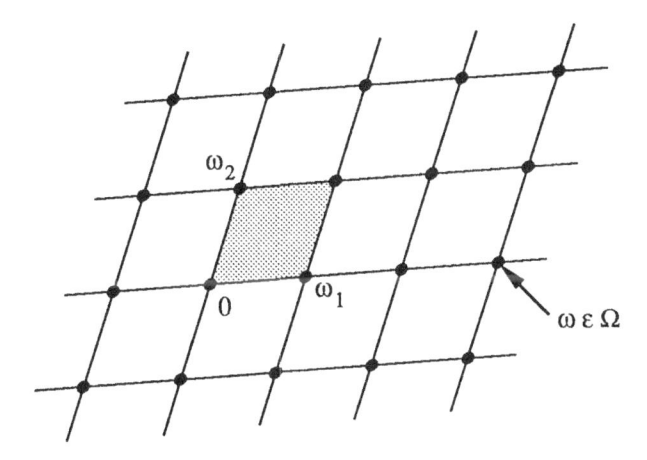

Figure 1.

simultaneously and confusion between their associated $\wp$-functions must be avoided.

We have not previously run into infinite series indexed by sets other than sets of integers, so (10.8) warrants some explanation. (It is not our intention, however, to get bogged down in a full-blown technical discussion of such series.) Since Ω^* is a discrete subset of $\mathbb{C}$, we are free to list its members in a sequence $z_1, z_2, z_3, \cdots$, where $z_n \neq z_m$ for $n \neq m$. We shall demonstrate that the series $\sum_{n=1}^{\infty} |z_n|^{-3}$ is convergent. Assume for a moment that this has already been done. An argument along the lines of those that established convergence in (10.1) and (10.2) can then be fashioned to show that the series of meromorphic functions $\sum_{n=1}^{\infty} \left[(z-z_n)^{-2} - z_n^{-2}\right]$ is normally convergent in $\mathbb{C}$ and absolutely convergent in $\mathbb{C} \sim \Omega^*$. Therefore, the formula

$$(10.9) \qquad \wp(z) = \frac{1}{z^2} + \sum_{n=1}^{\infty} \left[\frac{1}{(z-z_n)^2} - \frac{1}{z_n^2}\right]$$

is seen to define a meromorphic function in $\mathbb{C}$ whose set of poles is the lattice Ω and whose singular part at any element ω of Ω is $S(z) = (z-\omega)^{-2}$. Furthermore, the absolute convergence of the series implies that $\wp$ depends only on the set Ω, not on our particular way of arranging the members of Ω^*: if $w_1, w_2, w_3, \cdots$ is a second enumeration of the points of Ω^*, then

$$\wp(z) = \frac{1}{z^2} + \sum_{n=1}^{\infty} \left[\frac{1}{(z-w_n)^2} - \frac{1}{w_n^2}\right]$$

gives an alternate description of the function in (10.9). The last fact justifies writing

$$\wp(z) = \frac{1}{z^2} + \sum_{\omega \in \Omega^*} \left[\frac{1}{(z-\omega)^2} - \frac{1}{\omega^2}\right],$$

without any reference to the sequence $z_1, z_2, z_3, \cdots$. Notice, incidentally, that $-z_1, -z_2, -z_3, \cdots$ does represent another way of listing the elements of Ω^*, so we have

$$\wp(z) = \frac{1}{z^2} + \sum_{n=1}^{\infty} \left\{ \frac{1}{[z-(-z_n)]^2} - \frac{1}{(-z_n)^2} \right\} = \frac{1}{z^2} + \sum_{n=1}^{\infty} \left[\frac{1}{(z+z_n)^2} - \frac{1}{z_n^2} \right].$$

On the other hand, it is also true that

$$\wp(-z) = \frac{1}{(-z)^2} + \sum_{n=1}^{\infty} \left[\frac{1}{(-z-z_n)^2} - \frac{1}{z_n^2} \right] = \frac{1}{z^2} + \sum_{n=1}^{\infty} \left[\frac{1}{(z+z_n)^2} - \frac{1}{z_n^2} \right].$$

We deduce that $\wp(-z) = \wp(z)$ for every z in $\mathbb{C}$; i.e., $\wp$ is an even function. Of course, all of the above statements are contingent on the convergence of the series $\sum_{\omega \in \Omega^*} |\omega|^{-3} = \sum_{n=1}^{\infty} |z_n|^{-3}$, something we now confirm.

The function φ defined on the circle $K = K(0,1)$ by $\varphi(x+iy) = |x\omega_1 + y\omega_2|$ is non-negative and continuous. Were $\varphi(z) = 0$ to hold for some z on K, it would imply that ω_2/ω_1 is real, contrary to our assumption. We infer that $c = 2^{-1} \min\{\varphi(z) : z \in K\}$ is a positive number. We shall show that $\sum_{n=1}^{N} |z_n|^{-3} \leq A = 4c^{-3} \sum_{k=1}^{\infty} k^{-2}$ holds for every N, from which the convergence of $\sum_{n=1}^{\infty} |z_n|^{-3}$ follows. Fix N and choose M so that Ω_M^*, the set of points $n\omega_1 + m\omega_1$ in Ω^* with $|n|+|m| \leq M$, includes $z_1, z_2, \ldots, z_N$. Since the lower bound

$$|n\omega_1 + m\omega_2| = (n^2 + m^2)^{1/2} |n(n^2+m^2)^{-1/2}\omega_1 + m(n^2+m^2)^{-1/2}\omega_2|$$

$$\geq 2c(n^2+m^2)^{1/2} \geq c(|n|+|m|)$$

is valid for every element $n\omega_1 + m\omega_2$ of Ω^* and since the number of integer pairs (n,m) for which $|n|+|m| = k$ is exactly $4k$ when $k \geq 1$, we obtain

$$\sum_{n=1}^{N} \frac{1}{|z_n|^3} \leq \sum_{\omega \in \Omega_M^*} \frac{1}{|\omega|^3} = \sum_{k=1}^{M} \sum_{|n|+|m|=k} \frac{1}{|n\omega_1 + m\omega_1|^3}$$

$$\leq \sum_{k=1}^{M} \sum_{|n|+|m|=k} \frac{1}{c^3(|n|+|m|)^3} = \frac{4}{c^3} \sum_{k=1}^{M} \frac{1}{k^2} \leq A,$$

as desired.

We arrive at the derivative of the $\wp$-function by differentiating the series in (10.8) termwise,

(10.10) $$\wp'(z) = -\sum_{\omega \in \Omega} \frac{2}{(z-\omega)^3}.$$

The latter series converges normally in $\mathbb{C}$ (Theorem 1.1), and, due to the convergence of $\sum_{\omega \in \Omega^*} |\omega|^{-3}$, it converges absolutely in $\mathbb{C} \sim \Omega$. If we pick

1. The Theorem of Mittag-Leffler

a sequential listing $z_1, z_2, z_3, \cdots$ for the members of Ω, then we notice that $z_1 - \omega_1, z_2 - \omega_1, z_3 - \omega_1, \cdots$ provides another such listing. The absolute convergence in (10.10) dictates that

$$\wp'(z) = -\sum_{n=1}^{\infty} \frac{2}{(z-z_n)^3} = -\sum_{n=1}^{\infty} \frac{2}{(z+\omega_1-z_n)^3}$$

for every z in $\mathbb{C}$. Using the first of these two representations, we remark that

$$\wp'(z+\omega_1) = -\sum_{n=1}^{\infty} \frac{2}{(z+\omega_1-z_n)^3} \ .$$

We thus discover a noteworthy property of $\wp'$: $\wp'(z+\omega_1) = \wp'(z)$ for every z in $\mathbb{C}$. Similarly, $\wp'(z+\omega_2) = \wp'(z)$ in $\mathbb{C}$. (More generally, $\wp'(z+\omega) = \wp'(z)$ holds for every ω in Ω.) These identities bestow on $\wp'$ membership in an exclusive class of functions known as "elliptic functions," concerning which there is an extensive literature. (By definition a function f is an *elliptic function* provided it is meromorphic in $\mathbb{C}$ and *doubly periodic*. The last condition demands the existence of a lattice $\Omega = \Omega(\omega_1, \omega_2)$ generated by non-zero complex numbers ω_1 and ω_2 with non-real ratio $\tau = \omega_2/\omega_1$ such that $f(z) = f(z+\omega_1) = f(z+\omega_2)$ for every z in $\mathbb{C}$ or, equivalently, such that $f(z) = f(z+\omega)$ holds in $\mathbb{C}$ for every ω in Ω. It can be shown that, unless f is a constant function, the generators ω_1 and ω_2 here can always be chosen so that Ω includes every period of f, meaning every ω for which $f(z+\omega) = f(z)$ throughout $\mathbb{C}$. If so, we call Ω the *period lattice* of f and refer to ω_1 and ω_2 as a pair of *primitive periods* for f.) The $\wp$-function, too, is an elliptic function, although this fact is not quite as obvious as it was in the case of $\wp'$. To prove it, consider the function $g(z) = \wp(z) - \wp(z+\omega_1)$. Then $g'(z) = \wp'(z) - \wp'(z+\omega_1) = 0$ in $\mathbb{C} \sim \Omega$, making g constant in that domain. Because $\wp$ is an even function and $-\omega_1/2$ is not a point of Ω, $g(-\omega_1/2) = \wp(-\omega_1/2) - \wp(\omega_1/2) = 0$, so g vanishes identically in $\mathbb{C} \sim \Omega$; i.e., $\wp(z) = \wp(z+\omega_1)$ whenever z belongs to $\mathbb{C} \sim \Omega$. Since $\wp(z) = \wp(z+\omega_1) = \infty$ for every z in Ω, $\wp(z) = \wp(z+\omega_1)$ is true everywhere in the complex plane. The same argument shows that $\wp(z+\omega_2) = \wp(z)$ in $\mathbb{C}$. This makes $\wp$ an elliptic function, one which has ω_1 and ω_2 as a set of primitive periods. The function $\wp$ and its derivative play central roles in the general theory of elliptic functions, a subject we are unfortunately without space to probe more deeply here. (Some of the properties of elliptic functions, among them important properties of $\wp$, are explored in the exercises.)

2 The Theorem of Weierstrass

2.1 Infinite Products

Mittag-Leffler's theorem assures us that, given a discrete subset E of a plane open set U and given the assignment of a rational singular part to every point of E, we can construct a meromorphic function in U whose pole-set is E and whose singular part at each point of E is the one prescribed. Weierstrass's theorem answers in the affirmative a related question: Does there exist an analytic function $f: U \to \mathbb{C}$ that has the given discrete set E as its set of zeros, the order of each of these zeros being specified beforehand? As the usual method of creating such a function involves infinite products, it is to this topic that we first direct our attention.

Let $\langle z_n \rangle$ be a sequence of complex numbers. In giving meaning to the infinite product $\prod_{n=1}^{\infty} z_n$ we initially discuss the special case — as things turn out, it is not too special — where all of the factors z_n are non-zero. Mimicking the process that led from $\langle z_n \rangle$ via its corresponding sequence of partial sums s_n to the infinite series $\sum_{n=1}^{\infty} z_n$, we use $\langle z_n \rangle$ to generate an associated *sequence $\langle p_n \rangle$ of partial products:* $p_1 = z_1, p_2 = z_1 z_2, \ldots, p_n = z_1 z_2 \cdots z_n, \cdots$. It may happen that $p = \lim_{n \to \infty} p_n$ exists and that $p \neq 0$. Under these conditions we say that the infinite product $\prod_{n=1}^{\infty} z_n$ is *convergent* and that the value of the product is p, a state of affairs we symbolize by writing $p = \prod_{n=1}^{\infty} z_n$. If $\lim_{n \to \infty} p_n$ fails to exist or if $\lim_{n \to \infty} p_n = 0$, then $\prod_{n=1}^{\infty} z_n$ is said to be *divergent*. There are a number of reasons for insisting on the condition $p \neq 0$ as part of the definition of convergence here. Not the least of these is the simple desire to remain algebraically consistent with the situation for finite products: if $z_1, z_2, \ldots, z_n$ are all non-zero, then their product is non-zero as well. To illustrate the definition, let $z_n = 2^{2^{1-n}}$ for $n = 1, 2, 3, \cdots$. Then

$$p_n = 2 \cdot 2^{1/2} \cdots 2^{(1/2)^{n-1}} = 2^{1 + (1/2) + \cdots + (1/2)^{n-1}} = 2^{2 - (1/2)^{n-1}},$$

which yields

$$\prod_{n=1}^{\infty} 2^{2^{1-n}} = \lim_{n \to \infty} 2^{2 - (1/2)^{n-1}} = 2^2 = 4 .$$

As a second example, consider the sequence $z_n = n^{-1}$. In this instance $p_n = (n!)^{-1} \to 0$ as $n \to \infty$, so the product $\prod_{n=1}^{\infty} n^{-1}$ is by definition divergent. We remark that an infinite product $\prod_{n=1}^{\infty} z_n$ of non-zero complex numbers converges if and only if for each $N \geq 1$ the truncated product $\prod_{n=N+1}^{\infty} z_n = \prod_{n=1}^{\infty} z_{N+n}$ is convergent. Furthermore, it is then true that

(10.11)
$$\prod_{n=1}^{\infty} z_n = p_N \cdot \prod_{n=N+1}^{\infty} z_n$$

for every N.

2. The Theorem of Weierstrass

The transition from the special case in which the factors are non-zero to a more general infinite product $\prod_{n=1}^{\infty} z_n$ is accomplished as follows: $\prod_{n=1}^{\infty} z_n$ is called *convergent* if there exists an index N such that $z_n \neq 0$ holds whenever $n > N$ and such that the truncated product $\prod_{n=N+1}^{\infty} z_n$, whose factors are non-zero, converges according to the earlier definition, in which event we define

$$\prod_{n=1}^{\infty} z_n = p_N \cdot \prod_{n=N+1}^{\infty} z_n \ .$$

Otherwise, $\prod_{n=1}^{\infty} z_n$ is pronounced *divergent*. Because of (10.11) it is evident that the value of $\prod_{n=1}^{\infty} z_n$ does not depend on which N is used here, as long as it has the two properties stated, and that the generalized understanding of an infinite product reduces to the original one when all the factors z_n are non-zero. Formula (10.11) remains in effect for an arbitrary convergent product. We emphasize: $\prod_{n=1}^{\infty} z_n$ is automatically divergent if $z_n = 0$ for infinitely many n; under the assumption that $\prod_{n=1}^{\infty} z_n$ is convergent, $\prod_{n=1}^{\infty} z_n = 0$ only when $z_n = 0$ for one or more values of n. It follows almost immediately from these definitions that, if $\prod_{n=1}^{\infty} z_n$ and $\prod_{n=1}^{\infty} w_n$ are convergent infinite products, then the product $\prod_{n=1}^{\infty} (z_n w_n)$ also converges, with

(10.12)
$$\prod_{n=1}^{\infty}(z_n w_n) = \left(\prod_{n=1}^{\infty} z_n\right)\left(\prod_{n=1}^{\infty} w_n\right) \ .$$

A simple condition which is necessary (but far from sufficient) for the convergence of $\prod_{n=1}^{\infty} z_n$ is that $\lim_{n\to\infty} z_n = 1$: assuming that $z_n \neq 0$ for every $n > N$ and writing $q_n = \prod_{k=1}^{n} z_{N+k}$, we compute

$$\lim_{n\to\infty} z_n = \lim_{n\to\infty} \frac{q_n}{q_{n-1}} = \frac{\prod_{n=N+1}^{\infty} z_n}{\prod_{n=N+1}^{\infty} z_n} = 1 \ .$$

It is sometimes convenient to express a product $\prod_{n=1}^{\infty} z_n$ in the form $\prod_{n=1}^{\infty}(1+w_n)$ by writing w_n for $z_n - 1$. In this notation the necessary condition for convergence translates to $\lim_{n\to\infty} w_n = 0$.

The problem of testing an infinite product for convergence can be transformed to the task of checking whether a related infinite series converges or diverges.

Theorem 2.1. *Let $\langle z_n \rangle$ be a sequence of non-zero complex numbers. The infinite product $\prod_{n=1}^{\infty} z_n$ converges if and only if the associated infinite series $\sum_{n=1}^{\infty} \text{Log } z_n$ converges, in which case*

(10.13)
$$\prod_{n=1}^{\infty} z_n = \exp\left(\sum_{n=1}^{\infty} \text{Log } z_n\right) \ .$$

Proof. Write $p_n = z_1 z_2 \cdots z_n$ and $s_n = \operatorname{Log} z_1 + \operatorname{Log} z_2 + \cdots + \operatorname{Log} z_n$. Then $e^{s_n} = p_n$. If the series $\sum_{n=1}^{\infty} \operatorname{Log} z_n$ is convergent with sum s, then $p_n = e^{s_n} \to e^s$ as $n \to \infty$; i.e., $\prod_{n=1}^{\infty} z_n$ converges and (10.13) holds. The converse is slightly more delicate. Assume that the infinite product $\prod_{n=1}^{\infty} z_n$ converges, and call the product p. By definition, $p \neq 0$. We may further assume for the purposes of this proof that p is not a point of the negative real axis. (If p does happen to be in the interval $(-\infty, 0)$, we simply consider in place of $\langle z_n \rangle$ a new sequence $\langle w_n \rangle$; namely, $w_1 = -z_1$ and $w_n = z_n$ for $n \geq 2$. Then $\prod_{n=1}^{\infty} w_n = -p$ does not belong to $(-\infty, 0)$. The two-series $\sum_{n=1}^{\infty} \operatorname{Log} z_n$ and $\sum_{n=1}^{\infty} \operatorname{Log} w_n$, differing only in their first terms, either converge or diverge together.) Thus, we may assume that the principal logarithm function is continuous at p. Since $p_n \to p$, we conclude that $\operatorname{Log} p_n \to \operatorname{Log} p$. The relation $e^{s_n} = p_n$ stamps s_n as a logarithm of p_n. Unfortunately, s_n need not be the principal logarithm of p_n. We can, however, express this number in the manner $s_n = \operatorname{Log} p_n + 2k_n \pi i$ for some integer k_n. The convergence of $\prod_{n=1}^{\infty} z_n$ implies that $z_n \to 1$ as $n \to \infty$. Consequently,

$$\lim_{n \to \infty} (k_{n+1} - k_n) = (2\pi i)^{-1} \lim_{n \to \infty} (s_{n+1} - s_n - \operatorname{Log} p_{n+1} + \operatorname{Log} p_n)$$

$$= (2\pi i)^{-1} \lim_{n \to \infty} (\operatorname{Log} z_{n+1} - \operatorname{Log} p_{n+1} + \operatorname{Log} p_n)$$

$$= (2\pi i)^{-1} (0 - \operatorname{Log} p + \operatorname{Log} p) = 0 \ .$$

Because $k_{n+1} - k_n$ is an integer, there must exist an index N with the property that $k_{n+1} - k_n = 0$ once $n \geq N$; i.e., $k_n = k_N = k$ for every $n \geq N$. We infer that $s_n = \operatorname{Log} p_n + 2k_n \pi i \to \operatorname{Log} p + 2k\pi i$. In other words, the series $\sum_{n=1}^{\infty} \operatorname{Log} z_n$ converges, its sum being $\operatorname{Log} p + 2k\pi i$ for some integer k. We have already observed that the convergence of $\sum_{n=1}^{\infty} \operatorname{Log} z_n$ guarantees the validity of (10.13). ∎

If it is only known that $z_n \neq 0$ for $n > N$, then it is the convergence of $\sum_{n=N+1}^{\infty} \operatorname{Log} z_n$ that is equivalent to the convergence of $\prod_{n=1}^{\infty} z_n$.

There is a notion of absolute convergence for infinite products that is analogous to absolute convergence for infinite series. The definition is most easily stated for products of the type $\prod_{n=1}^{\infty} (1 + w_n)$: $\prod_{n=1}^{\infty} (1 + w_n)$ is *absolutely convergent* provided $\prod_{n=1}^{\infty} (1 + |w_n|)$ is convergent. Since

$$\lim_{w \to 0} \frac{|\operatorname{Log}(1+w)|}{|w|} = 1 \ ,$$

the comparison test implies that, for any complex sequence $\langle w_n \rangle$ tending to zero, the three series

$$(10.14) \quad \sum_{n=1}^{\infty} |\operatorname{Log}(1+w_n)| \ , \quad \sum_{n=1}^{\infty} \operatorname{Log}(1+|w_n|) \ , \quad \sum_{n=1}^{\infty} |w_n|$$

2. The Theorem of Weierstrass

either converge or diverge in unison. (In the first series ignore those terms — there are at most a finite number of them — in which $w_n = -1$.) According to Theorem 2.1 the infinite product $\prod_{n=1}^{\infty}(1+|w_n|)$ converges if and only if the infinite series $\sum_{n=1}^{\infty} \text{Log}(1+|w_n|)$ does. As a consequence, the absolute convergence of a product $\prod_{n=1}^{\infty}(1+w_n)$ is equivalent to the convergence of any (hence, all) of the series in (10.14). For example, the fact that the series $\sum_{n=1}^{\infty} n^{-2}$ converges implies that the infinite product $\prod_{n=1}^{\infty}(1+n^{-2}z)$ is absolutely convergent for every complex number z, whereas the divergence of $\sum_{n=1}^{\infty} n^{-1}$ means that $\prod_{n=1}^{\infty}[1+(-1)^n n^{-1}z]$ converges absolutely only if $z = 0$. For a product written in the straightforward fashion $\prod_{n=1}^{\infty} z_n$, with $z_n \to 1$, it is the first of the series in (10.14) that is the simplest to deal with when it comes to articulating a convenient criterion for absolute convergence: $\prod_{n=1}^{\infty} z_n$ converges absolutely if and only if $\sum_{n=N}^{\infty} |\text{Log } z_n|$ is convergent, where N is taken large enough to insure that $z_n \neq 0$ is satisfied whenever $n \geq N$. Note especially that the absolute convergence of a product $\prod_{n=1}^{\infty} z_n$ imparts absolute convergence — and, with it, convergence — to the series $\sum_{n=N}^{\infty} \text{Log } z_n$, as soon as N is suitably large. On the basis of Theorem 2.1 we can thus assert that an absolutely convergent infinite product is, in fact, convergent. Furthermore, the factors in an absolutely convergent infinite product can be rearranged at will without fear of upsetting the convergence or changing the value of the product: if $\prod_{n=1}^{\infty} z_n$ converges absolutely, then $\prod_{n=1}^{\infty} z_n = \prod_{n=1}^{\infty} z_{\sigma(n)}$ for every permutation σ of the positive integers. When all factors z_n are non-zero, this follows from (10.13) and the corresponding fact for absolutely convergent series,

$$\prod_{n=1}^{\infty} z_n = \exp\left(\sum_{n=1}^{\infty} \text{Log } z_n\right) = \exp\left(\sum_{n=1}^{\infty} \text{Log } z_{\sigma(n)}\right) = \prod_{n=1}^{\infty} z_{\sigma(n)} ;$$

when $z_n = 0$ for some n, it is not hard to see that $\prod_{n=1}^{\infty} z_n = \prod_{n=1}^{\infty} z_{\sigma(n)} = 0$.

2.2 Infinite Products of Functions

Assume that the domain-set of each function in a sequence $\langle f_n \rangle$ includes a subset A of the complex plane. We proclaim the infinite product $\prod_{n=1}^{\infty} f_n$ *pointwise convergent in A* if for each fixed z in A the infinite product $\prod_{n=1}^{\infty} f_n(z)$, whose n^{th} factor is the value of f_n at the given point, converges. This being the case, we can define a function $f: A \to \mathbb{C}$ through the rule of correspondence $f(z) = \prod_{n=1}^{\infty} f_n(z)$. Naturally, we write $f = \prod_{n=1}^{\infty} f_n$ in A. Owing to the way in which infinite products have been defined, the set of zeros of the product f is nothing but the union $\cup_{n=1}^{\infty} Z_n$ of the sets $Z_n = \{z \in A : f_n(z) = 0\}$.

As was true for infinite series of functions, pointwise convergence for infinite products of functions is just too weak a form of convergence to serve adequately our purposes. Once again the mode of convergence that best meets our needs is uniform convergence — or, to be more exact, normal convergence. An infinite product of functions $\prod_{n=1}^{\infty} f_n$ is said to *converge uniformly on a set A* provided (i) $\prod_{n=1}^{\infty} f_n$ is pointwise convergent in A and (ii) there exists an index N with the following two properties: f_n is free of zeros in A whenever $n > N$ and the partial product $f_{N+1} f_{N+2} \cdots f_{N+n}$ tends to the truncated infinite product $\prod_{n=N+1}^{\infty} f_n$ uniformly on A as $n \to \infty$. (N.B. Under this definition, which is a standard one, it is not automatically the case that the uniform convergence of $\prod_{n=1}^{\infty} f_n$ on a set A carries with it the uniform convergence on A of all the truncated products $\prod_{n=N+1}^{\infty} f_n, N = 1, 2, 3, \cdots$. Extra information is needed if such a conclusion is to be drawn. For instance, if each individual factor f_n happens to satisfy a condition $a_n \leq |f_n(z)| \leq b_n$ for every z in A, where a_n and b_n are positive constants — this is definitely true when the set A is compact and each factor f_n is both continuous and zero-free in A — then the uniform convergence of $\prod_{n=1}^{\infty} f_n$ on A does imply the same type of convergence for all truncated products.) The statement that an infinite product $\prod_{n=1}^{\infty} f_n$ *converges normally in an open set U* means that it converges uniformly on each compact subset of U. When the factors f_n are continuous functions in U, the normal convergence of $\prod_{n=1}^{\infty} f_n$ in this open set is implied by the uniform convergence of this product on every closed disk in U. The most frequently cited criterion for the uniform convergence of an infinite product is a carry-over of the Weierstrass M-test from series to products.

Theorem 2.2. *Suppose that each of the factors in an infinite product of functions $\prod_{n=1}^{\infty} f_n$ is defined on a set A. If there exists a sequence $\langle M_n \rangle$ of real numbers such that $|f_n(z) - 1| \leq M_n$ is satisfied for every z in A and such that the series $\sum_{n=1}^{\infty} M_n$ converges, then $\prod_{n=1}^{\infty} f_n$ converges absolutely and uniformly on A.*

Proof. Set $M = \prod_{n=1}^{\infty}(1 + M_n)$. The fact that $\sum_{n=1}^{\infty} M_n$ converges and $M_n \geq 0$ accounts for the convergence of this product. (Recall the discussion surrounding (10.14).) Moreover, on the basis of the comparison test we can assert that for each fixed z in A the series $\sum_{n=1}^{\infty} |f_n(z) - 1|$ is convergent, a detail that insures the absolute convergence of the product $\prod_{n=1}^{\infty} f_n(z)$. It follows that $\prod_{n=1}^{\infty} f_n$ is at least pointwise convergent in A. We now select and fix an index N with the property that $\sum_{k=N+1}^{\infty} M_k < 1/2$. If $n > N$, then the function f_n can have no zeros in A, for the reason that the inequality $|f_n(z) - 1| \leq M_n \leq \sum_{k=N+1}^{\infty} M_k < 1/2$ is in force there. To finish the proof it is enough to demonstrate that the convergence of $g_n = f_{N+1} f_{N+2} \cdots f_{N+n}$ to $g = \prod_{n=N+1}^{\infty} f_n$ is uniform on A. We shall certainly accomplish this if we can verify that the estimate

$$(10.15) \qquad |g(z) - g_n(z)| \leq 2eM \sum_{k=N+n+1}^{\infty} M_k$$

2. The Theorem of Weierstrass

is valid in A, for the right-hand side of (10.15) is independent of z and tends to 0 as $n \to \infty$. Fix $n \geq 1$. Since $|f_k(z) - 1| < 1/2$ holds everywhere in A once $k > N$ and since

$$|\operatorname{Log}(1+w)| = \left|\int_1^{1+w} \frac{dz}{z}\right| \leq \int_1^{1+w} \frac{|dz|}{|z|} \leq 2|w|$$

when $|w| < 1/2$, our choice of N determines that

$$\left|\sum_{k=N+n+1}^{\infty} \operatorname{Log} f_k(z)\right| \leq \sum_{k=N+n+1}^{\infty} |\operatorname{Log} f_k(z)|$$

$$\leq 2 \sum_{k=N+n+1}^{\infty} |f_k(z) - 1| \leq 2 \sum_{k=N+n+1}^{\infty} M_k < 1$$

for every point z of A. To repeat, the inequality

(10.16) $$\left|\sum_{k=N+n+1}^{\infty} \operatorname{Log} f_k(z)\right| \leq 2 \sum_{k=N+n+1}^{\infty} M_k < 1$$

holds throughout A. When $|w| \leq 1$ one obtains a simple bound for $|e^w - 1|$ via

$$|e^w - 1| = \left|\int_0^w e^z dz\right| \leq \int_0^w e^{\operatorname{Re} z} |dz| \leq e|w|,$$

so referring first to (10.13) and then to (10.16) we estimate

$$|g(z) - g_n(z)| = |f_{N+1}(z)| \cdots |f_{N+n}(z)| \left|\exp\left[\sum_{k=N+n+1}^{\infty} \operatorname{Log} f_k(z)\right] - 1\right|$$

$$\leq (1 + M_{N+1}) \cdots (1 + M_{N+n}) \cdot e \left|\sum_{k=N+n+1}^{\infty} \operatorname{Log} f_k(z)\right|$$

$$\leq 2e \prod_{n=1}^{\infty} (1 + M_n) \cdot \sum_{k=N+n+1}^{\infty} M_k = 2eM \sum_{k=N+n+1}^{\infty} M_k$$

for every z belonging to A. This confirms (10.15) and, in so doing, completes the proof of the theorem. ∎

2.3 Infinite Products and Analytic Functions

The analogue of Theorem VII.3.2 in the context of infinite products is the following proposition.

Theorem 2.3. *Suppose that each of the factors in an infinite product of functions $\prod_{n=1}^{\infty} f_n$ is analytic in an open set U and that $\prod_{n=1}^{\infty} f_n$ converges normally in U, with the function f as its product. Then f is analytic in U. Furthermore, under the assumption that f does not vanish identically in any component of U, the series of meromorphic functions $\sum_{n=1}^{\infty}(f_n'/f_n)$ is normally convergent in U, where its sum is the function f'/f.*

Proof. To prove that f is analytic in U we need only certify its analyticity in the interior of every closed disk that is contained in U. Let $K = \overline{\Delta}(z_0, r)$ be such a disk. The normal convergence of $\prod_{n=1}^{\infty} f_n$ in U demands the uniform convergence of this infinite product on K. We are thus able to fix an index N such that f_n is free of zeros in K as soon as $n > N$ and such that $g_n = f_{N+1} f_{N+2} \cdots f_{N+n}$ converges to the truncated product $g = \prod_{n=N+1}^{\infty} f_n$ uniformly on K as $n \to \infty$. It follows, in particular, that $g_n \to g$ normally in the open disk $\Delta = \Delta(z_0, r)$. Since each of the functions g_n is analytic in Δ, Theorem VII.3.1 attests to the fact that g is analytic in Δ. By (10.11) we are allowed to write $f = f_1 f_2 \cdots f_N g$ in Δ, which makes evident the analyticity of f there.

Assume now that f does not vanish identically in any component of U. Then plainly none of its factors f_n can be identically zero in a component of U. Accordingly, the functions f'/f and f_n'/f_n for $n \geq 1$ are all meromorphic in U. We show that, given a closed disk $K = \overline{\Delta}(z_0, r)$ in U, we can produce an index N such that no f_n with $n > N$ has a zero in K and such that the series $\sum_{n=N+1}^{\infty}(f_n'/f_n)$ converges uniformly on K. In the process we shall check that $f'/f = \sum_{n=1}^{\infty}(f_n'/f_n)$ in K. The obvious implication of these remarks is that, in the sense appropriate to series of meromorphic functions, $\sum_{n=1}^{\infty}(f_n'/f_n)$ converges normally in U, where its sum is f'/f. In order to carry out this program we are forced to work in a disk slightly larger than the given one. For this reason we fix $s > r$ with the property that the disk $K_0 = \overline{\Delta}(z_0, s)$ still lies in U. Due to the uniform convergence of $\prod_{n=1}^{\infty} f_n$ on K_0, we can fix an index N so that for every $n > N$ the function f_n is zero-free in K_0 and so that $g_n = f_{N+1} f_{N+2} \cdots f_{N+n} \to g = \prod_{n=N+1}^{\infty} f_n$ uniformly on K_0. Notice that the functions g_n and g are all free of zeros in K_0. As $g_n \to g$ normally in the disk $\Delta_0 = \Delta(z_0, s)$, Theorem VII.3.1 tells us that $g_n' \to g'$ normally in Δ_0. We make the further claim that in the present situation the convergence of g_n'/g_n to g'/g is also normal in Δ_0. Let us suppose for an instant that we can support this claim. An easy calculation reveals that

$$\frac{g_n'}{g_n} = \frac{f_{N+1}'}{f_{N+1}} + \frac{f_{N+2}'}{f_{N+2}} + \cdots + \frac{f_{N+n}'}{f_{N+n}},$$

the nth partial sum of the series $\sum_{n=N+1}^{\infty}(f_n'/f_n)$. We infer that this series is normally convergent in Δ_0 — hence, uniformly convergent on K — its

2. The Theorem of Weierstrass

sum there being g'/g. Because $f = f_1 f_2 \cdots f_N g$ in Δ_0, we thus obtain

$$\frac{f'}{f} = \frac{f_1'}{f_1} + \frac{f_2'}{f_2} + \cdots + \frac{f_N'}{f_N} + \frac{g'}{g} = \sum_{n=1}^{\infty} \frac{f_n'}{f_n}$$

in this disk.

All that is left to prove is that $g_n'/g_n \to g'/g$ uniformly on each compact set in Δ_0. Fix such a set, say A. Because the functions involved are continuous and nowhere zero in A, the numbers $a_n = \min\{|g_n(z)| : z \in A\}$ and $a = \min\{|g(z)| : z \in A\}$ are all positive. Also, the fact that $g_n \to g$ uniformly on A implies that $a_n \to a$. As a result, it is clear that $b = \min\{a, a_1, a_2, \cdots\} > 0$. Let $c = b^{-2} \max\{|g(z)| + |g'(z)| : z \in A\}$. Then for every z in A and every $n \geq 1$ we have

$$\left| \frac{g'(z)}{g(z)} - \frac{g_n'(z)}{g_n(z)} \right| = \frac{|g_n(z) g'(z) - g(z) g_n'(z)|}{|g(z)||g_n(z)|}$$

$$\leq \frac{|g'(z)||g_n(z) - g(z)| + |g(z)||g'(z) - g_n'(z)|}{|g(z)||g_n(z)|}$$

$$\leq c|g(z) - g_n(z)| + c|g'(z) - g_n'(z)| .$$

The last expression tends to zero uniformly on A, thereby forcing the uniform convergence of g_n'/g_n to g'/g on this set. ∎

The efforts we have invested in the last two theorems have an immediate payoff in the form of a lovely identity.

EXAMPLE 2.1. Show that

(10.17)
$$\sin(\pi z) = \pi z \prod_{n=1}^{\infty} \left(1 - \frac{z^2}{n^2}\right)$$

for every complex number z.

For $n = 1, 2, 3, \cdots$ set $f_n(z) = 1 - (z/n)^2$. If K is a compact set in $\mathbb{C}$ and if $c = \max\{|z|^2 : z \in K\}$, then

$$|f_n(z) - 1| = \frac{|z|^2}{n^2} \leq \frac{c}{n^2}$$

for every z in K. Since the series $\sum_{n=1}^{\infty} n^{-2}$ converges, Theorem 2.2 allows us to declare the product $\prod_{n=1}^{\infty} f_n$ uniformly convergent on K. It follows that $\prod_{n=1}^{\infty} f_n$ converges normally in the whole complex plane. We conclude by way of Theorem 2.3 that the formula $f(z) = z \prod_{n=1}^{\infty} [1 - (z/n)^2]$ defines an entire function, one with a simple zero at every integer and no other zeros. Theorem 2.3 also provides a little extra information about f — namely, that the meromorphic function f'/f admits the representation

$$\frac{f'(z)}{f(z)} = \frac{1}{z} + \sum_{n=1}^{\infty} \frac{2z}{z^2 - n^2}$$

in $\mathbb{C}$. In Example 1.1 we learned that for all complex numbers z

$$\pi \cot(\pi z) = \frac{1}{z} + \sum_{|n| \geq 1} \left(\frac{1}{z-n} + \frac{1}{n} \right)$$

$$= \frac{1}{z} + \sum_{n=1}^{\infty} \left(\frac{1}{z-n} + \frac{1}{n} + \frac{1}{z+n} - \frac{1}{n} \right)$$

$$= \frac{1}{z} + \sum_{n=1}^{\infty} \frac{2z}{z^2 - n^2}.$$

The identity of f'/f is thus revealed to us.

The function $g(z) = \sin(\pi z)$ is a second entire function whose only zeros are simple zeros at the integers. Therefore, the singularities of $h = f/g$ in $\mathbb{C}$ are removable. Upon their removal h becomes an entire function — in fact, an entire function without zeros. What is more,

$$h'(z) = \frac{h'(z)h(z)}{h(z)} = h(z) \left[\frac{f'(z)}{f(z)} - \frac{g'(z)}{g(z)} \right] = h(z)[\pi \cot(\pi z) - \pi \cot(\pi z)] = 0$$

throughout the complex plane, making h a constant function there. Since $h(0) = \lim_{z \to 0}[f(z)/g(z)] = \pi^{-1}$, the sole value that h assumes is π^{-1}. This leads to

$$\sin(\pi z) = g(z) = \frac{f(z)}{h(z)} = \pi z \prod_{n=1}^{\infty} \left(1 - \frac{z^2}{n^2} \right)$$

for every z.

Consider again a plane open set U and a discrete subset $E = \{z_1, z_2, \cdots\}$ of U. Given a sequence $\langle m_n \rangle$ of positive integers, we would like to construct a function f that is analytic in U and that has E as its set of zero there, the zero at z_n having order m_n. If $E = \{z_1, z_2, \ldots, z_p\}$ is a finite set, this creates no problem: we just take $f(z) = (z-z_1)^{m_1}(z-z_2)^{m_2} \cdots (z-z_p)^{m_p}$. One might expect that $f(z) = \prod_{n=1}^{\infty}(z - z_n)^{m_n}$ would do the trick in the case of infinite E, but such hopes are dashed by the fact that this infinite product does not, in general, converge normally in U. Something akin to the convergence inducing summands of Mittag-Leffler's construction is called for here. Fortunately, precisely the right thing is available, as the proof of the main result in this section makes apparent.

Theorem 2.4. (Weierstrass's Theorem) *Let $E = \{z_1, z_2, z_3, \cdots\}$ be a discrete subset of an open set U in the complex plane, and for $n = 1, 2, 3, \cdots$ let m_n be a positive integer. There exists an analytic function $f: U \to \mathbb{C}$ that has E as its set of zeros, the zero at z_n being one of order m_n for $n = 1, 2, 3, \cdots$. The quotient of any two functions fitting this description is both analytic and zero-free in U.*

2. The Theorem of Weierstrass

Proof. We assume that the set E is infinite, having already dealt with the situation for a finite set E in essentially trivial fashion. We also assume for the purposes of the construction in this proof that the origin is not a point of E. (If the origin does belong to E, we simply carry out the ensuing construction for the set $E \sim \{0\}$ and then multiply the result by z^m, where m is the order stipulated for the zero at the origin.) Let δ_n denote the radius of the largest open disk centered at z_n that is contained in U. Mirroring the proof of Mittag-Leffler's theorem, we initially consider two special cases.

Case 1: $|z_n|\delta_n \geq 1$ for every n. (N.B. If $U = \mathbb{C}$, then Case 1 is the general case.) We recall that $|z_n| \to \infty$ as $n \to \infty$ in this case. The function $L_n(z) = \text{Log}(1 - z_n^{-1}z)$ is analytic in the open disk $\Delta(0, |z_n|)$, where its Taylor series expansion reads

$$\text{Log}\left(1 - \frac{z}{z_n}\right) = -\sum_{k=1}^{\infty} \frac{1}{k}\left(\frac{z}{z_n}\right)^k.$$

For $n = 1, 2, 3, \cdots$ we choose and fix an index $d(n)$ with the property that

(10.18) $$m_n \sum_{k=d(n)+1}^{\infty} \frac{1}{k}\left(\frac{1}{2}\right)^k \leq \left(\frac{1}{2}\right)^n.$$

Our reason for doing this is that the bound

(10.19) $$\left| m_n \text{Log}\left(1 - \frac{z}{z_n}\right) + m_n \sum_{k=1}^{d(n)} \frac{1}{k}\left(\frac{z}{z_n}\right)^k \right| \leq \left(\frac{1}{2}\right)^n$$

then holds for every z in the closed disk $K_n = \overline{\Delta}(0, |z_n|/2)$, as follows from (10.18) via the computation

$$\left| m_n \text{Log}\left(1 - \frac{z}{z_n}\right) + m_n \sum_{k=1}^{d(n)} \frac{1}{k}\left(\frac{z}{z_n}\right)^k \right| = \left| -m_n \sum_{k=d(n)+1}^{\infty} \frac{1}{k}\left(\frac{z}{z_n}\right)^k \right|$$

$$\leq m_n \sum_{k=d(n)+1}^{\infty} \frac{1}{k}\left(\frac{|z|}{|z_n|}\right)^k \leq m_n \sum_{k=d(n)+1}^{\infty} \frac{1}{k}\left(\frac{1}{2}\right)^k \leq \left(\frac{1}{2}\right)^n.$$

We now define a function f_n by

$$f_n(z) = \left(1 - \frac{z}{z_n}\right)^{m_n} \exp\left[m_n \sum_{k=1}^{d(n)} \frac{1}{k}\left(\frac{z}{z_n}\right)^k\right].$$

This function is clearly an entire function whose only zero is one of order m_n at z_n. Referring to (10.19) and remembering that $|e^w - 1| \leq e|w|$ when

$|w| \leq 1$, we compute for any z in the set K_n

$$|f_n(z) - 1| = \left| \exp\left[m_n \operatorname{Log}\left(1 - \frac{z}{z_n}\right) + m_n \sum_{k=1}^{d(n)} \frac{1}{k}\left(\frac{z}{z_n}\right)^k \right] - 1 \right|$$

$$\leq e \left| m_n \operatorname{Log}\left(1 - \frac{z}{z_n}\right) + m_n \sum_{k=1}^{d(n)} \frac{1}{k}\left(\frac{z}{z_n}\right)^k \right| \leq 2^{-n} e \;;$$

i.e., the estimate

(10.20) $$|f_n(z) - 1| \leq 2^{-n} e$$

is valid throughout K_n. Let K be an arbitrary compact set in the complex plane. The fact that $|z_n| \to \infty$ as $n \to \infty$ means that K is a subset of K_n — hence, that (10.20) holds for every z in K — once n is sufficiently large. This implies that the series $\sum_{n=1}^{\infty} M_n$ is convergent, where we take $M_n = \max\{|f_n(z) - 1| : z \in K\}$. On the strength of Theorem 2.2 we conclude that the product $\prod_{n=1}^{\infty} f_n$ converges uniformly on K. Consequently, this product is normally convergent in $\mathbb{C}$. Theorem 2.3 informs us that $f = \prod_{n=1}^{\infty} f_n$ is an entire function, one whose zero-set is E and whose zero at z_n has order m_n for $n = 1, 2, 3, \cdots$. In Case 1, therefore, we have exhibited a function with all the desired features (and even more, for f is an entire function, not just a function that is analytic in U).

Case 2: $|z_n|\delta_n < 1$ for every n. Here, as we recall from the proof of Mittag-Leffler's theorem, $\delta_n \to 0$ as $n \to \infty$. Copying what we did in that argument, we select and fix for $n = 1, 2, 3, \cdots$ a point of ζ_n of ∂U for which $|z_n - \zeta_n| = \delta_n$. Then $|(z_n - \zeta_n)/(z - \zeta_n)| < 1$ holds for every point z of the annulus $G_n = \{z : \delta_n < |z - \zeta_n| < \infty\}$, which means that the function L_n defined in G_n by

$$L_n(z) = \operatorname{Log}\left[1 - \left(\frac{z_n - \zeta_n}{z - \zeta_n}\right)\right] = \operatorname{Log}\left(\frac{z - z_n}{z - \zeta_n}\right)$$

is analytic. If we represent L_n in G_n by its Laurent expansion centered at ζ_n, we learn that

$$\operatorname{Log}\left(\frac{z - z_n}{z - \zeta_n}\right) = -\sum_{k=1}^{\infty} \frac{1}{k}\left(\frac{z_n - \zeta_n}{z - \zeta_n}\right)^k$$

for every z in G_n. Also, if $d(n)$ is again selected to insure that (10.18) is true, we discover in the present situation that

(10.21) $$\left| m_n \operatorname{Log}\left(\frac{z - z_n}{z - \zeta_n}\right) + m_n \sum_{k=1}^{d(n)} \frac{1}{k}\left(\frac{z_n - \zeta_n}{z - \zeta_n}\right)^k \right| \leq \left(\frac{1}{2}\right)^n$$

2. The Theorem of Weierstrass

whenever z belongs to the set $A_n = \{z : 2\delta_n \leq |z - \zeta_n| < \infty\}$.
In Case 2 we introduce the function g_n:

$$g_n(z) = \left(\frac{z - z_n}{z - \zeta_n}\right)^{m_n} \exp\left[m_n \sum_{k=1}^{d(n)} \frac{1}{k}\left(\frac{z_n - \zeta_n}{z - \zeta_n}\right)^k\right].$$

Since ζ_n lies on ∂U, g_n is analytic in U, where its sole zero is a zero of multiplicity m_n at z_n. From (10.21) we conclude by means of calculations similar to those which produced (10.20) in Case 1 that the bound

(10.22) $\qquad |g_n(z) - 1| \leq 2^{-n} e$

holds for every element z of A_n. The fact that $\delta_n \to 0$ implies that any given compact subset K of D is contained in A_n as soon as n is suitably large. In conjunction with (10.22) and Theorem 2.2 this observation lets us know that the product $\prod_{n=1}^{\infty} g_n$ converges uniformly on each such K — hence, normally in U. Once more invoking Theorem 2.3 we find in Case 2 that the function $f = \prod_{n=1}^{\infty} g_n$ meets all the requirements of the theorem.

An infinite discrete subset E of U that does not already come under the umbrella of Case 1 or Case 2 can be written as a disjoint union $E = E_1 \cup E_2$, where E_1 either is finite or fits into Case 1 and where E_2 is either finite or covered by Case 2. We can thus construct analytic functions g and h in U whose respective zero-sets are E_1 and E_2, each zero being of the order stipulated in the specifications for E. Then $f = gh$ has all the properties demanded by the theorem. If $\tilde{f}$ is a second function with the same properties, then the quotient $\tilde{f}/f$ has only removable singularities in U, at each of which it has a non-zero limit. Upon removal of these singularities, $\tilde{f}/f$ becomes analytic and zero-free in U, which explains the final assertion of the theorem. ∎

Except for changing the word "analytic" to "holomorphic," Theorem 2.4 remains valid for an open set U in the extended complex plane $\widehat{\mathbb{C}}$, provided $U \neq \widehat{\mathbb{C}}$. (Since the only functions holomorphic in all of $\widehat{\mathbb{C}}$ are constant functions, Theorem 2.4 cannot be true for $U = \widehat{\mathbb{C}}$.) The proof is left as an exercise for the reader, with the following reminder: for a function f to have a zero of order m at ∞ means that $g(z) = f(z^{-1})$ has a zero of order m at the origin.

An immediate consequence of Theorem 2.4 is that an entire function f (to avoid trivial cases, assume that f is not identically zero but has an infinite number of zeros) admits representations of the type

$$f(z) = g(z) z^m \prod_{n=1}^{\infty} \left(1 - \frac{z}{z_n}\right)^{m_n} E_n(z),$$

where g is a zero-free entire function, m is a non-negative integer, the numbers $z_1, z_2, z_3, \cdots$ are the non-zero roots of f, m_n is the order of the zero that f has at z_n, and E_n is a function of the form

$$E_n(z) = \exp\left[\sum_{k=1}^{d(n)} \frac{1}{k}\left(\frac{z}{z_n}\right)^k\right]$$

for a suitably large non-negative integer $d(n)$. (We do not exclude the possibility of being able to choose $d(n) = 0$, in which case we would interpret this formula to mean $E_n = 1$.) Any such representation is called a *Weierstrass product expansion* of f.

Because we approached Theorem 2.4 primarily as an existence theorem, no serious effort was made in its proof to be efficient about the choice of "convergence inducing factors." In concrete situations these factors can often be selected much more economically than would be suggested by that proof, if not dispensed with entirely. As a case in point, just consider a discrete subset $E = \{z_1, z_2, z_3, \cdots\}$ of the complex plane with the property that the series $\sum_{n=1}^{\infty} |z_n|^{-p-1}$ is convergent, where p is a non-negative integer. If $p = 0$, then the function $f_n(z) = 1 - (z/z_n)$ satisfies

$$|f_n(z) - 1| = \frac{|z|}{|z_n|}$$

for every z in $\mathbb{C}$; if $p \geq 1$, then for all z belonging to the disk $K_n = \overline{\Delta}(0, |z_n|/2)$ we have

$$\left|\operatorname{Log}\left(1 - \frac{z}{z_n}\right) + \sum_{k=1}^{p} \frac{1}{k}\left(\frac{z}{z_n}\right)^k\right| = \left|-\sum_{k=p+1}^{\infty} \frac{1}{k}\left(\frac{|z|}{|z_n|}\right)^k\right|$$

$$\leq \frac{1}{2}\sum_{k=p+1}^{\infty}\left(\frac{|z|}{|z_n|}\right)^k = \frac{1}{2}\left(\frac{|z|}{|z_n|}\right)^{p+1}\left(1 - \frac{|z|}{|z_n|}\right)^{-1} \leq \left(\frac{|z|}{|z_n|}\right)^{p+1},$$

so for such z the function

$$f_n(z) = \left(1 - \frac{z}{z_n}\right)\exp\left[\sum_{k=1}^{p} \frac{1}{k}\left(\frac{z}{z_n}\right)^k\right]$$

is readily seen to obey the estimate

$$|f_n(z) - 1| \leq e\left(\frac{|z|}{|z_n|}\right)^{p+1}.$$

(Don't forget that $|e^w - 1| \leq e|w|$ when $|w| \leq 1$.) In either case the convergence of $\sum_{n=1}^{\infty} |z_n|^{-p-1}$ forces the uniform convergence of the product $\prod_{n=1}^{\infty} f_n$ on each compact set in $\mathbb{C}$. As a consequence,

(10.23) $$f(z) = \prod_{n=1}^{\infty}\left(1 - \frac{z}{z_n}\right)$$

2. The Theorem of Weierstrass

when $p = 0$ or

(10.24) $$f(z) = \prod_{n=1}^{\infty} \left(1 - \frac{z}{z_n}\right) \exp\left[\sum_{k=1}^{p} \frac{1}{k}\left(\frac{z}{z_n}\right)^k\right]$$

when $p \geq 1$ defines an entire function with a simple zero at each point of E and with no other zeros. Significant here is that in the case of (10.23) no convergence inducing factors are needed, while in (10.24) the limits on the sums appearing in those factors are independent of n.

An important consequence of the Weierstrass theorem is a clarification of the structure of meromorphic functions.

Theorem 2.5. *Suppose that a function f is meromorphic in an open subset U of the complex plane. Then f can be represented in U as $f = g/h$, where $g: U \to \mathbb{C}$ and $h: U \to \mathbb{C}$ are analytic functions that have no common zeros.*

Proof. If f is already analytic in U, we simply use $g = f$ and $h = 1$. If not, the set E of poles of f in U is a non-empty, discrete subset of U. Appealing to Weierstrass's theorem, we construct an analytic function h in U whose zero-set there is E and whose zero at any point of E has the same order as the pole of f at that point. The function $g = fh$ is then meromorphic in U, where its set of singularities is E. A moment's thought reveals that g possesses a finite, non-zero limit at each such singularity. These singularities are thus seen to be removable. Upon their removal g becomes an analytic function in U, one for which $f = g/h$ in this open set. By construction, the set of zeros of g is disjoint from E, the zero-set of h. ∎

In the representation $f = g/h$ of Theorem 2.5 the zeros of f in U are the zeros of g, while its poles are the zeros of h. Incidentally, this representation is not unique, since we can obviously write $f = (gk)/(hk)$ for any function k that is both analytic and free of zeros in U. On the other hand, it can be shown that this is the only departure from uniqueness in such a representation. Theorem 2.5 carries over to any open set U in $\widehat{\mathbb{C}}$, apart from $U = \widehat{\mathbb{C}}$, provided one transcribes "analytic" to "holomorphic."

Mittag-Leffler's theorem and the Weierstrass theorem can also be effective when used in concert, to which the proof of the next theorem bears witness. The theorem states that, given a discrete subset of a plane open set U, we can construct a function analytic in U whose Taylor series expansions about the points of E have predetermined initial segments.

Theorem 2.6. *Let $E = \{z_1, z_2, z_3, \cdots\}$ be a discrete subset of an open set U in the complex plane, and for $n = 1, 2, 3, \cdots$ let $p_n(z) = \sum_{k=0}^{d_n} a_k^{(n)} z^k$ be a polynomial of degree d_n. There exists an analytic function $f: U \to \mathbb{C}$ that for $n = 1, 2, 3, \cdots$ has a Taylor series expansion about z_n of the form*

$$f(z) = p_n(z - z_n) + O[(z - z_n)^{1+d_n}].$$

Proof. (We refer the reader to Example VIII.2.3. for a review of "big O notation.") Through application of the Weierstrass theorem we can create an analytic function $g: U \to \mathbb{C}$ that has E as its set of zeros, the zero at z_n being one of order $1 + d_n$ for $n = 1, 2, 3, \cdots$. The Taylor expansion of g about z_n thus has the appearance

$$g(z) = c_n(z - z_n)^{1+d_n} + O[(z - z_n)^{2+d_n}],$$

in which $c_n \neq 0$. The function h_n defined in U by $h_n(z) = p_n(z - z_n)/g(z)$ has a pole at z_n. Let S_n be the singular part of h_n at z_n. Using Mittag-Leffler's theorem we construct a meromorphic function $h: U \to \widehat{\mathbb{C}}$ whose pole-set is E and whose singular part at z_n is S_n for $n = 1, 2, 3, \cdots$. For z in any suitably small punctured disk centered at z_n we have

$$h(z) = h_n(z) + O(1) = \frac{p_n(z - z_n)}{g(z)} + O(1).$$

It follows that the function $f = gh$, which is clearly meromorphic in U and which has E as its singular set in U, shows a Laurent expansion of the form

$$f(z) = p_n(z - z_n) + g(z)O(1) = p_n(z - z_n) + O[(z - z_n)^{1+d_n}]$$

in such a punctured disk. We conclude that the singularities of f in U are all removable. Removing them turns f into a function that is analytic in U and has the specified structure near every point of E. ∎

2.4 The Gamma Function

One of the most famous special functions in mathematics comes up quite naturally in the context of infinite products of analytic functions. It is the "gamma function" of Leonhard Euler (1707-1783), a function denoted by Γ and defined by the formula

(10.25) $$\Gamma(z) = \frac{e^{-\gamma z}}{z} \prod_{n=1}^{\infty} \left(1 + \frac{z}{n}\right)^{-1} e^{z/n}.$$

Here γ is "Euler's constant,"

(10.26) $$\gamma = \lim_{n \to \infty} \left(1 + \frac{1}{2} + \cdots + \frac{1}{n} - \operatorname{Log} n\right).$$

(N.B. Since the sequence $\gamma_n = \sum_{k=1}^{n} k^{-1} - \operatorname{Log} n$ is easily seen to be decreasing and non-negative, $\gamma = \lim_{n \to \infty} \gamma_n$ exists. In rough approximation, $\gamma \approx 0.577$.) The fact that $\sum_{n=1}^{\infty} n^{-2}$ converges implies — recall (10.24) — that the function f given by

(10.27) $$f(z) = ze^{\gamma z} \prod_{n=1}^{\infty} \left(1 + \frac{z}{n}\right) e^{-z/n}$$

2. The Theorem of Weierstrass

is an entire function, one with a simple zero at each of the non-positive integers and with no other zeros. Therefore, the function $\Gamma = 1/f$ is meromorphic in the whole complex plane, it has simple poles at $0, -1, -2, \cdots$, and it is zero-free. The infinite product in (10.27) is normally convergent in the complex plane. The function defined by this product is bounded away from zero on each compact subset of $U = \mathbb{C} \sim \{-1, -2, -3, \cdots\}$. From these two pieces of information it follows without difficulty that the product in (10.25) converges normally in U. With justification provided by Theorem 2.3 we can state that in $\mathbb{C} \sim \{0, -1, -2, \cdots\}$

(10.28) $$\frac{\Gamma'(z)}{\Gamma(z)} = -\gamma - \frac{1}{z} - \sum_{n=1}^{\infty} \left[\frac{1}{z+n} - \frac{1}{n}\right].$$

As a matter of fact, in the sense fitting to series of meromorphic functions this series converges normally in $\mathbb{C}$, so (10.28) actually holds throughout $\mathbb{C}$, both sides taking the value ∞ at $0, -1, -2, \cdots$.

If $z \neq 0, -1, -2, \cdots$, then we obtain from (10.25) and (10.26)

$$\Gamma(z) = \frac{1}{z} \exp\left[z \lim_{n \to \infty}\left(\operatorname{Log} n - \sum_{k=1}^{n}\frac{1}{k}\right)\right] \lim_{n \to \infty} \prod_{k=1}^{n}\left(1+\frac{z}{k}\right)^{-1} e^{z/k}$$

$$= \lim_{n \to \infty}\left\{\frac{1}{z}\exp\left[z\left(\operatorname{Log} n - \sum_{k=1}^{n}\frac{1}{k}\right)\right] \prod_{k=1}^{n}\left(\frac{k}{z+k}\right) e^{z/k}\right\},$$

which simplifies to "Gauss's formula":

(10.29) $$\Gamma(z) = \lim_{n \to \infty} \frac{n!\, n^z}{z(z+1)\cdots(z+n)}.$$

Therefore, for $z \neq 0, -1, -2, \cdots$ we find that

$$\Gamma(z+1) = \lim_{n \to \infty} \frac{n!\, n^{z+1}}{(z+1)(z+2)\cdots(z+n+1)}$$

$$= \lim_{n \to \infty}\left[\frac{nz}{z+n+1} \cdot \frac{n!\, n^z}{z(z+1)\cdots(z+n)}\right] = z\,\Gamma(z).$$

In this way we arrive at the "functional equation" of the gamma function:

(10.30) $$\Gamma(z+1) = z\,\Gamma(z).$$

From (10.29) we also learn that

$$\Gamma(1) = \lim_{n \to \infty}\frac{n!\,n}{(n+1)!} = \lim_{n \to \infty}\frac{n}{n+1} = 1.$$

We conclude using (10.30) that $\Gamma(n) = n!$ for $n = 1, 2, \cdots$. Referring to (10.12) and Example 2.1, we notice that the function f in (10.27) satisfies

$$f(z)f(-z) = \left[ze^{\gamma z} \prod_{n=1}^{\infty} \left(1 + \frac{z}{n}\right) e^{-z/n}\right] \left[(-z)e^{-\gamma z} \prod_{n=1}^{\infty} \left(1 - \frac{z}{n}\right) e^{z/n}\right]$$

$$= -z^2 \prod_{n=1}^{\infty} \left(1 - \frac{z^2}{n^2}\right) = -\frac{z \sin(\pi z)}{\pi}.$$

In conjunction with the functional equation for Γ, this observation leads to another oft-cited identity satisfied by the gamma function,

(10.31) $$\Gamma(z)\Gamma(1-z) = \frac{\pi}{\sin(\pi z)}.$$

Indeed,

$$\Gamma(z)\Gamma(1-z) = -z\,\Gamma(z)\Gamma(-z) = -\frac{z}{f(z)f(-z)} = \frac{\pi}{\sin(\pi z)}.$$

Thus we see, for instance, that $[\Gamma(1/2)]^2 = \pi$. Because Γ is obviously a positive function on the interval $(0, \infty)$, we deduce that $\Gamma(1/2) = \sqrt{\pi}$.

There are an assortment of other ways to represent $\Gamma(z)$, most of them applicable for limited ranges of z. As one example, we point to

(10.32) $$\Gamma(z) = \lim_{n \to \infty} \int_0^n t^{z-1} \left(1 - \frac{t}{n}\right)^n dt,$$

valid when $\operatorname{Re} z > 0$. (N.B. When $0 < \operatorname{Re} z < 1$ the integral involved here is improper at its lower limit, but it is convergent.) We evaluate the integral in (10.32) by making the change of variable $t = ns$, $dt = n\,ds$ and then doing integration by parts repeatedly:

$$\int_0^n t^{z-1}\left(1 - \frac{t}{n}\right)^n dt = n^z \int_0^1 s^{z-1}(1-s)^n ds$$

$$= n^z \cdot \frac{n}{z} \int_0^1 s^z(1-s)^{n-1} ds = n^z \cdot \frac{n(n-1)}{z(z+1)} \int_0^1 s^{z+1}(1-s)^{n-2} ds$$

$$\cdots = \frac{n^z\, n!}{z(z+1)\cdots(z+n-1)} \int_0^1 s^{z+n-1} ds = \frac{n!\, n^z}{z(z+1)\cdots(z+n)}.$$

Statement (10.32) now follows from (10.29). A second well-known integral formula for the gamma function — in fact, a formula often used as the definition of $\Gamma(z)$ in a real-variable setting — is

(10.33) $$\Gamma(z) = \int_0^{\infty} t^{z-1} e^{-t} dt,$$

3. Analytic Continuation   507

valid under the assumption that $\operatorname{Re} z > 0$, which causes this improper integral to be convergent. In deriving (10.33) we make use of the elementary inequalities

(10.34) $$1 - \frac{t^2}{n} \leq e^t \left(1 - \frac{t}{n}\right)^n \leq 1,$$

which hold when $0 \leq t \leq n$. For $z = x + iy$ with $x > 0$, we determine that

$$\left| \int_0^n t^{z-1} e^{-t} \, dt - \int_0^n t^{z-1} \left(1 - \frac{t}{n}\right)^n dt \right| \leq \int_0^n |t^{z-1}| \left[e^{-t} - \left(1 - \frac{t}{n}\right)^n \right] dt$$

$$= \int_0^n t^{x-1} e^{-t} \left[1 - e^t \left(1 - \frac{t}{n}\right)^n \right] dt \leq \int_0^n t^{x-1} e^{-t} \left[1 - \left(1 - \frac{t^2}{n}\right) \right] dt$$

$$= \frac{1}{n} \int_0^n t^{x+1} e^{-t} \, dt \to 0$$

as $n \to \infty$, since $\lim_{n \to \infty} \int_0^n t^{x+1} e^{-t} \, dt = \int_0^\infty t^{x+1} e^{-t} \, dt$ is finite. Recalling (10.32), we infer that

$$\int_0^\infty t^{z-1} e^{-t} \, dt = \lim_{n \to \infty} \int_0^n t^{z-1} e^{-t} \, dt$$

$$= \lim_{n \to \infty} \int_0^n t^{z-1} \left(1 - \frac{t}{n}\right)^n dt = \Gamma(z)$$

when $\operatorname{Re} z > 0$ and in this manner obtain (10.33).

The gamma function has a multitude of fascinating properties which a shortage of space prevents us from going into here. Interested readers are urged to look at Emil Artin's beautiful little monograph *The Gamma Function* (Holt, Reinhart, and Winston, New York, 1964) to find out more about this function.

3 Analytic Continuation

3.1 Extending Functions by Means of Taylor Series

The final method we shall discuss for constructing analytic functions has a definite "organic" flavor to it. Much of the special vocabulary associated with the method would look perfectly at home in a biology text! Analytic continuation can be likened to growing a tree from a seedling. Here the "seedling" takes the form of a Taylor series $\sum_{n=0}^{\infty} a_n (z - z_0)^n$ with a positive radius of convergence ρ_0 or, to be more precise, the analytic function g_0 that this series defines in the disk $\Delta_0 = \Delta(z_0, \rho_0)$ through the rule of

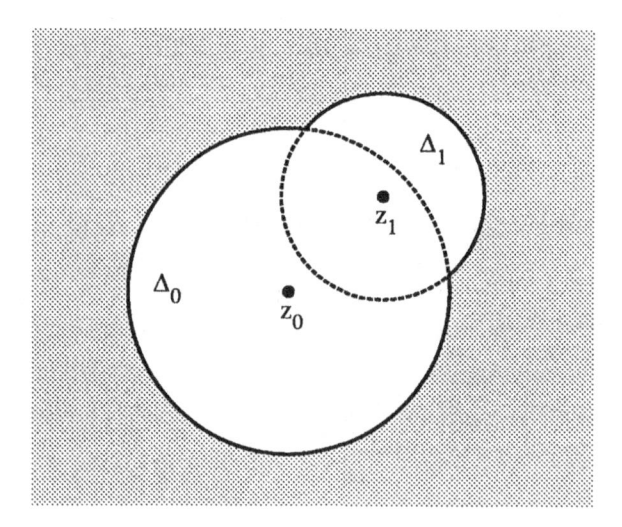

Figure 2.

correspondence $g_0(z) = \sum_{n=0}^{\infty} a_n(z - z_0)^n$. (In fact, one of the technical expressions frequently encountered in the context of analytic continuation describes g_0 not as a "seedling," but as an "analytic germ at z_0.") We can envision g_0 "growing" by a process we now sketch. Given a point z_1 of Δ_0, we can expand g_0 in a Taylor series centered at z_1 — say $g_0(z) = \sum_{n=0}^{\infty} b_n(z - z_1)^n$. We are assured by Theorem VII.3.4 that the radius of convergence ρ_1 of this series is no smaller than $\rho_0 - |z_1 - z_0|$. In particular, $\rho_1 > 0$. The function g_1 defined in $\Delta_1 = \Delta(z_1, \rho_1)$ by the formula $g_1(z) = \sum_{n=0}^{\infty} b_n(z-z_1)^n$ is analytic, and it agrees with g_0 in $\Delta_0 \cap \Delta_1$. If it happens, as well it may, that $\rho_1 = \rho_0 - |z_1 - z_0|$, then g_0 does not experience any "growth" at z_1. In this case g_1 is nothing but the restriction of g_0 to the disk Δ_1. (N.B. There exist situations in which $\rho_1 = \rho_0 - |z_1 - z_0|$ for every choice of z_1 in the disk Δ_0. Such severely "stunted growth" is exemplified by $g_0(z) = \sum_{n=1}^{\infty} z^{n!}$ in $\Delta_0 = \Delta(0, 1)$. We refer the reader to Exercise VII.5.64.) If, on the other hand, it is true that ρ_1 is larger than $\rho_0 - |z_1 - z_0|$, then the domain $D_1 = \Delta_0 \cup \Delta_1$ properly contains Δ_0 and the function f_1 given in D_1 by

$$f_1(z) = \begin{cases} g_0(z) & \text{if } z \in \Delta_0 \text{ ,} \\ g_1(z) & \text{if } z \in \Delta_1 \text{ ,} \end{cases}$$

is analytic (Figure 2). At the risk of working our analogy to death we might say that g_0 has sprouted a "branch" g_1 at z_1. In precise technical language we speak of g_0 being "analytically continued" to g_1 in Δ_1 (and also to f_1 in D_1).

Nothing prevents us from repeating the above construction; i.e., we can choose a point z_2 of D_1, expand f_1 in a Taylor series about z_2, denote by g_2 the analytic function defined by this series in its full disk of convergence

3. Analytic Continuation

Δ_2, and attempt to define a function f_2 in the domain $D_2 = \Delta_0 \cup \Delta_1 \cup \Delta_2$ by insisting that $f_2 = g_j$ in the disk Δ_j for $j = 0, 1, 2$. In other words,

$$f_2(z) = \begin{cases} f_1(z) & \text{if } z \in D_1, \\ g_2(z) & \text{if } z \in \Delta_2. \end{cases}$$

It is not difficult to see that, under these conditions, the set $D_1 \cap \Delta_2$ is still a domain. By construction, $f_1 = g_2$ in some small open disk centered at z_2. On the authority of Corollary VIII.1.6 we can say that $f_1 = g_2$ throughout $D_1 \cap \Delta_2$. As a consequence, f_2 is unambiguously defined in D_2 and is plainly analytic there. The possibility exists, of course, that Δ_2 is contained in D_1, in which event $D_2 = D_1$, $f_2 = f_1$, and this attempt to extend f_1 will have failed. If, however, Δ_2 does not lie completely in D_1, then we shall have succeeded in continuing g_0 analytically to a domain D_2 that properly includes D_1.

On the face of it we ought to be able to iterate the foregoing construction indefinitely, hoping thereby to continue g_0 analytically to larger and larger domains. Unfortunately, the process can develop a hitch. To see what that might be, let us suppose that we have successfully managed to carry out the extension procedure through $n (\geq 2)$ stages. Thus, we have produced functions $g_0, g_1, \ldots, g_n$, each the sum of a Taylor series in its disk of convergence. We label those disks $\Delta_0, \Delta_1, \ldots, \Delta_n$. Our assumption is that for $j = 1, 2, \ldots, n$ the center z_j of Δ_j is an element of the domain $D_{j-1} = \Delta_0 \cup \Delta_1 \cup \cdots \cup \Delta_{j-1}$, but that Δ_j is not contained in D_{j-1}. Moreover, the functions $g_0, g_1, \ldots, g_n$ are assumed to be "compatible," in the sense that $g_j = g_k$ throughout $\Delta_j \cap \Delta_k$ when Δ_j and Δ_k have non-empty intersection. Subject to these hypotheses, we obtain a well-defined analytic function f_n in $D_n = \Delta_0 \cup \Delta_1 \cup \cdots \cup \Delta_n$ by setting $f_n(z) = g_j(z)$ for z belonging to Δ_j, $0 \leq j \leq n$. We seek to push this construction one step further by picking a point z_{n+1} of D_n, expanding f_n in a Taylor series about z_{n+1}, and using the series obtained in this way to define an analytic function g_{n+1} in Δ_{n+1}, the disk of convergence of the series. So far, so good. We would like then to define a function f_{n+1} in the domain $D_{n+1} = \Delta_0 \cup \Delta_1 \cup \cdots \cup \Delta_{n+1}$ by requiring that $f_{n+1} = g_j$ in Δ_j for $0 \leq j \leq n+1$. It is here that things can go awry, for a real possibility exists that there is an index $j (0 \leq j \leq n)$ such that the intersection $\Delta_{n+1} \cap \Delta_j$ is not empty, but such that g_{n+1} does not coincide with g_j in this intersection. If that happens, then f_{n+1} is not a well-defined function in $\Delta_j \cap \Delta_{n+1}$ — and the continuation process grinds temporarily to a halt. We could, to be sure, try our luck with a different choice for z_{n+1}, but there is no guarantee that the same scenario wouldn't be repeated. We emphasize that this obstruction to the further analytic continuation of g_0 is likely to make its presence felt at any stage of the continuation process beyond the first two. (N.B. There is another circumstance in which the attempt to continue g_0 beyond D_n through the method described would end in failure — namely, it is conceivable that for every choice of z_{n+1} in D_n the disk Δ_{n+1} is contained in D_n! As previously noted, this is the fate that the function g_0 defined in $\Delta_0 = \Delta(0,1)$ by $g_0(z) = \sum_{n=1}^{\infty} z^{n!}$ suffers at the initial step: g_0 admits no analytic continuation beyond Δ_0. The latter kind of breakdown in the extension procedure is not nearly as traumatic as the former. It simply indicates that we have continued g_0 to what might be styled a "maximal domain of analyticity." We say no more about this case, even though it is not without interest.)

There are several options open to us for dealing with the above impediment to analytic continuation. One of these is to isolate conditions which prevent its occurrence and under which we are at least occasionally in a position to state with certainty that our original function g_0 is analytically continuable to a given domain D which contains Δ_0. (See Theorem 3.4.) This is to some extent a head-in-the-sand approach, for it does not confront the general issue of what the interruption of the continuation process is trying to tell us about the nature of analyticity. Nevertheless, it is the approach we intend to take. Our avowed purpose in the present chapter is, after all, to present various methods of constructing analytic functions — in this instance, returning to our earlier analogy, the "cultivation" of functions from g_0. A more sophisticated treatment of the phenomenon would incorporate it into a full-scale theory of "multi-valued functions" and would lead ultimately to the study of functions that are analytic not in plane domains, but on so-called "Riemann surfaces." Space limitations make it impossible for us to give more than an inkling of that theory in this book. We refer the reader to more advanced texts — e.g., *An Introduction to Riemann Surfaces* by George Springer (Addison-Wesley, Reading, Mass., 1957) — for a thorough discussion of such ideas.

In this section we implemented the process of analytic continuation by making use of Taylor series. When formalizing the concept it actually pays to adopt a slightly more general (and more flexible) point of view. That is what we do in the next section.

3.2 Analytic Continuation

By an *analytic function element* — we frequently abbreviate this expression to "function element," omitting the adjective "analytic" — is meant a pair (f, D), where D is a domain in the complex plane and f is an analytic function whose domain-set is D. The prototypical analytic function elements are those of the form (g, Δ), in which $\Delta = \Delta(z_0, \rho)$ is the disk of convergence of a Taylor series $\sum_{n=0}^{\infty} a_n(z - z_0)^n$ and g is the function given in Δ by $g(z) = \sum_{n=0}^{\infty} a_n(z - z_0)^n$. We reserve a special name for function elements of this type, calling them *analytic germs at* z_0. (We normally refer to such a germ (g, Δ) simply as the "germ g," it being implicit that the domain-set involved is the disk of convergence of the Taylor series which defines g.) An arbitrary function element (f, D) obviously determines an

3. Analytic Continuation

analytic germ at each point z_0 of D — to wit, the germ generated by expanding f in a Taylor series about z_0.

We say that one analytic function element (f_1, D_1) is a *direct analytic continuation* of another such element (f_0, D_0) provided $D_1 \cap D_0$ is nonempty and $f_1(z) = f_0(z)$ for every z in the intersection. (Since this definition is plainly symmetric in f_0 and f_1, it is correct to speak of (f_0, D_0) and (f_1, D_1) as direct analytic continuations of each other.) For example, if (g_0, Δ_0) is an analytic germ at z_0 and if z_1 is an arbitrary point of Δ_0, then the germ g_1 determined by g_0 at z_1 is a direct analytic continuation of g_0. Notice that, when (f_0, D_0) and (f_1, D_1) are direct analytic continuations of one another, we can house both of these function elements in a "larger" function element (f, D) by setting $D = D_1 \cup D_2$ and defining f in D through the requirement that f coincide with f_0 in D_0 and with f_1 in D_1.

Observe that the relation between function elements of being mutual direct analytic continuations respects the process of differentiation: if (f_0, D_0) and (f_1, D_1) are direct analytic continuations of each other, then this property persists for (f_0', D_0) and (f_1', D_1), (f_0'', D_0) and (f_1'', D_1), etc. There are other relationships involving functions or their derivatives that are also preserved under direct analytic continuation. To illustrate this point, suppose that $P(z, w) = a_0(z) + a_1(z)w + \cdots + a_n(z)w^n$ is a polynomial in the complex variable w with coefficients $a_0(z), a_1(z), \ldots, a_n(z)$ that are entire functions of z. Assuming it to be true of an analytic function element (f_0, D_0) that $P[z, f_0(z)] = 0$ for every z in D_0, then any direct analytic continuation (f_1, D_1) of (f_0, D_0) will automatically satisfy $P[z, f_1(z)] = 0$ for all z in D_1 (Exercise 4.36). Similarly, if $w = f_0(z)$ provides a solution in D_0 to an ordinary differential equation

$$c_n(z) \frac{d^n w}{dz^n} + c_{n-1}(z) \frac{d^{n-1} w}{dz^{n-1}} + \cdots + c_1(z) \frac{dw}{dz} + c_0(z) w = 0$$

whose coefficients $c_0(z), c_1(z), \ldots, c_n(z)$ are entire functions, then $w = f_1(z)$ will furnish a solution of the same equation in D_1 (Exercise 4.37).

To say that an analytic function element (f, D) is merely an *analytic continuation* of another such element (f_0, D_0) — again this relationship is a reciprocal one — means that it is possible to pass from (f_0, D_0) to (f, D) via a finite chain of direct analytic continuations; i.e., there exists a finite sequence $(f_0, D_0), (f_1, D_1), \ldots, (f_{n-1}, D_{n-1}), (f_n, D_n) = (f, D)$ of analytic function elements such that (f_{j+1}, D_{j+1}) is a direct analytic continuation of (f_j, D_j) for $j = 0, 1, \ldots, n-1$. (N.B. These conditions neither assume nor imply that $f_k = f_j$ in $D_k \cap D_j$ when $k \neq j+1$. Thus, unlike what happened in the case of a direct analytic continuation, it is not generally possible to consolidate the members of such a chain into a "super" function element $(\widetilde{f}, \widetilde{D})$ by taking $\widetilde{D} = \cup_{j=0}^n D_j$ and demanding that $\widetilde{f} = f_j$ in D_j for $j = 0, 1, \ldots, n$. The reason: $\widetilde{f}$ may fail to be a well-defined function.) Consider, by way of example, the analytic function elements (f_0, D_0) and (f, D), in which $D_0 = D = \{z : \operatorname{Re} z > 0\}$ and the two functions are defined in that set by $f_0(z) = \operatorname{Log} z$, $f(z) = \operatorname{Log} z + 2\pi i$. These function elements are manifestly not direct analytic continuations of each other. On the other hand, they definitely are analytic continuations of one another. They can, for instance, be connected by a string of direct analytic continuations (f_0, D_0), (f_1, D_1), (f_2, D_2), (f_3, D_3), (f, D) as follows: $D_1 = \{z : \operatorname{Im} z > 0\}$, $f_1(z) = \operatorname{Log} z$; $D_2 = \{z : \operatorname{Re} z < 0\}$, $f_2(z) = \operatorname{Log} |z| + i\theta(z)$, where θ is the branch of $\arg z$ in D_2 that takes its values in the interval $(\pi/2, 3\pi/2)$; $D_3 = \{z : \operatorname{Im} z < 0\}$, $f_3(z) = \operatorname{Log} z + 2\pi i$. We remark that the functional relationships preserved under direct analytic continuation are also preserved under its weaker cousin, "indirect" analytic continuation.

3.3 Analytic Continuation Along Paths

The important link between analytic continuation and plane topology is most conveniently expressed through the medium of "analytic continuation along a path." We shall operate with this concept exclusively on the level of germs, where the uniqueness of such a continuation accompanies its existence. Suppose then that g is an analytic germ at a point z_0 and that $\gamma : [a, b] \to \mathbb{C}$ is a path with z_0 as its initial point. We say that g can be *analytically continued along* γ if for every t in $[a, b]$ there is an analytic germ g_t at the point $\gamma(t)$ such that $g_a = g$ and such that the following compatibility condition is met: there exists a constant $\delta > 0$ with the property that g_s and g_t are direct analytic continuations of each other whenever the elements s and t of $[a, b]$ satisfy $|s - t| < \delta$. In these circumstances we describe the one-parameter family of germs $\{g_t\}$ as an *analytic continuation of* $g = g_a$ *along* γ *to* $\tilde{g} = g_b$. The germ $\tilde{g}$ really is an analytic continuation of g according to the strict meaning of that term, for we can move from g to $\tilde{g}$ through the chain of direct analytic continuations $g = g_a, g_{t_1}, g_{t_2}, \ldots, g_{t_{n-1}}, g_b = \tilde{g}$ simply by taking an arbitrary partition $a = t_0 < t_1 < \cdots < t_n = b$ of $[a, b]$ with the feature that $t_k - t_{k-1} < \delta$ for $1 \leq k \leq n$. What the idea of continuing g to $\tilde{g}$ along a path lends to this picture is the sense of $\tilde{g}$ developing from g as the end-product of a continuous evolution, rather than being the last stop in a sequence of jumps. (In actual fact analytic continuation and analytic continuation along a path are equivalent notions.) If g can be continued analytically along γ — this is by no means a foregone conclusion — then the continuation is unique, as the following result proclaims.

Lemma 3.1. *Suppose that* $\{g_t\}$ *and* $\{g_t^*\}$ *are analytic continuations of an analytic germ* g *along a path* $\gamma : [a, b] \to \mathbb{C}$. *Then* $g_t = g_t^*$ *for every* t *in* $[a, b]$.

3. Analytic Continuation

Proof. Define E to be the set of t in $[a,b]$ such that $g_s = g_s^*$ for every s satisfying $a \leq s \leq t$. Since $g_a = g_a^* = g$, the point a belongs to E. Let t_0 designate the least upper bound of this non-empty, bounded set. Then t_0 belongs to $[a,b]$. We shall be finished with the proof if we can only show that (i) t_0 is a member of E and (ii) $t_0 = b$. To verify (i) we first select a point t_1 from the set E such that $|t_1 - t_0| < \min\{\delta, \delta^*\}$, in which δ and δ^* are compatibility condition constants for the continuations $\{g_t\}$ and $\{g_t^*\}$, respectively. Because t_1 is an element of E, we know that $g_{t_1} = g_{t_1}^*$. By the definitions of δ and δ^* we are told that both g_{t_0} and $g_{t_0}^*$ are direct analytic continuations of g_{t_1}. If ρ and ρ^* denote the radii of convergence of the Taylor series which define g_{t_0} and $g_{t_0}^*$, then we know that $g_{t_0}(z) = g_{t_0}^*(z) = g_{t_1}(z)$ for every z in the non-empty open set $\Delta_0 \cap \Delta_1$, where Δ_0 is the open disk of radius $\min\{\rho, \rho^*\}$ centered at $\gamma(t_0)$ and Δ_1 is the domain-set of g_{t_1}. From the principle of analytic continuation (Corollary VIII.1.6) we infer that $g_{t_0}(z) = g_{t_0}^*(z)$ throughout the disk Δ_0. As a result, these functions determine the same analytic germ at $\gamma(t_0)$; i.e., $g_{t_0} = g_{t_0}^*$. The definition of E implies that $g_s = g_s^*$ when $a \leq s < t_0$. We have just added to $s = t_0$ to the list of points for which these germs coincide and so identified t_0 as a point of E.

We turn next to (ii). Assume, to the contrary, that $t_0 < b$. We can then choose t_2 satisfying $t_0 < t_2 < b$ and $|t_2 - t_0| < \min\{\delta, \delta^*\}$. For any s in the interval $[t_0, t_2]$ both of the germs g_s and g_s^* at $\gamma(s)$ are direct analytic continuations of g_{t_0}, which implies that g_s and g_s^* coincide with g_{t_0} in some non-empty open set. Appealing once again to the principle of analytic continuation, we conclude just as we did above that $g_s = g_s^*$ whenever $t_0 \leq s \leq t_2$. But this clearly puts t_2, a number larger than t_0, in the set E, an impossibility when one considers the definition of t_0. The only way to avoid such a contradiction is to have $t_0 = b$, which is the second thing we needed to prove. ∎

Lemma 3.1 entitles us to speak of "the" analytic continuation of a germ along a path, assuming that a continuation does exist in the first place. Let it be stated emphatically, however, that an analytic germ at a point z_0 need not be analytically continuable along every path originating at z_0. Consider, as an example, the germ g generated at the point $z_0 = 1$ by the function $f(z) = \sqrt{z}$. The function g is nothing more than the restriction of f to the disk $\Delta = \Delta(1,1)$. Its Taylor series representation in Δ reads

$$g(z) = \sum_{n=0}^{\infty} \binom{1/2}{n} (z-1)^n ,$$

in which we make use of the symbol $\binom{\lambda}{n}$ to indicate a generalized binomial coefficient: if λ is any complex number, then

$$\binom{\lambda}{n} = \frac{\lambda(\lambda-1)\cdots(\lambda-n+1)}{n!}$$

for $n = 1, 2, 3, \cdots$, while $\binom{\lambda}{0} = 1$. Let γ be the path defined by $\gamma(t) = 1-t$ for $0 \leq t \leq 1$. We claim that g does not admit an analytic continuation along γ. Were $\{g_t\}$ to be such a continuation, then the relation $[g(z)]^2 = z$, which is satisfied by $g = g_0$ in Δ, would be preserved by g_t for all t in the interval $[0,1]$, meaning that $[g_t(z)]^2 = z$ would hold for every z in Δ_t, the domain-set of g_t. This would, of course, make the germ g_1 a branch of the square root function in a disk containing the origin, an entity we know to be non-existent. (Recall Chapter III.4.2.) The consequence: no such continuation is possible. On the other hand, the given germ g can be continued analytically along the path β that is defined on $[0, 2\pi]$ by $\beta(t) = e^{it}$. (More generally, g can be continued along any path α starting at 1 whose trajectory stays away from the origin.) To present this continuation explicitly, write $\Delta_t = \Delta(e^{it}, 1)$ for $0 \leq t \leq 2\pi$ and define $g_t: \Delta_t \to \mathbb{C}$ as follows: when $0 \leq t \leq \pi/2$, $g_t(z) = \sqrt{z}$ for all z in Δ_t; when $\pi/2 < t < 3\pi/2$,

$$g_t(z) = \begin{cases} \sqrt{z} & \text{if } z \in \Delta_t \text{ and } \operatorname{Im} z \geq 0 \ , \\ -\sqrt{z} & \text{if } z \in \Delta_t \text{ and } \operatorname{Im} z < 0 \ ; \end{cases}$$

when $3\pi/2 \leq t \leq 2\pi$, $g_t(z) = -\sqrt{z}$ for all z in Δ_t. Each of the functions g_t is an analytic germ. To see this, notice that g_t is continuous in Δ_t and satisfies $[g_t(z)]^2 = z$ throughout Δ_t. It follows from discussions in Chapter III that g_t provides a branch of the square root function in Δ_t. In particular, g_t is an analytic function. A short computation provides the Taylor expansion of g_t about e^{it}:

$$g_t(z) = e^{it/2} \sum_{n=0}^{\infty} e^{-int} \binom{1/2}{n} (z - e^{it})^n \ .$$

The radius of convergence of this series is easily found to be 1, which implies that the series represents g_t throughout its disk of convergence, that disk being none other than Δ_t. By definition, this makes g_t an analytic germ at e^{it}. It is not difficult to check here that g_s and g_t are direct analytic continuations of one another as soon as $|s - t| < \pi$, so $\{g_t\}$ does supply an analytic continuation of g along β. As a matter of fact, it continues g to $\widetilde{g} = g_{2\pi} = -g$. In a similar way we can analytically continue $-g$ along β — the family $\{-g_t\}$ furnishes the continuation — and arrive back at g! (In general, when g gets continued along a closed path α in $\mathbb{C} \sim \{0\}$ beginning

3. Analytic Continuation

and ending at 1, the continuation terminates in the germ $(-1)^n g$, where $n = n(\alpha, 0)$ is the winding number of α about the origin.)

We make the observation that if g is an analytic germ at a point z_0, if g has domain-set $\Delta(z_0, \rho)$, and if $\gamma: [a, b] \to \mathbb{C}$ is any path with initial point z_0 whose trajectory lies in the disk $\Delta(z_0, \rho/2)$, then the analytic continuation $\{g_t\}$ of g along γ definitely exists: g_t is just the germ determined by g at $\gamma(t)$. That the necessary compatibility condition holds is essentially trivial here. Each germ g_t is a direct analytic continuation of g and the domain-set of g_t has radius at least $\rho - |\gamma(t) - z_0|$, which is larger than $\rho/2$. This means that g_t and g_s agree with g in some open set containing z_0 — hence, are direct analytic continuations of each other — for all t and s in $[a, b]$.

We record a few general remarks about the analytic continuation $\{g_t\}$ of an analytic germ g along a path $\gamma: [a, b] \to \mathbb{C}$. Fixing a compatibility condition constant δ for this continuation, we let $\Delta_t = \Delta[\gamma(t), \rho(t)]$ be the domain-set of the germ g_t. Our first comment is that the estimate

$$(10.35) \qquad \rho(s) \leq \rho(t) + |\gamma(s) - \gamma(t)|$$

is in force whenever $|s - t| < \delta$. If, namely, it were the case that $\rho(s)$ exceeded $\rho(t) + |\gamma(s) - \gamma(t)|$, then the disk Δ_t would be contained in Δ_s. As g_s is a direct analytic continuation of g_t, the former germ would have to agree with g_t throughout Δ_t. This would imply that g_t and g_s give rise to the same Taylor series at $\gamma(t)$. Since g_s is analytic in Δ_s, the Taylor series it generates at $\gamma(t)$ would have a radius of convergence no smaller than $\rho(s) - |\gamma(s) - \gamma(t)|$, a number that is by assumption larger than $\rho(t)$. This state of affairs would be inconsistent with the definition of $\rho(t)$. Accordingly, (10.35) must hold once $|s - t| < \delta$. By symmetry,

$$(10.36) \qquad \rho(t) \leq \rho(s) + |\gamma(s) - \gamma(t)|$$

is also true for such s and t. From (10.35) and (10.36) we extract the information that either $\rho(t) = \infty$ for every t in $[a, b]$ or the inequality

$$(10.37) \qquad |\rho(s) - \rho(t)| \leq |\gamma(s) - \gamma(t)|$$

is valid whenever s and t satisfy $|s - t| < \delta$. An immediate consequence of (10.37) and the continuity of γ is that $\rho(t)$ is a continuous function of t on $[a, b]$, unless $\rho(t) = \infty$ for all t in this interval. Because $\rho(t) > 0$ for every t in $[a, b]$, we are thus in a position to conclude that

$$(10.38) \qquad \rho = \min\{\rho(t) : a \leq t \leq b\} > 0 \, .$$

Incidentally, we enjoy a certain measure of flexibility in our choice of the constant δ, always having the freedom to replace our initial selection by any smaller positive number. Once aware of (10.38) we can reap the benefits of hindsight and, by taking advantage of the uniform continuity of γ on [a, b], shrink our original δ to a size that compels $|\gamma(s) - \gamma(t)| < \rho$ to hold whenever $|s - t| < \delta$. Under this added constraint on δ the statement

(10.39) $$|\gamma(s) - \gamma(t)| < \min\{\rho(s), \rho(t)\}$$

is true for all s and t in $[a, b]$ with $|s - t| < \delta$. The consequence of (10.39) for such s and t is this: not only are the germs g_t and g_s direct analytic continuations of one another, but g_t is actually obtained by expanding g_s in a Taylor series about $\gamma(s)$, and vice versa. We have no plans to exploit this fact, but we make note of it in passing for the simple reason that some treatments of analytic continuation include (10.39) as part of the compatibility condition in the definition of a continuation of g along γ. While we are on the subject, we should also point out that some authors prefer to state the aforementioned compatibility condition in a local version, phrasing it so: corresponding to each t in $[a, b]$ there exists a constant $\delta_t > 0$ with the property that g_s is a direct analytic continuation of g_t whenever s in $[a, b]$ satisfies $|s - t| < \delta_t$. We leave as an exercise the verification that this formulation is equivalent to the one we have given.

Let $\{g_t\}$ be the analytic continuation of a germ g to a germ $\tilde{g}$ along a path $\gamma : [a, b] \to \mathbb{C}$. Then $\tilde{g}$ can be analytically continued back to g along the path $-\gamma$, the germ corresponding to t in the reverse continuation being g_{b+a-t}. If $\tilde{g}$ admits a further analytic continuation — denote it by $\{\tilde{g}_s\}$ — along a path $\tilde{\gamma}: [c, d] \to \mathbb{C}$, then the germ g can be continued along the path $\gamma^* = \gamma + \tilde{\gamma}$. Recalling that γ^* has for its domain-set the interval $[a, b + d - c]$, we patch $\{g_t\}$ and $\{\tilde{g}_s\}$ together to form the continuation $\{g_r^*\}$ of g along the composite path as follows: $g_r^* = g_r$ if $a \le r \le b$ and $g_r^* = \tilde{g}_{r+c-b}$ if $b \le r \le b+d-c$. That $\{g_r^*\}$ obeys the necessary compatibility condition is not immediately obvious (by contrast, the validity of the local compatibility condition referred to above is instantly clear), but this fact is not hard to verify, especially if compatibility condition constants for $\{g_t\}$ and $\{\tilde{g}_s\}$ are selected in such a way that (10.39) and its counterpart for $\{\tilde{g}_s\}$ are satisfied. Finally, if a path $\gamma_1 : [a_1, b_1] \to \mathbb{C}$ is derived from γ by a change of parameter $h : [a_1, b_1] \to [a, b]$, then the assignment of the germ $g_{h(u)}$ to u defines an analytic continuation of g along γ_1. In this instance the compatibility condition can be checked by appealing to the uniform continuity of h on the interval $[a_1, b_1]$.

The last observation in this section can be summarized as follows: if a germ g can be analytically continued along a path γ and if $\tilde{\gamma}$ is a path that lies sufficiently "close" to γ, then g is continuable along $\tilde{\gamma}$ as well.

Lemma 3.2. *Suppose that $\{g_t\}$ is the analytic continuation of a germ g along a path $\gamma : [a, b] \to \mathbb{C}$ and that $0 < \rho < \min\{\rho(t): a \le t \le b\}$, where $\Delta_t = \Delta[\gamma(t), \rho(t)]$ is the domain-set of g_t. If $\tilde{\gamma} : [a, b] \to \mathbb{C}$ is any path with the properties that $\gamma(a) = \tilde{\gamma}(a)$ and that $|\gamma(t) - \tilde{\gamma}(t)| < \rho/4$ for every t in $[a, b]$, then g has an analytic continuation $\{\tilde{g}_t\}$ along $\tilde{\gamma}$. Furthermore, when $\gamma(b) = \tilde{\gamma}(b)$ the two continuations lead to the same terminal germ; i.e., $g_b = \tilde{g}_b$.*

3. Analytic Continuation

Proof. The hypotheses imply that $\widetilde{\gamma}(t)$ is an element of Δ_t for each t in $[a, b]$. Because g_t is analytic in Δ_t, it determines a germ at $\widetilde{\gamma}(t)$. We label that germ $\widetilde{g}_t$ and write its domain-set $\Delta[\widetilde{\gamma}(t), \widetilde{\rho}(t)]$ in abbreviated form as $\widetilde{\Delta}_t$. We can be certain that

$$\widetilde{\rho}(t) \geq \rho(t) - |\widetilde{\gamma}(t) - \gamma(t)| \geq \rho - \frac{\rho}{4} = \frac{3\rho}{4}.$$

Moreover, the functions g_t and $\widetilde{g}_t$ coincide in the intersection $\Delta_t \cap \widetilde{\Delta}_t$. The obvious candidate for the analytic continuation of g along $\widetilde{\gamma}$ is $\{\widetilde{g}_t\}$. Plainly $\widetilde{g}_a = g_a = g$, so only the compatibility condition for $\{\widetilde{g}_t\}$ needs to be verified. We choose $\delta > 0$ small enough to guarantee the validity of the following statements for all s and t in $[a, b]$ satisfying $|s - t| < \delta$: (i) $|\gamma(s) - \gamma(t)| < \rho/2$; (ii) $|\widetilde{\gamma}(s) - \widetilde{\gamma}(t)| < \rho/2$; (iii) g_s and g_t are direct analytic continuations of each other. For (i) and (ii) the selection of δ is made possible by the uniform continuity of γ and $\widetilde{\gamma}$ on $[a, b]$; to achieve (iii) we invoke the compatibility condition for $\{g_t\}$. We now claim that $\widetilde{g}_s$ and $\widetilde{g}_t$ are related by direct analytic continuation whenever $|s - t| < \delta$. Fix s and t subject to this constraint. The inequalities

$$|\gamma(t) - \widetilde{\gamma}(t)| < \frac{\rho}{4} \leq \widetilde{\rho}(t),$$

$$|\gamma(t) - \gamma(s)| < \frac{\rho}{2} \leq \rho(s),$$

$$|\gamma(t) - \widetilde{\gamma}(s)| \leq |\gamma(t) - \widetilde{\gamma}(t)| + |\widetilde{\gamma}(t) - \widetilde{\gamma}(s)| < \frac{\rho}{4} + \frac{\rho}{2} = \frac{3\rho}{4} \leq \widetilde{\rho}(s),$$

demonstrate that $\gamma(t)$ is a point of the set $U = \Delta_t \cap \widetilde{\Delta}_t \cap \Delta_s \cap \widetilde{\Delta}_s$, making U a non-empty open set. Moreover, by design $\widetilde{g}_t = g_t = g_s = \widetilde{g}_s$ in U. The principle of analytic continuation then forces $\widetilde{g}_t$ and $\widetilde{g}_s$ to agree throughout $\widetilde{\Delta}_t \cap \widetilde{\Delta}_s$; i.e., $\widetilde{g}_t$ and $\widetilde{g}_s$ are direct analytic continuations of each other. We have thus succeeded in certifying $\{\widetilde{g}_t\}$ as the analytic continuation of g along $\widetilde{\gamma}$. Finally, if $\gamma(b) = \widetilde{\gamma}(b)$, then the above construction and the uniqueness of analytic continuation along a path combine to yield $\widetilde{g}_b = g_b$ for any path $\widetilde{\gamma}$ with all the other stated properties. ∎

3.4 Analytic Continuation and Homotopy

In Chapter V.7.1 we introduced the notion of homotopy for paths in the complex plane. At the time our primary concern was the relationship between winding numbers and topology, so the focus was on free homotopy of closed paths. We did, however, pay lip-service to the idea of "homotopy with fixed endpoints," which is the brand of homotopy that comes into play in conjunction with analytic continuation. To recall the definition of the concept assume that $\alpha: [a, b] \to \mathbb{C}$ and $\beta: [a, b] \to \mathbb{C}$ are paths in a plane domain D and that these two paths have both the same initial point and the same terminal point. For α and β to be *homotopic with fixed endpoints in D* means that there exists a continuous function $H: R = \{(t,s): a \leq t \leq b, 0 \leq s \leq 1\} \to D$ which is blessed with the following two properties:

$$\begin{cases} \text{(i)} \quad H(t,0) = \alpha(t) \,, \; H(t,1) = \beta(t) \quad \text{for } a \leq t \leq b \,; \\ \text{(ii)} \quad H(a,s) = \alpha(a) \,, \; H(b,s) = \alpha(b) \quad \text{for } 0 \leq s \leq 1 \,. \end{cases}$$

(Any such H is called a *fixed-endpoint homotopy from α to β in D*.) With H we associate a one-parameter family of paths $\gamma_s: [a,b] \to D$ ($0 \leq s \leq 1$), each of which has the same endpoints as α and β: γ_s is defined by $\gamma_s(t) = H(t,s)$. One thinks of the homotopy H as a mechanism whereby the path $\alpha = \gamma_0$ can be continuously deformed to the path $\beta = \gamma_1$ in the domain D, the intermediate path γ_s representing the state of the deformation process at the instant s.

A fundamental connection between analytic continuation and plane topology is signalled by a result known as the "Monodromy Theorem."

Theorem 3.3. (Monodromy Theorem) *Let z_0 and z_1 be points of a plane domain D, and let g be an analytic germ at z_0 that can be analytically continued along every path in D with initial point z_0 and terminal point z_1. If two paths α and β of this description are homotopic with fixed endpoints in D, then the continuations of g along α and β produce the same terminal germ at z_1.*

Proof. Suppose that α and β are parametrized on the interval $[a,b]$. We choose a fixed-endpoint homotopy H from α to β in D, and let γ_s be the intermediate path associated with H for $0 \leq s \leq 1$. We denote by $\tilde{g}$ the germ at z_1 obtained as the end-product of the analytic continuation of g along α. Consider the set E consisting of all s in $[0,1]$ for which continuation of g along the path γ_s leads to $\tilde{g}$ as a final germ. Since $\gamma_0 = \alpha$, E is non-empty. This set has a least upper bound s_0, which clearly satisfies $0 \leq s_0 \leq 1$. To finish the proof we shall argue that (i) s_0 is a member of E and (ii) $s_0 = 1$. For this, designate by $\tilde{g}_0$ the germ at z_1 that results from the continuation of g along γ_{s_0}. The message of Lemma 3.2 is that we can produce a number $\epsilon > 0$ with the following property: if $\gamma: [a,b] \to \mathbb{C}$ is any path from z_0 to z_1 fulfilling the requirement that $|\gamma(t) - \gamma_{s_0}(t)| < \epsilon$ for every t in $[a,b]$, then not only is the analytic continuation of g along γ possible, but this continuation necessarily ends with the germ $\tilde{g}_0$. Next, because the continuous function H is uniformly continuous on the compact set R (Theorem II.4.8), we can fix $\eta > 0$ so as to make certain that

$$|\gamma_s(t) - \gamma_{s_0}(t)| = |H(t,s) - H(t,s_0)| < \epsilon$$

holds for every t in $[a,b]$ whenever $|s - s_0| < \eta$. By the definition of a least upper bound we can assert the existence of a number s_1 in E for which

3. Analytic Continuation

$s_0 - \eta < s_1 \leq s_0$. The fact that s_1 is in E means that the continuation of g along γ_{s_1} leads to $\tilde{g}$; on the other hand, the preceding remarks pin down the outcome of that continuation as $\tilde{g}_0$. We conclude that $\tilde{g}_0 = \tilde{g}$, which places s_0 in E. If s_0 were less than 1, then we could select a number s_2 in the interval $(s_0, 1)$ satisfying $s_2 < s_0 + \eta$. In continuing g along γ_{s_2} we would again arrive at $\tilde{g}_0 = \tilde{g}$; i.e., s_2 would be an element of E larger than s_0, an unacceptable situation. Consequently, $s_0 = 1$ and the continuation of g along $\beta = \gamma_1$ does terminate in the germ $\tilde{g}$, as claimed. ∎

The monodromy theorem is most readily applied in a simply connected domain D, where any pair of paths $\alpha, \beta : [a, b] \to D$ that share a common initial point and a common terminal point are homotopic with fixed endpoints: if $D = \mathbb{C}$, then $H(t, s) = (1-s)\alpha(t) + s\beta(t)$ gives a fixed-endpoint homotopy from α to β in D; if $D \neq \mathbb{C}$, such a homotopy is delivered by

$$H(t, s) = f^{-1}\{(1-s)f[\alpha(t)] + s f[\beta(t)]\},$$

in which f is a conformal mapping of D onto $\Delta(0, 1)$. A simply connected domain D is the setting for the next theorem, which formulates the third and final method of constructing analytic functions that we discuss in this book.

Theorem 3.4. *Let z_0 be a point of a simply connected domain D in the complex plane, and let g be an analytic germ at z_0 that can be analytically continued along every path in D which has z_0 for its initial point. Then there exists a unique analytic function element (f, D) whose germ at z_0 is g.*

Proof. Define a function $f: D \to \mathbb{C}$ as follows: given z in D, choose an arbitrary path γ in D with initial point z_0 and terminal point z, continue g analytically along γ to a terminal germ $\tilde{g}$ at the point z, and set $f(z) = \tilde{g}(z)$. The number $f(z)$ so obtained does not depend on the choice of the path γ. Indeed, after a preliminary change of parameter — such a change does not affect the terminal germ of the continuation — we may always assume that γ has for its domain-set the interval $[0, 1]$. Because D is simply connected, any two paths $\gamma : [0, 1] \to D$ and $\tilde{\gamma} : [0, 1] \to D$ with initial point z_0 and terminal point z are homotopic with fixed endpoints in D. By the monodromy theorem the continuations of g along γ and $\tilde{\gamma}$ produce the same terminal germ. As a result, the independence of $f(z)$ from the selection of γ is confirmed. In other words, f is a well-defined function in D.

To show that f is analytic in D, it suffices to verify that for each point z_1 of D this function is analytic in some open disk centered at z_1. Fix such a point z_1 and fix along with it a path $\gamma : [0, 1] \to D$ that starts at z_0 and ends at z_1. Let $\tilde{g}$ be the germ at z_1 arising from the continuation $\{g_t\}$ of g along γ, let $\Delta(z_1, \rho)$ be the domain-set of $\tilde{g}$, and let $\Delta = \Delta(z_1, r)$ be an open disk with $0 < r < \rho/2$ such that Δ is contained in D. If z belongs to Δ, the analytic continuation $\{g_s^*\}$ of g along the path $\gamma^* = \gamma + [z_1, z]$ is not hard to describe: $g_s^* = g_s$ when $0 \leq s \leq 1$ and g_s^* is the germ determined at $\gamma^*(s)$ by $\widetilde{g}$ when $1 \leq s \leq 2$. (Since for $1 \leq s \leq 2$ the germ g_s^* is obviously a direct analytic continuation of $\widetilde{g} = g_1^*$ and since the domain-set of g_s^* includes z_1 — its radius is at least $\rho/2$ — the compatibility condition for $\{g_s^*\}$ is easily checked.) By definition $f(z) = g_2^*(z) = \widetilde{g}(z)$. Such being the case for every z in Δ, the functions f and $\widetilde{g}$ agree in Δ, which fact makes evident the analyticity of f in this disk. We have therefore managed to construct an analytic function element (f, D). The argument just presented applies to the special case where we take $z_1 = z_0$ and $\gamma(t) = z_0$ for $0 \leq t \leq 1$ (then $\widetilde{g} = g$) and demonstrates that f and g coincide in some open disk centered at z_0, so g is the germ determined by f at z_0. The principle of analytic continuation vouches for the uniqueness of any function element (f, D) exhibiting this property. ∎

We stress that in using Theorem 3.4 to construct an analytic function it is necessary to prove that the original germ g is analytically continuable along all relevant paths. This is not going to be the case automatically.

3.5 Algebraic Function Elements

In this section we put Theorem 3.4 to work in showing how analytic functions arise through the solution of algebraic equations. The discussion starts with a polynomial function P of two complex variables z and w. We write P in the form

(10.40) $$P(z, w) = p_0(z) + p_1(z)w + \cdots + p_n(z)w^n,$$

where $p_0(z), p_1(z), \ldots, p_n(z)$ are polynomial functions of z. To avoid certain degenerate or trivial situations we shall always tacitly assume that $n \geq 1$ and that the leading coefficient p_n is not the zero polynomial. A simple example of what we have in mind here is the function $P(z, w) = w^n - z$. Another example that turns out to be significant in various contexts is the quadratic (in w) polynomial $P(z, w) = w^2 - 4(z-a)(z-b)(z-c)$, where a, b, and c are distinct complex constants. By holding the value of z in (10.40) fixed — say $z = z_0$ — we obtain a polynomial function φ of the variable w alone, $\varphi(w) = P(z_0, w)$. Without giving the matter much thought we might hope that φ would have degree n and possess n different roots. Inevitably, however, there is a set of points z_0 for which φ fails to have n distinct zeros. We name any point z_0 of this type an *exceptional point for P*. In case P is "irreducible" — this means that P cannot be expressed as the product of two non-constant polynomial functions of z and w — the set of exceptional points is necessarily a finite set, as we shall later see; in the "reducible" case the exceptional set may become infinite. For example, every complex number z_0 is exceptional for the polynomial $P(z, w) = z^2 + 2zw + w^2 = (z+w)^2$; if $P(z, w) = w^n - z$, then the exceptional set consists of the origin; for the polynomial $P(z, w) = w^2 - 4(z-a)(z-b)(z-c)$ the exceptional set is

3. Analytic Continuation

just $\{a, b, c\}$. In general, a point z_0 can qualify as exceptional for one of two reasons: (i) z_0 is a zero of the coefficient p_n in (10.40), in which event the degree of φ drops below n; (ii) $p_n(z_0) \neq 0$ — so φ is of degree n — but φ has a multiple root. In the latter instance there exists a point w_0 (any multiple zero w_0 of φ will have the property) such that both $P(z_0, w_0) = \varphi(w_0) = 0$ and $P_w(z_0, w_0) = \varphi'(w_0) = 0$; i.e., P and P_w, its partial derivative with respect to the variable w, have a common zero in $\mathbb{C}^2$ of the form (z_0, w_0).

The first result of this section is a version of the "Implicit Function Theorem." It informs us that at any non-exceptional point z_0 for P the equation $P(z, w) = 0$ implicitly defines a collection of n analytic germs $g_1, g_2, \ldots, g_n$ which obey the relationship $P[z, g_j(z)] = 0$ throughout their respective domain-sets.

Theorem 3.5. *Assume that z_0 is not an exceptional point for the polynomial $P(z, w) = p_0(z) + p_1(z)w + \cdots + p_n(z)w^n$. Let $w_1, w_2, \ldots, w_n$ be the solutions of the equation $P(z_0, w) = 0$. Corresponding to each w_j there exists a unique analytic germ g_j at z_0 such that $g_j(z_0) = w_j$ and such that $P[z, g_j(z)] = 0$ for every z in the domain-set of g_j. In fact, if D is any plane domain containing z_0 and $f: D \to \mathbb{C}$ is any continuous function satisfying $P[z, f(z)] = 0$ for every z in D, then there is an open disk centered at z_0 in which f coincides with the unique germ g_j among $g_1, g_2, \ldots, g_n$ that takes the value $f(z_0)$ at z_0.*

Proof. Since the nth-degree polynomial $\varphi(w) = P(z_0, w)$ finds in the points $w_1, w_2, \ldots, w_n$ a full complement of distinct roots, each of these is necessarily a simple root. Therefore, $\varphi'(w_j) \neq 0$ for every j. We use the continuity of φ' to select a number $s > 0$ with the following properties: the open disks $\Delta_1 = \Delta(w_1, s), \Delta_2 = \Delta(w_2, s), \ldots, \Delta_n = \Delta(w_n, s)$ are pairwise disjoint and $\varphi'(w) \neq 0$ holds for every w in the set $K = \cup_{j=1}^n \partial \Delta_j$. Naturally $\varphi(w) \neq 0$ for all w in K, so $|\varphi|$ has a positive minimum value on this compact set. Call that minimum value m. Appealing to the uniform continuity of the polynomial P on the compact set $E = \{(z, w) : z \in \overline{\Delta}(z_0, 1), w \in K\}$ in $\mathbb{C}^2$, we can choose and fix a radius r in $(0, 1)$ with the property that

$$|P(z, w) - P(z_0, w)| < m$$

whenever z belongs to $\Delta = \Delta(z_0, r)$ and w to K.

Consider a point z of Δ. Owing to the selection of r the polynomial ψ defined by $\psi(w) = P(z, w)$ satisfies

$$|\varphi(w) - \psi(w)| < m \leq |\varphi(w)|$$

for every w on $\partial \Delta_j$. According to Rouché's theorem the functions φ and ψ must have the same number of zeros in Δ_j. By construction φ has a simple zero at w_j as its one and only zero in this disk. As a result, ψ has exactly one root in Δ_j. Since this is the case for $j = 1, 2, \ldots, n$ and since

Δ_j and Δ_k are disjoint when $j \neq k$, we conclude that for each fixed z in Δ the equation $P(z,w) = 0$ has exactly n solutions, one in each of the disks Δ_j. This fact enables us to define functions $f_j: \Delta \to \Delta_j$, $1 \leq j \leq n$, by the following prescription: if z is in Δ, then $f_j(z)$ is the unique point of Δ_j for which $P[z, f_j(z)] = 0$. Notice especially that no point of Δ can be exceptional for P. It follows that, when $P(z,w) = 0$ and z lies in Δ, $P_w(z,w) \neq 0$.

For fixed z in Δ the formula $Q(w) = wP_w(z,w)/P(z,w)$ defines a rational function of w whose only singularities in $\mathbb{C}$ are located at the points $\zeta_1 = f_1(z), \zeta_2 = f_2(z), \ldots, \zeta_n = f_n(z)$, each being no worse than a simple pole. (N.B. If $\zeta_j = 0$ for some j, then the singularity of Q at ζ_j is actually a removable singularity, rather than a pole.) Through an application of the residue theorem we discover that

$$\int_{|w-w_j|=s} Q(w)\,dw = 2\pi i \operatorname{Res}(\zeta_j, Q) = 2\pi i \lim_{w \to \zeta_j}(w - \zeta_j)Q(w)$$

$$= 2\pi i \lim_{w \to \zeta_j}\left[wP_w(z,w)\frac{w-\zeta_j}{P(z,w)}\right] = 2\pi i\,\zeta_j P_w(z,\zeta_j) \lim_{w \to \zeta_j}\frac{w-\zeta_j}{P(z,w)}$$

$$= 2\pi i\,\zeta_j P_w(z,\zeta_j)\frac{1}{P_w(z,\zeta_j)} = 2\pi i\,\zeta_j = 2\pi i\,f_j(z)\,.$$

In this way we see that f_j admits an integral representation in Δ; namely,

(10.41) $$f_j(z) = \frac{1}{2\pi i}\int_{|w-w_j|=s}\frac{wP_w(z,w)\,dw}{P(z,w)}\,.$$

If z is a point of Δ and if $\langle z_k \rangle$ is any sequence in Δ such that $z_k \to z$, then we infer from the uniform continuity of the function P_w/P on the set $\{(\zeta, w): |\zeta - z| \leq (r - |z - z_0|)/2, w \in K\}$ that, as $k \to \infty$,

$$\frac{wP_w(z_k, w)}{P(z_k, w)} \longrightarrow \frac{wP_w(z, w)}{P(z, w)}$$

uniformly on $\partial \Delta_j$. This implies that

$$\lim_{k \to \infty} f_j(z_k) = \lim_{k \to \infty} \frac{1}{2\pi i}\int_{|w-w_j|=s}\frac{wP_w(z_k, w)\,dw}{P(z_k, w)}$$

$$= \frac{1}{2\pi i}\int_{|w-w_j|=s}\frac{wP_w(z, w)\,dw}{P(z, w)} = f_j(z)\,,$$

which reveals that f_j is a continuous function.

Suppose next that R is an arbitrary closed rectangle in Δ. Formula (10.41), combined with identity (V.5.20) and the observation that for fixed

3. Analytic Continuation

w in $\partial \Delta_j$ the rational function $\widetilde{Q}$ of z given by $\widetilde{Q}(z) = P_w(z,w)/P(z,w)$ is analytic in Δ, lets us conclude that

$$\int_{\partial R} f_j(z)\,dz = \frac{1}{2\pi i} \int_{\partial R} \left\{ \int_{|w-w_j|=s} \frac{w\,P_w(z,w)\,dw}{P_z,w)} \right\} dz$$

$$= \frac{1}{2\pi i} \int_{|w-w_j|=s} w \left\{ \int_{\partial R} \frac{P_w(z,w)\,dz}{P(z,w)} \right\} dw = \frac{1}{2\pi i} \int_{|w-w_j|=s} 0\,dw = 0 \,.$$

On the basis of Morera's theorem we can now assert that f_j is analytic in Δ. Let g_j be the analytic germ determined by f_j at z_0. Then $g_j = f_j$ in Δ, though the domain-set of g_j may actually be a disk of radius larger than r. Clearly $g_j(z_0) = f_j(z_0) = w_j$ and $P[z, g_j(z)] = P[z, f_j(z)] = 0$ for every z in Δ. The principle of analytic continuation sees to it that $P[z, g_j(z)] = 0$ throughout the domain-set of g_j.

Finally, assume that D is a plane domain which contains z_0 and that $f\colon D \to \mathbb{C}$ is a continuous function which satisfies $P[z, f(z)] = 0$ for every z in D. Because $P[z_0, f(z_0)] = 0$, it can only be the case that $f(z_0) = w_j$ for some j. The continuity of f then allows us to choose a t in $(0, r)$ for which the disk $\Delta(z_0, t)$ is contained in D and $f[\Delta(z_0, t)]$ is contained in Δ_j. The fact that $P[z, f(z)] = 0$ for all z in $\Delta(z_0, t)$, coupled with the definition of f_j, makes it evident that f coincides with f_j — hence, with the germ g_j — in $\Delta(z_0, t)$. ∎

We earlier made the statement that for an irreducible polynomial P of the type (10.40) there are at most a finite number of exceptional points. This important fact is obvious when $n = 1$. The key to confirming it for larger n lies in the realization that an irreducible polynomial P and its partial derivative P_w are related by identities of the form

(10.42) $\qquad A(z,w)P(z,w) + B(z,w)P_w(z,w) = C(z)\,,$

in which A and B are polynomial functions of z and w, while C is a non-zero polynomial function of the single variable z. (N.B. Among all non-zero polynomial functions $C(z)$ expressible in the manner (10.42) there is a unique monic polynomial — i.e., its highest coefficient is 1 — of smallest degree. It is referred to in the literature as the *discriminant* of P.) Given the identity (10.42) we observe that for a point (z_0, w_0) in $\mathbb{C}^2$ to obey $P(z_0, w_0) = P_w(z_0, w_0) = 0$ it is necessary that $C(z_0) = 0$. As a consequence, the exceptional points for P must appear among the finitely many zeros of C and the finitely many zeros of the coefficient p_n in (10.40) — hence, must themselves be finite in number.

We shall only sketch the derivation of (10.42), for it hinges on a non-trivial piece of algebraic information concerning the irreducible polynomial P: *if F is a non-constant polynomial function of z and w whose degree in w is less than n and if F is a factor of a product GP, where G is also a polynomial function of z and w, then F must be a factor of G.* A proof of this fact can be found in many elementary abstract algebra texts, e.g., *Topics in Algebra* by I.N. Herstein (Xerox College Publishing, Lexington, Mass., 1964). Assuming that $n \geq 2$, we arrive at (10.42) through an explicit construction. We start by dividing P_w into P (we are thinking here of old-fashioned long division with respect to the variable w) and express the result in the manner

(10.43) $$P(z,w) = Q(z,w)P_w(z,w) + R(z,w) .$$

The functions Q and R, the quotient and remainder that come out of the division process, have the general structure

$$r_0(z) + r_1(z)w + \cdots + r_\ell(z)w^\ell ,$$

in which the coefficients $r_0, r_1, \ldots, r_\ell$ are rational functions of z. We can multiply both sides of (10.43) by an appropriate non-zero polynomial function of z — label it A_0 — to clear the denominators from all the rational coefficients appearing on its right-hand side. We thereby transform (10.43) into an identity

(10.44) $$A_0(z)P(z,w) = B_1(z,w)P_w(z,w) + C_1(z,w) ,$$

where B_1 and C_1 are polynomial functions of z and w. Note that the degree of C_1 with respect to w is smaller than $n-1$, the degree of P_w in w. If the degree of C_1 in w is positive, we repeat the above procedure, this time dividing C_1 into P_w. By so doing we obtain an identity analogous to (10.44) relating P_w and C_1,

$$A_1(z)P_w(z,w) = B_2(z,w)C_1(z,w) + C_2(z,w) ,$$

in which the degree of C_2 in w is at most $n-3$. If that degree is again positive, we continue by dividing C_2 into C_1 to get a relation

$$A_2(z)C_1(z,w) = B_3(z,w)C_2(z,w) + C_3(z,w) .$$

Here the degree of C_3 with respect to w is at most $n-4$. This process must come to a halt after a finite number $m(1 \leq m \leq n-1)$ division steps. It terminates in an identity

$$A_{m-1}(z)C_{m-2}(z,w) = B_m(z,w)C_{m-1}(z,w) + C(z) ,$$

where C is a polynomial function of the variable z alone. The function C is not the zero polynomial. If it were, we would have $A_{m-1}C_{m-2} = B_m C_{m-1}$, making C_{m-1} a factor of $A_{m-1}C_{m-2}$. Thus, C_{m-1} would divide the right-hand side of the equation — to simplify the notation we henceforth suppress any mention of the variables z and w —

$$A_{m-1}A_{m-2}C_{m-3} = B_{m-1}A_{m-1}C_{m-2} + A_{m-1}C_{m-1} ,$$

3. Analytic Continuation

and so would also be a factor of $A_{m-1}A_{m-2}C_{m-3}$. Backtracking in this way we would find C_{m-1} among the factors of GP, with $G = A_{m-1}A_{m-2}\cdots A_0$. The irreducibility of P would force C_{m-1} to be a factor of G, which it clearly is not, for C_{m-1} has positive degree with respect to w and G doesn't. The conclusion: C is not the zero polynomial. Lastly, we can express C as

$$C = A_{m-1}C_{m-2} + (-B_m)C_{m-1}\ .$$

Since $C_{m-1} = A_{m-2}C_{m-3} - B_{m-1}C_{m-2}$, this leads to a second representation for C,

$$C = (-B_m A_{m-2})C_{m-3} + (A_{m-1} - B_m B_{m-1})C_{m-2}\ .$$

Through the repetition of this back-substitution process we eventually arrive at an identity of the form (10.42).

Modulo our ability to locate the roots of p_n and C, the preceding discussion furnishes us with an algorithm to help pin down the exceptional points for an irreducible polynomial P. (*Warning.* Unless C is actually the discriminant of P, there may be zeros of C that are not exceptional for P.) Another spin-off from this discussion and from Theorem 3.5 is the identification of a new situation in which analyticity can be inferred from continuity.

Theorem 3.6. *Let $P(z,w) = p_0(z) + p_1(z)w + \cdots + p_n(z)w^n$ be an irreducible polynomial in z and w. If D is a domain in $\mathbb{C}$ and if $f: D \to \mathbb{C}$ is a continuous function that satisfies $P[z, f(z)] = 0$ for every z in D, then f is an analytic function.*

Proof. Since P is irreducible, the set E of exceptional points for P that lie in D is a finite set. Theorem 3.5 implies that f is analytic in $D \sim E$. The continuity of f insures that it has no worse than removable singularities at the points of E — hence, is actually analytic in D. ∎

An analytic function element (f, D) with the feature that $P[z, f(z)] = 0$ for every z in D, where P is a non-zero polynomial function of two complex variables, is known as an *algebraic function element*. The crowning result of this section is an existence theorem for algebraic function elements.

Theorem 3.7. *Suppose that a simply connected domain D in the complex plane contains none of the exceptional points for a polynomial $P(z,w) = p_0(z) + p_1(z)w + \cdots + p_n(z)w^n$. If (z_0, w_0) is a point of $\mathbb{C}^2$ such that z_0 belongs to D and $P(z_0, w_0) = 0$, then there is a unique analytic function element (f, D) such that $f(z_0) = w_0$ and $P[z, f(z)] = 0$ for every z in D.*

Proof. By Theorem 3.5 there is a unique analytic germ g at z_0 about which is true that $g(z_0) = w_0$ and $P[z, g(z)] = 0$ throughout the domain-set $\Delta_0 = \Delta(z_0, \rho_0)$ of g. We shall prove that g can be analytically continued along every path in D originating at z_0. Once this has been shown, we can quote Theorem 3.4 and assert the existence of a unique analytic function element (f, D) whose germ at z_0 is g. Then $f(z_0) = g(z_0) = w_0$ and, since $P[z, f(z)] = P[z, g(z)] = 0$ for all z in some open disk centered at z_0, the principle of analytic continuation dictates that $P[z, f(z)] = 0$ everywhere in D. Furthermore, any analytic function element (f, D) with these two properties must, according to the last statement in Theorem 3.5, have g as its germ at z_0. The uniqueness aspect of the present result is then seen to be a consequence of uniqueness in Theorem 3.4. The proof thus comes down to demonstrating the continuability of g along the necessary paths in the domain D.

Let $\gamma: [a, b] \to D$ be a path with initial point z_0. For each c satisfying $a < c \leq b$ denote by γ_c the restriction of γ to the interval $[a, c]$. We consider the set E of all c in $(a, b]$ such that g admits an analytic continuation along the path γ_c. If c is close enough to a that $\gamma([a, c])$ is contained in the disk $\Delta(z_0, \rho_0/2)$, then, as previously noted, the continuation $\{g_t\}$ of g along γ_c exists: for $a \leq t \leq c$, g_t is just the germ determined by g at $\gamma(t)$. Therefore, the set E is not empty. Let d be its least upper bound — then $a < d \leq b$ — and set $z_0^* = \gamma(d)$. We verify (i) that d is a member of E and (ii) that $d = b$. Because z_0^* is not an exceptional point for P, the equation $P(z_0^*, w) = 0$ has n different roots $w_1, w_2, \ldots, w_n$. In view of Theorem 3.5 there exists for each w_j a unique analytic germ g_j at z_0^* satisfying the requirements $g_j(z_0^*) = w_j$ and $P[z, g_j(z)] = 0$ for every z in $\Delta_j = \Delta(z_0^*, \rho_j)$, the domain-set of g_j. We choose and fix a number r with the feature that $0 < r < \min\{\rho_j/2 : 1 \leq j \leq n\}$. Next we select a number c in E, $c < d$, enjoying the property that $\gamma([c, d])$ lies in $\Delta(z_0^*, r)$. If β is the restriction of γ to the interval $[c, d]$, then each of the germs g_j can be analytically continued along the path $-\beta$, for the trajectory of this path is a subset of $\Delta(z_0^*, \rho_j/2)$. We use $\widetilde{g}_j$ to indicate the germ at the point $[-\beta](d) = \gamma(c)$ obtained when g_j is so continued. Being in fact a direct analytic continuation of g_j, the germ $\widetilde{g}_j$ obeys the condition $P[z, \widetilde{g}_j(z)] = 0$ throughout its domain-set, a set that includes the point z_0^*. Since their respective values at the point z_0^* — namely, $w_1, w_2, \ldots, w_n$ — are different, the germs $\widetilde{g}_1, \widetilde{g}_2, \ldots, \widetilde{g}_n$ are distinct. Of course, $\widetilde{g}_j$ can be continued along $\beta = -(-\beta)$ to g_j. By the definition of E the germ g is analytically continuable along the path γ_c to a germ $\widetilde{g}$ at $\gamma(c)$. As an analytic continuation of g, the germ $\widetilde{g}$ satisfies $P[z, \widetilde{g}(z)] = 0$ for every z in its domain-set. A glance at the final sentence in Theorem 3.5 reveals that $\widetilde{g}$ must be one of the germs $\widetilde{g}_1, \widetilde{g}_2, \ldots, \widetilde{g}_n$. As such, $\widetilde{g}$ is analytically continuable along β. We infer that g can be continued along the path $\gamma_c + \beta = \gamma_d$. This makes d an element of E. Finally, if d were less than b, we could pick a number c_1 such that $d < c_1 \leq b$ and such that $\gamma([d, c_1])$ is contained in $\Delta(z_0^*, r)$. Then g^*, the end-product of the continuation of g along γ_d, could be further continued along α, the restriction of γ to the interval $[d, c_1]$. Therefore, g would admit an analytic continuation along the path $\gamma_d + \alpha = \gamma_{c_1}$, in conflict with the definition

3. Analytic Continuation

of d. It follows that $d = b$ and that g can be analytically continued along $\gamma = \gamma_b$, as insisted. ∎

3.6 Global Analytic Functions

If one starts with an analytic function element (f_0, D_0) and builds the collection of all function elements (f, D) that are analytic continuations of (f_0, D_0), one is faced with what appears to be a colossal hodgepodge of functions and domains. By no means is it evident that these function elements, these "patches" of function, fit together in any coherent and harmonious way to form a "whole" function. Nor should it be evident, for the very good reason that the construction which manages to consolidate all these elements into a single function generally forces one out of the familiar territory of the complex plane and requires one to do complex analysis on a "Riemann surface." In this section we afford the reader a glimpse of that construction and its implications. As indicated earlier, however, limitations of space prevent us from giving anything beyond an overview of the subject. Those who find their interest piqued by the ideas that follow should at this stage be adequately prepared to pursue them in a more advanced text.

The fundamental concept involved here is that of a "global analytic function." A *global analytic function* $\mathcal{F}$ (the expression *complete analytic function* is sometimes used to describe the same object) is a non-empty family of analytic function elements that enjoys the following two special properties: (i) any pair of function elements belonging to $\mathcal{F}$ are analytic continuations of each other and (ii) any analytic function element that is related through analytic continuation to any function element in $\mathcal{F}$ is itself a member of $\mathcal{F}$. In short, $\mathcal{F}$ is the totality of analytic function elements that can be obtained through the arbitrary analytic continuation of any one of its members. Each of the individual function elements that make up $\mathcal{F}$ is known as a *branch of* $\mathcal{F}$. The reader may be excused for looking with suspicion at the title "global analytic function" we have bestowed on $\mathcal{F}$. It would seem to be a serious misnomer: $\mathcal{F}$ is not a function at all, but a whole collection of functions! A justification for the terminology will be offered in due course.

Everyone's favorite example of a global analytic function is the succeeding one.

EXAMPLE 3.1. Let $\mathcal{L}$ denote the collection of all analytic function elements (f, D) with the property that $e^{f(z)} = z$ for every z in D. Show that $\mathcal{L}$ is a global analytic function.

If (f, D) belongs to $\mathcal{L}$ and if (f_1, D_1) is a direct analytic continuation of (f, D), then $e^{f_1(z)} = e^{f(z)} = z$ for every z in $D \cap D_1$. By the principle of analytic continuation $e^{f_1(z)} = z$ holds everywhere in D_1, so (f_1, D_1) is a member of $\mathcal{L}$. Since an arbitrary analytic continuation of (f, D) is linked to it by a finite chain of direct analytic continuations, any such continuation stays in $\mathcal{L}$. All that remains for us to check is that any two members of $\mathcal{L}$ are analytic continuations of each other. Because the relation here is a transitive one and because a function element is obviously an analytic continuation of every germ that it generates, it suffices to demonstrate that every germ in $\mathcal{L}$ is an analytic continuation of some fixed member of $\mathcal{L}$. For this member we pick the function element (f_0, D_0), in which $D_0 = \mathbb{C} \sim (-\infty, 0]$ and $f_0(z) = \text{Log } z$.

Suppose now that $\widetilde{g}$, an analytic germ at a point z_0, belongs to $\mathcal{L}$. Then $\widetilde{g}$ constitutes a branch of $\log z$ in its domain-set $\widetilde{\Delta} = \Delta(z_0, \rho)$, so $z_0 \neq 0$. A straightforward calculation reveals that

$$\widetilde{g}(z) = \widetilde{g}(z_0) + \sum_{n=1}^{\infty} \frac{(-1)^{n-1}(z-z_0)^n}{n \, z_0^n}$$

in $\widetilde{\Delta}$. In particular, we see that $\rho = |z_0|$. Next, we write $z_0 = |z_0|e^{i\theta_0}$ with $0 \leq \theta_0 < 2\pi$. Then

$$\widetilde{g}(z_0) = \text{Log } |z_0| + i(\theta_0 + 2k_0\pi)$$

for some integer k_0. We proceed assuming $k_0 \geq 0$, the case $k_0 < 0$ being handled similarly. Define a path γ by $\gamma(t) = |z_0|e^{it}$ for $0 \leq t \leq \theta_0 + 2k_0\pi$. (N.B. Should $\theta_0 = k_0 = 0$, then $\widetilde{g}$ would be the germ determined by f_0 at z_0, and none of what ensues would be necessary. For this reason we shall suppose that either $\theta_0 > 0$ or $k_0 > 0$. Incidentally, in the case where $k_0 < 0$ the path γ would be replaced in the following discussion by $\gamma(t) = |z_0|e^{-it}$, $0 \leq t \leq -\theta_0 - 2k_0\pi$.) We claim that $\widetilde{g}$ is obtained via the analytic continuation of g, the germ generated by f_0 at the point $|z_0|$ of the positive real axis, along the path γ. If this is true, it certainly marks $\widetilde{g}$ as an analytic continuation of g — hence, of (f_0, D_0). Observe that g is given by

$$g(z) = \text{Log } |z_0| + \sum_{n=1}^{\infty} \frac{(-1)^{n-1}(z-|z_0|)^n}{n|z_0|^n}$$

in the disk $\Delta = \Delta(|z_0|, |z_0|)$. For $0 \leq t < \infty$ let g_t designate the analytic germ defined by

$$g_t(z) = \text{Log } |z_0| + it + \sum_{n=1}^{\infty} \frac{(-1)^{n-1}(z-|z_0|e^{it})^n}{n|z_0|^n e^{int}}$$

in the domain-set $\Delta_t = \Delta(|z_0|e^{it}, |z_0|)$; i.e., g_t is the unique branch of $\log z$ in Δ_t whose value at the point $|z_0|e^{it}$ is $\text{Log } |z_0| + it$. Plainly $g_0 = g$ and $g_{\theta_0+2k_0\pi} = \widetilde{g}$. Furthermore, when $|s - t| < \pi$ the germs g_t and g_s are direct analytic continuations of each other. Perhaps the easiest way to make this clear is to express g_t in closed form: when $0 \leq t \leq \pi/2$, $g_t(z) = \text{Log } z$ for

3. Analytic Continuation

every z in Δ_t; when $\pi/2 < t < 3\pi/2$,

$$g_t(z) = \begin{cases} \text{Log}\, z & \text{if } z \in \Delta_t \text{ and } \text{Im}\, z \geq 0, \\ \text{Log}\, z + 2\pi i & \text{if } z \in \Delta_t \text{ and } \text{Im}\, z < 0; \end{cases}$$

when $3\pi/2 \leq t \leq 2\pi$, $g_t(z) = \text{Log}\, z + 2\pi i$ for every z in Δ_t; for t satisfying $2k\pi \leq t < 2(k+1)\pi$ with $k \geq 1$, $g_t = 2k\pi i + g_{t-2k\pi}$. This description leaves no doubt that $g_t = g_s$ in $\Delta_t \cap \Delta_s$ when $|s-t| < \pi$. It follows that $\{g_t\}$, where we now confine t to the interval $[0, \theta_0 + 2k_0\pi]$, is an analytic continuation of g to $\tilde{g}$ along γ. We have thus verified that $\mathcal{L}$ is a global analytic function.

The global analytic function $\mathcal{L}$ in Example 3.1 is called the *global logarithm function*. A similar discussion would reveal that $\mathcal{F}$, the family of all analytic function elements (f, D) that satisfy $[f(z)]^2 = z$ for every z in D, is a global analytic function, the *global square root function* (Exercise 4.43). More generally, if $P(z, w) = p_0(z) + p_1(z)w + \cdots + p_n(z)w^n$ is an irreducible polynomial function of z and w, then the collection $\mathcal{A}_P$ consisting of all analytic function elements (f, D) such that $P[z, f(z)] = 0$ throughout D is a global analytic function, the *global algebraic function corresponding to P*. The irreducibility of P is crucial to this pronouncement. The argument which shows that all members of $\mathcal{A}_P$ are related by analytic continuation is non-trivial. Because it involves a number of ideas unavailable in this book, we omit the proof.

Given a global analytic function $\mathcal{F}$ we carry out the following construction. For each fixed z in $\mathbb{C}$ we build what can be envisioned as a stack of points above z, the elements of which are in one-to-one correspondence with the collection of analytic germs at z that belong to $\mathcal{F}$. We write $p = (z, g)$ to signify that p is the point of the stack associated with the germ g at z. Of course, it is entirely possible that $\mathcal{F}$ contains no germs at z, in which event the stack above z is empty. This happens, for instance, in the case of the global logarithm function $\mathcal{L}$ at $z = 0$. Over any point z_0 other than the origin the stack determined by $\mathcal{L}$ comprises infinitely many points $p_0, p_1, p_{-1}, p_2, p_{-2}, \cdots$: for each integer k we set $p_k = (z_0, g_k)$, where g_k is the unique branch of $\log z$ in the disk $\Delta(z_0, |z_0|)$ whose value at z_0 is $\text{Log}\, z_0 + 2k\pi i$. If $P(z, w) = p_0(z) + p_1(z)w + \cdots + p_n(z)w^n$ is an irreducible polynomial in z and w, then the global analytic function $\mathcal{A}_P$ generates a stack of n points above every point z of $\mathbb{C}$ that is not exceptional for P, one stack-point corresponding to each solution w of the equation $P(z, w) = 0$ (Theorem 3.5). Over the exceptional points for P the stacks contain fewer than n points and may well be empty, as the one above the origin is when $P(z, w) = w^2 - z$. If we now assemble all the stacks to which a global analytic function $\mathcal{F}$ gives rise in this way, we obtain an object called the "Riemann surface" of $\mathcal{F}$. To formalize the definition: the *Riemann surface of the global analytic function $\mathcal{F}$* is the set $S(\mathcal{F})$ whose elements are the ordered pairs $p = (z,g)$, where z is a point of the complex plane and g is an analytic germ at z that belongs to $\mathcal{F}$. (N.B. To say that $p_1 = p_2$ for two such pairs $p_1 = (z_1, g_1)$ and $p_2 = (z_2, g_2)$ means that $z_1 = z_2$ and that g_1 and g_2 are the same germ at z_1. The latter condition demands more than just $g_1(z_1) = g_2(z_1)$. It insists that g_1 and g_2 be represented by the same Taylor series at z_1; i.e., it insists that $g_1^{(n)}(z_1) = g_2^{(n)}(z_1)$ for $n = 0, 1, 2, \cdots$.) The image which $S(\mathcal{F})$ is supposed to conjure up is that of a multi-tiered surface suspended over a portion of the complex plane. The technical name for what we have heretofore described as the "stack of points" generated by $\mathcal{F}$ above z is the *fiber of $S(\mathcal{F})$ over z.*

(*Remark.* As defined, the Riemann surface $S(\mathcal{F})$ of a global analytic function $\mathcal{F}$ is, properly speaking, a subset of the cartesian product $\mathbb{C} \times \mathcal{F}$. In many instances, however, $S(\mathcal{F})$ is realizable as an actual surface in the four-dimensional space $\mathbb{C}^2$. Take, for example, the case of the global logarithm function $\mathcal{L}$. The Riemann surface $S(\mathcal{L})$ can be legitimately identified with $\{(z, w) \in \mathbb{C}^2 : e^w = z\}$, a concrete two-dimensional surface in $\mathbb{C}^2$. If we regard the complex plane as sitting inside $\mathbb{C}^2$ disguised as the plane $\{(z, w) \in \mathbb{C}^2 : w = 0\}$, then the fiber of $S(\mathcal{L})$ over any non-zero point $(z, 0)$ of the complex plane — the preposition "over" is slightly misleading here — reduces to the set $\{(z, \mathrm{Log}\, z + 2k\pi i) : k = 0, \pm 1, \pm 2, \cdots\}$ in $\mathbb{C}^2$. Similarly, when $\mathcal{F}$ is the global square root function, the Riemann surface $S(\mathcal{F})$ can be thought of as $\{(z, w) \in \mathbb{C}^2 : z \neq 0, w^2 = z\}$. The fiber of $S(\mathcal{F})$ over $(z, 0)$ in this situation consists of the two points $(z, \sqrt{z})$ and $(z, -\sqrt{z})$.)

There are two functions of particular importance defined on the Riemann surface $S(\mathcal{F})$. The first of these is the function $\mathcal{P}: S(\mathcal{F}) \to \mathbb{C}$ with rule of correspondence $\mathcal{P}(z, g) = z$. It is termed the *projection of $S(\mathcal{F})$ to the complex plane*. We now force the symbol $\mathcal{F}$ to do double duty, in that we retain it to denote the second function in question here, for it is the function on $S(\mathcal{F})$ that incorporates into a single function the diverse analytic function elements which make up the global analytic function $\mathcal{F}$. The function $\mathcal{F}: S(\mathcal{F}) \to \mathbb{C}$ is defined by $\mathcal{F}(z, g) = g(z)$. (N.B. As g is a germ at z, the expression $g(z)$ makes perfectly good sense.) Let (f, D) be any of the analytic function elements that constitute $\mathcal{F}$. We can associate with (f, D) a replica of the domain D on the surface $S(\mathcal{F})$, the set $\widetilde{D}$ consisting of all points (z, g) on $S(\mathcal{F})$ such that z lies in D and g is the germ determined at z by f. (The function $\mathcal{P}$ projects $\widetilde{D}$ in a one-to-one fashion onto D, and this projection has every nice property imaginable.) Let (z, g) be an element of $\widetilde{D}$. Since g is the germ of f at z, we see that $\mathcal{F}(z, g) = g(z) = f(z)$. Thus, $\mathcal{F}$ assigns to a point of $\widetilde{D}$ exactly the same value that f assigns to the corresponding point of D. As things turn out, the mimicry of f that $\mathcal{F}$ displays in $\widetilde{D}$ is complete. For all intents and purposes we might just as well regard D as a subset of $S(\mathcal{F})$ and f as the restriction of $\mathcal{F}$ to that subset. It is in this sense that $\mathcal{F}$ combines all the function elements of its namesake global analytic function into one function. A price is exacted for this

3. Analytic Continuation

consolidation: we must abandon the complex plane and work instead on $S(\mathcal{F})$. Fortunately, at least as far as the function $\mathcal{F}$ is concerned, the transition to complex analysis on $S(\mathcal{F})$ is quite straightforward. For example, differentiation can be done as follows: $\mathcal{F}'(z,g) = g'(z)$, $\mathcal{F}''(z,g) = g''(z)$, etc.

Let $p_0 = (z_0, g_0)$ be a point of $S(\mathcal{F})$, where g_0 has domain-set $\Delta(z_0, \rho_0)$. For r satisfying $0 < r \leq \rho_0$ we designate as the *open disk on $S(\mathcal{F})$ with center p_0 and radius r* the set $\Delta(p_0, r)$ composed of all points (z, g) on $S(\mathcal{F})$ with the property that z is an element of $\Delta(z_0, r)$ and g is the germ generated at z by g_0. The disk $\Delta(p_0, r)$ is a carbon copy of the plane disk $\Delta(z_0, r)$ and lies above it on the surface $S(\mathcal{F})$. We emphasize that $\Delta(p_0, r)$ is defined not for arbitrary $r > 0$, but only for r in the interval $(0, \rho_0]$. If Δ_1 and Δ_2 are open disks on $S(\mathcal{F})$ and if p is a point of $\Delta_1 \cap \Delta_2$, it is not difficult to check that this intersection contains some open disk centered at p. Also, if p_1 and p_2 are distinct points of $S(\mathcal{F})$, then there exists an $r > 0$ for which $\Delta(p_1, r)$ and $\Delta(p_2, r)$ are disjoint.

With the notion of an open disk on $S(\mathcal{F})$ thus established, the way is cleared for the installation in $S(\mathcal{F})$ of the full apparatus of topology, embracing all of the concepts discussed previously in the contexts of the complex plane $\mathbb{C}$ and extended complex plane $\widehat{\mathbb{C}}$: open set, closed set, convergent sequence, connected set, domain, compact set, etc. Indeed, since the earlier definitions were couched in terms of open disks, they transfer to the present setting with little or no formal change. One important topological fact about $S(\mathcal{F})$ is that it is connected. Another is that open disks on $S(\mathcal{F})$ are domains; i.e., they are open, connected sets. We are also entitled to speak of continuity for functions of the type $\Phi: A \to S(\mathcal{F})$, where A is a subset of $\mathbb{C}$ (or $\widehat{\mathbb{C}}$), and those of the kind $\Phi: A \to \mathbb{C}$ (or $\Phi: A \to \widehat{\mathbb{C}}$), in which A is a subset of $S(\mathcal{F})$. Once again, the previous definitions, which were formulated with the aid of disks, carry over essentially verbatim. To cite two examples, both the projection $\mathcal{P}$ and the function $\mathcal{F}$ are readily seen to be continuous functions from $S(\mathcal{F})$ into the complex plane. As a matter of fact, the restriction of $\mathcal{P}$ to any open disk $\Delta(p, r)$ on $S(\mathcal{F})$ furnishes a homeomorphism between $\Delta(p, r)$ and the plane disk $\Delta[\mathcal{P}(p), r]$.

Paths on $S(\mathcal{F})$ enjoy a special significance. A path Γ on $S(\mathcal{F})$ is just a continuous function of the type $\Gamma: [a, b] \to S(\mathcal{F})$. For reasons that will be clear in a second, we express $\Gamma(t)$ in the manner $\Gamma(t) = (\gamma(t), g_t)$, where $\gamma: [a, b] \to \mathbb{C}$ is the path in the complex plane defined by $\gamma(t) = \mathcal{P}[\Gamma(t)]$. (We refer to γ as the *projection of Γ into the complex plane.*) Let $p_1 = (z_1, g)$ be the initial point of Γ, and let $p_2 = (z_2, \widetilde{g})$ be its terminal point. For each t in $[a, b]$ we are presented by Γ with a germ g_t at $\gamma(t)$. By definition, $g_a = g$ and $g_b = \widetilde{g}$. Furthermore, a moment's reflection should convince one that the definition of continuity for Γ at a point t_0 of $[a, b]$ is a restatement of the fact that $\{g_t\}$ satisfies at t_0 the local compatibility condition for an analytic continuation along γ. The result: $\{g_t\}$ gives the analytic continuation of g to $\widetilde{g}$ along γ. Conversely, if $\{g_t\}$ is known to be the analytic continuation of a germ g to a germ $\tilde{g}$ along a path $\gamma\colon [a,b] \to \mathbb{C}$, then $\Gamma(t) = (\gamma(t), g_t)$ defines a path on $S(\mathcal{F})$ starting at the point $p_1 = (\gamma(a), g)$ and terminating at $p_2 = (\gamma(b), \tilde{g})$. (The path Γ is called a *lift* of γ to $S(\mathcal{F})$.) One consequence of the definition of a global analytic function is that any two points p_1 and p_2 of $S(\mathcal{F})$ can be joined by a path on that surface.

It is possible to do contour integration on $S(\mathcal{F})$. If $\Gamma\colon [a,b] \to S(\mathcal{F})$ is a piecewise smooth path — this condition simply asks that the projection γ of Γ into $\mathbb{C}$ be piecewise smooth — and if Φ is a continuous, complex-valued function on the trajectory of Γ, then we make the definitions

$$\int_\Gamma \Phi(p)\, dp = \int_a^b \Phi[\Gamma(t)]\, \dot{\gamma}(t)\, dt$$

and

$$\int_\Gamma \Phi(p)\, |dp| = \int_a^b \Phi[\Gamma(t)]\, |\dot{\gamma}(t)|\, dt \ .$$

The first equation defines the *complex line integral of Φ along Γ*, the second gives meaning to the *integral of f along Γ with respect to arclength*. As an illustration, we remark that

$$\int_\Gamma \mathcal{F}'(p)\, dp = \mathcal{F}[\Gamma(b)] - \mathcal{F}[\Gamma(a)] \ ,$$

a formula whose verification is left to the reader as an instructive exercise. (See Exercise 4.49.)

Finally, we introduce the counterpart of analyticity for functions whose domain-sets are subsets of $S(\mathcal{F})$. Let $\Delta = \Delta(p_0, r)$ be an open disk on $S(\mathcal{F})$, and let $\mathcal{P}_\Delta$ designate the restriction of the function $\mathcal{P}$ to Δ. Then $\mathcal{P}_\Delta$ projects Δ homeomorphically onto the plane disk $\mathcal{P}(\Delta) = \Delta[\mathcal{P}(p_0), r]$. In particular, $\mathcal{P}_\Delta$ has an inverse function $\mathcal{P}_\Delta^{-1}\colon \mathcal{P}(\Delta) \to \Delta$. A complex-valued function Φ whose domain-set includes Δ is pronounced *holomorphic in Δ* provided the function $\varphi = \Phi \circ \mathcal{P}_\Delta^{-1}$, which is just an ordinary complex-valued function of a complex variable, is analytic. (The literature talks of $\mathcal{P}_\Delta$ as a "local coordinate chart" on $S(\mathcal{F})$ at p_0. The statement that $\Phi \circ \mathcal{P}_\Delta^{-1}$ is analytic is often rendered by the phrase "Φ is analytic in Δ with respect to the local coordinate system given by $\mathcal{P}_\Delta$.") To differentiate Φ at a point p of Δ we take derivatives of φ at the corresponding point of $\mathcal{P}(\Delta)$: $\Phi'(p) = \varphi'[\mathcal{P}(p)]$, $\Phi''(p) = \varphi''[\mathcal{P}(p)]$, etc. For example, both $\mathcal{P}$ and $\mathcal{F}$ are holomorphic functions in Δ: if $p_0 = (z_0, g_0)$, then we obtain

$$\mathcal{P} \circ \mathcal{P}_\Delta^{-1}(z) = z \ , \quad \mathcal{F} \circ \mathcal{P}_\Delta^{-1}(z) = g_0(z)$$

for every z in $\mathcal{P}(\Delta)$, which confirms that $\mathcal{P} \circ \mathcal{P}_\Delta^{-1}$ and $\mathcal{F} \circ \mathcal{P}_\Delta^{-1}$ are analytic there. A function Φ is said to be *holomorphic in U*, an open subset of $S(\mathcal{F})$, if Φ is holomorphic in every open disk on $S(\mathcal{F})$ that is contained in U. For such Φ it makes sense to take derivatives $\Phi'(p), \Phi''(p), \Phi'''(p), \cdots$

3. Analytic Continuation

at a point p of U: we merely choose an open disk Δ in U that contains p and compute the derivatives of $\Phi \circ \mathcal{P}_\Delta^{-1}$ at the point $\mathcal{P}(p)$. The results do not depend on which disk Δ is selected. Prime examples of functions that are holomorphic on the whole surface $S(\mathcal{F})$ are, as might be expected, $\mathcal{P}$ and $\mathcal{F}$. Other simple examples include $\mathcal{P}^2$, $\mathcal{P} + \mathcal{F}$, $\mathcal{P}\mathcal{F}^2 + i\mathcal{F}'$, and $\mathcal{F}'e^{\mathcal{F}}$. The concept of meromorphicity for a function defined in an open subset of $S(\mathcal{F})$, but taking values in $\widehat{\mathbb{C}}$, is treated in a similar fashion.

We have tried in this section to indicate how many of the concepts met throughout the book can be transported from the complex plane to the Riemann surface $S(\mathcal{F})$. Naturally, this whole procedure raises many questions. What changes does the theory undergo in the new setting? What form do fundamental theorems, such as Cauchy's theorem, take on $S(\mathcal{F})$? What is the significance of points of $\mathbb{C}$ whose fibers in $S(\mathcal{F})$ are empty or "deficient" in some way? And most importantly, is there any real purpose to generalizing complex analysis along these lines? The answers to these and other questions will have to be sought elsewhere. We end here by taking a parting look at two of the Riemann surfaces that have received occasional mention in the preceding pages.

A point p on the Riemann surface $S(\mathcal{L})$ of the global logarithm function $\mathcal{L}$ has the form $p = (z, g)$, where g is a germ of $\mathcal{L}$ at z. The value of g at z is $\mathrm{Log}\, z + 2k\pi i$ for some integer k. We can use this integer as a handy means of identifying p, writing $p = (z, k)$ in place of $p = (z, g)$. For a fixed integer k we refer to the set $S_k = \{(z, k) \in S(\mathcal{L}): z \neq 0\}$ as the kth *sheet of* $S(\mathcal{L})$. Of course, $S(\mathcal{L}) = \cup_{k=-\infty}^{\infty} S_k$. What do these sheets actually look like and just how do they fit together to form $S(\mathcal{L})$; i.e., how can we utilize them to picture this surface? To answer these questions we assume that we have at our disposal an infinite number of facsimiles of the complex plane, one copy for each integer k. We distinguish between the various copies by writing $\mathbb{C}_k$ to specify the copy corresponding to k. A model of the sheet S_k is produced by slitting $\mathbb{C}_k$ along the entire length of its negative real axis — i.e., forming $\mathbb{C}_k \sim (-\infty, 0]$ — and then restoring the points of the interval $(-\infty, 0)$ to the upper edge of the slit (Figure 3). (Think of physically cutting a piece of paper along a ray, but imagine doing it so deftly that all the paper molecules on the cut-line itself wind up on the upper edge of the cut!) To build $S(\mathcal{L})$ from these sheets we perform a massive gluing job: for each integer k we paste S_k and S_{k+1} together along half of their respective boundaries, each point on the upper edge (bold line) of the slit that borders S_k being matched with the corresponding point on the lower edge (dotted line) of the slit bordering S_{k+1}. In this way we form from the sundry sheets S_k a continuous surface $S(\mathcal{L})$, S_k blending smoothly with S_{k-1} below it and S_{k+1} above it for every integer k. In moving on $S(\mathcal{L})$ across a former slit line one makes a transition from one sheet to an adjoining sheet. Perhaps the most evocative image of $S(\mathcal{L})$ that emerges from this construction is that of an immense spiral ramp, as one sometimes finds in parking garages, winding endlessly upward and

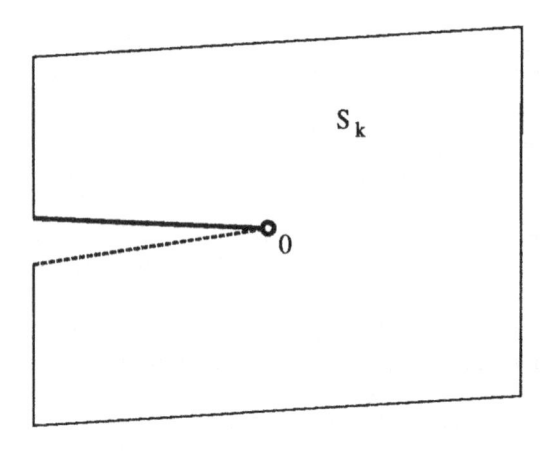

Figure 3.

downward around a central axis. In the above notation the holomorphic function $\mathcal{L}$ is given on $S(\mathcal{L})$ by $\mathcal{L}(z,k) = \text{Log } z + 2k\pi i$. Its derivative has the formula $\mathcal{L}'(z,k) = z^{-1}$. Furthermore, $\mathcal{L}$ provides a conformal mapping (i.e., a univalent holomorphic function) of $S(\mathcal{L})$ onto the complex plane. From a conformal point of view, $S(\mathcal{L})$ is thus the same as the complex plane!

If $\mathcal{F}$ is the global square root function, then a point p on the Riemann surface $S(\mathcal{F})$ has the structure $p = (z, g)$, where the germ g is a branch of the square root function in the disk $\Delta(z, |z|)$. At the point z itself the value of g is either $\sqrt{z}$ or $-\sqrt{z}$. If we relabel the point p for which the former is true $p = (z, +)$ and the point for which the latter holds $p = (z, -)$, then we achieve a decomposition of $S(\mathcal{F})$ into two sheets, $S_+ = \{(z, +) : z \neq 0\}$ and $S_- = \{(z, -) : z \neq 0\}$. To construct the Riemann surface $S(\mathcal{F})$ we start with two copies of the complex plane — call them $\mathbb{C}_+$ and $\mathbb{C}_-$ — we slit each along its negative real axis, just as we did in constructing $S(\mathcal{L})$, in order to obtain concrete realizations of S_+ and S_-, and we then fasten S_+ and S_- together according to the following prescription: the upper edge of the slit that borders S_+ is joined to the lower edge of the slit on the boundary of S_- — and vice versa. As things stand the result of this cut-and-paste operation is admittedly hard to visualize. If a correct topological picture of $S(\mathcal{F})$ is all that is desired, however, a simple one is available. We merely have to substitute for the complex plane slit along the negative real axis its topological equivalent, the Riemann sphere slit from north pole to south pole along the great semi-circle that corresponds to the interval $(-\infty, 0)$ under stereographic projection. In this model S_+ and S_- amount topologically to hemispheres which, when joined together in accordance with the above instructions, form the whole Riemann sphere $\widehat{\mathbb{C}}$, except for punctures at the points 0 and ∞ (Figure 4). Even more is true. The holomorphic function $\mathcal{F}$, whose rule of correspondence can be expressed by $\mathcal{F}(z,+) = \sqrt{z}$ and $\mathcal{F}(z,-) = -\sqrt{z}$, provides a conformal mapping of $S(\mathcal{F})$ onto $\mathbb{C} \sim \{0\}$.

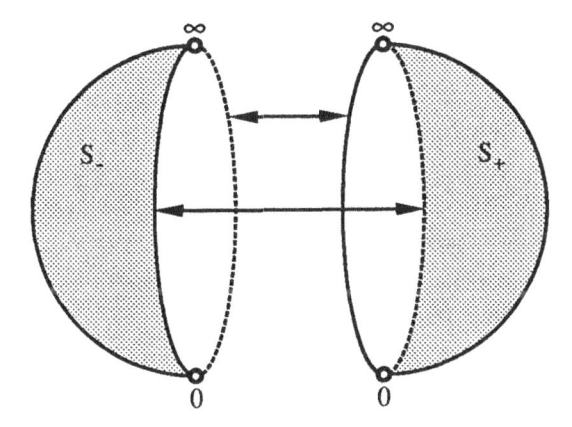

4 Exercises for Chapter X

4.1 Exercises for Section X.1

4.1. Prove Mittag-Leffler's theorem for an open subset U of $\widehat{\mathbb{C}}$ that contains the point ∞. (*Hint.* Let f_0 be the meromorphic function in $U_0 = U \sim \{\infty\}$ constructed in the proof of Theorem 1.2 for the set $E_0 = E \sim \{\infty\}$ and the given assignment of singular parts. Check that f_0 has a removable singularity at ∞. Remember: should the set E_1 that appears in the aforementioned proof be a finite set, then the contribution to f_0 arising from E_1 would require no convergence inducing summands.)

4.2. From (10.4) derive the identity

$$\pi \tan(\pi z) = -2 \sum_{n=-\infty}^{\infty} \left(\frac{1}{2z - 2n + 1} + \frac{1}{2n - 1} \right)$$

for z in $\mathbb{C}$. (*Hint.* At one point the fact that $\tan 0 = 0$ comes into play.)

4.3. Verify that

$$\pi \csc(\pi z) = \frac{1}{z} + \sum_{|n| \geq 1} (-1)^n \left(\frac{1}{z-n} + \frac{1}{n} \right)$$

for z in $\mathbb{C}$. (*Hint.* $2\csc(\pi z) = \tan(\pi z/2) + \cot(\pi z/2)$.)

4.4. Identify the meromorphic function f defined in $\mathbb{C}$ by $f(z) =$

$\sum_{n=-\infty}^{\infty}(-1)^n(z-n)^{-2}$.

4.5. Prove that $f(z) = \sum_{n=-\infty}^{\infty}(z^3-n^3)^{-1}$ defines a meromorphic function in $\mathbb{C}$. Determine this function explicitly.

4.6. If $\wp$ is the Weierstrass function associated with a lattice Ω, confirm that $\wp'(z) = 0$ for every "half-lattice" point z; i.e., for every z such that $2z$ belongs to Ω, but z itself does not.

4.7. Let Ω be the lattice generated by ω_1 and ω_2, and let $\wp$ be its associated Weierstrass function. Demonstrate that

$$\zeta(z) = \frac{1}{z} + \sum_{\omega \in \Omega^*} \left(\frac{1}{z-\omega} + \frac{1}{\omega} + \frac{z}{\omega^2} \right)$$

defines a meromorphic function in $\mathbb{C}$ and that $\zeta'(z) = -\wp(z)$. Conclude that $\zeta(z+\omega_1) = \zeta(z) + \eta_1$ and $\zeta(z+\omega_2) = \zeta(z) + \eta_2$ for every z in $\mathbb{C}$, where η_1 and η_2 are constants. Prove that these constants obey "Legendre's relation": $\eta_1\omega_2 - \eta_2\omega_1 = 2\pi i$. (*Hint.* For the last part integrate $\zeta(z)$ along the boundary of a certain parallelogram centered at the origin.)

4.8. Show that the only entire elliptic functions are the complex-valued constant functions on $\mathbb{C}$.

4.9. Let $\wp$ be the Weierstrass function associated with a lattice Ω. Since $\wp$ is an even function with singular part $S(z) = z^{-2}$ at the origin and since $\wp(z) - S(z) \to 0$ as $z \to 0$ (Why is this?), the Laurent expansion of $\wp$ in a small punctured disk $\Delta^*(0, r)$ has the form $\wp(z) = z^{-2} + a_2 z^2 + a_4 z^4 + O(z^6)$. Check that $a_2 = 3\sum_{\omega \in \Omega^*} \omega^{-4}$ and $a_4 = 5\sum_{\omega \in \Omega^*} \omega^{-6}$. Then show that the elliptic function $f = (\wp')^2 - 4\wp^3 + 20a_2\wp + 28a_4$ satisfies $f(z) = O(z^2)$ in $\Delta^*(0, r)$, whence $f(z) \to 0$ as $z \to 0$. Deduce from this information that $f(z) = 0$ for every z in $\mathbb{C}$. Conclude that $\wp$ and $\wp'$ are related by the equation $(\wp')^2 = 4\wp^3 - 20a_2\wp - 28a_4$.

4.10. If f is a non-constant elliptic function whose period lattice Ω is generated by ω_1 and ω_2, if P is the closed parallelogram with vertices $0, \omega_1, \omega_2$, and $\omega_1 + \omega_2$, and if E is the set (modulo Ω) of poles that f has in P, then E is non-empty and $\sum_{z \in E} \text{Res}(z, f) = 0$. Justify this statement. Conclude that it is not possible for f to have only a single (modulo Ω) simple pole in P. (*N.B.* In dealing with the zeros and poles of an elliptic function f — or, more generally, with solutions of equations of the sort $f(z) = w$ — in such a "period parallelogram" P, it is customary to work "modulo Ω," meaning that points z and $z+\omega$ with ω in Ω are considered to be one and the same point. For example, we would say that the Weierstrass function $\wp = \wp_\Omega$ has two poles in P, these accounted for by the double pole at the origin, for the actual poles which $\wp$ has at $0, \omega_1, \omega_2$, and $\omega_1 + \omega_2$ coalesce modulo Ω. We abide by this convention in forming the set E in the present problem — we take all the poles of f, we eliminate duplicates modulo Ω, and only then do we sum the residues — and in similar situations

4. Exercises for Chapter X

later. *Hint.* By replacing f with $g(z) = f(z+c)$ for a suitable constant c, one can reduce the proof to the case in which f is free of poles on ∂P. Notice that no two points in the interior of P coincide modulo Ω.)

4.11. Let f and P be as in the preceding exercise. Show that, modulo Ω, the number of zeros that f has in P is equal to its number of poles in this set, provided multiplicity is taken into account. (N.B. If points z_0 and $\tilde{z}_0$ coincide modulo Ω, then the multiplicity of f at z_0 is the same as its multiplicity at $\tilde{z}_0$. Thus the multiplicity of a zero or pole is independent of which representative we pick for it modulo Ω.)

4.12. Again taking f and P as in Exercise 4.10, prove that $f(P) = \widehat{\mathbb{C}}$. More precisely, show that for any w in $\widehat{\mathbb{C}}$ the number of solutions (modulo Ω) in P of the equation $f(z) = w$ is independent of w, assuming that solutions are counted with proper regard for multiplicity.

4.13. Let $\wp$ be the Weierstrass function associated with the plane lattice Ω that is generated by ω_1 and ω_2, and let P be the closed parallelogram with vertices $0, \omega_1, \omega_2,$ and $\omega_1 + \omega_2$. Certify that $\wp'$ has simple zeros at the points $\omega_1/2$, $\omega_2/2$, and $(\omega_1 + \omega_2)2$, but has (modulo Ω) no additional zeros in P. Use this information to verify that for given w in $\widehat{\mathbb{C}}$ the equation $\wp(z) = w$ has a unique solution (modulo Ω) in P when $w = \infty$, $\wp(\omega_1/2)$, $\wp(\omega_2/2)$, or $\wp[(\omega_1 + \omega_2)/2]$, and exactly two solutions (modulo Ω) in P otherwise. Conclude, in particular, that the complex numbers $\wp(\omega_1/2)$, $\wp(\omega_2/2)$, and $\wp[(\omega_1 + \omega_2)/2]$ are distinct. (*Hint.* Recall Exercises 4.6 and 4.12.)

4.14. Let $\Omega = \Omega(\omega_1, \omega_2)$ be a plane lattice, let $\wp$ be the corresponding Weierstrass function, let $a = 60 \sum_{\omega \in \Omega^*} \omega^{-4}$ and $b = 140 \sum_{\omega \in \Omega^*} \omega^{-6}$, and let C denote the locus of points (z, w) in $\mathbb{C}^2$ such that $w^2 = 4z^3 - az - b$. Show that for each element (z, w) of C there is a unique (modulo Ω) point ζ of the closed parallelogram P with vertices $0, \omega_1, \omega_2,$ and $\omega_1 + \omega_2$ for which $\wp(\zeta) = z$ and $\wp'(\zeta) = w$. (*Hint.* Utilize Exercises 4.9 and 4.13.)

4.15. If $\wp$ is the Weierstrass function corresponding to the plane lattice $\Omega = \Omega(\omega_1, \omega_2)$ and if $e_1 = \wp(\omega_1/2)$, $e_2 = \wp(\omega_2/2)$ and $e_3 = \wp[(\omega_1 + \omega_2)/2]$, verify that $(\wp')^2 = 4(\wp - e_1)(\wp - e_2)(\wp - e_3)$. Conclude with the help of Exercise 4.9 that $e_1 + e_2 + e_3 = 0$. (*Hint.* Consider the function $f = (\wp - e_1)(\wp - e_2)(\wp - e_3)/(\wp')^2$. Use Exercise 4.8.)

4.16. Let $\omega_1 = 1$ and $\omega_2 = i$. If $\wp$ is the Weierstrass function associated with the lattice that ω_1 and ω_2 generate, check that $\wp(iz) = -\wp(z)$ and $\wp(\bar{z}) = \overline{\wp(z)}$. Deduce that, in the notation of the preceeding exercise, $e_2 = -e_1$, e_1 is real, and $e_3 = 0$.

4.17. Suppose that D is the interior of a closed rectangle R in the complex plane with one of its vertices at the origin and that f is a conformal mapping of D onto a half-plane D'. Confirm that f is the restriction to D of an elliptic function F having R as one quarter of a period parallelogram.

Would the same be true if, rather than a rectangle, R were some other parallelogram?

4.2 Exercises for Section X.2

4.18. Compute the infinite product $\prod_{n=1}^{\infty}[(n+1)^2(n^2+2n)^{-1}]$.

4.19. Test the following infinite products for convergence; in case of convergence, decide whether it is absolute: (i) $\prod_{n=1}^{\infty}[(-1)^n n(2n+1)^{-1}]$; (ii) $\prod_{n=1}^{\infty}[1+(-1)^n n^{-3/2}]$; (iii) $\prod_{n=1}^{\infty}\cos[n^{-1}]$; (iv) $\prod_{n=1}^{\infty}\sqrt[n]{n}$; (v) $\prod_{n=2}^{\infty}[1+(-1)^n n^{-1}\operatorname{Log}^{-2} n]$.

4.20. For which values of z does $\prod_{n=1}^{\infty}(1+z^n)$ converge. What about $\prod_{n=1}^{\infty}\sqrt[n]{z}$?

4.21. Let $\langle x_n \rangle$ be a sequence such that $0 \leq x_n < 1$ for every n. Given that the infinite product $\prod_{n=1}^{\infty}(1-x_n)$ is convergent; prove that it is absolutely convergent, i.e., $\prod_{n=1}^{\infty}(1+x_n)$ converges. (*Hint.* Show that the sequence of partial products $\langle p_n \rangle$ corresponding to $\prod_{n=1}^{\infty}(1+x_n)$ is monotone and bounded. How do $1+x$ and $1-x$ compare when $0 < x \leq 1$?)

4.22. Let $\lambda > 0$. Show that the product $\prod_{n=2}^{\infty}[1+(-1)^n n^{-\lambda}]$ converges absolutely when $\lambda > 1$, merely converges when $1/2 < \lambda \leq 1$, and diverges when $0 < \lambda \leq 1/2$. (*Hint.* In the case $0 < \lambda \leq 1/2$ verify that $\prod_{n=2}^{2N}[1+(-1)^n n^{-\lambda}] \to 0$ as $N \to \infty$.)

4.23. Use Example 2.1 to determine $\prod_{n=2}^{\infty}(1-n^{-2})$.

4.24. Establish the identity $(1-z^2)^{-1} = \prod_{n=1}^{\infty}(1+z^{2^n})$ for $|z| < 1$.

4.25. Let $p_1 < p_2 < p_3 < \cdots$ be the sequence of prime natural numbers; i.e., $p_1 = 2, p_2 = 3, p_3 = 5, \cdots$. Demonstrate that the infinite product $\prod_{k=1}^{\infty}(1-p_k^{-z})^{-1}$ converges absolutely and normally in the half-plane $U = \{z: \operatorname{Re} z > 1\}$. Go on to prove that the function defined in U by this product is none other than Riemann's zeta-function! Through this process obtain the "Euler product formula" for $\zeta(z)$ in U: $\zeta(z) = \prod_{k=1}^{\infty}(1-p_k^{-z})^{-1}$. Infer from it that ζ is free of zeros in U. (*Hint.* To obtain the product formula make use of the fact that any integer $n \geq 2$ has a unique factorization of the type $n = p_{k_1}^{\alpha_1} p_{k_2}^{\alpha_2} \cdots p_{k_r}^{\alpha_r}$ where $k_1 < k_2 < \cdots < k_r$ and where $\alpha_1, \alpha_2, \ldots, \alpha_r$ are positive integers. Observing that $(1-p_k^{-z})^{-1} = \sum_{\ell=0}^{\infty} p_k^{-\ell z}$, try to get a feel for what's going on by checking what happens when one forms the products $(1-2^{-z})^{-1}(1-3^{-z})^{-1}$ and $(1-2^{-z})^{-1}(1-3^{-z})^{-1}(1-5^{-z})^{-1}$.)

4.26. Show that the infinite product $\prod_{n=1}^{\infty}\exp(n^{-z}\operatorname{Log} z)$ converges normally in the half-plane $U = \{z: \operatorname{Re} z > 1\}$, and identify the function that this product defines in U.

4.27. Let $\langle z_n \rangle$ be a sequence of non-zero complex numbers in the disk $\Delta = \Delta(0,1)$ with the property that the series $\sum_{n=1}^{\infty}(1-|z_n|)$ converges.

4. Exercises for Chapter X

Verify that the formula

$$f(z) = cz^p \prod_{n=1}^{\infty} \frac{\overline{z}_n}{|z_n|} \left(\frac{z_n - z}{1 - \overline{z}_n z}\right),$$

where p is a non-negative integer and c is a constant of unit modulus, defines an analytic function in Δ, one that satisfies $|f(z)| \leq 1$ for every z in Δ. Where are the zeros of f and what are their multiplicities?

4.28. Establish the Weierstrass product theorem for an open set U in $\widehat{\mathbb{C}}$ — $U \neq \widehat{\mathbb{C}}$ — that contains the point ∞.

4.29. Let $E = \{z_1, z_2, z_3, \cdots\}$ be a discrete subset of an open set U in the complex plane, let $\langle w_n \rangle$ be an arbitrary sequence of complex numbers, and let $\langle m_n \rangle$ be an arbitrary sequence of positive integers. Confirm the existence of an analytic function $f: U \to \mathbb{C}$ that for $n = 1, 2, 3, \cdots$ takes the value w_n with multiplicity m_n at z_n.

4.30. Let D be a domain in the complex plane. Construct a discrete subset E of D with the property that every point of ∂D is a limit point of E. Use E to prove: there exists an analytic function $f: D \to \mathbb{C}$ that admits no extension to an analytic function $F: \widetilde{D} \to \mathbb{C}$ with $\widetilde{D}$ a domain that properly contains D.

4.31. Find an explicit Weierstrass product expansion for the entire function f: (i) $f(z) = \cos z$; (ii) $f(z) = e^z - 1$; (iii) $f(z) = e^{a\pi z} - e^{b\pi z}$; (iv) $f(z) = \sqrt{z}\sin(\pi\sqrt{z})$.

4.32. With the aid of the functional equation (10.30) confirm that $\mathrm{Res}[-n, \Gamma(z)] = (-1)^n (n!)^{-1}$ for $n = 0, 1, 2, \cdots$.

4.33. Obtain the representation

$$\Gamma(z) = \sum_{n=0}^{\infty} \frac{(-1)^n}{n!(z+n)} + \int_1^{\infty} t^{z-1} e^{-t}\, dt$$

for every z in $\mathbb{C} \sim \{0, -1, -2, \cdots\}$.

4.34. The logarithmic derivative $\Psi = \Gamma'/\Gamma$ of the gamma function is known as the *digamma function* or, alternatively, as the *Gauss psi-function*. Derive the identity $\Psi'(z) = \sum_{n=0}^{\infty}(z+n)^{-2}$. Also, show that Ψ satisfies the functional equations

$$\Psi(1+z) - \Psi(z) = z^{-1} \quad , \quad \Psi(1-z) - \Psi(z) = \pi\cot(\pi z).$$

4.35. Define f by $f(z) = \Psi(z) + \Psi[z + (1/2)] - 2\Psi(2z)$ for z in the domain $D = \mathbb{C} \sim \{0, -1/2, -1, -3/2, \cdots\}$. By calculating f', show that f is constant in D. Remembering that Ψ is the logarithmic derivative of Γ, argue from the constancy of f to the fact that $\Gamma(z)\Gamma[z + (1/2)] = e^{az+b}\Gamma(2z)$ for every z in D, where a and b are constants. By comparing the values of both sides of this identity when $z = 1/2$ and $z = 1$, determine a and b. In this way arrive at the so-called "duplication formula" of Legendre:

$$\Gamma(2z) = \pi^{-1/2} 2^{2z-1} \Gamma(z) \Gamma[z + (1/2)] \ .$$

4.3 Exercises for Section X.3

4.36. Let $P(z,w) = a_0(z) + a_1(z)w + \cdots + a_n(z)w^n$ be a polynomial in w whose coefficients $a_0(z), a_1(z), \ldots, a_n(z)$ are entire functions of z. Given that an analytic function element (f_0, D_0) has $P[z, f_0(z)] = 0$ for every z in D_0 and that a second function element (f, D) is an analytic continuation of (f_0, D_0), confirm that $P[z, f(z)] = 0$ for every z in D.

4.37. If an analytic function element (f_0, D_0) has the property that $w = f_0(z)$ furnishes a solution in D_0 to the differential equation

$$c_n(z) \frac{d^n w}{dz^n} + c_{n-1}(z) \frac{d^{n-1} w}{dz^{n-1}} + \cdots + c_1(z) \frac{dw}{dz} + c_0(z) w = 0 \ ,$$

where $c_0(z), c_1(z), \ldots, c_n(z)$ are entire functions of z, and if (f, D) is an analytic continuation of (f_0, D_0), then $w = f(z)$ gives a solution to the same equation in D. Prove this.

4.38. Exhibit the analytic continuation $\{g_t\}$ of g, the analytic germ generated at the point 1 by the function $f(z) = \text{Log } z$, along the path γ given by $\gamma(t) = e^{it}$ for $0 \leq t \leq 2\pi$. How is the terminal germ $\tilde{g}$ of this continuation related to g? What terminal germ is obtained when g is continued along the path $\gamma + \gamma$? What about $-\gamma$? Or a general path $\alpha: [a,b] \to \mathbb{C} \sim \{0\}$ for which $\alpha(a) = \alpha(b) = 1$?

4.39. Let g be the analytic germ determined at the point 1 by the function $f(z) = z^{2/3}$. Find the analytic continuation $\{g_t\}$ of g along the path γ defined on $[0, 2\pi]$ by $\gamma(t) = e^{it}$. In what relation do g and the terminal germ $\tilde{g}$ of this continuation stand?

4.40. It is known that an analytic germ g at a point z_0 of a domain D can be continued analytically along all closed paths γ in D starting — hence, terminating — at z_0. Assuming that such a path γ is homotopic in D, with fixed endpoints, to a constant path at z_0, what is the end-product of the continuation of g along γ?

4.41. Determine all exceptional points z for the polynomial $P(z,w)$:
(i) $P(z,w) = w^3 - z^2 + z$; (ii) $P(z,w) = z^2 w^2 + (2z+2)w + 1$;
(iii) $P(z,w) = 2zw^3 + 3zw^2 + 1$.

4.42. Let $P(z,w) = w^3 - z^2$, a polynomial whose only exceptional point is at the origin, and let $D = \mathbb{C} \sim [0, \infty)$. According to Theorem 3.7 there is an analytic function $f: D \to \mathbb{C}$ such that $f(-1) = 1$ and $P[z, f(z)] = 0$

4. Exercises for Chapter X

for every z in D. Find a formula for f.

4.43. Show that the only exceptional points for the polynomial $P(z, w) = w^3 - z^2 w^2 + (4/27)$ are 1 and -1. By Theorem 3.7 there is a unique analytic function $f: \Delta(0, 1) \to \mathbb{C}$ such that $f(0) = -\sqrt[3]{4}/3$ and $P[z, f(z)] = 0$ for every z in $\Delta(0, 1)$. Verify that f is a zero-free, even function which is real-valued on the interval $(-1, 1)$, and check that the only point z of $\Delta(0, 1)$ for which $f'(z) = 0$ is $z = 0$. What is the image of the interval $(-1, 1)$ under f?

4.44. Consider a linear differential equation

$$(10.45) \qquad c_n(z) \frac{d^n w}{dz^n} + c_{n-1}(z) \frac{d^{n-1} w}{dz^{n-1}} + \cdots + c_0(z) w = 0 ,$$

say with entire functions for coefficients, in a simply connected domain D where the leading coefficient c_n is zero-free. Prove that for any given point z_0 of D and any preassigned complex numbers $w_0, w_1, \ldots, w_{n-1}$ there exists a unique analytic function $f: D \to \mathbb{C}$ such that $w = f(z)$ solves (10.45) in D and such that $f(z_0) = w_0$, $f'(z_0) = w_1, \ldots, f^{(n-1)}(z_0) = w_{n-1}$. (*Hint.* Apply Theorem 3.4. Feel free to use the following basic existence and uniqueness theorem from the theory of ordinary differential equations: if ζ_0 is a point of the complex plane for which $c_n(\zeta_0) \neq 0$ and if $\omega_0, \omega_1, \ldots, \omega_{n-1}$ are arbitrary complex numbers, then there is a unique analytic germ g at ζ_0 such that $w = g(z)$ satisfies (10.45) throughout the domain-set of g and such that $g(\zeta_0) = \omega_0$, $g'(\zeta_0) = \omega_1, \ldots, g^{(n-1)}(\zeta_0) = \omega_{n-1}$.)

4.45. Demonstrate that $\mathcal{F}$, the family of all analytic function elements (f, D) with the property that $[f(z)]^2 = z$ for every z in D, is a global analytic function.

4.46. Let $P(z, w) = z^2 w^2 - (z^2 + 1) w + 1$. Show that the collection of analytic function elements (f, D) with the property that $P[z, f(z)] = 0$ for every z in D is not a global analytic function by producing two such elements that are not analytic continuations of each other. What does this tell you about P?

4.47. Let $\Delta_1 = \Delta(p_1, r_1)$ and $\Delta_2 = \Delta(p_2, r_2)$ be open disks on the Riemann surface $S(\mathcal{F})$ of a global analytic function $\mathcal{F}$, and let p be a point of $\Delta_1 \cap \Delta_2$. Demonstrate that $\Delta(p, r)$ is contained in $\Delta_1 \cap \Delta_2$ when $r > 0$ is suitably small.

4.48. Suppose that p_1 and p_2 are different points of the Riemann surface $S(\mathcal{F})$. Confirm that there is a radius $r > 0$ for which the disks $\Delta(p_1, r)$ and $\Delta(p_2, r)$ are disjoint.

4.49. Let U be an open set on $S(\mathcal{F})$, let $\Phi: U \to \mathbb{C}$ be a continuous function, and let $\Psi: U \to \mathbb{C}$ be a primitive for Φ in U, meaning that Ψ is holomorphic in U and has $\Psi'(p) = \Phi(p)$ for every p in U. Show that $\int_\Gamma \Phi(p) dp = \Psi[\Gamma(b)] - \Psi[\Gamma(a)]$ for any piecewise smooth path $\Gamma: [a, b] \to U$.

(*Hint.* First treat the case in which the trajectory of Γ lies in an open disk on $S(\mathcal{F})$ that is contained in U.)

Appendix A

Background on Fields

1 Fields

1.1 The Field Axioms

The definition of a field starts with a set F that is equipped with two binary operations. These operations are conventionally called *addition* and *multiplication* and are denoted by $+$ and $\cdot$. Thus, to each pair of elements a and b from F they assign a *sum* $a+b$ and a *product* $a \cdot b$ (usually abbreviated to ab), also belonging to the set F. The definition goes on to require that the following axioms be satisfied:

F.1 (Associative Laws): $(a+b)+c = a+(b+c)$ and $(ab)c = a(bc)$ for all $a, b,$ and c in F;

F.2 (Commutative Laws): $a+b = b+a$ and $ab = ba$ for all a and b in F;

F.3 (Distributive Law): $a(b+c) = ab+ac$ for all $a, b,$ and c in F;

F.4 (Existence of Identities): there exist a pair of distinct elements of F, by tradition designated 0 and 1, that are neutral elements for the operations of addition and multiplication, respectively — i.e., $a+0 = a$ and $a \cdot 1 = a$ for every a in F;

F.5 (Existence of Inverses): corresponding to each a in F there exists an element of F, ordinarily represented by $-a$, with the property that $a+(-a) = 0$; corresponding to each a in $F^* = F \sim \{0\}$ there exists an element of F, normally symbolized by a^{-1} or $1/a$, such that $aa^{-1} = 1$.

When the above requirements are met the set F endowed with the given operations is called a *field*. The axiom system is known as the set of *field axioms*. These axioms imply that the *additive identity* 0 and the *multiplicative identity* 1 are uniquely determined. The requirement $0 \neq 1$ prevents the set $\{0\}$ from trivially satisfying the field axioms, and so forces any field to contain at least two elements. (There is a field with exactly two elements!)

Also implicit in the axioms is the uniqueness of the *additive inverse* $-a$ of an arbitrary element a of F and the *multiplicative inverse* a^{-1} of a non-zero element a of F. *Subtraction* and *division* in F are defined in the expected manner: $a - b = a + (-b)$ for a and b in F; $a/b = a \cdot b^{-1}$ for a in F and b in F^*. The meaning of a^n for a in F and n an integer is also the anticipated one: $a^n = a \cdot a \cdots a$ (n factors) if n is positive, $a^{-n} = (a^{-1})^n$ if n is positive and $a \neq 0$, and $a^0 = 1$ if $a \neq 0$.

The three fields that have a direct bearing on this book are the fields $\mathbb{Q}$ of rational numbers, $\mathbb{R}$ of real numbers, and $\mathbb{C}$ of complex numbers. The general study of fields is an important component of present-day algebra.

1.2 Subfields

Let F be a field. If a subset E of F is closed with respect to the operations of F — meaning that $a + b$ and ab belong to E whenever a and b do — if the elements 0 and 1 from F are members of E, and if E is itself a field with respect to the operations it thus inherits from F, then E is termed a *subfield* of F. This definition insures that, if a is an element of E, then its additive inverse $-a$ in the field F also belongs to E and, naturally, furnishes the additive inverse of a with respect to the field structure of E. Likewise, the multiplicative inverse in the field E of any non-zero member a of E coincides with its inverse a^{-1} from F. The field $\mathbb{Q}$, for instance, is a subfield of $\mathbb{R}$. In Hamilton's construction of the complex number field $\mathbb{C}$ the set $E = \{(x, 0) \colon x \in \mathbb{R}\}$ forms a subfield of $\mathbb{C}$.

1.3 Isomorphic Fields

Consider a pair of fields F and F'. It may happen that there exists a function $\varphi \colon F \to F'$ which is one-one, which has range F', and which satisfies

$$\varphi(a+b) = \varphi(a) + \varphi(b) \quad , \quad \varphi(ab) = \varphi(a)\varphi(b)$$

for all a and b in F. (N.B. The addition and multiplication on the left-hand sides of these equations are performed in F; the operations on the right-hand sides are carried out in F'.) If so, we say that F and F' are *isomorphic fields* and that φ is an *isomorphism of F onto F'*. The function φ supplies a dictionary for passing from algebraic statements made in the context of F to their exact translations in the language of F'. The existence of such a function implies that there is nothing intrinsic to the field structure of F to distinguish it from F'. For all practical purposes F and F' are the same field! Referring again to Hamilton's construction of the complex numbers, we note that $\varphi(x) = (x, 0)$ defines an isomorphism of $\mathbb{R}$ onto $E = \{(x, 0) \colon x \in \mathbb{R}\}$, a subfield of $\mathbb{C}$. It is this fact that we had in mind

2. Order in Fields 545

when in Chapter I we spoke of $\mathbb{R}$ and E as "structurally indistinguishable" fields.

2 Order in Fields

2.1 Ordered Fields

Roughly speaking, an "ordered field" is one in which there is established a sense of "positive" and "negative" that is compatible with the field structure. To make this precise, suppose that a field F has a subset P (for "positive elements") which exhibits the following two properties:

OF.1: $a + b$ and ab belong to P whenever both a and b do;

OF.2: P does not contain the element 0; if a is any non-zero element of F, then P contains either a or $-a$, but not both.

We then say that F is *ordered by* P. Indeed, we can define an order relation $>$ on F by declaring that $b > a$ if and only if $b - a$ lies in P. (N.B. The statement that $a < b$ is synonymous with $b > a$. The notation $b \geq a$ indicates that either $b > a$ or $b = a$.) Thus, asserting that $a > 0$ is equivalent to stating that a is a member of P. Using (OF.1) and (OF.2) one easily checks that the ordinary rules for operating with inequalities become valid in F. It is conceivable, of course, that there is more than one subset P of F satisfying (OF.1) and (OF.2). As a result, there may be many ways of turning a given field F into an *ordered field*, by which is meant a field in which an order has been specified through a choice of P. It is also possible that F has no subsets P enjoying these properties. When the latter situation occurs, F is called an *unorderable field*. In such a field one must live without an algebraically consistent notion of order.

The field $\mathbb{Q}$, when thought of as an ordered field, is typically assigned its *standard ordering*: a rational number $x = m/n$, where m and n are integers and $n \neq 0$, is declared positive if and only if mn belongs to $\mathbb{N}$, the set of natural numbers. The usual constructions of the real field $\mathbb{R}$ from $\mathbb{Q}$ show how to extend this standard ordering of $\mathbb{Q}$ to the *standard ordering* of $\mathbb{R}$. The field $\mathbb{C}$, on the other hand, is unorderable. To see this, suppose that P were a subset of $\mathbb{C}$ endowed with properties (OF.1) and (OF.2). Since $i \neq 0$, either i would belong to P or $-i$ would. In both cases (OF.1) would imply that $i^2 = (-i)^2 = -1$ is an element of P. Again by (OF.1), $1 = (-1)^2$ would have to be in P, placing both 1 and -1 there. This would violate condition (OF.2). The conclusion: no such P exists, so $\mathbb{C}$ cannot be ordered.

2.2 Complete Ordered Fields

Let F be an ordered field, say with order relation $>$, and let S be a non-empty subset of F. To say that S is *bounded above* means there exists an element b of F with the property that the statement $b \geq x$ is true for every x in S. Such an element b is called an *upper bound for S in F*. If S is bounded above, it may be the case — it does not happen automatically — that among all upper bounds for S in F there is a smallest one, call it u; i.e., u is an upper bound for S and $b > u$ holds for every other upper bound b. In these circumstances we refer to u as the *least upper bound* (or *supremum*) *of S in F*. We express this relationship between S and u by writing $u = \sup S$. In a similar fashion one can say what it means for S to be *bounded below* and for an element of F to be a *lower bound* for S. If S has a *greatest lower bound* (also called an *infimum*) ℓ in F, then we indicate this fact symbolically by means of $\ell = \inf S$. Take, for example, the set $S = \{x \in \mathbb{Q} : x^2 < 2\}$. If $F = \mathbb{Q}$ with its standard ordering, then S is bounded above but has no supremum in F. By contrast, if $F = \mathbb{R}$ with its standard ordering, then the same set S has $\sup S = \sqrt{2}$.

A field F ordered by $>$ is pronounced *complete* relative to the given ordering if every non-empty subset S of F that is bounded above has a supremum in F. This implies the dual property, the existence of an infimum in F for every non-empty subset of F that is bounded below. A fundamental statement about the real number system is:

Theorem 2.1. *The field $\mathbb{R}$ of real numbers with its standard ordering is a complete ordered field.*

Even more is true. The standardly ordered field $\mathbb{R}$ is up to an order-preserving isomorphism the only complete ordered field: if F is any complete ordered field, then there exists an isomorphism φ of $\mathbb{R}$ onto F with the added property that $\varphi(b) > \varphi(a)$ in the order in F whenever $b > a$ with respect to the standard ordering of $\mathbb{R}$.

2.3 Implications for Real Sequences

A sequence $\langle x_n \rangle$ of real numbers is said to be *monotone* if it is either non-decreasing ($x_1 \leq x_2 \leq x_3 \leq \cdots$) or non-increasing ($x_1 \geq x_2 \geq x_3 \geq \cdots$). Theorem 2.1 has the following implication for such sequences: *a monotone and bounded sequence $\langle x_n \rangle$ of real numbers is convergent.* In fact, it is a simple exercise to check that

$$\lim_{n \to \infty} x_n = \sup\{x_n : n = 1, 2, 3, \cdots\}$$

in the non-decreasing case, whereas

$$\lim_{n \to \infty} x_n = \inf\{x_n : n = 1, 2, 3, \cdots\}$$

2. Order in Fields

in the non-increasing case.

Consider now an arbitrary bounded sequence $\langle x_n \rangle$ of real numbers. Let us assume that $|x_n| \leq c$ for all n. We define a new sequence $\langle y_n \rangle$ by setting

$$y_n = \sup\{x_k : k = n, n+1, n+2, \cdots\}$$

for $n \geq 1$. Then $-c \leq y_n \leq c$ and $y_1 \geq y_2 \geq y_3 \geq \cdots$. According to the above remarks, $M = \lim_{n\to\infty} y_n$ exists. This number M is known as the *limit superior* of the given sequence and is denoted by $\limsup_{n\to\infty} x_n$. To repeat: for a bounded sequence $\langle x_n \rangle$ of real numbers,

$$\limsup_{n\to\infty} x_n = \lim_{n\to\infty} \sup\{x_k : k = n, n+1, n+2, \cdots\}.$$

Similarly, the *limit inferior* of such a sequence, denoted $\liminf_{n\to\infty} x_n$, is defined by

$$\liminf_{n\to\infty} x_n = \lim_{n\to\infty} \inf\{x_k : k = n, n+1, n+2, \cdots\}.$$

The important observation to be made about $M = \limsup_{n\to\infty} x_n$ and $m = \liminf_{n\to\infty} x_n$ is that these numbers are accumulation points of $\langle x_n \rangle$. As a matter of fact, M is the largest accumulation point of $\langle x_n \rangle$ and m the smallest. Furthermore, $\langle x_n \rangle$ is convergent with limit ℓ if and only if $m = M = \ell$. The proofs of the preceding observations are straightforward exercises. As one important spin-off from this discussion we record:

Theorem 2.2. (Bolzano-Weierstrass Theorem) *Any bounded sequence of real numbers has at least one real accumulation point.*

Appendix B

Winding Numbers Revisited

1 Technical Facts About Winding Numbers

1.1 The Geometric Interpretation

We furnish here the rigorous argument that justifies the geometric interpretation of winding numbers presented in Chapter V. Let us first recall the situation there. Given a closed, piecewise smooth path $\gamma:[a,b] \to \mathbb{C}$ and a point z of $\mathbb{C} \sim |\gamma|$, we gave geometric meaning to the winding number $n(\gamma, z)$ as follows: we fixed a circle $K = K(z,r)$ centered at z; we defined an auxiliary path $\beta:[a,b] \to \mathbb{C}$, the radial projection of γ on K, by $\beta(t) = z + r\{[\gamma(t) - z]/|\gamma(t) - z|\}$ and denoted by $v(t)$ the radial vector from z to $\beta(t)$; we decided what it should mean for $v(t)$ to perform one complete revolution over a subinterval $[c,d]$ of $[a,b]$; we characterized each such revolution as either positive or negative; we then made an assertion that we now turn into a theorem.

Theorem 1.1. *The winding number* $n(\gamma, z)$ *gives the net number of complete revolutions — meaning the number of positive revolutions minus the number of negative revolutions — executed by the vector* $v(t)$ *as t increases from a to b.*

Proof. For convenience we take $r = 1$. (The choice of r affects only the length of $v(t)$, not the number of revolutions it performs.) With a view toward minimizing notational difficulties, we also assume that $z = 0$ and that $\beta(a) = \beta(b) = 1$. (N.B. The general case can be reduced to this special one by considering in place of γ the path $\gamma_1:[a,b] \to \mathbb{C}$ defined via $\gamma_1(t) = e^{i\varphi}[\gamma(t) - z]$, where $\varphi = -\operatorname{Arg}[\beta(a) - z]$. Then $n(\gamma, z) = n(\gamma_1, 0)$, as a simple calculation confirms, and the radial projection β_1 of γ_1 onto the circle $K_1 = K(0,1)$ is given by $\beta_1(t) = e^{i\varphi}[\beta(t) - z]$. Thus, since we have chosen $r = 1$, $\beta_1(a) = \beta_1(b) = 1$. Finally, the revolutions that $v(t)$ performs exactly mirror those of its counterpart $v_1(t)$.) We may further suppose that β is a non-constant path, for the contents of the theorem are essentially

1. Technical Facts About Winding Numbers

trivial otherwise.

Because β is uniformly continuous on $[a,b]$, we can choose and fix $\delta > 0$ so as to insure that $|\beta(t) - \beta(s)| < 1$ holds whenever t and s are points of $[a,b]$ for which $|t-s| < \delta$. We remark next that the set $A = \{t \in [a,b]: \beta(t) \neq 1\}$ is a non-empty open set in the topology of the real line $\mathbb{R}$. As such, A is expressible as a disjoint union of open intervals, possibly an infinite number of them, and this breakdown of A into open intervals is unique. Consider an arbitrary component interval of A — call it (c,d). The definitions of A and (c,d) dictate that $\beta(t) \neq 1$ when $c < t < d$, whereas $\beta(c) = \beta(d) = 1$. In other words, $[c,d]$ is an interval over which $v(t)$ potentially executes a complete revolution. (Conversely, if $v(t)$ does perform a complete revolution over a subinterval $[c,d]$ of $[a,b]$, then (c,d) is easily seen to be one of the component intervals of A.) If the point -1 happens to belong to $\beta([c,d])$, then necessarily $d-c \geq \delta$. (If $d-c < \delta$, then the fact that $\beta(c) = 1$ would, by the choice of δ, force $\beta([c,d])$ to lie in the disk $\Delta(1,1)$, which does not contain -1.) It follows that the number of intervals (c,d) in the decomposition of A with the feature that -1 belongs to $\beta([c,d])$ cannot exceed $(b-a)/\delta$ — hence, is finite. Suppose that $(c_1,d_1), (c_2,d_2), \ldots, (c_p,d_p)$ is a complete list of the component intervals of A with this property, labeled so that $c_1 < d_1 \leq c_2 < d_2 \leq \cdots \leq c_p < d_p$. These intervals create a subdivision $a = d_0 \leq c_1 < d_1 \leq c_2 \cdots \leq c_p < d_p \leq c_{p+1} = b$ of the interval $[a,b]$. (Of course, if -1 is not in the trajectory of β to begin with, this becomes the trivial partition $a = d_0 < c_1 = b$.) For $0 \leq k \leq p$ we let β_k^* denote the restriction of β to the interval $[d_k, c_{k+1}]$ and for $1 \leq k \leq p$ let β_k designate the restriction of β to $[c_k, d_k]$. Then $\beta = \beta_0^* + \beta_1 + \beta_1^* + \cdots + \beta_p + \beta_p^*$. (In the not unlikely event that $d_k = c_{k+1}$ one can actually ignore the term β_k^* in this expression, for its deletion has absolutely no effect on the path-sum.) We use γ_k^* and γ_k to indicate the restrictions of γ to the corresponding intervals, so that $\gamma = \gamma_0^* + \gamma_1 + \gamma_1^* + \cdots + \gamma_p + \gamma_p^*$.

By construction the point -1 is not a member of $\beta([d_k, c_{k+1}])$, which fact implies that $\gamma([d_k, c_{k+1}])$ lies in $\mathbb{C} \sim (-\infty, 0]$, a domain where the function $f(\zeta) = \text{Log}\,\zeta$ is analytic. As $\text{Arg}[\gamma(d_k)] = \text{Arg}[\gamma(c_{k+1})] = \text{Arg}\,1 = 0$, we conclude that for $0 \leq k \leq p$

$$\int_{\gamma_k^*} \frac{d\zeta}{\zeta} = \text{Log}[\gamma(c_{k+1})] - \text{Log}[\gamma(d_k)] = \text{Log}\,|\gamma(c_{k+1})| - \text{Log}\,|\gamma(d_k)|\,.$$

In particular,

(B.1) $$\text{Im}\left(\int_{\gamma_k^*} \frac{d\zeta}{\zeta}\right) = 0$$

for $k = 0, 1, \ldots, p$. Next, set $D = \mathbb{C} \sim [0, \infty)$ and let Ψ be the branch of the argument in D whose values lie in the interval $(0, 2\pi)$. Then $g(\zeta) = \text{Log}\,|\zeta| + i\Psi(\zeta)$ defines a branch of the logarithm function in D. In Chapter V we defined revolutions of $v(t)$ in terms of the behavior of a certain reference angle $\theta(t)$. That angle is represented here by $\theta(t) = \Psi[\beta(t)] = \Psi[\gamma(t)]$, provided t belongs to any of the intervals (c_k, d_k). Writing $\ell(t) = g[\gamma(t)] = \text{Log}\,|\gamma(t)| + i\,\theta(t)$ and exploiting the fact that $\gamma[(c_k, d_k)]$ is contained in D — hence, that the derivative of ℓ is given by $\dot\ell(t) = g'[\gamma(t)]\dot\gamma(t) = \dot\gamma(t)/\gamma(t)$ at every point of differentiability t which γ has in (c_k, d_k) — we compute for $k = 1, 2, \ldots, p$:

$$\int_{\gamma_k} \frac{d\zeta}{\zeta} = \int_{c_k}^{d_k} \frac{\dot\gamma(t)\,dt}{\gamma(t)} = \lim_{\substack{s \to d_k^- \\ u \to c_k^+}} \int_u^s \frac{\dot\gamma(t)\,dt}{\gamma(t)}$$

$$= \lim_{\substack{s \to d_k^- \\ u \to c_k^+}} \int_u^s \dot\ell(t)\,dt = \lim_{s \to d_k^-} \ell(s) - \lim_{u \to c_k^+} \ell(u)$$

$$= \text{Log}\,|\gamma(d_k)| - \text{Log}\,|\gamma(c_k)| + i\left[\lim_{s \to d_k^-}\theta(s) - \lim_{u \to c_k^+}\theta(u)\right].$$

Recalling what it means for the vector $v(t)$ to perform a single revolution over $[c, d]$, we note that the imaginary part of the final expression is equal to 2π if $v(t)$ performs a positive revolution over $[c_k, d_k]$, -2π if $v(t)$ makes a negative revolution over $[c_k, d_k]$, and 0 if $\beta([c_k, d_k])$ is not the full circle $K(0, 1)$. These are the only three possibilities. Because $n(\gamma, 0)$ is a real number, we can combine this information with (B.1) to obtain

$$n(\gamma, 0) = \frac{1}{2\pi i}\int_\gamma \frac{d\zeta}{\zeta} = \text{Re}\left(\frac{1}{2\pi i}\int_\gamma \frac{d\zeta}{\zeta}\right) = \frac{1}{2\pi}\,\text{Im}\left(\int_\gamma \frac{d\zeta}{\zeta}\right)$$

$$= \frac{1}{2\pi}\,\text{Im}\left(\int_{\gamma_0^*}\frac{d\zeta}{\zeta} + \int_{\gamma_1}\frac{d\zeta}{\zeta} + \cdots + \int_{\gamma_p}\frac{d\zeta}{\zeta} + \int_{\gamma_p^*}\frac{d\zeta}{\zeta}\right) = P - N\,,$$

where P is the number of positive revolutions performed by $v(t)$ and N is the number of negative revolutions. ∎

Notice that Theorem 1.1 expresses $n(\gamma, z)$ in terms of quantities that make sense for an arbitrary closed path $\gamma\colon [a, b] \to \mathbb{C} \sim \{z\}$, even one that is not piecewise smooth. Thus, Theorem 1.1 can serve as a definition of $n(\gamma, z)$ in this more general setting. Most of the properties of winding numbers observed for piecewise smooth paths (e.g., those listed in Lemma V.2.1) remain valid in the generalized situation, although they require new confirmation.

1.2 Winding Numbers and Jordan Curves

We turn next to the proof of Lemma V.2.1(iii). We remind the reader that, in stating and proving this result, we accept the Jordan curve theorem as a given. Needless to say, a much shorter argument would be possible if we had

1. Technical Facts About Winding Numbers   551

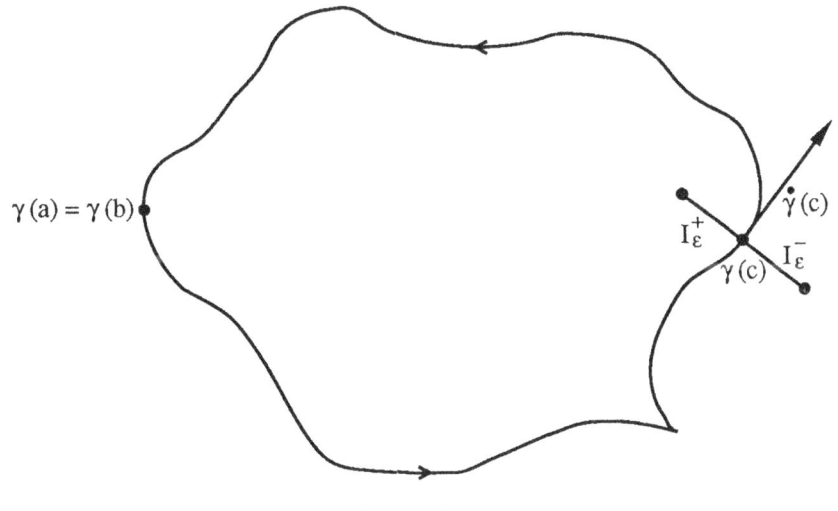

Figure 1.

access to more sophisticated topological machinery. The chief ingredient in the proof that we furnish is the ensuing lemma. (See Figure 1.)

Lemma 1.2. Let $\gamma: [a,b] \to \mathbb{C}$ be a simple, closed path, and let c be a point of (a,b) at which γ is differentiable with $\dot\gamma(c) \neq 0$. There exists an $\epsilon > 0$ for which it is true that the sets $I_\epsilon^+ = \{\gamma(c) + si\dot\gamma(c) : 0 < s \leq \epsilon\}$ and $I_\epsilon^- = \{\gamma(c) + si\dot\gamma(c) : -\epsilon \leq s < 0\}$ lie in different components of $\mathbb{C} \sim |\gamma|$.

Proof. Without loss of generality we may suppose that $\gamma(c) = 0$ and that $\dot\gamma(c)$ is both real and positive. (The proof can always be reduced to this case by subjecting γ to a preliminary translation and rotation.) With this normalization the sets under consideration are $I_\epsilon^+ = \{si\dot\gamma(c) : 0 < s \leq \epsilon\}$ and $I_\epsilon^- = \{si\dot\gamma(c) : -\epsilon \leq s < 0\}$, which are intervals on the positive and negative imaginary axis, respectively. We first note that there exists an $\epsilon > 0$ with the property that the set $I_\epsilon = \{si\dot\gamma(c) : -\epsilon \leq s \leq \epsilon\}$ meets the Jordan curve $J = |\gamma|$ only at the origin. If not, there would be a sequence of non-zero real numbers $\langle s_n \rangle$ such that $s_n \to 0$, but such that $s_n i \dot\gamma(c)$ lies on J. Thus, $s_n i \dot\gamma(c) = \gamma(t_n)$ for some t_n in $[a,b]$. After passing to a subsequence and relabeling, if necessary, one could assume that $t_n \to t_0$, a point of $[a,b]$. Now $\gamma(t_n) = s_n i \dot\gamma(c) \to 0$ by construction and $\gamma(t_n) \to \gamma(t_0)$ by the continuity of γ. Therefore $\gamma(t_0) = 0 = \gamma(c)$. From the simplicity of γ we would infer that $t_0 = c$. Because $\gamma(t_n) \neq 0$ for $n = 1, 2, 3, \cdots$, the simplicity of γ would also make sure that $t_n \neq c$ for such n. These facts would allow us to write

$$\dot\gamma(c) = \lim_{n \to \infty} \frac{\gamma(t_n) - \gamma(c)}{t_n - c} = \lim_{n \to \infty} \frac{\gamma(t_n)}{t_n - c} = \lim_{n \to \infty} \left[\frac{s_n i \dot\gamma(c)}{t_n - c} \right].$$

552   Appendix B. Winding Numbers Revisited

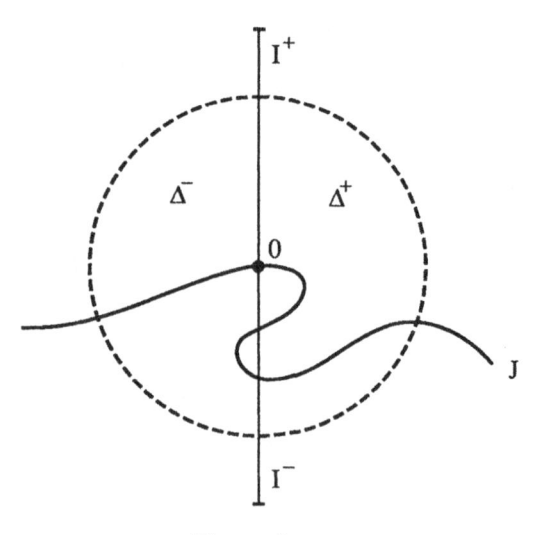

Figure 2.

By assumption $\dot\gamma(c) \neq 0$, so we would have $s_n/(t_n-c) \to -i$. But $s_n/(t_n-c)$ is real for all n, producing a contradiction. Accordingly, we are at liberty to choose and fix $\epsilon > 0$ for which $I_\epsilon \cap J = \{0\}$. We abbreviate $I = I_\epsilon$, $I^+ = I_\epsilon^+$, and $I^- = I_\epsilon^-$ for the rest of the proof. We claim that I^+ and I^- are contained in different components of $\mathbb{C} \sim J$. Since I^+ and I^- are connected sets and neither of them meets J, each of these sets does lie in a component of $\mathbb{C} \sim J$. We shall suppose that they are contained in the same component of $\mathbb{C} \sim J$ and derive another contradiction.

The critical step in obtaining a contradiction is to make a small observation. Let $\Delta = \Delta(0, r)$, where $0 < r < \epsilon\dot\gamma(c)$. (Don't forget: $\dot\gamma(c) > 0$.) If $\Delta^+ = \{z \in \Delta : \operatorname{Re} z > 0\}$ and $\Delta^- = \{z \in \Delta : \operatorname{Re} z < 0\}$, then we observe that J must have points in both Δ^+ and Δ^-. (See Figure 2.) Indeed, because γ is differentiable at c and because $\gamma(c) = 0$, we can express $\gamma(t)$ in the form

$$\gamma(t) = \dot\gamma(c)(t-c) + E(t) ,$$

where the function $E \colon [a,b] \to \mathbb{C}$ satisfies $\lim_{t \to c} |E(t)|/|t-c| = 0$. If $t \neq c$ is near enough to c that $|E(t)| < \dot\gamma(c)|t-c|$, then $\dot\gamma(c)(t-c)$ and the real part of $\gamma(t)$ must have the same sign. Since $\dot\gamma(c) > 0$, we conclude that $\operatorname{Re}[\gamma(t)] > 0$ must hold for all t near c with $t > c$, whereas $\operatorname{Re}[\gamma(t)] < 0$ for all t near c with $t < c$. This fact, together with the continuity of γ, implies that J meets both Δ^+ and Δ^-.

Assume now that I^+ and I^- lie in the same component of the set $\mathbb{C} \sim J$ — call that component D. An elementary argument shows that there is a simple polygonal path β in D with initial point $\epsilon i \dot\gamma(c)$, with terminal point $-\epsilon i \dot\gamma(c)$, and with $|\beta|$ otherwise disjoint from I. It follows that the path $\alpha = \beta + [-\epsilon i \dot\gamma(c), \epsilon i \dot\gamma(c)]$ is simple and closed, and that the

1. Technical Facts About Winding Numbers

Jordan curve $J_1 = |\alpha| = |\beta| \cup I$ meets J only at the origin. We next choose r with $0 < r < \epsilon \dot{\gamma}(c)$ and with the property that $\Delta = \Delta(0, r)$ does not intersect $|\beta|$. Employing the notation introduced in the last paragraph, we conclude that $\Delta \sim J_1 = \Delta \sim I = \Delta^+ \cup \Delta^-$. The Jordan curve theorem applied to J_1 demands that Δ^+ and Δ^- be subsets of different components of $\mathbb{C} \sim J_1$. (Otherwise the disk Δ would intersect only one component of $\mathbb{C} \sim J_1$, which would prevent the origin from being a boundary point of its second component, as required by the Jordan curve theorem.) The set $J \sim \{0\}$ is connected and does not meet J_1. This means that $J \sim \{0\}$ has to be contained in a single component of $\mathbb{C} \sim J_1$. As a consequence, J must fail to intersect one of the sets Δ^- or Δ^+. We have thereby arrived at the anticipated contradiction. The alternative is for I^- and I^+ to be contained in different components of $\mathbb{C} \sim J$. ∎

After this preparation, we are ready for the proof of:

Theorem 1.3. *Let $\gamma: [a, b] \to \mathbb{C}$ be a simple, closed, piecewise smooth path and let D be the bounded component of $\mathbb{C} \sim |\gamma|$. Then either $n(\gamma, z) = 1$ for every z in D or $n(\gamma, z) = -1$ for all such z.*

Proof. In light of Lemma V.2.1(i) it suffices to exhibit just one point z_0 of D for which $n(\gamma, z_0) = 1$ or $n(\gamma, z_0) = -1$. To accomplish this we begin by choosing c in (a, b) with the property that $\dot{\gamma}(c)$ exists and is not zero. As γ is piecewise smooth and non-constant, the existence of such a number c is evident. The previous lemma permits us to select $\epsilon > 0$ so that the sets I_ϵ^+ and I_ϵ^- described there lie in different components of $\mathbb{C} \sim |\gamma|$. In order to simplify notation in this proof we shall assume that $\gamma(c) = 0$, that $\text{Arg}[\dot{\gamma}(c)] = \pi/2$, and that $\epsilon = 1$. One can always reduce the proof to this situation through preliminary translation, rotation, and dilation of γ. With these normalizations $I_\epsilon^+ = [-1, 0)$ and $I_\epsilon^- = (0, 1]$. We proceed under the assumption that $[-1, 0)$ is contained in D and $(0, 1]$ is contained in D^*, the unbounded component of $\mathbb{C} \sim |\gamma|$. We shall locate a point z_0 of D for which $n(\gamma, z_0) = 1$. In the opposite case (i.e., when $(0, 1]$ is contained in D and $[-1, 0)$ in D^*) a similar argument would identify a point z_0 of D with $n(\gamma, z_0) = -1$.

To aid in the proof we select an open disk $\Delta = \Delta(0, R)$ of radius $R > 1$ that contains the bounded set $|\gamma|$ — hence, that encompasses the domain D — and apply Theorem II.3.6 to choose a polygonal arc A in D^* whose endpoints are 1 and R. As a final preparatory step we fix r in $(0, 1)$ for which the closed disk $\overline{\Delta}(-r, r)$ is contained in Δ and is disjoint from A. (See Figure 3.)

By construction, the point $z_0 = -r$ belongs to D. We claim that $n(\gamma, z_0) = 1$. For the proof of this let γ_1 and γ_3 designate the restrictions of γ to the intervals $[a, c]$ and $[c, b]$, respectively, and let $\gamma_2(t) = -r + re^{it}$ for $0 \le t \le 2\pi$. Then $\beta = \gamma_1 - \gamma_2 + \gamma_3$ is a closed and piecewise smooth

554   Appendix B. Winding Numbers Revisited

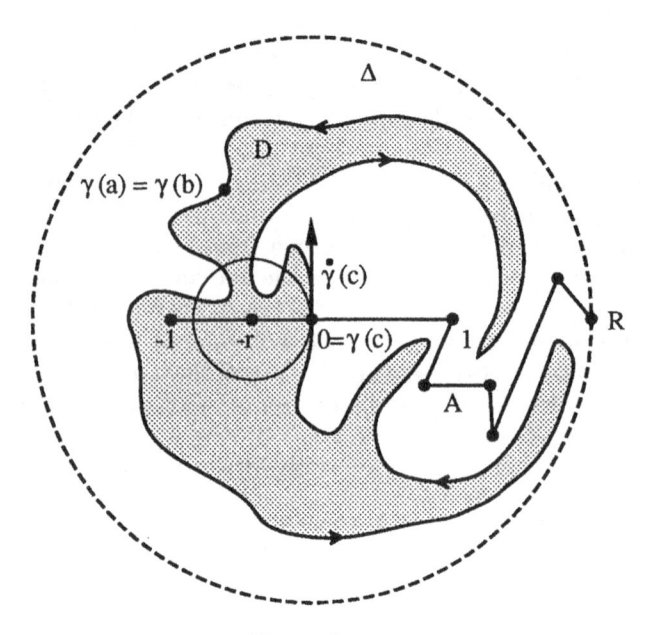

Figure 3.

path. We notice that

$$n(\beta, z_0) = \frac{1}{2\pi i} \int_{\gamma_1} \frac{d\zeta}{\zeta - z_0} - \frac{1}{2\pi i} \int_{\gamma_2} \frac{d\zeta}{\zeta - z_0} + \frac{1}{2\pi i} \int_{\gamma_3} \frac{d\zeta}{\zeta - z_0}$$

$$= \frac{1}{2\pi i} \int_{\gamma} \frac{d\zeta}{\zeta - z_0} - \frac{1}{2\pi i} \int_{\gamma_2} \frac{d\zeta}{\zeta - z_0} = n(\gamma, z_0) - n(\gamma_2, z_0) ,$$

while direct calculation yields $n(\gamma_2, z_0) = 1$. The path β is parametrized on the interval $[a, b + 2\pi]$, with $\beta(t) = 0$ only for $t = c$ and $t = c + 2\pi$. Since $\gamma(c) = 0$ and $\text{Arg}[\dot\gamma(c)] = \pi/2$, it follows that $\text{Im}[\gamma(t)] < 0$ for all t sufficiently close to c with $t < c$, whereas $\text{Im}[\gamma(t)] > 0$ for all t near to c with $t > c$. (Recall the verification of a similar fact in the proof of Lemma 1.2.) Also, $\text{Im}[\gamma_2(t)] > 0$ when $0 < t < \pi$ and $\text{Im}[\gamma_2(t)] < 0$ when $\pi < t < 2\pi$. These facts permit us to select intervals $[t_1, s_1]$ and $[t_2, s_2]$ with $t_1 < c < s_1 < t_2 < c + 2\pi < s_2$ concerning which the following statements are valid: $\text{Im}[\beta(t_1)] < 0$ and $\text{Im}[\beta(s_1)] < 0$; $\text{Im}[\beta(t_2)] > 0$ and $\text{Im}[\beta(s_2)] > 0$; $\beta([t_1, s_1])$ and $\beta([t_2, s_2])$ are subsets of a small disk $\Delta_0 = \Delta(0, r_0)$, so chosen that Δ_0 does not contain the point z_0, does not intersect A, and is itself a subset of Δ.

To complete the proof, we consider a closed, piecewise smooth path α that is constructed by modifying β on each of the intervals $[t_1, s_1]$ and $[t_2, s_2]$. Namely, the restriction of α to $[t_k, s_k]$ ($k = 1, 2$) is just some smooth parameterization on this interval of the directed line segment from $\beta(t_k)$ to

1. Technical Facts About Winding Numbers

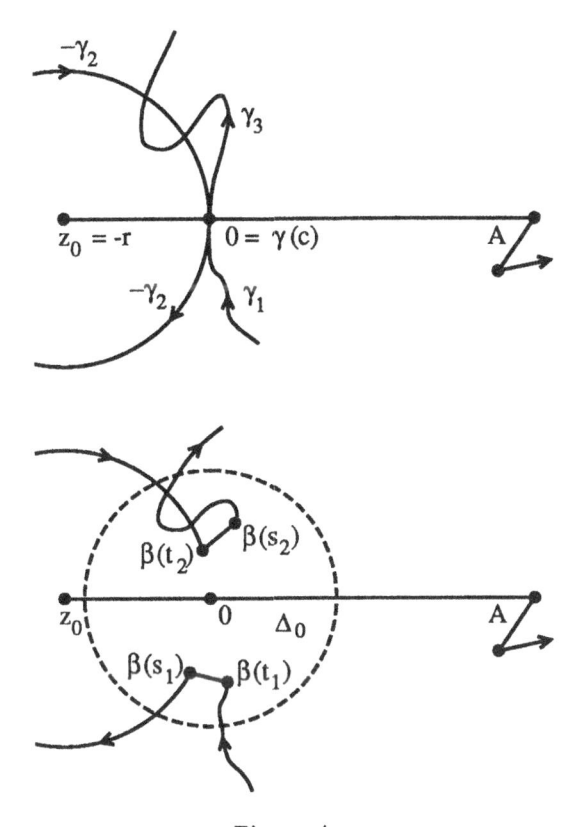

Figure 4.

$\beta(s_k)$. Otherwise, α agrees with β. (See Figure 4.) If β_k and α_k designate the restrictions of β and α to $[t_k, s_k]$ we see that

$$\int_\beta \frac{d\zeta}{\zeta - z_0} - \int_\alpha \frac{d\zeta}{\zeta - z_0} = \int_{\beta_1 - \alpha_1} \frac{d\zeta}{\zeta - z_0} + \int_{\beta_2 - \alpha_2} \frac{d\zeta}{\zeta - z_0} = 0 \ .$$

The last assertion is a consequence of the definition of α and of the local Cauchy theorem applied to the function $f(\zeta) = (\zeta - z_0)^{-1}$ and the paths $\beta_k - \alpha_k$ ($k = 1, 2$) in the disk Δ_0.

We infer that $n(\beta, z_0) = n(\alpha, z_0)$. On the other hand, since the whole construction of α was engineered to put z_0 in the unbounded component of $\mathbb{C} \sim |\alpha|$, Lemma V.2.1(ii) allows us to conclude that $n(\alpha, z_0) = 0$. Consequently,

$$n(\gamma, z_0) = n(\gamma_2, z_0) + n(\beta, z_0) = 1 + n(\alpha, z_0) = 1 \ ,$$

as asserted. ∎

Index

Abel's theorem, 294
absolute convergence of a series, 249
    of an infinite product, 492
absolute value, 5
accumulation point, 38, 354
admissible density, 429
algebraic function element, 525
analytic continuation, 510
    along a path, 512
analytic function, 67
analytic function element, 510
analytic modulo isolated singularities, 309
angle-preserving, 379
annulus of convergence, 269
anti-conformal mapping, 379, 390
anti-Möbius transformation, 400
argument, 6
argument principle, 340
Arzelà-Ascoli theorem, 282
attracting fixed point, 405

Bernoulli numbers, 293
Bolzano-Weierstrass theorem, 53, 547
boundary of a set, 35
bounded set, 52
branch of a global analytic function, 527
    of an inverse, 86
    of $\arg z$, 91
    of $\log f(z)$, 175
    of $\log z$, 91
    of the λ-power, 92, 182
    of the pth-root, 88, 183
branch covering principle, 344, 361

Cantor's theorem, 56
Carathéodory-Osgood theorem, 445
Casorati-Weierstrass theorem, 321
Cauchy convergence criterion, 53, 246, 248
    estimates, 167
    inequality, 27
    integral formula, 161, 192
    sequence, 53
    theorem, 148, 188
Cauchy-Riemann equations, 69
Cauchy-Schwarz inequality, 430
chain rule, 66
change of parameter, 116
circle in $\widehat{\mathbb{C}}$, 352, 398
closed path, 110
closed set, 34, 354
closure of a set, 35
compact set, 54, 356
complete analytic function, 527
complex line integral, 119
complex number, 2
complex plane, 4
component of an open set, 51, 356
composition of functions, 19
conditional convergence, 249

Index

conformal mapping, 379, 390
conformal modulus, 428
conjugate complex number, 5
    harmonic function, 215
    Möbius transformation, 394
connected set, 48, 356
continuous function, 40, 355
contractible path, 203
contractive sequence, 54
convergence of a sequence, 36, 355
    of a series, 248
    of an infinite product, 490
convex set, 206
cross-ratio, 396
curvilinear angle, 376
cycle, 186

De Moivre's formula, 9
dense subset, 207, 280
derivative, 63
diffeomorphism, 378
differentiability, 62
    in the real sense, 97
dilation, 391
directional derivative, 107
Dirichlet problem, 227
    series, 297
disconnected set, 47
discrete mapping, 306, 356
    mapping theorem, 306, 358
    subset, 306, 356
disk of convergence, 261
divergence of a series, 248
    of an infinite product, 490
domain, 50, 356
domain-set of a function, 18

Elementary Möbius transformation, 391
elliptic function, 489
    transformation, 405
entire function, 75
equicontinuity, 279
essential singularity, 309, 322
exceptional point, 520

extended boundary, 354
    closure, 354
    complex plane, 25, 350
    domain-set, 353
extremal length, 429

Factor theorem, 169
field, 2, 543
finitely connected along the boundary, 440
fixed point, 30, 394
fundamental theorem of algebra, 168

Gamma function, 504
Gauss-Lucas theorem, 207
Gauss's formula, 505
geometric series, 251
global analytic function, 527
Goursat's theorem, 193

Hadamard's theorem, 173
harmonic conjugate, 215
harmonic function, 102, 172, 215
Harnack's inequalities, 232
Hilbert transform, 334
holomorphic function, 67, 357
homeomorphism, 352, 424
homologous cycles, 187
homologous to zero, 187
homotopic paths, 197, 517
Hurwitz's theorem, 348
hyperbolic transformation, 405

Infinite product, 490
inner radius of convergence, 269
integral with respect to arclength, 119
interior of a set, 35
interior point, 34, 354
invariant circle, 408
inverse function, 20
inverse function theorem, 348, 360
inversion, 391
isogonal, 379
isolated point, 45

isolated singularity, 309

Jacobian, 107, 377
Jordan contour, 160
    curve theorem, 111
    domain, 445

L'Hospital's rule, 304
λ-powers of a complex number, 16
Lagrange's identity, 27
Laplace's equation, 102, 215
Laurent series, 269
Legendre duplication formula, 539
    relation, 536
length function, 132
length of a path, 120, 131
limit of a function, 43, 355
    of a sequence, 36, 354
limit point of a set, 43, 355
limit superior, 260, 545
Liouville's theorem, 167
locally bounded family, 282
locally conformal, 382
locally connected, 443
locally connected along the boundary, 440
logarithm, 15
loxodromic transformation, 405

Magnitude, 5
maximum principle, 170, 221, 360
mean value property, 220
meromorphic function, 318, 356
Mittag-Leffler's theorem, 480
Möbius transformation, 25, 390
modulus, 5
monodromy theorem, 518
Montel's theorem, 285
Morera's theorem, 165
multiplicity of a function value, 303, 361
multiplier, 404

Normal convergence of a sequence, 247
    of a series, 253, 477
    of an infinite product, 494
normal family, 278
normalized form of a Möbius transformation, 393
normalized matrix, 393

Open mapping, 347, 356
open mapping theorem, 75, 347, 359
open set, 34, 354
order of a function value, 303
    of a Möbius transformation, 405
    of a pole, 312, 322
ordered field, 545
orientation-preserving (-reversing), 378
oriented angle, 375
oriented path, 160, 447
outer radius of convergence, 269

Parabolic transformation, 402
path, 109
path sum, 115
Picard's theorem, 168, 299, 321
piecewise smooth path, 112
pointwise bounded family, 282
pointwise convergence of a sequence, 244
    of a series, 253
    of an infinite product, 493
Poisson integral, 229
    kernel, 228
polar representation, 6, 15
pole, 309, 322
polygonal arc, 49
    path, 116
polynomial function, 18
power series, 260
primitive, 125
principal arcsecant, 105
    arcsine, 83
    arctangent, 84
    argument, 8

Index 559

λ-power, 16
logarithm, 15
n^{th}-root, 10
principal part, 309, 322
principle of analytic continuation, 307
purely imaginary, 5

Radius of convergence, 261, 269
range of a function, 18
rational function, 18
rectifiable path, 131
reflection in a circle, 399
regular arc, 376
regular for the Dirichlet problem, 228
removable singularity, 309, 322
repelling fixed point, 405
residue, 310
    at infinity, 323
residue theorem, 323
reverse path, 115
Riemann extension theorem, 237, 310
    hypothesis, 258
    mapping theorem, 420
    sphere, 352
    surface, 529
    zeta-function, 258
right-inverse, 85
ring of convergence, 269
rotation, 391
Rouché's theorem, 341

Schwarz lemma, 172
    reflection principle, 453
    theorem, 230
Schwarz-Christoffel formula, 459
sense-preserving (-reversing), 378
sequence, 35
similarity transformation, 22, 30
simple path, 110
simply connected domain, 195, 422
singular part, 309, 322
    set, 309

singularity at infinity, 322
smooth path, 112
starlike set, 49
stereographic projection, 351
subfield, 2, 542
subharmonic function, 172, 220
subsequence, 36
superharmonic function, 220
symmetry with respect to a circle, 401

Target-set of a function, 18
Taylor series, 260
trajectory, 110
translation, 391
triangle inequality, 11
trigonometric functions, 77

Uniform Cauchy sequence, 246
uniform continuity, 57
uniform convergence of a sequence, 244
    of a series, 253
    of an infinite product, 494
univalent function, 19
unorderable field, 4, 545

Vitali's theorem, 364

Weierstrass M-test, 253
    $\wp$-function, 486
    theorem, 498
winding number, 153

z- and $\bar{z}$-partial derivatives, 98

Undergraduate Texts in Mathematics

(continued from page ii)

Malitz: Introduction to Mathematical Logic.
Marsden/Weinstein: Calculus I, II, III. Second edition.
Martin: The Foundations of Geometry and the Non-Euclidean Plane.
Martin: Transformation Geometry: An Introduction to Symmetry.
Millman/Parker: Geometry: A Metric Approach with Models. Second edition.
Moschovakis: Notes on Set Theory.
Owen: A First Course in the Mathematical Foundations of Thermodynamics.
Palka: An Introduction to Complex Function Theory.
Pedrick: A First Course in Analysis.
Peressini/Sullivan/Uhl: The Mathematics of Nonlinear Programming.
Prenowitz/Jantosciak: Join Geometries.
Priestley: Calculus: An Historical Approach.
Protter/Morrey: A First Course in Real Analysis. Second edition.
Protter/Morrey: Intermediate Calculus. Second edition.
Ross: Elementary Analysis: The Theory of Calculus.
Samuel: Projective Geometry. *Readings in Mathematics.*
Scharlau/Opolka: From Fermat to Minkowski.
Sigler: Algebra.
Silverman/Tate: Rational Points on Elliptic Curves.
Simmonds: A Brief on Tensor Analysis. Second edition.
Singer/Thorpe: Lecture Notes on Elementary Topology and Geometry.
Smith: Linear Algebra. Second edition.
Smith: Primer of Modern Analysis. Second edition.
Stanton/White: Constructive Combinatorics.
Stillwell: Elements of Algebra: Geometry, Numbers, Equations.
Stillwell: Mathematics and Its History.
Strayer: Linear Programming and Its Applications.
Thorpe: Elementary Topics in Differential Geometry.
Troutman: Variational Calculus and Optimal Control with Elementary Convexity. Second edition.
Valenza: Linear Algebra: An Introduction to Abstract Mathematics.
Whyburn/Duda: Dynamic Topology.
Wilson: Much Ado About Calculus.

Made in the USA
Las Vegas, NV
27 August 2023